The Potential of U.S. Cropland to Sequester Carbon and Mitigate the Greenhouse Effect

by
R. Lal
J.M. Kimble
R.F. Follett
and C.V. Cole

ISBN 1-57504-112-X

Ann Arbor Press
121 South Main Street
Chelsea, MI 48118
Ann Arbor Press is an imprint of Sleeping Bear Press, Inc.

Printed in the United States of America.

10 9 8 7 6 5 4 3 2 1

Executive Summary

This report, *The Potential of U.S. Cropland to Sequester Carbon and Mitigate the Greenhouse Effect*, assesses the potential of U.S. cropland to sequester carbon (C). It concludes that properly applied soil restorative processes and best management practices (BMPs) can mitigate the greenhouse effect both by decreasing the emissions of greenhouse gases (GHGs) from U.S. agricultural activities and by making U.S. cropland a major sink for C sequestration. Worldwide soil restoration and adoption of BMPs has a potential to mitigate effectively a large proportion of the annual increase in atmospheric concentration of CO_2.

Chapter 1 summarizes our objectives in preparing this report. Improved agricultural practices have great potential to increase carbon sequestration and decrease the net emission of carbon dioxide and other greenhouse gases, but policy makers have not widely recognized this potential. Since the 1980s, considerable scientific information has been collated about the potential of agricultural lands to sequester C (Johansson, et al. 1993; Lal et al., 1995a, b; 1998a, b). But the available information has not been synthesized in a form that policy makers and land managers readily can use to mitigate CO_2 emissions in relation to the potential greenhouse effect. This report therefore (i) collates and synthesizes the available information on the contribution of U.S. cropland agriculture to the mitigation of CO_2 emission, (ii) assesses the role of soil restorative processes and BMPs on C sequestration, (iii) evaluates the potential role of biofuel production as C offset for fossil fuels, and (iv) evaluates the impact of soil erosion on C emission.

Chapter 2 describes the greenhouse processes and global trends in greenhouse gas (GHG) emissions (including concentrations of CO_2, CH_4, and N_2O) and the three principal components of anthropogenic global warming potential (GWP) — fossil fuel combustion and transport, the chemical industry, and deforestation and agriculture. Chapter 3 presents data on U.S. emissions and agriculture's role in relation to them.

The atmospheric concentration of natural GHGs has increased steadily since the industrial revolution. For example, concentration of CO_2 has risen from 280 parts per million by volume (ppmv) in about 1850 to 365 ppmv in 1995 and is increasing at 0.5%/yr (DOE/EIA, 1996). At this rate of increase, CO_2 concentration is expected to reach 600 ppmv during the 21st century. The concentration of CH_4 has increased from 0.8 to 1.74 ppmv and is increasing at 0.75%/yr, and that of N_2O has increased from 288 parts per billion (ppbv) to 311 ppbv and is increasing at 0.25%/yr. The GWP of these gases relative to that of CO_2 for a 100-year time horizon is 21 for CH_4 and 310 for N_2O (DOE/EIA, 1996).

The greenhouse effect is essential to the existence of life on earth. The concern arises because of the excessive rate of global warming due to anthropogenic perturbations. Most climatologists and environmentalists believe that anything more than an increase of 0.1°C/decade or 1°C over the 21st century would be excessive in terms of adverse impact on ecosystems. When climate change exceeded this rate in the past, forest ecosystems in the northern hemisphere were devastated (IPCC, 1990 a, b).

Estimates of total U.S. emissions of GHGs range from 1442 million metric tons of carbon equivalent (MMTCE), including 66 MMTCE from agricultural activities (DOE/EIA, 1996), to 1666 MMTCE, including 80 MMTCE from agriculture (USEPA, 1995 b). This report estimates total emissions of 1600 MMTCE, including 116 MMTCE from agricultural activities. Agriculture thus plays an important role in GHG emissions and, with adoption of BMPs, also can be a major sink for C sequestration in soil and terrestrial/aquatic ecosystems. Agriculture's role should include both sequestration of C into the soil and the production of biomass fuel crops to substitute for fossil fuels.

Chapter 4 deals with the soil organic carbon (SOC) pool in soils of the U.S. and its loss due to cultivation and provides a reference point for the magnitude of the C sequestration potential. It also compares the SOC pool in U.S. cropland with that of world soils. The mass of soil C (1 m depth) in cultivated soils of the world (1727 Mha) is estimated at 1,670,000 MMTC (1670 GT), and IPCC (1995) estimates an historic soil C loss of 55,000 MMTC (55 GT). The atmospheric inventory of C was reported as 740,000 MMT (740 GT) in 1986 and is increasing at about 0.5% per year (DOE/EIA, 1996); the atmospheric inventory now probably is approaching 780,000 MMT (780 GT). The entire 55,000 MMT (55 GT) of historic soil-C loss from cultivated soils worldwide thus accounts for about 7% of the current atmospheric inventory.

Cropland soils potentially can sequester a considerable part of this lost C (55,000 MMTC), provided that recommended practices are adopted. U.S. cropland area is about 136.6 M ha or about 8% of the total world cropland area and 19.4% of U.S. land area. The SOC stock of U.S. soil is about 80,700 MMT in U.S. soils and 15,600 MMT in U.S. cropland, and cropland historically has lost about 5,000 MMT SOC from its preagricultural levels. One reasonably can assume that cropland potentially can sequester 4,000 to 6,000 MMT, with an average of 5,000 MMTC in cropland soils — potentially more, with new technologies and proper management.

Chapter 5 deals with principal processes governing GHG emission from the pedosphere and establishes a link between SOC content and soil quality. Agricultural activities that lead to GHG emissions from terrestrial and aquatic ecosystems in general, and from U.S. cropland soils in particular, include deforestation, biomass burning, plowing, applications of fertilizers and manures, cultivation of rice paddies, intensive use of organic soils, overgrazing and denuding of rangelands in semiarid and arid regions, and animal husbandry or raising cattle and other ruminants. The principal processes of GHG emission include soil degrading processes such as erosion, decline in soil structure leading to crusting and compaction, soil fertility depletion and acidification, salinization, and decline in soil biodiversity. Prevalence of these processes leads to decline in soil quality, reduction in biomass productivity, decline in SOC content, and emission of GHGs.

Chapter 6 outlines strategies for mitigating GHG emissions from U.S. cropland. These strategies include reversion of marginal agricultural land to natural or planted fallows, restoration of degraded soils, intensification of prime agricultural land through widespread adoption of BMPs, and production of biofuels on idle land. Chapter 6 also describes attributes of U.S. cropland and historical changes in the land area and outlines options for C sequestration.

Chapter 7 deals with soil erosion management, which is a high priority for enhancing soil quality, improving the environment, decreasing GHG emissions, and increasing C sequestration in soil. Controling soil erosion and restoring degraded soils and ecosystems are important strategies that deserve careful consideration. These strategies, through adoption of BMPs, can increase production of below-ground and above-ground biomass and increase SOC content. Soil erosion control can prevent emission of 12 to 22 MMTC/yr.

A vast potential for C sequestration also exists in the Conservation Reserve Program (CRP) and Wetland Reserve Program (WRP), as Chapter 8 describes, including conservation buffers, cropped wetland management, grazing land restoration, and conversion of marginal land to pasture land. Properly implemented, several proven land conversion and restorative measures also can increase SOC content to mitigate the greenhouse effect — that is, they can sequester it, as well as prevent its emission.

The strategy is to take marginal land out of agricultural production and apply restorative measures (e.g., afforestation, cover crops) to it. Land conversion through CRP or alternative agricultural uses, such as pasture, and WRP can sequester 6 to 14 MMTC/yr. Restoration of degraded soils (eroded lands, mineland and toxic soils, salt-affected soils) and soil fertility management can sequester 11 to 25 MMTC/yr.

Chapter 9, on biofuel offset, outlines the potential of using cropland to produce biomass for direct fuel to produce power. The biofuel offset potential for 10 Mha of cropland is estimated at 35 to 63 MMTC/yr. An additional potential of sequestering 3-5 MMTC/yr as SOC content can result from improved management of idle land.

Chapter 10 appropriately elaborates the potential for sequestering C by intensifying prime agricultural land, while Chapter 11 describes the overall potential to sequester C. The BMPs for agricultural intensification include conservation tillage (CT) and crop residue management, water management, and improved cropping systems. Adoption of CT and crop residue management in humid and sub-humid regions potentially can sequester 35-107 MMTC/yr. Management of irrigation land can sequester an additional 5 to 11 MMTC/yr. Adoption of improved cropping systems (e.g., soil fertility management, rotation, winter cover crops) can sequester 19-52 MMTC/yr. The total C sequestration potential of U.S. cropland through improved management thus is 75-208 MMTC/yr. At an average of 142 MMTC/yr, this potential is 9.5% (1485 MMT-CE/yr [DOE estimate]) or 8.3% (1709 MMTCE/yr [EPA estimate]) of the total U.S. annual emissions of all GHGs.

Agricultural practices with large C sequestration potential are CT and crop residue management (49% of the total 142 MMTC/yr), improved cropping systems (25%), land restoration (13%), land conversion and alternate land use (7%), and irrigation (6%). Biofuel offset through biomass produced on 10 mha of land is estimated at 32-38 MMTC/yr. Soil erosion prevention could reduce emissions by 12-22 MMTC/yr, and conversion to CT could produce an estimated savings in fossil fuel of 1-2 MMTC/yr. The overall potential of U.S. cropland for C sequestration, fossil fuel offset, and erosion control therefore is 120-270 MMTC/yr.

These estimates of the C sequestration potential of U.S. croplands show that adoption of BMPs is a partial solution to the environmental problem. Important agricultural practices with a vast potential to mitigate the greenhouse effect are: (i) CT, crop residues mulch, CRP, and conservation buffers, (ii) crop rotations based on frequent use of cover crops (legumes and grasses) in the farming cycle, (iii) fertility management by judicious use of fertilizers, manures, and urban/agricultural wastes to replenish soil fertility and replace nutrients harvested in crops and animals, (iv) irrigation and water management on drought prone soils in semiarid and sub-humid climates, (v) reclamation of salt-affected soils, and (vi) mineland restoration for growing row crops, with improved cropping systems based on rotations with pastures. In this regard, site-specific soil management techniques and other modern soil capability evaluation techniques are high priority technologies.

Projected global warming is a serious national and international issue — a major environmental threat which we can address through reduced use of fossil energy, increased C storage in agricultural soils, and expanded biofuel production to offset fossil fuel use. In addition to continuing such ongoing programs as CRP and WRP, we need to establish policies that encourage farmers to improve soil quality by sequestering C in the pedosphere. Such policies will enhance soil quality, increase productivity, enhance farm income, improve water quality, and mitigate the greenhouse effect.

As this report, *The Potential of U.S. Cropland to Sequester Carbon and Mitigate the Greenhouse Effect*, concludes, adoption of science-based and best management practices on U.S. cropland can indeed create a win-win situation. It is in our national interest to encourage widespread adoption of proper agricultural practices for sustainable use of our natural resources and improvement of our nation's environment.

Lal, Kimble, Follett, and Cole

Contents

Tables

Figures

Photos

About the Authors

Dr. R. Lal is a Professor of Soil Science in the School of Natural Resources at Ohio State University. Before joining Ohio State in 1987, he served as a soil scientist for 18 years at the International Institute of Tropical Agriculture, Ibadan, Nigeria. Prof. Lal is a fellow of the Soil Science Society of America, the American Society of Agronomy, the Third World Academy of Sciences, the American Association for Advancement of Sciences, and the Soil and Water Conservation Society.

Dr. Lal is a recipient of the International Soil Science Award, the Soil Science Applied Research Award of the Soil Science Society of America, and the International Agronomy Award of the American Society of Agronomy. He is past President of the World Association of Soil and Water Conservation and of the International Soil Tillage Research Organization.

Dr. J.M. Kimble is a Research Soil Scientist at the USDA Natural Resources Conservation Service National Soil Survey Laboratory in Lincoln, Nebraska. Dr. Kimble manages the Global Change project of the Natural Resources Conservation Service, Soil Survey Division, and has worked more than 15 years with the U.S. Agency for International Development projects, dealing with soil related problems in more than 40 developing countries. He is a member of the American Society of Agronomy, the Soil Science Society of America, the International Soil Science Society, and the International Humic Substances Society.

Dr. R.F. Follett is Supervisory Soil Scientist, USDA-ARS, Soil Plant Nutrient Research Unit, Fort Collins, CO. He previously served 10 years as a National Program Leader with ARS headquarters in Beltsville, MD. Dr. Follett is a Fellow of the Soil Science Society of America, American Society of Agronomy, and Soil and Water Conservation Society. He was twice awarded the USDA Distinguished Service Award (USDA's highest award).

Dr. Follett organized and wrote the ARS Strategic Plans both for *Ground-Water Quality Protection — Nitrates* and for *Global Climate Change — Biogeochemical Dynamics*. He has served as editor or coeditor of several books and as a guest editor for the *Journal of Contaminant Hydrology*. His scientific publications include topics about nutrient management for forage production, soil-N and -C cycling, ground water quality protection, global climate change, agroecosytems, soil and crop management systems, soil erosion and crop productivity, plant mineral nutrition, animal nutrition, irrigation, and drainage.

Dr. C.V. Cole is a Senior Research Scientist, Natural Resource Ecology Laboratory, Colorado State University. From 1950 until his retirement in 1993, he was a Research Soil Scientist, Agricultural Research Service, USDA. In the first years of his career, he contributed to advances in phosphorus chemistry; a practical soil test developed from this research is used worldwide to identify phosphorus deficient soils.

Dr. Cole's research expanded into plant responses to phosphorus and broader studies of phosphorus recycling in relation to organic carbon and nitrogen transformations. He was the Principal Investigator on interdisciplinary studies which embraced a range of biological and physical-chemical reactions in soils and which integrated ecological and pedological principles: "Organic Matter and Nutrient Cycling in Semiarid Agroecosystems," a cooperative effort between Colorado State University and the Agricultural Service, USDA, supported from 1979 to 1985 by Ecosystem Studies, National Science Foundation (NSF); and, with Dr. E.T. Elliott, "Organic C, N, S, and P Formation and Loss from Great Plains Agroecosystems," funded by NSF Ecosystem Studies from 1985-1989. This work led to significant advances in understanding management effects on long-term soil productivity and degradation, and it opened avenues of collaborative research into production and environmental concerns for a range of agroecosystems.

Dr. Cole served as a member of the Scientific Advisory Committee (on Biogeochemical Cycles) of the Scientific Committee on Problems ofthe Environment (SCOPE) and as chairman of the Scientific Advisory Committee on Global Phosphorus Cycles. He is a member of Working Group II of the Intergovernmental Panel on Climate Change (IPCC) Second Assessment Report and is convening lead author of Chapter 23, "Agricultural Options for Mitigation of Greenhouse Gas Emissions," in *Climate Change 1995 — Impacts, Adaptations, and Mitigation of Climate Change: Scientific-Technical Analyses*.

Acknowledgments

The authors have drawn heavily on material and data from numerous sources. We have taken the liberty of citing data from many colleagues and friends around the country. It is impossible to name all those who have very generously helped and contributed in one way or the other toward completing this book. Nonetheless, the names of few need a special mention.

Personnel from USDA-ARS, ERS, NRCS, USDOE, USGS, and several universities provided by the data and information used. We are particularly thankful to the following people for sharing information and time with us: Richard Arnold, Anita Daniels, Norm Fausey, William Fryrear, Mohinder Gill, Don Goolsby, Larry Hagen, Doug Karlen, Terry Logan, Greg Marland, Hank Mayland, Curtis Monger, Keith Paustian, James Rhoades, Mathew Romken, Carman Sandretto, Susan Sampson-Liebig, James Schepers, Don Suarez, Don Tanaka, Paul Unger, Sharon Waltman, Andy Ward, Carol Whitman, and Dan Yaalon. We gratefully acknowledge the generous help we received from Steve Stover of NRCS, especially his providing the data on land use, erosion, crop, and other information developed from the NRI.

Our special thanks are due to Ms. Brenda Swank of The Ohio State University, who did a meticulous job in preparing the volume through typing multiple drafts. We also thank Maria Lemon, Ph.D., of The Editor Inc., for her many comments and suggestions and her work editing the manuscript and creating the camera-readies for printing.

Special thanks also go to those who reviewed this manuscript — Richard Arnold, D. Fryrear, Robert Grossman, William Larson, Andrew Manale, Ronald Paetzold, David Pimentel, Bobby Stewart, and Paul Unger. Their time, effort, and comments helped us to improve the final manuscript.

CHAPTER 1

Objectives

Improved agricultural practices have great potential to increase carbon sequestration and decrease the net emission of carbon dioxide and other greenhouse gases, but policy makers have not widely recognized this potential. Since the 1980s, considerable scientific information has been collated about the potential of agricultural lands to sequester C (Johansson, et al. 1993; Lal et al., 1995a, b, 1998a, b). But the available information has not been synthesized in a form that policy makers and land managers readily can use to mitigate CO_2 emissions in relation to the potential greenhouse effect.

This report assesses the potential to sequester carbon and concludes that properly applied soil restorative processes and regionally or locally recommended best management practices (BMPs) can mitigate the greenhouse effect both by decreasing the emissions of GHGs (greenhouse gases) from U.S. agricultural activities and by making U.S. cropland a major sink for C sequestration. Worldwide adoption of such measures potentially can sequester a considerable part of this lost C back into cropland soils. In some ecoregions, adoption of BMPs may lead to C content in soils over and above that found in native ecoregions. In other regions, however, it may be difficult to improve over the existing C content because of other constraints.

Increasing the SOC content of degraded soils by .01%/yr (Lal, 1997) could lead to C sequestration equal to the annual increase in atmospheric CO_2-C (Cole et al., 1996). These efforts also will improve soil, water, and air quality and will increase agricultural productivity: a win-win situation, environmentally and economically. We therefore urge U.S. policy makers to implement governmental policies that will encourage farmers to adopt practices that enhance, through intensification, prime agricultural land production; release marginal land for restoration; and decrease C losses and increase SOC content — and we urge them to emphasize the importance of this strategy to the world community.

This report (i) collates and synthesizes the available information on the contribution of U.S. cropland agriculture to the mitigation of CO_2 emission, (ii) assesses the role of soil restorative processes and BMPs on C sequestration, (iii) evaluates the potential role of biofuel production as C offset for fossil fuels, and (iv) evaluates the impact of soil erosion on C emission. This report is prepared for policy makers and land managers as an aid to implementing appropriate land use and soil management strategies for sustainable management of the nation's soil and water resources. The information contained is also of interest to researchers, to help them identify priorities and knowledge gaps so that they can help policy makers, and land managers, to adopt environmentally friendly but profitable technologies.

Some discussion of processes involved is included, to facilitate identification of appropriate technological options for soil-specific situations. However, to simplify this report and focus on appropriate options for mitigating CO_2 emissions, we deliberately have omitted detailed information on soil taxonomic properties, ecosystem characteristics, and predominant degradative processes. For additional information on technical matters, including CH_4 and N_2O emissions, readers should consult the literature this report references.

CHAPTER **2**

Basic Processes

The Greenhouse Process

Temperature within a greenhouse is warmer than the ambient temperature because glass does not allow long-wave radiation to reflect out of the greenhouse. Some gases in the atmosphere possess properties analogous to glass, and they also absorb heat, much as the glass does in a greenhouse, and do not allow long-wave radiation to reflect back into space.

These gases are of both natural and synthetic origin. Natural gases include water (H_2O) vapors; carbon dioxide (CO_2); carbon monoxide (CO); methane (CH_4); nitrous oxide (N_2O), nitrogen oxide (NO), and nitrogen dioxide (NO_2), collectively called NO_x; and ozone (O_3). Some synthetic compounds which possess similar attributes have been introduced since the 1930s and include several chlorofluorocarbons and chlorofluorohydrocarbons (collectively referred to as CFCs in this report).

The atmospheric capacity to trap solar heat depends on the concentration of these natural and synthetic gases. Concentration of natural trace gases has been increasing steadily because of human activities (e.g., fossil fuel consumption, changes in land use). As a result of a concentrated effort to phase out CFCs because of their effect on the protective atmospheric ozone layer, which effort began with the 1987 Montreal Protocol, the CFC concentration in the 1990s has been decreasing. This volume therefore discusses sources, sinks, and implications of increasing tropospheric concentrations of only CO_2, CH_4, N_2O and the agricultural sources and sinks of CO_2.

The greenhouse effect is a natural process that has made the earth a habitable planet. The natural concentration of GHGs has rendered earth's temperature a tolerable 15°C rather than a frigid -18°C (Schlesinger, 1995). These gases' capacity to trap solar radiation within earth's atmosphere, which is similar to that of the glass in a greenhouse, lead to their being called, collectively, "greenhouse gases" (GHGs) or "radiatively active gases." Concern has arisen because anthropogenic factors have accelerated this natural process since the beginning of the era of industrialization. GHG enrichment of the atmosphere can cause the "greenhouse effect" of creating a substantial increase in the mean global temperature. Some climatologists believe that these emissions already have increased global temperature by 0.5°C ($\pm$ 0.2° C) in the past 100 years (Mahlman, 1997). In addition, further global warming of 1-5°C, due to human enrichment of the atmosphere with GHGs, is projected within the next 100 years (OSTP, 1997). This increase may lead to a global shift of vegetation or ecological zones.

The overall impact of the potential climate change on global C pools and fluxes is highly complex and difficult to predict, but for each 1°C increase in mean annual temperature, the vegetation zones may move poleward by 200 to 300 km (IPCC, 1990 a, b). To prevent warming from reaching this level, the world community must develop long-term goals for stabilizing the concentration of all GHGs. A "business-as-usual" scenario would lead to 700 ppmv enrichment of CO_2 by 2100 (OSTP, 1997). To slow or try to reverse this predicted rapid increase requires not only decreased emissions of CO_2, but also increased sequestration of CO_2-C into soil and the terrestrial/aquatic ecosystems.

The heat trapping potential of different GHGs differs widely, some being more effective and longer lasting than others. The effectiveness of these gases in causing global warming is usually expressed as radiative forcing.

Because of the differences in their GWP, radiative forcing of all gases is expressed in terms of MMTCE. Two convenient units of expressing gaseous emissions are teragram and petagram. One teragram, or Tg, is 10^{12} g and equals one million metric tons (MMT). One petagram, or Pg, is 10^{15} g and equals 1000 MMT.

DOE/EIA (1996) estimates the total U.S. emissions of GHGs is 1442 MMTCE per year, while the US-EPA (1995 a, b) estimates 1666 MMTCE, and this range shows the uncertainties of various estimates. The DOE/EIA estimate for U.S. agriculture includes 52 and 14 MMTCE from CH_4 and N_2O, respectively, while estimates by EPA for these same two gases are 61 and 19 MMTCE. Neither DOE/EIA nor US-EPA estimate agricultural emissions of CO_2 resulting from production input energy; our estimate in this report for agricultural production input energy is 27.9 MMTC of CO_2 per year. In addition, we estimated that erosion from cropland causes emission of 15 MMTC/yr.

U.S. emission thus may total 1485 MMTCE or 1709 MMTCE, or an average of both — 1600 MMTCE. Addition of the estimates of radiative forcing of CH_4 and N_2O to our estimate for CO_2 results in a total radiative forcing from U.S. cropland agriculture of about 109, or 123 MMTCE per year, depending on whether we use, respectively, the DOE/EIA or the EPA estimates. In either case, U.S. agriculture accounts for about 7.3% of the U.S. total.

Global Trends in Greenhouse Gas Emissions

The concentration of GHGs in the atmosphere has increased steadily since about 1850. A substantial part of the total increase so far has been attributed to deforestation, conversion to farmland, and other agricultural activities (Post et al., 1982).

Gas Concentrations

Carbon dioxide

CO_2 is the most important GHG, because increase in its concentration causes about 50% of the total radiative forcing (Rodhe, 1990). The concentration of CO_2 in the atmosphere was about 280 ppm in about 1850 and 365 ppm in 1996 (Fig. 1), and it is increasing at the rate of 0.5%/yr. If this trend continues, CO_2 concentration will be 600 ppmv during the 21st century (OSTP, 1997). A doubling of atmospheric CO_2 over pre-industrial levels is projected to lead to an equilibrium global warming in the range of 1.5° to 4.5° C (Mahlman, 1997).

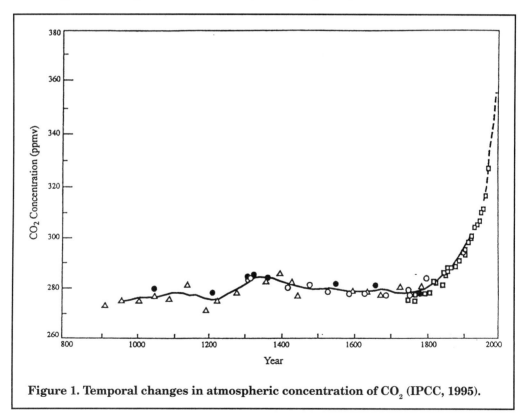

Figure 1. Temporal changes in atmospheric concentration of CO_2 (IPCC, 1995).

Seasonal variations in CO_2 concentration (Fig. 2) show that it reaches the maximum in spring and the minimum in autumn. A gradient in CO_2 concentration exists between the northern and the southern hemisphere, to the extent of about 3 to 5 ppmv. This gradient may be due to the existence of a major CO_2 sink (2-3 Pg C/yr) in the northern hemisphere (Enquete Commission, 1992; Tans et al., 1990; Keeling et al., 1989).

Methane (CH_4)

CH_4 concentration in 1996 was about 1.74 ppmv, which is about twice

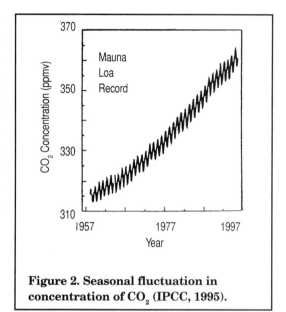

Figure 2. Seasonal fluctuation in concentration of CO_2 (IPCC, 1995).

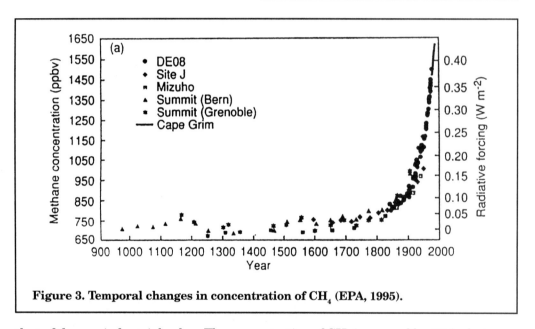

Figure 3. Temporal changes in concentration of CH_4 (EPA, 1995).

that of the pre-industrial value. The concentration of CH_4 increased by 1%/yr between 1960 and 1980, and the current rate of increase is about 0.8%/yr (Fig. 3).

An interhemispheral gradient exists, similar to that for CO_2, with about 0.08 ppmv higher concentration in the northern (1.78 ppmv) compared with the southern (1.7 ppmv) hemisphere. A pronounced annual variation also occurs in CH_4 concentration, with a maximum in spring and minimum in autumn. Seasonal fluctuations in sources and sinks cause these variations in concentration, and the magnitude of sink is related to the tropospheric concentration of OH radical. The CH_4

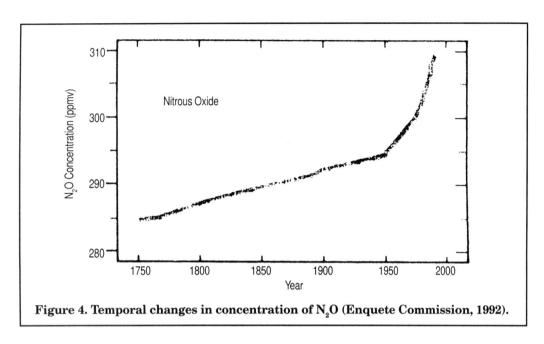

Figure 4. Temporal changes in concentration of N$_2$O (Enquete Commission, 1992).

production by ruminants ranges from 30 to 45 l/kg of dry matter consumed (Shibata, 1994). The CH$_4$ emission rates from rice paddies range from 0.01 to 1.4 g/m^2/day (Sass, 1994).

In addition to its direct effect on radiative forcing, CH$_4$ also has indirect GWP because of its role in photochemical ozone formation, which in turn affects the concentration of other radiatively active gases. Oxidation of CH$_4$ in the stratosphere leads to the formation of H$_2$O vapor at these altitudes, which also has a GWP. The indirect GWP of CH$_4$ is about 2/3 of its direct effect.

Nitrous oxide (N$_2$O)

N$_2$O concentration is 311 ppbv and is increasing at 0.8 ppbv/year (Fig. 4). Sources and sinks of N$_2$O are not very well understood, but the increase is due principally to production and evolution of natural and fertilizer-derived N$_2$O from the soil surface, biomass burning, and biotic processes in forest soils. Several estimates show that 0.17 to 3.52 kg of N$_2$O can be emitted to the atmosphere per 100 kg of N in the fertilizer applied (CAST, 1992). Total use of nitrogenous fertilizer is about 80 MMT/yr in the world and 11 MMT/yr in the U.S. (FAO, 1992).

Anthropogenic Sources of Greenhouse Gases

Three principal components comprise the anthropogenic global warming potential (radiative forcing): fossil fuel combustion and transport, the chemical industry, and agricultural and land use changes, including deforestation.

Fossil fuel combustion and transport

Fossil fuel combustion and transport is the major source of radiative forcing. It accounts for about 50% of the total GWP, of which 40% is due to CO_2 and 10% to CH_4 and O_3.

Chemical industry

Chemical products (CFCs, halogens, etc.) accounted for 20% of GWP in the 1980s. These are entirely synthetic compounds and did not exist before the 1930s.

The concentration of CFCs has decreased steadily during the 1990s, due to strict regulatory measures.

Agricultural and land use changes, including deforestation

The agriculture sector worldwide accounts for about one fifth of the annual anthropogenic increase in greenhouse forcing, producing about 50 and 75% of anthropogenic methane and nitrous oxide emissions and about 5% of anthropogenic carbon dioxide emissions (Cole et al., 1996). Deforestation, biomass burning, and other land use changes account for an additional 14% (Fig. 5).

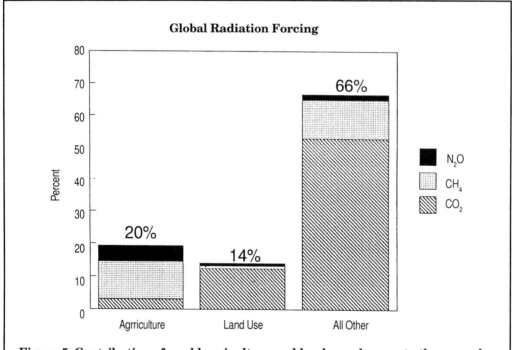

Figure 5. Contribution of world agriculture and land use changes to the annual increase in radiative forcing (Rosenberg, Cole, and Paustian, in press).

The total area of tropical forests is about 1.9 billion ha (19 M km^2), and the annual deforestation rate is 0.9% or 17 million ha (M ha). Tropical deforestation is occurring mostly in the Amazon Basin, Central America, the Congo Basin, and Sumatra (Photo 1). The resulting GHG emissions are due to biomass burning, biomass decomposition, and mineralization of soil organic carbon (SOC) content. The exact magnitude of emissions due to SOC mineralization is not known.

Agricultural activities affect radiative forcing both directly and indirectly. Direct effects result from emissions of CO_2, CH_4, and N_2O due to deforestation, biomass burning (Photo 2), ruminant animals (Photo 3), SOC decomposition from plowing, rice paddy cultivation (Photo 4), fertilizer application (Photo 5), use of manure (Photo 6), and degrading of wetlands. Plowing or soil turnover (Photos 7a, b) is a principal cause of CO_2 emission from cropland.

Within the U.S., the direct effects of deforestation on GHG emissions are minor to none. Indirect radiative forcing due to agricultural activities include emissions of N_2O, NO_x, and NH_3 by cattle feedlots and microbial activities in soils and water following applications of fertilizers and manures.

Excessive use of chemical fertilizers can exacerbate N_2O emissions. Further NO_x and NH_3 emissions lead to soil acidification and eutrophication of natural waters. Ecosystem or land degradation (e.g., decline in quality of soil, water, and vegetation) decreases biomass productivity and reduces the sink capacity for CO_2 assimilation.

Photo 1. Tropical deforestation is a principal source of greenhouse gas emissions.

Photo 2. Biomass burning leads to emissions of CO_2, CH_4, CO, and N_2O.

Photo 3. Ruminant animals are a major source of CH_4 emissions.

Photo 4. Cultivation of rice paddies leads to CH$_4$ emissions.

Photo 5. Application of nitrogenous fertilizers leads to N$_2$O emissions.

The Potential of U.S. Cropland to Sequester Carbon and Mitigate the Greenhouse Effect

Photo 6. Use of manures and other organic wastes on farmland leads to emissions of N_2O.

Photo 7a. Plowing causes oxidation of SOC and emission of CO_2.

Photo 7b. Plowing causes oxidation of SOC and emission of CO_2.

CHAPTER 3

The Role of Agriculture in U.S. Emissions of Three GHGs

Agriculture is a major industry in the U.S. In 1995, farming contributed about $63.2 billion (0.9%) to the national gross domestic product (GDP) and directly employed 1.6 million people (1.2% of the U.S. labor force). Farming plus agriculturally related industries (input supplying industries, processing, and distribution) contributed about $1 trillion, or 13.5% of GDP, and employed 23 million people (or 17.3% of the U.S. labor force). Agriculture contributes to radiative forcing through emissions of CO_2, CH_4, and N_2O. Estimates of the contributions of agriculture to U.S. emissions are tentative and difficult to obtain, because a considerable part of agricultural emissions are included in industrial and other sources and are from non-point sources. Because of the diffuse nature of the sources, aggregation and scaling up are challenging tasks.

Agricultural activities contribute CO_2 emissions through combustion of fossil fuel, SOC decomposition, and biomass burning. Although energy-intensive, total energy use in the farm sector is small compared to that of other major industries. Because there is little deforestation in the U.S., forests are a net sink for CO_2. Contribution of CH_4 from agricultural activities is primarily from enteric fermentation in ruminant animals, rice cultivation, and biomass burning. Principal sources of N_2O are soils, fertilizers and manures, and biomass burning.

The data in Table 1 show considerable U.S. contributions to global emissions and radiative forcing, with about 20% emissions of CO_2, 8% of CH_4, and 8% of N_2O. Total U.S. emissions of the three gases contribute 20% of the global radiative forcing. The estimate of total U.S. emissions in Table 1, expressed as MMTCE, is 1442 MMTCE, and contributions of CH_4 and N_2O from agricultural sources are 52 and 14 MMTCE, respectively, or 66 MMTCE total (DOE/EIA, 1996). Contributions of 42.9 MMTCE by U.S. agriculture includes 27.9 MMTCE from energy and fertilizer use and 15 MMTCE from CO_2 emission through erosion from cropland.

Another estimate of total U.S. emissions is 1666 MMTCE (USEPA, 1995b), which includes contributions of CH_4 and N_2O of 61 and 19 MMTCE, respectively, or 80 MMTCE total (USEPA, 1995 b). To this can be added the 15 MMTCE of CO_2

Table 1. Emission information about greenhouse gases, global and U. S. levels.

Global Atmospheric Concentrations, Natural and Anthropogenic
Sources, and Absorption of Five Greenhouse Gases

				PP trillion	
	Carbon Dioxide	Methane	Nitrous Oxide	CF11	CF12
Preindustrial atmospheric concentration (ppm)	278	0.700	0.275	0	0
1992 atmospheric concentration (ppm)	356	1.714	0.311	268	503
Average annual change (ppm)	1.6	0.008	0.0008	0	7
Average change per year (%)	0.4	0.6	0.25	0	1.4
Atmospheric lifetime (years)	50-200	12	120	50	102
Emissions from natural sources (MMT of gas)	550,000	110-210	6-12	0	0
Emissions from anthropenic sources (MMT of gas)	26,030	300-450	4-8	ND	ND
Absorption (MMT of gas)	564,670	460-660	13-20	ND	ND
Annual increase in the atmosphere (MMT of gas)	11,370-12,830	35-40	3-5	ND	ND

Emissions and Global Warming Potential of Three Greenhouse
Gases from the U.S. and U.S. Farming Activities

				PP trillion	
	Carbon Dioxide	Methane	Nitrous Oxide	CF11	CF12
Emissions from all United States sources (MMT of gas)	5287	31	0.471	<0.1	<0.1
Emissions from United States agriculture (MMT of gas)	157*	9	0.174	—	—
Global warming potential (100 years relative to CO_2)	1	21	310	1320	6650
Global warming potential from:					
Emissions from all United States sources (MMTCE)	1442	178	39	13320	128768
Emissions from United States agriculture (MMTCE)	42.9*	52	14	—	—

ND = data unavailable
MMTCE = million metric tons of carbon equivalent
Data in Table 1 are adapted from EIA (Energy Information Administration). 1996. Emissions of greenhouse gases in the United States.
DOE/EIA-0573(95) U.S. Dept. of Energy. Washington, D.C. 140 p.
*As per calculations made in this report.

emissions from U.S. cropland from accelerated erosion and 27.9 MMTCE from fertilizer use, discussed in the previous paragraph.

The overall contribution of U.S. cropland to radiative forcing resulting from CO_2 emissions from production inputs is rather small (Table 2). The direct input of fertilizers and pesticides is estimated at 12.9 MMTCE and the indirect input of energy at 15.0 MMTCE, or 27.9 MMTC total, and emission of CO_2-C by erosion at 15 MMTC. The total CH_4 and N_2O from agricultural activities and the 27.9 MMTC of production energy input contribute about 7% of the total U.S. emissions, regardless of whether the DOE/EIA (Table 1) or the USEPA estimates of total U.S. emissions are used —1485 (1457) or 1709 (1681) MMTCE, respectively. Gebhart et al. (1994) estimated that U.S. agriculture releases about 38.1 MMTC annually into the atmosphere. This report estimates soil erosion and energy emissions of 42.9 MMTC annually, which includes 15 MMTC/yr from erosion.

Table 2. Production inputs and farm energy uses, directly in operating machinery and equipment on the farm and indirectly in fertilizers and pesticides produced off the farm.

Farm Input	Quads	Joules*E15 kg	C/Joule*E9	MMTC
Fertilizer + Pesticides **	0.83 total *			
Nitrogen **	0.63	664.2	14.10 **	9.4
Phosphate **	0.06	63.24	15.74 **	1.0
Potash **	0.05	52.7	17.10 **	0.9
Pesticides **	0.09	94.86	17.12 **	1.6
			Subtotal	12.9
Gasoline *	0.180	189.72	18.41 ***	3.5
Diesel *	0.460	484.84	18.93 ***	9.2
LP gas *	0.065	68.51	16.36 ***	1.1
Natural Gas	0.085	89.59	13.78 ***	1.2
Totals:	1.62			27.9

* Quads of energy interpolated from: Figure 3.3.1 (Production inputs — Energy) ERS. 1997. Agricultural Resources and Environmental Indicators, 1996-97. U.S. Department of Agriculture, Economic Research Service, Natural Resources and Environmental Division. Agricultural Handbook 712. A quad is a quadrillion BTUs (British Thermal Units).

** Values estimated by R.F. Follett based upon partial energy input data in: Energy Use Survey, CY 1987. The Fertilizer Institute. 501 Second Street, N.E. Washington D.C.

*** Values based upon personal communication with Dr. Greg Marland (Oakridge National Laboratory, TN) and from: Marland, G. and Turhollow, A.F. 1991. CO$_2$ emissions from the production and combustion of fuel ethanol from corn. Energy 16: 1307-1316.

Precisely identifying sources of emissions in relation to agricultural activities is important for developing effective mitigation strategies. In addition to fossil fuel consumption, SOC and biomass decomposition/burning are the principal causes of GHG emissions from agricultural activities. A drastic reduction in the consumption of gasoline and LP gas has occurred since the late 1970s (Fig. 6), but diesel consumption has risen slightly in the 1990s (USDA-ERS, 1997). Conservation tillage (CT) and other energy-efficient farming methods can reduce net fuel consumption.

Fertilizer use is another principal source of GHG emissions. Total fertilizer consumption is about 20 MMT (FAO, 1996; USDA-ERS, 1994), and fertilizer use efficiency can be increased by decreasing losses due to erosion, leaching, and volatilization. Adoption of site-specific management (precision farming) may make fertilizer use more efficient. Agricultural sources of GHG emission, other than SOC and biomass, include ruminant animals and manure application for CH$_4$, and fertilizer use for N$_2$O.

Despite the relatively low contributions of U.S. agriculture to overall radiative forcing, soils and agriculture have a high potential as a C sink and for the production of biofuel energy to replace fossil fuel energy. Because of large historic emissions due to changes in land use, the potential for C sequestration is very large. Judicious soil

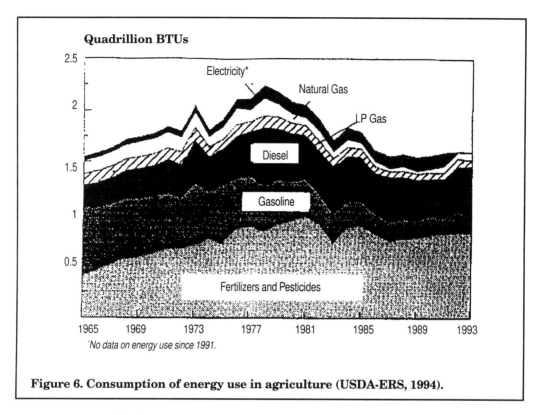

Figure 6. Consumption of energy use in agriculture (USDA-ERS, 1994).

management and adoption of appropriate farming/cropping systems can make soils an important sink for carbon, and soils and cropping systems can be effective tools to mitigate the radiative forcing of other nonagricultural activities.

The SOC Pool in U.S. Soils and SOC Loss from Cultivation

Land use changes in forests, grasslands, and wetlands have transformed large areas from relatively stable ecosystems to agroecosystems under extensive and intensive management. The introduction of agriculture, which involves land clearing, draining, sod breaking, cultivating, replacing perennial vegetation with annual crops, and fertilizing, has had major impacts on C pools and fluxes around the globe. In the initial phases of these transformations, major losses of CO_2 to the atmosphere occurred as soil C levels adjusted to reduced C inputs and increased soil disturbance.

Intense pressure for production also has led to serious soil degradation through erosion and nutrient losses. These trends continue in many areas of the world. In the U.S. and other industrialized countries, however, with available inputs of energy and technology, agricultural productivity has steadily increased, land degradation has slowed or reversed, and soil C pools have stabilized or increased (Cole et al., 1993).

Organic matter in the soils of the semiarid U.S. Great Plains was seriously depleted during intense cultivation by pioneer farmers, and soil productivity decreased 71% during the 28-year interval after sod breaking (Flach et al., 1997). The combination of low yielding varieties, complete straw removal at harvest, and intensive cultivation resulted in serious losses of SOC and available nutrients. After 1940, management practices were modified to include higher yielding varieties, application of N and P fertilizers, no straw removal at harvest, and stubble mulch tillage. These management changes arrested soil C and N losses, and soil C levels stabilized (Cole et al., 1990).

Figure 7, which is based on a combined data analysis and modeling effort (Donigian et al., 1994), shows historical changes in the SOC of agroecosystems of the Central United States, including the Corn Belt and a portion of the Great Plains. This diagram represents the aggregated results of thousands of model runs, including climatic subregions, soil distribution, major cropping systems, and soil management practices of the study area. The striking decline in SOC levels during the period from 1907 to 1940 parallels the losses recorded over the same

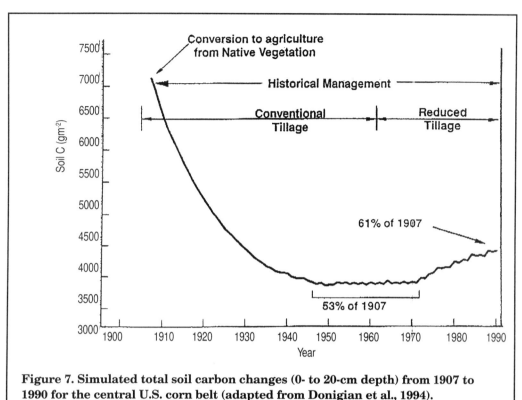

Figure 7. Simulated total soil carbon changes (0- to 20-cm depth) from 1907 to 1990 for the central U.S. corn belt (adapted from Donigian et al., 1994).

period in the semiarid Great Plains. The model runs indicate that SOC levels dropped to 53% of precultivation levels in the 1960s and increased subsequently with the adoption of conservation tillage (CT) practices and the advent of higher yielding varieties which produce more crop residues.

Kern (1994) estimated total SOC stock in soils of the 48 contiguous states at 80.7 ± 18.6 MMTC. The data in the map on the back cover show high SOC concentration in cropland soils of the midwestern region. Uncultivated soils were in equilibrium with native vegetation cover and accumulated large SOC reserves, especially the soils of the northern Great Plains, which reached their level of SOC equilibrium after the last glaciation. Cultivation of these soils disrupted this steady state.

Grassland and forest soils tend to lose from 20 to 50% of the original SOC content in the zone of cultivation within the first 40 to 50 years of cultivation (Campbell and Souster, 1982; Tiessen et al., 1982; Mann, 1985; 1986; Schimel, 1986; Johnson and Kern, 1991; Rasmussen and Parton, 1994; Houghton 1995). Changes in SOC content thereafter become a function of soil management and erosion (Rasmussen and Collins, 1991). Gebhart et al. (1994) estimated that mean SOC content in 3 states (Texas, Kansas, and Nebraska) is 59.2 MMTC/ha in the top 0.3 m of soil for cropland, compared with 90.8 MMT C/ha in native pastures and rangeland. Grigal and Ohmann (1992) estimated the SOC pool in cropland forest soils of the lake states and observed the effect of forest management on the SOC pool. Cultivation and management-induced changes in the SOC pool are mostly due to changes in the size of the slow pool (i.e., intermediate turnover rate; Parton et al., 1987; Camberdella and Elliott, 1992).

The data in Table 3 show that total SOC content of the lower 48 states is 59.4 GT (59,400 MMT). Kern (1992, 1994) estimated that the SOC pool of the major U.S. field cropland for the top 30-cm depth is 6969 MMT, with a range of 5304 to 8654 MMT (Table 4). The arable land area is about 19.4% of the total land area. However, cropland concentrated in the upper Mississippi valley has a high SOC content (back cover). We estimated that SOC pools in cropland to a 1-m depth is 15,600 MMT. From this, C lost to the atmosphere may be 5000 MMT.

Kern (1994) estimated a mean loss of 1329 MMT, with a range of 993 to 1670 MMT (Table 5). Kern's calculations of C pool and emissions may be underestimates. U.S. cropland's potential to sequester C is in the range of 5000 MMT, which may be achievable over a 50-year period, by the year 2050, through adoption of BMPs.

Table 3. Estimated soil organic carbon pool for North America (Waltman and Bliss, 1997).

Land Mass	SOC total soil profile depth (10^3 MMT)	Percent of World SOC (1550 Pg*)	SOC 100 cm soil depth (10^3 MMT)
North America	346.7	22%	266.8
U.S. (lower 48)	59.4	4%	52.0
U.S. (Alaska)	13.5	1%	13.5
Canada	262.3	17%	190.0
Mexico	11.5	1%	11.3

*Post et al., 1982

Table 4. The SOC pool in major field cropland in the contiguous U.S. (Kern, 1993).

Depth interval (cm)	Mean	SOC pool Minimum	Maximum
		——MMT——	
0-8	2270	1710	2821
8-15	1807	1383	2240
15-30	2892	2211	3593
0-30	6969	5304	8654

Table 5. Historic soil carbon losses in the surface 30 cm after cultivation, for major field cropland in the contiguous U.S. (Kern, 1994).

Status	Mean	SOC pool Minimum	Maximum
		—MMT—	
Precultivation SOC	8298	6297	10324
Current SOC	6969	5304	8654
Losses	1329	993	1670

CHAPTER 5

Processes Governing Emissions from the Pedosphere

A temperature increase of 0.1°C/decade or 1°C over the 21st century would have extremely adverse impacts on ecosystems and the global economy. When climate change has exceeded this rate in the past, forest ecosystems in the northern hemisphere have been devastated (Rijsberman and Stewart, 1990). Over the past century, the Earth's surface has warmed by about 0.5°C ($\pm$ 0.2°C). A doubling of atmospheric CO_2 over pre-industrial levels may lead to equilibrium global warming in the rage of 1.5° to 4.5°C (Mahlman, 1997). As the climate warms, the rate of evaporation will increase, leading to an increase in global mean precipitation of about 2% ($\pm$ 0.5%) per 1°C of global warming (Enquete Commission, 1992).

To prevent this unacceptable increase, we must limit the concentration of all GHGs to a maximum of < 550 ppmv for CO_2-equivalent at mid-century (2050), compared to the current level of about 430 ppmv CO_2-equivalent (Flavin and Tunali, 1996). The world community thus must develop long-term goals for stabilizing the concentration of all GHGs. Agriculture can play a major role in achieving the goals of sequestering C and mitigating radiative forcing.

A thorough understanding of the biotic/abiotic interactions between soil, plants, animals, and microbes is necessary if we are to identify options to mitigate radiative forcing by adopting BMPs that enhance C sequestration in soil and terrestrial/aquatic ecosystems and that reduce emissions and losses of C and N from the ecosystem. Plants and animals are the principal products of interest, and increasing their production requires soil inputs of fertilizers, manure and other organic amendments, and irrigation.

Plant Action

The primary producers of energy — plants — synthesize CO_2 and solar energy into carbohydrates through the uptake of essential nutrients and water from the soil. Most of the CO_2 they absorb reenters the atmosphere through respiration of plants and animals and through microbial decomposition. Microbes play a crucial role in the global carbon cycle, and understanding microbial processes may be crucial to identifying and adopting effective mitigation options.

Lal, Kimble, Follett, and Cole

If a fraction of the CO_2 which plants absorb can be immobilized within the terrestrial/aquatic ecosystems, overall CO_2 concentration in the atmosphere will decrease. Soil is the principal medium of all terrestrial biotic/abiotic linkages; soil interacts with the atmosphere, hydrosphere, biosphere, and lithosphere. It is to the soil that we add fertilizers, manure and other organic materials, and supplemental irrigation water. It is also the soil that emits GHGs and releases other elements when the ecological balance of the cycles of H_2O, C, and nutrient elements in the lithosphere are disturbed due to drastic anthropogenic (and sometimes natural) disturbances.

Cycles within an ecosystem are so interdependent that disturbing one may disturb the others drastically. For example, a disturbance in the hydrologic cycle, leading to an increase in runoff, causes accelerated soil erosion, which may lead to losses of C, N, and other elements. Deposition of sediments in depressions and aquatic ecosystems may lead to C sequestration. Increase in soil water storage leads to anaerobiosis and prevalence of reducing conditions. Denitrification and/or volatilization of nitrogen, whether inherent or applied as fertilizers and manures, is a major source of N_2O. Along with several ions, dissolved organic carbon (DOC) also can be released from the system, and little is known about its fate.

Depending on land use and management, soil can function as either a source of, or a sink for, atmospheric C. It can play an important role in sequestering C from the atmosphere and decreasing the buildup of GHGs in the atmosphere. The SOC pool plays a major role, both as a sink for atmospheric CO_2 and as a source of CO_2, CH_4, and N_2O emissions (Fig. 8).

Soil Processes

The "pedosphere" concept is a broad one which encompasses those entities we think of as "soils." It plays a crucial role in the global C cycle. It comprises the thin layer of soil or the uppermost layer of earth's crust that supports all terrestrial life and regulates environmental quality. The pedosphere is the seat of action of all interactive processes that link the atmosphere with the biosphere, hydrosphere, and geosphere.

Several important pedospheric processes can lead to C sequestration in soil and within an ecosystem. These processes include humification, aggregation, translocation within the pedosphere, deep rooting, and calcification (Appendix 1). The SOC content at any point is the balance between processes that determine emissions and sequestration. Emission of C from the soil to the atmosphere is accentuated by erosion, leaching, methanogenesis, volatilization, and mineralization (Appendix 1). In contrast, C sequestration in the soil is enhanced by humification, aggregation, calcification, and stratification or deep sequestration.

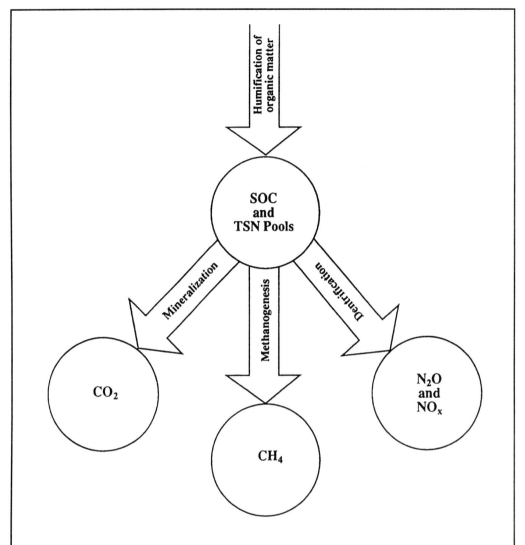

Figure 8. The SOC and total soil nitrogen (pools) serve as sinks for atmospheric CO_2 through humification. However, SOC is also a source of CO_2, CH_4, and N_2O emissions through several biotic/abiotic processes occurring in soil.

Soil Quality

The objective for agriculture therefore must be to adopt land use and management practices that increase C sequestration and reduce GHG emissions. The basic strategy for reaching that objective is to improve and maintain high soil quality. Soil quality is defined as "the capacity of the soil to function within ecosystem boundaries to sustain biological productivity, maintain environmental quality, and promote plant and animal health" (Doran and Parkin, 1994).

Lal, Kimble, Follett, and Cole

A strong relationship exists between both soil quality and C sequestration and soil quality and the greenhouse effect (Bezdicek et al., 1996). Numerous inherent soil properties and processes affect emissions of GHGs and sequestration of C in soil. Indicators of soil quality in relation to C sequestration are SOC and soil inorganic carbon (SIC) contents, aeration porosity, aggregation and mean weight diameter (MWD) of aggregates, available water capacity (AWC), cation exchange capacity (CEC), electrical conductivity (EC), soil bulk density, and soil biodiversity. The relative importance of these indicators varies among soils, and site-specific information is needed for quantitative assessment of soil quality.

A strong relationship exists between soil quality and SOC content, and soil structure and SOC content (Beare et al., 1994; Tisdall, 1996; Angers and Carter, 1996). Angers (1992) observed that SOC content strongly correlated with aggregation and the MWD of aggregates. The increase in SOC content under 5 years of alfalfa, from 26 g/kg soil to 30 g/kg soil, increased MWD from 1.5 mm to 2.3 mm. Hudson (1994) reported a strong positive correlation between SOC content and AWC, and a negative correlation between SOC and soil bulk density. Unger (1995a, b) observed improvements in soil surface conditions with an increase in SOC content. SOC-induced improvements in soil quality increase biomass productivity, which leads to further enrichment of SOC content.

Agriculture is linked closely to global climate change (Adams et al., 1990; Iserman, 1994) and often has been blamed for such modern environmental problems as contamination of surface and ground water, non-point source polution, emission of GHGs into the atmosphere, and loss of biodiversity. Land misuse, soil mismanagement, and adoption of poor agricultural practices obviously lead to accellerated soil erosion, high rates and amounts of runoff, transport of chemicals and pollutants into natural waters, and emissions of GHGs. The prevalance of such poor agricultural practices, especially but not only in developing coutries, has caused widespread soil and environmental degredation, a decline in soil quality, and low productivity. Low return agriculture perpetuates poverty and leads to low or no investment in agriculture and soil resources, over-exploitation of soil resources, mining of soil fertility, and continued decline of soil quality. This trend must be reversed.

Agricultural intensification through the adoption of scientifically proven BMPs can solve, rather than cause, numerous environmental problems, including CO_2 emission. BMPs can improve SOC content, enhance soil quality, restore degraded ecosystems, increase biomass production, improve crop yield, and encourage reinvestment in soil resources for soil restoration.

Improving soil quality by enhancing SOC content would reverse the trend in degradation, increase productivity, sequester C in soil, and help mitigate the greenhouse effect. Thus an important national objective and general agenda should be to use BMPs to improve soil quality, intensify agricultural use of prime lands, restore marginal lands, and restore degraded soils.

CHAPTER **6**

Strategies for Mitigating Emissions from Cropland

The soil is an ultimate storehouse of C, because CO_2 taken away from the atmosphere via photosynthesis can be stored in the soil in either living organisms (biomass carbon) or in their residues, in a form which resists further biological degradation. Soil humus is relatively resistant to biological degradation and has a long turnover time, ranging from several decades to several centuries. Long turnover time is related to a more resistant fraction and placement deeper within the soil profile.

Adopting best management practices (BMPs) can make croplands a net sink for C and decrease emissions of N_2O and CH_4 (CAST, 1992). The sink capacity of croplands for atmospheric CO_2 can be enhanced by offsetting losses of C (through oxidation, methanogenesis, and erosion) with C gains (through fixation and stabilization). Table 6 summarizes several options for mitigating GHGs from cropland (Cole et al., 1996).

The options related to CO_2 emission can be grouped in three categories, (i) land conversion and restoration, (ii) biofuels for offsetting fossil fuels, and (iii) intensification of prime agricultural land. The basic strategy for land conversion and restoration is to convert marginal agricultural land to nonagricultural restorative uses (e.g., grassland, forest, or wetland), restore lands drastically disturbed by different processes (e.g., erosion, salinization, fertility depletion), and restore wetlands and organic soils to their natural vegetation. This strategy allows some marginal land to be used to produce biomass for conversion into biofuel.

These approaches imply intensification of prime agricultural lands through widespread adoption of BMPs. The BMPs are soil and ecoregion specific and involve using conservation tillage (CT) and residue management, increasing fertilizer and water use efficiencies by decreasing losses, increasing cropping intensity by eliminating summer fallowing, and frequently using cover crops. The effective use of organic manures and other by-products is important to increasing SOC content, improving soil quality, enhancing biomass productivity, and reducing CO_2 emissions.

U.S. Cropland

Cropland includes land used to produce crops, e.g., corn, soybeans, wheat, barley, and millets. Because of the nature of their use, croplands have been influenced drastically by anthropogenic perturbations, including deforestation, biomass burning, tillage, fertilization, and vehicular and animal traffic. Landscape units constituting cropland thus have characteristics that differ from those of their undisturbed counterparts within the same landscape and ecoregions. The principal differences lie in the total carbon pool of the root zone, hydrologic and energy balance, slope characteristics, and stock and dynamics of plant nutrients.

The total arable land area in the U.S. is estimated at 185.7 Mha or 19.4% of the total land area. U.S. cropland area has been relatively constant during the 20th century. The cropland area was 134 Mha (330 M acres) in 1910, 152 Mha (375 M acres) during World Wars I and II, 154 Mha (380 M acres) during the export boom of the 1970s, and 134 Mha (330 M acres) since the mid-1980s. The decrease in cropland area since the 1980s is partly due to diversions from production by federal farm programs (see section on CRP).

U.S. cropland is of five distinct types. (1) Harvested cropland consists of land used continuously for growing crops and from which crops are harvested; it constitutes 63% (84.4 Mha) of the total cropland. (2) Cropland which was sown to crops but on which crops failed and were not harvested constitutes about 1% (1.3 Mha) of the total cropland. (3) Cropland maintained as summer fallow constitutes about

Table 6. Strategies for C sequestration in U.S. cropland soils.

Mitigation option	Relative potential
1. Soil erosion management	H
2. Land conversion and restoration	
a. Convert marginal agricultural land to grassland, forest or wetland	H
b. Restore degraded lands	H
c. Restrict use of organic soils	H
d. Restore wetlands	H
e. Reclaim mineland and toxic soils	L
f. Restore salt-affected soils	L
3. Intensification of prime agricultural land	
a. Erosion control (conservation tillage, buffers, CRP)	H
b. Supplemental irrigation	M
c. Soil fertility management	M
d. Improved cropping systems and winter cover	H
e. Elimination of summer fallowing	H
4. Biofuels	
a. Energy crops for fossil fuel	H
b. Biogas from liquid manures	H
c. Efficient grain drying	L
d. Soil C sequestration in lands for biofuel	L
5. Improving fertilizer use efficiency	
a. Crop rotations and tillage method	H
b. Nitrification inhibitors	M
c. Mixed farming	L
d. Recycling organic material	M
e. Biological nitrogen fixation	L
6. Management of rice paddies	
a. Residue management and tillage methods	M
b. Water management	M
c. Fertility management	M

Adapted from IPCC (1995)
H, M, and L refer to high, medium, and low potential, respectively.

7% (9.4 Mha) of the total cropland. (4) Idle cropland, which includes land in cover and soil improvement crops or left completely idle because of physical or economic reasons, constitutes about 15% (20.1 Mha) of the total cropland. (5) Cropland used for pastures grown in rotation with crops comprises 14% (18.8 Mha) of the total cropland (Heimlich and Daugherty, 1991). Thus some U.S. cropland could form a base for biofuel production (note: the idle land alone comprises about 20 Mha).

Sustainable Management Strategies

Just as the conversion of undisturbed land to annual cropping can disturb established equilibria and decrease SOC content, adoption of ecologically compatible practices (e.g., use of fertilizer and manures, CT, and forages and legumes in rotation) can shift the equilibrium SOC content upward by eliminating or minimizing the degrading processes. Soil erosion control is an important consideration for maintaining a favorable ecological balance. Principal strategies thus are to adopt land use and management systems which suppress mechanisms which exacerbate GHGs emission, and to enhance mechanisms that sequester C in soil (Fig. 9). It is important to adopt an ecologically sound and holistic approach to natural resources management (Peterson et al., 1993; Sindelar et al., 1995.

Strategies to prevent emission include: (i) soil erosion management, (ii) supplemental irrigation for drought management, (iii) reclamation of mineland and toxic soils, (iv) reclamation of salt-affected soils, (v) soil fertility restoration, and (vi) wetland restoration. Strategies to enhance mechanisms that sequester C include: (i) conservation tillage, (ii) management of crop residues and other biomass, (iii) the Conservation Reserve Program, (iv) conservation buffers, (v) improved systems of water management on drained cropland, (vi) improved cropping systems with winter cover crops, and (vii) restoration of eroded soils. Implementing these strategies requires identifying appropriate techniques of soil, water, and natural resources management and then identifying policies that can facilitate their adoption. High priority should apply to those processes/mechanisms that have a high potential either to reduce CO_2 emissions and/or to increase C sequestration (Cole, 1995).

The concept of agricultural sustainability needs broadening to include increasing agronomic productivity, improving resource conservation, and enhancing environmental quality. Such broadening makes C sequestration an important aspect of agricultural sustainability. Sustainable land use must be assessed in terms of its impact on the SOC pool. A nonnegative trend in the SOC pool would imply a sustainable land use/soil management system. All other factors remaining the same, a sustainable system would enhance SOC content and increase carbon sequestration within the solum.

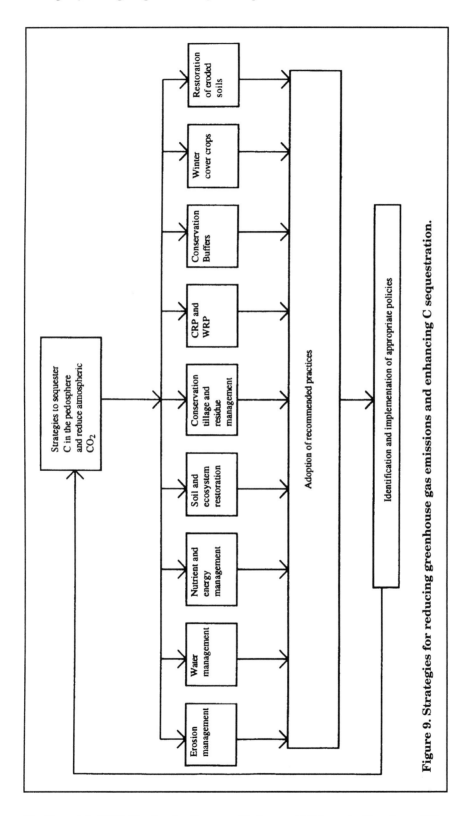

Figure 9. Strategies for reducing greenhouse gas emissions and enhancing C sequestration.

Agricultural sustainability and soil quality are interlinked intimately. Sustainable agricultural systems maintain and enhance soil quality, and vice versa [agricultural sustainability = f(soil quality), and soil quality =f(soil content, aggregation, and aggregate size distribution)]. An appropriate criterion for assessing the sustainability of an agricultural system thus must be based on SOC dynamics. Lal and Kimble (1997) proposed an index of soil quality based on SOC dynamics, mean diameter of stable aggregates, and percent aggregation. In this approach, a sustainable agricultural system would maintain a non-negative trend in soil quality over a long period of time, by maintaining or enhancing SOC content and soil aggregation. Managing SOC content to enhance soil quality is crucial to attaining sustainable agriculture.

While misuse and mismanagement of agricultural ecosystems can create severe environmental problems, adopting BMPs can provide a partial solution to environmental issues. Scientifically based agricultural practices have a strong potential to mitigate the greenhouse effect by sequestering C in soils and in biomass. A vast array of agricultural practices can facilitate C sequestration. These include soil management, water management, nutrient and crop management, and biofuel options—and the most appropriate options for natural resource management should be chosen based on their effects on agronomic productivity, environmental quality, and profitability.

CHAPTER **7**

Soil Erosion Management

A large proportion of SOC content is concentrated near the soil surface and therefore is highly vulnerable to the oxidative processes associated with soil erosion. About 400 million cubic meters of sediment are dredged each year in the maintenance and establishment of waterways and harbors (Sopper, 1993). Erosion may be severe on some of the 10 Mha of public roads and highways in the U.S., also, as well as on agricultural land.

Loss of topsoil SOC is a major on-site impact of accelerated soil erosion, which also increases GHG emissions by: (i) exposing carbon locked within aggregates through their disruption by the erosive forces of raindrops, runoff, and wind, (ii) mineralizing carbon thus exposed through microbial decomposition and other oxidative processes, and (iii) decreasing the soil's capacity to produce biomass by depleting soil fertility, losing water as runoff and through decreasing soil available water capacity (AWC), burying or inundating crops, and causing other numerous direct and indirect effects of accelerated soil erosion.

While soil erosion may accentuate carbon emissions, sedimentation and downslope deposition may lead to deep burial of carbon or its translocation into lakes and oceans, where it may be sequestered over geological periods. On the whole, however, accelerated soil erosion exacerbates carbon emissions (Clark et al., 1985; Lal, 1995a; Paul et al., 1997a; Flach et al., 1997). It is generally believed that 20% of the carbon dislocated by erosion may be released eventually into the atmosphere (Lal, 1995a).

Accelerated soil erosion is a severe problem on a large proportion of U.S. cropland (Table 7). Soil erosion by water is severe in Illinois, Indiana, Iowa, Kansas, Kentucky, Texas, and other states where row crops are grown. Wind erosion is also severe in arid and semiarid regions including Arizona, Colorado, New Mexico, Texas, and Wyoming. The erosion hazard is caused by a combination of erosive climate and highly erodible land.

Because of inherent soil characteristics, highly erodible land also exists in several states, including Texas, New York, Nebraska, Montana, Colorado, and Alabama. Phillips et al. (1993) estimated that projected increases in global temperature may accentuate the soil erosion hazard in the U.S. With all other factors held constant, change in the erosivity factor due to climate change may lead to changes in sheet and rill erosion by +2 to +16% in croplands, -2 to +10% in pasture land, and -5 to +22% in rangeland.

Cropland erosion significantly decreased for the 10-year period from 1982 to 1992. The sheet and rill erosion decreased by 30%, wind erosion by 35%, and total erosion by 32.1% (USDA NRCS, 1995). Lee (1990) estimated that the average soil erosion rate decreased from 9.6

Table 7. Soil erosion by wind and water (USDA-SCS, 1992).

Erosion Type	Land use	Erosion amount (MMT/yr)
I. Water erosion	Cropland cultivated	1038
	Cropland uncultivated	46
	Forest land	0
	Pasture land	114
	Range land	438
II. Wind erosion	Cropland cultivated	833
	Cropland uncultivated	12
	Forest land	0
	Pasture land	9
	Range land	1590

MT/ha/yr (4.3 tons/acre/yr) in 1982 to 8.5 MT/ha/yr (3.8 tons/acre/yr) in 1987, an overall decrease of 11%. Brown (1991) reported that total soil erosion (sheet and rill erosion) from cultivated cropland was 1926 MMT in 1982, 1620 MMT in 1985, 1070 MMT in 1990, and 670 MMT in 1995. This decrease in erosion is mostly the result of the increase in cropland area under conservation tillage (CT), the Conservation Reserve Program (CRP), and afforestation. Yet, some 6 Mha of highly erodible land exists in the contiguous U.S. (Gomez, 1995). The entrained material contains higher SOC content than the field soil because soil erosion leads to enrichment of C in sediments (Zobeck and Fryrear, 1986; Fryrear et al., 1994).

The C emission reduction potential through soil erosion management can be computed in three different ways. *The first approach* is based on estimating the C entrained in the displaced sediments. This approach involves estimating the amount of total sediments displaced. The total suspended sediment transport in U.S. rivers was about 400 MMT/yr in the 1980s (Meade and Parker, 1984). The load in 12 major U.S. rivers in 1991 was estimated at 336 MMT/yr for suspended load and 113.5 MMT/yr for dissolved load (Leeden et al., 1991). Assuming that 75% of the suspended load (mostly due to erosion) is contributed by cropland, sediment transport attributed to cropland is about 250 MMT/yr. Assuming a delivery ratio of 10% and SOC content of sediment of 3% (Lal, 1995 a), total SOC displaced by soil erosion from cropland is 75 MMT/yr (Table 8).

Of the SOC displaced, an estimated 15 MMT is emitted into the atmosphere due to rapid microbial decomposition/volatilization of the carbon exposed by the breakdown of aggregates (Eq. 1), 7.5 MMT is transported to the ocean and other aquatic ecosystems, and 52.5 MMT is relocated over the landscape, including by burial in some depositional sites (Fig. 10). The 15 MMTC emission is attributed to cropping/cultivation activities and should be added to the numbers in Table 1.

Table 8. Soil erosion and C dynamics on U.S. cropland.

Processes	Statistics	Total SOC (MMTC/yr)
1. Sediment transport by major rivers	336 MMT/yr	
a. Estimated contribution by cropland	75% or 250 MMT	
b. Sediment delivery ratio	10%	
c. SOC content in sediment	3%	
d. Total SOC displaced by erosion		75
2. Decomposition		
a. SOC displaced — biodegraded, mineralized, and released as CO_2	20%	15
b. C relocated over the landscape	70%	52.5
3. Sediment transport to the ocean from cropland		
a. Total sediments	250 MMT/yr	
b. SOC transported to the ocean		7.5
4. Total runoff		
a. Runoff volume	50×10^9 m³	0.3
b. Organic C in runoff	(6 mg/L)	

Sediment transport data from Meade and Parker (1984); Leeden et al. (1991).
For SOC content, see Lal (1995a).

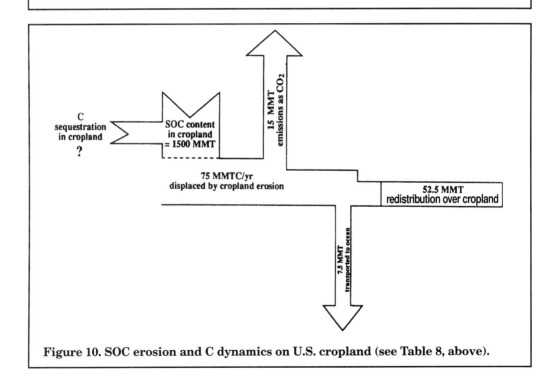

Figure 10. SOC erosion and C dynamics on U.S. cropland (see Table 8, above).

This is also the amount of C emission that can be mitigated through adoption of erosion control practices.

An estimate of the average amount of C emitted by sediments displaced by erosion =

$$\frac{250 \text{ MMT}}{\text{yr}} \times \left(\frac{100}{10}\right) \times \left(\frac{3}{100}\right) \times \frac{20}{100} = 15 \text{ MMT/yr} \text{ .... Eq.1}$$

The second approach is to estimate the C contained in the sheet and rill erosion from cropland. Total sediments displaced by wind and water erosion from cropland are 2000 MMT, which contain 60 MMT of SOC. If 20% of the C displaced is emitted into the atmosphere, it would lead to total emission to the atmosphere of 12 MMTC/yr.

The third approach is to estimate the C and N moved or translocated by eroded sediments in different ecoregions. Based upon Follett et al. (1987), we assumed a C:N ratio of 12:1, and the total amount of SOC displaced by erosion would be 115 MMTC/yr (Table 9). Assuming that 20% of this C is released eventually to the atmosphere as CO_2 (Lal, 1995 a), then the C emission resulting from soil erosion would be 0.2 x

Table 9. Total N and C displaced in eroded sediments (recalculated from Follett et al., 1987).

Region	Total N	Total C
	——MMT/yr——	
Northeastern	0.3	3.6
Lake	0.6	7.2
Corn Belt	4.4	52.8
Northern Plains	2.1	25.2
Appalachian	0.7	8.4
Southeastern	0.2	2.4
Delta	0.5	6.0
Southern Plains	0.5	6.0
Mountain	0.2	2.4
Pacific	0.1	1.2
Total	9.6	115.2

Assuming the C:N ratio of 12:1

113.9 = 22.8 MMTC/yr and the amount of C emitted through erosional processes may range from 12 to 22.8 MMTC/yr. Thus, erosion control can lead to mitigation of CO_2 emission at the rate of 12-23 MMTC/yr. While this is the estimated amount of net SOC lost, the gross C displaced by soil erosion over the landscape may be much larger, depending on the delivery ratio (Lal, 1995a) and the SOC content of sediments.

Further, these calculations do not account for the SIC contained in eroded sediments and they also underestimate the C emission from sediments transported by wind erosion. Therefore, the C emission reduction of soil erosion management at 12 to 23 MMTC/yr is a highly conservative estimate. The C thus emitted by erosion from cropland is added to the total emission from U.S. agriculture. Reduction of CO_2-C emission by soil erosion control is not considered as C sequestration in this report.

CHAPTER 8

Land Conversion and Restoration

Land conversion and restoration converts marginal agricultural land into ecologically compatible land use systems. Taking fragile lands out of agricultural production curtails soil degradation, sets soil restorative processes in motion, and leads to C sequestration. Several land conversion and restoration programs are important to these processes.

Conversion of Marginal Land

The Conservation Reserve Program (CRP)

The Conservation Reserve Program (CRP), as a provision of the 1985 and 1990 Farm Bills, was intended to convert highly erodible land from active crop production to permanent vegetative cover for a 10-year period. The 1996 Farm Bill made major changes in the CRP — for example, it makes highly erodible land, which best management practices (BMPs) cannot protect, targets for temporary land retirement. One of its objectives is to reduce transport of sediment and pollutants into natural waters (Dukes, 1996).

Lands eligible for inclusion within CRP were those with an annual erosion rate of 42.8 MT/ha (19.1 tons/acre), which is about 3 times the national average rate of soil erosion. Using this criteria, about 24% of all U.S. cropland is eligible for inclusion in CRP. Considering all other socioeconomic and political factors, however, eligible cropland that can be included in CRP is about 28.3 Mha (70 M acres).

Farmers are encouraged to participate in CRP; incentives include the government's sharing the cost of converting the cropland to alternate uses (e.g., tree plantations, growing cover crops) and making payments to the farmers throughout the 10-year CRP contracts. About 14.75 Mha (36.4 M acres) of land currently are enrolled in the CRP (Table 10), of which about 1 Mha are planted in trees; the remainder is in grasses. Table 11 shows the regional distribution of old and new

CRP and indicates large enrollment in Northern and Southern Plains states, Mountain states, and the Corn Belt.

Implementing CRP, like adopting a conservation tillage (CT) or residue management system, can lead to C sequestration in soil through erosion control, incorporation of biomass in the soil, etc. Osborn (1993) reported that the average erosion reduction from adoption of CRP on highly erodible land is 42.6 MT/ha/yr (19 tons/acre/yr), giving a total erosion reduction of 636 MMT/yr (700 M tons/yr) on 14.8 Mha (36.5 M acres) of enrollment. CRP also provides a great opportunity for U.S. agriculture to mitigate radiative forcing by sequestering atmospheric CO_2 as C in soil and trees (CAST, 1992). Follett (1993) estimated that about 13.8 Mha of CRP could sequester between 3 and 10 MMTC as SOC over a 10-year period.

Gebhart et al. (1994) assessed the impact of CRP on SOC content in soils of Texas, Kansas, and Nebraska. The mean SOC content of the 0- to 300-cm depth was 59.2 MTC/ha for cropland, 65.1 MTC/ha for CRP, and 90.8 MTC/ha for native pasture. The C sequestration rate of CRP was 1.1 MTC/ha/yr. The principal cause of the increase in SOC may be erosion control (Osborn, 1993; Davie and Lant, 1994) and

Table 10. Estimated eligible cropland under proposed CRP criteria.

Region	CRP areas contracted under signups 1-12	
	Total	Minimum eligible
	—Mha—	
Northeast	0.08	0.08
Appalachian	0.49	0.45
Southeast	0.69	0.45
Delta States	0.49	0.45
Corn Belt	2.27	2.06
Lake States	1.21	0.77
Northern Plains	3.93	2.83
Southern Plains	2.15	1.74
Mountain	2.71	2.47
Pacific	0.73	0.45
Total	14.75	11.75

USDA-ERS (1996)

Table 11. Regional distribution of old and new CRP.

Region	CRP enrollment	
	Old CRP	New CRP
	—%—	
Northeast	1	1
Appalachian	3	2
Southeast	5	4
Delta States	3	4
Corn Belt	15	14
Lake States	8	6
Northern Plains	27	31
Southern Plains	15	14
Mountain	8	21
Pacific	5	3

Note: Numbers may not add upto 100%, due to rounding.

subsequent increase in total soil nitrogen (Staben et al., 1996). Paustian et al. (1995) showed that CRP would sequester about 25 MMTC over a 10-year period and observed that the rate of SOC accumulation under CRP ranged from < 10 to more than 40 g $C/m^2/yr$, with the highest rate in more humid regions.

Follett and Kimble (unpublished data) recently have made experimental measurements of average rates of SOC accumulation (5 year average) — 60 g C/m^2/yr in the 0 to 5-cm soil depth (P $\geq$ 0.05) and 80 g C/m^2/yr in the 0- to 10-cm soil depth (P $\geq$ 0.10) — across 10 sites in 8 states in the Great Plains and western corn belt. They also observed that average amounts of SOC in cropped soils (0- to 10-cm depth) was about 60% of that observed for paired native grassland soils. After 5 years in CRP, the average amount of SOC in the 0- to 10-cm depth had increased to 70% of that observed in paired native grassland sites.

Based on experiments by Lee (1993a, b), USEPA conducted a study (1995a) to predict the impact of CRP on C sequestration in soil and observed that planting hardwood trees can sequester significant additional C. Lee (1993a, b) observed that maintaining the full 16.2 Mha (40 M acres) or CRP-40 could result in an additional C sequestration of almost 215 MMT in 2015 and 350 MMT by 2035. This accumulation is relative to the base case, full implementation of the current CRP plans (assuming that land begins to revert to agricultural production after 1995). Total C sequestration in the base case is about 1050 MMTC in 1995, increasing to 1160 MMTC in 2015 and 1360 MMTC in 2035 (Lee, 1993a, b).

If CRP land is planted with trees, the C sequestration potential can be enhanced greatly. Lee (1993a, b) predicted that C sequestration by implementing CRP then would be as follows:

Year	C sequestration (MMTC/yr)
2000	2.2
2010	4.4
2020	7.0
2030	4.6

Assuming a potential C sequestration rate of 30 to 70 g C/m^2/yr with an average rate of C accumulation of 50 g C/m^2/yr and 100% enrollment in 16.2Mha, the full potential for C sequestration of CRP is about 8.1 MMTC/yr (Eq. 2).

Therefore the C sequestration potential of CRP =

$$(16.2 \text{ Mha}) \times \left(\frac{10^4 \text{m}^2}{\text{ha}}\right) \left(50 \times 10^{-6} \text{ MT } \frac{C}{\text{m}^2\text{yr}}\right) = 8.1 \text{ MMTC/yr} \quad\text{....... Eq. 2}$$

Conservation Buffers

Conservation buffers are vegetative filter strips used in conjunction with recommended practices. Filter strips and riparian wetlands also can control flooding and minimize water pollution (Lant et al., 1995a, b).

Vegetated strips, ranging from 5 to 50 m wide, usually are installed along streams and on agricultural lands to minimize soil erosion and risks of transport of non-point source pollutants into streams. Such strips are called by different names, e.g., Filter Strips (Photo 8) to absorb run-on water and sediments from an upstream land, Riparian Buffers (Photo 9) established along a stream to absorb sediments and chemicals before the runoff enters a stream, and Grass Waterways (Photo 10) specifically designed for safe disposal of excess water from agricultural land.

Depending on the characteristics of the root and above-ground biomass, these conservation buffers can be extremely effective in absorbing water, sediments, and chemicals transported from agricultural lands. Filter strips are extremely effective in retaining surface-applied swine manure and other constituents (Chaubey et al., 1994), because they decrease the runoff rate and encourage sedimentation.

Aase and Pikul (1995) assessed the effectiveness of a double row of tall wheat grass established with 15 m between the rows. They observed a gradual increase in SOC content from upslope to downslope from 11 g/kg to 12 g/kg for 0 to 5-cm depth, 8 g/kg to 10 g/kg for 15 to 20-cm depth, and 6 g/kg to 6.5 g/kg for 25 to 30-cm depth. Robinson et al. (1996a) observed that the initial 3 m of the vegetative filter strip removed more than 70%, while 9 m removed about 85% of the sediments from runoff.

Photo 8. Filter strips are established to absorb runon water and sediments from upstream land.

Photo 9. Riparian buffers established along a stream.

Photo 10. Grass waterways facilitate safe disposal of excess runoff and trap sediments.

The USDA has a vountary program to develop 3.2 M km (2 M miles) of Conservation Buffers by the year 2002 (personal communication, Max Schnepf, 1997; CSU, 1997). Conservation Buffers have different widths, ranging from 6 to 30 m (20 to 100 ft) for Filter Strips, 15 to 45 m (50 to 150 ft) for Riparian Buffers, and 24 to 30 m (80 to 100 ft) for Grass Waterways. An average width of all types of Conservation Buffers is 10 m. Therefore, the total land area to be covered by Conservation Buffers by the year 2020 is (3.2 x 10^9 x 10 m) 3.2 x 10^{10} m^2 or 3.2 M ha.

It is assumed that the average rate of C accumulation in Conservation Buffers is similar to that of the land under CRP, with a range of 30 to 70 g c/m^2/yr and an average rate of 50 g C/m^2/yr. The total C sequestration potential of Conservation Buffers is 1.6 MMTC/yr (Eq. 3). This estimate is probably an underestimate, since riparian areas tend to be very productive and certainly more productive than the average CRP land, which tends to be on a slope and highly erodible.

Therefore the C sequestration potential of Conservation Buffers =

$$3.2 \times 10^{10} \ m^2 \times 50g \ \frac{C}{m^2 \, yr} = 1.6 \ \text{MMTC/yr} \ \dots\dots\dots\dots\dots\dots \text{Eq. 3}$$

Restoring Wetlands and Restricting the Use of Organic Soils

Wetlands are an important component of the overall environment. Approximately 15% of the world's wetlands occur in the U.S. (40 M ha), and nearly 30% of the nation's total wetlands are coastal (Rabenhorst, 1995). Depending on their origin, wetlands may be palustrine, lacustrine, riverine, estuarine, or marine. Wetlands include the swamps, bogs, marshes, mires, fens, and other wet ecosystems that cover about 6% of the land surface of the world.

Peat accumulation is a result of reduced oxidation of the biomass produced in wetlands. The principal processes involved in wetlands are fermentation, methanogenesis, and S reduction. Because of these processes, wetlands have an important effect on ecosystem health and the global C cycle. Their ability to accumulate peat lets them affect global CO_2 levels. They also can filter pollutants (agricultural chemicals) and store sediments transported from agricultural land through erosional processes. Because of their role as environmental filters, wetlands are designated as "kidneys" of the land. Therefore, wetland management and wetland restoration are important strategies for mitigating the GWP and improving the overall environment. Drainage and cultivation of wetlands causes the release of a considerable amount of C worldwide.

Most marsh or wetland soils are classified as Histosols or Entisols. Histosols predominantly are formed from organic soil materials. If organic materials comprise as much as 40 cm of the upper 80 cm of the soil, then the soil is classified as Histosol. Wetland ecosystems comprise about 6% of the world's land area, but they account for a disproportionately high percentage (14.5%) of the SOC pool, which in these soils may range from 20 to 200 kg/m² in the upper 1 meter (Rabenhorst, 1995). In comparison, most inorganic soils contain < 10 kg C m² in the upper 1 meter. A large proportion of organic soils occur in the cold regions, e.g., in Arctic and Tundra climates (Oechel and Vourlitis, 1995).

Histosols can be drained and cultivated, which leads to higher rates of decomposition. The degree of decomposition is closely related to bulk density and to the initial subsidence that takes place within a very few years (perhaps only 2-3 years) after drainage. When drained and cultivated, the soil warms, the rate of decomposition increases, and the SOC is lost rapidly as a result of oxidation (Photo 11). Armento and Menges (1986) estimated that, before human intervention, net global retention of C in wetland peats was 57-83 MMTC/yr. Gorham (1991) estimated global accumulation of 76 MMTC/yr in northern peatlands. However, drainage of wetlands over the last 100 years may have shifted the balance from a large sink to a net source of atmospheric CO_2. Currently, global wetlands are estimated to be a sink of about 80 MMTC/yr and a source of about 55 MMTC/yr. About one-fourth of C sequestered by wetlands is re-released into the atmosphere by CH_4 emissions.

Wetland management has a potential for C sequestration (Mitsch and Wu, 1995). Natural wetlands have a potential to accumulate peat at the rate of 43 g C/m²/yr (Mitsch and Gosselink, 1993). Gorham (1991) estimated a peat accumulation rate of 29 g C/m²/yr. Most data show that the net accumulation of C (peat accumu-

Photo 11. Organic soils decompose with drainage and cultivation. This post installed on a peat soil in Florida in 1924 shows loss of about 1 m of peat in about 70 years.

Table 12. U.S. wetland conversions, 1954-91 (USDA-ERS, 1994).

Wetland converstion to	1954-74	1974-83	1982-87	1987-91
		10^3 ha/yr		
Cropland	243	95	20	12
Urban use	22	6	23	24
Others	14	70	10	8
Total	279	171	53	44

lation minus CH_4 emission) in wetlands is about 25 g C/m^2/yr, with a range of 15 to 35 g C/m^2/yr.

A substantial conversion of natural wetlands to croplands and urban and other uses in the U.S. has occurred since the 1950s (Table 12)—a total of about 237,000 ha (585,390 acres) (Table 13). While wetlands converted to agricultural land use (crops, pastures, and managed forests) can be restored through incentives and subsidies, those converted to urban and other uses cannot. The Wetland Reserves Program (WRP) was established as a part of the 1990 Farm Bill to restore and protect wetlands that have been converted partly or fully to agricultural use (Robinson, 1993; Reese et al., 1993). The WRP places accepted land areas under 30-year or permanent easements that prohibit draining. The WRP pays compensation, based on the type of easement and the appraised land value, up to the full value of the land. The program goal was to enroll about 0.4 Mha by 1995, but the goal was not realized. The expanded WRP envisages the introduction of 2 Mha of reserve of wetlands, consisting primarily of drained bottomland previously planted to agricultural crops. That being the case, the C sequestration potential of 2 Mha of wetland restoration is about 0.5 MMTC/yr (Eq. 4).

Table 13. Average annual wetland loss due to agriculture (USDA-SCS, 1992).

Year	Area lost (ha)	Total area (10^6 ha)
1954-74	161,134	3.38
1974-83	63,562	0.64
1982-92	12,551	1.39
Total	237,247	5.41

The average C sequestration potential of 2 Mha restored wetlands =

$$(2 \text{ Mha}) \times (\frac{0.25 \text{ MT}}{\text{ha yr}}) = 0.5 \times 10^6 \text{ MT/yt} = 0.5 \text{ MMTC/yr} \dots\dots\dots\dots \text{Eq. 4}$$

The economic evaluation of WRP by Heimlich et al. (1997), which raised a question about the ownership of the converted wetlands, has led to suggestions for legislative changes about property rights. About 75% of the remaining wetlands in

Table 14. U.S. cropland moderately and severely eroded by wind and water erosion (calculated from USDA-SCS, 1992).

Land area	Moderate erosion (11.2-44.9 Mg/ha/yr)			Severe erosion (> 44.9 Mg/ha/yr)		
	Water	Wind	Total	Water	Wind	Total
			—10⁶ ha—			
Total	21.7	16.8	38.5	2.6	3.7	6.3
Sd	0.3	0.3	0.1	0.2		

the contiguous 48 states (36.84 Mha) are privately owned, which historically implied the right to convert wetlands to other uses. These rights were changed by legislation in the 1970s.

Restoration of Degraded Soils

Eroded Lands

Moderately eroded cropland is defined as land that erodes at a rate of 1 to 4 times the tolerable soil loss of 11.2 MT/ha/yr (5 ton/acre/year). In comparison, severely eroded land erodes at a rate > 4 times the tolerable soil loss (> 44.2 MT/ha/yr). Eroding at an excessive rate for several decades, if not centuries, some croplands are highly degraded and are characterized by low SOC contents and low natural fertility, resulting in poor quality soils.

Wind erosion is a severe hazard in arid and semiarid regions, water in humid regions, and both wind and water in semiarid and sub-humid regions. Inevitably, some croplands are subjected to both wind and water erosion. The available data do not permit separate estimates of such land. In the interest of simplification, the data in Table 14 show U.S. cropland subjected to moderate and severe erosion by wind and water. These areas are added for estimating C sequestration potential. Therefore, U.S. cropland area subjected to moderate levels of wind and water erosion is estimated at 38.5 Mha, and that subjected to severe erosion is estimated at 6.3 Mha (Gomez, 1995).

Some of these lands are enrolled under the Conservation Reserve Program (CRP), and areas of low to moderately eroded lands which previously were not enrolled under CRP are now, under the 1996 Farm Bill, eligible in some circumstances and can be diverted to soil restorative measures for C sequestration. Areas of such lands are estimated at 28.6 Mha (38.5 + 6.3 - 16.2 = 28.6 Mha). Assuming that the C sequestration potential of this land is similar to that of CRP at about 30 to 70 g/m²/yr with an average rate of 50 g/m²/yr, the total C sequestration through restoration of eroded cropland is 14.3 MMTC/yr (Eq. 5).

Average C sequestration potential through restoration of eroded soils =

$$28.6 \text{ Mha} \times 10^4 \left(\frac{m^2}{ha} \times \frac{50g}{m^2 \, yr} \right) = 14.3 \text{ MMTC/yr} \dots\dots\dots\dots\dots \text{Eq. 5}$$

These estimates are similar to those for adoption of a conservation tillage (CT) system. Chapter 10 explains a more thorough way to estimate the C sequestration potential of 22.5 MMTC/yr through crop residues. Based on the assumption that an average rate of C sequestration is 3778 kg C/ha/yr, the calculations show a C sequestration potential of 447 MMTC/yr and assume that 10% of the above-ground plus below-ground residue C can be converted to SOC. Thus the C sequestration rate is about 22.3 MMTC/yr for 50% of the residue produced.

Minelands and Toxic Soils

Mining activities lead to drastic disturbance in the topsoil that contains the most SOC content. The topsoil is removed, and the exposed subsoil is edaphologically inferior and supports little or no vegetative growth (Photo 12). Such drastically disturbed lands constitute a significant part of anthropogenically degraded soils. Surface mining of all mineral soils in the U.S. has affected over 1.6 Mha, and additional thousands of hectares will be disturbed each year (Sopper, 1993). A major part of the soils disturbance is due to strip-mining for coal in the eastern region of the country.

An estimated up to 0.63 Mha of land strip-mined for coal requires reclamation (Sutton and Dick, 1987). Unreclaimed lands are prone to severe erosion, often 100 times more than the adjacent undisturbed lands (USEPA, 1976; Photo 13). Reclaimed minelands can be used to grow row crops (Dunker and Janson, 1987; Schroder, 1995a) (Photo 14) and pastures (Photo 15). Several studies have shown that applications of sludges and other amendments enable growth of crops, grasses, and other vegetation (Pichtel et al., 1994) and lead to improvements in SOC contents (Schaller and Sutton, 1978). Bennett (1977) reported, from an experiment in West Virginia, that SOC in a strip-mined spoil was increased from 0.11 to 1.17% in the top 20 cm over a 4-year period. Peterson et al. (1982) observed changes in SOC content by 30 g/kg after 7 years of reclamative treatment. Hinesly et al. (1982) and Stucky et al. (1980) observed a similar magnitude of increases.

Improvements in SOC content affect soil quality as measured by the productivity index (PI) (Olson 1992; Barnhisel et al., 1992; Hammer, 1992). Improvements in soil quality are reflected in higher yields of grain crops (Underwood and Sutton, 1992), small grains (Pedersen et al., 1992), sorghum, soybeans, wheat (Barnhisel et al., 1992), and pastures.

The rate of SOC sequestration in reclaimed mineland depends on many factors, including post-reclamation management. Nitrogen deficiency frequently lim-

Photo 12. Abandoned mineland needs revegetation.

Photo 13. Unreclaimed minelands are prone to severe erosion.

Photo 14. Grain crops can be grown on reclaimed mineland.

Photo 15. Reclaimed mineland sown to pastures in Ohio.

Table 15. Plant species[1] recommended for strip mine reclamation (collated from Follett, 1980).

Plant	
Legumes	Alfalfa, white clovers, crimson clove, birdsfoot trefoil, Sericea lespedeza, red clover, crown vetch, hairy vetch, kura clover, zigzag clover and white and yellow sweet clover
Grasses	Weeping lovegrass, deer-tongue, little blue stem, tall fescue (Ky-31), bermuda grass (Midland and Tufcote varieties), millet, sudan grass, wheat grass, indiangrass, switch grass, redtop, bent grass, blue grama, needle grass
Woody plants	Winterfat, fourwing salt bush, silver sage brush, Maximowicz pea shrub, pygmy pea shrub, Siberian salt tree, Chinese wolfberry, matrimonyvine, trumpet gooseberry, Mugwort wormwood

[1] *Note: species selection depends on climate, resources available for vegetation, and many other factors. Thus, species adapted for use in the humid regions in the eastern U.S. usually will not be well adapted for use in the semi-arid Great Plains or more arid regions in the western U.S.*

---

its mine-spoil revegetation in the eastern U.S. Application of organic amendments to these systems may provide a long-term source of N and reduce the need for repeated N fertilization.

Schoenholtz et al. (1992) observed that the use of organic amendments provided a more stable source of N and increased the SOC content of the top 0- to 10-cm depth of an Appalachian mine soil. Application of wood chips along with inorganic fertilizers increased the SOC content of 0-10 cm from 23.0 g/kg in July, 1987, to 32.3 g/kg in October, 1989, an increase of 5.8 Mg/ha/yr, assuming a bulk density of 1.4 Mg/m^3. In North Dakota, Schroeder (1995b) observed in 2 years an increase from 1.4% to 3.0% in SOC content. Caspall (1975) reported an increase in SOC content in the surface 15 cm of Illinois spoils, from an initial level of 0.24% to 1.5% in 14 years. This amounts to an increase of about 2 MT/ha/yr of SOC in the top 15-cm layer (Gee et al., 1978).

For the entire rooting depth, however, increases in SOC may be much higher, especially when highly adapted plant species are grown. Table 15 shows appropriate species for reclaiming mineland, species with adaptation to acid soil of low inherent fertility. Growing such species in conjunction with appropriate reclamative measures may lead to SOC sequestration at a rate of 4 to 5 MT/ha/yr.

Assuming some 0.63 Mha of drastically disturbed land in the U.S. (Sutton and Dick, 1987), and that reclamation of these soils can lead to net SOC sequestration at the rate of 1 to 3 MT/ha/yr, the total SOC sequestration potential is 1.3 MMTC/yr (Eq. 6).

Average C sequestration in 0.63×10^6 of drastically disturbed land =

$$(0.63 \text{ Mha}) \times \left(\frac{2 \text{ MT}}{\text{ha yr}} \right) = 1.26 \times 10^6 \text{ MT/yt} = 1.3 \text{ MMTC/yr} \dots\dots\dots \text{ Eq. 6}$$

We suggest that the duration for which SOC sequestration can continue following reclamation is likely less than 30 years. Little experimental data show temporal changes in SOC content of reclaimed mineland or other drastically disturbed soils. Most available research information deals with the agronomic/biomass productivity and related aspects of mineland reclamation (Dunker et al., 1992).

Salt-affected Soils

Irrigated cropland and pastureland in arid and semiarid regions is prone to salinization (Photo 16). The total area of salt-affected soils in the U.S. is 19.6 Mha (Table 16). Salt-affected soils are located primarily in the following water resources regions: Missouri (44.2%), Souris-Red-Rainy (15.9%), California (8.1%), Texas-Gulf (7.2%), Pacific Northwest (4.6%), Great Basin (3.8%), Arkansas-White-Red (3.7%), and Colorado (3.7%) (USDA-SCS, 1992). Crop growth in severely salt-affected soils is poor. High salt content in soils can decrease biomass yield from 4% to 29%/mmho/cm, depending on the crop.

Some crops are more sensitive to salt concentration than others. The most sensitive crops include beans (*Phaseolus vulgare*), with a yield reduction at 19%/mmho/cm, cowpea (*Vigna unguiculata*) at 14.0%, sorghum (*Sorghum bicolor*) at 16%, and soybean (*Colycine max*) at 20%. Crops less sensitive to salt content include cotton (*Gossypium hirsutum*) at 5.2%, sugarcane (*Succharum officinarum*)

Photo 16. Salinization of rangeland in arid regions.

Lal, Kimble, Follett, and Cole

at 5.9%, wheat (*Triticum aestivum*) at 7.1%, barley (*Hordeum vulgare*) at 7%, corn (*Zea mays*) at 7% for forage and 12% for grains, and rice (*Oryza sativa*) at 12%. Crops vary widely in their threshold salinity levels for yield reduction. Sensitive crops have a lower threshold to salinity level than tolerant crops. Crops with a high salinity threshold level are barley (6.0-8.0 mmho/cm), cotton (7.7 mmho/cm), sorghum (6.8 mmho/cm), soybean (5.0 mmho/cm), sugarbeet (*Beta vulgaris*) (7.0 mmho/cm), rice (5.6 mmho/cm), and wheat (6.0 mmho/cm). Most sensitive crops have a low threshold salinity level, which is 1.0 mmhos/cm for beans, 1.8 for corn, 1.3 for cowpea, 1.9 for sugarcane, and 1.5 for sweet potato.

In addition to using amendments (e.g., gypsum, organic matter, and green manure) and leaching salts out of the root zone, choosing appropriate crops and cropping systems improves above- and below-ground biomass production and enhances soil quality by increasing SOC content. Limited data are available on the

Table 16. Cropland and pastureland affected by salinity and sodicity (USDA-SCS, 1992).

Water resources region	Total area (10^3 ha)
New England	0
Mid-Atlantic	0
South Atlantic-Gulf	15.1
Great Lakes	0
Ohio	13.2
Tennessee	0
Upper Mississippi	579.4
Lower Mississippi	294.1
Souris-Red-Rainy	3116.2
Missouri	8684.3
Arkansas-White-Red	722.1
Texas-Gulf	1412.1
Rio Grande	765.5
Upper Colorado	309.0
Lower Colorado	410.6
Great Basin	750.1
Pacific Northwest	910.4
California	1584.0
Hawaii	0
Caribbean	6.4
Total	19622.5

reclamative effects of amendments and improved cropping systems on SOC content. Sodic soils with a high sodium absorption ratio (SAR) have massive structure because the colloidal fraction is in a dispersed state (Gupta and Abrol, 1990). Replacement of Na^+ by Ca^{+2} and Mg^{+2} on the exchange complex can improve aggregation by forming stable organo-mineral complexes (microaggregates) that enhance soil physical quality and improve biomass productivity. Biomass production on reclaimed salt-affected soils may be increased by a factor of 2 to 5 (Szabolcs, 1998; Gupta and Abrol, 1990).

The microbial population that oxidizes SOC probably can adapt and be more salt-tolerant than the planted vegetation. In this instance, salinity under agricultural conditions would be likely to decrease the SOC content. Salinity *per se* does not affect the process of the formation of secondary carbonates, but saline water would require a higher leaching fraction and, in turn, allow for more carbonate dissolution and thus serve as a sink for CO_2.

Assuming that an increase in the SOC content of reclaimed soils is at the rate of 50 to 150 kg/ha/yr, the rate of C sequestration through reclaiming salt-affected soils is 2 MMTC/yr (Eq. 7).

Average C sequestration by reclamation of 20 Mha of salt-affected soils =

$$(2 \text{ Mha}) \times 0.1 \text{ MT/ha/yr} = 2 \text{ MMTC/yr} \quad \dots\dots\dots\dots \text{Eq. 7}$$

Soil Fertility Restoration

Soil fertility refers to a soil's ability to store and supply nutrients to meet a plant's requirements for achieving its maximum growth potential. The soil's inherent fertility is inadequate to meet the demands of intensive agricultural practices and high yield expectations. Soil fertility can be depleted rapidly unless it is replenished through discriminate, and judicious, use of chemical fertilizers and organic amendments. Soil fertility depletion is the principal cause of low biomass production that leads to a decline in SOC content, soil quality, and sustainability. Fertility mining practices also have led to severe depletion of the SOC content of the soils of Africa and South Asia.

In Africa, nutrient depletion is at the continental scale (Stoorvogel and Smaling, 1990) and is the principal cause of low yields and low SOC content. The prevalent low input and subsistence agricultural practices are responsible for soil fertility depletion throughout the tropics and subtropics. These practices are similar to those which led to severe soil degradation in the southeast U.S. during the 19th century.

Data from long-term fertility management experiments are needed to assess temporal changes in SOC content. Soil fertility depletion is not a major problem in U.S. cropland. However, increasing the efficiency of fertilizer use, reducing the use of chemical fertilizers, and increasing the use of organic materials and nutrient recycling are important considerations. We discuss the latter strategies in Chapter 11.

Carbon Sequestration Potential from Land Conversion and Soil Restoration

The discussion presented in this chapter shows that adoption of soil restorative measures on U.S. cropland has a potential to sequester 17 to 39 MMTC/yr with an average of 28 MMTC/yr. This potential includes a large component of C sequestration from land conversion (e.g., CRP, WRP, and conservation buffers). Restoration of degraded soils constitute a large share, most of which is due to restoration of eroded soils.

The estimate of 28 MMTC/yr represents a small yet important sink in comparison with total U.S. emissions of GHGs, at either 1485 (1.9%) or 1709 (1.6%) (Table 1), and total emissions from agricultural activities estimated at either 109 MMT (25.5%) or 123 MMT (22.6%) per year (Table 2). Realization of the potential of soil restorative measures, however, should be considered in the context of other soil management strategies (see Chapter 11).

CHAPTER 9

Biofuels for Offsetting Fossil Fuel

Agriculture requires energy, but since the mid 1980s, it has succeeded in decreasing its requirements for gasoline and LP gas and has kept its use of diesel fairly constant. In 1992, annual fuel use was 1.6 billion gallons of gasoline, 3.1 billion gallons of diesel, and 0.6 billion gallons of LP gas (Table 17). Even though energy use has been decreasing, several options remain for additional improvements in energy use on farms: (i) conservation tillage (CT), (ii) higher fertilizer and water use efficiencies, (iii) energy-efficient grain drying systems, and (iv) production of biofuel substitutes.

Biofuels are increasingly recognized as a feasible alternative to fossil fuel (Abelson, 1995; Giampietro et al., 1997; Paustian et al., 1998). Biofuels sometimes can be burnt directly. In other cases, a biofuel is any type of solid, liquid, or gaseous fuel that can be produced from biomass substrates and that can be used as a substitute for fossil fuel (e.g., ethanol, methanol, biodiesel, or direct combustion of biomass). Ethanol can be obtained from fermentation of sugar crops (e.g., sugarcane, sugarbeet, and sweet sorghum), starchy crops (e.g., corn and cassava), and woody plants containing cellulose. Methanol can be obtained

Table 17. Fuel purchased for on-farm use in the U.S. (USDA-ERS, 1994).

Year	Gasoline	Diesel	LP gas
	————————10⁹ gallons————————		
1974	3.7	2.6	1.4
1975	4.5	2.4	1.0
1976	3.9	2.8	1.2
1977	3.8	2.9	1.1
1978	3.6	3.2	1.3
1979	3.4	3.2	1.1
1980	3.0	3.2	1.1
1981	2.7	3.1	1.0
1982	2.4	2.9	1.1
1983	2.3	3.0	0.9
1984	2.1	3.0	0.9
1985	1.9	2.9	0.9
1986	1.7	2.9	0.7
1987	1.5	3.0	0.6
1988	1.6	2.8	0.6
1989	1.3	2.5	0.7
1990	1.5	2.7	0.6
1991	1.4	2.8	0.6
1992	1.6	3.1	NA

1 Mg of gasoline = 42.2 GJ
13.78 kg C/GJ of natural gas
19.94 kg C/GJ of petroleum liquids
1 Mg of straw = 3 x 10⁶ k cals

from wood or woody crops, and biodiesel fuels from oil crops (e.g., soybean, rapeseed, sunflower, and palms). Each gallon of biofuel produced could save roughly 0.5 gallons of oil. Marland and Turhollow (1991) calculated that burning ethanol releases 17.53 kg C/GJ (1 GJ = 10^9 J = 10^6 KJ) as compared with 13.78 for natural gas, 19.94 for petroleum liquids, and 24.12 kg C/GJ for coal. However, C from fossil fuel is being released from long-term storage and results in a net increase in atmospheric CO_2.

Several options can increase agricultural production of biofuels (Paustian et al., 1998). The land area for biofuel production can be increased by substituting it for other agricultural crops (e.g., those in surplus supply) and growing forage crops and tree plantations as agroforestry systems. Biofuel crops also can be grown as an integral component of soil and water conservation programs, e.g., buffer strips. Idle land also can be converted to biofuel production. Numerous species can be grown as dedicated biofuel crops.

Idle cropland constitutes about 15% of total cropland or about 20 Mha, and some of it could be used for production of dedicated biofuel crops. Part of the above-ground crop residues produced from some of the existing cropland in the U.S. also could be used for biofuel, provided that this is consistent with the maintenance of adequate levels of ground cover for soil and water conservation. Even if only 5% of the above-ground residues were used (about 25 MMT residues or 10 MMTC), 1093 MMT of above- and below-ground residues (437 MMTC) would remain available for soil-C sequestration (see Table 21, Chapter 10).

Growing short-rotation woody crops and herbaceous energy crops on 10 Mha of idle or other cropland could lead to a substantial amount of fossil C offset. Assuming a net C assimilation rate of 5 T/ha/yr, 10 Mha will assimilate 50 MMTC/yr. The energy substitution factor is usually 0.7, with a range of 0.65-0.75. Thus the net energy saving may equal 35 MMTC/yr, with a range of 32-38.

Average Biofuel C offset =

$$(10 \text{ Mha}) \times (5 \text{ MT/ha/yr}) \times (0.7) = 35 \text{ MMT/yr} \dots\dots\dots\dots\dots \text{Eq. 8}$$

Whereas the above-ground 50 MMT/yr of biomass produced is used to generate biofuel offset at 35 MMTCE/yr, the below-ground biomass of about 100 MMT/yr would lead to enriched SOC content. At 40% C contained in the below-ground biomass and with 20% efficiency of humification, the C sequestered in the soil is 4 MMTC/yr (Eq. 9), with a range of 3 to 5 MMTC/yr.

Average SOC sequestered by the below-ground biomass produced on idle land =

$$(100 \frac{\text{MMT}}{\text{yr}}) = (0.40)(0.10) = 4 \text{ MMTC/yr} \dots\dots\dots \text{Eq. 9}$$

Giampietro et al. (1997) argued that large-scale liquid biofuel production is not a viable option. Direct use of biomass (e.g., fuel in power production) is a better option.

In an attempt to assess the potential of biofuel solids for fossil fuel offset, several researchers (Hall et al., 1993; Sampson et al., 1993; Cole et al., 1995) assumed that a net C yield of dedicated energy crops in temperate regions can total 5-9 Mg C ha^{-1}/yr^{-1} through adopton of short-rotation woody crops and herbaceous energy crops. Therefore, net C accumulation for 10 Mha of cropland converted to dedicated biofuel cropland is 50-90 MMTC/yr^{-1}.

Estimating biomass energy production potentials requires assumptions not only about available land, productivity, plant species, and percent of the crop to be used but also about collection and transport, conversion efficiencies, and fuel substitution factors. Overall, agricultural biofuels (energy crops, agroforestry, and crop residues) have the potential to substitute for 0.40 to 1.50 MMT fossil fuel C per year (Cole et al., 1996).

CHAPTER **10**

Intensification
of Prime Agricultural Land

Accelerated soil erosion is a serious environmental and economic issue, both in the U.S. and around the world (Pimentel et al., 1995). Adoption of effective conservation measures is important to achieving a sustainable use of soil and water resources (Fig. 11). Soil erosion control can be achieved through widespread adoption of: (i) conservation tillage (CT), (ii) management of crop residue and other organic material, (iii) the Conservation Reserve Program (CRP), (iv) conservation buffers, (v) water management systems, including runoff management, drought management through supplemental irrigation, and drainage of seasonally wet

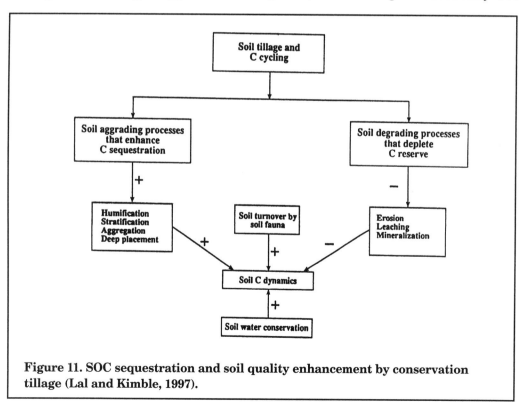

Figure 11. SOC sequestration and soil quality enhancement by conservation tillage (Lal and Kimble, 1997).

The Potential of U.S. Cropland to Sequester Carbon and Mitigate the Greenhouse Effect

lands to improve soil aeration and infiltration capacity, (vi) adoption of improved soil fertility management practices, including improved use of organic amendments, chemical fertilizers, and biological N fixation, and (vii) adoption of improved cropping systems.

This text discusses the carbon sequestration potential of these measures separately, for convenience and clarification. Because of their strong interactions, however, it is difficult to estimate separately the C sequestration potential of each practice. For example, one cannot adopt a CT system in isolation. Crop residue management, incorporation of a cover crop within a rotation cycle, and other mulch farming techniques are integral components of CT. Similarly, the impact of soil fertility management, cropping system, and erosion control may be included within that of the C sequestration potential of CT.

Conservation Tillage and Residue Management

The two practices of CT and residue management go hand in hand, and it is difficult to isolate the ecological impact of one from that of the other. A CT system includes residue management, crop rotation, and pest control measures.

Conservation Tillage

Basic concepts

A CT system is defined as "any tillage and planting system that maintains at least 30% of the soil surface covered by residue after planting to reduce water erosion; or where wind erosion is a primary concern, maintains at least 1000 kg/ha of flat, small grain residue equivalent on the surface during the critical wind erosion period" (CTIC, 1990; 1995; 1996; 1997).

In fact, CT is a generic term encompassing all tillage systems that reduce loss of soil and water from cropland, relative to conventional tillage (Lal, 1989; Lal and Kimble, 1997; Blevins and Frye, 1993). Conventional tillage includes plow-based methods, such as successive operations of plowing (soil turnover with a moldboard plow), mixing (with a disc plow), and pulverization (with a rotovator). CT eliminates one or several of these operations.

Crop-residue management is an integral part of any CT system and also includes selecting crops that produce sufficient quantities of residues (e.g., corn, wheat, small grains, sorghum) and sowing cover crops to provide an effective ground cover. Rather than turning under plant materials or crop residues following harvest, the residues are left on the soil surface to protect the soil against the erosive forces of rainfall, runoff, and wind. Some important CT variants include no till, ridge till, minimum till, and sod till (Appendix 1).

About 37% of the land farmed in the U.S. is now managed with a CT system (Table 18). Of the 119.2 Mha of cropland in 1997, 44.4 Mha used some form of CT, including no-till, minimum till, or ridge till (CTIC, 1997). Edwards et al. (1992) estimated that approximately 48% of arable land in the southeastern U.S. is managed under some form of CT, and this figure may increase to 65% by the year 2000. Application of CT to farmland in the U.S. is likely to continue to increase (Schertz, 1988). Estimates of adoption of a CT system by 2010 range widely, including 50-60% of the cropland (Crosson, 1981), 72% of the cropland (OTA, 1982), 63-82% of the cropland (Schertz, 1988), and 95% of the cropland (USDA, 1975). Table 19 shows the results of the 1997 National Crop Residue Management Survey conducted by the Conservation Technology Information Center (1997), which assessed the changes in CT from 1989 to 1997.

Increase in SOC content

Long-term use of a CT system leads to an increase in SOC content, enhancement of soil quality, and improvement in soil resilience (Grant, 1997; Black and Tanaka, 1997) (Fig. 11). Tillage accentuates

Table 18. Arable land area in conservation tillage in the U.S. (the data of 1968-1986 are those compiled by Schertz, 1988; those from 1987-1995 are those compiled by CTIC, 1997).

Year	Land area (10^6 ha)	% of planted area
1968	2.4	2.0
1969	3.2	2.0
1970	4.1	3.4
1971	4.5	3.6
1972	4.9	4.1
1973	6.1	4.7
1974	6.9	5.2
1975	7.3	5.6
1976	8.1	5.9
1977	9.7	7.0
1978	12.6	9.2
1979	13.4	9.5
1980	15.8	10.9
1981	17.4	11.8
1982	26.7	18.2
1983	28.3	22.6
1984	35.2	25.3
1985	38.5	27.8
1986	39.7	32.9
1987	34.8	31.6
1988	35.6	32.3
1989	29.0	25.7
1990	29.7	26.1
1991	32.0	28.1
1992	35.9	31.4
1993	39.3	34.9
1994	40.2	35.0
1995	40.0	35.5
1996	42.0	36.0
1997	44.4	37.0

A change in definition of conservation tillage, from 1989 onward, changes the estimates of area.

C oxidation by increasing soil aeration and soil residue contact, and accelerates soil erosion by increasing exposure to wind and rain (Grant, 1997). In contrast, CT can improve soil aggregation (Beare et al., 1994; Lal et al., 1994) and change the vertical distribution and retention of SOC content. Several experiments have shown more SOC content in soils of CT, compared with plow-till seedbeds (Doran, 1980; Doran et al., 1987; Rasmussen and Rohde, 1988; Tracy et al., 1990; Havlin et

Table 19. Rate of conversion of conventional tillage to conservation tillage for the 8-year period from 1989-1996, and the projected rate of change for the next 15 years until the year 2010 (calculated from CTIC, 1996).

Tillage method	1989	1990	1991	1992	1993	1994	1995	1996	Projected area in 15 years
					—10^6 ha—				
A. > 30% ground cover									
No-till	5.73	6.83	8.34	11.37	14.10	15.79	16.56	17.4	36.2
(%)	(5.06)	(6.0)	(7.33)	(9.92)	(12.52)	(13.73)	(14.69)	(14.78)	$\Delta 5$
Ridge-till	1.10	1.23	1.31	1.36	1.40	1.44	1.37	1.37	1.59
(%)	(0.97)	(1.08)	(1.15)	(1.19)	(1.24)	(1.26)	(1.22)	(1.17)	$\Delta 1$
Mulch-till	22.2	21.6	22.39	23.19	23.83	23.0	22.1	23.3	27.1
(%)	(19.62)	(18.98)	(19.66)	(20.24)	(21.16)	(20.0)	(19.6)	(19.8)	$\Delta 1$
B. <30% ground cover									
Reduced-till	28.6	28.7	29.3	29.7	29.6	29.6	28.4	30.3	35.2
(%)	(25.3)	(25.3)	(25.7)	(25.9)	(26.3)	(25.76)	(25.2)	(25.77)	$\Delta 1$

- *Figures in parenthesis are % of planted area under each category. The differences are planted by conventional till.*
- *Δ means change*

al., 1990; Wood et al., 1991a, b; Kern and Johnson, 1991, 1993; Power, 1994; Lafond et al., 1994; Reicosky et al., 1995). Increase in SOC content also leads to improvement in soil structure and aggregation under CT, compared with plow-till (Bruce et al., 1990; Lal et al., 1994).

Cambardella and Elliott (1992) observed, for the Duroc loam soil near Sydney, Nebraska, that total SOC content in the 0- to 20-cm depth was 3.1 kg/m² for bare fallow, 3.5 kg/m² for stubble mulch, 3.7 kg/m² for no-till, and 4.2 kg/m² for a native sod. They observed that conventional tillage practices can lead to losses of 40% or more of total SOC during a period of 60 years. Edwards et al. (1992) observed that conversion from plow till to CT increased SOC content in the 0- to 10-cm layer from 10 g/kg to 15.5 g/kg in 10 years, an increase of 56%. A summary of the available literature indicates that the SOC sequestration potential of conversion to CT ranges from 0.1 to 0.5 MT/ha/yr for humid temperate regions and from 0.05 to 0.2 MT/ha/yr for semiarid and tropical regions.

The increase in SOC with time, however, may follow a sigmoid response. Thus, the increase in SOC content may be none or slight in the first 2 to 5 years of conversion (Franzluebbers and Arshad, 1996) and large in the next 5 to 10 years. We estimate that the SOC increase may continue over a period of 25 to 50 years, depending on soil properties, climatic conditions, and management. The beneficial effects of CT on SOC content, however, may be short-lived if the soil is plowed even after a long time under CT (Gilley and Doran, 1997). In contrast to the SOC data available from long-term experiments conducted on research farms where CT is adopted on a continuous basis, farmers choose tillage practices from season to season. A farmer may use a CT system for some years and then convert back to conventional tillage. Thus, it is difficult to apply the SOC data from continuous CT experiments to farmer-managed cyclic tillage systems.

Mechanisms of increase in SOC content

Several mechanisms are responsible for the increase in SOC content from adoption of a CT system (Fig. 12). The changes include: (i) a net decrease in losses due to soil erosion, (ii) an increase in soil aggregation, and (iii) a decrease in mineralization losses. The SOC emission of 15 MMTC/yr due to erosion discussed in Chapter 10 does not cover all C affected by soil erosion. A considerable part of the 52.5 MMTC/yr redistributed over the landscape can be better managed through CT and residue management.

Energy savings also occur with adoption of a CT system. Kern and Johnson (1993) estimated that fossil fuel emissions from field manipulations and herbicide production are 53 kg C/ha/yr for plow till, 45 kg C/ha/yr for minimum till, and 29 kg C/ha/yr for no till. Therefore, conversion to CT involves both C sequestration in the soil due to enrichment of SOC and savings in fossil fuel.

The magnitude of enrichment in SOC content differs among soils and agroecosystems. In Ohio, Dick et al. (1986a, b) observed that total SOC content in the top 45 cm in soil managed with NT was significantly more than on soil under conventional till, by 13.3% for continuous corn (59.6 vs. 52.6 MT/ha), 16.7% for corn-soybean rotation (55.8 vs. 47.8 MT/ha), and 2.0% for corn-oats-meadow rotation (61.6 vs. 60.4 MT/ha) (Fig. 13).

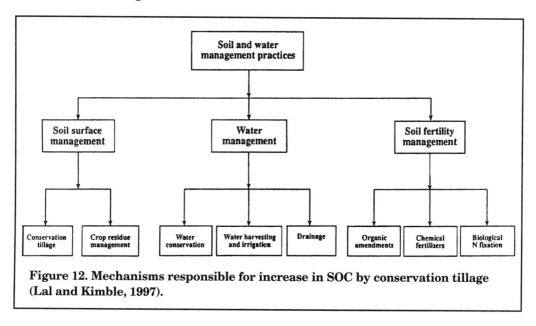

Figure 12. Mechanisms responsible for increase in SOC by conservation tillage (Lal and Kimble, 1997).

Increase in SOC content from a CT system may result from incorporation of a cover crop within a rotation cycle (Blevins et al., 1983; Frye et al., 1988; Utomo et al., 1990). Lee et al. (1993) assumed that adoption of a CT system would increase SOC content of the top 15-cm layer of soil by 0.2 kg/m^2 during the next 100 years. With incorporation of a cover crop within the rotation cycle, the increase in SOC content may be 0.4 to 0.8 kg/m^2. The time required for attainment of the steady

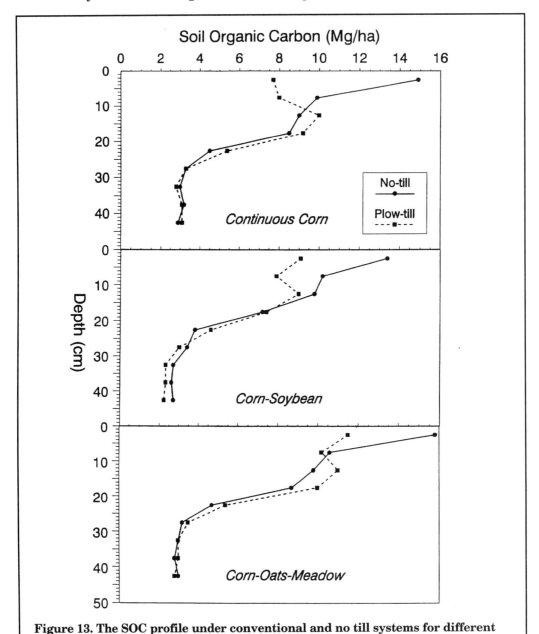

Figure 13. The SOC profile under conventional and no till systems for different crop rotations in NW Ohio (redrawn from Dick et al., 1986a).

state level of SOC after change from conventional to a CT system may be about 10 years (Kern and Johnson, 1991; 1993).

Lal (1997) estimated that adoption of a CT system in the U.S. has a total C sequestration potential of 350 MMTC to 1400 MMTC by the year 2020.

The data in Table 20 show that C sequestration through the adoption of CT systems varies among tillage methods. Conversion from conventional moldboard plow to CT caused a change in C sequestration of 975 kg/ha for chisel plow, 602 kg/ha for ridge till, and 525 kg/ha for no till (Karlen et al., 1998). The projected increase in land area for the next 15 years is 18.8 Mha under no till, 0.22 Mha under ridge till, 4.2 Mha under mulch till, and 4.9 Mha under reduced till (CTIC, 1997). These are conservative estimates in comparison with those made by Kern and Johnson (1993) and Lal (1997).

Table 20. The rate of C sequestration in soil under different tillage systems (adapted from Karlen et al., 1998).

Depth (cm)	SOC content g/kg	kg/ha	Change in SOC content kg/ha
(a) Moldboard plow			
(r_b = 1.3 Mg/m^3)			
0-5	23.7	1540.5	
5-10	23.5	1527.5	
10-20	22.3	2899.0	
Total		5967.0	—
(b) Chisel plow			
(r_b = 1.26 Mg/m^3)			
0-5	29.1	1833.0	
5-10	28.9	1820.7	
10-20	26.1	3288.6	
Total		6942.3	975.3
(c) Ridge till			
(r_b = 1.25 Mg/m^3)			
0-5	32.8	2050.0	
5-10	28.5	1593.8	
10-20	23.4	2925.0	
Total		6568.6	601.6
(d) No till			
(r_b = 1.19 Mg/m^3)			
0-5	37.3	2219.4	
5-10	26.0	1547.0	
10-20	22.9	2725.1	
Total		6491.5	542.5

The C sequestration potential

The following calculations are based on the assumption that the C sequestration potential of conversion from conventional to a CT system is 0.5 MT/ha/yr for no-till, 0.5 MT/ha/yr for mulch till, and 0.5 MT/ha/yr for ridge till. With these assumptions and the projected land area under different types of CT system, equations 10 to 12 show the C sequestration potential.

(i) No till — 18.8 Mha of additional land area over 15 years with C sequestration rate of 0.5 MT/ha/yr:

Average C sequestration potential with no till =

$$(18.8 \text{ Mha}) \times (\frac{0.5 \text{ MT}}{\text{ha yr}}) = 9.4 \text{ MMTC/yr} \dots\dots\dots \text{Eq. 10}$$

(ii) Mulch till — 9.1 Mha of additional land area over 15 years with C sequestration rate of 0.5 MT/ha/yr:

Average C sequestration potential with mulch till =

$$(9.1 \text{ Mha}) \times (\frac{0.5 \text{ MT}}{\text{ha yr}}) = 4.6 \text{ MMTC/yr} \dots\dots \text{Eq. 11}$$

(iii) Ridge till — 0.22 Mha of additional land area over 15 years with C sequestration rate of 0.6 MT/ha/yr:

Average C sequestration potential with ridge till =

$$(0.22 \text{ Mha}) \times (\frac{0.6 \text{ MT}}{\text{ha yr}}) = 0.13 \text{ MMTC/yr} \dots\dots \text{Eq. 12}$$

The total C sequestration potential for this scenario is about 14.1 MMTC/yr.

Another approach to calculating the potential of CT for C sequestration is to assume that 100 Mha of additional cropland area would be brought under CT, which would lead to an average rate of increase in SOC content of 0.0625%/yr to a 1-m depth for a soil with a mean bulk density of 1.4 MT/m^3. This approach gives the C sequestration potential over the next 25 year period as 875 MMTC (Eq. 13).

Average C sequestration potential over next 25 years =

$$(100 \text{ Mha}) \times 6.25 \times 10^{-4} \times 1.4 \text{ MT/m}^3 \times (1\text{m}) \times (\frac{10^4 \text{ m}^2}{\text{ha}}) = 875 \times 10^6 \text{ MT} = 875 \text{ MMTC/yr} \dots\dots \text{Eq. 13}$$

The C sequestration potential also may be computed on the assumptions of Lee et al. (1993), who proposed an average increase of SOC content by 0.8 kg/m^2 to a 1-m depth by conversion from a conventional to a CT system. Assuming a C sequestration potential of 0.6 to 1.0 kg/m^2 with an average rate of 0.8 kg/m^2, the C sequestration potential of a CT system is 32 MMTC/yr (Eq. 14).

Average C sequestration potential over next 25 years =

$$(100 \text{ Mha}) \times 10^4 (\frac{\text{m}^2}{\text{ha}}) \times 0.8 \frac{\text{kg}}{\text{m}^2} \times \frac{\text{MT}}{10^3 \text{ kg}} = 800 \text{ MMTC or } 32 \text{ MMTC/yr} \dots\dots \text{Eq. 14}$$

Most available data support the assumption by Lee (1993a, b). Their calculations are also similar to those by Kern and Johnson (1993) and Lal (1997).

Another approach to calculating the potential is to use a mean rate of SOC sequestration of 0.4 MT/yr of additional land that will be converted to CT from 1997 onward. Assuming that the additional cropland brought under CT after 1997 is 55.6 Mha (100 Mha-44.4 Mha; see Table 18), and SOC sequestration continues for 25 years, the total C sequestration potential of new land converted to CT is 22.2 MMT/yr. The land area converted to CT since 1990 (15.4 Mha) will sequester an average of 0.4 MT/yr for another 15 years, with an additional potential of 6.2 MMT/yr, and that converted in the 1980s (15.6 Mha) will sequester for another 10-year period, with C potential of 6.2 MMT/yr. Computed this way, the SOC sequestration potential of converting to CT is about 34.6 MMTC/yr.

Therefore, we estimate that the C sequestration potential of widespread adoption of CT on U.S. cropland for 25 years is 32 MMTC/yr.

Adoption of CT also saves fossil fuel. The data by Kern and Johnson (1993) showed that savings in fossil fuel from converting from a plow based system is 8 kg C/ha/yr for minimum tillage (e.g., mulch till and ridge till) and 24 kg C/ha/yr for no till. Therefore, the estimated savings in fossil fuel carbon from converting to minimum till (Eq. 15) and no till (Eq. 16) are 1.6 MMTC/yr, assuming that 50% of the 100 Mha will be under no till and 50% under minimum till.

Savings in fossil fuel from conversion to minimum till =

$$(50 \text{ Mha}) \times \frac{8 \text{ kg}}{\text{ha yr}} \times \frac{1 \text{ MT}}{10^3 \text{ kg}} = 0.4 \text{ MMTC/yr} \quad\text{.................... Eq. 15}$$

Savings in fossil fuel from conversion to no till =

$$(50 \text{ Mha}) \times \frac{24 \text{ kg}}{\text{ha yr}} \times \frac{1 \text{ MT}}{10^3 \text{ kg}} = 1.2 \text{ MMTC/yr} \quad\text{.................... Eq. 16}$$

Adoption of CT has a potential to save fossil fuel equivalent to 1.2 MMTC/yr.

Crop Residues and Other Biomass Management

The SOC pool is a function of the quantity of crop residues, plant roots, and other organic material returned to the soil and of the rate of their decomposition. Crop residues and other organic material constitute a major resource for soil surface management, energy production, and other uses. Organic materials of high importance to enhancing SOC content and C sequestration are the below-ground or root biomass and the total biomass produced by weeds.

An important use of crop residues, plant roots, and other organic materials is to improve soil quality by enhancing the SOC content. The latter is also the principal mechanism for sequestering carbon in soil. Because of its beneficial effects in improving soil and environmental quality, the USDA developed a Crop Residue Management Action Plan (Schertz and Bushnell, 1993). The goal of this plan was to implement some form of crop residue management on nearly 75% of the 54.7 Mha (135 M acres) of highly erodible land. Therefore, residue management is over and above the CT impact on C sequestration.

The increase in SOC content through crop residue application depends on the quantity and quality of the residue, soil properties, and management. Within a cropping/farming system, the equilibrium level of SOC content often is related linearly to the amount of crop residue applied to the soil (Larson et al., 1972; Rasmussen et al, 1980). Paustian et al. (1992) observed that lignin content of the residue has a strong positive effect on SOC accumulation. The maximum level of SOC achievable also depends on the retention capacity of the soil. Gregorich et al. (1996) and Hassink and Whitmore (1997) observed that the net rate of accumulation of SOC depends, not on the protective capacity of a soil *per se,* but on the extent to which the soil already is filled by SOC. Therefore, soils of different texture receiving identical amounts of fresh residues generally attain SOC content to the same level. The surface applied crop residues decompose more slowly than those that are incorporated by tillage, because they have less contact with soil microorganisms (Reicosky et al., 1995) and soil water (Grant, 1997). Angers et al. (1995) reported that conversion of corn residue C into SOC in the 0- to 24-cm layer was about 30% of the total input.

Several factors determine the quantity of crop residues and other biomass produced, including soil quality, ecoregional characteristics, cropping systems, soil and crop management practices, and weather conditions. There are several estimates of crop residue production in the U.S. (Larson et al., 1978; USDA, 1978; Lal, 1995b), which totals about 1118 MMT, including total above-ground residues (508.3 MMT) and below-ground biomass plus weeds (609.9 MMT) (Table 21).

Management of crop residues and other biomass with conventional tillage potentially sequesters C similarly to CT. However, included also within the estimates of CT on C sequestration is the impact of crop residue management. It is difficult to estimate separately the C sequestration potential of individual components. It is assumed that 50% of the residue produced is used with conversion to CT. Potentially the remaining 50% of the residue produced on cropland can be used efficiently. The average carbon content of crop residues is about 40%. If 10% of the C contained in 25% to 75% (average 50%) of the crop residues can be sequestered as humus, the C sequestration potential of crop residues is 22.5 MMTC/yr (Eq. 17).

Average C sequestration through C residues management =

$$(\frac{118}{2} \text{ MMT}) \times (0.40)\,(0.10/\text{yr}) = 22.5 \text{ MMTC/yr} \ldots\ldots\ldots \text{Eq. 17}$$

There is another option for using 50% of this crop residue. Assuming that 1/4 of the total above-ground residue produced can be used for direct biofuel purposes,

Table 21. Estimation of total C produced on crop residue in the United States.

Crop	Crop (1) Category	Area planted in 1996 (2)(3) 1000 ha	U.S. total above-ground crop residue 1000 tons	weight (4) tons/ha	Residue Carbon (5) above grnd MMTC	root&weed MMTC	Total (6) Residue C MMTC	Residue C sqstrd/yr range[4] MMTC		
Corn	1	35868	245345	6.84	98.1	117.8	215.9	10.8	to	21.6
Sorghum	2	6283	23264	3.70	9.3	11.2	20.5	1.0	to	2.0
Soybeans	3	24616	84412	3.43	33.78	40.5	74.3	3.7	to	7.4
Cotton	4	5767	4220	0.73	1.7	2.0	3.7	0.2	to	0.4
Wheat	6	30195	106587	3.53	42.6	51.2	93.8	4.7	to	9.4
Barley	7	2877	12769	4.44	5.1	6.1	11.2	0.6	to	1.1
Rice	—	1214	7755	6.39	3.1	3.7	6.8	0.3	to	0.7
Other row	5	5085	10278	2.02	4.1	4.9	9.0	0.5	to	0.9
Other field	8	6753	13649	2.02	5.5	6.6	12.0	0.6	to	1.2
Totals		118659.254	508279		203.3	2434.6	447.3	22.4	to	44.7

(1) USDA-ERS, 1997

(2) CTIC, 1997. 1996 total planted acres from CTIC web site, individual crop acres based on % of acres for crop categories 1-5 from ref (1). Rice acres estimated to have increased from 2,857 acres and from 5,617 lbs/ac ave yields reported in 1991 (USDA, 1991).

(3) USDA, 1978. Other row and other field crop acres from USDA (1987), their percent of the total residue production of the other categories plus rice is based upon the same percentage that was calculated from data reported in USDA 1978. Improving soils with organic wastes.

(4) The ratios of residue weight to harvested crop were obtained from Larson, Holt, and Carson (1978), Larson et al. (1983), and Banarjee et al. (1990).

(5) This assumes that all residues (i.e. above ground, below ground, and weeds) are 40% C by weight. A multiplier factor of 2.2, based on wheat/fallow data (Follett, et al. 1997) was used to estimate the weight of weeds plus roots.

*(6) The residue C produced = 1,118,214,000*0.40 or 447,286,000 tonnes of residue C.*

Note #1. Calculation of total above-ground crop residue produced for 1996 is 508,279 thousand tons compared to 1978 production of 391,009 thousand tons reported by USDA, 1978.

Note #2. The residue C produced is calculated as 447,286,000 tons of residue C produced on 118,659,300 ha. Thus, the overall average C produced/ha is 3.77 tons C/ha/yr.

Note #3. The efficiency of incorporation of residue C is calculated as weight of C sequestered divided by tons weight of residue C returned to the soil. It was assumed that the percent of C sequestered into soil organic C was the same for all crop types.

Note #4. Follett et al. (1997) reported the efficiency of incorporation of total residue C was 5.4% when measured in an 84-year study at Akron, CO and 10.5% when measured in a 20-year study at Sidney, NE. Both studies were with wheat-fallow. For a 5 or 10% C sequestration efficiency, amounts of total residue C sequestered would be 447,286,000 times 0.05 or 0.10 or 22,364,300 and 44,728,600 tons residue C, respectively.

the net C amount in 280 MMT of crop residue is 112 MMT. With an energy substitution factor of 0.6 to 0.7, C emission reduction through use of 25% of crop residue as biofuel is 67 to 78 MMTC/yr. If this were the option, the amount of C sequestration in soil (Eq. 17) would be reduced by 50%, to 11.2 MMTC/yr.

Multipurpose use of crop residue needs to be assessed carefully in view of several demands on this precious resource. Crop residue has an important role in erosion control, nutrient cycling, soil quality enhancement, animal fodder and bedding, composing, and potential biofuels use. Because of the multipurpose use of

Table 22. Irrigated land area under different crops (USDA-ERS, 1994).

Crop	1969	1974	1978	1982	1987	1992	1993	1994
				10^3 ha				
Corn	1336	2267	3522	3441	3239	4170	3968	4291
Wheat	810	1336	1215	1862	1498	1660	1579	1579
Rice	891	1053	1215	1296	972	1417	1255	1377
Soybeans	283	202	526	931	1053	1255	1296	1296
Cotton	1255	1498	1903	1377	1417	1498	1781	1741
Hay	3198	3239	3603	3441	3482	3401	3482	3603
Total	15830	16680	20405	19838	18785	20769	20486	21053

biofuels, there is a danger of double accounting, especially since there is a lack of experimental data on efficient use of residues for different purposes.

Irrigation Water Management

Surface Irrigation

Irrigated agriculture is important to crop production in arid and semiarid regions that receive total precipitation < 400 mm per year. Irrigated agriculture worldwide grows some 1/3 of the earth's total crops, and 1/2 the value of all crops harvested grow on 1/6th of the world's cropland. An estimated third of the earth's food crops grow on a mere 50 Mha (4% of the earth's cropland) of irrigated land. Water management on agricultural land has a strong influence on SOC dynamics (Kimmelshue et al., 1995).

Irrigated farming is practiced mostly in California, Idaho, Arkansas, Colorado, Kansas, Nebraska, and Texas. Many irrigable soils are inherently low in SOC content, ranging from 0.5 to 0.7% in the top 0- to 10-cm depth, or about 6 Mg SOC/ha at the 10-cm depth. Converting these less productive soils to economic cropland via irrigation can result in a drastic increase in both above- and below-ground biomass production. Irrigation has increased biomass production from 5 to 13 Mg C/ha to 23 to 32 Mg C/ha (Mayland, 1961; Mayland and Murray, 1978). Supplemental irrigation thus can lead to 2 to 5 times more C synthesis on irrigated than unirrigated soils. With irrigation, a large proportion of the biomass produced can be returned to the soil as SOC pool. Variation in the effect of irrigation on SOC depends on the soil type, temperature regime, and type of crops grown. In dry areas, SOC content is likely to increase when soil is irrigated, as most soils in dry areas have inherently low levels of SOC content.

Surface irrigation is used on some 21 Mha (USDA-SCS, 1992; Table 22; Photos 17 and 18) and has improved their biomass production vastly. Some calciferous irrigated soils also contain large concentrations of calcium carbonate or SIC content, as much as 1.2 Mg C/ha/10 cm of soil depth for each 1% of $CaCO_3$ present (Monger, 1993).

The acidifying effects of accelerated plant growth and chemical fertilizers may dissolve $CaCO_3$ and release some C as CO_2. Some dissolved SIC and SOC also may be leached and transported out of the ecosystem into other aquatic ecosystems. Irrigated soils also are plowed frequently, exposing them to accelerated soil erosion and to oxidation of SOC and release of C as CO_2.

Recently, Unger (personal communication) collected soil samples from paired fields near Bushland, TX, that had been cropped largely to sorghum for nearly 40 years, with one irrigated and the other dryland. However, the two fields' SOC content showed no measurable difference. Comparison of both to a native field on the same soil that had never been cultivated shows that the SOC is 1/3 less for the

cultivated fields. Thus, soil erosion and other SOC loss processes may likely mask the beneficial affects of irrigation for sequestering SOC on these fields. This also may be related to the higher soil temperature in Texas versus cooler temperatures in the North, where increases in SOC occur under irrigation.

Some positive and negative feedback mechanisms thus are set in motion by introducing irrigated agriculture to arid and semiarid regions. Research information on the dynamic of SOC and SIC contents under irrigated agriculture is scarce.

Photo 17. Center pivot irrigation.

Some data are available on the SOC content of irrigated soils. In Wyoming, Mayland (1961) reported that the SOC content of irrigated soils differed among crop rotations. He observed that, for the 40-year period from 1917 to 1956, the SOC content of the 0-20 cm layer was 2.0% for corn-wheat-oats rotation with manure, 1.35% for corn-wheat-oats without manure, and 0.82% for continuous corn. The data from Nebraska show that irrigated soil gained 1.66 Mg/ha-30 cm over a 15-year period (Table 23). Therefore, the rate of SOC sequestration in this irrigated soil was 110 kg/yr/ha-30 cm. Because of a wide range of soils, climate, cropping systems, and fertility management practices, we can justifiably assume that conversion of dryland farming to irrigated agriculture may increase SOC content in the soil profile by 50 to 150 kg/ha/yr with an average rate of 100 kg/ha/yr.

The total C sequestration potential of 21 Mha of irrigated soils thus is 2.1 MMTC/yr (Eq. 18).

Photo 18. Furrow irrigation.

Table 23. Increase in SOC content by central pivot irrigation in Nebraska sandhill soils (mixed mesic, Typic Ustipsamments) over the 15-year period (Lueking and Schepers, 1985).

Depth (cm)	Irrigated	Unirrigated	Changes due to irrigation
		kg/ha	
0-7.5	8600	8230	+370
7.5-15	6400	5650	+750
15-30	9190	8650	+540
Total 0-30	24190	22530	+1660

Average C sequestration in 21 x 10^6 ha irrigated land in U.S. =

$$(21 \times 10^6 \text{ ha}) \times \frac{0.1 \text{ MT}}{\text{ha yr}} = 2.1 \text{ MMTC/yr} \dots\dots\dots\text{ Eq. 18}$$

Judicious management of irrigated land is crucial, because there are high risks of soil erosion and salinization. Before designing and adopting an irrigation system, it is important to conduct a detailed feasibility study, based on soil profile analyses for total and soluble salt content, nature of salts (whether Ca, Mg, K, or Na), soil structure and water transmission properties, surface and subsurface drainage, and the quality of the irrigation water. Implementing irrigation schemes with disregard to these factors can lead to salinization at a rapid rate, which will deplete SOC content and accentuate C emission from the soil.

In addition to affecting SOC content, irrigation of soils in arid and semiarid regions also affects the SIC pool and its dynamics (Suarez and Rhoades, 1977; Suarez and Simunek, 1997). This is a complex issue, and all interacting processes involved are not understood clearly, especially with regard to solubilization of secondary carbonates, e.g., $CaCO_3$. Assuming a net sequestration rate of 50 to 150 kg/ha/yr and an average rate of 100 kg/ha/yr of C as secondary carbonates (personal communication, D. Yaalon, Israel), the potential of irrigation in sequestering SIC is 2.1 MMTC/yr (Eq. 19).

$$21 \text{ Mha} \times 0.1 \frac{\text{MTC}}{\text{ha yr}} = 2.1 \text{ MMTC/yr} \dots\dots\dots\text{ Eq. 19}$$

The total potential of irrigated cropland in C sequestration is thus 4.2 MMTC/yr (Eqs. 18 and 19). This amount of C sequestration would only partly offset the C emission by increasing the energy input required for crops grown on irrigated land.

Sub-irrigation on Poorly Drained Soils

Natural wetlands are areas that are inundated or saturated by surface or ground water often enough and for a long enough period of time to support a prevalence of vegetation typically adapted for life in saturated soil conditions (Snyder, 1995). There also are converted wetlands, farmed wetlands, and prior converted wetlands.

About 10% of the total surface area of the earth is too wet for arable land use (Driessen and Dudal, 1991). In the U.S., this land comes under the Wetland Reserve Program (WRP). Additionally, some agricultural lands are seasonally wet, due to impeded internal drainage resulting from natural or anthropogenic subsurface horizons having restricted movement of water.

Table 24. Crop land area drained in the North Central Region (USDA, 1987; Fausey et al., 1995).

State	Rank	Land area drained (10⁶ ha)	% of all drainage	% of all cropland
Illinois	1	3.97	90	35
Indiana	2	3.27	85	50
Iowa	3	3.15	90	25
Ohio	4	3.00	80	50
Minnesota	7	2.58	75	20
Michigan	11	2.32	70	30
Missouri	13	1.72	70	25
Wisconsin	16	0.91	45	10

Table 25. Drainage-induced differences in the SOC content of a Crosby-Kokomo soil association in Ohio (modified from Fausey and Lal, 1992).

Tillage	SOC content (%)		
	Drained	Undrained	t-test
No till	1.24	1.94	*
Raised beds	1.35	1.76	NS
Ridge till	0.59	1.47	*
Moldboard plow	0.94	1.41	*

Significant at 5% level of probability
NS = not significant

Drainage of cropland in the Midwest began after 1850, when the Swamp Land Acts of 1848 and 1850 released large amounts of swamps and wetlands still owned by the federal government (Fausey et al., 1995). These lands were released for private development, and organization of drainage districts was initiated in the early 1900s. The Reclamation Act of 1902 led to the creation of the Bureau of Agricultural Engineering within the USDA.

A large portion of the cropland area thus has received drainage in the eight states that make up the Great Lakes and Corn Belt states (Table 24) — Illinois, Indiana, Iowa, Michigan, Minnesota, Missouri, Ohio, and Wisconsin. In Ohio, over 60% of the cropland has an excess wetness limitation. Drainage of wetlands for agricultural production accounts for 87% of national wetland losses, and 2/3 of remaining wetlands are in agricultural areas (Prato et al., 1995). Dramatic losses in wetlands (from 87.1 Mha in the lower 48 states in the 1600s to 40.1 Mha in 1990s) led President Bush to adopt a no-net-loss policy for wetlands in 1990.

The process of anaerobiosis in undrained arable lands with seasonal wetness leads to emissions of radiatively active gases, e.g., CH_4 and N_2O (Lal et al., 1995). However, providing drainage to such lands improves aeration and enhances mineralization of SOC content. Draining seasonally wet agricultural land leads therefore to a net loss of SOC content, especially when drained upland is managed with conventional tillage. Fausey and Lal (1992) and Sullivan et al. (1998) observed that drained soil had lower SOC content than undrained soil for all tillage methods (Table 25). The loss of SOC due to drainage in the top 50-cm depth was 36.4% for no till, 23.3% for raised beds, 40.0% for ridge till, and 33.3% for moldboard plow. The average loss, mean of all tillage methods, was 33.3%. Improved management of such lands may restore SOC content, such as does conversion from a conventional to a CT system.

Drained agricultural lands are better managed through CT or through a sub-irrigation system, which involves recycling the drained water during summer. That improves soil moisture content and biomass yield and is projected to increase SOC content by 0.05% to 0.075%

Table 26. Cropland area drained and in need of drainage in the U.S. (USDA-ERS, 1987).

Item	U.S.	29 selected states	21 irrigation states
		10^6 ha	
Drained land	43.4	31.9	11.5
Irrigated cropland drained	7.8	1.1	6.7
Drainage treatment needed	1.0	0.7	0.3

with an average rate of 0.0625%/yr over the next 25 years.

Total cropland area that has been drained is about 43.4 Mha (Table 26). Potential C sequestration with improved management of these lands is 3.8 MMTC/yr (Eq. 20).

C sequestration potential of 43.4 Mha with increase in SOC at 0.0625%/yr =

$$(43.4 \text{ Mha}) \times (10^4 \frac{m^2}{ha})(1.4 \frac{MT}{m^3})(6.25 \times 10^{-4}) \times 1m = 3.80 \times 10^6 \text{ Mg} = 3.8 \text{ MMTC/yr} \ldots\ldots\ldots\ldots \text{Eq. 20}$$

Improved Cropping Systems

Adoption of improved cropping systems has a vast potential for C sequestration. An important component of BMPs (best management practices) is increasing the efficiency of fertilizer use. Improved cropping systems and more efficient fertilizer use are integral components of agricultural intensification.

Fertilizer Management

Soil fertility status affects the amount of biomass produced, and the SOC content is related *directly* to the quantity of crop residues returned to the land and *inversely* to the N deficit in the soil. Several experiments have shown that additions of fertilizers on a regular basis for many years often lead to an increase in SOC content (Rasmussen et al., 1980; Janzen, 1987a, b; Campbell et al., 1991a, b; Glendining and Powlson, 1991).

Paustian et al. (1992) reported that fertilizer N addition increased the SOC level by 15 to 19% by increasing net primary productivity (NPP) and the residue-C input. In southwestern Saskatchewan, Campbell and Zentner (1993) observed a strong positive correlation between soil organic nitrogen and the quantity of crop residue returned, and a strong negative correlation with apparent N deficit (e.g., N exported in grains-N applied as fertilizer). Franzluebbers et al. (1994) observed that the SOC content of the 0- to 50-mm depth was 62% more, in wheat cultivation, with fertilization than without. Robinson et al. (1996b) observed a 22% increase in SOC content due to application of NPK fertilizer to a soil in Iowa. Gregorich et al. (1996) observed that soil under continuous corn, fertilized for more than 30 years, had greater amounts of SOC than unfertilized systems. They estimated that, in fertilized soils, from 22 to 30% of SOC in the plow layer was derived from the corn residue. Application of fertilizers also may improve soil physical quality, e.g., increase the geometric mean diameter of aggregates (Darusman et al., 1991).

The long-term experiment on Sanborn plots at Columbia, Missouri, highlights the importance of soil fertility management on SOC content. Anderson et al. (1990) observed that, following 100 years of continuous cropping, SOC contents in unfertilized versus fertilized plots were, respectively, 0.9% and 1.3% under wheat, 0.5% and 1.1% under corn, 1.4% and 1.4% under timothy, and 1.2% and 1.5% under rotation.

The data in Table 27 show that soil fertility management practices followed over a century had an important impact on topsoil thickness for corn but little or no effect for soil under rotation or in timothy. The least topsoil thickness (under corn) is presumably due to the highest erosion rate and also shown by the highest clay content from exposed subsoil. For all three cropping systems studied, plots that received manure or inorganic fertilizers

Table 27. Average topsoil thickness and soil organic matter and clay contents after 100 years of cropping in Sanborn Field under 6-year rotation (corn-oat-wheat-clover-timothy-timothy) (Gantzer et al., 1991).

Management	Topsoil thickness (cm)	Soil organic carbon (%)	Clay (%)
Corn unfertilized	17.8	0.7	26.8
Corn manured	21.3	1.4	30.0
6-year rotation unfertilized	31.0	0.9	14.6
6-year rotation manured	28.5	1.5	18.4
6-year rotation fertilized	33.5	1.3	17.6
Timothy, unfertilized	45.0	1.3	16.7
Timothy, manured	43.5	2.1	16.7

had 60 to 100% more SOC concentration than those that received neither. The SOC concentration in continuous corn with application of manure (1.4%) was double that without application of manure (0.7%). Similarly, manure application maintained the SOC concentration of the 6-year rotation plot at 1.5%, compared with only 0.9% for unmanured treatment, a difference of 56%.

In pasture, SOC content of manured plots was 59% more than that of the unmanured treatment. Soil fertility maintenance, through manuring and inorganic fertilizer application, therefore maintained SOC concentration at high levels. In addition to the high production of below- and above-ground biomass, erosion control was an important factor in the high SOC content of manured plots. The conservation effectiveness of manuring and inorganic fertilizer treatments under different management systems appears in topsoil thickness and the clay content (Table 27).

Data (Table 28) from a 32-year experiment show that use of fertilizer on corn increased SOC content in the upper 42-cm layer by 8 MT/ha or at the rate of 0.25 MT/ha/yr. Varvel (1994) reported on the effects of nitrogen fertilization on SOC dynamics in Sharpsburg silty clay loam near Mead, Nebraska. The data in Table 29 show that the rate of N fertilizer application had a significant effect on SOC content.

The fact that SOC content declined drastically when no N application occurred has important policy implications. In contrast, optimal rates of N fertilizer (both organic and inorganic) application increase SOC content.

The net cropland area that receives chemical fertilizers and organic amendments is about 117.5 Mha, most of which is receiving adequate rates of fertilizers. Fertilizer use in the U.S. from 1989-90 to 1995-96 (Table 30) remained fairly stable, with fertilizer N, P, and K use changing by 1.1, 0.2, and 0.0 MMT of nutrients, respectively, during that period. Nonetheless, soil and crop management practices, better use of soil test information, site-specific applications of fertilizers, and precision farming techniques can make

Table 28. Amount of total SOC in a Brookston clay loam soil under fertilized and unfertilized corn after 32 years of cultivation (Gregorich et al., 1996).

Treatment	Depth	SOC content
	(cm)	(MT/ha)
Fertilized corn	0-10	28.5 ± 1.3
	10-26	46.2 ± 2.0
	26-42	14.6
	Total	89.3
Unfertilized corn	0-10	25.4 ± 2.2
	10-26	39.2 + 2.1
	26-42	16.7
	Total	81.3
Sod	0-10	62.7 ± 4.8
	10-22	44.0 ± 4.9
	22-40	25.0
	Total	131.7

Table 29. Total C sequestered in each rotation at different rates of N fertilizer application for 0- to 15-cm depth from 1984 through 1992 at Mead, NE (modified from Varvel, 1994).

SOC sequestered at different N rates			
Rotation	0	Low	High
	————kg/ha/yr————		
Continous corn	-534	186	1358
Corn-soybeans	-855	-203	-363
Continuous grain sorghum	259	1316	1538
Corn-sorghum	-283	186	-69
Grain sorghum-soybeans	333	228	892
Corn-Oats+clover-grain sorghum-soybeans	870	644	1420
Corn-soybeans-grain sorghum-oats+clover	454	1012	1261

fertilizer use far more efficient (Larson and Robert, 1991). Assuming that improved soil fertility management practices can enhance SOC content at a rate of 50 to 150 kg/ha/yr with an average rate of 100 kg/ha/yr, the C sequestration potential through soil fertility management on cropland is 11.8 MMTC/yr (Eq. 21).

Average C sequestration through soil fertility management =

$$(117.5 \text{ Mha}) \times 0.1 \ \frac{\text{MTC}}{\text{ha yr}} = 11.8 \text{ MMTC/yr} \ \dots\dots \text{ Eq. 21}$$

This is a very conservative estimate, and the actual C sequestration potential through soil fertility management may be 2 to 4 times higher. Even this conservative estimate shows an offset of the C-emission associated with fertilizer production. Total fertilizer use in the U.S. is about 20 MMT of plant nutrients, including 11 MMT of N (Table 30). Each MT of synthesized N causes emission of 1-2 MTC. Improving N use efficiency thus can reduce drastically the C emission synthesizing fertilizer causes.

Table 30. Fertilizer use in the U.S. (FAO, 1996; USDA-ERS, 1994).

Year	N	P	K	Total
	\-------MMT of nutrients\-------			
1989/90	10.0	3.9	4.7	18.6
1990/91	10.2	3.8	4.5	18.5
1991/92	10.4	3.8	4.6	18.8
1992/93	10.3	4.0	4.7	19.0
1993/94	11.5	4.1	4.8	20.4
1994/95	10.6	4.0	4.6	19.2
1995/96	11.1	4.1	4.7	19.9

Vast improvement in the efficiency of fertilizer use is possible in the U.S. and throughout the world. The GWP mitigation potential of soil fertility management, however, cannot be accurately assessed without considering the specific cropping system involved, because soil fertility management is linked closely with the cropping system.

Organic Manures and By-products

Applying plant nutrients and especially N to soil, either as inorganic fertilizers or in organic forms, influences the SOC content. Several long-term experiments support this conclusion by demonstrating that regular application of organic manures/composts increases SOC content (Table 27) (Glendining and Powlson, 1991). Jenkinson (1991) reported a 3-fold increase in the SOC pool in soil by applying 35 T/ha of farmyard manure, and SOC content presumably is increasing still after 150 years of the Rothamsted experiment in the United Kingdom. Over the same period, the unmanured plots lost some SOC, and those fertilized with 144 kg N, 35 kg P, and 90 kg K (about 50% of the N and K of the manured plots) gained some SOC until 1920 and then equilibrated at about 10 T/ha higher than the unfertilized plots.

Tester (1990) used high rates (33 MT/ha to 268 MT/ha) of manure on Evesboro loamy sandy soil and observed that 24 to 37% of the annually added organic matter from sewage-sludge and beef manure, respectively, decomposed during the first year. During the next 3 years, an additional 28 and 37% of the residual organic matter decomposed. Consequently, SOC content in the top 13-cm layer increased from 0.06% to 6.5% over the 50-month period.

Paustian et al. (1992) observed that lignin content of the organic residue applied had a strong positive effect on SOC accumulation. With the addition of organic amendments, the percent of active SOC content increased during the first 10 years. From a 58-year study, Collins et al. (1992) observed that SOC content was the highest in the manured treatment, compared to N-fertilized, non-fertilized, or burned residue treatments. Davis et al. (1997) evaluated the impact of the application of feedlot manure on the SOC content of sandy (primarily loamy sands and some sandy loams) and loamy soils. Sandy soils received 40-66 MT, and clayey soils received 44-77 MT manure/ha/yr. The SOC content of both soils increased, especially in the top 0- to 20-cm layer.

The U.S. produces a large amount of a wide range of organic materials. These materials include livestock wastes (Tables 31-34), biosolids (sewage sludge) and septage, food processing wastes, industrial organic wastes, logging residues, wood processing wastes, and yard clippings (Table 35). Agricultural composting is increasing in the U.S. (Kashmanian and Rynk, 1996). These anaerobically digested biosolids contain 1-6% organic N, 1-3% ammonium N, 1.5-5% total P, and 0.2-0.8% total K (Forste, 1997). Judiciously applied to cropland, these wastes are a valuable resource (Muchovej and Pacovsky, 1997).

Table 31. Animal manure and nutrients excreted by feedlot beef cattle in the U.S. (USDA-ARS, 1990).

State	No. animals (10⁶)	Manure 10⁶ Mg	N	P	K
				—10³ Mg—	
California	0.39	0.94	17.9	6.11	8.8
Colorado	0.90	2.16	41.0	14.0	43.2
Illinois	0.30	0.72	13.7	4.7	14.4
Iowa	1.02	2.45	46.6	15.9	49.0
Kansas	1.70	4.08	77.5	26.5	81.6
Minnesota	0.33	0.79	15.0	5.1	15.8
Nebraska	2.15	5.16	98.0	33.5	103.2
Oklahoma	0.32	0.77	14.6	5.0	15.4
South Dakota	0.27	0.6	12.4	4.2	13.0
Texas	2.11	5.06	96.1	32.9	101.2
U.S.	10.06	24.1	457.9	156.7	482.0

Table 32. Manure produced by the swine industry in the U.S. (Sweeten, 1992).

Pasticulture	Quantity
Number of head (10⁶)	55.3
Solid manure production (10⁶ Mg)	14.1
N (10⁶ Mg)	0.66
P (10⁶ Mg)	0.42
K (10⁶ Mg)	0.66

Table 33. Poultry manure produced on U.S. farms in 1990 (Moore et al., 1995).

Animal	Number (M)	Manure generated (10⁶ kg)
Broilers	5966	8948
Layers	272	1044
Turkeys	283	3085
Total	6520	13078

Table 34. Annual manure produced by various animals (equalized on 454 kg or 1000 lbs. of animal mass) (modified from Mikkelsen, 1997).

Animal species	Manure production (MT/yr)
Broiler	4.1
Sheep	5.4
Horse	7.3
Beef	7.7
Dairy	10.9
Swine	14.5

Table 35. Municipal solid wastes of organic origin produced in 1990 (USDA-ARS, 1993).

Material	Weight generated (10^6 Mg)	% total municipal solid waste
Wood	11.2	6.3
Food waste	12.0	6.7
Yard trimmings	31.8	17.9
Total	55.0	30.9

Table 36. Land use changes from 1982-1992 (USDA-SCS, 1992).

Use	1992	Change 1982-1992
	Mha	
Crop	154.8	-15.6
Pasture	51.0	-2.4
Range	161.5	-4.0
Forest	159.9	+0.24

Walker et al. (1997) estimated that, of the more than 1000 MMT of organic and inorganic agriculturally recyclable by-products generated each year in the U.S., about 40% (400 MMT) are crop residues, 5% dairy and beef manures (50 MMT), 3% poultry and swine manure (30 MMT), 15% municipal solid waste (150 MMT), and less than 1% (<10 MMT) biosolids. The municipal solid waste of organic origin is about 55 MMT/yr, including wood (11.2 MMT), food waste (12.0 MMT), and yard trimmings (31.8 MMT) (USDA-ARS, 1993). In addition to crop residue, the total organic material available for disposal on agricultural soils is about 145 MMT, containing about 58 MMTC. If 5% to 15% (average 10%) of this C can be sequestered in soil as SOC through judicious application on cropland, the C sequestration potential of this resource is 5.8 MMTC/yr (Eq. 22).

Average C sequestration potential of using organic materials =

$$145 \ \frac{MTC}{yr} \times 0.4 \times 0.1 = 5.8 \ MMTC/yr \ \dots\dots\dots\dots \ Eq. \ 22$$

Rotation and Winter Cover Crops

Croplands constitute an important component of major land use in the U.S. (Table 36). Principal crops include wheat and other close-grown cereals, corn and other widely spread row crops, and soybean and other short canopy row crops. Increasing SOC content in cropland requires the addition of biomass-C that compensates for or exceeds annual losses from oxidation, erosion, leaching, volatilization, and crop removal.

In addition to fertility management and erosion control, increased cropping intensity can lead to an increase in SOC by increasing the return of plant residues to the soil. Crop management strategies that alter the timing, placement, quantity, and quality of crop residue input can affect the size, turnover, and vertical distribution of both active and passive pools of SOC (Franzluebbers et al., 1994). Crops grown in rotation often produce more and higher quality plant dry matter than those grown in monoculture (Copeland and Crookston, 1992). Cropping systems and rotations affect SOC content through differences in tillage methods and fertility maintenance practices.

Salter and Green (1933) computed rates of SOC loss for different rotations from a 30-year experiment. Change in SOC content, as a percentage of the total amount present in the soil, ranged from -3.12%/yr for corn to +1.36% for hay. A high (r^2 = 0.71) correlation existed between SOC content and crop yield. Wood et al. (1991a, b) observed that 4 years of wheat-fallow rotation caused a decline of 620 kg/ha of SOC, compared with an increase of 280 kg/ha under wheat-corn-millet-fallow rotation. Hunt et al. (1996) observed that the magnitude of differences in SOC content due to tillage methods depended on the crop (corn, wheat, or cotton).

Experiments at the Sand Mountain Substation near Crossville, Alabama, showed that the effect of rotation on SOC content depended on the amount of residues returned (Edwards et al., 1992). The SOC content at 0- to 30-cm depth was 11.8 g/kg for continuous soybean, 12.1 g/kg for corn-soybean, and 13.5 g/kg for continuous corn. Wood et al. (1991a, b) monitered similar effects and also observed the highest SOC content under corn-wheat, compared with soybean-wheat or corn-wheat-soybean-wheat rotation.

The effects of crop rotation on SOC content, based on growing winter cover crops, have been documented widely in numerous long-term experiments conducted throughout the U.S. Planting winter cover crops after fall harvest increases cropping intensity and residue return and reduces the length of time the soil is left fallow. Growing cover crops can increase the SOC content in the soil (Rosswall and Paustian, 1984; Stivers and Shannon, 1991; Johnston, 1991; Wood et al., 1991a, b; Campbell et al., 1991; Collins et al., 1992; Campbell and Zentner, 1993; Robinson et al., 1996a, b; Hu et al., 1997). However, the magnitude of increase may be small in some instances (Kuo et al., 1997) and will vary among soils, ecoregional characteristics, and cover crops, depending on the biomass produced. For a Kamouraska clay soil, Angers (1992) observed that SOC content under alfalfa increased from 26 g/kg to 30 g/kg over a 5-year period. Krall et al. (1995) reported that complete removal of alfalfa was not necessary to grow the following no-till corn.

The data in Fig. 13 (page 60) from Dick et al. (1986a) show that the highest SOC content in soils was under the corn-oats-meadow rotation. The total SOC content in the top 45-cm depth was 51.8 MT/ha for corn-soybean rotation, 56.1 MT/ha for continuous corn, and 61.0 MT/ha for corn-oats-meadow rotation. Incorporation of cover crops (vetch or grass with a potential to produce a large amount

Table 37. Some legume cover crops for use in CT systems and to enhance C sequestration (compiled from Power et al., 1983).

Region	Cover crop
1. Southeast	Crimson clover, hairy vetch, common vetch, crown vetch, alfalfa, lupines, arrow leaf clover, red clover
2. Temperate Regions	Hairy vetch, big flower vetch, and alfalfa
3. Corn Belt and Great Plains	Hairy vetch (cultivar Madison), alfalfa, and sweet clover
4. Pacific Northwest	Several grain legumes can be grown in these regions, e.g., peas, lentils, dry beans, chick peas, and lupines
5. Northeast and Midwest	Hairy vetch, Korean lespedeza, alfalfa, sweet clover, red clover

of below-ground biomass) within a rotation can improve SOC content significantly (Frye et al., 1988; Utomo et al., 1990; Lal et al., 1998; Delgado et al., 1998).

In addition to providing a large amount of biomass, legume cover crops also are a principle source of N. Some legume crops listed in Table 37 can provide 90 to 120 kg of N/ha to the following grain crop, as well as reduce soil erosion and limit runoff from winter precipitation. The use of winter legume crops has been limited because of competition for water, especially in sub-humid and semiarid regions. Even if winter legume crops trap snow or enhance other water conservation measures, climatic variability in the drier regions can limit emergence and growth of the following grain crop.

This limitation also can restrict double cropping or multiple cropping even in regions where irrigation is available (Power et al., 1983). Even when legumes are grown as a green manure crop, such as in the Midwest and Great Plains, they often can deplete soil water to the extent that yield of the following grain crop is reduced. Therefore, selection of appropriate species, choice of an adapted cultivar, and development of management techniques are essential to successful incorporation of legume cover crops in a rotation cycle (Table 37).

Winter cover crops can be grown in 26 states for cropland areas sown to corn and soybeans (Table 38). Since winter wheat is itself a winter cover crop, the land in winter wheat cannot be sown to a different cover crop. Therefore, land area suitable for adopting improved cropping systems based on a rotation with winter cover crop is about 51 Mha.

With improved rotational and management systems adopted over the next 25 years, the SOC content of these lands can be increased at 100 kg/ha/yr to 300 kg/ha/yr, with an average rate of 200 kg/ha/yr. The total C sequestration potential of these lands is thus about 10.2 MMTC/yr (Eq. 23).

Average C sequestration potential of adopting improved rotations on 51×10^6 ha over 25 year period =

(51 Mha) (5 MT/ha)/25 years = 10.2 MMTC/yr...Eq. 23

Lal, Kimble, Follett, and Cole

Elimination of Summer Fallow

Summer fallowing often is practiced in semiarid regions. Land under summer fallow in the U.S. totals 9.4 Mha, or 7% of the total cropland. Summer fallowing reduces SOC content by decreasing inputs of plant residues, increasing decomposition rates, and increasing soil erosion (Photo 19) (Rasmussen et al., 1980; McGill et al., 1981; Hargrove, 1986; Havlin et al., 1990; Campbell et al., 1991). Angers (1992) reported that a fallow-based system resulted in an SOC loss. Wood et al. (1991a, b) observed that a wheat-fallow system for a 4-year period resulted in a loss of 620 kg/ha of SOC. Replacement of a fallow system with intensive cropping has been shown to increase SOC content (Collins et al., 1992; Wood et al., 1991a; b). Bremer et al. (1995) observed that, over a 25-year period, SOC content in the 0 to 15-cm depth was 11% higher in continuous wheat than in wheat-fallow rotation. Summer fallow accounts for about 9.4 Mha of cropland. The potential for C sequestration through increase in SOC of 100 kg/ha/yr to 300 kg/ha/yr with an average SOC sequestration rate of 200 kg/ha/yr is 1.9 MMTC/yr (Eq. 24).

Table 38. Land area under row crops that can be grown in rotation with winter cover crops in appropriate states (USDA-SCS, 1992).

State	Wheat	Corn	Soybeans
		—10^3 ha—	
Alabama	18.4	190.9	358.2
Arkansas	278.5	40.8	1358.2
Connecticut	0	23.1	0
Delaware	20.9	76.3	69.4
Georgia	77.7	394.9	455.7
Illinois	373.8	5286.6	3547.7
Indiana	193.0	2646.8	2054.0
Iowa	11.6	5415.9	3410.6
Kentucky	105.0	683.2	338.9
Maryland	87.8	294.4	119.9
Massachussettes	0	15.4	0
Michigan	255.1	1448.4	473.4
Minnesota	1103.1	3127.2	2086.3
Missouri	716.4	1050.3	1875.5
Nebraska	889.1	3351.4	899.9
New Jersey	13.4	54.7	60.8
New York	45.2	566.0	19.1
North Carolina	252.4	652.5	517.3
Ohio	681.2	1588.0	1554.7
Pennsylvania	111.8	779.8	31.7
Rhode Island	0	0.9	0
South Carolina	142.5	272.2	396.7
Tennessee	61.8	337.2	560.9
Virginia	108.9	217.9	153.8
West Virginia	3.4	50.2	7.4
Wisconsin	27.6	1790.2	223.1

Total land area under corn = 30.4 million ha
Total land area under soybeans = 20.6 million ha
Total land area suitable for winter cover crops = 51 million ha

Average C sequestration potential on 9.4 Mha =

$$(9.4 \text{ Mha}) \times \frac{0.2 \text{ MT}}{\text{ha yr}} = 1.9 \text{ MMTC/yr} \quad \text{............ Eq. 24}$$

Photo 19. Soil erosion from land under summer fallow.

Improvement in Crop Yields

Because of the increasing demand for food grains and oil seeds throughout the world, it is likely that crop production in the U.S. will increase (Table 39). The rate of increase may be 1.4%/yr for the 20-year period between 1990 and 2010. A comparison of the data on crop residue in Table 21 (page 65)with

Table 39. Projected increase in grains and soybean production in the U.S. (Flach et al., 1997; Crosson, 1992; USDA-FAS, 1991a; b).

Crop	Production (MMT) 1987-1990	2010	Av. annual increase %
Wheat	59.2	87	1.8
Coarse grain	204.7	264	1.3
Soybean	49.9	69	1.5
Total	313.8	425	1.4

the estimates of USDA for residue production in 1978 shows an increase in residue production from 319 MMT in 1978 to 508 MMT in 1996, or an increase of 10.5 MT/yr. Increased residue production is associated with increased grain yield. The increase in yield will be due to varietal improvement, efficient use of fertilizer and pesticides through site-specific management, widespread adoption of integrated pest management (IPM) and other advances in biotechnology. Increases in crop yield also will lead to corresponding increases in crop residue and biomass that may be returned to the soil.

On the basis of the data in Table 40, a conservative estimate of the additional rate of residue C produced will be 5 to 7 MMTC/yr, with an average of 6 MMTC/yr. Assuming that 5 to 15% (average 10%) of this C is sequestered in the soil through adoption of BMPs, the rate of C sequestration is 0.6 MMTC/yr (Eq. 25).

Lal, Kimble, Follett, and Cole

Average C sequestration through increase in biomass production =

$$(447.3 \text{ MMTC/yr}) \times \frac{1.24}{100} \times \frac{10}{100} = 0.6 \text{ MMTC/yr} \dots\dots \text{Eq. 25}$$

Management of Rice Paddies

Land area used for rice cultivation in the U.S. is about 1.3 Mha, which is 0.7% of the world rice area and less than 1% of the U.S. cropland. Rice is grown in two geographical regions of the country, the lower Mississippi River valley (Arkansas and Mississippi) and the Sacramento River Valley (California). Total rice production in the U.S. is 7.9 MMT (FAO, 1995). Total rice straw production is estimated at 12.5 MMT, which contains about 5 MMT of C. Improved management of this straw (through compost and mulch farming rather than burning) may lead to C sequestration at the rate of 1 MMTC/yr.

The Carbon Sequestration Potential of Arable Land

The discussion in this chapter shows the C sequestration potential of adopting improved cropping/farming systems (Table 40). The total potential is 75-208 MMTC/yr, of which about 50% is due to conservation tillage and residue management, 6% to supplemental irrigation and water table management, and 25% to adoption of improved cropping systems.

Table 40. Estimates of total C sequestration potential through improved management of U.S. cropland.

Scenario		MMTC/yr
A. Land Conversion and Restoration		
1. Land Use		6-14
a. Conservation Reserve Program	5-11	
b. Conservation Buffers	1-2	
c. Wetland Reserve Program	0.3-0.7	
2. Land/Soil Restoration		11-25
a. Eroded lands	9-20	
b. Mine land	0.6-2	
c. Salt affected soils	1-3	
B. Intensification of Prime Agricultural Land		
3. Conservation Tillage and Residue Management		35-107
a. Conservation tillage	24-40	
b. Residue management	11-67	
4. Irrigation/water management		5-11
a. Supplemental	2-6	
b. Sub-irrigation on poorly drained soils	3-5	
5. Improved Cropping Systems		19-52
a. Fertilizer management	6-18	
b. Organic manures and by-products	3-9	
c. Rotation & winter cover crops	5-15	
d. Summer fallow elimination	1-3	
e. Management of rice straw	0.5-1.5	
f. Idle land conversion to biofuel production	3-5	
Total Potential		75-208

MMTC = million metric tons of carbon

Priority strategies for C sequestration in this category are conservation tillage (23%), residue management (26%), fertilizer management (9%), organic manures and by-products (4%), rotations and winter cover crops (7%), summer fallow elimination (1%), improvement in crop yields (<1%), management of rice straw (1%), irrigation management (6%), and C sequestration in idle land (3%). In addition, land conversion (CRP, WRP, etc.) has the potential to sequester 7% and land restoration an additional 13% of the total potential (Table 40).

CHAPTER 11

U.S. Cropland's Overall Potential to Mitigate the Greenhouse Effect

The U.S. cropland's overall potential to sequester C sequestration is 75-208 MMTC/yr (Table 40, Chapter 10, page 81). This potential includes 7% due to land conversion (e.g., CRP, conservation buffers, WRP), 13% due to restoration of degraded soils, 49% due to conservation tillage (CT) and residue management, 6% due to irrigation water management, and 25% due to adoption of improved cropping systems.

Properly managed, U.S. cropland thus can be a major sink for C sequestration. The total C sequestration potential of 75-208 MMTC/yr represents a large sink in comparison with total U.S. emissions of GHGs at either 1485 MMT (8.5%) or 1709 MMT (7.3%). The total SOC sequestration potential is 1.16 or 1.03 times the emission from all agricultural activities, which are estimated at 109 MMT or 123 MMT/yr.

Techniques for Sequestration

Techniques with a large potential for C sequestration are erosion prevention and control through CT and residue management, improved cropping systems, land restoration, land conversion, and irrigation and water management (Fig. 14). This potential can be realized through appropriate policies that encourage farmers to adopt improved soil, crop, and water management practices.

Table 41 presents an overall summary of the potential of U.S. cropland for C sequestration in soil, fossil fuel offset, and savings in energy due to adoption of CT. In addition to the C sequestration potential of 75-208 MMTC/yr, fossil fuel offset through production of biofuel potentially totals 35-63 MMTC/yr. A savings of fuel consumption equal to 1-2 MMTC/yr results from conversion from conventional till to a CT system. Effective control of soil erosion can curtail C emission from displaced sediment, totaling an estimated 12-22 MMTC/yr (Table 41). The overall potential of U.S. cropland to mitigate CO_2 is thus 120 to 270 MMTC/yr.

These estimates do not consider the CO_2 fertilization effect due to potential climate change. A high degree of uncertainties exist in our estimates and about the level at which various mitigation options can be implemented.

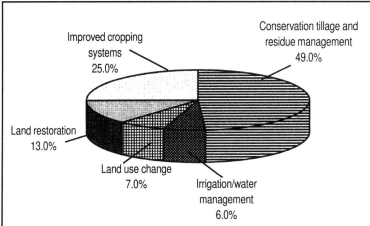

Figure 14. A pie diagram showing C sequestration potential of different components of improved management of U.S. cropland (see Table 41, below, for details).

Conversion of natural ecosystems to agriculture has caused an estimated total loss of 4000 to 6000 MMT SOC from U.S. cropland. Partial recovery of this loss, achievable through the range of alternatives Chapter 6 outlines (Table 6, page 27), can mitigate CO_2 emissions substantially.

The available data show clearly that SOC loss from cultivated land is very rapid, and the total loss may be 20 to 50% during the initial 20 to 50 years of cultivation (Tiessen et al., 1982; Mann, 1986;

Table 41. The overall potential of U.S. cropland for C sequestration and fossil fuel offset and erosion control.

Carbon sequstration/fossil fuel offset Scenario	MMTC/yr
1. C sequestration in soil	75-208
2. Carbon offset through biofuel production	32-38
3. Saving in fuel consumption	1-2
4. Reduction of C emission from eroded sediments	12-22
Total	120-270

Rasmussen and Parton, 1994). The data on restorative effects of various options on C sequestration are scarce. Further, the technology and practices needed to realize this potential are often difficult to implement. It may be easier to prevent the loss of SOC than to restore it.

Rates of SOC Sequestration

The quantity of carbon sequestration in soil has a practical upper limit. In addition, achieving this limit may require at least 50 years. In contrast, C offset through biofuel production and C accumulation in wetlands can be maintained, in principle, indefinitely. The maximum potential of 5000 MMT may not be realiz-

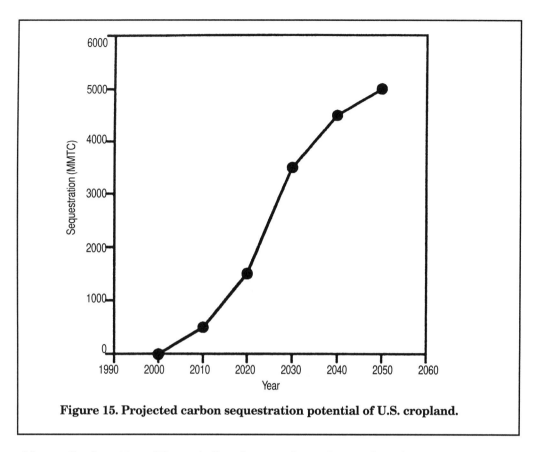

Figure 15. Projected carbon sequestration potential of U.S. cropland.

able until after 2050 (Fig. 15). Reaching it depends on identifying relevant policies, developing appropriate programs, and successfully implementing those policies and programs.

Possible Implementation Obstacles

The potential calculated in this report may be based on optimistic assumptions about the adoption of CT, CRP, winter cover crops, biofuel production, etc. With world population increasing and potential food shortages ahead, the land under CRP may have to be cropped. Further, sub-irrigation may be possible only in warmer parts of the country, and application of biosolids to farmland is subject to the logistical problems of distribution and application.

Other problems also may limit the realization of this potential. With the possible increase in mean global temperature, insect damage to crops and weed infestation on cropland may increase drastically. While reducing productivity, these pests may increase the use of insecticides and herbicides, thus increasing the use of fossil fuel and decreasing the potential of SOC sequestration in cropland.

Table 42. Estimates of land use and land use change in the U.S. (Alaska excluded; Puerto Rico and Virgin Islands included) (Kellogg et al., 1994).

Land use	Area (Mha) 1982	Area (Mha) 1992	Change	Percent change	Percent of land base in 1992
Total surface area	785.122	785.122	0.000	0.0	0.0
Water area	19.426	19.788	0.363	1.9	NA
Land Base	765.696	765.334	-0.362	0.0	100.0
a. Federal land	163.782	165.113	1.331	0.8	21.6
b. Non-federal land	601.914	600.221	-1.693	-0.3	78.4
c. Developed land	31.718	37.375	5.657	17.8	4.9
d. Non-federal rural land	570.196	562.846	-7.350	-1.3	73.5
e. Cropland	170.363	154.724	-15.639	-9.2	20.2
(i) Cultivated	148.201	131.714	-16.487	-11.1	17.2
Irrigated	19.494	19.298	-0.196	-1.0	2.5
Non-irrigated	128.707	112.416	-16.291	-12.7	14.7
(ii) Non-cultivated	22.162	23.009	0.847	3.8	3.0
Irrigated	5.494	5.869	0.376	6.8	0.8
Non-irrigated	16.668	17.140	0.472	2.8	2.2
f. Pasture land	53.375	50.963	-2.412	-4.5	6.7
g. Range land	165.482	161.455	-4.028	-2.4	21.1
h. Forest land	159.599	159.840	0.240	0.2	20.9
i. CRP land	0.000	13.776	13.776	NA	1.8
j. Other rural land	21.378	22.090	0.712	3.3	2.9

The potential of U.S. agriculture to mitigate the greenhouse effect also must be assessed in terms of the increasing and diverse demands on prime agricultural land. The nation's population at the beginning of 1998 is 269 million; it increased by 2.4 million during 1997. The overall growth rate is 0.9%, or a total increase of 8.1% during the 1990s. The U.S. population doubles every 60 years, and each person consumes a large quantity of energy, and this accentuates the problem of urbanization.

Prime agricultural land is being lost rapidly to urbanization and nonagricultural uses. The data in Table 42 show that, for the decade ending in 1992, cultivated cropland decreased by 16.5 Mha (11.1%), 84% (13.8 Mha) of which may be due to the CRP and the rest due to other land uses. The prime agricultural land area is shrinking. Therefore, intensified protection of existing prime agricultural land is a high priority.

Required Action

Through the implementation of appropriate governmental policies, farmers should be encouraged to adopt practices that decrease C losses and increase SOC content. Agricultural policy should be reformed to encourage flexible land use for C sequestration in soil, to raise the value of crops produced per unit input of fossil fuel consumed, to encourage the use of biofuels, to minimize the risks of soil degradation, and promote the restoration of degraded soils.

CHAPTER 12

Conclusions:
The Win-Win Strategy

Agriculture is believed to cause environmental problems, especially those related to water contamination and the greenhouse effect. Calculations presented in this volume amply demonstrate, however, that scientific agriculture also can be a solution to environmental issues in general and to mitigating the greenhouse effect in particular. In fact, agricultural practices have a potential to sequester more C in soil than farming emits through land use and fossil fuel combustion.

That potential points to another important criterion that must be considered in evaluating agricultural sustainability: the effectiveness and efficiency of agricultural practices in sequestering C in soil. Sustainable agricultural systems thus involve those soil, crop, and water management techniques that increase productivity while enhancing C sequestration in soil — crop residue management, conservation tillage (CT), nutrient management, precision farming, water management involving drainage and irrigation, and restoration of degraded soils through CRP/WRP, afforestation, and other proven technologies.

Agricultural Profits
from Environmental Improvements

Natural resources management and agricultural sustainability require maintaining high soil quality (e.g., biomass productivity and environmental quality). Soil quality, air quality, and water quality (Fig. 16) are strongly linked, and farmers should be encouraged to maintain soils of their cropland at high quality. Doing so is both economically profitable and socially responsible.

Increases in SOC content increase agronomic productivity. Bauer and Black (1994) observed that 1 MT/ha of SOM content in the upper 30-cm layer equals about 35 kg/ha of additional biomass and 16 kg/ha of wheat grain. Similar relationships between crop yield and SOC content exist for other crops. Pimentel et al. (1995) assumed the cost of soil erosion at the rate of $3/MT of soil for nutrients and $2/MT of soil for water loss.

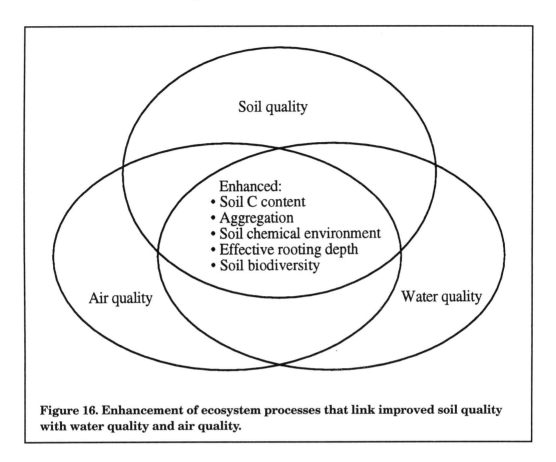

Figure 16. Enhancement of ecosystem processes that link improved soil quality with water quality and air quality.

Both nutrients and the water-holding capacity of soil are directly related to the SOC content. Therefore, indicators of high soil quality include high SOC content, low C:N ratio, high aggregation, high buffering capacity for pH changes, deep effective rooting depth, and high soil biodiversity. High quality soils lead to C sequestration and cause low levels of GHG emissions. Because of their high buffering and filtering capacity, soils of high quality maintain high water quality.

Soil quality management includes restoration of degraded soils and adoption of BMPs on croplands. Soil restoration involves reversing degrading trends through erosion control, drought management, rehabilitation of mineland and other drastically disturbed lands, reclamation of salt-affected soils, soil fertility enhancement and site-specific management, and wetland restoration (Fig. 17). Scientifically proven technologies can enhance overall soil quality, increase biomass productivity, and improve water and air quality. Restorative trends thus are win-win: they have a snowballing effect on improving productivity and enhancing the environment (Fig. 17).

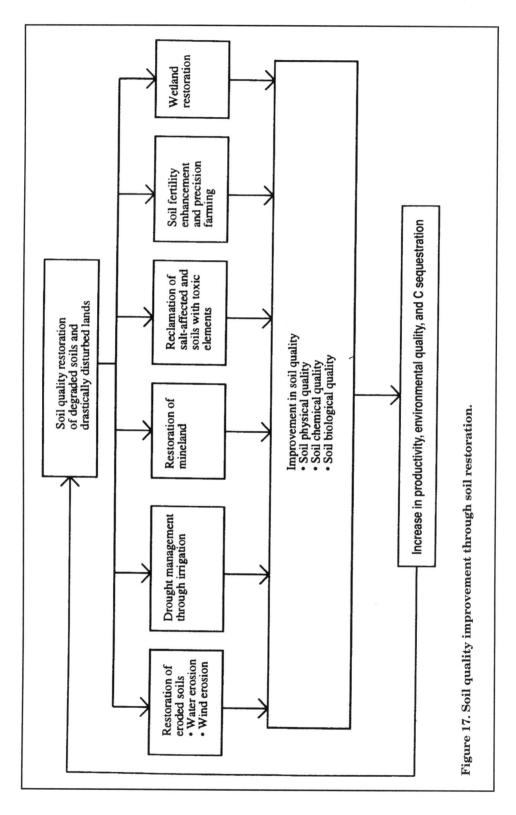

Figure 17. Soil quality improvement through soil restoration.

SOC's Monetary Value

By adopting a system/holistic approach and identifying and implementing appropriate policies, we can realize the vast potential of U.S. cropland for C sequestration at the rate of 75-208 MMTC/yr. Like the other components of an ecosystem (Costanza et al., 1997), the SOC is a valued commodity. Some agronomists (W.D. Kemper, personal communication) estimate the value of plant nutrients and available water in SOC at about $0.20/kg or $200/MT. If so, the SOC sequestered is equivalent to $15 to 42 billion.

Reduction of C emissions on a nationwide scale also would require a carbon tax estimated as ranging from $50 to $350/MT of C (OECD, 1994). Assuming an average tax of $50/MT of C, a total savings of 123 to 295 MMTC/yr is equivalent to $6 to 15 billion/yr. Therefore, total savings by C sequestration in soil and fossil fuel offset from proper land use and soil management is about $21-57 billion/yr or $80-220 per capita per year.

SOC's Environmental Value

The value of SOC is more than just its ability to hold water and nutrients for plant growth. Its hidden value lies in its ability to regulate the environment, especially to mitigate the greenhouse effect, and we must objectively assess the value of a unit quantity of SOC in cropland in terms of its total stock of SOC, and farmers' contributions to the environment in sequestering C in the soil. Part of the atmospheric enrichment of CO_2, CH_4, and N_2O is due to agricultural activities. Adoption of BMPs on croplands can reverse these trends while helping to mitigate the enrichment resulting from U.S. emissions outside of agriculture by sequestering C in soil.

Realizing this vast potential depends on society's willingness to foster BMPs to restore degraded soils and to enhance the soil quality of degraded lands and ecosystems. The win-win scenario will increase productivity, improve soil quality, and mitigate the greenhouse effect. It is important, therefore, that C lost from the pedosphere through historical land use and farming practices be restored to its natural potential.

Global Potential

Through implementation of a coordinated program on C management in soils, the U.S. also can provide an example to the world community that restoration of degraded soils on a global scale is an option to mitigate the greenhouse effect. About 2×10^9 ha of soils worldwide are degraded (Oldeman, 1994) and low in SOC content. Soil restoration, by establishing vegetative cover and improving soil nutrient supply, would enhance SOC and SIC contents, improve soil quality, and increase biomass productivity.

Lal (1997) estimated that increasing the SOC content of degraded soils at the rate of 0.01%/yr could lead to C sequestration at the rate of 3000 MMTC (3.0 Pg/yr) (Eq. 26). This rate of C sequestration is equal to the net annual increase in atmospheric CO_2. Herein lies an option for a truly win-win strategy, based on restoring degraded soils, enhancing soil quality through C sequestration, increasing agronomic productivity, improving water quality, and mitigating the greenhouse effect. U.S. policy makers should strongly emphasize the importance of this strategy to the world community (Eq. 26).

Average C sequestration potential through restoration of degraded soils in the world =

$$2 \times 10^9 \text{ ha} \times 10^4 \frac{m^2}{ha} \times 1m \times 1.5 \text{ MT} \times 10^{-4} \text{ g/g} = 3000 \text{ MMTC/yr} \quad \text{..................................... Eq. 26}$$

The potential for increased C sequestration also exists worldwide, through adoption of improved agricultural and forestry practices, better management of rice paddies and wetlands, and fertility management through nutrient cycling and application of organic by-products.

References

Aase, J.K., and J.L. Pikul, Jr. 1995. Terrace formation in cropping strips protected by tall wheatgrass barriers. J. Soil Water Cons. 50: 110-112.

Abelson, P.H. 1995. Renewable liquid fuels. Science 268: 955-.

Adams, R.M., C. Rosenzweig, R.M. Peart, J.T. Ritchie, B.A. McCarl, J.D. Glyer, R. Bruce Curry, J.W. Jones, K.J. Boote, and L.H. Allen, Jr. 1990. Global climate change and U.S. agriculture. Nature 345: 219-224 pp.

Anderson, S.H., C.J. Gantzer, and J.R. Brown. 1990. Soil physical properties after 100 years of continuous cultivation. J. Soil Water Cons. 45: 117-121.

Angers, D.A. 1992. Changes in soil aggregation and organic C under corn and alfalfa. Soil Sci. Soc. Am. J. 56: 1244-1249.

Angers, D.A., R.P. Voroney, and D. Côte. 1995. Dynamics of soil organic matter and corn residues affected by tillage practices. Soil Sci. Soc. Am. J. 59: 1311-1315.

Angers, D.A., and M.R. Carter. 1996. Aggregation and organic matter storage in cool, humid, agricultural soils. In M.R. Carter and B.A. Stewart (eds) "Structure and Organic Matter Storage in Agricultural Soils", CRC Press, Boca Raton, FL: 193-211.

Armento, T.V., and E.S. Menges. 1986. Patterns of change in the C balance of organic soil wetlands of the temperate zone. J. Ecology 74: 755-774.

Aulack, M.S., J.W. Doran, and A.R. Mosier. 1992. Soil denitrification: significance measurement and effects of management. Adv. Soil Sci. 18: 1-57.

Banerjee, N.K., A.R. Mosier, K.S. Uppal, and N.N. Goswami. 1990. Use of encapsulated calcium carbide to reduce denitrification losses from urea-fertilized flooded rice. Mitteilungen der Deutschen Bodenkundlichen Gesellschaft 60: 245-248.

Barnhisel, R.I., M. Hower, and L.D. Beard. 1992. Development of a soil productivity index for use in prime farmland reclamation. In: R.E. Dunker, R.I. Barnhisel and R.C. Darmody (eds) "Prime Farmland Reclamation", Univ. Illinois, Urbana, IL: 205-212.

Bauer, A., and A.L. Black. 1994. Quantification of the effect of soil organic matter content on soil productivity. Soil Sci. Soc. Am. J. 58: 185-193.

Beare, M.H., P.F. Hendrix, and D.C. Coleman. 1994. Water-stable aggregates and organic matter fractions in conventional and no-tillage soils. Soil Sci. Soc. Am. J. 58: 777-786.

Bennett, O.L. 1977. Strip mining-new solutions to an old but growing problem. Crops and Soils 12-14, January 1997.

Bezdicek, D.F., R.I. Papendick, and R. Lal. 1996. Importance of soil quality to health and sustainable land management. In J.W. Doran and A.J. Jones (eds) "Methods for Assessing Soil Quality", Soil Sci. Soc. Am. Special Pub. 49, ASA-SSSA, Madison, WI: 1-8.

Birdsey, R.A., and L.S. Heath. 1993. Carbon sequestration impacts of alternative forestry scenarios. USEPA, USDA-FS, Washington, D.C.

Black, A.L., and D.L. Tanaka. 1997. A conservation tillage cropping systems study in the northern Great Plains of the United States. In: E.A. Paul, K. Paustian, E.T. Elliott and C.V. Cole (eds) "Soil Organic Matter in Temperate Agroecosystems: Long-term Experiments in North America". CRC Press, Boca Raton, FL: 335-342 pp.

Blevins, R.L., M.S. Smith, G.W. Thomas, and W.W. Frye. 1983. Influence of conservation tillage on soil properties. J. Soil Water Conserv. 38: 301-305.

Blevins, R.L., and W.F. Frye. 1993. Conservation tillage: an ecological approach to soil management. Adv. Agron. 51: 34-77.

Bouwman, A.F. 1994. Estimated global source distribution of N_2O. In K. Minami, A. Mosier and R. Sass (eds) "CH_4 and N_2O: Global Emissions and Controls From Rice Fields and Other Agricultural and Industrial Sources". National Institute of Agro-Environmental Sciences, Tsukuba, Japan: 147-159.

Bremer, E., B.H. Ellert, and H.H. Janzen. 1995. Total and light-fraction carbon dynamics during four decades after cropping changes. Soil Sci. Soc. Am. J. 59: 1398-1403.

Bronson, K.F., A.R. Mosier, and S.R. Bishnoi. 1992. Nitrous oxide emissions in irrigated corn as affected by nitrification inhibitors. Soil Sci. Soc. Am. J. 56: 161-165.

Brown, L. 1991. The global competition for land. J. Soil Water Cons. 46: 394-397.

Bruce, R.R., G.W. Langdale, and A.L. Dillard. 1990. Tillage and crop rotation effect on characteristics of a sandy surface soil. Soil Sci. Soc. Am. J. 54: 1744-1747.

Cambardella, C.A., and E.T. Elliott. 1992. Particulate soil organic matter changes across a grassland cultivation sequence. Soil Sci. Soc. Am. J. 56: 777-783.

Cambardella, C.A., and E.T. Elliott. 1994. Carbon and nitrogen dynamics of soil organic matter fractions from cultivated grassland soils. Soil Sci. Soc. Am. J. 58: 123-130.

Campbell, C.A., D.R. Cameron, W. Nicholaichuk, and H.R. Davidson. 1977. Effects of fertilizer N and soil moisture on growth, N content and moisture use by spring wheat. Can. J. Soil Sci. 57: 289-310.

Campbell, C.A., and W. Souster. 1982. Loss of organic matter and potentially mineralizable N from Saskatchewan soils due to cropping. Can. J. Soil Sci. 62: 651-656.

Campbell, C.A., K.E. Bowren, M. Schnitzer, R.P. Zentner, and L. Townley-Smith. 1991a. Effect of crop rotations and fertilization on soil organic matter and some biochemical properties of a thick Black Chernozem. Can. J. Soil Sci. 71: 377-387.

Campbell, C.A., V.O. Biederbeck, R.P. Zentner, and G.P. Lafond. 1991. Effect of crop rotations and cultural practices on soil organic matter, microbial biomass and respiration in a thin, black Chernozem. Can. J. Soil Sci. 71: 363-376.

Campbell, C.A., and R.P. Zentner. 1993. Soil organic matter as influenced by crop rotations and fertilization. Soil Sci. Soc. Am. J. 57: 1034-1040.

Caspall, F.C. 1975. Soil development on surface mine spoils in western Illinois. In 3rd Symp. on Surface Mining and Reclamation, 21-23 Oct. 1975, Louisville, KY, Vol II. Natl. Coal Assoc., Washington, D.C.: 221-228.

CAST 1992. Preparing U.S. agriculture for global climate change. Council for Agricultural Science and Technology, Task Force Report No. 199, Washington, D.C., 96 pp.

Chaubey, I., D.R. Edwards, T.C. Daniel, P.A. Moore, Jr., and D.J. Nichols. 1994. Effectiveness of vegetative filter strips in retaining surface-applied swine manure constituents. Trans. ASAE 37: 845-850.

Clark, E.H. II, J.A. Haverkamp, and W. Chapman. 1985. Eroding soils: The off-farm impacts. The Conservation Foundation, Washington, D.C.

Cole, V. 1995. Agricultural options for mitigation of greenhouse gas emissions In "Climate Change 1995: Impacts, Adaptation and Mitigation of Climate Change: Scientific Technical Analyses". IPCC Working Group II. Cambridge Univ. Press, Cambridge, U.K.: 745- 771.

Cole, C.V., I.C. Burke, W.J. Parton, D.S. Schimel, D.S. Ojima, and J.W.B. Stewart. 1990. Analysis of historical changes in soil fertility and organic matter levels of the North American Great Plains. In: P.W. Unger, T.V. Sneed, and R.W. Jensen (eds) Proc. International Converence on Dryland Farming, Amarillo, TX, 15-19 Aug. 1988. Texas A & M University, College Station, TX: 436-438 p.

Cole, C.V., K. Paustian, E.T. Elliott, A.K. Metherell, D.S. Ojima, and W.J. Parton. 1993. Analysis of agroecosystem carbon pools. Water, Air and Soil Pollution 70: 357-371.

Cole, C.V., C. Cerri, K. Minami, A. Mosier, N. Rosenberg, D. Sauerbeck, J. Dumanski, J. Duxbury, J. Freney, R. Gupta, O. Heinemeyer, T. Kolchugina, J. Lee, K. Paustian, D. Powlson, N. Sampson, H. Tiessen, M. van Noordwijk, and Q. Zhao. 1996. Chapter 23. Agricultural Options for Mitigation of Greenhouse Gas Emissions. In: Climate Change 1995 — Impacts, Adaptations and Mitigation of Climate Change: Scientific-Technical Analysis. IPCC Working Group II, Cambridge Univ. Press, pp. 745-771.

Collins, H.P., P.E. Rasmussen, and C.L. Douglas, Jr. 1992. Crop rotation and residue management effects on soil carbon and microbial dynamics. Soil Sci. Soc. Am. J. 56: 783-788.

Colorado State University 1997. Agronomy News. Cooperative Extension, Fort Collins, CO, Sept. 1997, 4 pp.

Comfort, S.D., K.A. Kelling, D.R. Keeney, and J.C. Converse. 1990. Nitrous oxide production from injected liquid dairy manure. Soil Sci. Soc. Am. J. 54: 421-427.

Copeland, P.J., and R.K. Crookston. 1992. Crop sequence affects nutrient composition of corn and soybean grown under high fertility. Agron. J. 84: 503-509.

Costanza, R., R. D'Arge, R. de Groot, S. Farber, M. Grasso, B. Hannong, K. Limburg, S. Naeem, R.V. O'Neill, J. Paruelo, R.G. Raskin, P. Sutton, and M. Van den Belt. 1997. The value of the world's ecosystem services and natural capital. Nature 387: 253-260.

Crosson, P. 1981. Conservation tillage and conventional tillage: a comparative assessment. Soil Water Cons. Soc., Ankeny, IA.

Crosson, P.R. 1992. United States agriculture and the environment: perspectives in the next 20 years. US-EPA, Washington, D.C. (unpublished).

CTIC. 1990. National survey of conservation tillage practices. Cons. Tillage Inf. Center, West Lafayette, IN.

CTIC. 1995. National survey of conservation tillage practices. Cons. Tillage Inf. Center, West Lafayette, IN.

CTIC. 1996. National survey of conservation tillage practices. Cons. Tillage Inf. Center, West Lafayette, IN.

CTIC. 1997. National survey of conservation tillage practices. Cons. Tillage Inf. Center, West Lafayette, IN.

Darusman, L., R. Stone, D.A. Whitney, K.A. Janssen, and J.H. Long. 1991. Soil properties after 20 years of fertilization with different N sources. Soil Sci. Soc. Am. J. 55: 1097-1100.

Davidson, E.A. 1991. Fluxes of nitrous oxide and nitric oxide from terrestrial ecosystems. In: J.E. Rogers and W.B. Whitman (eds) "Microbiological Production and Consumption of Greenhouse Gases". Am. Soc. Microbiology, Washington, D.C.: 215-235.

Davie, D.K., and C.L. Lant. 1994. The effect of CRP enrollment on sediment loads in two southern Illinois streams. J. Soil Water Cons. 49: 407-412.

Davis, J.G., M. Young, and B. Ahnstedt. 1997. Soil characteristics of cropland fertilized with feedlot manure in the South Platte river basin of Colorado. J. Soil Water Conserv. 52: 327-331.

Delgado, J.A., R.T. Sparks, R.F. Follett, J.L. Sharkoff, and R.R. Riggenbach. 1998. Use of winter cover crops to conserve soil and water quality in the San Luis Valley of South Central Colorado. In R. Lal (ed) "Erosional Impact on Soil Quality". CRC Boca Raton, FL (In Press).

Despain, W. 1995. A summary of the SWCS-WRP survey. J. Soil Water Conserv. 50: 632-633.

Dick, W.A., D.M. Van Doren, Jr., G.B. Triplett, Jr., and J.E. Henry. 1986a. Influence of long- term tillage and rotation combinations on crop yields and selected soil parameters: results obtained for a Mollic Ochraqualf soil. OARDC Res. Bull. 1180, Wooster, OH, 30p.

Dick, W.A., D.M. Van Doren, Jr., G.B. Triplett, Jr., and J.E. Henry. 1986b. Influence of long- term tillage and rotation combination on crop yield and selected soil parameters. II. Results obtained for a Typic Fragiudalf soil. OARDC Res. Bull. 1181, Wooster, OH 34 p.

DOE/EIA. 1996. Emission of greenhouse gases in the United States 1995. Energy Information Administration, U.S. Dept. of Energy, Washington, D.C.

Donigian, A.S., Jr., T.O. Barnwell, R.B. Jackson, A.S. Patwardhan, K.B. Weinreich, A.L. Rowell, R.V. Chinnaswamy, and C.V. Cole. 1994. Assessment of alternative management practices and policies affecting soil carbon in agroecosystems of the central United States. Publication No. EPA/600/R-94/067. US-EPA, Athens, GA.

Doran, J.W. 1980. Microbial changes associated with residue management with reduced tillage. Soil Sci. Soc. Am. J. 44: 518-524.

Doran, J.W., D.G. Fraser, M.N. Culik, and W.C. Liebhardt. 1987. Influence of alternative and conventional agricultural management on soil microbial processes and N availability. Am. J. Alternative Agric. 2: 99-106.

Doran, J.W., and T.B. Parkin. 1994. Defining and assessing soil quality. In J.W. Doran et al. (eds) "Defining Soil Quality For A Sustainable Environment". Soil Sci. Soc. Am. Special Publ. 35, ASA-SSSA, Madison, WI: 3-21.

Driessen, P.M., and R. Dudal (eds) 1991. The major soils of the world. Agric. Univ. Wageningen, Katholieke Universiteit, Leuven, 310 pp.

Dukes, D. 1996. CRP: A wakeup call for agriculture. J. Soil Water Cons. 51: 140-141.

Dunker, R.E., and I.J. Jansen. 1987. Corn and soybean response to topsoil replacement and irrigation on surface mined land in western Illinois. J. Soil Water Conserv. 42: 277-280.

Dunker, R.E., R.I. Barnhisel, and R.C. Darmody (eds) 1992. Prime Farmland Reclamation. Univ. Illinois, Urbana, IL, 284pp.

Duxbury, J.M., and A.R. Mosier. 1993. Status and issue concerning agricultural emissions of greenhouse gases. In H.M. Kaiser and T. Drennen (eds) "Agricultural Dimensions of Global Climate Change", St. Lucia Press, Delray Beach, FL: 229-258.

Edwards, J.H., C.W. Wood, D.L. Thurlow, and M.E. Ruf. 1992. Tillage and crop rotation effects on fertility status of a Hapludalt soil. Soil Sci. Soc. Am. J. 56: 1577-1582.

Elliott, E.T., and C.V. Cole. 1989. A perspective on agroecosystem science. Ecology 70: 1597-1602.

Energy Use Survey. 1987. The Fertilizer Institute, Washington, D.C.

Enquete Commission. 1992. Climate Change: a threat to global environment. Economica Verlag, Bonn, Germany, 235 p.

Eswaran, H., E. Van Den Berg, and P. Reich. 1993. Organic carbon in soils of the world. Soil Sci. Soc. Am. J. 57: 192-194.

FAO 1992. Current world fertilizer situation and outlook. FAO, Rome, Italy.

FAO 1996. Production Year Book. FAO, Rome, Italy.

Fausey, N.R., and R. Lal. 1992. Drainage-tillage effects on a Crosby-Kokomo soil association in Ohio. III. Organic matter content and chemical properties. Soil Tech. 5: 1-12.

Fausey, N.R., L.C. Brown, H.W. Belcher and R.S. Kanwar. 1995. Drainage and water quality in Great Lakes and Cornbelt states. J. Irrigation & Drainage Eng. 121: 283-288.

Flach, K.W., T.O. Barnwell, Jr., and P. Crosson. 1997. Impact of agriculture on atmospheric CO_2. In: E.A. Paul, K. Paustian, E.T. Elliott and C.V. Cole (eds) "Soil Organic Matter in Temperate Agroecosystems: Long-term Experiments in North America". CRC Press, Boca Raton, FL: 3-13.

Flavin, C., and O. Tunali. 1996. Climate of hope: new strategies for stabilizing the world atmosphere. Worldwatch Paper 130, Worldwatch Institute, Washington, D.C. 84 p.

Follett, R.G. 1980. Reclamation and revegetation of land areas disturbed by man: an annotated bibliography of agricultural research, 1972-1980. USDA, Bibliography and Literature of Agric. No. 8, 55 pp.

Follett, R.G. 1993. Global climate change, U.S. agriculture and carbon dioxide. J. Prod. Agric. 6: 181-190.

Follett, R.F. 1998. CRP and microbial biomass dynamics in temperate climates. In R. Lal, J. Kimble, R.F. Follett and B.A. Stewart (eds) "Management of Carbon Sequestration", CRC Press, Inc., Boca Raton, FL: 305-322.

Follett, R.F., S.C. Gupta, and P.G. Hunt. 1987. Conservation practices: relation to the management of plant nutrients for crop production. In "Soil Fertility and Organic Matter as Critical Components of Production Systems", Soil Sci. Soc. Am. Spec. Publ. 19, Madison, WI: 19-51.

Follett, R.F., E.A. Paul, S.W. Leavitt, A.D. Halvorson, D. Lyon, and G.A. Peterson. 1997. Carbon isotope ratios of Great Plains soils in wheat-fallow systems. Soil Sci. Soc. Am. J. 61: 1068-1077.

Forste, J.W. 1997. Biosolids processing, products and uses. In J.E. Rechcigl and H.C. Mackinnon (eds) "Agricultural Uses of By-Products and Wastes", Am. Chemical Soc. Washington, D.C.: 50-61.

Franzluebbers, A.J., F.M. Hons, and D.A. Zuberer. 1994. Long-term changes in soil carbon and nitrogen pools in wheat management systems. Soil Sci. Soc. Am. J. 58: 1639-1645.

Franzluebbers, A.J. and M.A. Arshad. 1996. Soil organic matter pools during early adoption of conservation tillage in northwestern Canada. Soil Sci. Soc. Am. J. 60: 1422-1427.

Frye, W.W., R.L. Blevins, M.S. Smith, S.J. Croak, and J.J. Varco. 1988. Role of annual legume cover crops in efficient use of water and nitrogen. In: ASA-CSSA-SSSA Special Publication 51, Madison, WI: 129-154.

Fryrear, D.W., C.A. Krammes, D.Z. Williamson, and T.M. Zobeck. 1994. Computing the wind erodible fraction of soils. J. Soil Water Conserv. 49: 183-199.

Gantzer, C.J., S.H. Anderson, A.L. Thompson, and J.R. Brown. 1991. Evaluation of soil loss after 100 years of soil and crop management. Agron. J. 83: 74-77.

Gardner, G. 1997. Recycling Organic Waste: From Urban Pollutant to Farm Resource. Worldwatch Paper 135, Washington, D.C. 59pp.

Gebhart, D.L., H.B. Johnson, H.S. Mayeux, and H.W. Polley. 1994. The CRP increases soil organic carbon. J. Soil Water Cons. 49: 488-492.

Gee, G.W., A. Bauer, and R.S. Decker. 1978. Physical analysis of overburden materials and mineland soils. In "Reclamation of Drastically Disturbed Lands", ASA-CSSA-SSSA, Madison, WI: 665-672.

Giampietro, M., S. Ulgiati, and D. Pimentel. 1997. Feasibility of large-scale biofuel production. Bioscience 147: 587-600.

Gilley, J.E., and J.W. Doran. 1997. Tillage effects on soil erosion potential and soil quality of a former conservation reserve program site. J. Soil Water Cons. 52: 184-188.

Glendining, M.J., and D.S. Powlson. 1991. The effect of long-term applications of inorganic nitrogen fertilizer on soil organic nitrogen. In W.S. Wilson et al. (ed) "Advances in soil organic matter research: Impact on agriculture and the environment". R. Soc. Chem., Cambridge, England: 329-338.

Gomez, B. 1995. Assessing the impact of the 1985 farm bill on sediment-related non-point source pollution. J. Soil Water Conserv. 50: 374-377.

Gorham, E. 1991. Northern peatlands: role in the C cycle and probable response to climatic warming. Ecological Applications 1: 182-195.

Grant, F.R. 1997. Changes in soil organic matter under different tillage and rotations: mathematical modeling in ecosystems. Soil Sci. Soc. Am. J. 61: 1159-1175.

Gregorich, E.G., B.H. Ellert, C.F. Dury, and B.C. Liang. 1996. Fertilization effects on soil organic matter turnover and crop residue carbon storage. Soil Sci. Soc. Am. J. 60: 472-476.

Grigal, D.F., and L.F. Ohmann. 1992. Carbon storage in upland forests of the Lake States. Soil Sci. Soc. Am. J. 56: 935-943.

Gupta, R.K., and I.P. Abrol. 1990. Salt-affected soils: their reclamation and management for crop production. In R. Lal and B.A. Stewart (eds) "Soil Degradation". Adv. Soil Sci. 11: 187-221.

Hall, D.O., F. Rosillo-Calle, R.H. Williams and J. Woods. 1993. Biomass for energy supply prospects. In B.J. Johansson, H. Kelly, A.K.N. Redy and R.H. Williams (eds) "Renewables for Fuels and Electricity", Island Press, Washington, D.C.: 593-651.

Hammer, R.D. 1992. A soil based productivity index to assess surface mine reclamation. In R.E. Dunker, R.I. Barnhisel and R.C. Darmody (eds) "Prime Farmland Reclamation", Univ. Illinois, Urbana, IL: 221-232.

Hargrove, W.L. 1986. Winter legumes as a nitrogen source for no-till grain sorghum. Agron. J. 78: 70-74.

Harper, L.A., P.F. Hendrix, G.W. Langdale, and D.C. Coleman. 1995. Clover management to provide optimum nitrogen and soil water conservation. Crop Sci. 35: 176-182.

Hassink, J., and A.P. Whitmore. 1997. A model of the physical protection of organic matter in soils. Soil Sci. Soc. Am. J. 61: 131-139.

Havlin, J.L., D.E. Kissel, L.D. Maddux, M.M. Claassen, and J.H. Long. 1990. Crop rotation and tillage effects on soil carbon and nitrogen. Soil Sci. Soc. Am. J. 54: 448-452.

Heimlich, R.E., and A.B. Daugherty. 1991. America's Cropland: where does it come from? In "Agriculture and the Environment". The 1991 Yearbook of Agriculture, Govt. Printing Office, Washington, D.C.: 3-9.

Heimlich, R.E., K.D. Wiebe, R. Claassen, and R.M. House. 1997. Recent evolution of environmental policy: lessons from wetlands. J. Soil Water Cons. 52: 157-161.

Hinesly, T.D., D.E. Redborg, E.L. Ziegler, and I.H. Rose-Innes. 1982. Effect of chemical and physical changes in strip-mined spoil amended with sewage sludge on the uptake of metal by plants. In: W.I. Sopper, E.M. Speaker and R.K. Bastian (eds) "Land Reclamation and Biomass Production with Municipal Wastewater and Sludge", Pennsylvania State Univ. Press, Univ. Park, PA: 339-352.

Houghton, R.A. 1995. Changes in storage of terrestrial carbon since 1850. In R. Lal, J. Kimble, E. Levine and B.A. Stewart (eds) "Soils and Global Change". CRC/Lewis Publishers, Boca Raton, FL.

Hu, S., N.J. Grunwald, A.H.C. Van Bruggen, G.R. Gamble, L.E. Drinkwater, C. Shennan, and M.W. Demment. 1997. Short-term effects of cover crop incorporation on soil C pools and N availability. Soil Sci. Soc. Am. J. 61: 901-911.

Hudson, B.D. 1994. Soil organic matter and available water capacity J. Soil Water Cons. 49: 189-194.

Hunt, P.G., D.L. Karlen, T.A. Matheny, and V.L. Quisenberry. 1996. Changes in carbon content of a Norfolk loamy sand after 14 years of conservation or conventional tillage. J. Soil Water Cons. 51: 255-258.

Intergovernmental Panel on Climate Change. 1990a. Scientific Assessment of Climate Change. Report prepared for IPCC by Working Group I: New York: WMO and UNEP.

Intergovernmental Panel on Climate Change. 1990b. Potential Impacts of Climate Change: Report Prepared for IPCC by Working Group II: New York: WMO and UNEP.

Intergovernmental Panel on Climate Change. 1992. The supplementary Report to the IPCC Impacts Assessment. W.J. Tegart and G.W. Sheldon (eds), Australian Govt. Publishing Services, Canberra.

Intergovernmental Panel on Climate Change. 1994a. Radiative Forcing of Climate Change. WMO/UNEP, Washington, D.C.

Intergovernmental Panel on Climate Change. 1994b. Radiative forcing of climate change: the 1994 report of the scientific assessment working group of IPCC, summary for policy makers, WMO/UNEP, Geneva, Switzerland.

Iserman, K. 1994. Agriculture's share in the emission of trace gases affecting the climate and some cause-oriented proposals for sufficiently reducing this share. Environ. Pollut. 83:1-21.

Janzen, H.H. 1987a. Effect of fertilizer on soil productivity in long-term spring wheat rotations. Can. J. Plant Sci. 67: 165-174.

Janzen, H.H. 1987b. Soil organic matter characteristics after long-term cropping to various spring wheat rotations. Can. J. Soil Sci. 67: 845-856.

Jenkinson, D.S. 1991. The Rothamsted long-term experiments. Agron. J. 83: 2-.

Jenny, H. 1941. Factors of soil formation. McGraw Hill Book Co., NY.

Johansson, B.J., H. Kelly, A.K.N. Redy and R.H. Williams (eds) 1993. Renewables for Fuels and Electricity". Island Press, Washington, D.C.

Johnson, M.G., and J.S. Kern. 1991. Sequestering C in soils: A workshop to explore the potential for mitigating global climate change. USEPA Rep. 600/3-91-031. USEPA Environ. Res. Lab., Corvallis, OR.

Johnston, A.E. 1991. Soil fertility and soil organic matter. In W.S. Wilson et al. (ed) "Advances in soil organic matter research: Impact on agriculture and the environment". R. Soc. Chem., Cambridge, England: 299-314.

Karlen, D.L., A. Kumar, R.S. Kanwar, C.A. Cambardella, and T.S. Colvin. 1998. Ridge-, chisel-, moldboard-, or no-tillage effects on a 15-year continuous corn N balance. Journal Paper No. J-17223, Iowa Agric. & Home Economic Expt. Station, Ames, IA. Soil & Tillage Res. (In review).

Kashmanian, R.M., and R.F. Rynk. 1996. Agricultural composting in the United States: trends and driving forces. J. Soil Water Cons. 51: 194-201.

Keeling, C.D., R.B. Bacastow, A.F. Carter, S.C. Piper, T. P. Whorf, M. Heiman, W.G. Mook, and H. Roeloffzen. 1989. A three-dimensional model of atmospheric CO_2 transport based on observed winds: I. Analysis of observational data. In D.H. Peterson (ed) "Aspects of Climate Variability in the Pacific and Western Americas". Geophysical Monograph 55, American Geophysical Union, Washington, D.C.: 165-236.

Kellogg, R.L., G.W. Te Selle, and J.J. Goebel. 1994. Highlights from the 1992 National Resource Inventory. J. Soil Water Cons. 49: 521-527.

Kern, J.S. 1992. Global scale erosion and SOC displacement. Agronomy Abstract, ASA, Madison, WI, 328 p.

Kern, J.S. 1994. Spatial patterns of soil organic carbon in the contiguous United States. Soil Sci. Soc. Am. J. 58: 439-455.

Kern, J.S., and M.G. Johnson. 1991. The impact of conservation tillage use on soil and atmospheric carbon in the contiguous United States. USEPA Rep. EPA/600/3-91/056. USEPA Corvallis, OR, 28 pp.

Kern, J.S., and M.G. Johnson, 1993. Conservation tillage impacts on national soil and atmospheric carbon levels. Soil Sci. Soc. Am. J. 57: 200-210.

Khalil, M.A.K., and R.A. Rasmussen. 1992. The global sources of N_2O. J. Geophys. Res. 97: 14651-14660.

Kimmelshue, J.E., J.W. Gilliam, and R.J. Volk. 1995. Water management effects on mineralization of soil organic matter and corn residue. Soil Sci. Soc. Am. J. 59: 1156-1162.

Kliewer, B.A., and J.W. Gilliam. 1995. Water table management effects on denitrification and nitrous oxide evolution. Soil Sci. Soc. Am. J. 59: 1094-1701.

Krall, J.M., J.J. Nachtman, and S.D. Miller. 1995. Irrigated corn production in no-till and plowed alfalfa sod. J. Soil Water Cons. 50: 75-76.

Kuo, S., U.M. Sainju, and E.J. Jellum. 1997. Winter cover crop effects on soil organic carbon and carbohydrate in soil. Soil Sci. Soc. Am. J. 61: 145-152.

Lafond, G.P., D.A. Derksen, H.A. Loeppky, and D. Struthers. 1994. An agronomic evaluation of conservation tillage systems and continuous cropping in East Central Saskatchewan. J. Soil Water Cons. 49: 387-393.

Lal, R. 1989. Conservation tillage for sustainable agriculture: tropics vs. temperate environments. Adv. Agron. 42: 85-197.

Lal, R. 1995a. Global soil erosion by water and carbon dynamics. In R. Lal, J. Kimble, E. Levine and B.A. Stewart (eds) "Soils and Global Change", CRC/Lewis Publishers, Boca Raton, FL: 131-142.

Lal, R. 1995b. The role of residue management in sustainable agricultural systems. J. Sust. Agric. 5: 51-78.

Lal, R. 1997. Residue management, conservation tillage and soil restoration for mitigating greenhouse effect by CO_2 enrichment. Soil & Tillage Res. 43: 81-107.

95a. Soils and Global Change. CRC/Lewis Publishers, Boca Raton, FL, 440pp.

Lal, R., J. Kimble, E. Levine and B.A. Stewart (eds) 1995b

Lal, R., and N.R. Fausey. 1993. Drainage and tillage effects on a Crosby-Kokomo soil association in Ohio. IV. Soil physical properties. Soil Technology 6: 123-135.

Lal, R., A.A. Mahboubi, and N.R. Fausey. 1994. Long-term tillage and rotation effects on properties of a central Ohio soil. Soil Sci. Soc. Am. J. 58: 517-522.

Lal, R., N.R. Fausey, and D.J. Eckert. 1995. Land use and soil management effects on emissions of radiatively-active gases from two soils in Ohio. In R. Lal, J. Kimble, E. Levine and B.A. Stewart (eds) "Soil Management and Greenhouse Effect", CRC/Lewis Publishers, Boca Raton, FL: 41-60.

Lal, R., J. Kimble, E. Levine and B.A. Stewart (eds) 1995a. Soils and Global Change. CRC/Lewis Publishers, Boca Raton, FL, 440pp.

Lal, R., J. Kimble, E. Levine and B.A. Stewart (eds) 1995b. Soil Management and Greenhouse Effect. CRC/Lewis Publishers, Boca Raton, FL, 385pp.

Lal, R., and J. Kimble. 1997. Conservation tillage for carbon sequestration. Nutrient Cycling in Agroecosystems 49: 243-253.

Lal, R., P. Henderlong, and M. Flowers. 1998. Forage and row cropping effects on soil organic carbon and nitrogen contents. In R. Lal, J. Kimble, R. Follett and B.A. Stewart (eds) "Management of Carbon Sequestration", CRC Press, Boca Raton, FL (In Press).

Lal, R., J. Kimble, R.F. Follett and B.A. Stewart (eds) 1998a. Soil Processes and the C Cycle. CRC Press, Boca Raton, FL. 609pp.

Lal, R., J. Kimble, R.F. Follett and B.A. Stewart (eds) 1998b. Management of C Sequestration in Soil. CRC Press, Boca Raton, FL, 457pp.

Lant, C.L., S.E. Kraft, and K.R. Gillman. 1995a. Enrollment of filter strips and recharge areas in the CRP and USDA easement programs. J. Soil Water Cons. 50: 193-200.

Lant, C.L., S.E. Kraft, and K.R. Gillman. 1995b. Farm bill and water quality in corn belt watersheds: conserving remaining wetlands and restoring farmed wetlands. J. Soil Water Cons. 50: 201-205.

Larson, W.E., C.E. Clapp, W.H. Pierre, and Y.B. Morachan. 1972. Effects of increasing amounts of organic residue on continuous corn. II. Organic C, N, P and S. Agron. J. 64: 204-208.

Larson, W.E., R.F. Holt, and C.W. Carlson. 1978. Residues for soil conservation. In W.R. Oschwald (ed) "Crop Residue Management Systems". ASA Special Publ. 31, ASA, Madison, WI: 1-15.

Larson, W.E., F.J. Pierce, and R.H. Dowdy. 1983. The threat of soil erosion to long-term crop production. Science 219: 458-465.

Larson, W.E., and P.C. Robert. 1991. Farming by soil. In R. Lal and F.J. Pierce (eds) "Soil Management For Sustainability". Soil Water Cons. Soc., Ankeny, IA: 103-112.

Lee, L.K. 1990. The dynamics of declining soil erosion rates. J. Soil Water Cons. 45: 622-624.

Lee, J.J. 1993a. Cited in "Climate Change Mitigation Strategies in the Forest and Agriculture Sectors" EPA, 1995, Washington, D.C.

Lee, J.J. 1993b. Cited in USEPA (1995b). Climate change mitigation strategies in the forest and agriculture sector. Washington, D.C., 64p.

Lee, J.J., D.L. Phillips, and R. Liu. 1993. The effects of trends in tillage practices on erosion and carbon content of soils in the U.S. corn belt. Water, Air and Soil Pollution 70: 389-401.

Leeden, F., F.L.Van der Troise, and D.K. Todd. 1991. The Water Encyclopedia Second Edition, Lewis Publishers, Chelsea, MI.

Lindau, C.W., P.K. Pollich, R.D. DeLaune, A.R. Mosier, and K.F. Bronson. 1993. Methane mitigation in flooded Louisiana rice fields. Biol. Fertil. Soils 15: 174-178.

Lindau, C.W., R.D. Delaune, D.P. Alford, and H.K. Kludge. 1994. Methane production and mitigation in rice. In K. Minami, A. Mosier and R. Sass (eds) "CH_4 and N_2O: Global Emissions and Controls From Rice Fields and Other Agricultural and Industrial Sources". National Institute of Agro-Environmental Sciences, Tsukuba, Japan: 79-86.

Lueking, M.A., and J.S. Schepers. 1985. Changes in soil carbon and nitrogen due to irrigation development in Nebraska's Sandhill Soils. Soil Sci. Soc. Am. J. 49: 626-630.

Mahlman, J.D. 1997. Uncertainties in projections of human-caused climate warming. Science 278: 746-1417.

Manley, J.T., G.T. Schuman, J.D. Reeder, and R.H. Hart. 1995. Rangeland soil carbon and nitrogen responses to grazing. J. Soil Water Cons. 50: 298-306.

Mann, L.K. 1985. A regional comparison of C in cultivated and uncultivated Alfisols and Mollisols in central United States. Geoderma 36: 241-253.

Mann, L.K. 1986. Changes in soil C storage after cultivation. Soil Sci. 142: 279-288.

Marland, G., and A.F. Turhollow. 1991. CO_2 emissions from the production and combustion of fuel ethanol from corn. Energy 16: 1307-1316.

Mayland, H.F. 1961. A characterization of soil aggregate properties and their effects on aggregate stability. M.S. Thesis, Univ. of Wyoming, Laramie, WY.

Mayland, H.F., and R.B. Murray. 1978. Mineral cycling aspects within the sagebrush ecosystems. In: "The Sagebrush Ecosystem: A Symposium", Utah State Univ., Logan, Utah, 251 p.

McGill, W.B., C.A. Campbell, J.F. Dormaar, E.A. Paul, and D.W. Anderson. 1981. Soil organic matter losses. In "Agricultural Land: Our disappearing heritage. A Symposium". Proc. Annu. Alberta Soil Sci. Worksh. 18th, Edmonton, 24-25 Feb. 1981. Alberta Soil and Feed Testing Lab., Edmonton.

Meade, R.H., and R.S. Parker. 1984. Sediment in rivers of the United States. In "National Water Summary 1984", USGS, Water Supply Paper 2275: 49-60.

Mikkelsen, R.L. 1997. Agricultural and environmental issues in the management of swine waste. In J.E. Rechcigl and H.C. Mackinnon (eds) "Agricultural Uses of By-Products and Wastes", Am. Chemical Soc., Washington, D.C.: 110-119.

Minami, K. 1994. Effect of nitrification inhibitors and slow-release fertilizer on emission of N_2O from fertilized soils. In K. Minami, A. Mosier and R. Sass (eds) "CH_4 and N_2O: Global Emissions and Controls From Rice Fields and Other Agricultural and Industrial Sources". National Institute of Agro-Environmental Sciences, Tsukuba, Japan: 187-196.

Mitsch, W.J., and J.G. Gosselink. 1993. Wetlands. 2nd edition, Van Nostrand Reinhold, New York, 722 pp.

Mitsch, W.J., and X. Wu. 1995. Wetlands and global change. In R. Lal, J. Kimble, E. Levine and B.A. Stewart (eds) "Soil Management and Greenhouse Effect", CRC/Lewis Publishers, Boca Raton, FL: 205-230.

Monger, C.F. 1993. Inorganic carbon in desert soils of New Mexico: microbial precipitation and isotopic significance. In: Proc. Int'l Symp. "Soil Processes and Management Systems: Greenhouse Gas Emissions and C Sequestration", 4-9 April, 1993, Columbus, Ohio.

Moore, P.A., Jr., T.C. Daniel, A.N. Sharpley, and C.W. Wood. 1995. Poultry manure management. J. Soil Water Cons. 50: 321-327.

Mosier, A.R. 1994. Nitrous oxide summary. In K. Minami, A. Mosier and R. Sass (eds) "CH_4 and N_2O: Global Emissions and Controls From Rice Fields and Other Agricultural and Industrial Sources". National Institute of Agro-Environmental Sciences, Tsukuba, Japan: 135-139.

Mosier, A.R., K.F. Bronson, J.R. Freney, and D.G. Keerthisinghe. 1994. Use of nitrification inhibitors to reduce nitrous oxide emission from urea fertilized soils. In K. Minami, A. Mosier and R. Sass (eds) "CH$_4$ and N$_2$O: Global Emissions and Controls From Rice Fields and Other Agricultural and Industrial Sources". National Institute of Agro-Environmental Sciences, Tsukuba, Japan: 197-207.

Muchovej, R.M.C., and R.S. Pacovsky. 1997. Future directions of by-products and wastes in agriculture. In J.E. Rechcigl and H.C. Mackinnon (eds) "Agricultural Uses of By-Products and Waste", American Chemical Society, Washington, D.C.: 1-19.

Odum, E.P. 1969. The strategy of ecosystem development. Sci. 164: 262-270.

OECD. 1994. The economics of climate change: Proceedings of an OECD/IEA Conference. Organization of Econoimc Cooperation and Developmnet, Paris, France.

Oechel, W.C., and G.L. Vourlitis. 1995. Effects of global change on carbon storage in cold soils. In R. Lal, J. Kimble, E. Levine and B.A. Stewart (eds) "Soils and Global Change", CRC/Lewis Publishers, Boca Raton, FL: 117-129.

Oldeman, L.R. 1994. The global extent of soil degradation. In D.J. Greenland and I. Szabolcs (eds) "Soil Resilience and Sustainable Land Use". CAB International Wallingford, U.K.: 99-118.

Olson, K.R. 1992. Assessment of reclaimed farmland disturbed by surface mining in Illinois. In: R.E. Dunker, R.I. Barnhisel and R.C. Darmody (eds) "Prime Farmland Reclamation", Univ. Illinois, Urbana, IL: 173-176.

Osborn, T. 1993. The Conservation Reserve Program: status, future and policy options. J. Soil Water Cons. 48: 271-280.

OSTP. 1997. Climate Change.. State of Knowledge. Office of Science Technology Policy, Washington, D.C., 18 pp.

OTA .1982. Impacts of technology on U.S. cropland and rangeland productivity. Office of Technology Assessment, U.S. Congress, Washington, D.C. 266 p.

Parton, W.J., D.S. Schmil, C.V. Cole and D.S. Ojima. 1987. Analysis of factors controlling soil organic matter levels in Great Plains Grasslands. Soil Sci. Soc. Am. J. 51: 1173-1177.

Paul, E.A., R.F. Follett, S.W. Leavitt, A.D. Halvorson, G.A. Peterson, and D.J. Lyon. 1997. Radiocarbon dating for determination of soil organic matter pool sizes and dynamics. Soil Sci. Soc. Am. J. 61: 1058-1067.

Paul, E.A., K, Paustian, E.T. Elliott, and C.V. Cole (eds) 1997. Soil Organic Matter in Temperate Agroecosystems. CRC Press, Boca Raton, FL, 413 p.

Paustian, K., W.J. Parton, and J. Persson. 1992. Modeling soil organic matter in organic-amended and nitrogen-fertilized long-term plots. Soil Sci. Soc. Am. J. 56: 476-488.

Paustian, K., C.V. Cole, E.T. Elliott, E.F. Kelly, C.M. Yonker, J. Cipra, and K. Killian. 1995. Assessment of the contributions of CRP lands to C sequestration. Agron. Abs. 87: 136.

Paustian, K., E.T. Elliott, G. Bluhm, and T. Kautza. 1997a. C sequestration with agricultural conservation practices. National Resources Ecology Lab., CSU, Fort Collins, CO, 4 pp.

Paustian, K., H.P. Collins, and E.A. Paul. 1997. Management controls on soil carbon. In: E.A. Paul, K. Paustian, E.T. Elliott and C.V. Cole (eds) "Soil Organic Matter in Temperate Agroecosystems: Long-term Experiments in North America". CRC Press, Boca Raton, FL: 15-49.

Paustian, K., C.V. Cole, D. Sauerbeck and N. Sampson. 1998. CO_2 mitigation by agriculture: an overview. Special Issue, Climate Change (In press).

Pedersen, W.L., R.E. Dunker, and C.L.Hooks. 1992. Evaluation of small grains on mine soils. In R.E. Dunker, R.I. Barnhisel and R.C. Darmody (eds) "Prime Farmland Reclamation", Univ. Illinois, Urbana, IL: 267-270.

Peterson, J.R., C. Lue-Hing, J. Gshwind, R.I. Pietz, and D.R. Zeng. 1982. Metropolitan Chicago's Fulton County sludge utilization program. In: W.I. Sopper, E.M. Speaker and R.K. Bastian (eds) "Land Reclamation and Biomass Production with Municipal Wastewater and Sludge", Pennsylvania State Univ. Press, Univ. Park, PA: 332-338.

Peterson, G.A., D.G. Westfall, and C.V. Cole. 1993. Agroecosystem approach to soil and crop management research. Soil Sci. Soc. Am. J. 57: 1354-1360.

Phillips, D.L., D. White, and B. Johnson. 1993. Implications of climate change scenarios for soil erosion potential in the USA. Land Degradation & Rehabilitation 4: 61-72.

Pimental, D., L.E. Hurd, A.C. Bellotti, M.J. Forster, I.N. Oka, O.D. Sholes, and R.J. Whitman. 1973. Food production and the energy crisis. Science 182: 44-.

Pimental, D.C., P. Harvey, K. Resosudarmo et al. 1995. Environmental and economic costs of soil erosion. Science 267: 1117-1120.

Pichtel, J.R., W.A. Dick, and P. Sutton. 1994. Comparison of amendments and management practices for long-term reclamation of abandoned mine lands. J. Env. Qual. 23: 766-772.

Post, W.M., T-H. Peng, W.R. Emanuel, A.W. King, and D.L.De Angelis. 1982. The global carbon cycle. Am. Sci. 78: 310-.

Power, J.F. 1994. Understanding the nutrient cycling process. J. Soil Water Cons. 49: 16-23.

Power, J.F., R.F. Follett, and G.E. Carlson. 1983. Legumes in conservation tillage systems: a research perspective. J. Soil Water Conserv. 38: 217-218.

Prato, T., Y. Wang, T. Haithcoat, C. Barnett, and C. Fulcher. 1995. Converting hydric cropland to wetland in Missouri. J. Soil Water Cons. 50: 101-106.

Prinn, R.D., R. Cunnold, R. Rasmussen, P. Simmonds, F. Alyea, A. Crawford, P. Fraser, and R. Rosen. 1990. Atmospheric emissions and trends of N_2O deduced from 10 years of ALEGAGE data. J. Geophys. Res. 95: 18369-18385.

Rabenhorst, M.C. 1995. Carbon storage in tidal marsh soils. In R. Lal, J. Kimble, E. Levine and B.A. Stewart (eds) "Soils and Global Change", CRC/Lewis Publishers, Boca Raton, FL: 93-103.

Rasmussen, P.E., R.R. Allmaras, C.R. Rohde, and N.C. Roager, Jr. 1980. Crop residue influences on soil carbon and nitrogen in a wheat-fallow system. Soil Sci. Soc. Am. J. 44: 596-600.

Rasmussen, P.E., and C.R. Rohde. 1988. Long-term tillage and nitrogen fertilization effects on organic N and C in a semi-arid soil. Soil Sci. Soc. Am. J. 44: 596-600.

Rasmussen, P.E., and H.P. Collins. 1991. Long-term impacts of tillage, fertilizer and crop residue on soil organic matter in temperate semi-arid regions. Adv. Agron. 45: 93-134.

Rasmussen, P.E., and W.J. Parton. 1994. Long-term effects of residue management in wheat- fallow: I. Inputs, yield and soil organic matter. Soil Sci. Soc. Am. J. 58: 523-530.

Reese, R., A. Bouzaher, and J. Shogren. 1993. "Long-term economic consequences of Alternative Carbon Reducing Conservation and Wetlands Reserves Programs: A BLS Analysis". Draft paper prepared for the USDE, OEA, Washington, D.C.

Reicosky, D.C., W.D. Kemper, G.W. Langdale, C.L. Douglas, Jr., and P.E. Rasmussen. 1995. Soil organic matter changes resulting from tillage and biomass production. J. Soil Water Cons. 50: 253-262.

Rice, C.W., P.E. Sierzega, J.M. Tiedje, and L.W. Jacobs. 1988. Stimulated denitrification in the microenvironment of a biodegradable organic waste injected into soil. Soil Sci. Soc. Am. J. 52: 102-108.

Rijsberman, F.R., and R.J. Stewart (eds) 1990. Targets and Indicators of Climate Change. German Bundestag Enquete Commission, Bonn, Economica Verlag, Germany.

Rixon, A.J. 1968. Oxygen uptake and nitrification at various moisture levels by soils and mats from irrigated pastures. J. Soil Sci. 19: 55-66.

Robinson, A. 1993. Wetland protection: what success. J. Soil Water Cons. 48: 267-271.

Robinson, C.A., M. Ghaffarzadeh, and R.M. Cruse. 1996a. Vegetative filter strip effects on sediment concentration in cropland runoff. J. Soil Water Cons. 50: 227-230.

Robinson, C.A., R.M. Cruse, and M. Haffarzadeh. 1996b. Cropping system and nitrogen effects on Mollisol organic carbon. Soil Sci. Soc. Am. J. 60: 264-269.

Rodhe, H. 1990. A comparison of the contribution of various gases to the greenhouse effect. Science 248: 1217-1219.

Rolston, D.E., A.N. Sharpley, D.W. Toy, and F.E. Broadbent. 1982. Field measurement of denitrification. III. Rates during irrigation cycles. Soil Sci. Soc. Am. J. 46: 289-296.

Rosenberg, N., C.V. Cole, and K. Paustian. 1998. Mitigation of Greenhouse Gas Emissions by the Agriculture Sector, a Special Editorial. Climatic Change. In press.

Rosswall, T., and K. Paustian. 1984. Cycling of nitrogen in modern agricultural systems. Plant Soil 76: 3-21.

Sahrawat, K.L., D.R. Keeney, and S.S. Adams. 1987. Ability of nitrapyrin, dicyandiamide and acetylene to retard nitrification in a mineral and an organic soil. Plant Soil 101: 179-182.

Salter, R.M., and T.C. Green. 1933. Factors affecting the accumulation and loss of nitrogen and organic carbon from cropped soils. J. Am. Soc. Agron. 25: 622-.

Sampson, R.N., L.L. Wright, J.K. Winjum, J.D. Kinsman, J. Benneman, E. Krusten and J.M.O. Surlock. 1993. Biomass management and energy. In J. Wisniewski and R.N. Sampson (eds) "Terrestrial Biospheric C Fluxes: Quantification of Sinks and Sources of CO_2", Kluwer Academic Publishers, Dordrecht, The Netherlands: 139-162.

Sass, R.L. 1994. Short summary chapter for methane. In K. Minami, A. Mosier and R. Sass (eds) "CH_4 and N_2O: Global emissions and controls from rice fields and other agricultural and industrial sources". NIAES, Tsukuba, Japan: 1-7.

Sass, R.L., and F.M. Fisher. 1994. CH$_4$ emission from paddy fields in the United States Gulf Coast area. In K. Minami, A. Mosier and R. Sass (eds) "CH$_4$ and N$_2$O: Global Emissions and Controls From Rice Fields and Other Agricultural and Industrial Sources". National Institute of Agro-Environmental Sciences, Tsukuba, Japan: 65-77.

Schaller, F.W., and P. Sutton (eds). 1978. Reclamation of drastically disturbed lands. ASA, Madison, WI.

Schertz, D.L. 1988. Conservation tillage: an analysis of acreage projections in United States. J. Soil Water Conserv. 43: 256-258.

Schertz, D.L., and J.L. Bushnell. 1993. USDA crop residue management action plan. J. Soil Water Cons. 48: 175-177.

Schimel, D.S. 1986. Carbon and N turnover in adjacent grassland and cropland ecosystems. Biogeochemistry 2: 345-357.

Schlesinger, W.H. 1995. An overview of the global carbon cycle. In R. Lal, J. Kimble, E. Levine and B.A. Stewart (eds) "Soils and Global Change". CRC/Lewis Publishers, Boca Raton, FL: 9-25.

Schoenholtz, S.H., J.A. Burger, and R.E. Kreh. 1992. Fertilizer and organic amendment effects on mine soil properties and revegetation success. Soil Sci. Soc. Am. J. 56: 1177-1184.

Schroeder, S.A. 1995a. Topographic influence on soil water and spring wheat yield on reclaimed mineland. J. Env. Qual. 24: 467-471.

Schroeder, S.A. 1995b. First estimation of cover for reclaimed mineland erosion control. J. Soil Water Conserv. 50: 668-671.

Shepherd, M.F, F. Barzetti and D.R. Hastie. 1991. The production of atmospheric NO$_x$ and N$_2$O from a fertilized agricultural soil. Atmos. Environ. 25A: 1961-1969.

Shibata, M. 1994. Methane production in ruminants. In K. Minami, A. Mosier and R. Sass (eds) "CH$_4$ and N$_2$O: Global emissions and controls from rice fields and other agricultural and industrial sources". NIAES, Tsukuba, Japan: 105-115.

Sindelar, B.W., C. Montagne, and R.R.H. Kroos. 1995. Holistic resource management: An approach to sustainable agriculture on Montana's Great Plains. J. Soil Water Cons. 50: 45-49.

Snyder, D. 1995. What farmers should know about wetlands. J. Soil Water Conserv. 50: 630-632.

Sopper, W.E. 1993. Municipal sludge use in land reclamation. Lewis Publishers, Boca Raton, FL, 163 pp.

Sotomayer, D., and C.W. Rice. 1996. Denitrification in soil profiles beneath grassland and cultivated soils. Soil Sci. Soc. Am. J. 60: 1822-1828.

Staben, M.L., D.F. Bezdicek, J.L. Smith, and M.F. Fauci. 1996. Assessment of soil quality in conservation reserve program and wheat-fallow soils. Soil Sci. Soc. Am. J. 61: 124-130.

Stivers, L.J., and C. Shannon. 1991. Meeting the nitrogen needs of processing tomatoes through winter cover cropping. J. Prod. Agric. 4: 330-335.

Stone, E.L., W.G. Harris, R.B. Brown, and R.J. Kuehl. 1993. Carbon storage in Florida Spodosols. Soil Sci. Soc. Am. J. 57: 179-182.

Stoorvogel, J.J., and E.M.A. Smaling. 1990. Assessment of soil nutrient depletion in sub-Saharan Africa, 1983-2000. Report 28, DLO Winand Staring Center for Integrated Land. Soil and Water Research (SC-DLO), Wageningen, 137 p.

Stucky, D.J., J.H. Bauer, and T.C. Lindsey. 1980. Restoration of acidic mine spoils with sewage sludge. Reclamation Rev. 3: 129-139.

Suarez, D.L., and J.D. Rhoades. 1977. Effect of leaching fraction on river salinity. J. Irrigation and Drainage Div. ASCE. IR2: 245-257.

Suarez, D.L., and J. Simunek. 1997. UNSATCHEM: Unsaturated water and solute transport with equilibrium and kinetic chemistry. Soil Sci. Soc. Am. J. (In press).

Sullivan, M.D., N.R. Fausey, and R. Lal. 1998. Long-term effects of sub-surface drainage on soil organic carbon content and infiltration in the surface horizons of a lakebed soil in northwest Ohio. In R. Lal, J. Kimble, R. Follett and B.A. Stewart (eds) "Management of Carbon Sequestration", CRC Press, Boca Raton, FL: 73-81.

Sutton, P., and W.A. Dick. 1987. Reclamation of acid mined lands in humid areas. Adv. Agron. 41: 377-405.

Sweeten, J.M., and J.R. Miner. 1992. Odor intensities at cattle feedlots in nuisance litigation. Paper #924514. Winter Meeting, ASAE, St. Joseph, MI.

Szabolcs, I. 1998. Salt buildup as a factor of soil degradation. In R. Lal, W.H. Blum, C. Valentine and B.A. Stewart (eds) "Methods for Assessment of Soil Degradation", CRC Press Inc., Boca Raton, FL: 253-264.

Tans, P.P., I.Y. Fung, and T. Takahashi. 1990. Observational constraints on the global atmospheric CO_2 budget. Science 247: 1431-1438.

Tester, C.F. 1990. Organic amendment effects on physical and chemical properties of a sandy soil. Soil Sci. Soc. Am. J. 54: 827-831.

Thompson, R.B., J.C. Ryden, and D.R. Lockyer. 1987. Fate of N in cattle slurry following surface application or injection to grassland. J. Soil Sci. 38: 689-700.

Thornton, F.C., and R.J. Valente. 1996. Soil emissions of nitric oxide and nitrous oxide from no-till corn. Soil Sci. Soc. Am. J. 60: 1127-1133.

Tiessen, H., J.W.B. Stewart, and J.R. Betany. 1982. Cultivation effects on the amounts and concentration of C, N and P in grassland soils. Agron. J. 74: 831-835.

Tisdall, J.M. 1996. Formation of soil aggregates and accumulation of soil organic matter. In M.R. Carter and B.A. Stewart (eds) "Structure and Organic Matter Storage in Soils", CRC Press, Boca Raton, FL: 57-95.

Tracy, P.W., D.G. Westfall, E.T. Elliott, G.A. Peterson, and C.V. Cole. 1990. Carbon, nitrogen, phosphorus and sulfur mineralization in plow and no-till cultivation. Soil Sci. Soc. Am. J. 54: 457-461.

Underwood, J.F., and P. Sutton. 1992. In R.E. Dunker, R.I. Barnhisel and R.C. Darmody (eds) "Prime Farmland Reclamation", Univ. Illinois, Urbana, IL: 1-10.

Unger, P.W. 1995a. Residue management for continuous winter wheat. J. Soil Water Cons. 50: 317-321.

Unger, P.W. 1995b. Organic matter and water-stable aggregate distribution in ridge-tilled surface soil. Soil Sci. Soc. Am. J. 59: 1141-1145.

USDA .1975. Office of planning and evaluation. Minimum tillage: A preliminary technology assessment, Washington, D.C.

USDA. 1978. Improving soils with organic wastes. Report to Congress in response to Section 1961 of the Food and Agric. Act of 1977 (P.L. 95-113), Washington, D.C.

USDA. 1987. Farm drainage in the United States: history, status and prospect. Misc. Pub. No. 1455, Washington, D.C.

USDA-ARS. 1990. Agricultural Statistics. 1990. U.S. Govt. Printing Office, Washington, D.C.

USDA-ARS. 1993. Agricultural Statistics. 1993. U.S. Govt. Printing Office, Washington, D.C.

USDA-ERS. 1987. Farm drainage in the United States: History, Status and Prospects. Miscellaneous Publication No. 1455, 170 pp.

USDA-ERS. 1994. Agricultural Resources and Environmental Indicators, Agricultural Handbook No. 705, Washington, D.C., 205 pp.

USDA-ERS. 1995. Tillage and cropping on HEL, AREI Updates, No. 6.

USDA-ERS. 1996. Cropping patterns of major field crops and associated chemical use. AREI Updates No. 18, Dec. 1996.

USDA-ERS. 1996. Conservation and the 1996 Farm Act. Agricultural Outlook, Special Article, November, 1996, 8 pp.

USDA-ERS. 1996. 1995 nutrient use and practices on major field crops. Agricultural Resources and Environmental Indicators (AREI) Updates, No. 2, May 1996.

USDA-ERS. 1997. Agricultural resources and environmental indicators, 1996-97. An economic research service report. Agric. Handbook No. 712, Washington, D.C., USA.

USDA-FAS. 1991a. World grain situation and outlook. Foreign Agricultural Service, FGS-91, Washington, D.C.

USDA-FAS. 1991b. World soil seed situation and outlook. Foreign Agricultural Service, FOP 10-91, Washington, D.C.

USDA-FS. 1979. User guide to soils: mining and reclamation in the west. USDA-Forest Service, General Technical Report INT-68, Ogden, Utah, 80 pp.

USDA-NASS. 1997. Agricultural Statistics, USDA, National Agricultural Statistics Service, Washington, D.C.

USDA-NRCS. 1995. National Resource Inventory, Data Base, Washington, D.C.

USDA-SCS. 1992. National Resource Inventory, Washington, D.C.

USEPA. 1976. Erosion and Sediment Control, Surface mining in Eastern U.S. vol. 1 EPA-625/3-76-006, U.S. Govt. Printing Office, Washington, D.C.

USEPA. 1995a. Climate change mitigation strategies in the forest and agriculture sectors. Washington, D.C., 64 pp.

USEPA. 1995b. Inventory of U.S. greenhouse gas emissions and sinks: 1990-94, Washington, D.C.

Utomo, M., W.W. Frye, and R.L. Blevins. 1990. Sustaining soil N for corn using hairy vetch cover crop. Agron. J. 82: 979-983.

Valente, R.J., and F.C. Thornton. 1993. Emissions of NO from soil at a rural site in central Tennessee. J. Geophys. Res. 98: 16745-16753.

Varvel, G.A. 1994. Rotation and nitrogen fertilization effects on changes in soil carbon and nitrogen. Agron. J. 86: 319-325.

Walker, J.M., R.M. Southworth, and A.R. Rubin. 1997. U.S. Environmental Protection Agency regulations and other stakeholder activities affecting the agricultural use of by-products and wastes. In J.E. Rechcigl and H.C. Mackinnon (eds) "Agricultural Uses of By-Products and Waste", American Chemical Society, Washington, D.C.: 28-47.

Waltman, S.W., and N.B. Bliss. 1997. Estimates of SOC content. NSSC, Lincoln, NE.

Wiebe, K.D., A. Tegene, and B. Kuhn. 1995. Property rights, partial interests, and the evolving federal role in wetlands conversion and conservation. J. Soil Water Cons. 50: 627-630.

Williams, S.T., M. Shameemullah, E.T. Watson, and C.I. Mayfield. 1972. Studies on the ecology of actinomycete in soil. IV. The influence of moisture tension on growth and survival. Soil Biol. Biochem. 4: 215-225.

Wood, C.W., J.H. Edwards, and C.G. Cummins. 1991a. Tillage and crop rotation effects on soil organic matter in a Typic Hapludalt on northern Alabama. J. Sust. Agric. 2: 31-41.

Wood, C.W., D.G. Westfall, and G.A. Peterson. 1991b. Soil C and N changes on initiation of no-till cropping systems. Soil Sci. Soc. Am. J. 55: 470-476.

Zobeck, T.M., and D.W. Fryrear. 1986. Chemical and physical characteristics of windblown sediment. II. Chemical characteristics and total soil and nutrient discharge. Trans. ASAE 29: 1037-1041.

APPENDIX 1

Definitions

Expression of Global Warming Potential

Radiative forcing

The term "radiative forcing" is defined as a change in the average net radiation at the top of the troposphere. A radiative forcing perturbs the balance between incoming and outgoing radiation. Over time, climate responds to the perturbation to re-establish the radiative balance. Annual increases in anthropogenic radiative forcing are calculated by figuring the sum of the net changes in emissions of each greenhouse gas, multiplied by its GWP, and expressed in terms of MMTCE. This allows a direct comparison of the relative contribution of each of the GHGs to the total annual change in radiative forcing.

Global warming potential (GWP)

The GWP of a greenhouse gas is the ratio of global warming or radiative forcing from one kilogram of a greenhouse gas to one kilogram of carbon dioxide over a period of time. This book uses the 100-year GWPs the IPCC recommends. Using this index provides a way to calculate the contributions of each of the GHGs to the annual increases in radiative forcing. This book represents all gases in units of million metric tons of carbon equivalent, or MMTCE. The GWP value used is 1 for CO_2, 21 for CH_4, and 310 for N_2O.

Million metric tons of carbon equivalent (MMTCE)

The term MMTCE is based on the conversion of all gases to equivalent GWP expressed on a weight basis (Eq. 27).

MMTCE = (MMT of gas) (GWP of gas) (12/44) .... Eq. 27

where MMT is million metric ton (1 MMT = 10^{12} g = 1 Tg = 1 tera gram).

Million metric tons of carbon (MMTC)

The unit MMT is based on the total weight of carbon expressed as million metric tons (10^6 Mg). For CO_2, MMTCE is numerically the same as MMTC and equals 1 Tg = 1 teragram = 10^{12} g = 1 MMTC. One gigaton (GT) equals 1000 MMT or 1 Pg (petagram = 10^{15}).

The Potential of U.S. Cropland to Sequester Carbon and Mitigate the Greenhouse Effect

Important Pedospheric Processes that Lead to C Sequestration in the Soil

Humification

This is the conversion of biomass into humic substances which are relatively resistant to microbial decomposition and have a very long (hundreds to thousands of years) turnover time.

Aggregation

This refers to formation of stable organo-mineral complexes in which C is bonded with clay colloids and metallic elements to form aggregates. The C enclosed within microaggregates is protected against microbial decomposition.

Translocation within the pedosphere

This implies eluviation and deposition of humus and stable micro-aggregates within the soil profile so that they are less disturbed by biotic and abiotic processes and are not exposed to climatic elements.

Deep rooting

Some species have deep rooting characteristics, and appropriate management can be applied to encourage root growth which can enhance SOC content in the sub-soil horizons (Photo 20).

Calcification and formation of secondary carbonates

This implies the formation through abiotic processes of secondary carbonates, carbonates, and other solid and insoluble compounds containing inorganic carbon. This process is extremely important in arid and semi-arid regions (Photo 21).

Photo 20. Use of appropriate genotypes and soil management techniques can increase root system development in sub-soil (root distribution to 2 m).

Photo 21. Some soils of the arid regions are characterized by a calciferous (caliche) horizon.

Soil Processes that Lead to Emissions of GHGs

Some soil processes also lead to emissions of CO_2, CH_4, and N_2O from soil into the atmosphere. Predominant among these are soil erosion, leaching, anaerobiosis, denitrificatin and volatilization, and mineralization.

Soil Erosion

Soil erosion by water (Photo 22) and wind (Photo 23) are major processes that exacerbate the greenhouse effect. Soil erosion is a selective process and involves preferential removal of SOC, SIC, and clay fractions. The C content of eroded sediments varies depending on the antecedent soil properties, climate, and parent material. However, enrichment of C by the erosional process occurs, with a mean C content of depositional material often being 2 to 3 times higher than that of the undisturbed soil.

Illuviation/Leaching

Cations and dissolved organic carbon (DOC) can be translocated to a lower horizon or leached out of the soil, especially in a humid climate. The magnitude of the loss and the fate of DOC, once it enters the aquatic ecosystem, are not known.

Anaerobiosis

Emission of CH_4 from soil occurs under reducing conditions. Natural wetlands (Photo 24), poorly drained agricultural lands (Photo 25), rice paddies, and organic soils (Photo 26) are the principal sources of CH_4 emissions from soil. The process of soil-CH_4 emission is called *methanogenesis*.

Photo 22. Soil erosion by water can be serious on cropland.

Photo 23, at left. Wind erosion in arid and semi-arid regions leads to displacement of C-rich sediment. Photo 24, below. Natural wetlands can sequester carbon.

The Potential of U.S. Cropland to Sequester Carbon and Mitigate the Greenhouse Effect

Photo 25. Poorly drained agricultural lands and wetlands are also a source of CH$_4$ emissions.

Photo 26. Mineralization of organic soils by cultivation is a source of CO$_2$.

Denitrification and volatilization

Soil emission of N_2O occurs through mineralization and denitrification of N in SOC, fertilizers, manures, and other organic materials.

Mineralization

Decomposition of complex organic compounds into simple inorganic compounds occurs through several biotic and abiotic reactions. As also is the case with denitrification, volatilization, and leaching, it is difficult to obtain reliable estimates of C emission through mineralization.

Some Important Variants of Conservation Tillage

No till

All pre-planting seedbed preparation is eliminated, weed control is primarily through herbicides, and crop residue is left on the soil surface as mulch (Photos 27-28).

Ridge till

Crops are planted on previously made raised beds, but the soil also is left undisturbed from harvest to planting (Photo 29). Ridges are made once every 2 or 3 years but repaired every spring, and planting on ridges is facilitated by the use of disk openers, coulters, or row cleaners.

Photo 27. Crop residue mulch is an important component of a no till system.

Photo 28. Crop residue mulch is an important component of a no till system.

Photo 29. Ridge till is a widely used conservation tillage system.

Photo 30. Chisel plowing is a common farm practice of a minimum till system.

Minimum till

This refers to any system of seedbed preparation that involves fewer tillage operations than conventional tillage and leaves 15-30% of residue mulch after planting or 500 to 1000 kg/ha of small grain residue equivalent throughout the critical wind erosion period. Commonly used minimum tillage practices include chisel-plant (Photo 30), disc plant, or sowing after para plowing (Photo 31). The family of minimum till systems also are referred to as mulch tillage, because crop residue mulch is maintained on the soil surface (Photo 32).

Sod till

A cover crop specifically is grown to produce a protective ground cover, and crops are seeded through the chemically killed sod (Photo 33).

**Photo 31, right.
Paraplowing is done to
alleviate subsoil
compaction while
leaving residue mulch
on the soil surface.
Photo 32, below. Mulch
farming reduces soil
erosion and conserves
water.**

Photos 33a, 33b. A cover crop improves soil quality and sequesters carbon.

The Potential of U.S. Cropland to Sequester Carbon and Mitigate the Greenhouse Effect

Abbreviations

AWC = available water capacity
BMP = best management practices
BTU = British Thermal Unit
CEC = cation exchange capacity
CFCs = chlorofluorocarbons
C:N = carbon:nitrogen ratio
CRP = Conservation Reserve Program
DOC = dissolved organic carbon
DOE = Department of Energy
EC = electrical conductivity
EIA = Energy Information Administration
EPA = Environment Protection Agency
FAO = Food and Agriculture Organization of the United Nations
g Clm^2/yr = gram of carbon per meter square per year
GDP = gross domestic product
GHG = greenhouse gases
GJ = gigajoule
GT = gigaton = petagram = 1000 MMT
GWP = global warming potential
ha = hectare
IMP = integrated pest management
IPCC = inter-governmental panel on climate change
kg = kilogram
m = meter
Mha = million hectares 5 f
mmhos/cm = millimhos/cm
MMT = million metric tons
MMTC = million metric tons of carbon
MMTCE = million metric tons of carbon equivalent
MT = metric ton
MTC/ha = metric ton of carbon per hectare
MWD = mean weight diameter
NPP = net primary productivity
OSTP = 0ffice of Science Technology Policy
Pg = Petagrom = 10^{15} g
ppbv = parts per billion by volume
ppmv = parts per million by volume
ppptv = parts per trillion by volume
SIC = soil inorganic carbon
SOC = soil organic carbon = C

Tg = Teragram = 10^{12} g
WRP = Wetland Reserve Program

Units of Measurement

Metric (SI) multipliers

Prefix	Abbreviation	Value
exa	E	10^{18}
peta	P	10^{15}
tera	T	10^{12}
giga	G	10^{9}
mega	M	10^{6}
kilo	k	10^{3}
hecto	h	10^{2}
deka	da	10^{1}
deci	d	10^{-1}
centi	c	10^{-2}
milli	m	10^{-3}
micro	μ	10^{-6}
nano	n	10^{-9}
pico	p	10^{-12}
femto	f	10^{-15}
atto	a	10^{-18}

Conversion factors

Energy and work

1 J = 10^{7} ergs = 0.738 ft lb
1 ft-lb = 1.36 J = 1.29×10^{-3} Btu = 3.24×10^{-4} kcal
1 kcal = 4.18×10^{3} J = 3.97 Btu
1 Btu = 252 cal = 778 ft-lb = 1054 J
1 eV = 1.602×10^{-19} J
1 kWh = 3.60×10^{6} J = 860 kcal
1 quad = 1 quadrillion Btus = 10^{15} Btu

Researchable Topics

To realize of the full potential of cropland management for carbon sequestration in soil, terrestrial, and aquatic ecosystems, we must have strong database and an understanding of the biophysical processes within the pedosphere that determine the SOC pool and its dynamics. Soil scientists traditionally have been involved in soil resource management more for agronomic productivity than for C sequestration to mitigate the greenhouse effect. Now that the C sequestration potential of soil is widely recognized, however, we must strengthen the database and understand the mechanisms, principles, soil carbon pools, and potentials involved. Scientific knowledge must be strengthened to identify sources of GHG emissions in relation to agricultural practices, and sinks of C in soils and other components of terrestrial and aquatic ecosystems.

Data and Information Base

The databases needed to provide credible estimates of the C pool and their dynamics are weak. The quality of existing field data is doubtful because of knowledge gaps, missing or incomplete information, and lack of resources to complete the collection of field data.

Land Use and Management

Data exist in great quantity on soil surveys, soil physical and chemical properties, land use, crop rotation, input used in crop production, and crop yield and factors affecting it. The major need at this point is for data on soil processes that affect SOC and SIC pools and their dynamics with respect to land use, cropping/farming systems, and soil and crop management practices. Specific research information is needed for C pool and dynamics, and their spatial and temporal changes, with regard to soil erosion management and to land conversion and restoration.

Soil erosion management

Data exist on potential risks of soil erosion and its impact on loss of water, nutrients and crop yield. However, there is a strong need to strengthen the database about the:

1. Impact of erosional processes by wind and water on SOC and SIC pools and on the fate of C carried by the sediments
2. Enrichment ratio of eroded sediments for C and N
3. Fate of C distributed over the landscape through eroded sediments.

Land conversion and restoration

There is a need to strengthen the database on the impact of soil restorative measures on soil quality enhancement and C sequestration in soil. Soil and site-specific information is needed about the impact of:

1. CRP, WRP, and conservation buffers, on the magnitude and mechanisms of C sequestration in soil
2. Restoration of eroded lands, on the rate and magnitude of C sequestration
3. Mineland reclamation, on C sequestration for different restorative techniques
4. Reclamation of salt-affected and polluted soils, on quality and quantity of SOC content
5. Soil fertility restoration, on SOC dynamics
6. Wetland restoration, on net C sequestration (the balance between SOC accumulation and CH_4 emission) for different aquatic ecosystems.

Biofuels

Little research information is available on the potential of using idle and marginal cropland to produce biofuels. What are the pros and cons of using organic material as direct fuels rather than converting those into liquid fuel products?

Intensification of Prime Agricultural Land

There is a scarcity of research information about the impact of:

1. Tillage methods (including cyclic tillage), residue management techniques, and interaction between them, on SOC dynamics
2. Water management (e.g., irrigation, drainage, sub-irrigation), on SOC and SIC pools and dynamics
3. Fertilizer practices and nutrient management, on C and N dynamics, GHG emissions, and N use efficiency
4. Use of organic materials and crop residues, on C sequestration efficiency

5. Cropping systems, crop rotations, and cover crops, on C and N dynamics
6. Management of rice paddies, on C dynamics.

Soil Degradation and C Dynamics

There is a paucity of data on the impact of soil degradative processes on SOC and SIC pools and their dynamics. In addition to erosional processes, there is also a need to strengthen the data about the:

1. Impact of the decline of soil structure (e.g., aggregation), on SOC and SIC dynamics
2. Salinization and SIC/SOC dynamics
3. Impact of mining and drastic soil disturbance, on SOC pool and fluxes
4. Impact of cultivation of organic soils, on the C pool and on C and N dynamics
5. Impact of intensive use of calciferous soils, on the SIC pool, formation of secondary carbonates, and the C balance (Photo 34)
6. Quantification of root biomass-C produced by various crops and cropping systems.

Photo 34. Inorganic or carbonate carbon is a major C pool in soils of arid and semi-arid climates.

Problems with Methodology

The existing data were collected with a wide range of methods for determining SOC and SIC contents. The supporting data on soil bulk density and textural properties are often not available. The SOC data are also not available for soil layers below the plow depth. The information on CT methods needs to be updated especially with regard to the cyclic tillage systems whereby farmers change from CT to plowing and vice versa because of soil compaction, weed pressure, or other site-specific reasons.

There is also a problem with the method of aggregating the data from a point or plot scale to a regional and national scale. There is a need to develop and standardize methods of monitoring SOC and SIC pools, and emissions of GHGs from soil. Scaling procedures need to be developed for aggregating the data.

Value of Soil Carbon

The actual value of SOC, both on-site and off-site, is not known. Economists, working with soil scientists/agronomists and water quality specialists, need to establish the actual value of SOC. Only then will its importance to policy makers be fully understood. Although its intrinsic value to soil quality is known, policies are linked with economic considerations.

Implementation of a Priority Research Program

The USDA (ARS and NRCS) in collaboration with universities, DOE, EPA, and other organizations should develop and implement a comprehensive research and evaluation program to fill the knowledge gaps this Appendix identifies. A systematic analysis of soil samples, obtained from on-going cropland management experiments in different ecoregions, is a high priority.

Numerous gaps and uncertainties exist related to obtaining reliable estimates of GHG emissions and the C sequestration potential of agricultural activities. Three major unknowns in C budgeting in relation to agricultural practices are: (i) understanding C dynamics in cropland soils with regard to management (e.g., irrigation, fertilizer and manure use, crop rotation), (ii) improving the knowledge base with regard to the impact of soil degradative and restorative processes on C pool and dynamics, and (iii) assessing the capacity of cropland soils as a C sink for different management options.

This volume has not addressed the issue of SIC and secondary carbonates. Soils of the arid region contain a major pool of carbonates. The dynamics of this vast pool, in relation to land use and management, are not known. It is important to quantify the magnitude of this pool and assess its dynamics in relation to cropping systems, fertilizer use, irrigation, and tillage methods.

Extensive analysis is needed about the net impact of agriculture in the western U.S. on the soil inorganic carbon budget and the net impact of various management changes. This would require that we run extensive realistic simulations for actual basins and for a variety of scenarios.

We have the necessary knowledge of the dynamics of carbonate dissolution and precipitation and the production and transport of CO_2 in the rootzone. This knowledge has been incorporated into a detailed chemical transport model which includes plant biomass production and salinity response. Analysis should result in a calculation of the present effect of irrigated agriculture on the inorganic carbon balance as well as the effect that several management options might have on the short to medium time scales.

Limited information exists about the effect of salinity on microbial respiration. Studies need to be conducted on the effect of salinity on the net change in SOC. These experiments should be conducted with several nitrogen treatments and should be run for several soils to account for differences in salt tolerance among microbial populations. Information is already available on the effects of salinity on plant and root biomass.

FRAN BAUM

THE NEW PUBLIC HEALTH

FOURTH EDITION

OXFORD

UNIVERSITY PRESS

AUSTRALIA & NEW ZEALAND

OXFORD
UNIVERSITY PRESS

Oxford University Press is a department of the University of Oxford.
It furthers the University's objective of excellence in research, scholarship,
and education by publishing worldwide. Oxford is a registered trademark
of Oxford University Press in the UK and in certain other countries.

Published in Australia by
Oxford University Press
253 Normanby Road, South Melbourne, Victoria 3205, Australia

First edition published 1998
Second edition published 2002
Third edition published 2008
Fourth edition published 2016
Reprinted 2016, 2018

National Library of Australia Cataloguing-in-Publication data

> Creator: Baum, Frances, author.
> Title: The new public health / Fran Baum.
> Edition: 4th edition.
> ISBN: 9780195588088 (paperback)
> Notes: Includes bibliographical references and index.
> Subjects: Public health—Australia.
>   Health promotion—Australia.

Dewey Number: 362.10994

Reproduction and communication for educational purposes
The Australian *Copyright Act 1968* (the Act) allows a maximum of one chapter
or 10% of the pages of this work, whichever is the greater, to be reproduced
and/or communicated by any educational institution for its educational purposes
provided that the educational institution (or the body that administers it) has
given a remuneration notice to Copyright Agency Limited (CAL) under the Act.

For details of the CAL licence for educational institutions contact:

Copyright Agency Limited
Level 15, 233 Castlereagh Street
Sydney NSW 2000
Telephone: (02) 9394 7600
Facsimile: (02) 9394 7601
Email: info@copyright.com.au

Edited by Julie Irish, Biotext
Cover image by Stocksy/Tom Tomczyk; Shutterstock/Pedo Pinto (walking man)
Text design by Denise Lane
Typeset by diacriTech, Chennai, India
Proofread by Vanessa Lanaway
Indexed by Bruce Gillespie
Printed by Sheck Wah Tong Printing Press Ltd

BRIEF CONTENTS

EXTENDED CONTENTS

LIST OF BOXES, FIGURES AND TABLES

BOXES

FIGURES

TABLES

ACKNOWLEDGMENTS

The author and the publisher wish to thank the following copyright holders for reproduction of their material:

AAP/AP, 369 /Greenpeace, 579; Adelaide Thinkers in Residence Office, 470; Extract, 'The Growing Impact of Globalization for Health and Public Health Practice' by R. Labonte, et al., Annual Review of Public Health, 2011, 296; Copyright © Australian Institute of Health and Welfare, 292–3, 295–6, 300–1; City of New York/Iwan Baan, 464; © Commonwealth of Australia Creative Commons Attribution 2.5 Australia licence, 299, 304, 305, 371–2; Department of Health (2009) National Mental Health Policy 2008, Canberra, Commonwealth of Australia, 73; Extract, When Corporations Rule the World by D. Korten, Earthscan, London, 1995, 125; Fairfax/Paul Matthew, 582; Getty Images/Handout, 114; Fernando M Gonçalves, 265, 375, 376, 472, 565, 574, 577, 581, 607; Department for Health and Ageing, SA Health, Government of South Australia, 631; Health Equity Knowledge and Health Equity, Victorian Health Promotion Foundation, 69; Inner South Community Health, 59; Joint United Nations Programme on HIV/AIDS (UNAIDS) (2014) Fast track: Ending the AIDS epidemic by 2030. Joint United Nations Programme on HIV/AIDS, Geneva, 273–4, 274–5; Simon Kneebone, 663; Paul Laris, 466, 477; Arthur Mostead, 576; National Health and Medical Research Council, 205; Navdanya Office, 119; Newspix, 654 /Amos Aikman, 120 /Fiona Hamilton, 365; Extract, Making Health Policy, by K. Buse, N. Mays & G. Walt, Open University Press, Maidenhead, UK, 2005, 163; The material on pages 115, from Working for the Few: Political capture and economic inequality, 2014 is reproduced with the permission of Oxfam GB, Oxfam House, John Smith Drive, Oxford OX4 2JY, UK www.oxfam.org.au. Oxfam GB does not necessarily endorse any text or activities that accompany the materials, 115; Panos/Piers Benatar, 262 /Stefan Boness, 578 /Tom Pilston, 121 /Vlad Sokhin, 298 /Penny Tweedie, 504 /Tuen Voeten, 381; Extract, Beyond the Welfare State, by C. Pierson, Polity Press, London, 1994, 101; http://repowerportaugusta.org, 483 (top), 483 (bottom); Letter dated 08 July, 2014 from the Premier Jay Weatherill RE reform of S.A. Boards and Committees, 534; Extract, 'Competing paradigms in qualitative research' by E. Guba & Y. Lincoln, Handbook of Qualitative Research, SAGE Publications, Thousand Oaks, CA, 1994, 158; Extract, The Art of Case Study Research by R.E Stake, SAGE Publications, Thousand Oaks, CA. 1995, 208; Jan Swasthya Abhiyan—Peoples Health Movement, India, 580; Extract, Quantity and Quality in Social Research by A. Bryman © 1988, Routledge, London, reproduced by permission of Taylor & Francis Books UK, 157; World Bank (2014) World Development Indicators. http://data.worldbank.org/data-catalog/world-development-indicators, 259, 264; Extract, The Health of Nations: Towards a New Political Economy, by G. Mooney, Zed Books, London, 2012, 98.

Every effort has been made to trace the original source of copyright material contained in this book. The publisher will be pleased to hear from copyright holders to rectify any errors or omissions.

ABOUT THE AUTHOR

Fran Baum is a public health researcher, teacher and advocate. She is the Matthew Flinders Distinguished Professor of Public Health and Foundation Director of the Southgate Institute for Health, Society and Equity at Flinders University, Adelaide, Australia. She was instrumental in developing both the Master of Public Health and Doctor of Public Health at Flinders University, and this book draws on that experience. Fran is one of Australia's leading researchers on the social and economic determinants of health. She has held many grants from the National Health and Medical Research Council and the Australian Research Council, including a Centre for Research Excellence on Policies for Health Equity and a prestigious Australian Research Council Federation Fellowship.

As well her academic work, Fran is a passionate health advocate. She is a member and past Chair of the Global Steering Council of the People's Health Movement—a global network of health activists (www.phmovement.org). She is a past national president of the Public Health Association of Australia. From 2005 to 2008 she served as a Commissioner on the World Health Organization's Commission on Social Determinants of Health. Her work has been recognised by her appointment as a Fellow of the Academy of the Social Sciences in Australia and of the Australian Health Promotion Association, and Life Member of the Public Health Association of Australia.

THANKS AND APPRECIATION

Just as it takes a village to raise a child, a book and its author need rich nourishment from many places. I am lucky to have many friends, family and colleagues who create my global village and I'm very happy to acknowledge the support, encouragement, friendship and love I receive from so many different sources.

Revisions for this and the third edition of this book were meant to be minor. In neither case did that prove possible, given the pace of changes in the world and the accumulating knowledge about both the threats to our collective health and the range of solutions. While being multidisciplinary and broad in scope are key strengths of the new public health, those characteristics make producing a comprehensive textbook on the subject challenging! I am very grateful to my publisher at Oxford, Debra James, and Development Editor Camha Pham, for accommodating the longer time needed to do justice to the growing knowledge this edition encompasses. Julie Irish has been a great copyeditor to work with.

My writing has been aided by the direct and indirect support of friends and colleagues including Julia Anaf, Pat Anderson, Lori Baugh-Littlejohns, Michael Bentley, Danny Broderick, Toni Delany, Tiana Della Putta, Judith Dwyer, Evelyne de Leeuw, Matt Fisher, Denise Fry, Liz Harris, Ilona Kickbusch, Michael Kidd, Ross and Libby Kalucy, Toby Freeman, Gwyn Jolley, Ron Labonte, Angela Lawless, Pamela Lyons, Dennis McDermott, Tamara MacKean, Lareen Newman, Clare Phillips, Jennie Popay, Christine Putland, Sue Richardson, Clare Shuttleworth, Melissa Sweet, Louise Townend, Carmel Williams, Paul Worley and Anna Ziersch. Each of you has helped by commenting on sections of the book, engaging in discussion about an aspect of the book that has enriched and deepened my understanding or developed my thinking about public health. Special thanks to Colin MacDougall for all our discussions about public health, reminders of the increasing absurdity of life and the inspiration to embark on the first edition.

Also thanks to the wonderful administrative staff I work with at Flinders University— Helen Scherer's quiet efficiency is much appreciated and my Executive Assistant Paula Lynch guards my time and provides me with amazingly professional support that makes my life easier and less stressful. Anna Lane has been my research assistant for just over a year and has been superb in that role. She has chased facts, figures, references and photos doggedly and with good humour. Her attention to detail has reduced the errors in the book. She has been a delight to work with, which has made the process of completing the book as pleasurable as possible.

The photos in this edition have been greatly enhanced by the photographic work of Fernando M. Gonçalves. His work so often captures the details of inequities and social injustice and I am very grateful to him for producing high-quality images for this edition.

From 2005 to 2008 I served as a Commissioner on the Commission on Social Determinants of Health (established by the Director-General of the World Health Organization) and in that process made lifelong friends who have greatly enriched the understanding I have brought to this edition of the book. These friends include Michael Marmot, who has been the global leader of the social determinants of health movement; Mirai Chatterjee, whose work with the Self-employed Women's Association in India is deeply inspiring; Sharon Friel, who has become a good friend and with whom I have the delight and privilege of co-directing our NHMRC Centre of Research Excellence on the Social Determinants of Health Equity; and Monique Bégin, a dear friend who has taught me how powerful the combination of charm, wisdom, intellect and passion is to advocacy for greater health equity. I owe a deep gratitude to all those involved in the Commission on Social Determinants of Health for deepening my understanding of and ability to act on social determinants of health.

My comrades in the People's Health Movement have ensured that I have not forgotten for a moment that the new public health is, above all, about changing our world for the better. Special thanks to Chiara Bodini, Anneleen De Keukelaere, Prem John, David Legge, Bridget Lloyd, Dave McCoy, Jihad Marshal, Ravi Narayan, Delen de la Paz, David Sanders, Claudio Schuftan, Amit Sen, Hani Serag, Mira Shiva, Ruth Stern, David Werner, Maria Zuniga—you all inspire me with your dedication and commitment to the struggle for health and provide a wonderful solidarity in the conviction that a better world is possible and that our movement will help that world eventuate.

I also have two very special groups of women with whom dining is a delight— Rubies and the Christies Beach Women's Shelter group—we share discussion, debate and laughter.

Academic life has become more pressured and competitive over the past decade as the pressures to win grants and to publish create considerable stress. My long-term work–life mentor Robin Maslen sadly died in 2008. I have missed his wise counsel very much but thank him for his many years of support. A great gift from him in the final months of his life was to introduce me to Dr Stefan Neszpor, who has helped me cope with a terrible bout of depression, provided me with ongoing life-enhancing support in dealing with the black dog and introduced me to the Haven, a Centre for Transformative Learning on Gabriola Island, BC, Canada. The Haven has become a very special place for me and helped me learn so much more about myself and the meaning of life. So heartfelt thanks to Stefan and all my Haven family with whom I have shared so much. Hugh Kearns has also offered me very useful support in balancing work with the rest of my life and taught me the dangers of the shiny balls that steal away my leisure time.

The book has stolen time away from family so thanks to you all for your love and support. Above all I want to thank Paul Laris, who totally supports me in my academic and advocacy endeavours. Always on hand to read, discuss and critique and to remind me that a good life is also love, fun, relaxation and living in the moment.

Fran Baum
Adelaide, March 2015

INTRODUCTION

> We are challenged to develop a public health approach that responds to the globalized world and its political, social and economic ramifications. The challenge is as large as when public health was first developed.
>
> *Kickbusch, 2005*

Since the publication of the first edition of *The New Public Health* in 1998 much has changed in the world. Neo-liberal globalisation has proceeded apace. A new century has dawned and people around the world have expressed huge hopes for it. Yet peace is appearing as elusive as it was in the past century, inequities are increasing more rapidly and terrorism is becoming a greater threat. Much hope appears to be vested in the advances in biomedicine as a means of raising world health standards, yet simple and proven public health measures, such as the provision of clean water, sanitation and universal education, could improve health for the world's poor just as dramatically. Threats to environmental sustainability continue to increase and there is no effective global treaty to address climate change. Despite the setting of the Millennium Development Goals by the United Nations, designed to improve the health of the world's poorest, only a few of these goals have been met. If anything, the world is becoming a less just place. For a moment, imagine a world where there are no major differences between the health experienced by people in different groups, where prejudice is unheard of, where no children live in poverty, where the wisdom and rights of those of us who are indigenous are totally respected, and where social and environmental considerations always balance economic decisions. Does that world appeal to you? Well, let your imagination wander a little further to a world with all those characteristics in which countries vary because of their culture and geography, rather than because of deep gulfs in wealth and poverty, and where cultural difference delights rather than frightens people.

Cloud cuckoo land, of course. Just think of the numerous political, economic and social barriers to those imaginings. Yet, there aren't any physical reasons why they couldn't become reality. Our problem is not insufficient food but uneven distribution; there is enough wealth to go around, but it is concentrated in the hands of a few; social and environmental considerations can be high priorities in decision making but only if we choose for them to be; health could be more evenly distributed if only the resources that create it were. The new public health strives for a fairer, more just, healthier, kinder world and recognises that it is human action rather than physical constraint that prevents us achieving it. One of the major constraints is the compartmentalised nature of our knowledge. This book integrates knowledge from a range of disciplines, methodologies and perspectives as they relate to public health. It also examines the role of public health in achieving a better world.

This book is written for students of public health, public and primary health care workers, health and environment planners, and those people who are interested in creating communities that maximise health for people and the environment. The book is deliberately broad in scope, in the belief that the health and environmental challenges facing humanity require cross-disciplinary perspectives. Narrow areas of knowledge certainly have their place but public health is, by its very essence, an integrative discipline that hunts and gathers theories and ideas from many other disciplines and professions. Key tasks for the expert public health worker are to be capable of integrating complex areas of knowledge and applying them to public health issues. This book intends to provide a basis for public health workers to do this.

This book concentrates on the social side of public health. Areas of public health pertaining to traditional laboratory science, such as microbiology, have a valuable role but are not covered here. Other textbooks cover that territory. This book integrates knowledge and methodologies from the social sciences, environmental sciences and humanities.

The term 'the new public health' is controversial to some as it implies that previous public health activity was not effective. However, new directions in public health are required to address the reality of our globalised and fast-changing world. In response, this book presents a package of ideas and directions for public health that explicitly focus on achieving a fairer, more sustainable and equitable world

My starting point is that no theory offers an entire solution to understanding the world and providing a basis for public health action. Consequently, I use theories eclectically, selecting bits here and there as they help me to understand the world and explain it to others. Sometimes I will veer towards a Marxist or neo-Marxist perspective. At other times I will use the postmodern technique of deconstruction of once-taken-for-granted truth. The one certainty I cling to is that our world is complex—far too complex to be neatly understood through any single theoretical lens. I have chosen the lenses that seem to best illuminate the practice of the new public health and provide a basis for action. For me that is the essential element— how to use theory to understand what forms of public health practice work and why they do.

One of my firmly held beliefs is that public health is nothing if not multidisciplinary and holistic. There is a role for specialists, of course, but the generalists are the ones who will take the overview and separate the wood from the trees. This book includes many methodologies (epidemiology, demography, surveys, interviews, observation) and perspectives from even more disciplines and professions (including sociology, medicine, psychology, anthropology, ecology, urban planning, architecture, engineering, social work, political science, economics). I have deliberately chosen not to play it safe and stick to firm ground on which I believe myself to be one of the experts. If I had done so, the book could not describe the essence of the new public health. To work in public health requires that you are something of an experimenter and risk-taker, which often means reaching out beyond the territory where you know you will be safe.

We live in dangerous times—there are so many threats to our individual health and, far more significantly, to that of our planet and collective life-support system. Public health should be contributing to reducing the danger. Its real strength is that it offers a holistic and varied view of the world—it is a planetarium rather than a microscope. Achieving the aims of the new public health will require commitment from creative, intelligent and dedicated people who can span disciplines and combine action and intellectual work. I hope this book will be a resource to inspire people to work in public health, as the rewards that could come from a truly new public health are evident. My aim has been to offer an integration of perspectives and methodologies so that students of public health can gain a sense of the scope of the new public health with sufficient detail to demonstrate the excitement and importance of the challenges it offers. I hope I have succeeded in ordering the cacophony of voices that make up the new directions in public health into some kind of harmony.

THE STRUCTURE OF THE BOOK

The book has eight parts. Part 1 provides a context for understanding health and the new public health. Various definitions of health are given, including biomedical, lay, outcome-focused and critical perspectives. This is followed by a history of public health and a description of the rise of the new public health internationally and within Australia.

Part 2 considers the political economy of public health. It argues that public health is deeply and inevitably rooted in social, economic and political circumstances and that understanding and reflecting on these is a prerequisite to effective action for public health. It considers ethical dilemmas faced by public health. It also stresses the political nature of public health and the ways in which its practice reflect social and political values. A chapter is devoted to globalisation, which considers in particular the impact and spread of neo-liberalism. The effect of global trading treaties and the growth of transnational corporations and their health impacts is also covered.

Part 3 provides a guide to the research methods used to study public health. Debates about types of methodologies and methods are reviewed, and a range of methods applied in public health described and analysed, including epidemiology, survey and qualitative methods. This chapter also demonstrates the importance of appropriate research as an underpinning of the new public health. Participatory approaches to research are described, and evaluation methods for the new public health are investigated.

Patterns of health, illness and mortality and international comparisons in health status are provided in part 4. The significant inequities in health status between different groups in Australia as a case study of inequities in the population are described, followed by a detailed consideration of the explanation for the existence of these inequities.

The health of the environment and its impact on human life is the focus of part 5. The deteriorating state of the physical environment and its questionable ability to support people in the future is described. Urbanisation and population growth are considered as threats to public health.

A vision of a healthy, sustainable and equitable future is presented in part 6. The changes needed to achieve this from the current economic system are described. The ways in which environmental sustainability and socially successful and equitable communities might be achieved are considered.

The variety of ways in which public health has attempted to improve the health of individuals is considered in part 7. These include medical strategies, behavioural change strategies, community and organisational development. Healthy settings initiatives (including Healthy Cities and Local Agenda 21) are described and the organisational changes required to make these settings-based projects effective are assessed. Public health advocacy and activism are described and analysed as a key strategy of the new public health. The final chapter in this section is devoted to healthy public policy, which examines the theory and provides examples. The discussion throughout the chapters is illustrated with many examples of public health initiatives. A critical approach is taken to each of the strategies and initiatives described.

The book concludes in part 8 by assessing what changes are required if we are to achieve the vision of a sustainable, healthy and equitable future.

Structure of the book

Part 1: Approaches to Public Health
What is health?
History of public health and health
  service development
Recent developments in public health
  policy

Part 2: Political Economy of Public Health
Politics and ideology
Neo-liberalism
Globalisation

Part 3: Researching Public Health
Epidemiology, qualitative methods,
action research, evaluation

Basic understandings
for the new public health

Part 4: Health Inequities: Profiles, Patterns and Explanations
Data on health status and inequities,
causes of inequities and the social
determinants of health

Part 5: Unhealthy Environments: Global and Australian Perspectives
What makes environments, societies
and individuals unhealthy

Analysing the causes of
poor health and health
inequities

Part 6: Creating Healthy and Equitable Societies and Environments
Visions of what a healthy, sustainable
and equitable society would be like

The vision of society that the
new public health wants to
achieve and practical
examples of how this society
might be achieved

Part 7: Health Promotion Strategies for Achieving Healthy and Equitable Societies
Medical and behavioural strategies
  and their limits
Participation, empowerment,
  community development and advocacy
Healthy settings and cities
Health public policy

How to achieve the vision
through action in
communities, through
advocacy, policy and practice
change

Part 8: Public Health in the Twenty-first Century
Consideration of where to next

PART 1

APPROACHES TO PUBLIC HEALTH

This part provides a context and history for understanding the new public health. Chapter 1 explores different definitions and perspectives on health, showing how these reveal the links between understandings of health and social and cultural factors. It makes it clear that 'health' is a contested concept that cannot be easily defined. It also considers the difference between individual health and population health, an essential distinction for students of public health to grasp. Chapter 2 places public health in a historical context, while chapter 3 examines the evolution of the new public health internationally and within Australia. It describes recent developments, examining them critically in light of the philosophy of the new public health. A discussion of definitions of public health, the new public health and other keywords is given in appendix 1.

1

UNDERSTANDING HEALTH: DEFINITIONS AND PERSPECTIVES

Health is a social, economic and political issue and above all a fundamental human right. Inequality, poverty, exploitation, violence and injustice are at the root of ill-health.

People's Health Movement (PHM), 2000

A "toxic combination of bad policies, economics and politics is, in large measure, responsible for the fact that a majority of people in the world do not enjoy the good health that is biologically possible".

Commission on Social Determinants of Health (CSDH), 2008, p. 26

KEY CONCEPTS

Introduction
Health: the clockwork model of medicine
Health as the absence of illness
Measuring health
Health: ordinary people's perspectives
Public and private lay accounts
Health in cultural and economic contexts
Spiritual aspects
Health: critical perspective
Health as 'outcomes'
Health and place: defining collective health
Population versus individual health: the heart of public health
Conclusion

INTRODUCTION

The above quotes sum up the approach to health taken in this book. Through this book we will explore the underlying social and economic determinants of health in detail. But first it is important to understand the many ways in which health is understood and used. The word 'health' carries considerable cultural, social and professional baggage, and its contested nature suggests that it is a key to our culture and a word that involves important ideas and strongly held values (Williams, 1983). Using it in different ways gives rise to particular ways of seeing the world and behaving. Definitions of health structure the ways in which the world is viewed and how decisions are made. Health policies, for example, are shaped by policy makers' assumptions about what health is.

Most public health workers see health as central to their work and often assume that everyone sees their world revolving around the pursuit of health. The blinkered view this can lead to was brought home to me forcefully when I was speaking to a community audience in Adelaide about the new public health and waxing lyrical about its virtues. I was stopped in my tracks by an older woman in the audience who raised her hand in order to ask me, 'Excuse me dear, what are we allowed to die of?' I had not only assumed health to be central but also that life and health were limitless! Health is a preoccupation of modern society. Crawford (1984, p. 63) views health as a cultural factor and comments: 'Health is a particularly important concept in the modern West. In disenchanted, secular and materialist cultures, health acquires a greater symbolic importance. Health substitutes for salvation and becomes a salvation of its own.'

The cultural importance attached to health and illness in Western society has been well illustrated by Susan Sontag (1979), who noted that not being healthy and being ill have often been seen as undesirable and as states that imply adverse moral and psychological judgments about the ill person. Being healthy is viewed as so important that it affects both the way people experience illness and the way they regard those who are ill.

Understanding the place and role of public health in our society requires an understanding of health and its manifestations. This chapter presents five main perspectives on health: health within the medical 'clockwork' model, health as defined by ordinary people, critical understandings of health, health as an 'outcome' and health as a characteristic of place or environment. It concludes with a consideration of the crucial difference between population and individual perspectives on health.

HEALTH: THE CLOCKWORK MODEL OF MEDICINE

For much of the past two centuries the discourse of Western health has been overshadowed by a biomedical perspective (Foucault, 1973). This operates from a clockwork definition of medicine in which the body is studied through its component parts (Underwood, 1986). Health is defined as the body operating efficiently like a machine. Any breakdowns in the body system mean that it is not healthy. The isolation, labelling and systematic classification of specific diseases by Linnaeus in the eighteenth century was an important part of the development of the clockwork model, later consolidated by an increasingly sophisticated understanding of the specific causes of diseases. What biomedicine has not done well is to consider disease within the context of the lives of people with disease.

HEALTH AS THE ABSENCE OF ILLNESS

Biomedicine does, however, distinguish between disease and illness (Curtis and Taket, 1996). Disease involves a set of signs and symptoms and medically diagnosed pathological abnormalities. Illness is primarily about how an individual experiences the disease. Illness can be culturally specific and may have social, moral or psychological aspects. Disease is viewed as more objective, involving professional rather than lay diagnosis. But both disease and illness can detract from health.

However, other perspectives on health have existed alongside the biomedical view. Traditional midwives, herbalists, indigenous forms of healing, Ayurvedic and Chinese medicine all operated from a significantly different view of health. There has always been a tradition of social medicine that has been more concerned with social and economic factors that affect health (Underwood et al., 1986), but it has consistently been the poor relation of the clockwork model.

Behavioural psychology added another dimension: the need to protect and maintain the body by appropriate lifestyle behaviours that minimise risk of disease. So, as is discussed in chapter 20, behavioural change and the promotion of healthy lifestyles have become major factors in the professional perspective on health over the past three decades.

The limitations of the clockwork model of health have been widely recognised. It has been accused of being too mechanistic and ignoring the social, psychological and spiritual aspects. It suggests that if a body is not diseased then it must, by implication, be healthy. This conception of health is dependent upon there being some idea of what constitutes normal functioning so that abnormal, diseased states can be identified. However, Litva and Eyles (1994, p. 1083) point out that standards of normality are 'almost impossible to discern', even for physiological phenomena.

The biomedical model of health assumes a mind/body dichotomy, and it does not place much emphasis on how an individual's mental health might affect physical health status. The ways in which Western medicine defines, diagnoses and treats mental illnesses have been much disputed, with debates between the psycho-analytical and the biological schools about definition and treatment. The whole definition of mental illness has been severely criticised by the anti-psychiatry movement (Laing, 1982) and others (Cochrane, 1983; Seedhouse, 2002), who have suggested that it is socially and culturally defined and far from being as 'scientific' as psychiatry makes it appear.

Curtis and Taket (1996) point out that the biomedical model of health has less legitimacy than it had in the past. Building on the work of Foucault, an increasing number of critics are demonstrating that the medical definitions of health reflect its culture. They (Curtis and Taket, 1996, chapter 3) give three examples that show how medical diagnoses have reinforced aspects of the existing status quo. Their first example is 'hysteria', which was a common medical diagnosis for women in the eighteenth and nineteenth centuries. It has since been contested by feminists. They discuss the medical labelling of homosexuality as a mental disorder and show that this reflected the contemporary values of society. Finally, they discuss how HIV has been presented (for example, as a 'gay plague') and consider the social consequences. They conclude by stating that accurate definitions of 'health', 'disease' and 'illness' are not important. More important are the 'multiple and complex ways in which these terms are used discursively' (Curtis and Taket, 1996, pp. 72–3).

The biomedical model has also been critiqued for extending the definition of disease. This has been described as 'disease mongering' and linked to pharmaceutical companies seeking new markets (Moynihan and Henry, 2006). Risk factors are also defined as a disease; for example, obesity itself is now often conceptualised as a disease.

Blaxter (2010) notes that the concepts of health and ill health are asymmetrical and not simply opposites. The absence of disease may be part of health, but health is more than the absence of disease.

HEALTH AND WELL-BEING

The limitations of health being defined as 'the absence of disease' led to the World Health Organization (WHO) defining it as the 'complete state of physical, mental and social well-being, and not merely the absence of disease or infirmity' (WHO, 1948). This has been criticised as being too Utopian and unachievable (Nutbeam, 1986; Sax, 1990, p. 1), but has provided a vision of health beyond that suggested by the biomedical model. It has been an important impetus for broadening health activity beyond disease prevention. However, the definition of a state of well-being has proved tricky. Aitkin (1996, p. 14) explained the problem thus:

> Throughout my early life the word that was commonly used was 'health' and its antonym of course was 'disease'. And if we weren't diseased by implication we were healthy.
>
> In the past 20 years or so we have begun to realise that we need something more positive than both of those words. We need something which I suppose was captured by the Victorian novelists who talked about a 'rude health'—that is a great state of gruntle, where the world feels a very good place, where we feel we ourselves are very good people, that problems and responsibilities are within our compass and do not exceed it.

A recent focus on mental health has led to definitions that go beyond the concentration on physical factors. VicHealth offers the following positive definition: 'Mental health is the embodiment of social, emotional and spiritual well-being. Mental health provides individuals with the vitality necessary for active living, to achieve goals and to interact with one another in ways that are respectful and just' (VicHealth, 1999, p. 4).

Indigenous definitions of health focus on the whole person within their context of land, community and culture. The National Aboriginal Community Controlled Health Organisation (NACCHO, 2014) says Aboriginal health is 'not just the physical well-being of an individual but refers to the social, emotional and cultural well-being of the whole Community in which each individual is able to achieve their full potential as a human being thereby bringing about the total well-being of their Community'.

MEASURING HEALTH

Measuring this state of well-being has not proved easy. Many instruments have been developed (for example the Nottingham Health Profile, SF36, McMaster Health Index), but none adequately captures a positive health state measure. They are also static and so do not express the dynamic picture of health that appears to more accurately represent the ways in which health can be interpreted. Bowling (2005) has reviewed the range of measures that have been developed to view health in relation to quality of life, dividing them into six categories: measures of functional ability, broader measures of health status, measures of psychological well-being, measures of social networks and support, and measures of life satisfaction and morale. Bowling notes that all the measures she reviews have serious limitations in terms of reliability, validity and techniques of analysis, and that most are developed from professional definitions. Overall, health has defied any straightforward quantitative measurement, reflecting both the limitations of questionnaire surveys and the actual complexity of health. It is easier to measure disease or its absence than to measure a more positive state of health or well-being.

Even in health promotion, there is a tendency for a disease and risk factor orientation to continue. This was noted by Antonovsky (1996), who called for a salutogenic orientation to the study of health, as this would operate from a continuum rather than a dichotomous model and would incorporate notions of health and disease. Building on this notion, public health researchers are now trying to discover how lay people define health.

HEALTH: ORDINARY PEOPLE'S PERSPECTIVES

> ... if we are in the business of health promotion in the widest sense we should always remain sensitive to the enormous number of ways of defining health and disease which are held by ordinary people.
>
> *Maclean, 1988, p. 43*

Health literature recognises that ordinary people may not see health in the same way as health professionals. Research findings indicate that health is a complex concept that combines a number of different dimensions.

People find it harder to define health than illness, probably because illness presents as a problem to which societies have to respond (Locker, 1981, pp. 98–101). Being healthy requires no action (unlike being ill) and may be taken for granted (Pill, 1988). Three main domains relating to the definition of health are found: health is not being ill, it is a necessary prerequisite for life's functions and it is a sense of well-being expressed in physical and mental terms. WHO's positive definition of health given previously reflects how ordinary people define health more accurately than do medical perspectives. In her study with working-class mothers in South Wales, Pill (1988) found that health was seen as an absence of illness, in terms of functional capacity and as a positive condition of mental and physical well-being. She found that the mothers who were more aware of the effects of lifestyle factors on health were also those who could perceive the dynamic relationship between individuals and their environments. Herzlich (1973), working with a French middle-class sample, found that health was defined as the simple absence of disease, a positive state of well-being and as having a reserve to cope with life and illness.

One of the most thorough delineations of the lay understanding of health has come from Blaxter (2010). Combining a survey with detailed follow-up of a sample of the survey group, she set out to define what people mean when they talk of health. She defined eight main perspectives, which elaborate on the three key categories listed earlier. Her findings suggest that health is variously viewed and is a complex and dynamic concept involving a number of different perspectives. The health definitions that she identified are shown in box 1.1.

BOX 1.1   DEFINITIONS OF HEALTH (BLAXTER'S SURVEY OF BRITISH SAMPLE)

- **Health as not ill/diseased**: typical comments were Health is when you don't have a cold or Health is when you don't feel tired and short of breath. Some responses indicated a view that people could be healthy even if they did have a disease: I am very healthy apart from this arthritis.
- **Health as a reserve**: some people saw health as a reserve—if someone becomes sick, they are able to recover quickly.
- **Health as behaviour, health as 'the healthy life'**: primarily used when describing the health of other people as opposed to the respondent's. I call her healthy because she goes jogging and doesn't eat fried food. She walks a lot and doesn't drink alcohol.
- **Health as physical fitness**: particularly popular with young men and less favoured by older people. Men tended to express health in terms of physical strength and fitness. Typical quotes were: There's tone to my body, I feel fit; I can do something strenuous and not feel that tired after I've done it. Women were more likely to define health in terms of outwards appearance, such as being slim, a good complexion, bright eyes and shining hair.
- **Health as energy, vitality**: seen in terms of both physical and psycho-social energy to do things, signified by being able to get up easily, not feeling tired and getting on with activities, having energy and enthusiasm for work and generally feeling good.
- **Health as social relationships:** defining health in terms of relationships with other people, and more likely to be expressed by women. Younger people saw this as being able to have good relationships with their families: having more patience with them, and enjoying the family. Older people saw it as being able to help others and enjoying doing so: You feel as though everyone is your friend.
- **Health as function**: health is the ability to do things, which overlaps with the association between health and energy and vitality. More older people mentioned this,

> possibly because they no longer took doing things for granted: She's 81 and she gets her work done quicker than me, and she does the garden.
> * **Health as psycho-social well-being**: some people defined health solely in terms of their mental state, typified by the statement: I think health is when you feel happy. When I'm happy I feel quite well.

Source: Blaxter, 2010.

Blaxter asked her respondents to differentiate between health in themselves and in others. She found that being healthy for oneself was to be unstressed and able to cope with life. For other people, health was fitness, the ability to work and perform normal roles and simply 'not being ill'. Litva and Eyles (1994), based on work in an Ontario community, report similar findings. They argue that it is useful to distinguish between 'being healthy' and 'health'. They define health as an abstract state of being—not being ill. Being healthy is having the resources for everyday life and 'a social construction that helps us understand our place in the world and that of others. It affects the ways in which we see the "causes" of illness. It makes being not ill very important unless we are willing to be viewed as deviant. It becomes a moral code' (Litva and Eyles, 1994, p. 1084). Popay et al. (2003) found a similar moral view of health among the working-class respondents they spoke with. When asked to account for the existence of health inequities between richer and poorer areas their accounts suggested they saw being healthy as an important part of their moral identity and the view was stressed that they should not 'give in to illness'. Blaxter (1997) in examining the strength of moral framework in lay accounts of health and illness suggests that to acknowledge health inequalities 'would be to admit an inferior moral status for one's self and one's peers'. However, Davidson et al. (2006) found in Scotland and the north of England that people from deprived backgrounds were very aware of the ways in which their health and well-being was affected by their circumstances, especially in terms of feelings of shame, anger, frustration, rejection, injustice and alienation they had in relation to other people. This was seen as being related to sleeplessness, fear, anxiety and stress. Thus while the cause of inequality was rooted in material circumstances the disadvantage was perceived as being compounded by comparison with the circumstances of others.

Another feature of lay definitions of health is that people with some disability or disease often describe themselves as healthy, especially on days when their disabilities seem less severe. People assess their own health subjectively and in terms of a reasonable expectation for their age and disability.

PUBLIC AND PRIVATE LAY ACCOUNTS

Another useful perspective has come from the work of Cornwell, who interviewed working-class men and women in the East End of London. She found that they offered public and private accounts of what they understood by health, and that she only heard the private accounts when she knew them reasonably well. This implies that studies based on one-off interviews may not get to people's private accounts of health. Cornwell (1984) reports that her respondents, in their public accounts, presented a view of health that conformed to a biomedical model, tended to have a moral component and divided causes of illnesses into those that were or were not the individual's fault. The private theories, by contrast, were based on their own experiences or those of people they knew. These explanations brought out a more complex view of health that involved the interplay between individual and structural factors on health. Health was now seen as intimately tied up with an individual's overall life circumstances.

HEALTH IN CULTURAL AND ECONOMIC CONTEXTS

Crawford's (1984) work involved interviews with 60 adults in the Chicago metropolitan area. He found that interpretations of health reflect the cultural and economic context of people's lives. For most people health represents a status, socially recognised and admired. He discovered two main discourses of health—health as a means of exerting self-control and health as a release mechanism—suggesting that health fulfils different functions for different people. He expanded on these two notions of health as follows:

- *Health was seen by some as self-control* and a set of related concepts that include self-discipline, self-denial and willpower. This was primarily the view of middle-class professional people, but also of some blue-collar workers. Health was something to be achieved through healthy behaviour. Health was not seen as something that springs from normal everyday activity. People felt they did not have time to be healthy and sometimes decided to pursue goals other than health. Judgment of others and self-blame are themes running through Crawford's interviews. He found a strain of moral judgments about being able to keep healthy, and commented: 'thinness is believed to be an unmistakable sign of self-control, discipline and willpower. The thin person is an exemplar of mastery of mind over body and virtuous self-denial … Conversely, fat is a confirmation of the loss of control, a moral failure, a sign of impulsiveness, self-indulgence, and sloth.' Crawford interprets this as internalisation of medical knowledge so that the body is seen as an object of rational control. The values of self-control, self-discipline and self-denial fit with modern individualism (see part 2 of this book) and the Protestant work ethic, and so attitudes to health become one expression of dominant values. Crawford also noted that his respondents appeared to feel that the macro-conditions that affect health were out of control (toxic environment), so self-control over the range of personal behaviours that also affect health was the only remaining option. People did not dwell on environmental hazards. A typical comment was 'Why worry about something you can't do anything about?'
- *Health was seen by some respondents as a release mechanism*, who equated it with feeling good as distinct from following rules of medical authority. Life is seen as a series of pressures, anxiety, frustration and worry, and as leaving no time for

health-promoting activity. Health is not rejected as a value but is often repudiated as a goal to be achieved through instrumental action. The working-class males Crawford spoke to were particularly likely to see leisure as free from concern about health, and there was some resentment of a public discourse that instils fear and demands more controls over behaviour in order to achieve health. There appears to be both interplay and contradiction between the two perspectives on health. Crawford believes this is understandable, given the contradiction between production and consumption in our culture—managers of labour want disciplined work forces but advertising encourages hedonism. Culturally, release is a means by which societal tensions are managed and is important in keeping consumption growing.

SPIRITUAL ASPECTS

Lay definitions of health may also include a spiritual dimension. Stainton-Rogers (1991) reports that some people saw their health as dominated by external religious or supernatural powers. Healing could result from intervention by God or some other supernatural power, as could falling ill in the first place. Indigenous people are particularly likely to have a belief system that is related to health and illness, which emphasised spiritual dimensions. The position of traditional healers may often rest on their perceived ability to call on external forces.

The growing literature on lay definitions of health presents a picture of complexity and cultural and social embeddedness. Health makes sense within the context of people's everyday lives—it is not something that can be neatly defined in static categories. The following perspective on health builds on these insights from ordinary people's definitions, but adds a political explanation for the existence of culturally dominant definitions of health.

HEALTH: CRITICAL PERSPECTIVE

Critical perspectives on health are those that seek to explain the purposes that are achieved through particular means of defining health. They are critical in the sense that they look beneath the surface appearance of a concept or phenomenon and offer an explanation as to why it is this way.

One such perspective on health that has been particularly influential is that which maintains that health is defined in such a way by the dominant forces in a capitalist society that it becomes a defining and controlling mechanism. Writers adopting this perspective use a Marxist analytical framework (Doyal, 1979; Navarro, 1979, 2002). Central to this view is the idea that capitalist societies are structured in such a way that they produce illness. The system is geared up to maximising profit rather than protecting the health of workers and their families. Health is affected by practices such as shiftwork, overtime, monotonous work tasks and dangerous chemicals in the workplace. It is defined in terms of the ability of people (particularly workers) to function and carry on with their normal activities. Doyal (1979, p. 34) comments: 'The defining of health and illness in a functional way is an important example of how a capitalist value system defines people primarily as producers—as forces of production.' She goes on to say that the functional definition of health does not concern itself with people's fears, anxieties, pain or suffering, and that this may limit expectations about health.

The People's Health Movement (www.phmovement.org), whose Charter was quoted at the start of this chapter, takes a critical view on health. The People's Health Assemblies held in Bangladesh in 2000, Ecuador in 2005 and Cape Town in 2012 have been grounded entirely in a critical understanding of health whereby the dominant global economic structures in the world are seen to have a massive effect on shaping ordinary people's health experiences. From a postmodern perspective, Petersen and Lupton (1996) also take a critical look at definitions of health and conclude that health maintenance has become an important aspect of being a 'good' citizen. They say that in contemporary Western society the pursuit of good health is both a right and an obligation. Individuals are obliged to remain healthy because being ill means they cannot be good citizens, and may become an economic burden.

The political economy perspective on health also criticises the individualistic definition of health that it sees as prevalent under capitalism. Doyal (1979, p. 35) comments that this 'emphasis on the individual origin of disease is of considerable social significance, since it effectively obscures the social and economic causes of ill health'. The issue of individualism and health is discussed in more depth in part 2 and is of central importance in understanding public health. It underpins a notion of health that stresses personal (and even moral) responsibility for maintaining health. This perspective that defines health in terms of individual responsibility absolves other factors from responsibility. McKinlay (1984, p. 12) noted:

> the emerging emphasis of personal responsibility for health mystifies the social production of disease and undermines demands for rights and entitlements for health care. Beneath the rhetoric about the cost of medical care and the obligation of the individual to remain healthy lies a political problem to shift the burden of costs back to labour and consumers and to paralyse regulatory efforts undertaken to control environmental and occupational hazards.

The political economy view sees health in terms of its distribution in society (and so focuses on inequities in health status, especially those resulting from class differences) and in terms of the structural factors that create or detract from health, such as environmental, housing and occupational conditions. From this perspective, studying the health of individuals is less valuable than studying the collective health of societies and the social and economic forces that affect collective health. Increasingly the political economy view (see chapter 5) stresses the connections between the health of peoples in rich countries and those in poor countries as the processes of economic globalisation continue apace.

HEALTH AS 'OUTCOMES'

Around the world health departments and ministries are seeking evidence that their efforts result in health outcomes. Almost always this search reflects a clockwork view of health in which short-term improvements brought about by clinical interventions are able to produce an outcome that can be measured by a randomised control trial. Rarely does the use of the term imply an understanding of the complexity of health or the social and economic context in which poor health is produced.

In order to be able to attribute a change in health status to any particular intervention it is necessary to exclude the contribution of all other factors, which is generally very difficult to achieve as it requires a research design that controls for all other possible factors. Generally a randomised controlled design (see chapter 7) can only be used to study clinical interventions. Ethical and practical grounds restrict their use in most other circumstances. 'Health' in clinical trials is invariably reduced to an absence of the particular ailment the clinical intervention was designed to cure. More sophisticated definitions of health defy the type of simple measurement required in experimental designs. Many of the measures used to measure health within the health care system present a health service provider rather than user perspective. For people themselves (and for their families, friends, employers and others) crucial factors not covered might be their ability to function, their quality of life and the extent to which they can live their lives normally.

The focus on health outcomes is laudable insofar as it requires health care providers to be accountable for expenditure and to demonstrate that the work they do is beneficial—requirements that have been less pressing in the past. Often, however, the concept is dealt with naively and risks becoming no more than an unreasonable demand for evidence that particular health services can demonstrate an impact on population health status in a short period of time rather than evidence concerning a broader aspiration for the well-being of society as a whole.

In practice, most of the outcomes measured relate to individuals and not populations. For many, health promotion and public health outcomes are crucial, but it is often more feasible to measure outcome in terms of capacities rather than health status (Baum, 1998). The value of health promotion and public health interventions over clinical care should be in their capacity to improve health in the longer term. For public health, intervention to improve health is akin to an investment in the future.

However, 'health outcomes' in the current health system discourse refer to accountability mechanisms and measures of the effectiveness and efficiency of particular (predominantly sickness-care) interventions. The term is rarely used within bureaucracies to refer to a broader project of improving health in a social or environmental sense. The Commission on Social Determinants of Health (2008) stressed the importance of measuring health outcomes according to health equity, which concerns its distribution within and between countries. Both the World Bank and WHO have used economic measures of health as a yardstick for the success of their programs. This reflects an underlying assumption that economic productivity is paramount. This is nowhere better illustrated than in the use of disability adjusted life years (DALYs) or DALEs (disability adjusted life expectancy) to determine the value of a health intervention. DALYs are calculated by assigning values to years of life lost at different ages. The value for each year of life lost rises from zero at birth to a peak at age 25 and then gradually declines with increasing age. As the very young, the elderly and people with disabilities do not contribute much to economic development, treatment aimed at them would result in fewer DALYs than treatment aimed at people in their early twenties. The DALY or DALE measures make the ethically questionable assumption that a year of life for a person with a disability is of less value than a year of life for a person without a disability.

HEALTH AND PLACE: DEFINING COLLECTIVE HEALTH

Most literature that defines health does so in terms of what it means to individuals, but in recent times health promotion has given more attention to what constitutes health in terms of a place or a population as a whole. For example, 'Healthy Cities', 'Healthy Schools' and 'Healthy Workplace' projects have attempted to define what would constitute health for each of these contexts. A concern of the Healthy Cities movement has been to move beyond a deficit model (for example, for a city, how many unemployed, how many households without running water) to one that captures the more dynamic and positive aspects of health (the number of trusting people, the availability of community meeting spaces). WHO's list of the qualities of a healthy city appear in box 1.2 and put as much emphasis on the processes within the city as they do on the physical features.

BOX 1.2   QUALITIES OF A HEALTHY CITY

A city should strive to provide:
1. a clean, safe, physical environment of high quality (including housing quality)
2. an ecosystem that is stable now and sustainable in the long term
3. a strong, mutually supportive and non-exploitative community
4. a high degree of participation in and control by the citizens over the decisions affecting their lives, health and well-being
5. the meeting of basic needs (food, water, shelter, income, safety and work) for all the city's people
6. access by the people to a wide variety of experiences and resources, with the chance for a wide variety of contact, interaction and communication
7. a diverse, vital and innovative city economy
8. connectedness with the past, with the cultural and biological heritage of city dwellers and with other groups and individuals
9. a form that is compatible with and enhances the preceding characteristics
10. an optimum level of appropriate public health and sick care services accessible to all
11. high health status (high levels of positive health and low levels of disease).

Source: WHO Regional Office for Europe, 2015a.

The raison d'être for WHO's Healthy Cities and Healthy Settings initiatives is the view that the collective structures of a community form the crucial determinants of a population health status. In this view health is not only a characteristic of individuals but also of a city or community. A WHO document on urban health, for instance, noted:

> Physical, economic, social and cultural aspects of city life all have an important influence on health. They exert their effect through such processes as population movements, industrialisation and changes in the architectural and physical environment and in social organisation. Health is also affected in particular cities by climate, terrain, population density, housing stock, the nature of the economic activity, income distribution, transport systems and opportunities for leisure and recreation.

WHO, 1993, pp. 10–11

Lay and health professional definitions of health rarely encompass these wide-ranging social, physical and economic factors, perhaps because people take them for granted. A critical perspective, based on an analysis of structural factors, leads to a broader view of health as does a perspective that takes as its starting point a consideration of collective entities (such as workplaces, schools, hospitals, cities, villages, country towns). Defining health in such terms is useful for the new public health because it appears more likely to keep a focus on positive definitions and on structurally rather than individually driven factors that affect people's health.

In recent years the concept of 'ecosystem health' has been used by ecologists. Healthy ecosystems are characterised by diversity, vigour, effective internal organisation and resilience. This approach to health integrates an overall consideration of the environment and the interdependence of systems with the overall ecosystem. This approach is characterised by holism and stresses that the health of people is dependent on the health of the biosphere, which is increasingly under strain and threat. Brown et al. (2005) argue strongly that such an approach to health is essential in the face of the threatened collapse of these systems and their ability to support human health.

POPULATION VERSUS INDIVIDUAL HEALTH: THE HEART OF PUBLIC HEALTH

The distinguishing feature of public health is its focus on populations rather than individuals. Public health studies the distribution of disease and positive attributes of health in whole populations. Clinical work is based on work with individuals who are either at high risk for a disease (e.g. who have high blood pressure or genetic susceptibility for a particular disease such as breast cancer) or who have a disease. Table 1.1 illustrates the difference between an individual and a population perspective by showing how the two approaches give rise to different questions. Understanding this difference is vital. Both are important but the new public health primarily concerns itself with population issues rather than those designed to cure individuals when they are sick.

TABLE 1.1   INDIVIDUAL AND POPULATION HEALTH PERSPECTIVES: THE DIFFERENCES EXPLAINED

	Individual (clinical) questions	Population questions
Smoking	How can we stop individuals smoking?	How can we change the social and economic environment so it discourages smoking?
Childhood obesity	How can we encourage children to lose weight?	What social and economic trends contribute to higher rates of childhood obesity in our society than in the past?
Diabetes	How do we encourage people with diabetes to self-manage their disease?	How do we alter food supply systems so they help prevent rates of diabetes going up?
Depression	How do we best counsel teenagers with depression?	Why have rates of teenage depression gone up in the last ten years? What can be done to prevent depression?
Homelessness	How can we provide homes to homeless young people?	How can we design a housing system that ensures no one is homeless?
Drug use	How can we educate people to use drugs responsibly?	What legislation and policies can we adopt to reduce harm from drugs?

Treating high-risk or diseased individuals does not have much impact on population health levels overall, but changing a risk factor across a whole population by just a small (and often clinically insignificant) amount can have a great impact on the incidence of a disease or problem in the community. This paradox makes it very hard for public health to be newsworthy. The changes that affect population health are usually not dramatic, but spread thinly across a population. Yet they can make significant improvements to health and well-being (see box 1.3 for further explanation).

BOX 1.3   HOW DOES A POPULATION HEALTH PERSPECTIVE DIFFER FROM AN INDIVIDUAL OR CLINICAL ONE?

Understanding this distinction is fundamental to good public health practice. The distinction is neither intuitive nor obvious. But it is crucial that every student of public health grasp this understanding.

Changing a risk factor across a whole population by just a small (and usually clinically insignificant) amount can have a great impact on the incidence of a public health problem in the community. This creates the prevention paradox expressed as follows by Geoffrey Rose:

> A preventive measure which brings much benefit to the population offers little to each participating individual. (Rose, 1985, p. 38)

EXAMPLES

Seat belts

If everyone in a population wears a seat belt while driving, the burden of mortality and morbidity from road accidents will reduce. However, very few of the individuals doing so will benefit directly—only the few who are involved in a life-threatening accident.

Body mass index

While a small reduction in the mean body mass index of a population will make very little difference to any one individual it would be significant in terms of the disease burden across the population.

Rose (1992) considered the distribution of a range of health risk factors in 32 societies—factors such as obesity, high blood pressure, heavy drinking. He found that the proportion of people with these risk factors was a reflection of the society's average behaviours in relation to these risks. Thus the proportion of heavy drinkers was a function of the society's average alcohol consumption, obesity of the average body mass index, high blood pressure of average blood pressure. Thus those with dangerously high levels were not minorities behaving very differently from the rest of their society but were part of a behavioural shift to which the norms of the whole of the rest of society seemed to contribute. This demonstrates how individual behaviour is strongly influenced by social norms.

Analysis on an individual level may be appropriate for understanding how individuals may be affected by a disease or some other problem, but may miss the influence of broad structural factors on health. Marmot (2001) illustrates this by quoting Sen's argument that famines do not occur in countries with well-functioning democracies, for a range of structural reasons. Comparing individual starving children in a refugee camp could never lead to this conclusion. The relevant level of analysis is social, political and economic. Similarly, Durkheim's (1979 [1897]) famous sociological analysis of suicide rates in the late nineteenth century determined that it was the characteristics of a society at large (such as the types of social relations and the ways in which the society understood suicide and related it to other social phenomena) that determined the rate rather than a simple aggregate of individual factors in relation to suicidal tendencies. He concluded from this that responses to suicide should be collective rather than focused on individuals: 'The only possible way, then, to check this current of collective sadness is by at least lessening the collective malady of which it is a sign and a result' (Durkheim, 1979 [1897], p. 391).

So viewing health and disease from a public health perspective means taking a view of the health of populations, not just of individuals within them.

CONCLUSION

Comprehending the various ways in which health is understood is an important background to appreciating the change in thinking about health that is called for by the new public health movement with its emphasis on the social, environmental and economic determinants of health. Health is viewed as a complex outcome that results from a range of genetic, environmental, social, political and economic factors. The next two chapters demonstrate through a history of public health that such broad interpretations have been accepted before but have rarely assumed a dominant and driving position in public policies.

CRITICAL REFLECTION QUESTIONS

1.1   How is it that a person with disabilities can experience health?

1.2   To what extent do the qualities of a Healthy City listed in box 1.2 apply to the city or community you live in?

1.3   How do an individual clinical and a population health response to a health issue of your choice differ?

Recommended reading

Blaxter, M. (2010) *Health*. Provides a detailed guide to definitions and constructions of health and how it relates to culture, social systems and medicine.

Rose, G. (1992) *The strategy of preventive medicine*. Provides the epidemiology argument for the importance of public health approaches over clinical approaches in maintaining and improving health in populations.

Useful websites

www.naccho.org.au
    National Aboriginal Community Controlled Health Organisation

www.who.int/social_determinants/en
    World Health Organization Social Determinants of Health

2

A HISTORY OF PUBLIC HEALTH

May the rich remember during the winter, when they sit in front of their hot stoves and give Christmas apples to their little ones, that the ship hands who brought the coals and the apples died from cholera. It is so sad that thousands always must die in misery, so that a few hundred may live well.

Rudolf Virchow during the 1848–49 cholera epidemic in Berlin, quoted in Waitzkin, 2006, p. 7

KEY CONCEPTS

Introduction

Era of Indigenous control

Colonial legacy

Theories of disease causation

Public health legislation and sanitary reforms

Australian responses

Status quo or radical change?

Relearning the nineteenth-century lessons: McKeown and Szreter

Nation-building era

Affluence, medicine, social infrastructure

Conclusion

INTRODUCTION

Contemporary public health approaches in Australia and many other countries reflect practices that came from nineteenth-century Europe and were spread around the globe through the processes of colonisation. An appreciation of their history, and of the crucial philosophies and practices that have been representative of public health at different times, are important to understanding why the new public health was labelled as such and how it is both a continuation of the past and a departure from it.

There have been seven distinct periods in the development of public health thinking and practice in all countries. These are summarised in table 2.1 based on Australian history, but the history has resonance for other countries too. In practice there is some overlap between the different eras. This chapter describes the first four eras—the era of Indigenous control, the colonial era, the nation-building era and the promise of medicine era. The final three eras are discussed in chapter 3.

TABLE 2.1   HISTORY AND DEVELOPMENT OF PUBLIC HEALTH IN AUSTRALIA

Period	Dominant policies and ideologies	Typical intervention models
Era of Indigenous control (estimated to be in excess of 40 000 years)	Strong links with land, traditional healers, emphasis on spirituality and integration of health and life.	Practice part of accepted culture handed on through oral tradition.
Colonial era (from white invasion until 1890s)	Control of infectious disease main aim. Strongly influenced by British practices. Emphasis on sanitary measures.	Quarantine Acts. Public Health Acts in colonies. Provision of clean water and sanitation.
Nation-building era (1890–1940s)	State action to improve the health of the nation. Seeking to 'improve the race'. Health linked to ideas of vitality, efficiency, purity and virtue.	Formation of Commonwealth Department of Health. Organised exercise programs to improve national physique, medical inspection of children, hygiene advice to the population.
Affluence, medicine and infrastructure (1950s–early 1970s)	Economic affluence and interventionist governments committed to improving quality of life. Considerable developments in clinical medicine, which led to a belief that finally medicine would conquer disease.	Considerable state intervention in areas that have an impact on health, such as housing and education. Health services associated with more and more sophisticated medical technology (e.g. organ transplants). Growth of hospitals and expanding health service budgets—little focus on public health.
Lifestyle era (late 1960s–mid-1980s)	Focus on effects of affluence in terms of chronic disease. Rediscovery of philosophy of prevention reflecting a desire to control costs of health services. Focus on individual behaviour. Epidemiological methods developed.	Lifestyle programs modelled on North American Heart Health programs, such as the North Coast lifestyle program, 'Life—be in it' campaign. Population surveys of risk factors. Some challenge to this era and foreshadowing of new public health by Community Health Program and women's and Aboriginal health movements.
New public health era (mid-1980s–mid-1990s)	Influenced by World Health Organization policies, especially the Alma Ata Declaration of Health for All (1978) and the Ottawa Charter (1986). Focus on collective measures, especially policy. Emphasis on poverty and social justice in public health policies. Economic recession and cutbacks in state expenditure. Imposition of structural adjustment policies in low income countries.	Development of healthy public policy (e.g. legislation to control sale and use of tobacco, drink-drive legislative controls). Policy support for community involvement in health promotion. 'Settings' approaches to health promotion (e.g. Healthy Cities, Healthy Schools, Healthy Worksites, Healthy Hospitals).

(continued)

TABLE 2.1   HISTORY AND DEVELOPMENT OF PUBLIC HEALTH IN AUSTRALIA
(*CONTINUED*)

Period	Dominant policies and ideologies	Typical intervention models
Global new public health (mid-1990s to twenty-first century)	Continued development of the settings approach but increasing recognition that the progress these might make is limited by the powerful forces of economic globalisation. This era is characterised by increased recognition of the impact of the policies and practices of international financial institutions on health, by the shrinking of the state and subsequent privatisation in so many parts of the world. The revolution in communications has led to a vibrant civil society (for example the People's Health Movement) that is opposing many aspects of economic globalisation. Calls for public health to be seen as a global public good and to be protected by international treaties and laws.	Increased focus on measures against terrorism including bioterrorism. Increased fear and preparation for pandemic disease including bird flu, SARS and Ebola. Calls for interventions to ensure trade treaties support health. Health impact assessment of a wide range of issues including infrastructure developments, welfare policies and transnational corporations.

ERA OF INDIGENOUS CONTROL

While there is little firm historical evidence relating to the public health practices of Australia's Indigenous peoples, there is enough to know that health was a concern, but that concepts of health and illness differed significantly from those of Europeans. Hunter (1993) describes traditional indigenous culture as understanding illness in terms of intrusion—'active intervention by someone or something as a consequence of the sufferer's actions within a social or sacred sphere' (p. 54). Traditional healers were an integral part of society, using a range of natural products such as plants and animals for healing. The Australian Indigenous people were hunter gatherers, who did not plant crops but lived off the available plants and animals in defined areas, moving around to take advantage of seasonal availability of food and water. Consequently, they did not have the public health problems associated with permanent settlements.

The notion of public health, in the Western sense, would make little sense to indigenous peoples. Health appears to have been a concept that was literally part of life. Behaviour within societies was prescribed by kinship and all individuals were able to place themselves in relation to any other individual. They believed that law derived from Dreamtime legends would ensure the continuity of people and nature (Saggers and Gray, 1991). The societies were based on intense cooperation and intricately linked relationships, possibly reflecting the needs of survival in a hostile physical environment. While Indigenous Australians do seem to have suffered from a number of illnesses prior to colonisation (including trachoma, yaws, endemic syphilis, skin lesions and hepatitis B) these appear to have been fairly mild compared with the epidemics after colonisation (Mitchell, 2006).

COLONIAL LEGACY

Histories of public health show that some form of collective public health measures has always been implemented by societies (Rosen, 1958; Brockington, 1975; Wilkinson and Sidel, 1991; Lewis, 2003). Examples are the Roman public baths, Roman laws governing burial of the dead and regulating dangerous animals and unsound goods, the regulation of prostitution in Ancient Rome and Greece, inoculation against smallpox in India and China before the Christian era, the isolation of people with leprosy in Europe in the middle ages and the quarantining of ships by the Venetians. Some teachings of major religions may also be seen as public health measures, such as those encouraging sobriety, cleanliness, isolation of people with infectious disease and the ritual abstention from food likely to convey parasites (Brockington, 1975, p. 1).

British responses to major nineteenth-century public health problems influenced the development of responses in its colonies. European societies were the first to focus considerable public effort on controlling disease and attempting to create healthier living environments. The nineteenth-century public health reforms were a response to the dislocation and disease brought about by rapid industrialisation, especially when two classic waterborne sanitation diseases, cholera and typhoid, unknown in Britain before the nineteenth century, became major causes of death. The following history of the development of public health provides a potted version. A fuller version tracking experiences in the UK, USA and Australia, is available in the two volumes by Lewis (2003).

THEORIES OF DISEASE CAUSATION

In the nineteenth century there were a number of rival theories as to how infectious disease was spread (Tesh, 1982). The 'miasma' theory held that disease resulted from inhaling bad smells from filth. The 'germ' or contagion theory held that pathogens (air or waterborne) were responsible for disease. Supernatural theories, such as those that saw disease as a reflection of God's wrath, were also common. Other theories involved the unsanitary habits of individuals. Public health consequences were that the contagion theory supported quarantining of people and goods, while the 'miasma' theory advocated cleaning up cities.

PUBLIC HEALTH LEGISLATION AND SANITARY REFORMS

The prime tool of the nineteenth-century public health movement in Britain and Australia was legislation (Reynolds, 1989; Lewis, 2003). Edwin Chadwick, author of the Report on the Sanitary Condition of the Labouring Population of Great Britain (1842), was the main driving force behind the first public health reforms. His efforts resulted in the 1848 Public Health Act, which gave local authorities the powers to remedy unsanitary conditions and to require adequate drainage and sanitation in towns (Kearns, 1988).

A defining moment in public health history was when the London physician John Snow removed the handle from the water pump in Broad Street in 1854 because he was convinced, on the basis of limited epidemiological evidence, that the water was the source of the current cholera epidemic. The data supporting his intervention (shown in chapter 7) were a direct challenge to the miasma theory of disease causation. Wilkinson and Sidel (1991) note that, despite data indicating that particular sources of water supply were the cause of cholera, Snow's attempts to persuade the water companies to move their intake upstream away from the pollution were only successful because people were concerned about the aesthetic qualities of the water.

Nonetheless, experiences such as Snow's, together with public health legislation, led to the appointment of medical officers of health by local authorities to enforce public health legislation and advise on appropriate measures. In 1845 a Health of Towns Association was formed with the aim of bringing 'the subject of sanitary reform under the notice of every class of the community' (Warren and Francis, 1987, p. 28). The 1872 Public Health Act compelled every statutory authority to appoint a medical officer of health. A Diploma of Public Health was introduced in the 1870s and this group of doctors grew steadily (Szreter, 1992), proving influential in bringing about sanitary, food

The Broad Street Pump has become an icon of public health history. It represents the first recorded time when epidemiological methods were used to control a disease. (Copyright 2001 by Larry J. Clark. Used with permission.)

and hygiene public health reforms, which led to improved health status and a dramatic reduction in deaths from infectious diseases (Warren and Francis, 1987).

The history of public health in nineteenth-century Britain suggests that the 1848 Public Health Act led to public health measures being enthusiastically taken up around the country. Szreter (1995) warns against this assumption, believing that the 1840s were really a false dawn of the public health movement that failed to capture the hearts and minds of the emerging governing classes. He argues that it was only after 1866 that an effective public health movement could be said to have evolved, and then it resulted from two main factors. First, an able municipal leadership, affected by the emergence of a 'civic gospel' movement that focused less on saving souls for the next life and more on the social and collective enterprise of improving the minds and bodies of the poor in this life, was committed to urban improvement despite the cost. From this movement came a sense of civil society and the importance of investment to improve society.

Another crucial factor was the electoral reform of 1867, which significantly extended the franchise, and provided an electorate in favour of public spending to control the excesses of the free market. The increase in female literacy at the end of the nineteenth century was important in contributing to declining rates of infant mortality (Lewis, 2003).

AUSTRALIAN RESPONSES

In Australia, as in Britain, public health measures were partly in response to a series of epidemics. In the last two decades of the nineteenth century all Australian colonies passed comprehensive Public Health Acts that were closely modelled on the British Acts (Woodruff, 1984; Curson and McCracken, 1989), except for one important respect: the responsibility for administering public health lay with central Boards of Health rather than local government (Curson and McCracken, 1989). This set the scene for continuing state government involvement in public health following the establishment of the Commonwealth of Australia in 1901. For the European city populations of the Australian colonies, the health problems were similar to those in British cities, as were the unsanitary conditions. This can be seen from these observations of the president of the Central Board of Health in South Australia during the 1840s:

> In Adelaide, so far as my observation goes, sanitary conditions are either utterly disregarded, or are left to the caprice of the inhabitants. The city may be correctly described as a 'city of stenches', and these are of the most disgusting kind. It is impossible to walk through any of the streets (especially after sunset) without being sickened by the smells from closets, stagnant water, and decomposing matter in the water table.
>
> *Woodruff, 1984, p. 36*

Adelaide had high rates of diarrhoeal disease compared with other areas of Australia. The infant mortality rate was 140 per 1000 live births in South Australia in the 1840s (higher than today's highest—Sierra Leone at 107) and in Melbourne between 1885–89 it reached as high as 179 (Lewis, 2003, p. 60). Typhoid was a serious health problem, and similar situations existed in all the Australian colonies. In 1900 Australia experienced a bubonic plague, which was particularly bad in Sydney. Responses to this outbreak of plague (Curson and McCracken, 1989) show how public health developed in direct response to disease threats, social interpretations of the threats and the ways chosen to cope with them. The plague was met with fear and hysteria, many seeing it as divine retribution for people's sinfulness. The Chinese were accused of spreading the disease. People with the disease were treated less than humanely, being shipped off to compulsory quarantine. The plague did, however, lead to improved sanitary inspection of houses, better monitoring of the city's housing and health situation and an evidence-based understanding that the plague was carried by fleas from infected rats (Lewis, 2003).

Another feature of Australian public health history was racism against Chinese migrants. The fear of cholera led to the introduction of a Quarantine Act in New South Wales in 1832 (and soon afterwards in the other colonies), which required new arrivals who were suspected of having come in contact with cholera to be isolated (Reynolds, 1995, p. 161). There is some evidence that these controls were imposed more rigorously on Chinese arrivals, and the fear of disease from this group of immigrants formed part of the anti-Chinese debate in the nineteenth century (Reynolds, 1995).

The nineteenth century was very different for Aboriginal people. Saggers and Gray (1991) indicate that Aboriginal people enjoyed a reasonable lifestyle and had generally adequate health before the arrival of Europeans in 1788. The next two centuries saw them become the sickest group in Australia. In the nineteenth century, violence from the invading Europeans was a significant cause of mortality, while newly introduced infectious diseases, such as smallpox and measles, took a considerable toll. The traditional hunter-gatherer lifestyle of the Indigenous population was severely disrupted and they lost their basic means of production and so, of good nutrition, appropriate shelter, safety and a healthy environment. Introduced products like sugar, tobacco and alcohol helped disrupt the traditional ways. White Australia did not offer any response to a deadly public health situation other than a belief, propped up by the new Darwinism, that in the survival of the fittest, the Australian Indigenous population was clearly not destined to survive. The legacy of this still casts a shadow over contemporary Indigenous health issues.

STATUS QUO OR RADICAL CHANGE?

There are at least two broad traditions of public health activism in the history of nineteenth-century British and Australian public health (Ross, 1991). One is typified by a desire to control disease and the poor who were seen to be the cause of it, rather than by a more altruistic desire to make society a fairer place. In Britain, Chadwick was greatly influenced by thinkers such as Malthus and Bentham, and was one of the authors of the notorious Poor Law Amendment Act (1834), which was based on the notion of the 'undeserving' poor. Benefits were no longer available to poor people except in workhouses where the conditions were so miserable that only the completely destitute would go to them. Ross (1991) argues that the utilitarian thinking that inspired reformers such as Chadwick was implicitly conservative. Epidemic disease represented a threat to social order and productivity and so warranted attention from society. Tesh (1982) points out that it was no accident that waterborne diseases (such as typhoid and cholera), which could more easily cross boundaries from poor to more affluent neighbourhoods, were the subject of more action than tuberculosis, the classic disease of poverty.

This analysis does not detract from the effectiveness of Chadwick's measures, but it does explain why none of his writings called for redistribution of income and wealth or better living conditions for the poor. Kearns (1988) argues that Chadwick was always careful to couch his proposals for sanitary reform so that they appeared important for an efficient capitalist economy. Tesh (1982) claims that the 'miasma' theory was the basis of Chadwick's arguments for sanitary reforms, and that this theory of disease causation suited the new industrial classes in Britain very well. The contagion theory of health with its implication of quarantine interfered with trade and so threatened profits.

Engels and Virchow provide good illustrations of a more progressive tradition of thought in nineteenth-century public health. Both recognised that disease generally affected the poor more than the rich. Engels provided a clear picture of the relationship between people's working and living conditions and specific disease when he reported on the situation of the working classes in Manchester in 1844 (Engels, 1993 [1845]). He was convinced that the ability to resist disease was a reflection of an individual's class and social position, which meant that changes to working and living conditions were likely to be influential in preventing disease. Similar conclusions were reached by Rudolf Virchow in the Prussian region of Upper Silesia in 1848 when he reported on a typhus epidemic, noting that the underlying social and working conditions were important causes (Waitzkin, 1981). His recommendations called for improved nutrition, more employment, better housing and free public education. As it turned out, the public health reforms of the mid-nineteenth century were less radical and comprehensive than those recommended by Engels and Virchow.

The environmental and sanitary reforms followed the recommendations of Chadwick to the Poor Law Commissioners in 1842 and reflected liberal rather than more revolutionary changes. The 1848 European revolutions were followed by a period of repression in which radical reforms to social and economic structures were unlikely. The progressive tradition of public health, however, remained an influence, as can be seen in some of the statements of the medical officer of health of that period. For instance, Warren and Francis (1987, p. 153) quote Sir John Simon, reflecting in 1890 on his annual reports as medical officer of health of the City of London from 1849 onwards:

> I did my best to make clear to the commission, what sufferings and degradation were incurred by masses of the labouring population through the conditions under which they were so generally housed in courts and alleys they inhabited: not only how unwholesome were those conditions, but how shamefully inconsistent with reasonable standards of civilisation; and how vain it must be to expect good social fruits from human life running its course under such conditions.

Lewis (2003) in his analysis of the nineteenth- and early twentieth-century public health movements in industrialised countries notes that a philosophy of 'blame the victim' was evident and dominant throughout the period. The different motivations behind the public health reform traditions are mirrored today, as can be seen in chapter 4. Perhaps the two most important lessons the nineteenth-century public health movement can teach public health practitioners at the turn of the twentieth and twenty-first centuries are that public health is inherently a political activity that is likely to be controversial and disputed, and that patterns of disease and health are a reflection of broader social inequities.

RELEARNING THE NINETEENTH-CENTURY LESSONS: MCKEOWN AND SZRETER

The successes of the nineteenth-century public health movement have been celebrated by the new public health movement since the 1980s, but the lessons had been forgotten in the decades following World War II. McKeown (1979) has been one of the most significant modern voices to remind us of the importance of non-medical factors in improving the health of populations in industrialised societies. He concluded that, with the exception of vaccination against smallpox, immunisation or medical therapies are unlikely to have had a significant impact on mortality in the nineteenth and early twentieth centuries. He argues that mortality was declining before effective medical interventions were available (see figure 2.1).

McKeown's analysis has been used by Australian public health activists keen to convince policy makers of the value of interventions. Hence, it is important to understand his arguments and their importance in promoting public health in health policy debates.

FIGURE 2.1 RESPIRATORY TUBERCULOSIS: MEAN ANNUAL DEATH RATES (STANDARDISED TO 1901 POPULATION, ENGLAND AND WALES)

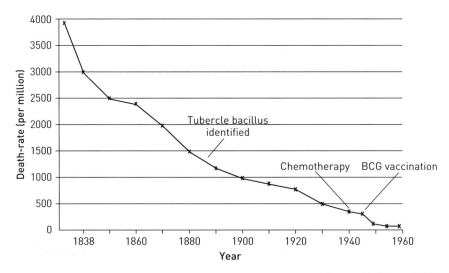

Source: McKeown, 1979, p. 92.

McKeown (1979) believed that improved living standards, especially nutrition, were responsible for the decline in mortality, but he has been questioned by Szreter (1988) who, after re-analysing McKeown's demographic data, concluded that the public health movement, with local interventions through the Ministry of Health and the actions of local authorities, was the most relevant factor. Szreter does not challenge McKeown's conclusion that the progress of modern scientific medicine was not responsible for the historical fall in mortality, but he does challenge his interpretation of what actually did account for the decline in mortality. Szreter recognises that rising living standards contributed to longer life expectancy, but argues that interventions from government authorities were the crucial factors that enabled health to be improved. He maintains that the argument proposing improved nutrition as the primary causal factor in the mortality decline was put forward by default rather than because of any convincing positive evidence (Szreter, 1988, p. 10).

In Szreter's view, McKeown's analysis that the 'invisible hand' of rising living standards led to better health only tells part of the story and dramatically plays down the role of hard-won improvements in working conditions, housing, education and various health services. Szreter argues convincingly that economic growth itself does not guarantee improved health but rather it all depends on how the fruits of that growth were deployed. This, in turn, depended on the cumulative net outcome of a 'rich history of political, ideological, scientific and legal conflicts and battles at both national and local levels' (Szreter, 1988, p. 35). In a later paper Szreter (1995) describes in detail how growing municipal activity and civic pride were essential components of the improvements in health status.

Both McKeown's thesis and Szreter's revision are crucial for the new public health. They agree that medical intervention played only a small part in the dramatic extension of life expectancy in the nineteenth century. Collectively, they establish the importance of general standards of living and of state intervention in improving the health of populations. Szreter's analysis, in particular, supports the crucial importance of the state's role in redistributing the benefits of economic growth, indicating that the invisible hand of the market does not automatically lead to improvements in health status. Further it is a powerful reminder that public health reforms require legislation, political commitment and popular support. Establishing the conditions for the confluence of these factors is a key aim of the new public health and one that it learnt from a close study of the process by which reform was achieved in the nineteenth century.

NATION-BUILDING ERA

The third period in public health history, lasting from the first decade of the twentieth century until the 1930s, saw public health promoted and used for nation building. Public health was typical of the growing state intervention in what had been civil society activities including education, social services, regulation of industry and labour relations (Lewis, 2003, p. 139). Each of these areas played a role in improving health through action on the social determinants. Across industrialised countries this period was characterised by a concern with strengthening the nation by improving the health and fitness of white citizens in particular and the quality and quantity of the population (Powles, 1988, p. 292; Lewis, 2003). Maintaining health was seen as part of a citizen's duty, to be encouraged by the state through school medical examinations and open-air exercise. This comment was made by the doctor responsible for establishing school inspections in Tasmania: 'I look forward to the day when the serious acceptance of a doctrine of national physical morality will cause preventable disease to be regarded as somebody's crime and when the preservation and protection of health will occupy a place in the daily round of unquestioned duty to the state and to one's neighbours' (quoted in Powles, 1988, p. 295).

Eugenics formed a major part of this nation-building phase, the pursuit of a 'pure race' being very much part of the agenda, once again demonstrating how public health at any particular time reflects dominant political and social attitudes. This was when many white Australians believed that Aboriginal people would soon 'die out' and that the role of 'civilised' whites was to 'smooth the pillow of the dying race'. That this was inevitable was supported by arguments derived from Darwinian notions of evolution that superior races would prosper as a result of the survival of the fittest. Similar eugenic movements were leading to forced sterilisations of people with mental illnesses or intellectual disabilities in Germany and calls for the introduction of similar measures in other countries (Lewis, 2003).

Public health services for infants, mothers and schoolchildren developed in this period. They generally emphasised teaching hygiene skills and saw an important part of their role as developing the health and fitness of the population in order to strengthen the state (Lewis, 2003). Being healthy came to be seen as the duty of a good citizen and the emphasis of the period was on individual rather than collective responsibility.

In this period economic progress was seen as leading to improvements in health. At least one Australian specialist in public health anticipated the arguments to be made by McKeown 40 years later when, at an Australian and New Zealand Association for the Advancement of Science (ANZAAS) meeting in 1930, he linked the decline in tuberculosis to rising living standards:

> The remarkable fall in the death rate from tuberculosis began even before the discovery of the tubercle bacillus by Koch, and its curve of descent corresponds to the curve of the rise in value of wages … Public health is purchasable. The Gordian knot of poverty and poor food, ignorance and infection, can be cut by a sword of gold. Food comes before Education, and Education precedes Health.
>
> *Quoted in Powles, 1988, p. 297*

This attitude towards public health demonstrates that the social and economic understanding of factors that create health was well and truly alive. The awareness of the importance of the social and economic determinants of health was very clear for some members of the medical profession as shown by the quotes from Dr E.P. Dark in box 2.1. The emphasis on nutrition, education and health was motivated by the desire to build up and strengthen the new Australian nation.

BOX 2.1   E.P. DARK: NEW PUBLIC HEALTH COMMENTATOR IN THE 1930s AND 1940s

The following quotes from the work of E.P. Dark published in the *Medical Journal of Australia* show that the 'new' public health thinking was evident in earlier periods of history.

> In this attempt to examine the relationship of poverty to disease, and to show that its abolition is an essential part of preventive medicine, it is impossible to avoid a brief discussion of politics and economics and a review of some of the workings of a system which accepts poverty of destitution as the inevitable lot of nearly half the people (Dark, 1939, p. 345).
>
> If capitalism cannot learn how to distribute what it can so abundantly produce, it must change or perish (Dark, 1939, p. 351).
>
> But give us decent housing, economic security, work for every man and woman who wants it, an optimum diet for every man, woman and child in the community, and then a socialised medicine could raise the health of the nation to an undreamed of level (Dark, 1941, p. 526).

In Australia an important part of the nation-building effort at this time was the formation of the Commonwealth Department of Health in 1921 (the first director-general was J.H.L. Cumpston) and the National Health and Medical Research Council in 1936. Both represented a commitment towards improving the health of the nation. It was also the period in which public health training grew around the world and the idea of public health as a profession developed. Lewis (2003) notes that it was at this time that the medical dominance of public health was entrenched.

Powles (1988) reports that this period of nation building was characterised by two main strands of thought. One was the ideology of progressivism, which reflected both European modernism and the ideas of the American President, Theodore Roosevelt. In health this ideology was linked to notions of vitality, efficiency, purity and virtue. Progressivism was not a radical movement that sought to bring about major change, but rather reflected a bourgeois concern to modify the more excessive effects of capitalism. It was also motivated by Utopian views, which had roots in the nineteenth century, typified by the views of a London physician, Benjamin Richardson, whose ideas Hetzel (1976) sees as influential on the development of Australian cities. Hetzel (1976, p. 27) quotes Richardson as advocating an ideal city, Hygeia, comprising 'no more than 25 persons per acre and homes under four storeys in height. Every home had an ample garden, as did also public buildings. The streets were wide and spotlessly clean, being washed daily. There was to be an underground railway system to eliminate noise.' Such a city was to be achieved by 'principal sanitary officers' working with medical officers and inspectors. Public homes with trained nurses would contribute to a decline in infant mortality. This vision was based firmly on an interventionist state committed to improving the health and well-being of its citizens.

The other strand was that of 'national efficiency', which stemmed from the belief that strong nations were essential for national protection and advancement. Strong nations were seen to result from reformed education systems, linking science and government, and more business-like government. This period consolidated the health of citizens as a legitimate concern of governments, and established some key legacies that still influence public health.

AFFLUENCE, MEDICINE, SOCIAL INFRASTRUCTURE

The postwar period was one of considerable affluence for industrial countries including Australia. Unemployment was low, immigration high, per capita income had never been higher, and successive governments were prepared to invest in social infrastructure. Education services were expanded in this period, state housing trusts and commissions provided social rather than welfare housing, and the provision of health services expanded considerably. Public health services were, however, in an in-between period. Infectious diseases were seen as less threatening, the provision of clean water and sewerage was extended to nearly all Australians, with the significant exception of many Aboriginal communities, and chronic lifestyle diseases were not recognised as the problem they were to become. Public health services existed in each state, typically staffed by medical officers with military or colonial experience. Their remit was mainly concerned with policing (together with local government) standards for clean air, water and food, and providing immunisation services. The tuberculosis and polio campaigns of the 1950s were major events and examples of the focus shifting from a structural and social approach (through better housing, more jobs) to a medical one based on immunisation, screening and treatment. Nonetheless, public health was seen as a poor cousin to medicine. The affluence of the times and the security of the Menzies era appear to have led to a belief that things would go on getting better, and that medicine would play a key role in this, rather than public health.

Ironically, the immediate postwar period did see much action that today might well be labelled 'new public health'. Federal and state governments invested in social and physical infrastructure. Nation building was the preoccupation of the day and some of the fruits of economic growth were ploughed into this activity. The growth in state educational services, housing, welfare provisions and community services was supportive of health. The formal public health sector in this period, however, was concerned with infectious diseases (which had already declined significantly) and had not yet appreciated the potential for a more holistic approach to public health. So it could be argued that, in a period when the public health profession is generally assessed as having been in decline, the broader public health agenda was, in fact, being advanced significantly.

The period from World War II until the 1970s was one in which available medical therapies mushroomed. After the 1950s, new drugs were developed, diagnostic techniques became more and more sophisticated and surgery opened up many new areas for medical intervention (including organ transplants). The period was a golden age for medicine because, in Western countries, these medical developments came at a time when economies were expanding, so there was finance for medical research and services to utilise and expand the new discoveries. Additionally, the growth of medicine coincided with a period of affluence, and rising living standards and life expectancy. While this correlation implied no causality, medicine was seen as one of the keys to the generally brightening social and health prospects for Australians.

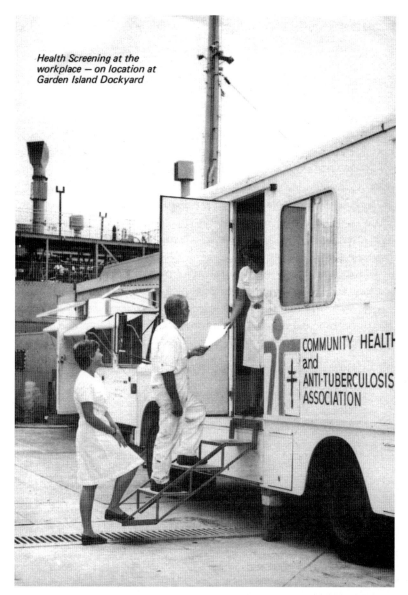

Health Screening at the workplace — on location at Garden Island Dockyard

COMMUNITY HEALTH and ANTI-TUBERCULOSIS ASSOCIATION

The 1950s were the first era of mass screening. Tuberculosis was common. Here the screening van is on location at the Garden Island Dockland in Sydney. (Community Health and Tuberculosis Association)

A detailed consideration of the life expectancy figures indicates that most significant gains were made in an earlier period before medical therapies were available. Between 1920–22 and 1953–55, men gained 13.5 years and women 14.8 years. The gains in the period of affluence were less dramatic—between 1953–55 and 1960–62 they were

0.8 years for men and 1.9 years for females (Hetzel, 1976, p. 31). These figures were similar in other developed countries, such as the USA and Europe, and led Hetzel (1976, p. 31) to conclude that the impact of the medical therapies had been in terms of quality of life rather than in extending life. Changes to social, economic and environmental conditions appear to have had a much greater effect on mortality rates (McKeown, 1979).

CONCLUSION

The 1973 oil crisis signalled the end of taken-for-granted affluence by Western countries. Recession was widespread in the 1970s in developed countries, and unemployment increased around the world. In many countries governments were elected with a mandate to undo the considerable advances made towards the establishment of welfare states. Australia was no exception, and from the 1980s onwards the language of economic rationalism (see chapter 5) became a central discourse of Australian political life. This environment was far less conducive to investment in medical therapies and technologies than the expansionary era and the costs of medicine were increasingly questioned. At the same time (and possibly as a consequence) public health began a resurgence that started with a consideration of what were dubbed 'lifestyle' chronic diseases (most notably diabetes, cancers including lung cancer, cardiovascular disease) and the role of individual behaviour in these. Simultaneously, other developments were encouraging a focus on the environmental causes of ill health. Together, these trends combined to set the stage for the new public health.

CRITICAL REFLECTION QUESTIONS

2.1   To what extent do you think the quote from Virchow at the start of the chapter applies today?

2.2   To what extent do you agree with the views of E.P. Dark in box 2.1?

Recommended reading

Lewis (2003) *The people's health. Public health in Australia 1799–1950*. Provides an excellent historical perspective.

Waitzkin, H. (1981) The social origins of illness: A neglected history. *International Journal of Health Sciences* 11, 77–103. Insightful classic piece on social determinants.

Useful website

http://sphtc.org/timeline/flash%20timeline/player.html
This US site provides a slide show timeline of key events in US public health history, which does include social aspects of public health.

3

THE NEW PUBLIC HEALTH EVOLVES

It is the same and at the same time it
is not the same
It is different and it is not different.

Zen saying

KEY CONCEPTS

INTRODUCTION

This chapter analyses recent developments in public health and related movements and describes the second revolution in public health, known as the new public health. The lifestyle, new public health and global new public health eras from table 2.1 will be discussed simultaneously, as the periods are not discrete. An overview of international milestones in the development of the new public health is provided in figure 3.1.

FIGURE 3.1   INTERNATIONAL MILESTONES IN THE DEVELOPMENT OF THE NEW PUBLIC HEALTH

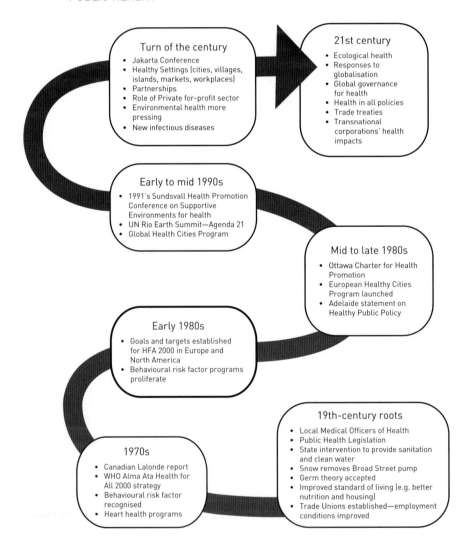

International developments in the new public health are discussed and then Australia's contribution and response to these as a case study of one country's response to the new public health.

The new public health is innovative because:

- it puts the pursuit of equity at the centre of public health endeavours
- it is based on the assumption (supported by considerable evidence) that social and environmental factors are responsible for much ill health
- it argues for health-promoting health services that are based on a strong system of primary health care
- it stresses the role of all sectors in impacting on health and the importance of health in all policies
- it stresses the importance of participation and involvement in all new public health endeavours.

INTERNATIONAL DEVELOPMENTS IN THE NEW PUBLIC HEALTH

THE 1970s: MEDICINE QUESTIONED AND LIFESTYLES TO THE FOREFRONT

By the 1970s the idea that medical advances had been responsible for extending average life expectancy in Britain was being challenged (Powles, 1973; McKeown, 1979). In all rich countries the costs of medicine were increasing and the returns per dollar appeared to be decreasing. All this contributed to changing ideas about the production and protection of health and set the scene for the development of the new public health.

The general social climate of the 1960s and 1970s was also important in these developing ideas: radicalism and social discontent was in the air in North America, Europe and Australia. The social revolution created a greater generation gap than had previously existed, fashions were more radical and protest was being popularly expressed through music, art and other forms. Change was the order of the day, imbued with an air of optimism that suggested life was likely to get better. In this period the environment movement was becoming more prominent and, while much of the message was one of doom, especially in regard to population, there was a sense that change could be achieved. The Vietnam War had divided US and Australian society and created a generation (now labelled the baby boomers) committed to social change by questioning the status quo. The women's movement was gaining momentum and in Australia Indigenous peoples were becoming more assertive about the historical neglect of their rights.

THE DISCOVERY OF 'LIFESTYLE'

Internationally, the 1970s saw the discovery of lifestyle and its impact on health. Canada was the first industrialised country to embrace the notion of healthy lifestyles within its health policy (Hancock, 1986) with the publication in 1974 of *A New Perspective on the*

Health of Canadians (known as the Lalonde report, after the then Health and Welfare Minister, Marc Lalonde). This report describes medicine and health care services as one of four 'health fields' that influenced health and illness, the others being human biology, the environment and lifestyle. It was significant in broadening the international health debate beyond medicine and treatment.

The 1970s brought about many different approaches to creating change in individual lifestyles (discussed in chapter 19). These almost exclusively drew on psychological theory and had little regard for individuals' social and economic circumstances. Theories developed in this decade included the health belief model (Becker, 1974), the theory of reasoned action (Ajzen and Fishbein, 1980), social learning theory (Bandura, 1977) and (in the next decade) the stages of change model (Prochaska and DiClemente, 1984). Disease-focused associations (including the National Heart Foundation and the Diabetes Association) were active in their campaigns to change risk factors. In many industrialised countries the 1970s saw the adoption of community programs designed to persuade people to change their health-related behaviour, including the Finnish North Karelia Project, the Pawtucket Heart Health program in the USA and the North Coast lifestyle program and the 'Life—Be in It' campaign in Australia (see chapters 18 and 19 for more details). Government policy statements stressed a behavioural approach to health, as seen in the US Healthy People Report, which directly reflected growing concern about the burgeoning US health budget. The UK *Prevention and Health: Everybody's Business*, published in 1976, is an example of a behavioural approach to health promotion.

In industrialised countries the behaviour modification approaches to health were developing alongside growing concern about inequities in health and the failure of health services (even deliberately egalitarian services such as the British National Health Service) to do much about them. The British Black Report on inequities in health was compiled in the second half of the 1970s (Townsend et al., 1992) and concluded very firmly that, while behavioural factors did play a role in health, they were not primarily responsible for the differences in health status between Britons in different social classes. They favoured explanations that lay in the social and economic support available to people and foreshadowed the World Health Organization (WHO) Ottawa Charter's emphasis on policy as a key strategy for health promotion. So the 1970s saw highly individual, behavioural approaches to health being developed, while social structural causes of illness and health were coming to be understood in more detail.

HEALTH FOR ALL BY THE YEAR 2000

The watershed of the 1970s was undoubtedly the goal set by WHO to achieve Health for All by the Year 2000 (HFA 2000), preferably by the promotion of primary health care (WHO, 1981). The key elements of this strategy as outlined in the Alma Ata Declaration were:

- an emphasis on global cooperation and peace as important aspects of primary health care
- recognition that primary health care should be adapted to the particular circumstances of a country and communities within it

- recognition that health status reflects broader social and economic development
- primary health care as the backbone of a nation's health strategy with an emphasis on health promotion and disease prevention strategies
- achievement of equity in health status
- participation in the planning, organisation, operation and control of primary health care, supported by appropriate education
- involvement of all sectors in the promotion of health
- the call for a 'new economic order'.

The initial application of the Health for All package was to developing countries where it was, at first, interpreted as a comprehensive package that tied health improvement to overall social and economic development. Many governments embraced the notion and primary health care was introduced in many communities. Fairly rapidly, however, the notion of selective primary health care (Walsh and Warren, 1979) was floated and reasserted the concept of focusing on a particular disease. This has been criticised by those advocating a more comprehensive approach (Rifkin and Walt, 1986; Werner and Sanders, 1997; Baum, 2007b) as it puts emphasis on eradicating and preventing disease through the actions of specialists. This contrasts with more holistic approaches that attempt to deal with the root causes of illness.

Selective primary health care encourages the view that medical interventions are the most crucial, usually to the detriment of other measures such as housing, education and nutrition (Sanders, 1985; Rifkin and Walt, 1986, pp. 561–2). A WHO document (Tarimo and Webster, 1994) suggests that a selective approach to primary health care often means that while a particular disease problem may be resolved, this will simply be replaced by another problem as the underlying causes of ill health have not been dealt with. Participation tends to be defined as a means of helping specialists tackle specific diseases, rather than a comprehensive community development process.

Of course, these criticisms do not mean that selective primary health care work is not valuable; they rather suggest that a comprehensive approach is likely to be more sustainable and effective in the long term (Baum, 2003, 2007b). Some believe both approaches can be used to advance public health (Taylor and Jolly, 1988), but they do have different philosophical bases, which affect how primary health care and health promotion are implemented (Baum and Sanders, 1995; Wass, 2000). The application of social psychology to health promotion has attracted much criticism because of the evidence that behaviour is not the most important determinant of health. Programs that focus on individual behaviour change have been widely criticised for ignoring the lack of opportunities for disadvantaged people to be healthy (Crawford, 1977; Tones, 1986, 1992; French and Adams, 1986; Tesh, 1988; Naidoo and Wills, 1994; Rodmell and Watt, 1986; Baum and Fisher, 2014). The individualism underlying behaviourism and its impact on public health is examined in detail in chapter 4. From the 1970s the individual approach to health promotion was really taking root. The WHO Ottawa Charter challenged that focus and continued to provide a blueprint for the new public health until the twenty-first century.

THE 1980s: DEVELOPING A NEW PUBLIC HEALTH

THE NEW PUBLIC HEALTH BIBLE: THE OTTAWA CHARTER

There were two driving forces behind the Ottawa Charter. It was clear that the Health for All by the Year 2000 strategy was not being adopted by industrialised countries, and the limitations of the lifestyle and behavioural approaches meant a new conceptualisation for health promotion was required. Also the time was opportune for a major health promotion statement. Health care budgets continued to expand in the 1980s and the prospect of cutting these by improving the health of populations became increasingly attractive.

Green and Raeburn (1988, p. 30) characterise the two approaches to health promotion as the 'individual versus the system', observing that these two views lead to divided ideological and theoretical perspectives on health promotion. Despite this, they note that, in practice, viewpoints are more integrated. Perhaps the genius of the Ottawa Charter lay in the fact that it managed to integrate many of the different perspectives on health promotion. While being seen as the foundation of the new public health, it did not reject behavioural and lifestyle approaches, but saw them as part of the acquisition of personal skills for health. Its five key strategies have become something of a mantra for health promotion (Baum, 1990), but, like most mantras, it has served a useful function in directing the task of health promotion towards a multipronged and multilevel strategy.

The Ottawa Charter rather cleverly builds on a number of social and public health movements, including nineteenth-century public health, feminism, the green and consumer movements and experiments in community development from the 1950s. Thus the Ottawa Charter did not emerge from a vacuum in the 1980s, but reflected numerous social and health movements of the previous century. Its claim to be 'new' derives from the way it pulled together numerous and diverse movements to present a package that gave public health a more radical and cohesive direction than had been the case for some time. It also served to make health promotion a legitimate and respectable aspect of the health scene.

The charter is based on the belief that health requires peace, shelter, education, food, income, a stable ecosystem, social justice and equity as prerequisites. Its five strategies are shown in box 3.1.

BOX 3.1   FIVE STRATEGIES OF THE OTTAWA CHARTER FOR HEALTH PROMOTION, 1986

- The development of healthy public policy, which recognises that most of the private and public sector policies that affect health lie outside the conventional concerns of health agencies. Rather they are in policies such as free and universal education, environmental protection legislation, progressive taxation, welfare, occupational health and safety legislation and enforcement, land rights legislation and control of the sale and distribution of substances such as alcohol and tobacco. Health becomes, therefore, a concern and responsibility of each sector of government.

- The creation of supportive environments in which people can realise their full potential as healthy individuals. The Charter recognises the importance of social, economic and physical environmental factors in shaping people's experiences of health.
- Strengthening community action refers to those activities that increase the ability of communities to achieve change in their physical and social environments through collective organisation and taking of action.
- The development of personal skills acknowledges the role that behaviour and lifestyles plays in promoting health. The skills called for are those that enable people to make healthy choices. It also extends the skills base for health to those associated with community organisation, lobbying and advocacy, and the ability to analyse individual problems within a structural framework.
- Reorientation of health services is a call for health systems to shift their emphasis from (in most industrialised countries) an almost total concentration on hospital-based care and extensive technological diagnostic and intervention to a system that is community based, more user-friendly and controlled, which focuses on health.

Source: WHO, 1986.

The Ottawa Charter stresses the importance of, and recommends:

- advocacy for health
- enabling people to achieve their full health potential
- mediation between different interests in society for the pursuit of health.

TABLE 3.1   CONTRASTS AND SIMILARITIES BETWEEN THE 'OLD' AND 'NEW' PUBLIC HEALTH

Old public health	New public health
Focus on improving physical infrastructure, especially in order to provide adequate housing, clean water and sanitation.	Focus on physical infrastructure, but also on social support, social capital, behaviour and lifestyles.
Legislation and key policy mechanisms, especially in the nineteenth century.	Legislation and policy rediscovered as crucial tools for public health. Stewardship responsibility for health systems to monitor health and argue for health perspective in all policies.
Medical profession has central place.	Recognition of intersectoral action as crucial. Medicine only one of many professions contributing.
In the nineteenth century public health was one of a series of social movements that worked to improve living conditions. Primarily expert-driven but some legitimation of community movement. Progressively more expert-dominated in the twentieth century.	Philosophy places strong emphasis on community participation, but in practice this is not often achieved, despite some real successes. Citizen (as opposed to consumer) interests dominant.

(continued)

TABLE 3.1   CONTRASTS AND SIMILARITIES BETWEEN THE 'OLD' AND 'NEW'
PUBLIC HEALTH (*CONTINUED*)

Old public health	New public health
Epidemiology is a legitimate research method.	Many methodologies recognised as legitimate.
Focus on disease prevention and health is seen as absence of illness.	Focus on disease prevention, health promotion and a positive definition of health.
Primary concern with the prevention of infectious and contagious threats to human health.	Concern with all threats to health (including chronic disease and mental health). Rapidly growing concern with sustainability and viability of the physical environment and the impact of environment deterioration on human health and focus on ecological aspects of public health.
Concern with improving the conditions of the poor and special-needs groups.	Equity and social justice an explicit aim of new public health philosophy.
Little attention paid to the political and economic context.	Political and economic context of health seen as crucial to addressing health issues.

THE 1990s: IMPLEMENTING THE OTTAWA CHARTER STRATEGIES

By the late 1980s a new style of public health theory and practice was emerging, with both continuities with the past and distinct differences, which are summarised in table 3.1. From the mid-1980s the term 'new public health' was taken up enthusiastically. Ashton and Seymour (1988) in their *The New Public Health* provided one of the first texts to examine the new ideas in detail. They built on Thomas McKeown's thesis, the UK Black Report, the Ottawa Charter and their own health promotion experience in Liverpool to set out an agenda for public health action that moved medicine out of the centre stage, introducing more social interventions, community participation and policy change.

A central strategic direction for the new public health was WHO's Healthy Cities program (Ashton, 1992; Tsouros, 1995), which in essence took cities as the units for public health program planning, rather than individuals. The program was spearheaded by the European WHO office and has grown from strength to strength since its launch in 1986. Despite some questioning about aspects of its implementation, it has signified new directions in public health. Healthy Cities is discussed further in chapter 23. A further crucial concept was that of healthy public policy, which has developed into the Health in All Policies approach (discussed in chapter 24).

Through the 1980s and subsequently, research methods in public health were progressively broadened so that the position of epidemiology as the only valid method of public health research was challenged. Qualitative methods have developed and become more widely used in public health research, indicating the need for methods that were able to reflect the complexity of social perspectives on health (see part 3 for details of this development).

THE 1990s TO THE TWENTY-FIRST CENTURY: INTERNATIONAL DEVELOPMENTS IN THE NEW PUBLIC HEALTH

Some of the key developments relevant to the new public health are shown in table 3.2. The focus on policy as a powerful tool of public health was consolidated at the Second International Health Promotion Conference held in Adelaide in 1988, where the theme was healthy public policy. Mahler (1988, p. 8), then Director-General of the World Health Organization and passionately committed to HFA 2000, explained the main aim of healthy public policy as being to create the preconditions for healthy living through:

- closing the health gap between social groups and between nations
- broadening the choices of people to make the healthy choices the easier and most possible
- ensuring supportive social environments.

Mahler also stressed that community participation and collaboration between all sectors of government were crucial aspects of healthy public policy.

TABLE 3.2 MEETINGS AND DOCUMENTS KEY TO THE NEW PUBLIC HEALTH, 1970s–2013

Meeting/document	Relevance to new public health
1978: Alma Ata Declaration on Primary Health Care	Promoted a social view of health and advocated the importance of intersectoral action to achieving health for all.
1986: 1st International Conference on Health Promotion, Ottawa	Produced the Ottawa Charter, which included among the five strategies of health promotion 'promoting healthy public policy' and 'creating supportive environments for health'.
1988: 2nd International Conference on Health Promotion: Healthy Public Policy, Adelaide	Produced the Adelaide Recommendations on Healthy Public Policy, defined as 'an explicit concern for health and equity in all areas of policy and by an accountability for health impact'.
1991: 3rd International Conference on Health Promotion: Supportive Environments, Sundsvall	Produced the Sundsvall Statement on Supportive Environments for Health, which 'recognized that everyone has a role in creating supportive environments for health' and stressed the importance of community empowerment.
1997: 4th International Conference on Health Promotion, Jakarta	Produced the Jakarta Declaration on Health Promotion into the 21st Century, which focused more on low- and middle-income countries than previous declarations. It advocated that public and private sectors should promote health, that health development requires a multisectoral approach, and that health promotion partnerships are necessary, including empowerment.
2000: 5th Global Conference on Health Promotion, Mexico City	Produced the Mexico Ministerial Statement for the Promotion of Health: From Ideas to Action, including key actions 'to position the promotion of health as a fundamental priority in local, regional, national and international policies and programs' and also 'to advocate that UN agencies be accountable for the health impact of their development agenda'.

(continued)

TABLE 3.2   MEETINGS AND DOCUMENTS KEY TO THE NEW PUBLIC HEALTH, 1970s–2013 (*CONTINUED*)

Meeting/document	Relevance to new public health
2005: 6th Global Conference on Health Promotion, Bangkok	Produced the Bangkok Charter for Health Promotion in a Globalised World, which reinforced the basic strategies of the Ottawa Charter and extended them as relevant for a globalised world. It made health promotion central to the global development agenda, a core responsibility of all governments and a requirement for good corporate practice; and called for global governance to address harmful impacts of trade, products, services and marketing strategies.
2008: Final Report of Commission on Social Determinants of Health	Provided extensive evidence on the impact of the social determinants of health and therefore the impact of activities in multiple sectors on health. It recommended the use of health equity impact assessments and endorsed the Health in All Policies approach.
2009: 7th International Conference on Health Promotion, Nairobi	Produced the Nairobi Call to Action for Closing the Implementation Gap in Health Promotion, which called for governments in low-, middle- and high-income countries to make health promotion integral to the policy and developmental agenda, including by implementing the recommendation of the Commission on the Social Determinants of Health.
2010: WHO–SA Health Meeting on Health in All Policies, Adelaide	Produced the Adelaide Statement on Health in All Policies, which 'emphasizes that government objectives are best achieved when all sectors include health and well-being as a key component of policy development'.
2011: United Nations high-level meeting on non-communicable disease prevention and control, New York	Reinforced cross-sectoral approaches and the use of regulation, including to discourage 'the production and marketing of foods that contribute to unhealthy diet'. Reinforced existing WHO strategies on diet, exercise and tobacco, and recognised growing incidence of non-communicable diseases as a threat to human progress.
2011: World Conference on the Social Determinants of Health, Rio de Janeiro	Produced the Rio Political Declaration on Social Determinants of Health, which states: 'Health in All Policies, together with intersectoral cooperation and action, is one promising approach to enhance accountability in other sectors for health, as well as the promotion of health equity and more inclusive and productive societies'.
2013: 8th International Conference on Health Promotion: Health in All Policies, Helsinki	Produced the Helsinki Declaration on Health in All Policies, which calls for all government sectors to implement a Health in All Policies approach with a social justice framework to monitor and evaluate public policies, and to determine the extent to which they support health-related human rights that promote health for all.

Source: Adapted from Baum et al., 2013d, p. 26.

The Ottawa and Adelaide conferences have been followed by a series of others (see table 3.2). Sundsvall in 1991 focused on environmental issues and sustainability. This conference was significant because human health promotion was explicitly linked with the health of the physical environment.

The 1997 Jakarta conference focused on partnerships in health and began to examine the processes of globalisation that were coming then to the forefront. The conference was controversial at the time because the organisers sought to involve representatives of large corporations. The argument in favour of this was that their activities have a very significant influence on health. Others expressed disquiet about the lack of opportunity to discuss the ethical and other dilemmas raised by involving players from the for-profit sector (Durham, 1997), and about the manner in which these 'new players' were involved and the lack of 'robust scientific and ethical debate about if, for what purpose, when, how and with what anticipated results, and under what conditions' the private sector should be involved in health promotion. These new players included multinational companies such as Coca-Cola, Guinness and SmithKline Beecham. The Fifth Global Conference on Health Promotion was held in Mexico City in 2000. Very little was new at this conference and it served to reinforce the directions taken in the previous conference, although it did not emphasise the involvement of the private sector as did the Jakarta conference. Most significantly, the Mexico Ministerial Statement on Health Promotion stressed the need for stronger human and institutional capacity building in order to ensure effective implementation of health promotion and called on nations to prepare country-wide plans of action for promoting health. In what will perhaps prove to be an ominous note, the Ministerial Statement warned that new and re-emerging diseases are threatening the very real gains in health that have been made. Certainly the experience with HIV in sub-Saharan Africa, where years are being wiped off life expectancy for the first time, suggests new diseases are a very real threat to global health. The sixth WHO health promotion conference was held in Bangkok in 2005 and focused on globalisation more than any previous conference. The output was the Bangkok Charter 2005, which called for global governance to address the harmful impact of trade, products, services and marketing strategies. Strident critiques of the pre-conference draft were made by civil society voices, for example:

> Not only does it take a 'neutral' view on globalization, but it takes an uncritical view of private–public partnerships, many of which advance corporate interests at the expense of people's health. Worst of all, the new charter takes the corporate line that the interests of the powerful corporations are basically (or at least potentially) pro-people, and that their commitment to equity, public health, and sustainable environment should be voluntary rather than through strong regulation and democratic process. In essence, it lets the crook off the hook! The Bangkok charter is typical of corporate and World Bank double-speak: all the progressive rhetoric with faulty analysis and unworkable solutions.
>
> *Werner, 2005*

It is certainly true that the new public health agenda does not offer much promise for corporate shareholders. Corporations invest in health services because they want to return a profit to their shareholders. In order to do this they may be keen to advance individualised health promotion practices among their members (and so subsidise gym memberships and encourage health checkups, for example). But they are very unlikely to engage in the community-building work that challenges the underlying inequities in health status or act on the social and economic determinants of health. These activities might concern controversial activities, such as opposing the introduction of a new factory into a residential neighbourhood, lobbying for the rights of garment industry outworkers, opposing coal seam gas exploration or working in coalition with police, welfare workers and educators to reduce child sexual abuse. None of these activities return the short-term profits that corporations are primarily concerned with.

The following conference in 2009 in Nairobi brought a focus on low- and middle-income countries and emphasised the importance of bringing health promotion in to the development agenda and instituting action on the social determinants of health, endorsing the work of the Commission on Social Determinants of Health. Most recently, the 2013 Helsinki conference was devoted to the topic of Health in All Policies (see below).

None of the subsequent WHO conferences has had the impact that the Declaration of Alma Ata and the Ottawa Charter have had on thinking about the new public health. Each has, however, added issues to the debates about health promotion, which have set the scene for the key directions for the new public health in the current century. Environmental issues, the role of transnational corporations and the globalisation of economic life and communications are now all vital issues for the new public health and these are debated and discussed in the following chapters of this book, especially chapter 5. Many of these issues were at the heart of the deliberations of the Commission on Social Determinants of Health.

NEW CENTURY: COMMISSION ON SOCIAL DETERMINANTS OF HEALTH—STRONG REINFORCEMENT FOR THE NEW PUBLIC HEALTH

The Commission on Social Determinants of Health worked from 2005 to 2008, amassing the evidence on the social determinants of health and what policies, initiatives and programs can best address them. Its work was assisted by nine knowledge networks, engagement with civil society organisations, and a scientific secretariat under the leadership of the world-renowned Sir Michael Marmot and eighteen other commissioners who brought policy, political, academic and civil society insights. Its final report was launched in 2008 and its main recommendations are provided in box 3.2. Subsequent to the Commission on Social Determinants of Health, WHO has taken some action: organising the World Conference on Social Determinants of Health in Rio de Janeiro in 2010, which resulted in a political declaration (see box 3.3 for extracts) that reinforced some of the recommendations of the Commission on Social Determinants of Health, and a resolution on the importance of social determinants passed at the 2012 World Health Assembly.

In 2013 the 8th International Conference on Health Promotion in Helsinki focused on Health in All Policies and reinforced the Commission's call for systematic approaches from all sectors of government to promote health through addressing social determinants (see chapter 24).

The Millennium Development Goals (MDGs) are one of the ways in which the United Nations systems support action on the social determinants of health. These have been revised and will be agreed in 2015 as Sustainable Development Goals (http://sustainabledevelopment.un.org/post2015.html). They promise to be broader and more inclusive of a social determinants perspective than the MDGs.

BOX 3.2   RECOMMENDATIONS OF THE COMMISSION ON SOCIAL DETERMINANTS OF HEALTH

IMPROVE DAILY LIVING CONDITIONS

- *Invest in early childhood development* including social/emotional and language/cognitive development.
- *Healthy places create healthy people*: place health and health equity at the heart of urban governance and planning, invest in rural areas, and take health equity into account in responses to climate change and other environmental degradation.
- *Fair employment and decent work* should be a central goal of national and international social and economic policy making; work should support health.
- *Provide social protection across the lifecourse* that can support a level of income sufficient for healthy living for all.
- *Ensure universal health care* with health systems based on principles of equity, disease prevention and health promotion, and workforce with capacity to act on social determinants of health.

Tackle the inequitable distribution of power, money and resources

- *Health equity in all policies, systems and programmes*: ensure that responsibility for action on health and health equity is at the highest level of government. Health ministries should adopt a social determinants framework across policy and programmatic functions and strengthen their stewardship role in supporting a social determinants approach across government.
- *Fair financing*: build national capacity for progressive taxation, assess potential for new national and global public finance mechanisms, increase global aid to 0.7% of GDP, and finance cross-government action on social determinants.
- *Market responsibility*: minimise the adverse effects of markets through use of health and health equity impact assessments in national and international agreements and policy making.
- *Gender equity*: create and enforce legislation that promotes gender equity and makes discrimination on the basis of sex illegal, promote education and economic participation for women, and invest in sexual and reproductive health services and programs.
- *Political empowerment*: ensure fair representation in decision making, and enable civil society to organise and act in a manner that promotes and realises the political and social rights affecting health equity.

- *Good global governance:* make health equity a global development goal, base strategies on a social determinants framework and institutionalise social determinants as a guiding principle across WHO departments and country programs.

MEASURE AND UNDERSTAND THE PROBLEM AND ASSESS THE IMPACT OF ACTION

- *Monitoring:* establish global health equity surveillance systems.
- *Research:* enable research funding bodies to create a dedicated budget for generating and global sharing of evidence on social determinants, including health equity intervention research.
- *Training:* increase training on social determinants for all health professionals and others including urban planners and educators. Increase capacity for health equity impact assessment among policy makers.

Source: Derived from CSDH, 2008, p. 2.

Commission on Social Determinants of Health meeting, Kobe, January 2008. (World Health Organization)

BOX 3.3   EXTRACTS FROM THE RIO DECLARATION ON SOCIAL DETERMINANTS OF HEALTH

'We understand that health equity is a shared responsibility and requires the engagement of all sectors of government, of all segments of society, and of all members of the international community, in an "all for equity" and "health for all" global action.'

'Good health requires a universal, comprehensive, equitable, effective, responsive and accessible quality health system. But it is also dependent on the involvement of and dialogue with other sectors and actors, as their performance has significant health impacts. Collaboration in coordinated and intersectoral policy actions has proven to be effective. Health in All Policies, together with intersectoral cooperation and action, is one promising approach to enhance accountability in other sectors for health, as well as the promotion of health equity and more inclusive and productive societies. As collective goals, good health and well-being for all should be given high priority at local, national, regional and international levels.'

The Declaration included pledges to:

- 'Work across different sectors and levels of government, including through, as appropriate, national development strategies, taking into account their contribution to health and health equity and recognizing the leading role of health ministries for advocacy in this regard'
- 'Adopt coherent policy approaches that are based on the right to the enjoyment of the highest attainable standard of health, taking into account the right to development as referred to, inter alia, by the 1993 Vienna Declaration and Programme of Action, that will strengthen the focus on social determinants of health, towards achieving the Millennium Development Goals'
- 'Support social protection floors as defined by countries to address their specific needs and the ongoing work on social protection within the United Nations system, including the work of the International Labour Organization'

Source: WHO, 2011.

GLOBAL HEALTH SYSTEMS TO PROMOTE THE NEW PUBLIC HEALTH

WHO's aspirational goal of Health for All by the Year 2000 was not met, and in sub-Saharan Africa, life expectancy even went backwards in response to the HIV epidemic (Sanders, 2006). Health inequities between and within countries have been tending to increase rather than decrease in the past two decades. Since the 1990s increasing doubts have been raised about the power of WHO as an effective voice in international health (Banerji, 2002; McCoy et al., 2006; Clift, 2014). The People's Health Movement was formed in 2000 in response to a perception that WHO was out of touch with the health concerns of people at the grassroots. This sentiment has been expressed in its three People's Health Assemblies held in Bangladesh in 2000, Ecuador in 2005 and South Africa in 2012 and expressed in the People's Charter for Health. The Global Health Watch (People's Health Movement et al., 2014) says WHO's weakness reflects the fact that its core contributions have been frozen and so it is reliant on tied activities that funders want to prioritise. It also notes its budget is completely inadequate and only equivalent to the budget of a large hospital in Europe or the USA. It has been noted that WHO's position as the global leader in health has also been challenged by a range of philanthropic bodies.

These initiatives (foremost among them the Gates Foundation; the Global Fund to Fight AIDS, Tuberculosis and Malaria; and the GAVI Alliance) attract huge funding and have the potential to provide significant support to good health systems in poor countries. However, their focus is on initiatives for specific diseases such as malaria, HIV and tuberculosis, which often creates a coordination nightmare for national ministries of health and results in a local brain-drain away from the public health system because the non-government organisations pay much higher salaries. Their focus on single diseases has meant approaches have been less comprehensive and national health systems have been weakened as staff flow to better-paid jobs in the new global non-government organisations. As these problems were recognised, in recent years global campaigns have developed to work towards universal health coverage and to strengthen health systems within countries.

Health systems in rich countries also have a tendency to focus on specific diseases and direct their strategies to lifestyle responses rather than those that tackle the underlying causes of ill health (Hunter et al., 2010; Baum and Fisher, 2014). This is despite the lack of evidence (reviewed in chapters 18 and 19) for lifestyle interventions in the absence of more systematic policy change.

In all settings, health systems are associated with the provision of curative health services to individuals. This is a crucial function to support population health, and ensuring that all groups in a population have access to health services is vital to health equity. But there is a role for health systems beyond this that concerns what the WHO Regional Office for Europe (WHO, 2007) has defined as 'stewardship', which involves 'influencing policies and actions in all the sectors that may affect health'. Here we consider what form of health system appears to best enable a nation to offer all its citizens access to appropriate and affordable health care, while simultaneously enabling the system to play a stewardship role for health and support the production of equitable population health. The fundamental features are:

- high proportion of public expenditure
- adoption of comprehensive primary health care as the backbone of the health system
- resistance to growing medicalisation
- commitment to a stewardship function for total population health.

We have seen earlier in this chapter that Health in All Policies approaches are being advocated by the European Union and WHO. Typically, health systems lead these interventions, as we see in chapter 24 on Healthy Public Policy. Implementation of these measures is vital in high-, middle- and low-income countries if health systems are to be sustainable, equitable and effective.

DOES SPENDING MORE ON CARE DETERMINE HEALTH OUTCOMES?

Evidence suggests that the amount of overall expenditure on health is not the key determinant of population outcomes but that the proportion of public expenditure (as part of total expenditure on health care) is more important. Table 3.3 shows that the

US health system (which has the highest health expenditure in the world) does not perform well on the key health outcome indicators of infant and maternal mortality when compared with the average of the top ten countries measured by the Human Development Index.

TABLE 3.3   HEALTH EXPENDITURE AND MORTALITY OUTCOMES, USA AND OECD COUNTRIES

	Health expenditure		Mortality	
	Per capita (at current prices and PPPs), 2011	As % of GDP, 2011	Infant (deaths per 1000 live births), 2012	Maternal (deaths per 100 000 live births), 2010
USA	8 482.70	17.9	6	21
Average of top ten Human Development Index countries	5 105.13	10.7	3.7	9.8
Average of very high Human Development Index countries	Not available	12.2	5	16

GDP = gross domestic product, PPP = purchasing power parity

Note: Purchasing power parities (PPPs) are currency conversion rates that both convert to a common currency and equalise the purchasing power of different currencies. In other words, they eliminate the differences in price levels between countries in the process of conversion.

Source: OECD, 2014d, 2014e; UNDP, 2014.

Anderson (2006) noted that OECD member states with high public expenditure on health (consistently more than 80 per cent) perform better in minimising infant mortality than those that rely on mixed public–private systems (consistently more than 40 per cent private). Anderson (2006) also cites data from developing countries to show that more privatised health systems are associated with generally worse health outcomes. The Health Systems Knowledge Network of the Commission on Social Determinants of Health similarly reviewed health system financing and concluded that a high proportion of public expenditure encourages equity in provision and outcome (Gilson et al., 2007).

Table 3.4 compares key indicators for the USA, Australia and Costa Rica and similarly illustrates the lack of relationship between both gross national income, and health care expenditure, life expectancy and infant mortality.

TABLE 3.4   HEALTH OUTCOMES AND EXPENDITURE, USA, COSTA RICA AND AUSTRALIA

Indicator	USA	Costa Rica	Australia
Life expectancy at birth, 2012	78.7	79.7	82.0
Infant mortality rate, 2013	5.9	8.4	3.0
Happy Planet Index, 2012	37.3 (rank 105th)	64.0 (rank 1st)	42.0 (rank 76th)
Gross national income (per capita, US$), 2013	53470	9550	65390
Health expenditure (per capita, PPP constant 2005 international $), 2013	8895	1311	4068

PPP = purchasing power parity

Note: Happy Planet Index from Abdallah et al., 2012

Note: Purchasing power parities (PPPs) are currency conversion rates that both convert to a common currency and equalise the purchasing power of different currencies. In other words, they eliminate the differences in price levels between countries in the process of conversion.

Source: Abdallah et al., 2012; World Bank, 2014.

There is strong evidence in favour of universal access and public funding of health care. For example, Australia, Canada, New Zealand and Britain spend less on their predominantly publicly funded systems, but all of these countries have longer average life expectancy than the USA, where access to health care is largely determined by insurance status, (figure 3.2). The USA, as well as being an outlier in terms of its total spending on health (16.9 per cent of GDP (OECD, 2014d)), also has the lowest public expenditure on health, just 47.6 per cent compared with 68.4 per cent in Australia and higher (in the high eighties percentile) for the Scandinavian countries and Luxembourg (OECD, 2014c). Most OECD countries have more than doubled the proportion of GDP spent on health since 1960.

Public health has contributed to much soul-searching in industrialised countries about the extent of health care expenditure. How much of the GDP should be spent on health? Should all services be available to all people regardless of cost? If not, which services should not be available? Who should be involved in these decisions? What should be the balance between spending on curative, preventive and promotive services? Would an investment in more preventive and promotive services save money in the longer term? These questions involve economic, ethical and medical considerations, and have no clear answers.

FIGURE 3.2 COMPARATIVE LIFE EXPECTANCY AND HEALTH EXPENDITURE

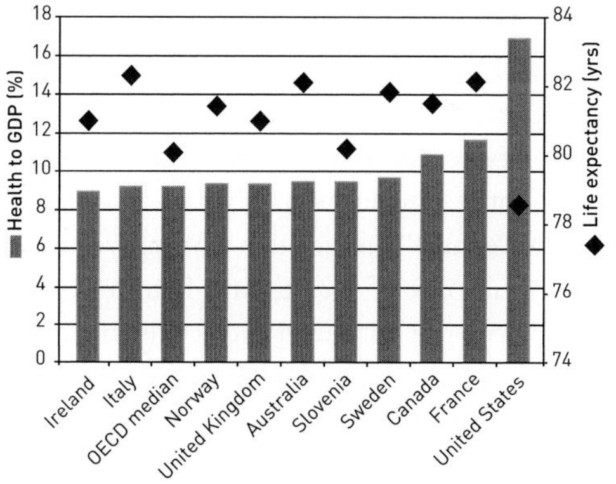

Sources: Adapted from AIHW, 2014b, Figure 2.12; OECD, 2014a, 2014d.

Health systems based on the principle of universal coverage achieved through public financing are generally cheaper, more efficient and more equitable (Gilson et al., 2007). This finding runs counter to the general belief that production and distribution of services is more efficient in the private sector, based on market principles. The reasons for this are complex, but arise from the fact that health is essentially a public good, where market principles do not work (Duckett and Wilcox, 2011, pp. 43–5).

Despite the evidence that high public expenditure on health is effective, public–private partnerships are being pursued in rich and poor countries alike. Within these systems public health and health promotion find it harder to gain funding and to be the drivers of the health system. For-profit medicine does not have a strong incentive to pursue disease prevention and health promotion, especially in relation to tackling the underlying social and economic determinants of health. Privatisation and the push to restrict the size, function and influence of the state are dynamic processes that have been shaping the world for the past two or more decades (see chapter 5). In middle- and low-income countries the effect of the imposition of a neo-liberal agenda (for fuller explanation see chapter 5) on health systems by the World Bank and the global funds has been to weaken the infrastructure of public health systems and leave them unable to cope with the pressing health issues. This has been particularly true in sub-Saharan Africa in the midst of the HIV epidemic. Sanders et al. (2005) point to the deterioration of African health systems and the need to confront this urgently with the re-establishment of strong public health systems. It is also one of the underlying causes of the 2014–15 Ebola epidemic in West Africa, where the weak public health infrastructure has intensified the epidemic because of the absence of adequate health sector response and the extreme poverty in the region (People's Health Movement, 2014). It is clear that the situation in Africa reflects a failure to deal with extreme poverty. In sub-Saharan Africa, almost half the population live on less than $1.25 a day. Sub-Saharan Africa is the only region that saw the number of people living in extreme poverty rise steadily, from 290 million in 1990 to 414 million in 2010, accounting for more than a third of people worldwide who are destitute (United Nations, 2013). In poor countries, through mechanisms such as structural adjustment packages (SAPS) and subsequently poverty reduction strategy papers (PRSP), pressure has been intense to privatise and introduce market measures such as user fees. SAPS and PRSP have imposed strict ceilings on government spending on health, limited public sector recruitment and advanced trade liberalisation, usually to the detriment of African countries (Sanders, 2006). Add to this the brain-drain of health personnel from poor countries and the fact that so many have died of AIDS in Africa. These reforms have had the impact of weakening health systems and reducing health service accessibility for poor people with user fees (Gilson et al., 2007) (see box 3.4). In such a position of crisis there is little hope of such overstretched health systems investing in disease prevention or health promotion.

BOX 3.4   DYING OF AIDS AND EBOLA IN AFRICA

It was 2002. I was visiting the Lilongwe Central Hospital in the capital of Malawi. The adult medical wards, male and female, presented a picture right out of Dante. There were two people to every bed, head to foot and foot to head, and in most instances, someone under the bed on the concrete floor, each in an agony of full-blown AIDS. With demonic, rhythmic regularity, another aluminium coffin would be wheeled into the ward to cart away the body of the person who had most recently died.

Every patient was a near cadaver. The wards rumbled with low, almost-inaudible moans, as though those who were ill could not summon the strength to give voice to the pain. The smell was awful: a room of rotting feces and stale urine. And the eyes, so sunken and glazed and pleading.

Source: S. Lewis, 2005. When Stephen Lewis wrote this he was the UN Secretary-General's Special Envoy on HIV/AIDS.

TWELVE YEARS ON AND A NEW EPIDEMIC HITS AFRICA—EBOLA

'I went and visited one clinic that only had one pair of rubber gloves that the whole staff had to share to care for patients,' Oliphant said. There are only about 1,200 people licensed to practice medicine in Liberia and so far 170 of them have died.

'It's just kind of mass despair everywhere people go,' Oliphant said, 'nobody is working, nobody has money, nobody has access to food—everyone is just kind of isolated and scared.'

Ebola is so infectious that patients need to be treated in isolation by staff wearing protective clothing.

People were drenched, but they carried on waiting because they had nowhere else to go. The first person I had to turn away was a father who had brought his sick daughter in the trunk of his car. He was an educated man, and he pleaded with me to take his teenage daughter, saying that while he knew we couldn't save her life, at least we could save the rest of his family from her.

Source: WHAM ABC 13, 2014.

There is an urgent need to reverse this trend and ensure that health services are accessible and affordable for all people through the provision of a comprehensive service. Ensuring an equity-based and population-based health system is not just a result of a technical process but it is also based on a values debate where there are likely to be competing interests. So, as Gilson et al. (2007) point out, political, health sector and civil society leadership will be essential. They suggest that one of the most urgent and important reforms is that health systems move towards a universal entitlement of service rooted in citizenship. The Australian Medicare model is a good example of a universal health insurance model that is based on a progressive levy (that is, those earning more income pay more) and provides free access to hospital care based on need, and free (for low-income earners) and heavily subsidised access to primary medical care for all. It also does well in health promotion measures. The British National Health Service is a further model of universal access. Both systems are under privatisation pressures. Gilson et al. (2007) produce strong evidence that a pro-equity health system would be based on a form of universality. Of course in resource-poor countries the major issue is lack of finance. The health systems of these countries are chronically underfunded and Gilson et al. (2007) call for a massive investment in health systems. More recently there has been a global call for the extension of Universal Health Coverage (UHC) globally. The resources are available in the world and fairer distribution of these could provide the massive investment needed for UHC. The example of the 'low income, high health' countries (Sri Lanka, China, Kerala State in India, Cuba and Cost Rica) demonstrate that primary health care and social investment (especially in women's education) can result in very high life expectancies without high economic development (Irwin and Scali, 2005). The Global Health Watch 4 (People's Health Movement et al., 2014) warns that UHC could be a means of extending private health care globally rather than a system based on public provision and that the private option is unlikely to be equitable. A commitment to social justice and willingness to use state mechanisms to provide universal access and supportive social environments is vital to equitable improvement in health.

COMPREHENSIVE PRIMARY HEALTH CARE AS THE BASIS OF HEALTH SYSTEMS

A massive global investment in a comprehensive primary health care strategy will go a long way towards making health care accessible. There is now a weight of evidence that health systems should be based on the Alma Ata principles of primary health care, which have withstood the test of time. Starfield et al. (2005) have shown that health systems based on primary health care are more cost-efficient and cost-effective, and countries with strong primary health care systems also have more equitable health outcomes. This lesson is crucial for resource-poor countries. The People's Health Movement advocates strongly for the revitalisation of comprehensive primary health care as the basis for health systems and sees that WHO should play a key role in advocating for this (McCoy et al., 2006; Sanders et al., 2011). It further suggests that WHO should assist countries to integrate fragmented pools of private and public finance, reverse privatisation and strengthen public health infrastructures so that they can plan effectively, deliver services and play the stewardship function detailed below (McCoy et al., 2006).

WHO reinforced the importance of primary health care in the 2008 World Health Report, *Primary Health Care: Now More Than Ever*. This report identified three disturbing trends that have a negative influence on population health outcomes and equity and undermine primary health care:

- a disproportionate focus on specialised curative care
- a command-and-control approach to disease control, focused on short-term results
- in some health systems, an unregulated commercialisation of health, which has been allowed to flourish.

Its key points are summarised in box 3.5. The implementation of comprehensive primary health care is threatened by the vertical nature of the global health initiatives discussed above. Their focus on providing funding for specific diseases undermines attempts to develop local and community-driven primary health care that responds to all diseases.

BOX 3.5   PRIMARY HEALTH CARE: NOW MORE THAN EVER

RECOMMENDED REFORMS

Access and equity

Ensure that health systems contribute to health equity, social justice and the end of exclusion, primarily by moving towards universal access and social health protection – *universal coverage reforms*;

People-centred

Reorganize health services as primary care, i.e. around people's needs and expectations, so as to make them more socially relevant and more responsive to the changing world while producing better outcomes – *service delivery reforms*;

Secure healthier communities

Integrating public health actions with primary care and by pursuing healthy public policies across sectors – *public policy reforms*;

Inclusive leadership

Replace disproportionate reliance on command and control on one hand, and laissez-faire disengagement of the state on the other, by the inclusive, participatory, negotiation-based leadership required by the complexity of contemporary health systems – *leadership reforms*.

Source: WHO, 2008.

Strong leadership from WHO, which supports effective national health systems that can coordinate the multitude of unilateral initiatives and persuade them to invest their resources in strong comprehensive primary health care, could result in better and more equitable access. It would also be the basis for citizen involvement in health systems and effective stewardship of the population's health through assessment of social and economic causes of illness and advocacy for action on them from all sectors.

WHO (2007) has noted that many countries fall short of their performance potential because weak health systems prevent the implementation of existing knowledge. This is especially true in the fragile states of Africa and Eastern Europe. The positive news is that the technical knowledge about what is needed is clear and the challenge is one of marshalling the social and political will to implement a comprehensive system. A major equity issue that will have to be addressed before resource-poor countries will be able to strengthen their health systems is stemming the brain-drain of health professionals from poor to rich countries. Poor countries pay for the cost of training health professionals, and currently many leave for rich countries and so further deplete the ability of those countries to provide effective health services. Mills et al. (2011) estimated the loss to sub-Saharan Africa as $2 billion. The brain-drain represents a subsidy from poor to rich countries. This pattern of migration needs to be reversed urgently. In 2010, the World Health Assembly adopted the first Code of Practice on the International Recruitment of Health Personnel (WHO, 2010c) that calls on wealthy countries to provide financial assistance to source countries affected by health worker losses.

RESISTING GROWING MEDICALISATION

The health and pharmaceutical industry is one of the fastest growing industries in the world. This means that there are strong global incentives to provide more drugs and health services. The extent of this has been such that the large companies involved have been accused of 'disease mongering'. Moynihan and Henry (2006) define this as 'the selling of sickness that widens the boundaries of illness and grows the markets for those who sell and deliver treatments'. Examples include male baldness, female sexual dysfunction, restless leg syndrome and a range of social phobia. Thus many conditions are given a medical diagnosis and so become amenable to services and products from the growing health sector. This tendency is one of the factors driving the fact that even in rich countries, despite increasing amounts of national revenue being spent on health services, the systems are struggling to meet people's needs and expectations.

In Australia, for example, it has been estimated that reducing the public subsidy for inappropriate prescriptions of several high-profile drugs to people with milder health problems could save hundreds of millions of dollars per year (Moynihan and Murphy, 2002). Most health systems are described as being in 'crisis' and the growing expenditure on health is a central political issue. The disease mongering is underpinned by the expectation created by contemporary medicine that disease can be conquered if only enough money is spent. The promise of genetic medicine makes this expectation even greater, despite the lack of evidence that its promise can be realised (O'Sullivan et al., 1999; Petersen and Bunton, 2002). In fact, death is inevitable and while good-quality health services are an essential aspect of society, they only make a minor contribution (estimates vary between 10 and 25 per cent) to health creation alongside the broader determinants of health (McKinlay and McKinlay, 1997; Standing Senate Committee on Social Affairs, Science and Technology, 2001). Politicians still base efforts to solve the 'health crisis' on more expenditure on health services. The 'disease-mongering' companies have multiple strategies to encourage this trend. Yet it appears a doomed strategy that may be very healthy for those who benefit from the provision of health care (from privatising companies through to the growing number of health professionals) but the crisis will never be solved because modern health care, backed by a powerful international industry, is insatiable.

In most countries' hospitals, medicines and doctors' consultations are growing as a proportion of health budgets. In Australia this is clearly at the expense of community-based and preventive activities (AIHW, 2014b). Factors fuelling this are the ever-expanding range of screening, diagnostic and treatment options, the above-average salary increases in the health workforce, population growth and the rising number of people living with chronic disease. Thus in most rich countries an overwhelming proportion of the health sector budget is devoted to medicine as opposed to public health and disease prevention (Baum and Dwyer, 2014). Primary health care receives the crumbs from the table and in every way is the poor relation of the hospital system. That this is the case reflects a complex mix of power relations and vested interests. Shifting this balance will take strong political leadership and, in democratic societies, a willingness to engage in citizen debate about how health sector resources should be deployed. Even in such debates, disease-based consumer movements, often receiving support from medical groups and the pharmaceutical industry (Moynihan and Henry, 2006), may assert their right to expensive health care even if the opportunity cost is less funding for population-wide strategies that will produce greater health benefit. Means need to be found of having a citizen debate based on a community-wide view of priorities. In poor countries the same pressures are evident in different ways. One example is the tendency to exclude traditional healing methods and downplay their value rather than exploring how they might be incorporated within a national health system. A further example is the way that infant formula has been marketed at the expense of breastfeeding, or oral rehydration solution rather than using local products (Werner and Sanders, 1997). Health systems have to learn to prioritise service provision and consider the balance between investment in heroic medicine, which benefits a few individuals, and in comprehensive services, which make more efficient use of scarce resources in delivering overall population health gains.

HEALTH SECTOR STEWARDSHIP FUNCTION

Health systems equipped to promote health in the twenty-first century must be prepared to take on a stewardship role for the health of the population they are responsible for. This means monitoring the state of population health and the extent of equity, and being a facilitator and advocate for all sectors to see health as an outcome of their actions. The importance of this approach was recognised by WHO when it established the Commission on Social Determinants of Health (see box 3.2). Yet despite the weight of evidence presented by the Commission, evidence of health systems taking concerted and systematic action on social determinants is very rare. A study of how former Australian state and territory health ministers accounted for this lack of action during their tenure suggests that the pressure of the medical lobby ensures that the acute care sector dominates health policy discussions and crowds out considerations of health equity and social determinants (Baum et al., 2013b).

The challenge for the twenty-first century is for states to invest in a health system and governance system that sees health as a measure of the outcome of all activities of government and the private sector. The Health in All Policies approach discussed earlier in this chapter exemplifies this approach and practical examples of the approach in action are provided in chapter 24. The health sector has the primary responsibility to advocate for this approach and to point to the ways in which improved and equitable population health is a result of not only health service provision, but also health promoting policies in all sectors. This book is concerned with presenting the evidence for such a system of public health and health promotion: it explains the importance of the social and economic determinants of health and elaborates on the theoretical and practice foundations of the new public health.

AUSTRALIA AND THE NEW PUBLIC HEALTH: 1970s TO THE PRESENT

NEW TYPE OF HEALTH SERVICES: THE AUSTRALIAN COMMUNITY HEALTH PROGRAM

> Health is a community affair. Communities must look beyond the person who is sick in bed or who needs medical attention …
> *Prime minister Gough Whitlam in 1973, quoted in Sax, 1984, p. 103*

The extent to which public health reflects the social and economic climate is well demonstrated by the evolution of the new public health in Australia. The roots of an understanding of the new public health were evident in earlier periods in Australian history (for example, see box 2.1 for E.P. Dark's 1930s view of health). The first Aboriginal-controlled health service was established in 1971 and anticipated many of the features of the subsequent WHO Health for All strategy. The early 1970s was a period of considerable social change in Australia and this manifested politically with the election of the progressive and reforming Whitlam Labor government in 1972. It introduced a Community Health Program (CHP) (Hospitals and Health Services Commission, 1973) that created energy and optimism for change in the way health and health services were approached in Australia. The CHP, set in place in 1973, was an innovative program designed to complement the new Medibank (now called Medicare), a universal public health insurance system, to ensure that all Australians had access to basic health services, and to develop a variety of new ways of delivering those services. The main feature of the program was that multidisciplinary health centres were to be responsible for the health of a given area (Owen and Lennie, 1992), offering such services as specialist child health, mental health, family planning, dental services, health education, immunisation, social work, domiciliary care and rehabilitation (Sax, 1984, pp. 106–7). The program also emphasised prevention of illness. The program review in 1976 (Hospitals and Health Services Commission, 1976) showed that it had developed more than 700 projects covering a large range of topics. While the program led to innovation, new services and new ideas about health delivery, it was difficult to change the mainstream health system. After the fall of the Whitlam government, responsibility for community health services devolved back to the states. The Commonwealth initiative had stimulated considerable innovation in some states, but others were hostile to the CHP's central ideas, leading to a period of uneven development, with Victoria and South Australia developing more innovative services than the other states. A further review in 1986 (Australian Community Health Association, 1986) found that, despite no direct Commonwealth involvement, there was still a reasonably coherent community health program. It also found that the core of community health was 'an illness-focused, residual service providing mainly tertiary and some secondary prevention' (p. 90).

Inner Southern Community Health Service, Melbourne

Despite this, Owen and Lennie (1992) point out that the CHP and the community health movement did offer locally available and managed services, and implemented community participation and prevention. Raftery (1995), while generally pessimistic about the impact of the CHP, acknowledged that it laid the foundation for the new public health in Australia in three ways: the development of a cadre of committed health professionals, some minor inroads into the prevailing biomedical culture, and the provision of a greater range of services to the community. A network of community health centres that were not necessarily managed by doctors was established with the idea of community participation central to their operation. The community health movement meant that, when the ideas of Health for All and the new public health were raised internationally in the mid-1980s, Australia had a tradition of such practices and a cadre of workers who readily adopted the new public health because it was already familiar.

The seedbed laid by the CHP meant Australia rapidly contributed to, incorporated and built on the new ideas in public health in its policy rhetoric and, by the late 1980s, Australia was regarded as one of the world leaders of innovation in public health practice. This was despite the fact that behavioural measures in health promotion were also very popular and meant that there were two approaches developing in parallel in Australia in the 1980s: one based on behaviour change and the notion that people would change to healthier habits if given information, and the other believing that structural change in the underlying causes of ill health (such as employment opportunities, housing and income support) were crucial to improving health and equity. The Federal Minister for Health from 1983 to 1990 was Dr Neal Blewett, a political scientist with a sophisticated understanding of health. He appreciated the need for health promotion, understood the importance of social determinants of health and was certainly a public health sympathiser. The Australian world-leading work undertaken to combat the HIV epidemic is reviewed in chapter 20 (box 20.3) and demonstrates the progressive nature of health policy in this period. The progressive policy approach was also evident when he established the Better Health Commission, whose three-volume report, *Looking Forward to Better Health*, identified cardiovascular disease, nutrition and injury as three priority areas (Better Health Commission, 1986).

A subsequent report, *Health for All Australians*, put the emphasis on the significant health inequities between different population groups. Unfortunately, the report did not suggest how these should be tackled and fell back on health service responses (McPherson, 1992, pp. 126–7). It set 28 national health goals and isolated five of these as central: hypertension, nutrition, injury prevention, older people's health and preventable cancers of the lung, skin and breast. So, despite a supportive minister, the final policies still gave most focus to disease. The program established to implement the recommendations, the National Better Health Program (NBHP), funded a broader range of activities than those linked to lifestyle (see, for example, the evaluation of the South Australian component of the NBHP in Baum et al., 1996). The Australian Healthy Cities Pilot Program was funded from the NBHP and aimed to develop a variety of health promotion programs that embraced the range of strategies in the Ottawa Charter (Worsley, 1990; Baum and Cooke, 1992). Another important initiative at the federal level in the early 1990s was the National Health Strategy. This inquiry reviewed the health system and the need for reform. It did this in a way that considered the importance of primary health care and health promotion and even produced a specific report on health inequities that was firmly grounded in an appreciation of social, environmental and economic factors (National Health Strategy, 1992).

STATE VARIATION IN COMMUNITY HEALTH AND HEALTH PROMOTION IN THE 1980s

The extent to which the Commonwealth's lead in health promotion was taken up in the states varied considerably. New South Wales and Western Australia focused on developing statewide health promotion units. The Queensland health system paid little attention to primary health care and health promotion in the 1980s, while Victoria and South Australia developed community-based initiatives, and their (then) Labor governments enthusiastically took up the ideas of the new public health as expressed in the Ottawa Charter and Alma Ata, and can be said to have been world leaders in these areas.

In Victoria the network of community health centres continued to expand (see, for example, Jackson et al., 1989). This state also established 43 District Health Councils with a mandate to focus on the health of their local communities, giving rise to innovation in community-based health promotion. There were significant advances towards implementing the strategies of the new public health in South Australia in the 1980s, many of which have been documented in a collection of essays (Baum, 1995a). Raftery (1995) sees that the commitment and vision of the Labor Health Minister John Cornwall were crucial to this process. Under his leadership, the state adopted a Primary Health Policy (South Australian Health Commission, 1989) and a Social Health Strategy (South Australian Health Commission, 1988), which made the WHO Health for All strategy and the Ottawa Charter relevant to South Australia. These policies created an environment in which experimentation with the concepts of the new public health could flourish. Community health centres and women's health centres (Shuttleworth and Auer, 1995) were expanded and the state experimented with community development strategies (Tesoriero, 1995); a successful Healthy Cities project was launched in an outer suburban area (Baum and Cooke, 1992); nutrition (Smith, 1995), drug and alcohol (Cormack et al., 1995), child and adolescent health services (Wigg, 1995) and mental health services (Martin and Davis, 1995) all struggled with making the rhetoric of the new public health relevant to their particular work settings, with some degree of success. The state was one of the first communities to pass legislation to control tobacco (Reynolds, 1995). More traditional public health activities were maintained and extended, including in local government (Kirke, 1995; Weston and Putland, 1995). Much was achieved but even so, when Minister Cornwall reflected on the attempts to implement the new public health, he noted accurately that the approach and the political agenda it implied were likely to meet with difficulties:

> the magic bullet approach … is much simpler than a necessarily complex approach based on the more accurate notion that health is the consequence of many and varied public policies interacting with the individual … At a political level the public policy approach lacks support because it produces results in the long term and less visibly than the short-term crisis intervention of heroic medicine. Coronary bypass surgery and level three intensive care for very low birthweight babies are newsworthy. Addressing questions of poverty, education, housing, nutrition and

income maintenance to overcome the problem of very low birthweight babies is not possible in a 60-second television news segment. Nor will it boost ratings or sell newspapers. It is a longer-term and less dramatic intervention. It also implies a consensus on equity of important health-producing goods and services which we have not yet achieved.

Cornwall, quoted in Raftery, 1995, p. 35

Minister Cornwall's comments are relevant to any country, region or institutions where the implementation of the new public health approach is an aim. Another significant feature of the 1980s was the establishment of Health Promotion Foundations in Victoria, South Australia (disbanded in 1999) and Western Australia, financed from tobacco taxes. VicHealth, in particular, has a broad perspective in keeping with the new public health agenda.

1990s: NEO-LIBERALISM TAKES HOLD IN AUSTRALIA

The period from the 1990s onwards has seen neo-liberalism, or economic rationalism, really take hold over Australian public life. The Hawke–Keating Labor governments imposed some aspects of this economic philosophy (see chapter 5 for extended discussion of neo-liberalism). The public health direction from 1990 has to be seen within the context of rapid changes in the management, administration and focus of health services in general, heightened by the Council of Australian Government's (COAG) wide-ranging reform agenda. COAG hoped to improve the efficiency and effectiveness of health and community service delivery by restructuring the planning, organisation and funding relationships between Commonwealth and state governments (Duckett and Willcox, 2011), and this aim has remained until the present day. Neo-liberalism has also been evident in the agenda of state governments, most significantly in Victoria's Kennett government in the 1990s, and Campbell Newman's Queensland Liberal National Coalition in power for three years until 2015. Both followed a policy of aggressive privatisation and cutbacks of public services. The Rudd, Gillard and Abbott governments have pursued aggressive health reform agendas in the face of increasing health sector costs. State governments have responded to these reforms with cuts to their community health and public health services, which have not been good for the new public health.

HOWARD'S AUSTRALIA AND THE IMPACT ON THE NEW PUBLIC HEALTH

In 1996 the Howard Coalition government was elected with a political philosophy unsympathetic to the new public health. Internationally this government was more in sympathy with the governments of Thatcher in the UK, and Reagan and G.W. Bush in the USA. The emphasis on individualism, personal responsibility and privatised health care has not been an environment in which the new public health flourished. The main concern of the Howard Coalition government was to strengthen the private sector's role in the provision of services. Its health promotion agenda focused on smoking and immunisation—both important—but did very little to address the social determinants of health. In the later Howard years emphasis shifted to a concern with biosecurity and preparedness for epidemics of newly emerging diseases such as SARS and avian flu.

NATIONAL, STATE AND LOCAL PUBLIC HEALTH RESPONSIBILITIES

The main public health roles of the state and Commonwealth health authorities are described in box 3.6. The description of these roles demonstrates the need for partnerships in public health. Activity is divided between Commonwealth, state, local government and non-government bodies. The National Health and Medical Research Council (NHMRC) and the Australian Institute of Health and Welfare (AIHW) play key roles, the former through research and development and the provision of expert advice, and the AIHW by providing national systems for collecting public health data. The Australian Population Health Development Principal Committee is a subcommittee of the Australian Health Ministers' Advisory Council. It coordinates national effort towards an integrated health development strategy that includes primary and secondary prevention, primary care, chronic disease, and child health and well-being. It has functioned as an intergovernmental committee and has not shown much strategic leadership or innovation.

BOX 3.6   KEY PLAYERS IN AUSTRALIAN PUBLIC HEALTH AND SUMMARY OF MAIN ROLES

FEDERAL GOVERNMENT

Role: coordination of public health activity, provision of regulatory framework and policy and vision leadership, funding of national public health activity, quarantine, immigration, nuclear safety, specific programs (including HIV, illicit drugs, cancer screening), national immunisation program, development, implementation and review of national health programs (e.g. women's health, injury, diabetes, mental health), Aboriginal health. All federal government departments have some impact on public health and so are potential partners, even if public health potential is not currently recognised. Key agencies include:

- Australian Government Department of Health (particularly the Population Health Division, the Indigenous and Rural Health Division, and Office of Health Protection, which includes emergency management, the Ebola Taskforce, immunisation, and chemical safety)
- Food Standards Australia New Zealand (develops and administers the Australia New Zealand Food Standards Code, which lists requirements for foods such as additives, food safety, labelling and genetically modified foods)

- Therapeutic Goods Administration (assessment and monitoring to ensure therapeutic goods available in Australia, including pharmaceuticals, are of an acceptable standard and keep up with international advances)
- Australian Radiation Protection and Nuclear Safety Agency (responsibility for protecting the health and safety of people and the environment from the harmful effects of ionising and non-ionising radiation)
- Safe Work Australia (leads the national effort to promote best practice in workplace health and safety, improve workers' compensation arrangements, and improve rehabilitation and return to work of injured workers)
- Australian Institute of Health and Welfare (AIHW) (national agency providing regular and reliable information and statistics on Australia's health and welfare)
- National Health and Medical Research Council (NHMRC) (funding body for health and medical research and principal independent advisory body on public and individual health)
- Primary Health Networks (funded by the Australian Government Department of Health from 2015, replacing the short-lived Medicare Locals)

STATE AND TERRITORY GOVERNMENTS

Role: delivery of public hospital and community health services, planning of and leadership for public health, maintaining up-to-date public health legislation, population health surveillance, regulating local government public health functions, promoting health and well-being, including working with other sectors to do so. Key activities include:

- epidemiological surveillance, identification of public health issues, intervention and monitoring of outcomes
- policy development and statutory responsibilities in regard to communicable diseases, environmental health, immunisation, food safety, radiation, workplace risk, water quality, drugs and poisons, emergency responses
- establishment of preventive and early detection programs including cancer screening, maternal and child health, school health and dental screening
- health promotion for specific population groups
- support and information to health care providers about disease control
- strategy development to meet new public health challenges
- evaluation of health services and programs
- collaboration with other organisations, including non-government organisations and the public, on shared public health concerns
- coordination with educators and providers to ensure an appropriately skilled public health workforce
- establishment of Health in All Policies initiatives (South Australia)
- establishment of health promotion foundations (e.g. VicHealth in Victoria).

LOCAL GOVERNMENT

Role: monitoring of food safety, environmental hazards, community, cultural and recreational development, maintaining roads, land use planning, provision of community services, local economic development. Some jurisdictions (South Australia and Victoria) have a legislative requirement to produce local public health plans.

NON-GOVERNMENT ORGANISATIONS, PROFESSIONAL AND COMMUNITY ORGANISATIONS

Role: lobbying, information provision, research, advocacy, policy development, fundraising. Key groups include:

- disease-focused groups (e.g. National Heart Foundation, Cancer Councils, Asthma Foundation)
- specific group–focused (e.g. Australian Women's Health Network; National Rural Health Alliance)
- consumer groups (e.g. Consumers Health Forum)
- National Association of Aboriginal Community Controlled Health Organisations
- Health Promotion Foundations (funded by tobacco taxes)
- professional associations (e.g. Public Health Association of Australia, Environmental Health Australia, Australasian Epidemiological Association).

PRIMARY HEALTH CARE PROVIDERS

Role: screening, individual and group health education, community development, advocacy, involvement in public health planning. Key groups include:

- Primary Health Networks (regional coordinating function)
- general practitioners
- community health centres.

UNIVERSITIES AND RESEARCH INSTITUTIONS

Role: teaching and research.

These institutions provide undergraduate and graduate education in health promotion and public health, and conduct research on a wide variety of public health topics.

COAG established bilateral Public Health Outcomes Funding Agreements (PHOFAs), which enabled 'broad banding of funding for public health activities' between 1997 and 2009. These no longer exist and public health activities are included in other COAG agreements, such as Closing the Gap in Indigenous Disadvantage. The Australian Government Department of Health also has less focus on its public health activities than in the past, and in 2014 the Abbott government cut funding for some significant health promotion programs.

RESEARCH FOR THE NEW PUBLIC HEALTH

The NHMRC, which funds the majority of public health and biomedical research in Australia, has increased the amount of funding allocated to public health and health services research through schemes such as the Centres of Research Excellence funding scheme. This scheme has a public health stream, but it is still a fraction of the funding given to biomedical research. Despite these improvements there is still much scope to improve the support given to research on the social and economic determinants of health (Baum et al., 2013a). One of the five recommendations of the Senate Community Affairs References Committee looking at Australia's domestic response to the Commission on Social Determinants of Health report was that the NHMRC should 'give greater emphasis in its grant allocation priorities to research on public health and social determinants research' (Senate Community Affairs Committee, 2013).

PREFERENCE FOR SELECTIVE PRIMARY HEALTH CARE AND LIFESTYLE HEALTH PROMOTION

The Howard Coalition government introduced a series of measures to strengthen general practice in Australia. While this led to some positive developments, these were at the expense of the development of a more comprehensive primary health care sector. A General Practice Reform Strategy (Weller and Dunbar, 2005) was established in the early 1990s, which injected considerable resources into one part of the primary health care sector. This was a missed opportunity to establish a stronger and more effective primary health care system with a new public health mandate that emphasised community participation, multidisciplinary team work, a focus on communities rather than individuals, work across sectors and health promotion. The most significant development was the establishment of Divisions of General Practice. Their establishment has meant that, for the first time, general practice had effective, resourced, regionally based organisations that could interact with other health care providers funded by the Commonwealth or the states.

For the first time in Australia the Divisions of General Practice encouraged GPs to come together and take a population view of the community they serve, opening opportunities for them to integrate more effectively with hospitals and community-based health services. The Phillips review of the Divisions program (Phillips, 2003) concluded the Divisions had helped coordination of care but also called for them to:

- address broader primary health care issues
- be more involved in population health including the reduction of health inequities
- engage more with Aboriginal community-controlled health services
- maintain a focus on supporting general practitioners and their practices
- become more accountable to the community for their performance (Weller and Dunbar, 2005, p. 8).

The next change to primary health care has come in the wake of the National Health and Hospitals Reform Commission established by the Rudd Labor government to make recommendations about the future of Australia's health system. One hope was that this commission would go some way to improving the fragmented governance and funding of the system between federal and state governments and so reduce the cost- and blame-shifting that has been the focus of concern for decades. This has not eventuated, but the Gillard Labor government established Medicare Locals, which replaced the Divisions of General Practice with a more comprehensive approach to primary health care. By July 2012, 61 Medicare Locals were established (Department of Health and Ageing, 2012). They were responsible, in collaboration with Local Health Networks (which manage hospitals and state-funded primary health care services), for coordinating the disjointed primary health care services to ensure patients have access to quality services in a timely and affordable way. Medicare Locals were expected to develop a population health plan by identifying needs, developing and implementing strategies, and monitoring and evaluating outcomes. These 'Healthy Communities Plans' had the potential to provide a regional focus for new public health planning and activities.

However, one unfortunate outcome was for some state governments to withdraw from the provision of health promotion and other community health services. Thus, in Queensland, very significant cuts were made to community nutrition programs and other health promotion services (Sweet, 2012). In South Australia, the government accepted recommendations to cut a range of community health services on the basis that the Medicare Locals would replace them, and abolished the health promotion branch within the Department of Health (Hughes, 2013). The Medicare Locals turned out to be short-lived, as in April 2014 the Abbott Coalition government announced they were to be replaced from June 2015 with Primary Health Networks, which appear to have a more clinical focus and little role in population health.

In Victoria, Primary Care Partnerships (PCPs) aim to improve the health of Victorians by engaging consumers, carers and communities in the planning and evaluation of services; improving health promotion, early intervention and continuity of care; and reducing the use of hospital services; all underpinned by a social model of health that begins to address social determinants. In 2012 there were 31 PCPs linking 1200 organisations, resulting in a more integrated and coordinated sector (Victorian Government Department of Health, 2010). Community health plans are based on population needs: they identify the strategic objectives, and map the partnerships that need to be forged, and the service systems and infrastructure that need to be better integrated.

The performance of regional Primary Health Care organisations such as the Medicare Locals and Primary Health Networks (Department of Health, 2014a) that have replaced them will depend in large part on how successful they are at getting GPs and state-funded primary health care services working together. A review of general practice and community health concluded that if such collaboration is to be encouraged (and the reviewers saw this as desirable), then it would be important to fund programs that explicitly encourage multidisciplinary teamwork (Fry and Furler, 2000). The collaboration will not happen without such incentives.

The community health sector, because of its philosophy and funding model, has been able to take a broader population view of health and put more emphasis on health promotion and disease prevention. While the types of health promotion in community health vary, very few emphasise behaviour change in isolation as a key strategy for improving health, and many have reflected the philosophy of the new public health (Baum, 1995a; Legge et al., 1996). However, these services have come under increased pressure to provide care for chronic conditions and this means that their ability to undertake broader health promotion is being restricted (Baum et al., 2013c). Victoria appears to be the only state that is increasing its investment in health promotion (as is shown by the range of services on offer at the Inner South Community Health Service in Melbourne: www.ischs.org.au).

Three main initiatives shaped health promotion strategy under the Rudd and Gillard governments: COAG's *National Partnership Agreement on Preventive Health* (NPAPH) (COAG 2008b); the report of the National Preventative Health Taskforce (2009a); and the Australian Government's response to the Taskforce report (Australian Government, 2010). COAG's NPAPH aimed to reform 'Australia's efforts in preventing the lifestyle risks that cause chronic disease' (2008b, p. 1). This included two programs, the 'Healthy Workers Initiative' ($289.4 million) and the 'Healthy Children Initiative' ($325.5 million), to promote improved health behaviours (especially diet- and exercise-related behaviours) in workplaces and among children. In 2011 a new Australian National Preventive Health Agency (ANPHA) was launched. These NPAPH agreements and ANPHA were cut by the Abbott government in 2014.

The Rudd–Gillard efforts in health promotion were welcome but have been criticised as being within a framework that primarily relates to behaviour rather than social health (Baum, 2011; Baum and Fisher 2011). Popay et al. (2010) note that health policy often starts by recognising the importance of the social determinants of health, but then a 'lifestyle drift' occurs whereby strategies are primarily concerned with behaviour change. This certainly appears true of recent Australian health promotion policy.

Where Australia was leading the world in terms of the new public health in the 1980s, this position can no longer be securely claimed. Perhaps one exception is VicHealth (see box 3.7).

BOX 3.7   VICHEALTH: HEALTH PROMOTION THAT TAKES NOTE OF SOCIAL DETERMINANTS

Since its establishment in the 1980s VicHealth has moved health promotion away from a focus on providing individuals with health messages in the hope that this would lead to behaviour change, to an approach based on changing communities and policies so that they are more supportive of health. This approach is well illustrated by its physical activity work, which has clearly emphasised creating supportive environments (for example, advocating the value of public transport) and working with planners and policy makers from a range of sectors. In recent years VicHealth has sponsored significant research on aspects of social determinants including an Aboriginal Health Research Unit, the effect of place on health and the development of health promotion programs for blue-collar workers (see www.vichealth.vic.gov.au for details). It also has campaigns on social connectedness, economic participation, preventing violence against women and reducing race-related discrimination. VicHealth also emphasises health equity (www.vichealth.vic.gov.au/Programs-and-Projects/Health-Inequalities/HI-Programs. aspx) and has developed a 'Fair Foundations' framework to guide its work in this area (VicHealth, 2013).

The social determinants of health inequities: the layers of influence

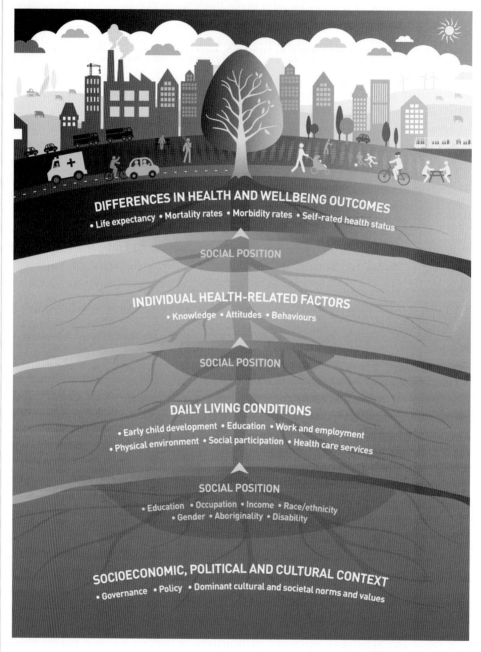

DIFFERENCES IN HEALTH AND WELLBEING OUTCOMES
• Life expectancy • Mortality rates • Morbidity rates • Self-rated health status

SOCIAL POSITION

INDIVIDUAL HEALTH-RELATED FACTORS
• Knowledge • Attitudes • Behaviours

SOCIAL POSITION

DAILY LIVING CONDITIONS
• Early child development • Education • Work and employment
• Physical environment • Social participation • Health care services

SOCIAL POSITION
• Education • Occupation • Income • Race/ethnicity
• Gender • Aboriginality • Disability

SOCIOECONOMIC, POLITICAL AND CULTURAL CONTEXT
• Governance • Policy • Dominant cultural and societal norms and values

VicHealth

SPECIFIC POLICY AREAS IN THE PAST 25 YEARS AND THEIR FIT WITH THE NEW PUBLIC HEALTH

Perhaps also reflecting a selective approach to health provision, a feature of Australian Government health policy since 1990 has been a series of policies relating to specific groups or issues. These are shown in table 3.5, which also indicates the extent to which they reflect new public health principles, including attention to the social determinants of health. The states and territories also make an important contribution to public health, as we saw in box 3.6. Examples of the programs and initiatives from Australian, state and territory governments are given in parts 6 and 7.

TABLE 3.5   KEY AUSTRALIAN GOVERNMENT HEALTH POLICIES, 1989–2014, AND THEIR FIT WITH THE NEW PUBLIC HEALTH

Policy area and policies	Fit with new public health
Women's health National Women's Health Policy and Program (1989) Australian Longitudinal Study on Women's Health (ALSWH) National Women's Health Policy (2010) (www.health.gov.au/internet/main/publishing.nsf/Content/national-womens-health-policy)	Australia was the first country in the world to develop such a policy and it led to a wide range of programs. It recognised equal pay, equal opportunity, child care and violence against women as important issues in women's health. The policy lapsed under the Howard government. The new policy in 2010 acknowledges social determinants and addresses some of them to a limited extent in terms of employment and education. It also focuses on groups of women with worse health status, including Aboriginal women and women with disabilities.
Men's health National Male Health Policy (2010) Ten to Men is the first national longitudinal study in Australia focusing exclusively on male health and well-being (www.tentomen.org.au/The-Study/the-study.html)	The first men's health policy in Australia, it acknowledges the social determinants of male health. 'Men's sheds', a program under the policy, addresses the mental health needs of men.
Aboriginal health National Aboriginal Health Strategy (1989) National Indigenous Reform Agreement (Closing the Gap) (2008) National Aboriginal and Torres Strait Islander Health Plan 2013–2023	Aboriginal health policy in Australia has paid more attention to new public health principles than many other areas, as shown by its focus on social determinants, including education and employment.
Mental health National Mental Health Policy (1992) National Action Plan for Promotion, Prevention and Early Intervention for Mental Health (2000) National Mental Health Policy (2008)	These policies focus on improving treatment and de-stigmatising mental health, and recognise that prevention of mental illness is important. They acknowledge the important role of carers and the significant emotional, social and economic burden they may bear. However, there is hardly any emphasis on maintaining a mentally healthy population.

(continued)

TABLE 3.5   KEY AUSTRALIAN GOVERNMENT HEALTH POLICIES, 1989–2014, AND THEIR FIT WITH THE NEW PUBLIC HEALTH (*CONTINUED*)

Policy area and policies	Fit with new public health
Rural health National Rural Health Strategy (1994) Healthy Horizons: A Framework for Improving the Health of Rural and Remote Australians National Strategic Framework for Rural and Remote Health (2012)	These strategies and frameworks outline a national approach to policy, planning, design and delivery of health services in rural and remote communities. They note the inequities between the health of rural and remote residents compared with urban dwellers. There is an acknowledgment that social determinants affect health and health access, but no strategies to deal with these issues. Overall, there is a strong focus on health services and little attention to population health or new public health perspectives.
Dental health National Oral Health Plan 2004–13	Recognises that poor oral health is most evident among Aboriginal and Torres Strait Islander peoples, people on low incomes, rural and remote populations, and some immigrant groups from non-English speaking backgrounds, particularly refugees. The plan is designed to increase access and emphasise the role of prevention, as well as the social impact of oral disease.
Legal and illegal drugs National Campaign Against Drug Abuse (started in 1985) National Drug Strategy 2010–15	The current strategy includes three pillars: demand reduction, supply reduction and harm minimisation. The main strategies are workforce capacity, evidence-based and evidence-informed practice and evaluation, and partnerships, as well as plain packaging for tobacco products. The strategy acknowledges the impact of social determinants on drug use.
Communicable diseases The Second National Hepatitis B Strategy 2014–2017 The Third National Sexually Transmissible Infections Strategy 2014–2017 The Fourth National Hepatitis C Virus Strategy 2014–2017 The Fourth National Aboriginal and Torres Strait Islander Blood Borne Viruses and Sexually Transmissible Infections Strategy 2014–2017 The Seventh National HIV Strategy 2014–2017	Reasonably comprehensive strategies to deal with these major infectious diseases but they each give much more attention to treatment than to prevention.

Policy area and policies	Fit with new public health
Non-communicable diseases National Preventive Health Strategy (2009) A Healthy and Active Australia website (www.healthyactive.gov.au)	There has been a much greater focus on non-communicable diseases in the twenty-first century as rates of obesity, diabetes and cardiovascular disease increased. However, most policies take a primarily behavioural or medical approach and pay little attention to the social determinants of health. One exception was the plain paper packaging for tobacco legislation.
Injury prevention National Injury Prevention and Safety Promotion Plan, 2004–14	The plan focuses on prevention across the life course from intentional (assault, suicide) and unintentional injury.

Because these policies are focused on one issue or one population group they also have the potential of distracting from a more comprehensive primary health care and health promotion sector. These programs usually mean that community health services or other agencies apply for funding for projects which generally have a limited life, and this means they have less potential to plan for services over a longer time scale. The area of health policy that best incorporates new public health principles has been that of Aboriginal health. The National Aboriginal Health Strategy, which was based on extensive consultation with Aboriginal and non-Aboriginal groups and communities, recognised the inextricable link between issues of dispossession, land rights and the history of colonial domination, and argued strongly for Indigenous control of health services and research. This document provided excellent assessment of Aboriginal health needs, and while the changes it has brought about have not been anything near as significant as they should have been, it has signalled a period in which Aboriginal health has attracted more attention from policy makers than it did previously. Aboriginal community-controlled health services are now well established and there is increasing evidence they are effective (see box 18.3 for more details). Both the Public Health Association of Australia and the Australian Medical Association have been particularly vocal in relation to Indigenous health issues, and their advocacy has been for a public health approach rather than a purely medical one. A major focus on Aboriginal and Torres Strait Islander health was made with the Australian Government's commitment to a national strategy to close the gap in life expectancy between non-Aboriginal and Aboriginal Australians. Most recently, a new National Aboriginal and Torres Strait Islander Health Plan 2013–2023 (Department of Health, 2013c) has also adopted a new public health approach. Its implementation is continuing under the Abbott Coalition government but the funding to support the implementation is under question.

One of the few areas of initiative and growth in the 1990s and early twenty-first century has been rural health. Drought, the downturn in rural industries, recession and depopulation focused attention on the needs of rural people. Health units, national rural health conferences and a series of policy frameworks for rural health (most recently the National Strategic Framework for Rural and Remote Health (Department of Health, 2012)) have increased national focus on this issue. The National Strategic Framework sets out as its vision that 'People in rural and remote Australia are as healthy as other Australians'. The strategy acknowledges the rapid change in social and economic circumstances of rural communities and the importance of a public health and primary health care perspective. However, there are particular difficulties, including the need for improved communications, means to overcome isolation, specialist training for health professionals and strategies to improve the recruitment of health professionals to rural and remote areas. The National Rural Health Alliance (http://nrha.ruralhealth.org.au) has proved very effective as an advocate for rural health and in keeping the issue on the political agenda, as well as advancing practice knowledge in the area.

Mental health has also been a significant policy focus in the last 25 years with a series of policies since the first in 1992. Most recently the National Mental Health Policy 2008 (Department of Health, 2009, p. 2) aims to:

- promote the mental health and well-being of the Australian community and, where possible, prevent the development of mental health problems and mental illness
- reduce the impact of mental health problems and mental illness, including the effects of stigma on individuals, families and the community
- promote recovery from mental health problems and mental illness
- assure the rights of people with mental health problems and mental illness, and enable them to participate meaningfully in society.

Beyondblue (www.beyondblue.org.au) has been a major national initiative that has played a very significant role in reducing the stigma associated with mental illness and helping it be understood, acknowledged and addressed by the wider community. It has focused on the prevention of mental illness and supported initiatives such as a campaign to create mentally healthy workplaces.

The National HIV Strategy, implemented since the 1980s, has been recognised as one of Australia's most successful public health initiatives, and is discussed in chapter 20. The National Drug Strategy is discussed in chapter 24.

HOW MUCH DOES AUSTRALIA SPEND ON PUBLIC HEALTH?

Australian community and public health activists have long complained about the imbalance in expenditure on hospital and diagnostic services and that on community-based caring, rehabilitation, disease prevention and health promotion.

Most commentators on the historical and current health systems in Australia agree that more emphasis should be placed on public health and primary health care activities (Sax, 1990; Bates and Linder-Pelz, 1990; National Health Strategy, 1992; Palmer and Short, 1994; Swerissen and Duckett, 1997; Wass, 2000; National Preventative Health Taskforce, 2009a; Duckett and Willcox, 2011). Currently, of the total expenditure on health goods and services, just 1.7 per cent is spent on public health and only 5 per cent on community health and other services, compared with 35 per cent on hospitals (see figure 3.3).

FIGURE 3.3 RECURRENT HEALTH EXPENDITURE, BY AREA OF EXPENDITURE AND SOURCE OF FUNDS, CURRENT PRICES, 2012–13

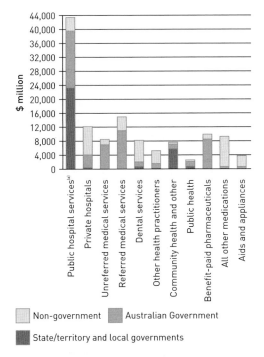

(a) Public hospital services exclude certain services undertaken in hospitals.
Can include services provided off-site, such as hospital in the home and dialysis.

Source: AIHW, 2014b, p. 38, figure 3.2.

There is practically no disagreement in the literature that the public and community figures should be increased, yet putting it into practice is proving difficult in the face of an intransigently curative health system. The tight fiscal environment has meant that departments have concentrated on core business (curing people) and shown less interest in the broad agenda established by the new public health. The preoccupation with cost-cutting and reorganisation within the federal and state health departments means that public health's agenda has not been significantly advanced and increasingly is receding. Opportunities have generally not been taken up by policy makers. While there are some instances of innovative public health practice within health services, these have had a marginal impact on the health system as a whole. Similarly, while successive reviews of health services have recommended a reorientation of the system to primary health care, none have made much progress towards achieving this (Dwyer, 2004). Boxall and Leeder (2006) and the National Health and Hospitals Reform Commission (2009) note that, if the system is to deal with increasing rates of chronic diseases such as diabetes, cancer and arthritis, there needs to be a radical reorientation of health service organisation in Australia in a direction that encourages better coordination of community-based care, while also encouraging a focus on health promotion and disease prevention.

Pressures towards privatisation are not abating (The Lancet, 2014). The rebate for those holding private health insurance, supported by successive Australian governments, is evidence of a policy direction that encourages less emphasis on strengthening the public sector and more on supporting private provision of health care. The AIHW (2005, 2014b) reports that this subsidy has risen from a cost of $2.7 billion in 2004–05 to $5.1 billion in 2012–13. Each time the private insurance companies put their prices up, so the cost of the government's rebate rises too. These are funds that could create an effective comprehensive primary health care system that could play a major role in implementing a new public health agenda. The subsidy enjoys bipartisan support and is evidence that the two dominant political parties in Australia accept the international trend towards increasing the privatisation of health care. In 2014 the idea of extending private health insurance cover to general practice was raised in a number of press reports, and four of the Primary Health Networks established in 2015 included private health insurers in their management.

Protests against privatisation of the Australian health system. (Fernando M. Gonçalves)

For the new public health, the move to privatised health care poses many difficulties. In particular, privatised provision of health care is likely to lead to a two-tiered system whereby people who are more affluent can afford private services, and poorer people generally rely on the public system, which would be starved of funds because health spending would be directed to the for-profit sector. In this environment, public services move away from the principle of universalism and become residual services. Almost inevitably, this affects the quality of the services, equity of access to them and, in time, equity of health outcome. Population health planning and policies would also be downgraded by a for-profit system that stands to make most profit for the provision of health services, not by keeping the whole population well and promoting equity.

CONCLUSION

Public health in the twenty-first century has to build on the significant developments of public health policies, practices, research and training that have achieved notable progress in promoting health, extending life and improving the quality of life in the past. These have been achieved through whole-of-government action in improving living standards. Now that the threats to public health are global—including a global obsession with reductions in the role of government and emphasis on privatisation of core services, and a deteriorating physical environment—the response must be increasingly global. The Industrial Revolution in the nineteenth century had a dramatic effect on the health of populations undergoing the process: states and civil society movements were forced to act in defence of health. They did this very effectively, as shown by the history of public health described earlier. A similar process now has to happen for the new public health. The contemporary movement will be increasingly global as well as national in nature because the threats to health and well-being are global.

The importance of a social perspective on health has been reinforced by the report of the Commission on Social Determinants of Health (2008) and subsequent documents from WHO (2011, 2013b). The imperative will be to keep this focus of public health strong in the face of a renewed emphasis on behavioural and genetic approaches to public health. Public health problems are increasingly complex, especially the environmental and social threats to health. Global growth in human population and economic activity seems paradoxically both unstoppable and unsustainable. International economic trends are less supportive of health, especially since the Global Financial Crisis, and inequity both within and between countries appears to be increasing. Governments, with few exceptions, are committed to reducing expenditure, and privatising and contracting out services.

These factors pose huge challenges for public health, but they also make the new public health agenda even more relevant. Part 2 examines the political economy of public health and demonstrates that the overall policy frameworks within which governments operate and the ideologies they hold are crucial to the new public health. It also examines neo-liberalism and considers the impact it is having on health. The challenge for the future is for individual countries to develop effective broad agendas to promote health and well-being internally and to work together to establish effective international governance to manage the increasingly manufactured and global risks to public health.

While the new public health has met some rough terrain in recent years, its basic philosophy, style of operation and commitment to social justice are likely to stand it in good stead in the coming years. The rest of this book is dedicated to understanding why its implementation is so important and what positive steps can be taken to ensure that its positive vision of a healthy, sustainable, peaceful and equitable society is met.

CRITICAL REFLECTION QUESTIONS

3.1   In what ways have politics influenced public health in the past three decades?

3.2   Do you think that the WHO Ottawa Charter for Health Promotion is still relevant to your country today?

3.3   To what extent do you agree that health systems should be based on primary health care?

3.4   Why do you think that comprehensive primary health care is likely to be more effective than a selective approach?

Recommended reading

Beaglehole and Bonita (2004) *Public health at the crossroads* provides an overview of the history of public health and a critique of developments. Part 1 provides an overview of global health status and determinants; part 2 a detailed consideration of the contribution of epidemiology; and part 3 a description of public health in poor and wealthy countries with speculation about its future.

Duckett and Willcox (2011) *The Australian health care system* provides a detailed guide to the structure and operation of the Australian health system.

Germov (2014) *Second opinion: An introduction to health sociology* provides a series of chapters on aspects of the health system, including on the social organisation of health care and issues of power and politics.

Useful websites

World Health Organisation, www.who.int/en
  The United Nations' health agency has led the development of international policies for the new public health.

Australian Institute of Health and Welfare, www.aihw.gov.au
  A great source of health sector data for Australia

World Bank database, www.worldbank.org
  Great source of comparative data on health systems and health statistics

VicHealth, www.vichealth.vic.gov.au
  A good example of an agency with an orientation towards the new public health

PART 2

POLITICAL ECONOMY OF PUBLIC HEALTH

> There is much merit in economic progress, but there is also an overwhelming
> role for intelligent and equitable social policies.
>
> *Amartya Sen, Nobel Prize-winning economist, 2001a, pp. 343–4*

Public health exists to make people and their communities healthier through change. Few people would disagree with the goal of promoting health but most would argue about how it should be achieved. Part 2 considers the political economy of public health. Interpretations of political economy are informed by people's values, experiences and ideologies, which are defined by Evans and Newnham (1992, p. 135) as 'sets of assumptions and ideas about social behaviours and social systems'. Consideration of competing ideologies and values can help explain why public health strategies do or do not succeed, which can reduce the frustration of public health practitioners when a strategy that made perfect sense from their ideological perspective is not accepted or does not work in a particular setting.

Tesh (1988) has argued that political beliefs and values have a defining influence on people's often implicit notions of disease prevention policy, and that this influence is exerted through 'hidden arguments'. She sees these implicit assumptions as fundamental: 'What is the legitimate source of knowledge? What is the nature of human beings? And what is the ideal structure of society?' She comments: 'Firmly, but often unconsciously, held answers to these questions guide scientists, policy makers, and ordinary citizens alike to different constellations of facts about the causes of disease and, hence, to different preferences for prevention policy' (Tesh, 1988, p. 3).

This idea about differing pathways leading to differing understandings affects all public health activity. The values and politics within a society help people to interpret and make sense of seemingly objective facts. Recognition of these values is important and their role in public health policy should be openly debated. Indeed, their central importance to public health suggests that practitioners would benefit from clarifying their own values, determining how they affect their world view and by being aware of the values and motives driving other players. Useful questions to pose in the analysis of public health situations are: 'who gains and who loses by any particular action?' and 'whose knowledge base is being used to support a particular course of action?'

Some students may express frustration with the relativities and uncertainties expressed in this part of the book. It is clear that public health is as much an art as a science. Some public health practitioners have approached their work as though it were a value-free activity based on proven scientific 'fact' and tried and tested practice. Others, particularly those with some training in the social sciences, approach their public health work recognising that knowledge is usually culturally and temporally specific; that it reflects gender, class, cultural and ethnic perspectives; that often 'accepted wisdom' is such because it reflects the practices and views of powerful groups within society. Students of ethical theory will recognise a number of distinct and often contrasting ethical stances underpinning common ideological and political positions in relation to public health. Ethics is essentially about what is the right thing to do: a deceptively simple question that we all know to be infinitely complex. It is immediately obvious that the 'right' thing to do depends on who we are considering.

Empirically, however, as a very crude rule of thumb it seems that those groups with the most power are the least likely to question or challenge the basis of their knowledge and understanding. Thus, women challenged male domination, the colonised challenged colonialism, non-medical practitioners challenged biomedical domination of public health, and Indigenous peoples challenge 'whiteness' as a dominant world view.

The history of public health and the evolution of the new public health demonstrated that public health has always been a contested and disputed area of practice and knowledge. Most significantly there has been a tension (not always creative) between medical understandings of health and those based on a more social interpretation. While the medical–social debate in public health has been relatively transparent, the tensions within public health policy and practice described in the next chapter are less visible and require more analysis to understand and lay bare. They are hidden and invisible, yet operate to influence public health policy and practice as surely as many more evident factors.

This part of the book is designed to provide an introduction to some of the key invisible factors that drive public health policy and practice. Chapter 4 examines the societal balancing act involving the often-conflicting forces of individualism and collectivism. This includes a discussion of the tendency of health and welfare policies to be based on a victim-blaming mentality, and touches on some of the principal ethical arguments relevant to public health. Chapter 5 considers the impact of neo-liberal globalisation on public health. The impact of the current system of regulating world trade through the World Trade Organization (WTO) and the International Monetary Fund (IMF) is described and its effects on health assessed. The themes from these chapters are revisited throughout the rest of the book and run as invisible streams through much public health policy and practice.

4

ETHICS, POLITICS AND IDEOLOGIES: THE INVISIBLE HANDS OF PUBLIC HEALTH

Medicine is a social science, and politics is nothing else but medicine on a large scale. Medicine, as a social science, as the science of human beings, has the obligation to point out problems and to attempt their theoretical solution: the politician, the practical anthropologist, must find the means for their actual solution.

Rudolf Virchow (1848), in his weekly medical newspaper Die Medizinische Reform, *quoted in Sigerist, 1941, p. 93*

KEY CONCEPTS

Introduction
Political systems and ideologies
Types of political systems
Growth of welfare states
Egalitarianism, socialism and capitalism
Ethical issues in public health
Roots of individualism
The dialectic between individualism and collectivism
Consequentialist and non-consequentialist ethics
Rights arguments
Victim blaming
Public health policies and individualism
Social-structural and communitarian perspectives
Individualism and the welfare state
Conclusion

INTRODUCTION

Public health is a political activity because it is about change, and its history shows that public health actions are expressions of prevailing political ideologies, the beliefs of those in government and the extent to which formal power holders are influenced by interest groups. Decisions to control harmful substances, to restrict individuals' behaviour, or take away their freedom of movement are invariably political and reflect the underpinning ideologies of those making the decisions. This chapter discusses:

- political systems and ideologies and their impact on public health
- ethical issues including the balance between the rights of individuals and those of society
- the impact of individualism on public health and welfare policies.

POLITICAL SYSTEMS AND IDEOLOGIES

Political ideologies and forms of government have varied considerably in the last few hundred years. European societies and their colonies have moved from forms of monarchy to a variety of more democratic governments. Political scientists analyse political systems and ask questions about society such as: Which groups hold power? How do they maintain legitimacy? How democratic is the system of government? To what extent are the rights of individuals protected? How much responsibility do governments take for the health and well-being of their populations? How equitable is the society? How does the political arrangement affect economic well-being? Which groups benefit most from the political arrangement? Answers to these types of questions are hotly debated by political scientists, who offer very different perspectives. It is not possible here to do more than alert the reader to the importance of political ideology. Political science should be a key public health discipline as it provides a framework for understanding these forces.

TYPES OF POLITICAL SYSTEMS

Walt (1994) cites Blondel's (1990) three basic criteria for the classification of political systems as reflecting points on the following dichotomies:

- democratic/undemocratic
- liberal/authoritarian
- egalitarian/inegalitarian.

It is always risky to offer generalisations about such complex entities as political systems, so please treat the typologies offered here with care. The new public health philosophy appears to assume a society that conforms to the left-hand side of Blondel's dichotomies: democratic, liberal and egalitarian. It envisages a society in which citizens can express their views through responsive institutions.

The potential for implementing the new public health differs over time and between societies. The political and economic context of countries plays a major role in shaping opportunities for health—some countries are much more committed to state intervention to promote health and reduce inequities than others, and the opportunities for the new public health are greatly shaped by the political systems countries have. The range of systems and their characteristics is shown in figure 4.1, which plots the ideological variation. The social democratic model (e.g. Sweden, Finland and Norway) is the most in tune with the aspirations of the new public health with its emphasis on universal rights and state intervention to reduce inequities. More right-wing systems (e.g. Australia, Canada and the USA) are more committed to a market model and less committed to intervention. Australian political orientation has moved between the right wing and centre over the past three decades without great variation in the ideologies of the parties, although the Australian Labor Party has generally shown more sympathy with intervening to protect health than the more conservative Coalition.

FIGURE 4.1 POLITICAL AND ECONOMIC SPECTRUM

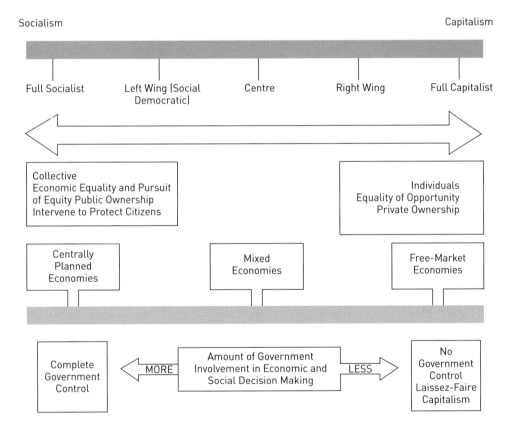

GROWTH OF WELFARE STATES

Social democratic and Liberal states societies have distinct ideological differences between political parties. Broadly speaking, the political ideologies on the right of the spectrum (exemplified by the US Reagan and Bush governments, the UK Thatcher and Cameron governments, the Australian Howard and Abbott Coalition governments and the Canadian Harper Conservative government) believe in the power of the market to meet the needs of people, and stress individual responsibility. Those move to the left (social democratic Labour governments) are more likely to believe that the state should intervene to ensure that the capitalist system does not ignore the needs and rights of those who are not powerful in the market. They are also more likely to believe in the value of an institutional welfare state to redistribute income and protect the interests of the poor and vulnerable.

After World War II all social democratic and some liberal states agreed that state provision of services and intervention to curb the excesses of capitalism were desirable or pragmatic. As a result, welfare states of various sorts were developed in most of these countries, including the UK, Australia, New Zealand and Canada. The most advanced social democratic states were in Scandinavia. Social-democratic welfare regimes redistribute extensively and are by far the most successful in achieving long-term reductions in socioeconomic inequality, particularly across generations (Korpi and Palme, 1998).

The 1980s, however, saw this consensus questioned in the wake of the oil crisis and economic recession of the mid-1970s and the spread of neo-liberal policies. Unemployment was rising, the population ageing and budgets for welfare states were increasing. The USA, Britain and New Zealand all elected political leaders who gave more commitment to market economics and less to the aims of equity. Australian political parties in the postwar period have mainly been in the centre of the left–right spectrum.

The one exception was the Whitlam Labor administration (1972–75), which was more reformist, but was controversially dismissed by the Governor-General. The most striking feature of the Australian political scene has been the convergence of, rather than the difference between, mainstream political parties. The two main parties (Labor and Liberal–Nationals) have broadly similar policies, and so the Hawke and Keating Labor governments adopted elements of the neo-liberal economic policies but combined these with some commitment to social policy to alleviate the effects of the market. The Howard (Liberal–National) Coalition government (1996–2007) was a more radically new right government, borrowing its agenda from Thatcher and Reagan. Its policies conformed to a standard neo-liberal agenda of cutting and privatising public services, reducing welfare payments and increasing their policing, reducing the protection of workers and deregulating as much state activity as possible. The Australian Rudd and Gillard Labor governments showed more concern with equity, especially in terms of their determination to close the gap in life expectancy between Aboriginal and non-Aboriginal Australians, and extended the welfare state by introducing Disability Care in 2013 to provide for the needs of people with disabilities (but at the same time reducing welfare payments to single parents). The unfairness of its proposed policies in the 2014 Budget has been seen as a hallmark of the Abbott Coalition government, including a co-payment for GP services, and harsh welfare provisions including a six-month period in which those under 30 could not claim income support in the case of unemployment. Marston et al. (2013) assess the changes in Australia and note that there has been a move from a system where Australians were seen to have unconditional welfare rights based on the fact of their citizenship to one in which moral judgments about people's right to welfare have become prominent.

Several studies have compared the impact of different political regimes on health. Navarro and Shi (2001) looked at the impact of the major political traditions in the advanced OECD countries from 1945 to 1980—social democratic, Christian democratic, liberal and ex-fascist—and compared them on determinants of income inequality, levels of public expenditure and health care benefits coverage, public support of services to families and the level of population health as measured by infant mortality. They concluded that the social democratic political tradition that was more committed to redistributive economic and social policies and full employment was generally more successful in improving the health of populations, such as by reducing infant mortality rates. Similarly Coburn (2004) concluded that social democratic forms of welfare regimes (that is, those that are less neo-liberal) have better health than those that are more neo-liberal. He also found that neo-liberal globalisation is associated with increasing inequalities within rich nations. Bryant (2012) assessed the commitment to tackling inequity between different developed countries. She concluded that the Scandinavian countries had the most systematic and comprehensive policy response and that Britain, Northern Ireland and Australia had 'mid-level responses', while Canada and the USA had an apparent absence of government policy responses to health inequities.

The Commission on Social Determinants of Health (see box 3.2) recommended the establishment of 'universal social protection policies' that would support a level of income sufficient for healthy living for all citizens. It also noted that such schemes are instrumental to achieving development goals in low- and middle-income countries and would benefit local economies. Most crucially, it noted that the world has enough resources to guarantee such social protection to all citizens (see box 4.1 for further views on this point).

BOX 4.1   MORAL GIANTS ON THE ETHICS OF WELFARE AND EQUITY

The world has enough for everyone's need, but not enough for everyone's greed.

Mahatma Gandhi

Poverty and inequity are 'not a preordained result of the forces of nature or the product of a curse of the deities … [but] the consequences of decisions which men and women take or refuse to take.'

Nelson Mandela, quoted in Heywood and Altman, 2000, p. 173

William Beveridge (one of the founders of the welfare state) spoke of the moral absolutes that should govern government policy above considerations of cost. He argued for 'bread and health for all before cake and circuses for anyone' and stressed that it was better to incur debt than 'to let children go hungry or sick and old unattended.' (1944, pp. 26–7)

EGALITARIANISM, SOCIALISM AND CAPITALISM

Socialist and communist governments promote centralised control over the economy and many other aspects of society, and ban private, capitalist enterprises. They aim to create egalitarian societies in which the differences between rich and poor are minimised. The dissolution of the Soviet Union, the loss of political control of most communist parties in eastern Europe and the adoption of a market economy in China have meant widespread disenchantment with all forms of socialism. These political events of the early 1990s have been portrayed as signalling 'the end of history' (Fukuyama, 1992), because the battle of political ideologies that characterised much of twentieth-century history appeared to some to have been won by capitalism and the market. However, most political scientists present a more complex picture in which the debates about fundamental political philosophical issues concerning the ways in which societies are organised and the relative balance of democracy, egalitarianism and authoritarianism are still very much open. The 'end of history' view implies that there were two 'pure' forms of government—communist and capitalist. Liberal democracies have, in fact, been mixed economies with a varying level of intervention from governments, and have featured significant debates about the best mix between private and public ownership, and which sector performs which functions best (Stretton, 1987). Korten (1995, p. 88) reminds us that prosperity in Western industrial nations in the postwar period was achieved by a democratic pluralism in which there was 'a pragmatic, institutional balance among the forces of government, market and civil society'.

Questioning of capitalism as a dominant economic form has increased since the Global Financial Crisis of 2008. The Occupy movement has offered a civil society face of this questioning and more academic texts are critiquing the basis of capitalism and its impact on population health (Quiggin, 2012; Stiglitz, 2012). Despite this questioning of capitalism, most of the world's major economic powers are governed by political parties on the right or right of centre in 2014. There are now few national governments who have a strong commitment to social democracy and which see reducing inequity as a central business of government. There are also few signs that governments are linking economic growth to the risks it poses to climate change. As we see in chapter 14, the evidence linking human activity to increasing carbon emissions and then to a warming climate grows stronger and stronger. Climate change is increasingly seen as the greatest moral challenge of our age and the most important public health issue we face (McMichael, 2013). Thus the vital political question is what policies will create conditions for human flourishing within the ecological limits of a finite planet? The urgency of the potential for climate disaster makes this the central political question for governments of all political persuasions and for all public policies, public health included.

Having outlined the nature of political systems we now move on to examine some of the ethical principles underpinning the new public health.

ETHICAL ISSUES IN PUBLIC HEALTH

INDIVIDUALISM

The drift of liberal-democratic states towards the right and the widespread adoption of neo-liberal policies since around 1980 has made individualistic ideologies more influential in the shaping of public policy. Individualism holds individuals totally responsible for their actions and the consequences, including health. In philosophical and ethical terms, its key principle is autonomy.

AUTONOMY AND PATERNALISM

'Autonomy' is derived from the Greek autos (self) and nomas (rule), and originally referred to self-rule or self-government in Greek city-states (Beauchamp and Childress, 2013, p. 57). In ethical terms, it refers to an individual's capacity to make free choices and ability to control the direction of his or her own life. Paternalism, on the other hand, refers to 'practices that restrict the liberty of individuals, without their consent, where the justification for such actions is either the prevention of some harm they will do to themselves or the production of some benefit for them that they would not otherwise secure' (Beauchamp, 1988). Many actions, rules and laws are justified by appeal to the paternalistic principle. Benign examples are laws that enforce seat belt legislation and speed control to protect drivers, and the restriction of the sale of tobacco to children. A less benign example was the Australian Government policy of removing Aboriginal children from their families so they could be brought up in a 'civilised' manner in white Australian homes. In the wake of terrorist attacks the USA, UK and Australian governments have passed laws that curtail the rights of individuals in what some consider to be quite draconian ways: for instance, the increased length of time in which people can be held without charge, and the increased powers to tap phones and monitor email communication. These laws are justified in terms of protecting the population from terrorist attacks but raise significant issues about restricting the rights of individuals. The revelations from Edward Snowden (former US computer professional) who leaked classified information from the National Security Agency showed the extent of US surveillance on phone and internet communications.

When diphtheria and typhoid were common, it was usual to isolate whole households in which even a single case was found (Last, 1998, p. 369). Thus, the freedom of the individual was curtailed for the perceived benefit of the wider community. It is also common practice to trace the sexual contacts of people with a sexually transmissible disease, even though this may infringe their right to privacy. At the start of the HIV epidemic it was not uncommon to hear calls for the isolation of people living with HIV, despite the obvious violation of their autonomy. In the past, this was commonly the practice applied to people with leprosy. The 2014–15 Ebola epidemic in West Africa has seen controversy over quarantine of doctors and nurses returning from Africa to other countries. In the USA, a returning nurse defied a state-imposed quarantine, saying that the science did not support the measure (Bidgood and Philipps, 2014). Autonomy is not always easy to define. Moore (1992) argues that, while conventional health promotion would see 'alcohol misuse' as a problem for Aboriginal people, it may also be seen as an act of rebellion against white paternalism. Drinking, in this view, can in part be interpreted as a means of asserting autonomy and 'an essentially political act' (p. 187) that challenges white attempts at health promotion.

A central concern of public health and health promotion has been the process of describing and quantifying risks to health. Empowerment and autonomy, in the view of the new public health, are increased when individuals have information about the risks to their health from their environment and their behaviour. Lupton (1995) deconstructs the discourse of risk in contemporary public health and argues that the experience of being labelled as 'at risk' may be detrimental to people's health status, and thus affect their autonomy because of the stigma associated with being a member of an 'at risk' group. Certainly people who are obese report feeling stigmatised and blamed for their weight, which Patterson and Johnston (2012) see as resulting from a 'moral panic' over obesity. Finally, recent developments have seen initiatives in which people have been paid to adopt healthy habits, such as paying women to give up smoking in pregnancy (Wolff, 2014), which represents a strong intervention by the state to influence behaviour. It is possible such measures enable people to resist peer-group pressure by saying they are changing their behaviour 'just for the money'.

ROOTS OF INDIVIDUALISM

Kingdom (1992) finds the roots of individualism in the writings of Thomas Hobbes, the English seventeenth-century political thinker who wrote in the context of a post-feudal society in which individual rights were being asserted against those of a powerful monarch. Hobbs portrayed people as acting in accordance with certain psychological principles, and especially that the instinct to avoid death supersedes all others. He saw individuals continually weighing up the costs and benefits of particular actions to their own well-being. In his view of the world, life was about individual self-interest. Locke, a seventeenth-century philosopher whose writings had an important influence on the drafters of the US Constitution, also strengthened the cause of individualism. He maintained that government's role was to protect people's natural rights to life, limb and liberty, but to leave people alone as much as possible. Kingdom (1992) pointed out that Locke saw capitalism as natural, and the unequal possession of property as a natural right that people bring to society, not one created by the state. He laid the path for the classical economics of Adam Smith, Thomas Malthus and David Ricardo, whose basic argument was that in a marketplace individuals will act in their own interests. Eventually, however, the market would make everyone better off. Collectively, the writings of these three philosophers made a case for the importance of allowing 'natural law' to dictate social and economic development. In chapter 3 the recent expression of these philosophies through neo-liberalism are discussed.

THE DIALECTIC BETWEEN INDIVIDUALISM AND COLLECTIVISM

The dialectic between individualism (autonomy) and collectivism (with its implications of paternalism) as a basis for understanding social and community organisation is one of the most fundamental to grasp and explore in the development of public health concepts and strategies. Each public health issue highlights expression of this dialectic. Should we control gun ownership or is the choice of owning a gun an individual matter? Should we restrict or ban substances such as tobacco that are dangerous to human health or can individuals make up their own minds about the risks? Are the causes of occupational stress located in individuals or do they reflect poor job conditions? Are many Australians overweight because of too much unhealthy food, or because high fat, high sugar food is consistently marketed by food companies?

How can we ensure that the measures we take will do no harm (non-maleficence) and preferably benefit others (beneficence)? In some circumstances public health measures may be injurious to individuals—such as the risks to a small number of children from pertussis and measles vaccinations—yet benefit the population as a whole. Last (1998, p. 354) estimates the risk of adverse effects of measles immunisation as 30 per 100 000 for convulsions and between 0.0 and 0.3 per 100 000 for encephalitis. These risks (maleficence towards some unknown individuals) are judged to be acceptable. Immunisation is voluntary in Australia, but there have been calls to follow the example of some US states where children are not admitted to school unless they have been immunised, making it almost mandatory.

Public health practitioners in Australia tend to override the principle of autonomy in relation to immunisation. Thus the Public Health Association of Australia's policy on immunisation (PHAA, 2012) stresses the benefits of vaccination but does not mention the possible harms to a small proportion of those immunised. The key duty defined for public health officials is to ensure that all who agree to have their children immunised are aware of the risks—informed consent. There has been concern in Australia from public health doctors about the increasing number of parents who are opting not to have their children immunised. Lupton (1995, p. 86) points out that they are often represented as neglectful, ignorant, overly anxious or not fully aware of the risks to their children of contracting childhood illnesses. She quotes an interview survey of women in the north-west of England who had not taken their children for the full series of immunisation. This found that the participants' decision was often made for fully considered and 'rational' reasons relating to their everyday knowledge of the risks of vaccination. For public health practitioners, parents such as these women are frustrating because the practitioners see the women's desire for autonomy and non-maleficence as conflicting with their desire as public health practitioners to protect the health of the population as a whole.

Since the threat of a bird flu pandemic has been evident, discussions have been held about the powers that governments might use to quarantine people even if they resist in the interests of the broader public's health. Australia, for instance, has the *Quarantine Act of 1908*, which is still remarkably powerful. It can be used to close borders, ban public gatherings, commandeer a school for a fever hospital, and quarantine people against their will. In January 2007 a national radio program in Australia considered what is likely to happen in the event of a pandemic. They quoted a speech by the then federal health minister, Tony Abbott, in which he predicted that there could be up to 40 000 Australians dead, hospitals with insufficient beds or equipment to deal with the sick, people fleeing cities and even riots (ABC, 2007). The program further quoted Professor Alison Bashford, a public health historian, who warned that lessons have to be learnt from history of past epidemics as governments are faced with a delicate balance between questions of habeas corpus over questions of detention and the extent to which the state can intervene in individual lives for the sake of public good (ABC, 2007). Thus governments face achieving a delicate balance between autonomy and collective rights in the face of highly infectious pandemics and so in an atmosphere of considerable fear and hysteria.

Debate about drug control has also struggled with the issue of the maleficent consequences of public policy. Until recently, strategies to combat illegal drugs were based on prohibition, and sanctions were imposed on those who contravened the prohibition. Increasingly, drug and alcohol experts in Australia have argued that prohibition can be harmful and counter-productive to the aim of reducing the harm caused by drugs. As a consequence, the concept of harm minimisation attempts to balance the needs of the population for protection from harmful drugs with the needs of the already addicted. The US policy provides a contrast in that it is far more prohibitionist and penalises and demonises users (see chapter 24 for a fuller discussion of drug policy).

This dialectic between individualism and collectivism has strong political implications. Those promoting individualism tend to be on the political right and argue that state intervention is promoting a 'nanny state' that excessively restricts the rights of individuals. Those on the left are more likely to argue for measures to protect the collective good even when they impinge on individual rights. However, neo-liberalism (see chapter 5 for definition) has been accompanied by a growing paternalism in the provision of social services so that commentators say 'Neoliberal paternalism is transforming the human services into a disciplinary regime for managing poverty populations' (Schram and Silverman, 2012, p. 129). This indicates that a more right wing perspective can also lead to nanny state–style interventions.

CONSEQUENTIALIST AND NON-CONSEQUENTIALIST ETHICS

Discussions about autonomy and paternalism should be considered within a broader ethical context. There are two broad groups of ethical theories: consequentialist and non-consequentialist, or deontological. Consequentialism holds that most ethical decisions are based on a calculation of the good that derives as a consequence of a given decision. Traditionally, this is known as teleology, and is summarised as 'the end justifies the means'. Conversely, a non-consequentialist or deontological position is based on the view that decisions should be guided by a set of inherent moral principles that one has a duty to follow, regardless of context or consequence. Thus, the process rather than the outcome determines the rightness of an action.

Utilitarianism is a consequentialist theory that is summed up as 'the greatest good for the greatest number'. It has been used to support public health measures, including immunisation. However, while utilitarianism, as developed by John Stuart Mill and currently propagated by Peter Singer and others, is seen as a driving force behind liberal and reformist social policies, it is also criticised as being insensitive to the needs of the disadvantaged individual. For example, utilitarians are likely to approve of public health research that may infringe on the autonomy and privacy of individuals if they believe that the research is likely to benefit society as a whole. Deontologists would oppose it because of actual or potential violation of autonomy and right to privacy.

RIGHTS ARGUMENTS

'Natural' or 'human' rights arguments are essentially deontological, and are based on the concept of a set of natural rights being the birthright of every human being. It is argued that any action that violates an established human right is, ipso facto, immoral. However, as there is no agreement, even among those who hold a human rights position, as to what those rights might be, there is considerable criticism of this perspective. Liberal democracies have always valued the protection of individual rights, only introducing public health measures that threaten individual rights when no alternative exists. By contrast, regimes and cultures operating on a more collective basis are more easily able to adopt policies that restrict individual behaviour. An example is China's one child policy, which has led to a series of measures that influence fertility behaviour—one of the more extreme examples of paternalism. All societies juggle the competing forces of individualism and collectivism, as Bayer (1986, p. 172) demonstrated when discussing the AIDS epidemic: 'These two great abstractions, liberty and communal welfare, are always in a state of tension in the realm of public health policy.' In the past few years appeals to the human right to health have become more common. The People's Health Movement has developed a Health for All Campaign on strong arguments about the human right to health (www.phmovement.org/en/campaigns/145/page). A campaign has been launched to argue for the adoption of a Framework Convention on Global Health and is strongly rooted in a human rights framework. Its aim is to 'create a right to health governance framework. It would be a global health treaty based on the right to health and aimed at closing national and global health inequities … It would establish a transformative understanding of the right to health to create the accountability now missing and adapt the right to our globalized world' (www.globalhealthtreaty.org). The World Health Organization has a cross-cutting program that is dedicated to human rights and health (www.who.int/hhr/en). At a national level the Australian Human Rights Commission contains an Aboriginal and Torres Strait Islander Social Justice Commissioner who has led the campaign for Indigenous health rights and to close the gap in life expectancy between Indigenous and other Australians (www.humanrights. gov.au/close-gap-indigenous-health-campaign). The right to health can be used to assert the collective right to good health and provision of health care but it can also be used to argue individuals' rights to medical treatment that might come at the expense of the collective good. Thus a consumer group (possibly financed by a pharmaceutical company) might argue for an expensive end-of-life treatment whose cost–benefit ratio is questionable and has a high opportunity cost in terms of the necessary forgoing of alternative treatment or preventive measures that bring much wider benefits to a population.

VICTIM BLAMING

One of the direct consequences of individualism for public health is a tendency to blame victims for their ill health, seeing people as totally responsible for things that happen to them. Thus, the success of well-off people in the employment market is attributed to their particular efforts, rather than to the advantages of having affluent parents who were able to buy them the best education, provide a materially secure upbringing, ensure top-class health care and introduce them to the culture of the professions. By contrast, the ill health and inferior social position of poor people is seen as a reflection of their lack of effort and inability to succeed. The tendency to blame victims in the USA was noted by Ryan (1972), who commented (p. 5):

> The generic process of Blaming the Victim is applied to almost every American problem. The miserable health care of the poor is explained away on the grounds that the victim has poor motivation and lacks health information. The problems of slum housing are traced to the characteristics of tenants who are labelled as 'Southern rural migrants' not yet 'acculturated' to life in the big city. The 'multiproblem' poor, it is claimed, suffer the psychological effects of impoverishment, the 'culture of poverty' and the deviant value system of the lower classes; consequently, though unwittingly, they cause their own troubles.

Crawford (1984) points to an increasing individualism in health services through the 1970s, explaining it by the contradictions arising from the threat of high medical costs, political pressures for the extension of health services entitlement and the politicisation of environmental and occupational health issues. Crawford (1984, p. 75) argues that victim blaming means 'The emphasis on individual responsibility for health mystifies the social production of disease and undermines demands for rights and entitlements to medical care'. Health becomes an issue of individual responsibility. A review of studies of lay conceptions of cause of inequality in health status indicates that most people see that they lie in individual causes rather than the economic structure. Other commentators (Brown and Margo, 1978; Naidoo, 1986) note that individualism has been an important philosophy behind the health education and health promotion movements, which often concentrate on changing individual behaviour rather than the conditions that create ill health in the first place. Based on an analysis of medical journals, Skolbekken (1995) has noted 'a trend, resembling an epidemic' that has changed the explanation for illness from external to individualised factors. This means that the focus is shifted away from the social and economic causes of illness. Blaxter (1997) points out the impossibility of transferring population risk factors to individuals (for example, only 7 per cent of men at 'high risk' of heart disease actually develop trouble in the following five years). Rose (1992) points out that most smokers will not die from smoking-related causes even though across a population higher rates of smoking increases death rates. Individuals see that so-called 'high risk' candidates do not get the disease predicted by population data, and use this to support individualistic interpretations. Common to all processes of victim blaming is the distraction from the social construction of the problem.

Victim blaming inherent in much health education and health promotion creates a political smokescreen that masks a host of factors that are fundamental to the creation of illness: poverty, gender inequality, racism, occupational hazards and environmental pollution (Labonté and Penfold, 1981; Baum, 2011). Not surprisingly, victim blaming and individualism is particularly popular with governments whose philosophical roots are based on free markets unfettered by government control. The then Australian Government health minister, Tony Abbott, provided a good example of this when on a national television program on the increase in overweight and obese people in the community he blamed parents for the fact their children are obese (see quote from Abbott in box 4.2).

Victim blaming is usually simplistic. It is probably true that if people were to eat less fat, exercise more, buy safer cars, lead less stressful lives and avoid violence they would be healthier. The beguiling simplicity of the logic, however, ignores many extraneous factors that make change difficult to achieve, and ignores the social, cultural and economic context in which decisions are taken. Victim-blaming approaches to health promotion may be emphasised by health professionals whose main training, skills and techniques are based on individualistic perspectives, such as psychologists, doctors, speech pathologists and physiotherapists.

Clearly such a position is unjust, and, indeed, arguments based on the concept of social justice have been central to critiques of the victim-blaming approach by the new public health movement. Social justice is concerned with an equitable distribution of resources and respect for the rights of individuals, especially those vulnerable to discrimination. It also stresses that people's behaviour usually reflects the physical, economic and social environment in which they find themselves. Philosophically, social justice has links to Rawles' (1971) Theory of Justice and to communitarian ideas (see below).

Victim blaming in regard to health also assumes that health has a centrality to people's lives that may not be accurate. Health educators and promoters, understandably, see health (especially physical health) as a most desirable state, scoring above many others, but lay people may not see health as central. A stressed single mother may make a balanced choice to smoke because it calms her nerves and is one luxury she can afford. A businessman who knows he drinks too much alcohol may continue to do so because it is good for his business lunches and provides him with an easy and accessible way to relax. A young man may risk his life driving dangerously because he feels suicidal and does not much care whether he lives or dies. In each of these cases, health considerations are not part of the 'weighing up' process that people make in their everyday lives. Adler and Stewart (2009) propose the concept of 'behavioral justice', by which they mean that individuals are maintaining healthy lifestyles but should be held accountable only when they have adequate resources to do so.

Another feature of victim blaming is that those defined as victims may themselves identify with the label. Seabrook's (1984) perceptive and poignant account of life in a poor working-class community in the British Midlands noted this tendency and commented that the people he observed saw their problems 'not as socially produced, but as a personal visitation' (p. 33). The poor believe they are in difficult situations because of their lack of merit and worth in a society that assumes success can be put down entirely to individual effort. Thus, pervasive victim blaming contributes to low self-esteem, which in turn contributes to poor health (see chapter 13 for further discussion).

PUBLIC HEALTH POLICIES AND INDIVIDUALISM

Tesh (1988) has argued that individualism is the principal ideology affecting disease prevention policy in the USA. It seems that this is true, to a greater or lesser degree, for most industrial societies. Box 4.2 highlights ways in which ideologies can affect political decision-making by contrasting three politicians who put emphasis on structural factors with three who place the focus on individual factors.

BOX 4.2   INDIVIDUALISM VERSUS COLLECTIVISM: COMPARING DIFFERENT POLITICIANS

The perspectives of a range of politicians from political parties in different periods with variant ideologies in different countries highlights the key role ideology plays in public health policy. Consider what type of policy each politician is likely to pursue as you read their words.

> Health status is one of the most telling indicators of disadvantage in our community. People's opportunities and access to important resources, goods and services such as housing, employment and education can be reflected in the health or illness they experience ... a social view of health is one that recognises the impact (both direct and indirect) which physical, socio-economic and cultural aspects of the environment have on the health of the community. A social view of health implies that we must intervene to change those aspects of the environment which are promoting ill health, rather than continue to simply deal with illness after it appears, or continue to exhort individuals to change their attitudes and lifestyles when, in fact, the environment in which they live and work gives them little choice or support for making such changes.
>
> *John Cornwall, South Australian Minister of Health (Cornwall, 1988)*

> We must loyally declare that all those medical measures taken will only produce benefits if they are accompanied by economic and financial resolutions that permit a rise in the standard of living of our citizens. It can be said that the fundamental bases that determine the welfare and progress of nations are precisely a good standard of living, adequate sanitary conditions and a widespread dissemination of culture.
>
> *Salvador Allende in 1939, President of Chile 1970–73 (Allende, 2006).*

> No one is in charge of what goes into my mouth except me. No one, no one is in charge of what goes into kids' mouths except their parents. It is up to parents more than anyone

else to take this matter in hand ... if their parents are foolish enough to feed their kids on a diet of Coca Cola and lollies well they should lift their game and lift it urgently.

Tony Abbott, when he was the Australian Government health minister (ABC, 2005).

I don't think we can put people in cotton wool. I don't think we can cover our population in cling-wrap. I think people need to return substantial authority over how they live their lives. I think people need to be allowed to make mistakes. Sometimes we have a right to be wrong.

Tony Abbott, when he was the Australian Government health minister (Abbott, 2006)

Our approach is about taking preventative action to reduce the likelihood of people becoming dependent on costly health treatment later in life, when it can often be too late. However, to make a difference to Wales' health, people also have to take responsibility for their own lifestyle choices - government cannot make individuals be healthy.

Mark Drakeford, Health Minister, Welsh Government (2014)

We know that health outcomes will be affected by housing educations, educational outcomes, employment outcomes. And we have committed to working across portfolios within the Commonwealth and across Commonwealth and state boundaries. It's really about us having a concerted effort. We can't fix our rheumatic heart fever if we don't fix people's housing conditions. We're going to do both of those, and it means we have to do a lot of talking across portfolios

Nicola Roxon, when she was the Australian Government health minister, talking about Aboriginal health (ABC, 2009)

That's why today, I want to talk about troubled families ... we've known for years that a relatively small number of families are the source of a large proportion of the problems in society ... Drug addiction. Alcohol abuse. Crime. A culture of disruption and irresponsibility that cascades through generations ... We're talking about behaviour – the behaviour of individuals, the failures of families ... and the consequences of that behaviour for society.

You can't fully address that without a debate about the codes of behaviour people choose to live by ... I have said this many times, but I will say it again. We will not fix these problems without a revolution in responsibility ... a recognition that we need in our country a massive step change in accepting personal responsibility, parental responsibility, and social and civic responsibility too.

David Cameron, British Prime Minister (Cameron, 2011)

Minister Abbott, Drakeford and, Prime Minister Cameron focus on changing individual behaviour and Minister Cornwall, Roxon and President Allende on altering people's social and economic circumstances. Since 1980, policy agendas in many governments and international organisations have reflected a strong belief in individualism. The underlying philosophy was well expressed in Margaret Thatcher's now infamous statement that 'there is no such thing as society'.

Grace (1991) observed that the British political climate in the 1980s enabled the ruling elite to target health promotion resources almost entirely on individual behaviour change and recast the citizen as consumer within a model of health as a product of consumer capitalism. In a similar vein two decades later, Popay et al. (2010) noted that while health policies may start off with an acknowledgment of the social determinants of health, the policy solutions are cast in terms of behavioural change—a process they labelled 'lifestyle drift'. This focus on individualism is despite the increasing evidence (summarised by the Commission on Social Determinants of Health, 2008) that illness and health are produced by social, economic and environmental structures (see chapter 12), which make a focus on individual responsibility for health status less feasible. Improved equity and better overall health status require strategies that tackle the underlying causes of illness. Policies that emphasise victim blaming and stress strategies aimed at encouraging individuals to change their behaviour are extremely unlikely to challenge structural inequities (Evans et al., 1994a; Wilkinson, 1996). This is most evident in the world's least developed countries, where the absence of the basic facilities of clean water, adequate housing, sanitation and sufficient food make the structural nature of illness and health starkly evident. The advocates of the focus on individual behaviour are usually those who wish to argue for reduced responsibility by the state for improving health status, or professionals whose techniques and skills are biased towards individual interventions.

A strong tenet of neo-liberalism is that what is in the interests of private individuals ends up being for the good of society. This view was explicitly expressed by Gordon Gecko, the ruthless New York stockbroker portrayed in the 1980s film *Wall Street* as: 'Greed is good'. If accepted, it reconciles the ethical dilemma of the conflict between autonomy and paternalism, since what is good for the individual is also good for the population and vice versa. Beauchamp (1988) sees this as flawed. Instead he promotes the view that 'The good of society and the sum of private goods of the individuals who make up society are not necessarily the same thing' (pp. 83–4). He cites Geoffrey Rose's prevention paradox (see chapter 1 for details), which argues that many public health measures with considerable potential for populations as a whole have little benefit to individuals, especially in the short term. In the case of most public health risks, many people will need to change their behaviour but very few will gain a direct benefit. Public health requires a perspective that moves beyond the individual to consider populations as a whole, yet objections to public health measures are often made in terms of the rights of individuals. Beauchamp illustrates this by reference to gun control. He cites the slogan of the US National Rifle Association: 'Guns don't kill people, people kill people'. Yet, while it is true that most gun owners will be innocent or innocuous, the aggregate level of gun ownership contributes to thousands of injuries or deaths. Similarly, most people are rarely involved in car accidents and so may not think they benefit by wearing seat belts, but if nobody wears them the population suffers greater injury among those that do have car accidents.

A utilitarian view of these examples would see the benefit of a few lives saved as outweighing the inconveniences to the many. However, there is an alternative view that may help to reconcile the apparent conflict between individual autonomy and state paternalism. Contractarian theorists argue that any rights or responsibilities held by an individual stem from an implicit contract with society (as represented by the state). Within this context, justice arguments are made. Rawles (1971) argues that the only ethical social contract must come from behind a 'veil of ignorance' where the designer of the contract does not know the outcome. Within such a society, inequalities of wealth, power and status can only be justified if they are of greatest benefit to the least advantaged.

The philosophy of individualism, however, has a powerful effect on people's interpretations as to why disease and illness occur. The tendency to focus on individual analysis means the social, structural and epidemiological perspectives on health are, at best, a confusing background to explaining why individuals have particular health problems.

Figure 4.2 shows the various perspectives used to assign causality in public health. Biomedical and psychosocial perspectives focus almost exclusively on individuals, their risk factors and their response to these. Issues of exposure to physical, economic, social or cultural factors tend to be left to epidemiologists, who zoom in on factors that individuals have been 'exposed' to (tobacco smoke, low income, environmental pollutants, for example). Most typically, their analyses assign individuals to specific risk categories, tracing biological pathways to health outcomes (Walsh et al., 1995).

FIGURE 4.2 ALTERNATIVE PERSPECTIVES FOR ASSIGNING CAUSALITY IN PUBLIC HEALTH

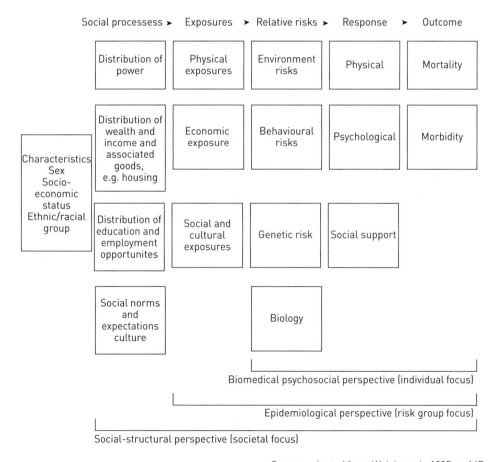

Source: adapted from Walsh et al., 1995, p. 147.

While this risk factor approach appears to follow the utilitarian path of calculating the risk and benefit to enable a redistribution of risk to benefit the population as a whole, there are major assumptions inherent in the identification and analysis of those risks. So, while they may study the effect of diet on individuals' health status, their risk factors and their exposure to particular foods, epidemiologists do not typically look at more structural questions concerning the advertising of food substances and the availability of healthy foods. The fundamental question as to who gains and who loses is usually left unasked.

SOCIAL-STRUCTURAL AND COMMUNITARIAN PERSPECTIVES

By contrast, a social-structural perspective on health brings to the fore the underlying social, economic, cultural and power issues, which tend to be hazy in the other perspectives. It is concerned with how society is organised and its impact on health and illness. Key issues for public health from this perspective involve the construction of social risk, the exercise of power on participation in society and the effects of income and wealth distribution on the pattern of health. Contractarian and social justice theories see these issues as central. In particular, communitarian theory has become prominent in recent decades in response to the dominance of rights-based arguments and the limits of neo-classical economic theory and practice. Box 4.3 explains its central principles. Its dominant themes are that individual rights need to be balanced with social responsibilities and that autonomous selves do not exist in isolation, but are shaped by the values and cultures of communities.

BOX 4.3   WHAT IS COMMUNITARIANISM?

Communitarianism is based on the belief that humans are social animals: 'Their being is composed in part by the community in which they exist; being in and of a community matters, and they take their identity partly from that' (Mooney, 2012, p. 124).
Key principles are that:

- the community is more than the sum of its individual parts
- sharing and reciprocity are core values
- community participation is out of both self-interest and community interest, and there is a merging between the two
- people are citizens before being consumers
- justice and social values are embedded in a society (compared to universalism of liberalism)
- it is only a good thing if the community it draws on is good.

Source: drawn from Mooney, 2012.

The Commission on Social Determinants of Health (see box 3.3) stressed a structural perspective on health that saw the distribution of power and resources in society as the most significant determinant of health and health equity. The Commission pointed strongly to the social and economic causes of disease and examined structural issues such as the nature of trade between countries and the ways these arrangements might impact on health. The Ottawa Charter for Health Promotion (see chapter 3) takes the position that health promotion activity has to work at the collective level to change environments and policies as well as educating individuals about health risks. The quotes provided in box 4.1 each suggest an ethic committed to redistributive social justice and stressing core values of justice, equity and the rights of all citizens to a reasonable standard of living.

However, such perspectives on health are less evident in public health literature than are the epidemiological, biomedical and psychosocial. Three reasons have been advanced to explain this (Walsh et al., 1995, p. 150):

1   Social structures are abstract and elusive while biological and psychological evidence is more tangible and obvious.
2   Western societies have a bias towards explaining social events in terms of personal characteristics. Walsh et al. (1995, p. 150) point out that in the West people 'have an analytic bias in favour of reductionism at the expense of integrative, intuitive, and convergent styles of knowing'. This supports Tesh's arguments that claim public health policy in the USA (and, by extension, other advanced capitalist economies, including Australia) is based on philosophies of individualism.
3   Recognition of the importance of social-structural factors in the creation of health and illness can lead to a sense of powerlessness. Explanations located in individual behaviour can lead to far more manageable policies and plans. When explanations are broadened to consider the range of social processes shown in figure 4.1, the extent of change implied can be overwhelming to the point of paralysis. This is even more so given the power of entrenched interests that may oppose change because it will threaten their position.

Smoking provides a good example. Early public health responses concentrated on changing people's behaviour by giving them information about the damaging effects of tobacco smoke. When this approach did not work, attention switched to legislation to control the sale and use of tobacco products. Some public health advocates have analysed the advertising and promotion practices of the tobacco industry as well as its enormous political and economic power in order to undermine its position. The public health battle against tobacco has used the range of perspectives shown in figure 4.2, but only after the individual-based ones were seen to be ineffective. Governments were very reluctant to take legislative action and only did so after considerable pressure from advocacy groups.

INDIVIDUALISM AND THE WELFARE STATE

The tendency towards individualism and victim blaming is not just a reflection of individual beliefs about why people from particular groups are sicker or suffer more social misfortune. These ideologies also express themselves in the forms of health and welfare provision available. These differences are shown in figure 4.3, in which

FIGURE 4.3   THE CHARACTERISTICS OF WELFARE REGIMES IN INDUSTRIALISED SOCIETIES

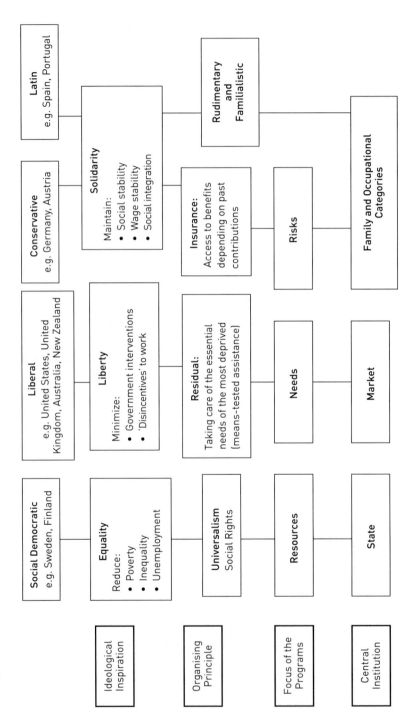

Source: Saint-Arnaud and Bernard, 2003, p. 503, Figure 2.

Saint-Arnaud and Bernard (2003) indicate how differences in political and economic structures and ideologies create different attitudes and policies towards welfare in industrialised countries.

The liberal approach views social welfare as a stopgap measure when economic market forces do not support people. It should support people until the market again provides for people's needs. In terms of ethical theory, a liberal model is based on consequentialist (possibly utilitarian) assumptions, while the social democratic model is based on a belief in the inherent value of its humanitarian ideals and it can therefore be seen as deontologically based. The social democratic approach sees welfare as a normal provision in industrial society and a humanitarian response to people in need. The conservative model is based on a social insurance model whereby people have a right to welfare based on past contributions. In the Latin model (the historic model in Latin America) the family is seen as the main support and citizenship confers no rights. Muller and Ventriss (1985) maintain that welfare has never been seen as a right in the USA, but as an act of charity for those who cannot provide for themselves. They also see a 'blame the victim' mentality prevailing. Such ideas stem back to the early white settlers who believed in individualism, hard work and self-help. Those in need were seen as being morally deficient. The Protestant work ethic did not believe in supporting people, except in the most extreme circumstances.

Australia's history has been far more influenced by the acceptance of government intervention, possibly reflecting Australia's history as a penal colony. From the 1950s until the 1970s welfare provision in Australia and the UK was less stigmatised than in the USA. Its role was disputed, but the consensus has seen welfare and other social provision as a right rather than charity. In the 1980s in the UK and in the 1990s in Australia this consensus shifted under the influence of neo-liberal economic policies. The criticisms against the institutional welfare state have been summarised by Pierson (1994) as:

1 it does not make economic sense because it replaces the 'natural' discipline of the market for capital to invest and labour to work
2 it does not encourage production because it displaces labour from the market-disciplined private sector to the undisciplined public sector, which encourages wage inflation
3 it is inefficient at delivering programs that meet the needs of individuals (now often referred to as 'customers') and is more responsive to the political claims of organised interest groups
4 it has not achieved a reduction in poverty or social inequities through its aim of redistribution
5 it tends to over-rule the lives of individuals with regulations and bureaucracy
6 the taxation of individuals to pay for the universalist services of the welfare state is presented as undermining individual liberties and denies individuals freedom of choice.

When political parties are driven by such beliefs about the welfare state, the general political environment is unlikely to be conducive to strong public health policies. Undermining of welfare provisions were a hallmark of the UK Thatcher government, the US Reagan and Bush administrations, the Howard government in Australia and the Conservative Cameron government in the UK. Even under the Australian Gillard Labor government welfare payments to single parents were restricted so that once a parent's youngest child had reached six, the parent was required to work. The Conservative Howard government introduced the Northern Territory Intervention which, among other measures, included compulsory 'income management' (or income quarantining), which dispensed payments in ways that ensured that money could not be spent on prohibited goods, such as alcohol or cigarettes (Bray et al., 2012). The austerity agenda that has formed part of the response to the Global Financial Crisis has also resulted in welfare cutbacks and tightening of eligibility and reduction in benefits. Similar policies tend to be promoted to resource-poor countries under structural adjustment packages and poverty reduction strategy papers promoted by the World Bank as part of their funding agreements (see chapter 5 for details). In many countries of the world there has been a shift from universal provision to notions of a residual 'welfare safety net' only for those in most need (Townsend, 2004). In Australia in 2014, the Treasurer Joe Hockey spoke of an 'entitlement culture' when referring to welfare recipients, and around the world welfare is becoming less of a universal provision for those in need and increasingly tied to entitlements (Schram and Silverman, 2012; Marston et al., 2013). Examples are welfare-to-work programs (framed as 'mutual obligations'), quarantining of welfare payments through the provision of vouchers instead of money, and the withdrawal of the single parent pension when children were over six years. These policies are based strongly on a model that sees the individual as primarily responsible for their life circumstances, regardless of the structural economic and social factors that have come to bear on them. Such a view is quite different to those expressing the importance of distributional equity shown in box 4.1. The example of different welfare regimes is evidence that ethics and values are much more than ideas, as they translate through policy into the way we shape and create our societies.

CONCLUSION

This chapter has examined the invisible hands of political systems, values and ethics, and shown that these have a considerable effect on public health policies and practice. It has highlighted the range of ethical dilemmas faced by public health practitioners. The new public health is not value-free, as it reflects a distinct set of ideologies and values that stress:

- the importance of balancing individual and collective needs in a way that is sensitive to ethical dilemmas and allows the promotion of health and equity to the greatest extent possible in a given setting
- the need for social-structural perspectives to be the driving force behind public health policy and practices.

We have seen that some political systems (e.g. social democratic and communist) place a greater value on collective well-being, while others (e.g. liberal democratic) place most emphasis on promoting market principles and the rights of individuals. We have seen how a philosophy of individualism is often at odds with a new public health approach, which rests on the importance of collective action to promote the public good even though in some cases the right of individuals are infringed (e.g. the right to smoke in public places).

This chapter has also shown that the ethical choices we make about the rights of individuals and the responsibilities we have to fellow citizens have a profound impact on the public policies that are developed to promote and protect health and well-being.

The ideas in this chapter will recur throughout the book as we examine the ways in which governments and others make and enact policies.

CRITICAL REFLECTION QUESTIONS

4.1   Which types of governments do you think are best placed to advance the aim of the new public health of improving population health and reducing inequities?

4.2   What health and welfare policies flow from a government committed to the values of individualism?

4.3   To what extent do you think governments should restrict individual liberty in order to protect health? For instance, should smoking be banned in restaurants? Should someone with an infectious disease such as Ebola be forcibly quarantined?

4.4   Should people be paid by the state to adopt healthy behaviours? For instance, should women be paid to quit smoking during pregnancy?

Recommended reading

Marston et al. (2013) *The Australian welfare state: Who benefits now?* provides an insightful account of changes to Australia's welfare system in the past three decades. It also sets the welfare debate in the context of debates about neo-liberalism and climate change.

Tesh (1988) *Hidden arguments: Political ideology and disease prevention policy.* A very important book with sharp analysis about the role of ideology in public health policy.

Useful website

The *Public Health Ethics* journal provides lists of latest articles debating current public health issues: http://phe.oxfordjournals.org

5

NEO-LIBERALISM, GLOBALISATION AND HEALTH

As human beings, it is in our power to take a correct turn, which would make the world safer, fair, ethical, inclusive and prosperous for the majority, not just for a few, within and between countries. It is also in our power to prevaricate, to ignore the road signs, and let the world we all share slide into further spirals of political turbulence, conflicts and war.

World Commission on the Social Dimension of Globalization, 2004, p. vii

KEY CONCEPTS

Introduction
What is globalisation?
What is neo-liberalism?
Key institutions
World trade system and health
International agreements that threaten global health
TRIPS and TRIPS-Plus
Trade in Services Agreement (TISA)
The impact of transnational corporations
The impact of neo-liberalism on health
Consumerism
The voices of dissent: civil society movements
Bringing the voice of ordinary people from the grassroots
Protest, advocacy and lobbying against international financial and trade institutions
'Watching' the global institutions
Conclusion

INTRODUCTION

Twenty years ago, if I had been writing this book and I wanted to gain access to the latest report on health inequities in London, I would have had to wait for the report to come through an inter-library loan service by sea mail. Today I can download it to my home computer in Adelaide the minute it is released in London. Now I can go to almost any capital city in the world and buy food from an American-based multinational fast-food chain or a brown drink that is the real thing. Every day I am in email and Skype contact with friends, relatives and colleagues around the world. I tweet about issues that concern me and find out what's happening in the world through the twittersphere. I post on Facebook to keep in contact with friends and family. I worry as more Australian companies are taken over by offshore transnational corporations (TNCs) that increasingly have no allegiance to any national social values. They exist as powerful entities in their own right, often with turnovers and budgets that exceed those of many countries. I hear on the radio that, while we have some fantastic opportunities for electronic communication, in many country towns big business interests are closing down banks and other services. I am concerned about the health effects of the closure of local manufacturing industries as the jobs move offshore to China where wages are lower and laws governing working conditions are less exacting. In my home town of Adelaide, two major car manufacturing plants that provided many jobs will have been closed down by TNCs by 2017 because it is cheaper to make cars elsewhere. Health services are also becoming big business and more and more likely to be part of a profit-making industry. When I read in the financial pages of the newspaper (also owned by a TNC) that the provision of health services is going to be one of the big opportunities for profit in the next decade, I wonder what that means for users of health services. I notice that there are more and more people on the move around the world and the numbers of refugees and asylum seekers increase. Terrorism has become a global concern. The Global Financial Crisis of 2008 and the austerity policies that followed it have had huge impacts on people's health, and the instability of financial systems is an ever-present threat. In the second decade of the twenty-first century we have not made the world fairer and more inclusive, even though the World Commission on the Social Dimension of Globalization quoted at the start of this chapter notes it is in our power. This chapter considers how the rapid changes to our world that make it more connected economically, socially and culturally have affected health and equity.

Underpinning this globalisation since the 1970s has been an economic theory—neo-liberalism—which has grown in dominance and influence around the world. In order to understand the position of public health in the twenty-first century it is necessary to understand the nature and impact of neo-liberalism and the extent to which it has colonised institutions and minds around the world. This chapter will examine the processes of globalisation and neo-liberalism through the prism of public health and provide the opportunity to reflect on and assess their impact on health.

WHAT IS GLOBALISATION?

It is not settled or secure, but fraught with anxieties, as well as scarred by deep divisions. Many of us feel in the grip of forces over which we have no control.

Giddens, 1999, p. 19

There is extensive debate about the meaning and implications of globalisation (see Held and McGrew, 2007, for summaries of these debates). For public health, globalisation has far-reaching implications, potentially both positive and negative. The process of globalisation has been continuing for some centuries. For indigenous peoples around the world, globalisation started when European powers invaded their lands and destroyed their traditional cultures, took away land and introduced new diseases that devastated populations. The resultant health effects are described in chapter 11.

Giddens (1999) sees globalisation as economic, political, technological and cultural, and Lang (1999) adds ideological to this list:

- **Economic globalisation** refers to the process of trade liberalisation, tariff reduction, standards harmonisation and deregulation. One of the most important results of this process has been that the growth in international trade has accelerated—for instance, by nearly 8.6 per cent per year over the period 1990–99 (Woodward et al., 2001, p. 876). This process has also seen greater mobility of capital and a massive increase in transnational investment.
- **Political globalisation** refers to the creation of global institutions that are establishing global forms of governance. Examples are the World Trade Organization, the General Agreement on Tariffs and Trade (GATT), and the General Agreement on Trades and Services (GATS). Most opinions argue that this process has weakened the capacity of national governments.
- **Technological globalisation** concerns the rapid breakthrough in communication technologies, such as satellites and the internet, which have made worldwide communication so much more rapid.
- **Cultural globalisation** makes cultural exchange easier; the processes result in breaking established social orders, but also may establish new social movements. It involves the fast transmission of ideas, images, fashion and information through new communication media.
- **Ideological globalisation** concerns the way in which political and corporate leaders sell a view that there is no alternative to the neo-liberal package of reforms. The argument is that citizens, companies and whole societies have no option but to accommodate these reforms, despite the negative consequences.

Economic globalisation is the aspect of the phenomenon over which there has been most debate and discussion in relation to health. Some have seen economic globalisation as an automatic consequence of the collapse of communism and the seeming triumph of free-market capitalism (Fukuyama, 1992).

Gray summed up the dreams of economic globalisers thus: 'The entire world was to be remade as a universal free market. No matter how different their histories and values, however deep their difference or bitter their conflicts, all cultures everywhere were to be corralled into a universal civilisation' (Gray, 2001, p. 26). To understand the economic dimensions of globalisation it is necessary to comprehend the nature of neo-liberal economics and the ideologies that surround them.

WHAT IS NEO-LIBERALISM?

> Neoliberalism … proposes that human well-being can best be advanced by liberating individual entrepreneurial freedoms and skills within an institutional framework characterized by strong private property rights, free markets and free trade.
>
> *Harvey, 2005, p. 2*

Neo-liberalism is a term used to describe trends since the 1970s that have seen a progressive reliance on economic factors as the basis for organising society and making government decisions. It describes a school of economic thinking that traces its origins back to Adam Smith's *Wealth of Nations*. This style of economic thinking is also known as *laissez-faire* or neo-classical economic thought. Modern proponents include Milton Friedman, F.A. Hayek and many academic economists. In Australia, neo-liberalism is also known as 'economic rationalism', a term that was coined in the 1990s (Pusey, 1991).

Neo-liberalism's basic tenet is that the free market should determine all economic transactions. Open competition in a free market will provide the greatest efficiency, intervention being seen as a distortion that results in efficient industry unfairly supporting inefficient industry. Neo-liberalism purports to ensure a 'level playing field' and promotes deregulation. Together these two concepts are seen as giving free reign to Adam Smith's notion of the 'invisible hand'. This assumes that there is a force behind the scenes that would guide the free market to ensure that outcomes are efficient and just. Neo-liberalism has come to dominate the policies of international agencies and national governments around the world in the last 40 years. Harvey (2005, p. 3) says it has become a hegemonic form of thought and 'has become incorporated into the common-sense way many of us interpret, live in and understand the world'. The policies associated with neo-liberalism have been the lowering of trade barriers, deregulation of the labour market, 'roll back' of state activities, privatisation or contracting out of public services, the use of private sector management techniques within public service departments, and the cutback of state funding for a range of activities including education, health, welfare, housing, arts and culture, and transport (Kelsey, 1995; Manne and McKnight, 2010). In Australia, Pusey's research (1991, 2010) has shown how the national policy apparatus was captured by neo-liberal ideology, which led to a redefinition of the fundamental tasks and priorities of the state so that public policy was reduced to 'a narrowly conceived ideology of maximum feasible market and business penetration into as many areas of the nation's life as possible' (Pusey, 2010, p. 135). The three stages of neo-liberalism are shown in box 5.1.

BOX 5.1   THREE STAGES OF NEO-LIBERALISM

NEO-LIBERALISM EMERGES

Neo-liberalism's dominance in political and economic decision making began to emerge in the early 1970s. This was a decade marked by an increasing pace of economic recessions, oil embargoes, and oil price rises that quadrupled the cost of capitalism's crude energy source.

In 1971 the USA permanently unpegged the US dollar from the gold standard. This set financial exchanges adrift, allowing money to be made through currency speculation, and entrenched the US dollar as the world's 'reserve currency' held by governments' central banks and other financial institutions 'in reserve' as a means to pay off international debt obligations and to stabilise the value of their own currency when needed.

In 1973 the Pinochet government elected in Chile was the first government to implement Hayek's and Milton Friedman's neo-liberal economics. The late 1970s and early 1980s saw the election of Reagan in the USA and Thatcher in UK, both with strong neo-liberal agendas.

Low- and middle-income countries experienced a debt crisis, partly in response to rising oil prices. This worsened when US-led monetary policy to control inflation led to huge increases in interest rates, rising from 11 per cent in 1979 to more than 20 per cent in 1981.

Structural adjustment policies meant countries borrowing money had to conform to these 'conditionalities':

- privatisation of state assets, in part to help governments pay off international loans
- deregulation to enable rapid private sector–led economic growth
- tax reform to attract foreign investment through lower corporate and marginal rates, or tax holidays for foreign investments
- public deficit (the shortfall between revenues and expenditures in any one fiscal year) and debt (total accumulated amount owed to creditors), in part to help governments pay off international loans
- rapid trade and financial market liberalisation, based on the theory that liberalisation leads to economic growth, which it sometimes but not always does.

NEO-LIBERALISM: REAGAN TO THE GLOBAL FINANCIAL CRISIS (GFC)

Declining profits for corporations saw them introduce new strategies:

- lowering production costs (labour-saving technology and outsourcing to low-cost countries)
- opening new markets (new treaties that reduced tariff and non-tariff barriers to goods and investment)
- increasing the financialisation of the economy (digital technologies, bank deregulation and removal of capital controls).

    More frequent financial crises culminated in the GFC of 2008. This was also affected by the burst of the housing bubble in the USA and the 'credit crunch'. Governments bailed out the banks, using public money to save private institutions. Direct stimulus spending in the early crisis years (2008–09) is estimated at around US$2.4 trillion across 50 countries, most of them G20 or OECD members (Ortiz and Cummins, 2013). These aspects resulted in growing income inequities.

NEO-LIBERALISM POST-GFC: GROWING INEQUITIES

Growth in income and wealth inequities continues and intensifies. While the 2008 GFC wiped out trillions of dollars in paper wealth, affecting the pensions and savings of many of the world's middle and working classes, the 24 million whom investment banks refer to as 'high and ultra-high net worth individuals' saw their balance sheets decline for a year or two, but then increase by more than 20 per cent after receiving trillions of dollars of public money (Baxter 2011).

The world's 85 wealthiest people own as much wealth as the bottom half (more than three billion) of the world's entire population (Oxfam, 2014b). Austerity agendas introduced new forms of structural adjustment:

- reductions in social protection spending and public sector employment
- increased VAT (consumption) taxation
- reduction or elimination of public deficits
- reduction of public debt
- increase in 'user pays' systems in public programs (co-payments)
- privatisation of state assets
- increased public–private partnerships characterised by the public absorbing most of the risk and little of the gain of private-sector financing for public goods or services.

Source: Labonté, 2012; Ortiz and Cummins, 2013; People's Health Movement et al., 2014.

Many governments around the world, especially in the USA, Canada, UK and Australia, now base their economic policy on neo-liberalism, which means they put economic considerations ahead of those of social policy and health and well-being of citizens.

This adherence to neo-liberalism often seems to fly in the face of evidence about the impact of the policies on health (see the discussion below about the health impacts of neo-liberalism). Davis (2014) shows how the arguments and interests shaping neo-liberalism are carefully constructed. He demonstrates that neo-liberal ideas and assumptions have found their way into everyday language and expectations. Central to these is the idea that the market is the most appropriate organising mechanism for society and this idea has led to free-market thinking in all areas. The philosopher Sandel (2012) calls this market triumphalism and notes the reach of markets 'into aspects of life traditionally governed by non-market norms is one of the most significant developments of our times' (see more in box 5.2).

BOX 5.2   INSIGHT INTO MARKET TRIUMPHALISM

Michael J. Sandel (2012) examines the 'moral limits of markets' and asks if there is something wrong with a world in which everything is for sale. Or 'are there certain moral and civic goods that markets do not honour and money cannot buy?' (p. 203)

Do you think the following activities should be open to market forces?

- Donating blood: in the USA you can sell blood; in most countries, giving blood is a donation.

- Paying children to read books: second-graders in Dallas, USA, are offered $2 for each book they read.
- Allowing a surrogate mother to carry a baby for another couple: couples from rich countries pay Indian women around $7000 to carry a baby for them.
- Allowing prisoners to have 'cell upgrades': prisoners in some US cities can pay for a clean, quiet cell away from cells of non-paying prisoners.
- Privatising prisons: private prisons were unheard of in the 1980s but are now increasingly common.
- Permitting corporate marketing at hospitals: McDonald's advertises at children's hospitals through Ronald McDonald houses.

In order to assess the impact of neo-liberalism it is necessary to understand the role of key financial institutions and the system of world trade and global finance regulation, as well as the roles of the most powerful actors.

KEY INSTITUTIONS

> Globalisation's rules favour the already rich (both countries and people within them) because they have greater resources and power to influence the design of those rules.
>
> CSDH Globalisation Knowledge Network, 2007, p. 9

Key institutions shaping neo-liberalism are the World Bank, the International Monetary Fund (IMF), the World Trade Organization (WTO), the Group of Eight (G8) and the Group of Twenty (G20). The World Bank and the IMF were created in the wake of World War II. The two institutions were given two distinct functions when they were formed following the United Nations Monetary and Financial Conference at Bretton Woods, New Hampshire, in July 1944—a meeting usually referred to as 'Bretton Woods':

- The World Bank (the full name is the International Bank for Reconstruction and Development) was designed to assist the rebuilding of Europe following the devastation of the war.
- The IMF was assigned the more difficult task of ensuring global economic stability and to save the world from future economic depression like that of the 1930s.

The creation of the World Bank and the IMF was heavily influenced by the British economist John Maynard Keynes, who believed that government intervention to direct and control markets was both desirable and necessary to economic prosperity and stability. To avoid another depression he advocated that government policy could help stimulate aggregate demand and that the pursuit of full employment was crucial both economically and socially. When it was formed, the IMF was based on recognition that markets often do not work well and that global collective action would be necessary to ensure economic stability. This recognition has been undermined to the point where it 'now champions market supremacy with ideological fervour' (Stiglitz, 2002, p. 12) and strongly reflected the 'Washington Consensus' of the 1980s that advocated the neo-liberal economic policies. The IMF appears to have lost all sense of its founding mission and appears more concerned with the needs of the developed countries than those of the poor (Potter, 1988; MacDonald, 2005; Ortiz and Cummins, 2013).

The World Bank has been heavily criticised in recent years on the grounds that it has funded many environmentally damaging projects; that it runs undemocratically for the benefit of high-income countries rather than for the benefit of low-income ones; that it pursues the interests of TNCs; that it has a damaging effect on the health of people in low-income countries; and that it is a tool of US foreign policy (Beder, 1993; Stiglitz, 2002; Korten, 2006). It has also been seen as responsible for the crippling burden of debt under which so many low-income countries have struggled since the 1970s. The burden of debt results in a net outflow of resources from rich to poor countries. MacDonald (2005) reports that the poorest countries in Africa had transferred $167 billion into debt service to their creditors in the rich world by the end of 2002. He contrasts this with the US$9 billion that UNICEF (the United Nations Children's Fund) estimated was needed for health and nutrition in all of Africa! The Globalisation Knowledge Network of the Commission on Social Determinants of Health (2007) reported that over the 1993–2005 period, the developing countries as a whole moved from being net recipients of financial flows to being net losers, to the tune of close to half a trillion dollars per year.

The World Bank's role also changed from its original mission of European reconstruction to focusing on poverty reduction in the de-colonising developing world. It is financed by the richest countries lending money to the poorest. It has made a profit every year since 1947. Voting power depends on the amount of money each country contributes. The USA has the largest say and, together with the UK, Germany, France and Japan, controls about 45 per cent of the votes. As with the IMF, the World Bank also adopted the Washington Consensus and focused on giving prescriptions (in the 1980s structural adjustment packages [SAPs], then poverty reduction strategies, and most recently austerity policies) to these countries that maintained that the way to solve poverty was to free markets from government controls and reduce the size of government. Generally these prescriptions have been seen to have disastrous results, especially in Africa. A United Nations study (UNRISD, 1995) describes the considerable social dislocation that has come with SAPs. Most people in countries experiencing SAPs have become more vulnerable economically. This has forced people to adopt multiple survival strategies because they cannot survive on one source of income. Most workers in the informal sector face increasing competition and struggle to survive. The net result of these dislocations is increasing polarisation, fragmentation and instability. Fried (1994) also claims that SAPs had a bad effect on the environment, as there has been a spread of chemical-intensive export agriculture, the extraction of more and more natural resources and a tendency for multinational industries to evade the stricter environmental controls imposed in developed countries. There is even some evidence that such practices are encouraged by the World Bank. In 1992 a memo was leaked from the Bank that argued that it was better policy to pollute areas where poor people lived and so the Bank should encourage dirty industries to move to less-developed countries where wages were lower, and the mortality and morbidity costs less (Beder, 1993, p. 181).

The SAPs have brought severe and sustained criticism from many groups around the world, including non-government organisations. In *Adjustment with a Human Face*, Cornia et al. (1988) document the effects of economic dislocation on people, especially children. In some countries, declining economic conditions have meant a reversal of earlier gains in child health status. In Zimbabwe, for example, the idealism and initial success of a nationwide primary health care program was undermined and restructured as a result of economic recession and SAPs in the 1980s (Sanders, 1993). A longitudinal study in Zimbabwe has indicated the deleterious effects on health of SAPs in that country (Bijlmakers et al., 1998). The strong criticism of the SAPs led to them being replaced with poverty reduction strategy papers (PRSPs). At first these seemed to be more benign but their impact is remarkably similar to SAPs. An analysis by WEMOS, a Dutch non-government organisation (Verheul and Cooper, 2001), suggests that PRSPs had similar effects to SAPs because the measures being advocated through them are much the same—for example, privatisation and selective approaches to disease prevention. Essentially, SAPs and PRSPs called on poor countries to deregulate their economies, privatise services wherever possible and cut back public services to the bare bones. Even relatively conservative commentators such as Joseph Stiglitz (2002), a former vice president of the World Bank, consider that the impact of these policies on newly developing nations with weak economies was devastating.

While the World Bank and IMF supported greater public investment in the immediate aftermath of the Global Financial Crisis (see box 5.3) this prescription very quickly reverted to the more normal neo-liberal focus under the guise of austerity policies. This policy advice has been very rapidly adopted by countries around the world and, by 2012, 94 countries (according to IMF reports) were cutting expenditure. This cutting was most severe in the developing world, which could least afford it. Of the 68 developing countries cutting public expenditure, the average was by 3.7 per cent compared with 2.2 per cent in the high-income countries that were cutting (Ortiz and Cummins, 2013). These cuts will have the greatest impact on people living in poverty and vulnerable circumstances. This is ironic given that the GFC was caused not by blow-outs in public spending but by financial market deregulation and subsequent reckless behaviour by financial institutions. An increasing number of economists are critical of neo-liberal austerity and argue that increased public investment should be used to boost employment and improve the living standards of the poor, together with measures to promote equity such as progressive taxation (Stiglitz, 2012; Stuckler and Basu, 2013) (also see chapter 16 for further discussion of alternative economic models to neo-liberalism).

BOX 5.3   THE GLOBAL FINANCIAL CRISIS: BAIL-OUT FOR THE BANKS, AUSTERITY FOR MOST

The Global Financial Crisis (GFC) has had and is having a dramatic impact on the health of people in countries around the world. It shows the way in which political economic issues shape our health and chances for more equitable health outcomes.

The GFC's immediate causes have been analysed by the conservative magazine *The Economist*. It notes 'The collapse of Lehman Brothers, a sprawling global bank, in September 2008 almost brought down the world's financial system' (The Economist, 2013a). It goes on to cite multiple causes: excessive risk taken by bankers including a flood of irresponsible mortgage lending, credit rating agencies (which are beholden to the banks!) that had far too generous credit ratings, and too great a willingness to take on debt. All this led to a sharp decline in trust, which is 'the ultimate glue of all financial systems'. Complex chains of debt and their collapse revealed that the system was built on flimsy foundations. The article also cites a failure of the central regulatory agencies to handle the crisis effectively.

This analysis, however, does not consider the system underpinning the GFC. Harvey (2010, p. 6) notes the GFC should be 'seen as the culmination of a pattern of financial crisis that had become both more frequent and deeper over the years' since the 1970s. There were very few crises between 1945 and the early 1970s. He sees neo-liberalism as a means of restoring and consolidating capitalist class power that had diminished in the immediate postwar years. He notes that one of the basic pragmatic principles of this that emerged in the 1980s was 'that state power should protect financial institutions at all costs' (p. 10). This is despite the fact that it flies in the face of the non-interventionism prescribed in other settings.

The response to the GFC has been to protect and bail out the banks that failed. This has happened around the world and with very few strings attached. By 2014 many bank CEOs are again earning huge salaries, even though these are built on the back of the public bail-outs.

Meanwhile, after a brief period of greater public investment, austerity policies have been introduced with significant health effects (Stuckler and Basu, 2013).

The World Trade Organization came into formal effect in 1995. It is quite different from the World Bank and the IMF as it does not set rules itself but provides a forum in which trade negotiations occur and then ensures the agreements are kept. Its negotiations are complex in the extreme and the round of negotiations continues for years. These are also done behind closed doors, making the influence brought to bear on the negotiations by corporate and other special interests invisible.

The G8 (originally the G7) was formed from the world's leading industrialised nations in the mid-1970s in response to the oil crisis of that decade. The seven original members were France, USA, UK, Germany, Italy, Japan and, joining slightly later, Canada. Russia attained full membership in 2003. These countries exert considerable power over the IMF and World Bank and together account for more than 40 per cent of the world's economic activity. This means they are also very powerful in the WTO negotiations. Like the other institutions the G8 uses neo-liberal economic assumptions and sees these as the solution to both global poverty and maintaining economic stability. The influence of the G8 has reduced somewhat as the G20 gains greater ascendancy. The G20 members represent around 85 per cent of global gross domestic product, more than 75 per cent of global trade, and two-thirds of the world's population.

The IMF, the WTO and the World Bank are seen as agents of rich countries and TNCs in their promotion of neo-liberalism. This promotion also happens through the policy prescriptions of the G8 and increasingly through the G20, both of which are groups of countries promoting neo-liberal economic growth (see box 5.1).

While these institutions are overwhelmingly neo-liberal in their policy orientation there are some chinks in the armour. Examples are:

1   the endorsement of public spending in the immediate aftermath of the GFC
2   countries that are investing in the public sector and increasing social and health spending (e.g. Iceland, Thailand, China, Brazil, Ecuador)
3   Lagarde's (Managing Director of the IMF) speech at the Davos Forum in January 2013 in which she noted that 'Excessive inequality is corrosive to growth, it is corrosive to society' and went on to say that 'a more equal distribution of income allows for more economic stability, more sustained economic growth, and healthier societies with stronger bonds of cohesion and trust'
4   the G20 meeting in Brisbane in November 2014, which called for more effective taxation on TNCs to reduce their tax avoidance activities.

We must note, however, that Marxist analysis shows in Harvey's words (2014, p. 172) that 'Distributional equality and capital are incompatible'. This is because, socially and historically, capital is constructed as a class in dominance over labour. Distribution of income and of wealth has to be lopsided in the favour of capital if capital is to be reproduced. This lopsided distribution is shown in box 5.4.

G20 meeting in Brisbane, November 2014. (Andrew Taylor/G20 Australia)

WORLD TRADE SYSTEM AND HEALTH

In the globalised world, trade is one of the most powerful forces linking our lives. It is also a source of unprecedented wealth. Yet billions of the world's poorest people are being left behind. Increased prosperity has gone hand in hand with mass poverty and the widening of already obscene inequalities between rich and poor (see box 5.4)

BOX 5.4   UNEQUAL WORLD

- Almost half of the world's wealth is now owned by just 1 per cent of the population.
- The wealth of the 1 per cent richest people in the world amounts to $110 trillion. This is 65 times the total wealth of the bottom half of the world's population.
- The bottom half of the world's population owns the same as the richest 85 people in the world.
- Seven out of ten people live in countries where economic inequality has increased in the last 30 years.
- The richest 1 per cent increased their share of income in 24 out of 26 countries for which we have data between 1980 and 2012.
- In the USA, the wealthiest 1 per cent captured 95 per cent of post-financial crisis growth since 2009, while the bottom 90 per cent became poorer.

Source: Oxfam, 2014b, pp. 2–3.

Understanding how such economic inequities come about requires an understanding of the ways in which the world's trade regime works to support the growth of capital and its concentration in a few hands. This regime is complex. It includes a range of structures and processes that shape the way in which world trade is conducted. The following factors are keys to considering the impact of globalisation on health:

- international treaties and agreements
- the increased power and size of TNCs.

INTERNATIONAL AGREEMENTS THAT THREATEN GLOBAL HEALTH

The adoption of neo-liberal policies by the World Bank and IMF led to a series of international agreements that liberalised trade and investment intensively from the 1990s. A series of international agreements have followed the establishment of the WTO, including the Agreement on Trade-Related Aspects of Intellectual Property Rights (TRIPS) and the General Agreement on Trade in Services, which have codified and consolidated the unequal terms of trade. In addition many bilateral and multilateral trade agreements have been negotiated in the past ten years or are in the process of being negotiated. Together, they establish a transnational regulatory framework that overrides national, regional and municipal jurisdictions and laws. Box 5.5 summarises the ways in which these agreements might threaten health.

BOX 5.5   HOW INTERNATIONAL TRADE AND INVESTMENT TREATIES THREATEN HEALTH

- The agreements reflect neo-liberal economic orthodoxy and promote unregulated financial and trade markets. They leave the poor unprotected and reduce government expenditure on social welfare and health provision. There is no evidence that neo-conservative policies (such as structural adjustment packages) promote equity, while there is increasing evidence that they do not. Equitable societies appear to be healthier (Wilkinson and Pickett, 2009), but the international agreements are likely to increase rather than decrease inequities.
- The Agreement on Trade-Related Aspects of Intellectual Property Rights (TRIPS) takes away the rights of indigenous people to biological resources they have controlled for generations (see box 5.6).
- The application of the TRIPS agreement to medicines has grave consequences for public health because it ensures high prices for medicines through enforcement of patents.
- The Trade in Services Agreement (TISA) threatens the universal provision of health and other services and supports privatisation of publicly funded services, which in turn is likely to reduce access for those on low incomes.
- The agreements favour the rights of corporations and investors over those of citizens and governments, and so reduce the sovereignty of nation states, which are prohibited from actions that restrict trade or investment even if they are for the common good of citizens.
- The agreements do not protect the rights of workers, maintain existing social welfare provisions or protect the environment, and may preclude national governments from enacting legislation that does.
- The agreements are negotiated in secret through processes that are not transparent. No opportunities for participation from NGOs or citizen groups are built into the draft process. The same applies to the dispute resolution procedures established under the agreements.
- The agreements are complex and almost impenetrable, making it difficult for governments (especially from poor countries) to respond to challenges under the agreements. This will increase the power of corporations.
- There are no global mechanisms to ensure the accountability of the international financial and trade bodies.

Source: McMurtry, 1997; Labonté, 1999; Lipson, 2001; Oxfam, 2002;
People's Health Movement et al., 2005; Gleeson, 2013.

TRIPS AND TRIPS-PLUS

An example of the way in which these international treaties can have an impact on health is provided by the TRIPS agreement. Labonté (2001) notes that TRIPS is an exception to other WTO agreements that liberalise trade. TRIPS requires that protections are extended, specifically extending corporate monopolies over drugs, foods (seeds) and other 'intellectual property'. Labonté further explains that under TRIPS and national patent laws, a private company can change less than 0.05 per cent of an organism (a part of a gene) and then claim proprietary rights to both the 0.05 per cent and the 99.95 per cent that remains essentially 'common'.

BOX 5.6   TRIPS: A NEW FORM OF COLONIALISM

Vandana Shiva, physicist, ecologist and activist, has written extensively on the threat that the Agreement on Trade-Related Aspects of Intellectual Property Rights (TRIPS) poses to the traditional knowledge and economies of low-income countries. She likens the imposition of intellectual property rights to a new form of colonialism by Western countries. She says the freedom that transnational corporations are claiming through the intellectual property rights protection in TRIPS is 'the freedom the European colonisers have claimed since 1494' (Shiva, 1998, p. 8). She points out that the system of intellectual property rights is heavily weighted in favour of transnational corporations and against citizens in general. This is because only the intellectual property rights of scientists and corporations who seek to patent genes and microorganisms in order to make a profit are recognised. Innovations of indigenous people and traditional societies are not. She describes this biopiracy thus:

Through patents and genetic engineering, new colonies are being carved out. The land, the forests, the rivers, the oceans, and the atmosphere have all been colonised, eroded and polluted. Capital now has to look for new colonies to invade and exploit for its further accumulation. These new colonies are, in my view, the interior spaces of the bodies of women, plants and animals. (Shiva, 1998, p. 11)

Her arguments are very well illustrated by the issue of 'seed freedom'. She notes that, for thousands of years, farmers—especially women—have evolved and bred seed freely with the help of nature to increase the diversity of nature and adopt it to the needs of different cultures, so that biodiversity and cultural diversity have mutually shaped one another (Shiva, 2014). However, intellectual property provisions under trade treaties have granted rights over seeds to transnational corporations like Monsanto so that 'seed as a common good became a commodity of private seed companies, traded on the open market'. She notes the adverse health, ecological and community well-being outcomes of this move.

Large corporations such as Monsanto are pressuring governments all over the world to pass laws that take control of seeds out of the hands of small farmers. 'La Via Campesina' ('The Peasant Way') is a network of more than 160 rural peoples' organisations in approximately 80 countries in the Global North and South (Mann, 2014). La Via Campesina coined the term 'food sovereignty' to advocate a model of peasant-based, sustainable, agroecological farming. Zacune describes this as 'Food sovereignty puts those who produce, distribute and consume food at the heart of food systems and policies, rather than forcing those systems to bend to the demands of markets and corporations. It defends the interests and the inclusion of the next generation' (Zacune, 2012, p. 5). Achievement of food sovereignty is threatened by TRIPS (Zacune, 2012).

This is shown in the TRIPS regime that, among other things, allows patenting of seeds. TRIPS poses a threat to genetic resources, sustainable agriculture, food security and the well-being of farmers (see box 5.6). Increased patent protection will lead to increasing prices and reduced access to medicine, and supports monopoly control. A typical example of the impact of TRIPS on small farmers in poor countries is a large US corporation such as Monsanto patenting crops that are protected from disease but that are seedless. This means the farmer may not be able to sow traditional seeds, yet will not be able to afford to buy seeds from the patent holder every year. The patent may be held on crops that the farmer's ancestors used for centuries (Shiva, 2014). Similarly, the General Agreement on Trade in Services results in the use of science and standards-setting as a mechanism for maintaining unfair trading practices. Lang (1999) notes that the United Nations Codex Alimentarius Commission (the international food standards body) is subject to undue influence from industry representation, which also results in discrimination against poorer nations.

The TRIPS agreement has also been used to ensure that low-income countries offer patent protection for pharmaceuticals. Implications are that access to medicine in poor countries will be difficult and the costs are likely to impose a strain on poor households. In November 2001 at a meeting of the WTO in Doha, Qatar, the issue of how the WTO's rules on patents may harm public health in poor countries was discussed for the first time, and at the Cancun meeting in 2003, a limited deal offering access to cheaper generic drugs for some of the very poorest countries was provided. This indicates that the extensive lobbying by non-government organisations may be having some limited effect. The fact that the pharmaceutical companies are making large profits and at the same time objecting to the use of generic drugs by poor countries has greatly assisted the lobbying efforts. The Cancun meeting was also the first where the less affluent countries organised among themselves to resist the demands of the more powerful countries. In the past decade an increasing number of middle- and low-income countries have challenged this ruling. They have allowed local manufacture of drugs that have undercut the prices of TNCs. The South African Government, facing a huge HIV epidemic, decided to allow the sale of generic antiretroviral drugs despite the protestations of the pharmaceutical companies. In Brazil, the government has produced its own generic versions of antiretrovirals; the cost of annual treatment has tumbled from $15 000 to less than $3000 (Médecins Sans Frontières et al., 2003). Governments have also issued compulsory licences, which allow the use of a patented innovation that has been licensed by a state without the permission of the patent titleholder. Compulsory licences are allowable under TRIPS provided a period of negotiation with the patent holder has occurred. Beall and Kuhn (2012) note that while there was a reasonable use of compulsory licences in the period 2003–06, they have declined since. They also note that many regional or bilateral trade agreements now include so-called TRIPS-Plus provisions that expand patent rights or limit generic production. The limited capacity of low-income countries

Indian civil society has been active in opposing the practice of seed patenting by multinational companies. (Navdanya)

to actually implement the available flexibilities introduced after the Doha agreement has been noted by Feldbaum et al. (2010), especially given the stricter protections found in many TRIPS-Plus measures, demonstrating the power of economic over public health considerations. Provisions in these agreements further narrow circumstances in which a compulsory license can be justified—virtually nullifying such opportunities—and extend periods of data exclusivity, enabling large pharmaceutical companies to prevent or delay generic competition (Dávila, 2011).

Akaleephan et al. (2009) attempted to quantify the impact of the US–Thai Free Trade Agreement on medicines access. According to their model, the TRIPS-Plus provisions of this agreement were estimated to increase medicine expenses by a minimum of US$806.4 million to US$5.2 billion and also delay generic accessibility. Michael Kirby (2014), a former Australian High Court judge, has called for reform of the TRIPS-Plus protection on intellectual property on drugs, saying 'Without changing the global laws on intellectual property, people will die needlessly'.

One of the early free trade agreements—the North American Free Trade Agreement (NAFTA)—has been assessed as privileging 'the rights of investors above social welfare and environmental concerns to ensure that barriers to trade are minimised' (White, 2010, p. 87). As a result of this treaty, the Mexican municipality of Guadalcázar had to pay the US Metalclad Corporation more than $16 million in compensation because it prevented them from creating a toxic waste landfill in an ecologically protected zone. The ability of TNCs to sue governments has the effect of making governments scared of legislating or decision making in favour of health and the environment. Free trade agreements have also been used by tobacco companies in an attempt to restrict public health legislation (see box 5.7) and it is feared that these agreements could prevent governments from enacting other public health measures such as restricting the advertising of fast food to children.

BOX 5.7   HOW TRADE AGREEMENTS MAY PREVENT GOVERNMENTS PASSING PUBLIC HEALTH LEGISLATION

Australia was the first country to pass legislation mandating plain packaging of tobacco, beginning on 1 December 2012. This means that all cigarette packs sold in Australia look similar to this:

Source: Cancer Council Victoria.

In November 2011, Philip Morris Asia Limited filed a claim against Australia under the Australia–Hong Kong bilateral investment treaty, alleging that the plain packaging law expropriates intellectual property. Mitchell and Studdert (2012) note that this Hong Kong–based subsidiary of the Philip Morris conglomerate purchased a 100 per cent stake in Philip Morris [Australia] Limited only months before the legislation was introduced and say they assume that this was solely so they were able to file this claim (p. 262). Gleeson and Friel (2013) see such actions as part of a global strategy by the tobacco industry to use international trade and investment dispute mechanisms to undermine tobacco control measures.

These investor–state dispute issues were the subject of an ABC Radio National *Background Briefing* episode that provides an excellent guide to the public health threats of these disputes, including the fear that their very existence will deter governments from passing legislation that protects public health (ABC, 2014).

An example of one of the TRIPS-Plus agreements is the Trans-Pacific Partnership Agreement. Negotiations are held in secrecy and draft texts are not publicly available. Considerable concern has been expressed about the public health impact of this agreement (Gleeson and Friel, 2013). Inequities are likely to be increased as wealthier countries can impose conditions that are more advantageous to themselves at the expense of poorer countries. The dispute mechanisms have also been shown to favour large corporations so the way is open for them to challenge public health legislation (see box 5.7). It is significant that the emerging trade treaties do very little to protect the health of the populations, and their main purpose is to increase the amount and freedom of trade.

Farmers such as these are under threat because of international trade agreements that protect TNCs at the expense of farmers. (Tom Pilston/Panos)

TRADE IN SERVICES AGREEMENT (TISA)

A further international agreement that will have an impact on health is the Trade in Services Agreement (TISA). Negotiations about this agreement started in secret in 2012 and seek to convert all forms of services across the world into tradeable commodities. The TISA negotiations need to be understood in the context that the economies of the rich countries are critically dependent on the services sector—services comprise 75 per cent of the economy in both the European Union and the USA. While manufacturing has continued to shift to countries other than the elite rich countries, economic growth in the high-income countries is dependent on the growth of the services sector, a large part of which is mediated through TNCs operating in services such as finance, insurance, media and management consultancy. The WTO attempted a similar move through the General Agreement on Trade in Services (GATS). Both agreements seek to liberalise trade, but the main objective of TISA is to open up all service sectors to international competition. Health is one of the targets of TISA because so much of gross domestic product spending in OECD countries is on health (around 10 per cent in Australia and nearly 17 per cent in the USA). TISA could mean that a public scheme such as Medicare could be challenged at the WTO because it does not provide the least trade-restrictive policies. Koivusalo and Tritter (2014) discuss the implications of trade agreements for the UK National Health Service and conclude that they will reduce the policy space in which member states can operate to manage and regulate health. They also note that the impetus for the agreement is trade and commercialisation, and the interests that are shaping the provisions are those of TNCs rather than member states in the European Union or their citizens.

Price et al. (1999) maintained that GATS posed a risk to the following features of European public health care systems: universal coverage, solidarity through risk-pooling, equity, comprehensive care and democratic accountability. TISA is now the mechanism by which these same risks exist. Koivusalo and Tritter (2014) note that the pressure to privatise basic services has continued on low- and middle-income countries. This is despite the fact that private provision of health services tends to focus on the most affluent people because this is more profitable. By contrast, public services are able to cross-subsidise and so are more likely to be equitable. Fundamentally, TISA makes trade liberalisation the aim of the provision of services rather than any public health considerations. TISA also threatens to reduce the sovereignty of national governments.

THE IMPACT OF TRANSNATIONAL CORPORATIONS

> When public health policies cross purposes with vested economic interests, we will face opposition, well-orchestrated opposition, and very well-funded opposition.
> *Margaret Chan, Director-General, World Health Organization, 2013*

A main factor underpinning the rapid growth of globalisation is TNCs. Aided by innovations in transportation and communication, and as a result of competition between developing countries for foreign investment, TNCs have been able to extend operations and take advantage of favourable regulatory and financial environments that seek to attract the investment of such corporations, particularly in low-income countries (Westaway, 2012). The United Nations Conference on Trade and Development (UNCTAD) stated that there were 82 000 TNCs worldwide in 2007, compared with 7000 in 1970 (Kale, 2012, p. 152). The UNCTAD estimates that TNCs account for one-third of total world exports of goods and services, and the number of people employed by them worldwide totalled about 77 million in 2008—more than double the total labour force of Germany (p. 154).

Trade agreements are widely seen to favour the position of TNCs, and these corporations have massive lobbying power through which they can influence these agreements to create a favourable operating environment for them, often at the expense of human rights or public health concerns (see box 5.8). The web of influence and the ways in which they exert power are shown in figure 5.1.

FIGURE 5.1   THE CORPORATE CONSUMPTION COMPLEX

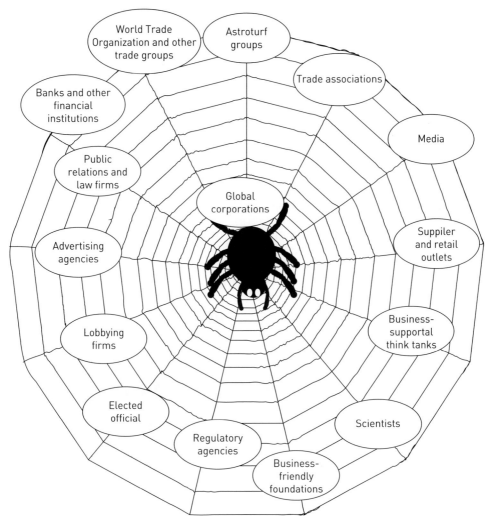

Source: Freudenberg, 2014, Figure 4.1, p. 96.

BOX 5.8   NEO-LIBERALISM AND TRANSNATIONAL CORPORATIONS

The profits of corporations were falling in the 1970s and this was put down to Keynesian economics. Pusey (2010, p. 126) says neo-liberalism was able to get rid of the institutional obstacles and 'open the field for turbo-capitalism and corporations on steroids'. Transnational corporations have five major cost inputs: capital, technology, land, wages and taxes. It is easiest to reduce the last two, and this is exactly what neo-liberal reforms have done. The prescription desired by transnational corporations was to 'deregulate the labour market, neutralise union power, get rid of collective wage bargaining, engage state power on the side of the corporations, strike down Keynesian economics, choke off the welfare state and reduce taxes' (Pusey, 2010, p. 126).

TNCs have an increasing number of critics (Korten, 1995; Estes, 1996; Shiva, 1996; Korten, 2006; Wiist, 2010; Freudenberg, 2014) who claim that TNCs are beyond the control of governments, are accountable to no external power, assume an increasing amount of control over the global economy and have little interest in goals other than economic ones. Klein (2001, p. xxi) describes these TNCs as 'select corporate Goliaths that have gathered to form our de facto global government'. In addition they are seen to exert great influence over international trade treaties that work in their favour. This situation makes it unlikely that the corporations will consider environment protection or human health promotion over short-term profit taking. TNCs have been criticised for their practices of manufacturing goods wherever they can most reduce costs (Klein, 2001). Often this means sweatshops in poor countries. For richer countries it means the withdrawal of manufacturing industries and so a decline in areas in which these industries have been the main employers. The size of TNCs is increasing and their revenue often exceeds that of nations, as shown in figure 5.2.

FIGURE 5.2   THE WORLD'S LARGEST CORPORATIONS' REVENUE COMPARED WITH SELECTED COUNTRIES

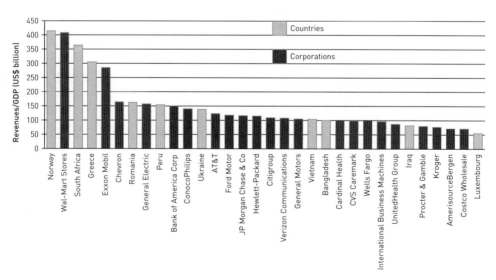

Source: Global Policy Forum, 2010.

The process of opening up so many countries to free trade and external institutions has assisted the growth and success of the TNCs. Korten (1995, p. 12) says of the corporations:

> These forces have transformed once beneficial corporations and financial institutions into instruments of a market tyranny that is extending its reach across the planet like a cancer, colonising ever more of the planet's living spaces, destroying livelihoods, displacing people, rendering democratic institutions impotent, and feeding on life in an insatiable quest for money … The problem is not business or the market per se but a badly corrupted global economic system that is gyrating far beyond control. The dynamics of the system have become so powerful and perverse that it is becoming increasingly difficult for corporate managers to manage in the public interest, no matter how strong their moral values and commitment.

Freudenberg (2014, p. 3) notes that 'decisions made by executive and managers in the food, tobacco, alcohol, pharmaceutical, firearm, automobile and other industries have a far greater impact on public health than the decisions of health officials, hospital directors and doctors'. There has been competition between developing countries to win investment from TNCs. Stiglitz (2007, p. 196) notes 'Globalization has compounded the problems arising from the misalignment of incentives to modern corporations. Competition among developing countries to attract investment can result in a race to the bottom, as companies seek a home with the weakest labor and environmental laws'. TNCs also pay low rates of taxation and have developed a series of measures to avoid doing so (Tax Justice Network, 2013).

Currently the terms of world trade are extremely favourable to TNCs and the richer countries where they are based. Their impact on low-income countries has been described by Erunke and Hafsat (2012), who maintain that the incursion of TNCs into Nigeria has worsened poverty in the country. They note that some multinational companies have sold products that have been proven to cause health problems in the population (they cite the example of a skin lightening cream that contains mercury). They also note that inappropriate consumption patterns have been encouraged where people swap to more expensive foreign products that are often not as healthy as the traditional products. For example, many Nigerians are now eating processed breakfast cereal instead of porridge made from local grains. They note that the TNCs, together with the domestic elite class, 'connive to siphon the nation's resources abroad for metropolitan developments' (p. 4).

Significant questions have been raised about the ways in which TNCs may have an adverse impact on human rights. While it is true that their operation may bring benefits in terms of the provision of jobs and income, much of the profit in low-income countries is taken offshore. The TNCs are also not governed by the United Nations convention on human rights and there have been many documented cases of abuse (Westaway, 2012). The lack of responsibility or accountability for human rights underlines the fact that the sole responsibility of TNCs is to produce a profit for their shareholders. Unlike governments, they have not signed up to human rights or other conventions, and while they may at times market themselves as showing corporate social responsibility, this is generally done with half an eye towards the likely impact of this seeming benevolence on their profit bottom line (Freudenberg, 2014).

Two groups of TNCs illustrate the impact these companies can have on public health. The first example is of 'Big Pharma', the collective term used to describe the world's major pharmaceutical corporations. Public health activists see them as very influential in the control of the trade in medicines and in dictating global trade rules and regulations (People's Health Movement et al., 2005). These companies have massive wealth (it is calculated that the combined worth of the five top drug companies is twice the combined GNP of all sub-Saharan Africa) and work directly with the US Government and European Commission to shape international rules on patents (Drahos and Braithwaite, 2004). The other example is the global food industry, which has massive and growing effects on diets. Chopra (2005, p. 49) describes the massive growth in a few TNCs that now dominate the entire food chain globally. Six corporations account for 85 per cent of world trade in grain, eight for 60 per cent of global coffee sales, seven for 90 per cent of tea consumed in the West, three for 80 per cent of the bananas. Moodie et al. (2013, p. 671) add that in the USA, the ten largest food companies control more than half of all food sales. Most of the profits from food are made through manufacturing the food, much to the frustration of farmers, who receive little of the profit that TNCs make from food. The food industry creates markets for its products. In 2002 more than 11 300 new food products were introduced in the USA alone, in a world where the number of undernourished people is increasing (Chopra, 2005). The food industry is the largest global investor in advertising and promotions and is ruthless in its search for new markets. Many of the new products are high in fat and sugar and are being blamed for the increase in obesity and related chronic diseases globally. Moodie et al. examine 'the unhealthy commodities industries', which they conceive of as the 'major drivers of [non-communicable disease] epidemics worldwide' and note that the potential for growth for the food industry is in low- and middle-income countries. They note that this growth is through 'ultra processed products' that are very high in fat and sugar, and that these products are fuelling epidemics of obesity, diabetes, and other diet-related chronic diseases. Koivusalo et al. (2013) note that trade in tobacco and alcohol products is associated with higher levels of consumption of these products and health-related problems. They also note that trade and investment treaties appear to be eroding the policy space for governments to intervene through restrictions on advertising, points of sale, taxation, and other measures likely to reduce the adverse health impact of these products.

THE IMPACT OF NEO-LIBERALISM ON HEALTH

If you are a stockbroker living in the USA or the Chief Executive of a TNC, you will probably feel that neo-liberal globalisation has been very good for your health. On the other hand, if you are raising a family and hold an insecure job in a US, European, Australian or Canadian city, or work in a factory in a low-income country in unsafe and extremely stressful working conditions, then you would see a primarily negative impact on your well-being. More extremely, if you are a poor person living in an African country that has been subject to the World Bank's and IMF's neo-liberal policy prescriptions, or you live in Greece or Spain and have been subjected to austerity in the wake of the GFC, you would feel that neo-liberalism has had a very poor effect on your health.

Neo-liberalism is the quintessential upstream variable and, as such, evidence of patterns of causality will be hard to prove in ways accepted by epidemiology, as it comprises 'multiple, interacting policy dynamics or processes, the effects of which may be difficult if not impossible to separate' (Labonté and Schrecker, 2006). Thus, trade liberalisation may reduce the incomes of some workers or shift them into the informal economy while reducing tariff revenues (and therefore funds available for public expenditures on health or education) before the benefits of any revenue gains from income and consumption taxes are felt. Simultaneously, the need to conserve funds for repaying external creditors may create a further expenditure constraint. Labonté et al. (2011) note that causal pathways linking globalisation with changes in health status are rarely linear, do not operate in isolation from one another, and may involve multiple stages and feedback loops

The model shown in figure 5.3 portrays the complex driving and constraining forces introduced by neo-liberal globalisation. It highlights five means (three direct and two indirect) by which globalisation influences health. The direct means are through impacts on the health system and the direct impact of policies (for example, policies in favour of privatisation) and through international markets (for example, the effect on pharmaceutical prices of the WTO agreement on TRIPS or the deregulation of financial markets that was the major cause of the GFC). The indirect effects include those operating through the national economy on the health sector (for example, IMF's prescriptions from 2010 for austerity policies and the subsequent contraction of government spending in 94 countries) and the impact on nutrition and living conditions resulting from changes in household income from changing labour market conditions or reduction in social security support. These relationships are complex and demonstrate the complexity of determining the impact of neo-liberalism on health.

The principal effects on health of neo-liberal globalisation are discussed under the following topics:

- assumption that economic growth is necessary to improve public health
- privatisation of public services and the transformation of services delivered to 'products'
- private sector ethos into public sector services and subsequent lack of commitment to broader social and environmental goals
- growing inequities evident under neo-liberal policies
- communication and consumer culture
- lack of global governance to regulate health and well-being impacts of neo-liberalism. A summary of the arguments are provided in table 5.1.

FIGURE 5.3   CONCEPTUAL FRAMEWORK LINKING GLOBALISATION TO HEALTH OUTCOMES

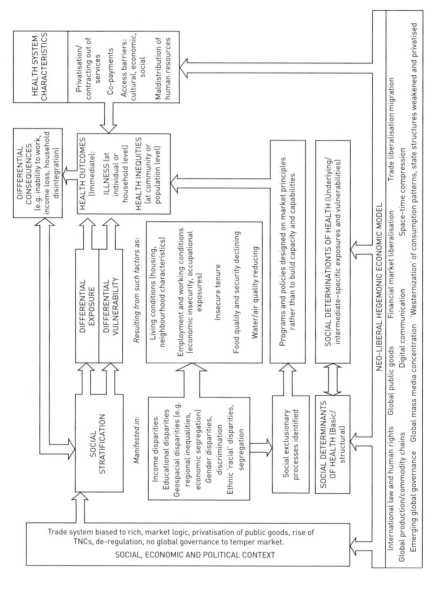

Source: developed by Labonté and Schrecker (2007), based on Diderichsen et al. (2001), as refined by Solar and Irwin (2005), for the Commission on Social Determinants of Health.

TABLE 5.1   NEO-LIBERAL GLOBALISATION, HEALTH AND EQUITY: THE ARGUMENTS
SUMMARISED

Neo-liberalism is good for health and equity	Neo-liberalism is bad for health and equity
Neo-liberalism encourages economic growth, which will ultimately benefit everyone through a trickle-down effect	Structured unfairness of the world trade system ensures an unfair distribution of wealth and is strongly biased in favour of the highest income strata of rich countries and transnational corporations
	Evidence suggests that the key issue is how governments invest the fruits of economic growth that determine health outcomes
Neo-liberalism encourages economic growth and so is good for the incomes of the poor and therefore for their health	Foreign direct investment and deregulation of financial markets has introduced economic vulnerabilities and insecurities globally
Privatisation and deregulation will improve the efficiency of the provision of a wide range of services that are vital to health, including housing, water, transport and health services	Neo-liberalism for poor countries has meant the imposition of structural adjustment packages and poverty reduction strategy papers, and private provision of services, which has affected the level of services and protection for local industries and resulted in rundown of public infrastructure and less services for the poor
	Privatisation is concerned more with opening up markets for transnational corporations than health and well-being
	Privatised services are concerned with profits more than public health
Public services are transformed by neo-liberal ideology to operate as more efficient markets do	Reformed bureaucracies do not stress nation building and public health is not a commodity that can be bought and sold
	Long-term outcomes are neglected
Opening up markets and deregulating trade will benefit all populations, increase wealth and reduce poverty	Neo-liberalism increases inequities both within countries and between them and, while there has been a reduction in poverty by some measures, levels of poverty continue to be high and unjust when wealth gaps are widening
Globalisation opens up communication around the world and creates a global village. This will result in more understanding and less conflict	Globalisation threatens to swamp the variety of cultures around the world and impose a common Americanised McCulture aided by the spread of transnational corporations. This domination fuels resentment and contributes to the growth of fundamentalism
Neo-liberalism provides a variety of consumer goods that enrich people's lives	Growth of advertising and marketing from transnational corporations imposes unhealthy diets (resulting in obesity and higher burden of chronic disease) and makes consumption an unhealthy obsession, as well as encouraging levels of consumption that are environmentally unsustainable
	Consumerism creates pressures for individuals and is a threat to mental health
	There is a digital gradient between and within countries, which means the benefits of the communications revolution are not shared equally
The current system of global governance provides sufficient safeguards for health	The current system of global governance is dysfunctional and needs strengthening so that it is able to regulate the market in favour of health

ECONOMIC GROWTH IS ASSUMED NECESSARY FOR IMPROVED PUBLIC HEALTH

The justification for neo-liberal policies lies in the belief that, ultimately, policies that encourage economic growth will be beneficial for human health and well-being. Proponents of this viewpoint refer to the experience of Western industrial societies, where economic growth appears to have been associated with better standards of living and longevity. They argue that wealth will trickle down the gradient so that eventually everyone benefits from economic growth. A British study suggests that this view represents a very partial reading of the historical evidence. Szreter (1995), a Cambridge historian, analysed the impact of rapid economic growth on the British population in the nineteenth century, and concluded that health and living standards did not improve significantly until economic growth was combined with significant state intervention (by national and local government) to control the excesses of the unfettered market, redistribute resources and organise collective facilities such as clean water and sewerage. Szreter concludes that without this state intervention, economic growth brings with it 'the four Ds' of disruption, deprivation, death and disease. He comments, 'The British historical case therefore confirms that without politically-managed continual and significant redistribution of resources towards the urban poor, there is no escape from the insecurity of "the four Ds" generated by rapid urbanising economic growth' (Szreter, 1995, p. 55). This lesson is vital for the urban poor of today's rapidly industrialising countries because it suggests that the way the fruits of development are used by governments is essential in determining health and equity outcomes.

The experience of countries that have achieved high standards of health without achieving high incomes is also relevant here. The fact that countries such as Cuba, China, Costa Rica, Sri Lanka and Kerala State in India have achieved high life expectancy while remaining low- or middle-income countries adds further evidence that wealth is not essential to achieve high levels of population health (see box 13.2 for further details).

So economic growth alone is likely to detract from, rather than contribute to, health. It is only when growth is combined with state action to ensure redistribution of resources and the direction of the benefits of economic growth to public projects for the communal good that health improves. Regulation of private sector activity is vital to protect health, and in the twenty-first century that means TNCs, as seen in the earlier section.

Stiglitz's (2002) account of the imposition of the free market mantra of the 1980s and 1990s makes it clear that the market liberalisation imposed on so many poor countries left havoc in its wake because there had been no protection or preparation, just the imposition of an ideology. He notes (2002, p. 17) that the small developing countries are like small boats launched on a voyage of the rough seas of market liberalisation 'before the holes in their hulls had been repaired, before the captain has received training, before life vests have been put on board'. The World Bank acknowledges that poverty and health are linked, but sees encouraging economic growth as a necessary condition for health improvement. Consistently the Bank advocates policies designed to assist the poor, but it does not consider the health consequences of unfettered economic growth. This has led to greater inequalities between countries and within countries, and health gain has been patchy around the world, with sub-Saharan Africa experiencing declining life expectancies (McDaid and Oliver, 2005). The focus on economic growth is often adopted without considering the impact of economic growth on the environment or people's lives. This means it is often at the expense of other goals of government such as protection of the environment or progressive, redistributive social policy (see chapter 16).

Those arguing that neo-liberalism is good for health base their case on the economic benefits of globalisation and the benefits that, it is argued, this brings to the poor. This view was summed up by Feacham (2001, p. 504), who said, 'globalisation, economic growth and improvements in health go hand in hand. Economic growth is good for the incomes of the poor and what is good for the incomes of the poor is good for the health of the poor.' Economic growth is seen to come from openness to trade and the inflow of capital, technology and ideas. But despite Feacham's very optimistic view, the review of evidence provided by Labonté et al. (2011) and Ortiz and Cummins (2013) paint a much less rosy picture. It is very hard not to come to the conclusion that world trade is organised in such a way that it systematically favours richer nations and, in particular, is geared to supporting TNCs. The net result is that, for the global population as a whole, the processes of economic globalisation (as described above) are not a positive force for poorer people in rich countries and for the vast majority of the population in developing countries. The beneficiaries from globalisation are the wealthiest in all societies and those with an investment in TNCs.

Feacham also argued that globalisation has been a means of reducing corruption in poor countries. Others have argued that the evolution of corruption has been intimately tied to colonialism. The interventions that Feacham sees as reducing corruption have involved the imposition of neo-liberal policies.

The key reason economic growth is pursued as a goal of government also relates to neo-liberalism. Harvey (2010) provides a Marxist analysis of the 'enigma of capital' and explains that capitalists are always looking for new markets. He sees that the capitalist system has lurched from crisis to crisis and after each one looks for new markets to provide profits. This is where neo-liberalism has been so helpful. Its advocacy of economic growth as the path to well-being and the need for a reduced public sector expand the space in which the private sector is able to operate and make profits. Capitalism is not concerned about well-being for the whole population, however, but rather for a few people who control the capital. Piketty (2014) (a French economist) has shown that the capital-owning class is highly concentrated and is currently further intensifying its wealth at the expense of the mass of the world's population. So long as economics are organised around the needs of this capital-owning class and they have a strong influence on political agenda, the focus on economic growth at all costs is not likely to change despite its adverse social, environmental, population health and equity impacts.

PRIVATISATION AND DEREGULATION

Neo-liberalism has led to a pandemic of privatisation or contracting out of public services, justified by the rhetoric that the role of government 'is steering not rowing' (Osborne and Gaebler, 1992). This appears to be happening without much regard for the impact on public health. The impact on the privatisation of housing, transport and education in the UK was described by Hutton (1995), who sees that it has had particularly bad consequences for the poorest part of the population. Ferreira (1997), a former World Bank economist, argues that even when privatisation is designed to be egalitarian, it may lead to increases in inequality and possible poverty. Privatisation implies short-term profit-taking, whereas public health is concerned with the long term and the preventable, especially improvements in well-being and decline in disease. Private sectors are not good at ensuring equity of access or concerned about overall equitable population health gain. Institutions have been weakened by the rapid privatisation of services and decreasing government control and accountability. While these policies have had an adverse impact on some sections in rich countries, their impact on poor countries has been more devastating. Carpenter (2000, p. 339), commenting on the impact of neo-liberalism, notes that it has 'made serious inroads against all forms of collectivism, fostering the expansion of the market and the erosion of state regulation of social life'.

Governments around the world that followed neo-liberal policy prescriptions have pursued policies of privatisation of many core services including utilities (gas, electricity, water supplies) and, increasingly, health services. Under the Cameron Coalition government in the UK, even the National Health Service is being contracted out and privatised by stealth. This process, however, started under the previous Labour government. Talbot-Smith and Pollock (2006) describe how the market reforms mean that a new National Health Service is being created and they question whether the original principles of the National Health Service of universality, comprehensiveness and free services at the point of delivery will be maintained. They also doubt whether the efficiency claimed for the new system will actually be more efficient. They note the high costs of the system such as 'making and monitoring contracts. Paying for capital, invoicing and accounting for every completed treatment, marketing services and dealing with fraud (which invariably increases with more and more complex chains of market exchanges, as is rife in US health care)' (Talbot-Smith and Pollock, 2006, p. 181). Pollock and Price (2013) say that the 2012 changes to the NHS amounted to an 'assault on universal health care' and have undermined the democratic and legal basis of the National Health Service in England, opening the way for privatisation. In Australia the pressure to privatise has been evident over the last two decades and has been intensifying under the Abbott Coalition government with their strong encouragement to state governments to sell off assets to rein in deficits, despite the obvious short-term nature of this strategy.

The effect of privatisation will take time to filter through to health outcomes, but the evaluation that has occurred suggests that it will have a negative impact on health. There is no evidence that privatisation of health services will increase access for poor people. In fact the contrary is true, as private, for-profit health services are interested in profits before access and equity. There is good evidence that the impact of the introduction of user charges falls most heavily on the poor and will result in their using services less (Creese, 1991; Gilson et al., 2007). Another face of the pressure for privatisation is through the campaign for universal health coverage. Sen (2013) notes that, in India, the policies promulgated to achieve universal health coverage have been to promote the private sector and to scale up health insurance schemes. He says the private sector is growing rapidly and developing into a corporate hospital-based system of care. Stiglitz (2002) notes that privatisations often happen before governments have time to put in place a regulatory framework that might have controlled some of the worse excesses of rapid marketisation. He notes that, in many countries, privatisation is jokingly referred to as 'briberization', and that often the process was accompanied by corruption. This was particularly the case in the former Soviet Republics.

Paddon's (1996, p. 17) assessment of the outcomes of privatisation of utilities in the Asia–Pacific region made two decades ago are still highly relevant:

* higher costs
* potentially significant environmental costs
* overstated benefits of revenue and debt reduction
* social costs of privatisation in terms of rising unemployment and loss of productive assets not being fully taken into account
* growing inequality.

The potential for shorter-term health effects from privatisation was shown in Britain by the muzzling of the report to the Chief Medical Officer on the health effects of the increased number of water disconnections that occurred after water supply services were privatised (Editorial, *The Lancet*, 1995, p. 598). Disconnections became more common as the price of water increased, and more people were unable to pay their bills.

Deregulation may also threaten public health. In the case of the labour market, it tends to undermine work conditions, such as penalty rates, occupational health and safety standards and permanency of employment. The labour market reforms introduced in Australia by the Coalition government in March 2006 were strongly opposed by the Australian Labor Party and the trade union movement. Both predict the reforms will have an adverse effect on family and social life and eventually on the health of workers and their families. A campaign against these reforms ('Your Rights At Work') by the Australian Council of Trade Unions was credited with the defeat of the Howard Coalition government in 2007.

The monitoring and surveillance role of public health may also be undermined if the powers of regulatory authorities are reduced, affecting the safety of substances and products, including food. Privatisation and deregulation are global movements, driven by multinational companies and international consulting firms, and strongly supported by the World Bank and IMF. The WTO Agreement sets limits on the use of clauses to give preference to local suppliers or contractors and makes it more difficult to put social and employment objectives into outsourcing contracts. Privatisation, then, is one facet of neo-liberal globalisation and the growth of TNCs.

PRIVATE SECTOR ETHOS INTO PUBLIC SECTOR SERVICES, INCLUDING PUBLIC HEALTH

Bureaucracies around the world, including in the health sector, have been transformed and restructured so that they operate more like private sector companies. Considerable reservations about the trends towards this new style of public management have been expressed (Bryson, 1987; Yeatman, 1987, 1990; Considine, 1990; Rees and Rodley, 1995; Beresford, 2000; Miller and Orchard, 2014). The appropriateness of applying an economic logic to a non-market public sector has been questioned and the trend has been criticised for being inconsistent with the expectation that public services ought to pursue the social goal of equity and be accountable for running equitable and accountable services to the public as a whole. Yeatman's (1987) observations, made when these processes were just beginning, remain true. She noted that the distinctiveness of public management is submerged by inappropriate private sector models and issues reduced to economic ones to the detriment of 'people and processes' approaches. Public health, and particularly the new public health, is fundamentally about people and processes. Economic considerations are important but should not be uppermost in a society's plans for public health. Davis (1995, pp. 132–3) comments of the new style of private sector management in the public sector that it 'is machismo management style that cuts, burns and slashes while demanding commitment to the new corporate culture and a health care market that is governed by a logic of cost rather than care and compassion'. When this logic is applied to public health policies, it is does not promote equity.

Privatised systems means more staff are on contracts and work-to-performance agreements and do not have the security of employment that was the hallmark of public services in earlier decades. Contract employment is likely to mean that public servants are more willing to please their political bosses than in the past because their continued employment depends on this. One effect of this change is that public servants' lives are characterised by far greater uncertainty, change and job insecurity than in the past. The British Whitehall II study looked at the impact of these public service changes on civil servants by comparing the health status of those affected by change and insecurity with those who were not (Ferrie et al., 1998). The study concluded that the anticipation of major organisational change results in significant increases in self-reported morbidity and small increases in clinically measurable outcomes. The authors conclude that public service changes have repercussions for health service use, employee well-being and organisational efficiency. They suggest that public health professionals should urge governments to take these costs into account when the returns of a flexible labour market are counted.

Self (1997) sees risks to the democratic working of bureaucracies as a consequence of their operating more like private companies. He foresees that protection against corrupt governments is likely to be weakened, noting that public policy problems are usually more complicated than those facing business, and that the expert and impartial advice so important to the operation of a liberal democracy may not be present.

A semantic indication of the adoption of market principles is the increasing use in many Australian bureaucracies of the term 'customer' instead of 'citizen'. The notion of 'customers' for public health services highlights the inappropriateness of the term for public services. Few people would 'shop' for public health services to prevent something they assume will never happen to them! Costello makes the point thus: 'Even the Police Commissioner, Neil Comrie, speaks of Victorians as "customers". That rather shocked me when I suddenly thought of the old adage "the customer is always right".

If we are all customers, how are the police going to now arrest anybody?' (Costello, 1996, p. 13)

Most crucially, the concept of customer negates that of citizen, and with it the notion of the rights and responsibilities that are associated with citizenship. Costello (1996) argues that 'customer' undermines the notion of a social contract between governments and citizens and changes the moral basis of government to a market-driven one. Customers are conceived as having a concern with their own needs only; citizens, by contrast, are also concerned with general social goals—important aspirations in a democratic society.

Public health is an activity that exemplifies the problems of applying a market logic to public sector activities. It is not a commodity that can be 'sold' like any other good. Its preventive nature means that it is not in 'demand', except when there is a threat to public health, when it is generally not possible to 'buy' public health measures to deal with a problem. Prevention requires planning, foresight, long-term investment and taking into account not just the potential for profit but also the potential to improve health and living conditions. In addition, most public health measures are best provided on a collective rather than an individual basis, so the market mechanisms are not appropriate as a means of organising them and the imposition of market models poses significant issues for public health. This is shown by the tendency for marketised models of public service to be dominated by short-term rather than long-term goals and purchaser–provider split.

Neo-liberalism and the managerialism that comes with it puts great emphasis on measuring the immediate output of expenditure, which led to the criticism that it elevates the quantifiable over the worthwhile. Bryson (1987) suggests that, based on her experience in the Victorian public service in the mid-1980s, the private sector ethos introduced into public sector management tends to introduce innovation with no discernible improvement for the clients of the system, despite the supposed focus on 'outcomes'. There appears to be a contradiction. While the rhetoric is concerned with making things better and improving outcomes, in practice energy is focused on processes and reorganisations of structures, rather than on improving the outcomes for citizens and their communities. In Australia, Pusey's (2010, p. 135) work has shown that the top public service has been focused on economic efficiency at the expense of other substantive long- and mid-term goals, which has come 'at great cost to our collective intelligence, to the historical creative, national building role of the Australian state'. This focus on short-term economic outcomes poses particular problems for public health, which is, by its very essence, concerned with long-term outcomes. A good public health outcome is something that does not occur, such as a disease epidemic, food poisoning or deaths from lung cancer. Such outcomes are difficult to account for in a world dominated by economic reckoning.

Another hallmark of reformed bureaucracies is that they distance themselves from service provision through mechanisms such as purchaser–provider splits (Ovretveit, 1995). Contracts have to be devised between purchasers and providers of agencies, often based on agreed outputs. The contracting culture has a number of implications for services with a health promotion or public health function (Hughes, 1996; South Australian Community Health Research Unit, 1996; Lewis and Walker, 1997):

- relationships between central agencies and service provision agencies are based on legalistic arrangement rather than trust. Contracting may result in a decline in trust and make effective relationships harder to maintain
- competition between providers does not encourage the collaboration that is so essential to health promotion and public health
- direct service provision is far easier to fund under a contract than disease prevention or health promotion activity. This may mean important public health strategies such as community development are not supported
- monitoring of service performance under contracting tends to favour quantitative measures of services provided, rather than assessment of quality or actual or potential impact on health status. Health promotion and public health work are difficult to measure in a reductionist way
- the service agreement system does not encourage community participation or local control.

Public service goals are about much more than the horizons set by neo-liberal philosophies. Public health is an essential element of nation building that is threatened by undue emphasis on economic considerations. In this vein, Saul (1997) argues that management has itself become an end, to the detriment of the public good, and that this is true for management in large corporations as well as within public services. He argues that managers have narrowed their focus and lost the ability to be creative or forward thinking. They have no commitment to broader social goals such as increasing equity within society. He comments (1997, p. 37) that in a corporatist society 'there is never any money for the public good because the society is reduced to the sum of the interests. It is therefore limited to measurable self-interest'.

Pusey (1991) and Stretton (1987) see that senior Australian public servants have been trained in a very narrow version of economics, which fails to set the discipline in its broader social and political context. Saul (1997) suggests that managers are typically taught a narrow range of management skills that does not provide them with the thinking and problem-solving skills necessary for the complex problems now facing major corporations and public services. What they do learn, in his view, is the importance of conformity. Part of a healthy society should be an insistence that its managers and bureaucrats have a commitment to broader social goals, including those of public health, and that private corporations are rewarded for social responsibility and their capacity for critical reflection, and penalised for a self-interested pursuit of profit. Korten (2006) suggests that unless there is a 'Great Turning' in values towards a far greater emphasis on social and environmental goals and less on economic goals, making profits and competition, then the future for all societies is grim.

THE GROWING INEQUITIES EVIDENT UNDER NEO-LIBERAL POLICIES

> … the resurgence of inequality after 1980 is due largely to the political shifts of the
> past several decades … the history of inequality is shaped by the way economic,
> social and political actors view what is just and what is not, as well as by the relative
> power of those actors and the collective choices that result.
>
> *Piketty, 2014, p. 20*

Neo-liberalism did not create inequities. Inequities have been present since they have
been measured and have been characteristics of many societies throughout history.
Think of slaves in the Roman Empire and in Ancient Egypt, and in medieval Europe with
its aristocrats and peasants. Yet in the period from World War II until the 1970s, evidence
indicates that economic inequities were in decline (Piketty, 2014). Neo-liberalism has
reversed that trend and the evidence is that the increase in inequities is intensifying in
the wake of the GFC. Inequities between rich countries and other countries have been
starkly evident since the nineteenth century when colonialism enabled the industrialising
countries to remove resources (including people as slaves) for their own benefit.

This pattern of colonial exploitation has continued under neo-liberalism in various
guises over the last 30 years. In the 1980s and 1990s, the rise of heavily indebted poor
countries created the situation where the outflow of funds through debt repayments was
greater than the inflow of aid. Between 1980 and 1987, the proportion of government
budgets allocated to interest payments increased from 9 per cent to 19 per cent in Latin
America, and from 7.7 per cent to 12.5 per cent in Africa (Korten, 1995, pp. 164–5). This
situation came about because of the expansion of TNCs into low-income countries,
which were attractive because of their low labour costs financed by an expansion of
commercial loans. There was little regard for the borrowers' ability to afford the interest
payments. The oil crises of 1973 and 1979 in members of the Organization of the
Petroleum Exporting Countries resulted in increased oil costs. The developing countries
continued to import the more expensive oil but had to borrow to pay for it. In 1979,
the governments of most developed countries began to impose monetary policies that
slowed both inflation and growth. The results for developing countries were that markets
declined and prices dropped. So they were faced with increasing debt, falling export
earnings and higher oil prices all at the same time (Potter, 1988), and neo-liberalism
meant that no rescue packages or relief were found.

The situation of the heavily indebted poor countries became the subject of a global
advocacy campaign in the 1990s and 2000s (see chapter 22) and the G8 took some
steps to relieve the situation in response. However, in order to be eligible for the IMF
and World Bank debt relief scheme, countries had agree to 'sound reform' and agree to
develop and implement a poverty reduction strategy paper (IMF, 2014). This imposed
neo-liberalism on the domestic policy, resulting in cutbacks to and lack of investment
in public services, which led to significant disadvantage and lack of progress for the
poor. Debt remains an issue, but where debt relief has been enacted, more resources are
available for health and social spending. Economic growth in some of these countries
has meant that they can now service debt more easily. Debt has been compounded

by illegal and commercial capital flight, which results in net resource transfers away from low-income countries. Henry (2012, p. 5) calculates that, since the 1970s, 'with eager (and often aggressive and illegal) assistance from the international private banking industry' the group of 139 countries he considers to be private elites accumulated $7.3 to $9.3 trillion of unrecorded offshore wealth in 2010. This was while structural adjustment packages were being imposed and governments urged to cut public spending and sell off assets, which had very adverse impacts on poor people and the general public infrastructures of the countries. Basically, assets of many countries have been privatised and the debt socialised. Henry (2012, p. 6) says 'in terms of tackling poverty it is hard to imagine a more pressing global issue to address'.

Wade (2001), Professor of Political Economy at the London School of Economics and ex-World Bank economist, notes the complexities of determining whether global inequalities are increasing or decreasing. He considers eight alternative measures of income distribution and concludes that none suggests that the distribution has become more equal over the past 20 years. Seven measures show varying degrees of increasing inequality. Evidence accumulating since then points to the widening of inequities (Oxfam, 2014b; Piketty, 2014). (Chapter 13 further details the growing inequity between the world's rich and poor.) This trend adds further weight to the argument that economic globalisation has not been good for equity. The ways in which the current distribution of resources adversely affect health and health equity were analysed by the Commission on Social Determinants of Health (2008). It noted that 'social norms, policies, and practices that tolerate or actually promote unfair distribution of, and access to, power, wealth, and other necessary social resources' create systematic inequalities in people's everyday lives.

One of the benefits of globalisation for many people in Western countries is the access to rapid communication and information. However, for poor people in rich countries (Baum et al., 2014c) and the vast majority of people in low-income countries, access to the new methods of communication are very limited. In Africa in 2011, only 16 in 100 people have access to the internet (and this had doubled since 2008). If this rate continues to grow in a similar manner, the notion of the world as a global village may be realised in the coming decade.

An important aspect of the argument about the overall health impact of neo-liberalism in the past 30 years concerns the impact on levels of poverty. There is little doubt that poverty rates are now increasing in many rich countries as the impact of the GFC (largely a result of the deregulation of financial markets that resulted from neo-liberal policies) comes home to roost. The United Nations is reporting rising rates of hunger, poverty and unemployment following the GFC and these appear to be resulting from the austerity policies (Ortiz and Cummins, 2013). This is most spectacularly shown in Greece, where the impact of the GFC and the policy response to it have had a dramatic effect on poverty and well-being (Ifanti et al., 2013). In Spain, the GFC has resulted in an 8 per cent increase in suicide rates and these are concentrated among those who lost their jobs (Bernal et al., 2013). Global poverty rates were declining in the decades prior to the GFC, according to some measures (Labonté et al., 2015). The absolute number of people worldwide who are living below the poverty line has also fallen, but there are important regional variations. Poverty rates and absolute numbers fell most dramatically in eastern Asia and China. Poverty rates and absolute numbers rose in Latin America and the Caribbean during the 1980s and 1990s—the 'lost decades' that many attribute to structural adjustment programs imposed by the World Bank and IMF—before beginning to fall in the 2000s. The recent decline in poverty in this region is a result of economic growth combined with redistributive social policies and programs (Birdsall et al., 2011). Poverty rates in sub-Saharan Africa fell only very slightly over the period 1980–2010, and failed completely to keep pace with population growth, as there was a continuing rise in the number of poor people. Globally, there were only modest gains in decreasing poverty at the $2-a-day level, and a substantial rise in the number of people subsisting on between $1.25 and $2 a day, which the World Bank notes 'points to the fact that a great many people remain vulnerable' (Chen and Ravallion, 2012, p. 3).

These data mean that the not very ambitious Millennium Development Goal of halving the number of people living in extreme poverty ($1 or $1.25 per day) between 1981 and 2015 has been achieved. But Labonté et al. (2015) note that the value of the global economy more than quadrupled during this same time period—from US$18 trillion in 1980 to US$80 trillion in 2011, suggesting that very little 'trickle down' of the benefits of that growth reached the 'bottom billion' (Collier, 2007). They further point out that halving these rates says nothing about the adequacy of the poverty lines themselves. Some have proposed an ethical poverty line established by working backwards from countries with an average life expectancy at birth of 74 years (considered an ethical minimum) to the average level of consumption associated with such a life expectancy (around $3 per day) (Edward, 2006). Using this $3-a-day poverty line increases the number of global poor by 1.3 billion to around 3.7 billion, or roughly half the planet's total population. This is the situation at a time when the accumulation of wealth by the richest people on the planet is increasing dramatically (Oxfam, 2014b; Piketty, 2014) and some redistribution of power and resources could realise the ethical poverty line. Analysis by The South Centre (Ortiz and Cummins, 2013) comes to the conclusion that the recent policy prescriptions from the IMF to contract government spending has resulted in the most vulnerable households bearing the brunt of these austerity measures.

CONSUMERISM

The last three decades have seen an increasingly rampant consumerism, much of which has promoted a North American culture. Shiva comments that globalisation is not a process that encourages cross-cultural interaction but rather 'is the imposition of a particular culture on all of the others' (Shiva, 1998, p. 104). She goes on to note that this means that 'global' does not represent a universal human interest but rather a 'particular local and parochial interest and culture that has been globalised through its reach and control, its irresponsibility and lack of reciprocity'. Important symbols of US-style capitalism are the chains of McDonald's restaurants and the soft drink Coca-Cola around the world. There are few countries where these products are not sold. In many countries there are signs that the Americanisation and corporatisation of culture is resented. Many indigenous peoples and other groups feel threatened by this Americanisation and, as a result of it, perceive their own culture as under threat. A sense of culture and belonging appears to be an important aspect of well-being, so promotion of culture is a health-promoting act and the imposition of a uniform consumer culture can be seen as a threat to health and well-being.

Neo-liberalism has been presented as spreading the benefits and choices of consumer society to an increasing number of people. Others, however, see that the choice is illusory and does not necessarily contribute to increased happiness or well-being. Consumerism is encouraged by clever marketing and the often artificial creation of demand (Klein, 2001). It also gives rise to competition between individuals to acquire the latest gadget or fashion. This has been noted to be unhealthy for individuals (James 2008; Stavropoulos, 2008) and the environment (Hamilton and Denniss, 2005). Some of these effects are shown in box 5.9. In some cases the demand created may be for unhealthy products. Thus Moodie et al. (2013) note that Western-style high-fat and high-sugar diets, which are increasing chronic diseases around the world, are strongly promoted by TNCs such as McDonald's, Coca-Cola and Nestlé, assisted by the free trade policies of the WTO. The WHO response has been its global strategy on diet, physical activity and health, which was watered down following intensive lobbying from the food industry (Chopra, 2005). Lee (2005) talks of the rise of the global consumer, who is the target of marketing by TNCs. She suggests that Western ideals of diet, body shape and looks are becoming widespread in non-Western countries and creating such markets as, for instance, people in South Korea having eye surgery to make their eyes more Western-looking, or Africans purchasing skin-lightening products.

BOX 5.9   CONSUMERISM: A THREAT TO HEALTH AND WELL-BEING?

Psychotherapist Pam Stavropoulos (2008) notes that depression is prevalent throughout Western society. She argues that this reflects the nature of individualistic, consumerist neo-liberal society and that living in such a society is itself a risk factor for depression.

Psychologist Oliver James (2008) sees materialism as a threat to health because it promises that consuming will make people happy, whereas in practice it doesn't. Neo-liberalism depends on expanding markets and increased consumption—a recipe that adds up to more mental illness.

Sir Michael Marmot, who chaired the Commission on Social Determinants of Health, describes a 'status syndrome' (Marmot, 2004) that drives people to compare how well-off they are. Consumer society intensifies such comparison and explains why it is likely to create mental health problems. Advertising that is at the heart of creating and increasing demand feeds off status comparisons.

LACK OF GLOBAL GOVERNANCE REGIME TO REGULATE HEALTH AND WELL-BEING IMPACTS OF NEO-LIBERALISM

> Global governance for health is achieved when we obtain a fair and equitable global governance system, based on a more democratic distribution of political and economic power that is socially and environmentally sustainable.
>
> *The Lancet–University of Oslo Commission on Global Governance for Health*
> *(Ottersen et al., 2014, p. 633)*

Finally, perhaps the most crucial issue in terms of neo-liberalism and health has been the failure for an adequate global governance system to develop to meet the demands of an increasingly globalised economy. The Lancet–University of Oslo Commission on Global Governance for Health reviewed the system of global governance in light of health concerns and determined that the system was dysfunctional in a number of respects, summarised in box 5.10. The Lancet Commission's analysis suggests that certain actors and institutions control too much power and that power needs to be redistributed.

BOX 5.10   DYSFUNCTIONS OF THE GLOBAL GOVERNANCE ACCORDING TO A LANCET COMMISSION

1   Democratic deficit: participation and representation of civil society, health experts, and marginalised groups, are insufficient in decision-making processes.

2   Weak accountability mechanisms: inadequate means to constrain power and poor transparency make it difficult to hold institutions to account for their actions.

3   Institutional stickiness: norms, rules, and decision-making procedures are often impervious to changing needs and can sustain entrenched power disparities, with adverse effects on the distribution of health.

4   Inadequate policy space for health: inadequate means exist at both national and global levels to protect health in global policy-making arenas outside of the health sector, such that health can be subordinated under other objectives.

5   Missing or nascent institutions: in a range of policy-making areas, there is a total or near absence of international institutions (e.g. treaties, funds, courts, and softer forms of regulation such as norms and guidelines) to protect and promote health.

Source: Ottersen et al., 2014, pp. 655–7.

Attention to each of these deficits will be required if the global system of economic and political governance is to be more supportive of health.

The transnational organisations that are central to globalisation (the World Bank, WTO and IMF) have no system of democratic governance, conduct their operations behind closed doors and do not engage in any meaningful way with voices of criticism (O'Keefe, 2000; Ottersen et al., 2014). Yet as Stiglitz (2002) points out, the IMF is actually a public bank, even though it does not act like it is. He notes that even though the IMF's actions affect the lives and livelihoods of billions of people throughout the developing world, they have little say in its actions. He sees that the IMF is always more concerned with getting creditors repaid than with maintaining full employment. Of course this reflects the fact that finance ministers and central bank governors have most say in the IMF and that the USA is most powerful in directing it. The World Bank is also not accountable for its policies. While there is now widespread agreement that its structural adjustment policy had devastating impacts on millions of people in poor countries, there has been no forum in which it has had to assess or report on the impact.

Some mechanism has to be found that will make the powerful international financial institutions accountable. Stiglitz (2002), based on his inside knowledge, concludes that there has to be a change in the governance of these financial institutions, including the veto held by the USA, and control over voting rights held by the USA and other rich countries. O'Keefe (2000) argues that deliberative democracy is needed at the heart of transnational decision-making bodies, especially those regulating business and finance. She describes the process thus: 'Deliberative democracy involves setting up processes and mechanisms that would allow decision-making to take place in a public sphere. This would be a step to transparency. For a start we would know what the decisions are. Decision-making would be deliberative. This means that decision-making would be based on reason and that these [sic] would be open to review' (O'Keefe, 2000, p. 168). A process of democracy should ensure that the interests of developing countries and vulnerable populations are represented. Woodward et al. (2001, p. 880) suggest this will mean 'international institutional reform, including changes in voting structures and negotiation processes, an increased role for civil society organisations and definition of the appropriate role of private companies'. Similar conclusions were reached by the Lancet Commission (see box 5.10). These democratic reforms are crucial to shaping globalisation processes so that they are more likely to promote health than they are currently, but the impact of India, China and Brazil developing more powerful economies is changing the balance of power, and the exact impact is as yet unclear.

THE VOICES OF DISSENT: CIVIL SOCIETY MOVEMENTS

> Health is a social, economic and political issue and above all a fundamental human right. Inequality, poverty, exploitation, violence and injustice are at the root of ill health and the deaths of poor and marginalised people. Health for all means that powerful interests have to be challenged, that globalisation has to be opposed, and that political and economic priorities have to be drastically changed.
>
> *People's Charter for Health, People's Health Movement, 2000*

The past decade has seen the growth of civil society movements directly challenging the tenets and goals of neo-liberalism. The People's Health Movement has led the way among health civil society with its work guided by the People's Charter for Health (quoted above). More broadly, the Occupy movement started in New York in 2011 (www.occupytogether.org) and has spread around the world. This movement protests against wealth inequity, TNCs and the impact of neo-liberalism on working people and their communities.

The power of civil society to change the behaviour of capitalism was shown during the industrial revolution in the UK. Szreter (1992) points out that, in the first stage of the industrial revolution, life expectancy in cities such as Manchester was as low as 28, and it was only when various groups in society (for example, trade unions and social reformers) began to advocate and lobby for legislative control over working conditions and living conditions that life improved for the new urban populations. This lesson suggests that current civil society protests against the exploitations that are accompanying globalisation are a very healthy sign and an essential part of a global movement to create a healthier and more equal world. Protests against neo-liberal globalisation have been growing over the past decades.

Health advocacy and activism are discussed further in chapter 22. Here, examples are given of the ways in which activists have opposed neo-liberalism on health grounds.

BRINGING THE VOICE OF ORDINARY PEOPLE FROM THE GRASSROOTS

The People's Health Movement (www.phmovement.org) was formed in 2000 at the first People's Health Assembly held in Savar, Bangladesh. This Assembly was conceived as a people's alternative to WHO's World Health Assembly, which is perceived to have become increasingly out of touch with ordinary people and susceptible to lobbying from powerful interests such as big business and the World Bank. A People's Charter for Health was drafted at the meeting and it reflects the ways in which the People's Health Movement has directly linked the political economy of the world with people's health status (see box 5.11).

Since its formation the People's Health Movement has become increasingly influential as a global voice of conscience. The Second People's Health Assembly, held in Cuenca, Ecuador, in July 2005, made similar calls for reform of the pattern of economic globalisation and documented these in the Cuenca Declaration (see box 5.11). The People's Health Movement has an active WHO Circle, which lobbies WHO to take a stance that is more protective of people's health status and less protective of the interests of those with power. In 2006 it launched a Right to Health campaign, which aims to ensure every human has access to health care.

BOX 5.11 THE PEOPLE'S HEALTH MOVEMENT'S OPPOSITION TO ECONOMIC GLOBALISATION

The People's Health Movement (PHM) (a network of health-related non-government organisations and a growing number of individuals), formed in December 2000 at the first People's Health Assembly, explains its philosophy in the People's Charter for Health

(2001) and the subsequent Cuenca Declaration (2005) and Cape Town Call to Action (2012). These documents list a series of demands that would roll back and reduce the impact of economic globalisation. They clearly show the ways in which a civil society network sees the World Bank, International Monetary Fund and World Trade Organization having an impact on health. Examples of these are as follows:

- Demand transformation of the World Trade Organization and global trading system so that it ceases to violate social, environmental, economic and health rights of people and begins to discriminate positively in favour of middle- and low-income countries in order to protect public health. Such transformation must include intellectual property regimes such as patents and the Trade-related Aspects of Intellectual Property Rights (TRIPS) agreements.
- Demand the cancellation of Third World debt.
- Demand radical transformation of the World Bank and International Monetary Fund so that these institutions reflect and actively promote the rights and interests of developing countries.
- Demand effective regulation to ensure that TNCs [transnational corporations] do not have negative effects on people's health, exploit their workforce, degrade the environment or impinge on national sovereignty.
- Demand that national governments act to protect public health rights in intellectual property laws.

Source: People's Charter for Health, People's Health Movement, 2000.

- PHM will defend health workers in their opposition to the privatisation of health services by building broad multi-sectoral alliances.
- PHM will campaign to end TRIPS, remove them from the WTO, and oppose bilateral Free Trade Agreements and TRIPS+.
- We call upon governments to use the Doha Agreement to provide people with affordable generic drugs.
- We oppose public–private partnerships because the private sector has no place in public health policy making.
- We call for a worldwide campaign for a UN Treaty on the Right to Water, ensuring that commodification and privatisation of this vital resource—life itself—is both reversed and prevented. Guided by evidence of devastating damage and by the precautionary principle, we demand a moratorium on extractive mining and petroleum exploration/extraction, a ban on patenting of life forms and processes, research on nanotechnology, release into the environment of GMOs, and on development and use of all biochemical weapons. Governments are accountable to people not transnational corporations and must guarantee rights relating to health and the environment through enforceable laws and regulations. Governments, international financial institutions (IFIs) and the WHO must cease to be accomplices to TNCs and imperialism. Dow, Monsanto and other companies must be forced to provide reparations to the thousands of uncompensated victims of disasters such as Bhopal and Agent Orange.

Knowledge and science must be reclaimed for the public good and freed from corporate control.

Source: Cuenca Declaration, People's Health Movement, 2005.

The Cape Town Call to Action (2012) sets out an alternative vision based on these four points:

- A reformed economic system that values every individual, not every dollar;
- Just, fair and democratic political and economic processes and institutions;
- Better and transformed global health governance that is free from corporate influence and the influence of unaccountable private actors;
- Equitable Public Health Systems that are universal, integrated and comprehensive, as well as provide a platform for appropriate action on social determination of health

Source: The Final Cape Town Call to Action, People's Health Movement, 2012.

The World Social Forum also provides an opportunity for discussion of issues associated with economic globalisation but with a focus on human rights. Its forums are conceived as an alternative to the World Economic Forum and provide an annual venue for non-government actors from around the world to meet and share their concerns about the direction of global economic policy. The slogan of the World Social Forum is 'Another World is Possible'.

PROTEST, ADVOCACY AND LOBBYING AGAINST INTERNATIONAL FINANCIAL AND TRADE INSTITUTIONS

In the past decade just about every meeting of the World Bank, IMF, the WTO, the G8 and G20 has attracted considerable protests from civil society concerned about the impact of the actions of these largely unaccountable international bodies. The potential power of civil society alliances was seen in Seattle in late 1999 when 50 000 citizens protested during the WTO's Third Ministerial Conference. The protesters were met with tear gas and batons, and violent images were beamed around the world. The protests were successful in stopping the new round of trade negotiations. Similar protests have happened during WTO meetings since that time. G8 and G20 meetings are also attracting similar protests. Austerity programs in Europe in the wake of the GFC have seen widespread protests at the imposition of measures and at the growing economic inequities since the GFC.

These protests reflect the serious concerns that civil society groups have about the impact of economic globalisation on health and especially the suite of international trade and investment treaties negotiated through the WTO agreements. They are seen to pose dramatic threats to health and well-being. Most recently there has been widespread protest concerning the Trans-Pacific Partnership Agreement and the Trade in Services Agreement. These proposed agreements follow others in attempting to free trade from

People's Health March during 3rd People's Health Assembly, Cape Town, July 2012. (Fran Baum)

restrictions including those relating to protection of people's health. Public health associations have been very vocal in opposing these treaties, warning of the health risks if TNCs can challenge national legislation designed to protect health.

Oxfam has an ongoing campaign for fairer rules of trade (www.oxfam.org.au/ whats-wrong-with-world-trade) and document on their website how the current rules disadvantage poor people and low-income nations.

There are also many examples of grassroots action from people directly affected by neo-liberal policies. Examples of these are provided in box. 5.12.

BOX 5.12   EXAMPLES OF GRASSROOTS ACTION AGAINST NEO-LIBERAL GLOBALISATION

A Bangladeshi peasants' movement, Nayakrishi Andolan, has focused 'primarily on the seed preservation, conservation, sharing and exchange among farmers'. Women have traditionally controlled seeds and the dependence of the farmers on the market for seeds had meant 'the displacement of women from the control of a crucial technology … once women lost the control they were disempowered and felt dispossessed'. Nayakrishi Andolan recognises that food production is the main activity of most communities in developing countries. One of its central aims is to keep the control of seeds with women. They have established a seed wealth network to help preserve biodiversity and to organise sharing of seeds. One of the slogans of Nayakrishi farmer women is

'Sisters, keep seeds in your hands ... that is fundamental to ensure food security' (Akhter, 2001). In 2014 grassroots groups and small farmers in India won a victory when the government put an indefinite moratorium on commercial cultivation of the genetically modified Bt brinjal, or eggplant. Concerns were also about the fact that commercial cultivation of Bt brinjal would put the food supply largely in the hands of the corporations that produce the seeds, and threaten biodiversity (Global Greengrants Fund, 2010).

Korean farmers are an organised group who protest against the impact of free trade on their livelihoods. They point out that their traditional methods will not survive unprotected against international business cartels. At the Cancún round, one of the farmers, Lee Kyong-hae, killed himself as a protest. At the Hong Kong meeting in 2005 there were noisy protests by the farmers (Watts, 2005a).

In San Jose del Golfo, a rural community in south-central Guatemala, local residents have been successful in halting the construction of a gold mine. The company started construction without any planning permission. Despite this, the community had to use a campaign of protest and blockades (which included community members being shot in one community). Finally on 26 February 2013, after almost two years of resistance and protracted negotiations with the communities, the contractor operating the machinery for the mining company pulled out all of the machinery and trucks and left. This was a much celebrated victory for the community (Leaning, 2014)

'WATCHING' THE GLOBAL INSTITUTIONS

A final role played by global civil society is that of watchdog. This role is well illustrated by the Global Health Watch, first published in 2005. The Global Health Watch is subtitled 'An alternative world health report'. The fourth edition was produced by the People's Health Movement (see p. 143), and involved activists and academics from many networks around the world. The People's Health Movement also organises a WHO Watch, which attends the annual WHO World Health Assembly and Executive Board meetings as observers to monitor the proceedings and evolve positions on key issues. One of the activities of the WHO Watch is the publication of a comprehensive commentary on the agenda items of the World Health Assembly (www.ghwatch.org).

CONCLUSION

Chapter 5 has hopefully convinced you that no student of public health can afford to ignore the powerful impact that political and economic arrangements have on the health of populations. These invisible hands are shaping every public health issue.

This chapter has:

- documented the threat posed by neo-liberal economic thought to the achievement of healthy and sustainable societies
- described the architecture of global neo-liberalism

- stressed the need for public services that are committed to supporting civil society and that see community well-being and the achievement of equity as their central goals
- argued the need for a change in the existing balance of global economic and political power away from TNCs and the institutions that support their activities at the expense of public well-being
- described the role of civil society in opposing the unhealthy aspects of neo-liberalism.

Neo-liberal globalisation has positive and negative impacts but the overall balance indicates that neo-liberalism is proving unhealthy for populations. Neo-liberalism will shape public health throughout the twenty-first century. Unless humans find means to regulate the activities of transnational companies and international capital, bring a more equitable system of distributing resources across global communities, and conduct economic life in a way that does not threaten the health of the ecosystem, the outlook for our collective health is bleak. Significant reform is required to ensure that neo-liberalism does not continue to have a detrimental effect on health and equity. In order for these reforms to come about there needs to be an ethical sea change. Currently the world is able to live with and accept the massive inequities that exist as if they are part of some natural order. Yet these inequities are social, political and economic—not biological—in origin. From a public health perspective they can be tackled if we can only imagine and then create the political, economic and social arrangements that would make trade fair, reduce inequities and so improve health.

These alternative arrangements are reviewed in parts 6 and 7 of this book. Chapter 16 particularly addresses the Keynesian alternative and more radical economic thinking coming from green economics, recognising that, despite the almost total dominance of neo-liberal thinking, there are viable and essential alternatives if the environment is to be protected and human health improved.

CRITICAL REFLECTION QUESTIONS

5.1   Compare the experience of a stockbroker in New York, an unemployed automotive worker in Adelaide or Detroit, and small-scale farmers in a country of your choice. How do you think they would see neo-liberalism affecting the health of themselves and their family?

5.2   Why might privatisation of water and health services have an adverse impact on health?

5.3   To what extent do you think national governments should regulate the activities of transnational corporations? In your country, which regulatory measure do you think would be most effective to promote health?

5.4   Select one of the civil society websites listed on pages 144–5, or on page 148, examine its content and determine what aspects of neo-liberalism it is protesting against. How effective do you think the movement is?

Recommended reading

Freudenberg (2014) *Lethal but legal: Corporations, consumption, and protecting public health* provides an excellent guide to the impact TNCs have on public health, and argues the need for regulation to control their practices in favour of public health.

People's Health Movement et al. *Global Health Watch 3* (2011) and *Global Health Watch 4* (2014): *An alternative world health report.*

Harvey (2005) *A brief history of neoliberalism* presents a concise and well-argued economic history with a succinct account of what neo-liberalism is, where it came from and where it is heading. In Chapter 5, 'Neoliberalism on Trial', he considers its impact on people's lives.

Harvey (2014) *Seventeen contradictions and the end of capitalism* is for those who want to gain a detailed understanding of the internal contradictions of capitalism and how the flow of capital has brought about crisis, including the Global Financial Crisis. The book stresses the inevitability of inequities that flow from capitalism.

Held and McGrew (2007) *Globalization/Anti-globalization: Beyond the great divide* provides 43 chapters on a variety of aspects of globalisation including conceptualisation, politics, impact on national culture, issues of governance and maintaining world order in the future. It provides a very good introduction to the topic.

Korten (1995) *When corporations rule the world* is recommended by Archbishop Desmond Tutu, who is quoted on the back cover: 'This is a "must-read" book— a searing indictment of an unjust international economic order, not by a wild-eyed idealist left-winger, but by a sober scion of the establishment with impeccable credentials. It left me devastated but also very hopeful. Something can be done to create a more just economic order.'

Stiglitz (2002) *Globalization and its discontents* provides an insider's perspective on the workings of the World Bank, IMF and WTO, and painstakingly deconstructs the ways in which their economic policies have had such a crippling effect on poor countries.

Wiist (2010) *The bottom line or public health* contains 20 chapters that provide a guide to the tactics used by corporations to increase their profits, and shows how this is detrimental to health.

Useful websites

Institutions

The World Bank www.worldbank.org

The International Monetary Fund www.imf.org/external

World Trade Organization www.wto.org

Civil society movements

The Occupy Movement www.occupytogether.org

The People's Health Movement www.phmovement.org

Oxfam's Fair Trade Campaign www.oxfam.org.au/whats-wrong-with-world-trade

PART 3

RESEARCHING PUBLIC HEALTH

All public health practitioners will engage in research at some time, or at least use the results of other people's research. Consequently, it is helpful to understand the strengths and weaknesses of different research methods and approaches. Information about health, disease, health services, people's lifestyles and the organisation of their societies is the lifeblood of public health, and research is needed to both describe and explain how these factors are related. Practitioners are often wary of research, and researchers are not always good at making their research relevant to practice.

Recently there has been a good deal of innovative thinking in public health research. This has increased its sophistication, which can make it more difficult for researchers, and particularly practitioners, to keep abreast of developments in research practice and theory. It is now likely to be multidisciplinary and combine qualitative and quantitative methodologies.

This part provides an overview of developments in public health research and a guide to the key elements of the most common methods and methodologies. Even if you will not be conducting much research yourself, you need to know how to evaluate research that may inform your practice or the policies you develop.

6

RESEARCH FOR A NEW PUBLIC HEALTH

The importance of doing away with the inappropriate and unnecessary conflict between quantitative and qualitative methodological approaches needs greater recognition. It is extremely dysfunctional when these research approaches are viewed as competing or mutually exclusive.

Dean et al., 1993, p. 229

KEY CONCEPTS

Introduction

Limits to epidemiology

Other forms of knowledge generation

Need to change focus of health research

Reflective research practice

Using previous research findings: systematic reviews

Ethical issues in research

Do no harm

Methodological soundness

Informed consent

Privacy, confidentiality and anonymity

Being an ethical researcher

Research with Indigenous Australians

Conclusion

INTRODUCTION

Two decades ago a chapter on research methods in a public health textbook would probably only have looked at epidemiology. Now public health is becoming increasingly methodologically eclectic and uses a range of methods from a variety of social science disciplines and epidemiology.

Despite this eclecticism, the debate about public health methodologies has become polarised. Epidemiologists often maintain that their set of methods is superior and more 'scientific', while those pointing out its limitations have tended to overlook its value in the quest to promote alternative research methods. This is understandable at a time of change and questioning of accepted practice, but it is necessary to examine the contribution of a variety of methods to understanding public health issues. I start from the position that all research can only lead to a partial understanding, but some methods are better suited for particular purposes than others. Good public health research involves interdisciplinary cooperation from colleagues to encourage dialogue across methodological divides. Collaborative and multidisciplinary approaches to research should be encouraged as the hallmark of the new public health.

This chapter sets the scene by discussing the debates about epistemology and methodology in public health. It argues that public health practitioners should use those methods and research tools that are best suited to the particular public health problem at hand and not be wedded to any particular methodology. The importance of participation in research will be discussed, and the mechanisms by which public health researchers can involve people in their research will be reviewed.

LIMITS TO EPIDEMIOLOGY

Epidemiology offers much to public health. It is particularly well suited to tracking down the causes of disease and to describing the patterns of disease in populations. It is not, however, a sufficient methodology to answer all public health questions. Indeed, any research, like all knowledge, will be conditional and bounded by time and circumstance. So long as epidemiology dominated the public health research imagination, the scope of public health was unnecessarily limited, and practitioners were like the man in the fable who lost his key. He arrived home with a friend, both somewhat the worse for drink, but dropped his key in a densely vegetated garden bed from where there was little chance of retrieving it. The friend went to see if there was a spare key in the car, but when he returned the man was busily looking under the lamp post in the street, well away from where the key had been dropped. When asked why, he said that that was where the light was!

For a number of decades, epidemiology was the lamp of public health. Other methods were 'unscientific' because they did not allow the rigorous control that was possible with many epidemiological methods. Yet, as in the fable, many of the problems sought by public health to research are out in the dark, beyond the light that can be shed by epidemiology. Realisation of this has led to an increasing recognition and use of many methods in public health research. Even the Lancet has recognised the limitations of epidemiology: 'Research on the health of populations is still dominated by experimental designs based on simplistic notions of causality that try to remove the variation and complexity of real-life health and disease process' (Anon., 1994, p. 429).

Why has method or methodology become a contentious issue in public health? If, for instance, you were to read a chemistry, physiology, or anatomy textbook, there would be little debate about the choice of method. The focus would be on explaining and understanding a research method that was well accepted by the field. By contrast, public health has seen more and more questioning of methodologies. The discipline evolved from medical science, and, until recently, most practitioners were doctors or nurses, trained within the biomedical sciences and therefore expecting to apply that paradigm of research to public health problems. Epidemiology is basically modelled on laboratory research and operates by establishing and testing hypotheses through carefully designed research methods—essentially a deductive process. This approach has had some remarkable successes in discovering the pattern and aetiology of disease, but it is less well equipped for understanding the complexities of many aspects of health. The growing realisation that health and illness reflect the structure, culture, power relationships, economy and politics of a society has resulted in public health seeking to understand more about health and disease than the immediate cause of any particular disease (Krieger, 1994).

Criticisms of epidemiology from social scientists centre on its almost total focus on controlled measurement to the exclusion of other forms of knowledge and analysis. Social scientists are often exasperated that epidemiologists do not go beyond describing the possible causes of disease to consider in more detail the social, economic and political factors that shape the disease. The Latin American social medicine movement has also called for a far more politically and socially engaged epidemiology movement (Breilh, 2003). Krieger (1994) points out that the metaphor and model for epidemiology has been the 'web of causation' through which it has become more and more concerned with modelling complex relationships among risk factors. She maintains that this view is atheoretical and includes a hidden reliance on biomedical individualism, to the neglect of social and environmental factors. She traces the atheoretical development of epidemiology and urges epidemiologists to become more critical in their approach to the cause of disease, proposing an eco-social framework for developing epidemiological theory (Krieger, 2011). She also stresses the importance of taking into account historical context and the contexts within which people live their lives. McMichael (2001) stresses that epidemiology reflects the sort of individualism that was described in chapter 4. This individualism has been shown in recent decades by the way epidemiology has focused on the study of the risks incurred by individuals because of their behaviour rather than the methodologically elusive but more crucial social influences on the health of populations. Ironically McMichael's earlier book *The LS Factor* (Hetzel and McMichael, 1989) is almost entirely focused on lifestyle risk factors. His later book delves much deeper into causality of diseases and provides a sweeping view of the social, biological, economic and political factors that account for disease patterns over time and across cultures. The limitations of epidemiology become very evident when such broad views are adopted.

Shy (1997) accuses academic epidemiology of serving clinical medicine more than public health. He supports his contention by arguing that epidemiology has limited itself to a narrow biomedical perspective that has dealt with risk factor and disease associations, rather than a population level of understanding. To support his case, he uses Rose's (1985) work on the difference between sick individuals and sick populations and the consequent need to study populations, not individuals. He points out that most epidemiological studies treat race, social class and economic status as potentially confounding factors rather than as potentially causative factors in their own right. He urges epidemiology to attempt to understand disease as a 'consequence of how society is organised and behaves, what impact social and economic forces have on incidence rates, and what community actions will be effective in altering incidence rates' (Shy, 1997, p. 480). If epidemiology were to be redefined in this way, it would be better positioned to contribute to public health's central aim of improving the health of populations rather than to understanding disease in individuals.

Internationally, the Health for All Strategy (WHO, 1981) and the Ottawa Charter (WHO, 1986) reinforced and legitimised the growing acceptance of the complexity of producing health and preventing disease, and made clear the need for research tools to complement epidemiology. This became contentious as epidemiologists typically had no experience of social science or qualitative methods, perceiving laboratory-based science as the gold standard for research. In terms of professional power, the 'new' public health research could also be seen as threatening. Typically, older male medically trained epidemiologists were being criticised and asked to broaden their research repertoires by younger, often female, researchers trained in the social sciences. Given that, until recently, epidemiologists have been predominantly male and medical, the feminist critiques of both medicine and positivism may have created a defensiveness that has made the adoption of innovation more difficult. It is therefore hardly surprising that there has been some resistance to change in public health research practices.

In defence of epidemiology, it should be noted that it has made important contributions to understanding disease patterns and factors that cause disease (most famously the link between tobacco smoking and lung cancer). It is good at establishing causal links, but there is a need to encourage a broader range of methods in public health and to accept that epidemiology may not be able to measure and make straightforward causal links between crucial determinants of health, such as social class, the quality of social relations and complex ecological systems.

OTHER FORMS OF KNOWLEDGE GENERATION

> the characterisation of the debate as an irresolvable one between positivism and interpretivism is disingenuous in our view. It is a device that obscures more than it reveals.
>
> *Kelly and Swann, 2004, p. v*

Since around 1970 there has been a growing body of criticism against conventional science. Feminist researchers (Keller and Longino, 1996) claim that science has been dominated by male values and reflected a very skewed view of the world. These critical voices have been strengthened by those of critical theorists, led by Habermas (Cheek et al., 1996, pp. 168–73), who have strongly challenged the view that science is the only form of research through which knowledge can be developed. They maintain that conventional science is also imbued with values, and that scientists, far from being objective observers, are continually making judgments based on their values in terms of what and how they will research. This has been illustrated in studies of medical textbooks that demonstrated that the discourses about women contained a strong hidden curriculum, which portrayed women in a conservative way, assuming heterosexuality, marriage and universal aspirations to have children (Koutroulis, 1990).

Social science postmodernist thinking over the past two decades has argued that knowledge is relative, its understanding depending on a range of social and cultural factors. Thus, people's positions in society (class and power position, gender, culture) play a crucial role in their interpretation of events and facts. This thinking has encouraged the view that discourses differ and that dogmatism governed by incontrovertible facts is often not helpful. Postmodernist theory suggests that 'discourses' determine how people view the world and are a mechanism for maintaining power within society. This theory has been strongly influenced by Michel Foucault, a French sociologist, who examined the development of the modern institutions of health care (Foucault, 1973) and prisons (Foucault, 1979). He concluded that power in modern society was influenced by the knowledge claimed by groups and how this knowledge comes to dominate discourses. Postmodern analysis offers the possibility of examining those discourses that dominate everyday life, and therefore the operation of activities such as public health. A postmodern perspective argues that all bodies of knowledge (including all forms of public health), and the network of power they support, should be treated critically.

The debate about knowledge generation in public health can also be seen as medical sciences versus social sciences. Typical views from medical practitioners are that social scientists, especially sociologists, speak a language they cannot understand and do work that has no practical relevance. Social scientists are often convinced that doctors are simply body technicians who are unable to communicate effectively with their patients but who believe they have god-like powers. They see doctors as empiricists with little understanding of the complexities of life. Of course, these polarised views are far from accurate—many social sciences operate within a positivist tradition and methodological debates are common in most social science faculties. Medical practice is not solely based on evidence derived from clinical trials, but also relies on clinical judgment and qualitative assessments of previous experience.

There has been debate about whether the differences between quantitative and qualitative research methods are a matter of fundamental epistemological issues or, more simply, those of a technical nature. This debate has been succinctly summarised by Bryman (1988, pp. 104–5):

> By an epistemological issue is meant a matter which has to do with the question of what is to pass as warrantable, and hence acceptable knowledge. In suggesting that quantitative researchers are committed to a positive approach to the study of society (Filmer, Phillipson et al., 1972), the view is being taken that they subscribe to a distinctive epistemological position, since the implication is that only research which conforms to the canons of scientific method can be treated as contributing to the stock of knowledge. Similarly, by subscribing to positions, such as phenomenology, verstehen, and naturalism, which reject the imitation of the natural scientist's procedures and which advocate that greater attention be paid to actors' interpretations, qualitative research can also be depicted as being underpinned by an epistemological standpoint.

Bryman (1988) goes on to demonstrate, by quoting antagonists from both research camps, that there have been strong views that the differences between the two approaches to research are based on epistemology, and equally strong ones that they are simply differences in technique. He is an advocate for methodological pluralism. This debate has also been evident in public health. Some biomedical and epidemiological researchers have asserted the superiority of knowledge that has been generated by 'scientifically sound' research, especially randomised controlled trials (Christie et al., 1987). Some qualitative researchers have suggested, for their part, that knowledge is only valid if based on constructivist methodology because there is a fundamentally different level of understanding between the two approaches to research (Guba and Lincoln, 1994). There is, however, an increasing call in public health for recognition of the need for the use of a greater diversity of methods (Phillimore and Moffatt, 1994; Black, 1994; Baum, 1995b; Popay and Williams, 1996; Dixon-Woods et al., 2004).

The debate about the desirability of different research methods is underpinned by beliefs about the nature of knowledge and understanding. Guba and Lincoln (1994), who have been at the forefront of thinking about qualitative methodology, defined four research paradigms, of which they believe two represent the received view and two challenge it:

- positivism aims for definition of objective truth and reduces all relationships to a statistical level
- postpositivism retains the basic beliefs of positivism but accepts some of the criticisms of the search for absolute truth and seeks hypothesis falsification rather than verification. It incorporates qualitative techniques to add a subjective perspective to the otherwise objective one
- critical theory focuses on critiquing and understanding inequities in society, seeking to change them as a result of research
- constructivism is the joint creation of knowledge between the researcher and the researched. In this view there is no static truth, but instead multiple and shifting realities.

In the discussion of research paradigms, it is common for textbooks to discuss two of these approaches: the conventional and the constructivist. The former is based on a positivist and reductionist approach to science—on the belief of a single truth that holds, regardless of time and place. The researcher studies a phenomenon objectively, removing values and contaminating factors. Gold standard research methods are experimentation and, in epidemiology, randomised control trials. Deviations from these methods are viewed as second-best options. The aim of science in this paradigm is to test hypotheses to discover the objective truth about the world and so make predictions. The notion of falsification or refutation is central (Popper, 1972). Conventional science has undoubtedly enjoyed hegemony in medical science and public health.

By contrast, the constructivist paradigm believes that truths are socially constructed and that reality is specific to time, place and culture. The researcher is seen as part of the reality being researched. The existence of objective knowledge is denied. The research process is one of enquiry, relying on a continuous process of iteration, analysis, critique, reiteration, reanalysis, synthesis (Reason, 1988).

I suspect that few public health practitioners will find themselves rooted in any one of these paradigms, but will shift between them, adopting approaches that could be classified under each. Perhaps the most comfortable position for new public health practitioners is within the critical theory perspective (see Crotty, 1998 for description). Public health research aims not just to understand but to use that understanding to bring about change.

The methodological and epistemological debates between aficionados of these two paradigms are often fierce, with little room for dialogue. Public health may be one of the few arenas where a more constructive and respectful dialogue is developing, based on the recognition that both approaches need to understand the complexities of public health problems, and the essential necessity of knowing the extent and pattern of disease and health.

Quantitative and qualitative methods have very different strengths. Quantitative research is essential for describing the extent and pattern of disease and the factors that are related to it within a community. Qualitative research can describe the meaning of disease, poverty or caring and can help us understand how public health strategies can assist in solving the problems (see box 6.1 for examples)

BOX 6.1   VALUE OF USING QUALITATIVE AND QUANTITATIVE METHODS IN PUBLIC HEALTH

WHY WOMEN SMOKE

Using a combination of methods, Graham (1987, 1994) demonstrated that women living in stressful situations may use cigarettes as a means of coping with the strains of their lives, even though they know the associated risks. Her early work on this topic focused on understanding the context in which women persisted with smoking. Her subsequent work confirmed that women smoke most when they have heavy caring responsibilities and low incomes (Graham, 1987, 1994).

EFFECTS OF REDUNDANCIES ON AUTOMOTIVE WORKERS

A longitudinal study of workers being made redundant from the Mitsubishi Motors plant in Adelaide showed that the workers' mental health took a drop at the time of leaving the plant, but for most recovered over the next two years. Those who continued to show signs of stress were the men who were finding it hard to meet their financial commitments (Ziersch et al., 2014). The qualitative analysis was able to tease out the ways in which the agency of the workers to respond to the job loss was constrained by the policy environment and the low performance of the privatised job networks (Anaf et al., 2013).

EFFECTS OF RACISM ON HEALTH

Research among Aboriginal people living in Adelaide showed that the more people had been subjected to racial assault, the worse their mental health was (Ziersch et al., 2011a). The qualitative data enabled exploration of the detailed experience of being subjected to racist assaults and the range of coping strategies that people use to mitigate its effects (Ziersch et al., 2011b).

The value of methodological pluralism was demonstrated in Australia by F research. Epidemiology was essential to track the spread of the epidemic and to d how the virus was distributed among different groups in the population. It was monitoring of disease that first alerted scientists to the existence of a new disease. Reports to the Centers for Disease Control in Atlanta, Georgia, suggested a rise in a rare form of cancer, Kaposi's sarcoma, leading to investigations that led to the eventual identification of the HIV virus. The evaluation of the National HIV/AIDS Strategy 1993–94 to 1995–96 (Feacham, 1995) reviews the course and impact of the HIV epidemic in Australia, and demonstrates the sophisticated Australian HIV surveillance strategy. The report is able to provide:

- observed AIDS incidence
- predicted incidence of HIV and AIDS
- estimate of number of people living with HIV (by state and territory of residence)
- number of reported cases of newly diagnosed HIV infection by sex, year of diagnosis and exposure category, state and territory
- international comparisons of HIV and AIDS incidence
- prevalence of HIV in different populations, including gay men and injecting drug users
- cross-sectional data on the behaviour of groups of people who are known to have a high prevalence of HIV (e.g. the percentage of men engaging in unprotected anal intercourse with regular or casual partners).

Epidemiological research has been essential to describing the HIV/AIDS epidemic in Australia and overseas, but it could not explore the reasons for the behaviours of people who acquired HIV, study social reactions to the disease or develop programs to prevent its spread. Australia funded a significant research program into social and behavioural research to increase understanding of responses to the epidemic, leading to a sophisticated program of prevention that appears to have been one of the most successful in the world (Feacham, 1995). Increasing qualitative research is used to understand the dynamics of the epidemic in African and Asian countries and is seen as essential to complement epidemiological information.

The need for epidemiology to be combined with social science methodologies has also been recognised by the North Karelia project researchers, who implemented a community-wide program to reduce the incidence of cardiovascular disease in central Finland (Tuomilehto and Puska, 1987). They recognised that classic epidemiological studies were insufficient to either mobilise the community in support of the program or bring about the social changes necessary to support individual behaviour change. Behavioural and social sciences were necessary to understand these processes. Popay and Williams (1996) argue that social science research has produced valuable insight into lay knowledge of health and illness. They contend that this knowledge is rarely taken seriously by public health researchers, and that they ignore it at their peril. It offers, they maintain, a means of understanding more about the relationship between social circumstances and individual behaviour, and may provide insights into the aetiology of disease (for example, wives of victims of asbestosis are reported as complaining to coroners' courts that their husbands' deaths resulted from exposure at work long before epidemiologists documented the link). Subjective reports of ill health in the absence of any definable physical illness may indeed be early warnings of illness in the future.

Lundy (1996) points out that quantitative methods are restricted in developing countries because of a lack of reliable data. Aside from this, she argues that in relation to the study of the impact of structural adjustment policies, qualitative data offers a grassroots view of their day-to-day impact that is missing from quantitative analysis. She says that, while data are not available to prove a causal relationship between structural adjustment and health, the evidence from those working in the health system and experiencing adjustment at first hand leaves no doubt that it has had a negative impact on health care provision, social welfare and environmental health. She describes how, as part of a structural adjustment program in Jamaica, the World Bank required the National Water Commission (NWC) to run at a profit. The senior health officials in the Jamaican Ministry of Health are convinced this impacted negatively on health. Lundy (1996, p. 322) reports one of these officials explaining:

> ... the head of the NWC actually went on television and made this statement during a debate on typhoid, he said quite clearly that he is not able to provide clean drinking water to all that need it, because some people can't pay for it ... So for example, I attribute the typhoid epidemic in Savanna-la-Mar fair and square on the deterioration of the environmental situation: the quality of the water and the efficiency of sewerage disposal. That's what causes typhoid. It's a breakdown in your social environmental structure.

> Q: And are you saying that this is related to structural adjustment?
> A: Oh yes. Oh yes absolutely. The water and sanitation situation is because the World Bank requires the Water Commission to run at a profit.

Lundy argues that the accumulation of such qualitative data is crucial in providing an insight into local perceptions of the impact of structural adjustment knowledge. Such insights may be an early sentinel of future health problems that will not be detected by quantitative research for many years.

The need for a new focus for public health research was recognised by the Commission on Social Determinants of Health (2008). Its final report highlighted the importance of broadening the scope of public health research to include a stronger emphasis on the social determinants of health. It requires that funding bodies—including national health, medical, and social research councils—invest more in research on the social determinants of health. This comprises epidemiological, and multi- and interdisciplinary research, including qualitative approaches, rather than the disease-specific biomedical focus of research funding that currently prevails (Global Forum for Health Research, 2004; Baum et al., 2013a).

NEED TO CHANGE FOCUS OF HEALTH RESEARCH

The Global Forum for Health Research has raised a series of problems with the focus of global research on health (Global Forum for Health Research, 2004). First, worldwide only 10 per cent of health research funds are allocated to the problems responsible for 90 per cent of the world's burden of disease. Second, greater emphasis should be placed on research on the social, economic and political determinants of ill health, relative to clinical and biological research. Third, there are significant barriers in terms of translating research into knowledge. In essence the critique of global health research is that it focuses on diseases of the rich world. Members of the People's Health Movement have commented of research priorities 'Despite substantial sums of money being devoted to health research, most of it does not benefit the health of poor people living in developing countries' (McCoy et al., 2004, p. 1630). Others have pointed out that there is very little research that focuses on efficacy research (testing interventions in a controlled setting) or implementation research (the 'how' of translating current research knowledge into practice within existing health and social systems) (Sanders et al., 2004). A 2004 World Health Report focused on the importance of producing better knowledge to support the strengthening of health services (WHO, 2004b). It called for more investment in innovative research on health systems and argued that researching how to implement health services within different health systems, population groups and diverse political and social contexts was vital. It also called for close interaction between health systems and health research systems so that there was more mutual learning, problem solving and innovation. Ågren (2003, p. 20) notes that public health research comes a very poor second to biomedical research and comments that:

> Research policy reflects both an over-confidence in the medical care services' ability to solve fundamental health problems and the strong economic interests that exist in the field of medical treatment. An individual and often deep-rooted biological approach dominates within the field of medicine, resulting in socially determined health discrepancies being studied relatively seldom or in many cases being ignored completely.

The reasons for the biases in research funding reflect a myriad of social, political and economic forces including the influence of pharmaceutical companies, and the entrenched power of the biomedical research establishment in rich country health and medical research bodies. Changing the balance of the research conducted is an important aspect of the new public health agenda. There needs to be more research on the ways in which social and economic factors affect health and what social, educational, housing and health interventions most improve health and health equity, as well as more applied health system research. It is vital that governments invest in this research to a greater degree than they do currently (Baum et al., 2013a).

The past few years have also seen a much greater call for effective transfer of research findings into practice so that there is a bridging of the gap between what is known and what is actually done (WHO, 2004b). One of the major blocks to this happening is the gap between the intentions, motivations and rewards that are the work experience of university researchers and policy makers. These are compared and contrasted in table 6.1.

TABLE 6.1   THE 'TWO COMMUNITIES' MODEL OF RESEARCHERS AND POLICY MAKERS

	University researchers	Government officials
Work	Discrete, planned research projects using explicit, scientific methods designed to produce unambiguous, generalisable results (knowledge focused); usually highly specialised in research areas and knowledge	Continuous, unplanned flow of tasks involving negotiation and compromise between interests and goals, assessment of practical feasibility of policies and advice on specific decisions (decision focused). Often required to work on a range of different issues simultaneously
Attitudes to research	Justified by its contribution to valid knowledge; research findings lead to need for further investigations	Only one of many inputs to their work; justified by its relevance and practical utility (e.g. in decision making); some scepticism of findings versus their own experience
Accountability	To scientific peers primarily, but also to funders	To politicians primarily, but also to the public, indirectly
Priorities	Expansion of research opportunities and influence of experts in the world	Maintaining a system of 'good governance' and satisfying politicians
Careers/rewards	Built largely on publication in peer-reviewed scientific journals and peer recognition rather than practical impact	Build on successful management of complex political processes rather than use of research findings for policy
Training and knowledge base	High level of training, usually specialised within a single discipline; little knowledge about policy making	Often, though not always, generalists expected to be flexible; little or no scientific training
Organisational constraints	Relatively few (except resources); high level of discretion, e.g. in choice of research focus	Embedded in large, interdependent bureaucracies and working within political limits, often to short timescales
Values/orientation	Place high value on independence of thought and action; belief in unbiased search for generalisable knowledge	Oriented to providing high quality advice, but attuned to a particular context and specific decisions

Source: Buse et al., 2005, p. 163.

REFLECTIVE RESEARCH PRACTICE

Most published research presents a sanitised view of the research process. A newcomer to research would gain the impression from published accounts that research was generally a smooth, logical process in which little goes wrong and which is immune from the vagaries and politics of everyday life. In practice it is rare for such immunity to operate. Public health research, like most other, is subject to the setting in which it is conducted and the researchers who conduct it. Social scientists have some tradition of reflection in their research practice and opening up their processes to take an honest look at them.

Feminist researchers have argued for the value of reflexivity in research (Stanley and Wise, 1990; Shakespeare et al., 1993), particularly in regard to how the researcher influences research. The feminists have questioned the claims of objectivity in traditional science and suggested that all forms of research reflect the biases and values of the researcher. Postmodern thinking similarly questions the notion of the existence of truth. The intellectual traditions stemming from postmodern thinking make all knowledge questionable. These perspectives pose such stark contrasts to medical science, which is quintessentially a modernist movement founded on the idea of seeking out the ultimate truth about bodies, disease and cures. A useful exploration of the impact of postmodern perspectives and the increasing lack of certainty about understanding the world is provided in a series of reflective essays by health and social welfare researchers (Shakespeare et al., 1993). The introduction to the collection sums up the challenge posed by reflective research:

> ... arguments between quantitative and qualitative research methodologies were about how you could best gain access to the truth. If we no longer search for the 'truth', or even some approximation of it, what are we doing? Many of our chapters reflect this uncertainty. We have found it both liberating and chastening at the same time. It has allowed us to challenge 'objectivity', to let ourselves in on the act, to be partial and to put forward the view from the standpoint of women, older people, people with learning disabilities and so on.
>
> *Shakespeare et al., 1993, p. 9*

Reflection is an important skill for a researcher and, while the tradition is most common among those using broadly qualitative measures, all researchers (and the quality of their research) are likely to benefit from it. Epidemiology does not have a tradition of reflective research practice, and could benefit from one. An unkind observer has defined much epidemiological research output as 'data untouched by human thought'! While extreme, this aphorism has a grain of truth, and encouraging a greater epidemiological imagination could improve the thoughtfulness and applicability of public health research.

Fook (1996), in her conclusion to a collection of essays, *The Reflective Researcher*, argues that postmodern and poststructuralist theory provide frameworks that encourage reflection on practice. Postmodernism stresses the importance of 'narratives' or 'discourses' in understanding how sense is made of the world. Poststructuralism is based on the notion that meaning is open to numerous interpretations, is not fixed and changes in different contexts. While most public health researchers do not see themselves as postmodernists or poststructuralists, they increasingly accept the importance of relativities, shifting meanings and individual interpretations, and use these to inform their practice while having a clear vision of what needs to change to create a more just and healthy world.

USING PREVIOUS RESEARCH FINDINGS: SYSTEMATIC REVIEWS

Increasingly emphasis has been placed on evidence-based practice in both medicine and public health (Chalmers et al. 1997). The Cochrane and Campbell Collaborations have established a worldwide movement that aims to produce a sound evidence base for clinical and health promotion practice. The Cochrane Collaboration has online learning resources that provide a guide to conducting systematic reviews and meta-analysis. Obtaining evidence is easier to do for most medical procedures than is often the case for health promotion and public health because the criteria and parameters of the review are typically more limited in scope (Oakley, 2001). Systematic review of research evidence is important for researchers to ensure they build on existing knowledge and research, and for practitioners so that they can find out what evidence there is for what works. Peersman et al. (2001) provide a guide to these reviews, and the steps they suggest are in box 6.2. Each of these stages is crucial and must be done with great rigour if the review is to be useful and accurate. It is worth noting that systematic reviewing is a time-consuming enterprise and the time required is often grossly underestimated by policy makers or those commissioning such reviews. Harden (2001, p. 113) notes that the resources allowed for a systematic review of the effectiveness and appropriateness of peer-delivered health promotion for young people were two full-time researchers working for nine months.

BOX 6.2   STEPS IN CONDUCTING A SYSTEMATIC REVIEW

- Formulating the review question
- Identifying relevant primary research
- Assessing identified studies for inclusion in the review
- Critically appraising studies meeting inclusion criteria
- Incorporating assessment of study quality in reviews
- Extracting relevant data
- Analysing and presenting results
- Interpreting result.

Source: Peersman et al., 2001.

The UK National Health Service has established the National Institute for Health and Care Excellence (NICE; www.nice.org.uk), which contains within it the Centre for Public Health Excellence. The Centre produces 'guidance' on the promotion of good health and prevention of illness. Most of its guidance relates to behaviour change (smoking, weight, alcohol use, breastfeeding) but guidance on community engagement and transport planning are in preparation for publication and reflect a stronger new public health perspective.

Increasing attention has been paid to qualitative evidence synthesis and a variety of methods have been developed to conduct such syntheses (Hannes and Lockwood, 2011). Examples include meta-ethnography, critical interpretive synthesis and realist reviews. Reviews now sometimes integrate quantitative and qualitative data to determine the weight of the aggregated data.

Evidence from systematic reviews is important for informing policy but will only ever be one of the factors taken into account in policy making (Glasziou et al. 2004). In public health, culture, human behaviour and social difference in populations play a much more significant role than in clinical medicine. This means external validity from research is a problem when conducting systematic reviews and that extrapolation from evidence to policy inevitably involves matters of judgment (Kelly et al., 2006). Added to this is the fact that population-wide interventions take considerable time to have a discernable impact (Briss, 2005) so the application of systematic review will be as much an art as a science.

ETHICAL ISSUES IN RESEARCH

Researchers are required to have ethical approval for their research from an Institutional Ethics Committee (IEC) or like body. The task of these IECs has been growing more complicated as the nature of health research has diversified to include that based on social science methodologies. Their work is based on guidelines issued by the National Health and Medical Research Council (NHMRC). The NHMRC updated the guidelines for all research involving people (NHMRC et al., 2007 [updated 2014]) and these guidelines are in clear language and invaluable to public health researchers. It is also important to consult the guidelines for research in Indigenous Australian communities (NHMRC, 2003) and an information paper on ethical aspects of qualitative methods in health research (NHMRC, 1996; NHMRC et al., 2007 [updated 2014]). McNeill et al. (1992) found that researchers were generally supportive of the IECs, even though they thought that the process of review was time consuming and demanding, and sometimes interfered with the progress of the research. Lumley (1996), for example, has described the laborious process she had to go through to gain permission for her statewide research into the effectiveness of diagnostic ultrasound in pregnancy. This process included having to approach 150 hospitals for permission and discovering different forms, processes and standards between the committees. While there have been some voices critical of the role of the IECs (Crotty, 1996), they are now accepted as part of the research scene and as arbiters on the ethical standards of research.

DO NO HARM

Medical practice has been based on the principle that no harm should be done to a patient (sometimes known as the principle of non-maleficence). This principle has not always been adhered to and some medical 'experiments' have caused considerable harm—see, for example, Coney's (1988) story of the treatment of women with cervical cancer at Auckland's National Women's Hospital. Most public health research is not physically invasive, with the potential for harm being more likely to be psychological, such as a respondent to an interview survey being asked sensitive questions about sexual behaviour or a study of dying that wishes to interview people in the last few months of their lives.

METHODOLOGICAL SOUNDNESS

Conducting badly designed research is considered to be unethical. Difficulties arise, however, in determining what constitutes sound design. IECs in medical settings have been most familiar with laboratory research based on a traditional positivist design. The growth in social science research, and especially that based on qualitative methods, has been challenging for many IECs as the assumptions underlying many qualitative methodologies are not familiar to some committee members. Daly (1996, p. 91) has commented: 'Qualitative research methods are the most difficult to describe and are commonly misunderstood by both funding bodies and ethics committees.' The misunderstanding relates to both the scientific credibility of the methods used and the issues of ethics raised by invasion of people's social lives to collect data. A common methodological difficulty is understanding the different assumptions about reliability and validity. Sampling techniques are usually not based on random selection and may be quite small. Methods of analysis are based on textual data rather than numbers. Exact details of the methodology cannot be specified in advance as they have to be responsive to the social setting in which they are applied. These differences have meant that some IECs have not easily given permission to research based on qualitative methodology (NHMRC et al., 2007 [updated 2014]).

Daly (1996, pp. 93–4) points out that social research using less structured research procedures is dependent on the integrity of the researcher to collect data in an ethical and responsible manner. She advises that all researchers (not just those using qualitative methods) 'might do worse than to cultivate human qualities like prudence, honesty, humility and caring and bring these to bear on their research task'.

Daly also suggests that ethical research should be based on appropriate research methods, defined as those that are most likely to address the research problem fully, given the constraints in the field. She warns that methodological prejudice is unethical, but this injunction, like the NHMRC's against poorly designed research, relies on a definitional consensus that may not exist.

INFORMED CONSENT

The informed consent of participants in all forms of research is a basic ethical right intended to protect the autonomy of the participants. The National Statement on Ethical Conduct in Human Research (NHMRC et al., 2007 [updated 2014], p. 16) states that informed consent in research normally requires that:

> participation be the result of a choice made by participants—commonly known as 'the requirement for consent'. This requirement has the following conditions: consent should be a voluntary choice, and should be based on sufficient information

and adequate understanding of both the proposed research and the implications
of participation in it.

In most public health research the process of gaining informed consent involves the
provision of an information sheet to participants and asking them to sign a consent form.
The information sheet should explain how the individual's privacy and confidentiality
are to be maintained and assure people that they have the right to withdraw from the
study at any point, without any consequences for them. The situation is complicated
when people have reduced capacity to provide informed consent, such as children
or people with a psychiatric or intellectual disability. In these cases consent is usually
sought from the legal guardians.

Some forms of qualitative research pose particular problems in regard to informed
consent. For instance, what should a participant observer do? One view holds that
participant observation does not have research participants in the sense that a survey
does. In this view the researcher interacts with people under ordinary conditions of life,
much like other participants. Consequently, the participant observer has no different
an ethical obligation to the people encountered in the course of research than she
or he would under other everyday circumstances (Jorgensen, 1989, p. 28). This view
means the researcher is not necessarily obliged to inform people of research intentions,
and sometimes success may depend on the researchers not revealing that they are
undertaking research—for example, research on criminal behaviour such as illegal drug
use. Jorgensen (1989) advises that research ethics have to be a constant concern of
participant observers as they participate, interact and develop relationships. The history
of social research shows that studies that have produced significant findings have
sometimes been based on covert research. Holman (1991) provides 10 case studies
of such research (pp. 96–104), and maintains that such research is justified in some
circumstances.

Considerable epidemiological research is based on existing databases. It is very often
impracticable for informed consent to be given for access to routine medical records,
yet data from routine records and the reporting of them provide important information
about disease patterns. Cancer registries, for instance, have been compiled without
obtaining the consent of each person with a diagnosis of cancer.

There have been instances of social research based on covert methods. The NHMRC
(2007 [updated 2014]) notes that variation from informed consent may be justified in
rare circumstances. Researchers seeking IEC approval for covert or deceptive research
should show why their research should be the exception to the general ethical principle.
Punch (1994) discusses this issue in detail and suggests that the likely social benefit of
the research has to be balanced against the rights of the people likely to be affected
by covert research. That is, does the utilitarian principle outweigh any violation of
individual autonomy. He suggests, for instance, that covert research may be justified
when studying groups with social and economic power as they would never agree to
being studied or to exposing their lives to scrutiny. Some sociologists have expressed
concern that most social research is conducted among the less advantaged members of
society and that studies of the rich and powerful are extremely rare. One reason may be

because poor people are less likely to object to the research process because of their relatively low social power. This raises the question of whether public health research has an inherent paternalism, which means that the researchers seek individuals' approval for their research enterprise but there is little public discussion about research priorities. Is there a means whereby communities of interest can be given a stake in determining research agendas—a form of collective informed consent for research priorities, thus reconciling utilitarian benefit with autonomy?

A major argument in favour of participative research is that it is less exploitative and therefore more ethical because it respects the rights and autonomy of participants and attempts to reduce the power of the researcher. The rights of participants to informed consent may be extended to informed participation in the future.

PRIVACY, CONFIDENTIALITY AND ANONYMITY

Participants in research are entitled to protection of their privacy, which means they should not be identifiable in research reports and that any identifying information should be removed from data as soon as possible. This concurs with the principle of non-maleficence. Informed consent forms generally assure research participants of confidentiality. NHMRC et al., (2007 [updated 2014]), however, point out that researchers should be aware that they cannot promise absolute confidentiality because they are not legally protected against testifying in court, nor to mandatory reporting if this is relevant to their profession.

Researchers using questionnaires and interviews have to ensure that their data is stored in such a way that particular individuals cannot be identified. Tapes should only be lodged in an archive if the research participant has agreed to this. Research reports do not normally identify individuals or provide information that would suggest an individual's identity.

Social, epidemiological and medical research can be intrusive, but should be planned so that it intrudes as little as possible. This includes ensuring that the empirical research is necessary and that the information sought cannot be gained from other sources.

Epidemiologists use existing data sets to study the causes of disease. The association between rubella and birth defects, cigarette smoking and cancer, ionising radiation and cancer, many adverse drug reactions including birth defects due to thalidomide, has been shown from such records (Last, 1998). Use of these statistics is considered ethical as individual privacy is not at stake. Last (1998) warns, however, that various interest groups may have an investment in censoring access to routinely collected data for epidemiological purposes. For example, industrial or commercial interest groups may oppose the collection of health statistics that can be used to identify occupational or environmental hazards.

Epidemiology also benefits by linking information from different data sets, but this raises particular issues relating to privacy, confidentiality and anonymity. The issue of privacy in public health relates closely to that of individualism. Societies that put a high premium on the rights of individuals are more likely to also protect individuals' privacy and autonomy. Rights-based ethical arguments stem largely from the liberal and humanist traditions, and while they had a prominent place in forming the values of many of our institutions, there are other approaches. Indeed, some writers on ethics argue that there is no rational basis for 'rights' at all. Public health researchers will frequently find themselves in the position of having to argue the ethics of placing community benefit apparently ahead of individual rights.

BEING AN ETHICAL RESEARCHER

These ethical concerns are focused on research being seen to be ethical, and convincing an ethics committee that it is. In this, ethical considerations are never resolved simply by a claim of legitimacy. The arguments and the various, usually conflicting, positions need to be explicitly analysed and articulated. Frequently the most obvious tension is that between individual and community good, but beneath the surface may be assumptions of consequentialist or deontological ethics or of principles of autonomy or paternalism. It is noticeable that writers on qualitative research matters devote more attention and soul-searching to ethical issues than do most other scientists, which may reflect differences in the paradigms within which the researchers are working, with postpositivist researchers typically showing more concern for the detail of ethical issues. Or it may reflect that qualitative research often raises more ethical issues. Miles and Huberman (1994) suggest a number of reasons for this. Most qualitative studies accept there will be multiple realities in idiosyncratic local context and that they 'tend to obscure general principles and make for situation-specific coping' (p. 289). It is also true that fieldwork and analysis are often unpredictable and specific to each situation. Because their exact nature cannot be specified in advance, neither can the ethical issues they give rise to, which means that qualitative researchers are more likely to encounter new ethical problems when they plan and conduct their research. Punch (1994) and Miles and Huberman (1994) provide thoughtful and full accounts of the ethical implications of qualitative research, and the reader is referred to these for more detailed consideration. The latter authors provide a list of issues that should be attended to before, during and after qualitative studies. Any public health researcher would benefit from considering them. They are given in table 6.2, and questions relevant to public health are posed.

TABLE 6.2   ETHICAL QUESTIONS FOR PUBLIC HEALTH RESEARCHERS

Issue	Questions raised for researcher
Worthiness of the project	Is the study worth doing? Will it make a significant contribution to public health?
Competence boundaries	Does the research team have the expertise to carry out a study of good quality? If not, is the necessary expertise available elsewhere?
Informed consent	Do the people involved have full information about the study? Can their consent be given freely and without coercion?

(continued)

TABLE 6.2   ETHICAL QUESTIONS FOR PUBLIC HEALTH RESEARCHERS (*CONTINUED*)

Issue	Questions raised for researcher
Benefits, costs and reciprocity	What will each group involved in the study gain from it? What do they have to invest in time, energy or money? How do the benefits to the researchers compare with those for the study participants? What will be the benefits to public health? Is there potential to make the research more participative?
Harm and risk	What might this study do to hurt the people involved? How likely is that harm to occur? Is any harm worth the potential benefits?
Honesty and trust	What is our relationship with the people we are studying? Do we trust each other? Are we telling the truth? Can any covert behaviour be justified?
Privacy, confidentiality and anonymity	In what ways will the study intrude and come closer to people than they may want? How will information be guarded? How identifiable are the individuals and organisations being studied? Can we keep the assurances of confidentiality that we make?
Intervention and advocacy	What will we do if we see harmful, illegal or wrong behaviour on the part of others during the study? If we believe an individual or group is being wrongly treated, will we advocate on their behalf?
Research integrity and quality	Is the study being carefully, thoughtfully and soundly conducted?
Ownership of data and conclusions	Who owns the data? Has this issue been resolved before the research starts? Who controls distribution of the research? What rights do the research participants have to comment on or contribute to the conclusions?
Use and misuse of results	How can we ensure the results are used to good effect? What is the best way for the research to have beneficial effects on public health? What if our results are used by others to bad effect?

RESEARCH WITH INDIGENOUS AUSTRALIANS

The history of colonialism and injustice to Indigenous Australians means that the ethics of public health research that affects them are particularly sensitive. Indigenous Australians have been the focus of anthropological research ever since white invasion, not always with benign effects. There has been growing attention to Indigenous health research since the 1980s. Anderson (1996, p. 154) reports that 'there is a growing, though tentative, recognition that research can be a valuable tool if deployed appropriately.' A sign of this recognition has been the willingness of the National Aboriginal and Islander Health Organisation (NAIHO) to collaborate with the NHMRC (2003) to develop a set of ethical guidelines for research in Indigenous communities, designed to protect Aboriginal people from exploitation by researchers. Commonwealth funding bodies require research they fund to comply with these guidelines, whose key principles include:

1 Research should be based on full consultation and negotiation with Aboriginal communities, providing them with the opportunity to understand and assess the proposal and its likely impact on the community.

2   Community members should be involved in the research wherever possible and researchers should endeavour to include members of the community in the research team. The involvement should be such that the community can continue to negotiate about the process of the research as it proceeds.
3   Issues associated with the ownership and publication of data should be negotiated in advance. They suggest that a plain English community report be made available.
4   Values important to Indigenous peoples should be acknowledged including respect for culture, acknowledgment of the history of colonisation and marginalisation and the responsibilities that Indigenous people have to Country, for kinship bonds, caring for others, maintenance of harmony and balance within and between the physical and spiritual realms.

Anderson (1996) raises a number of important issues relating to ethics and Aboriginal health research:

• How can the tensions between what the research community believe constitutes 'good science' be reconciled with what the community believe to be acceptable?
• How can the benefits of research to Aboriginal communities be assessed? Will the communities or the researchers be the primary beneficiaries? How can the process of the research operate to increase the skills of the Aboriginal people involved?
• How can the potential dangers of the research be anticipated?
• How can non-Indigenous researchers develop skills in working in a cross-cultural context and learn to negotiate difference?
• What are the ethical implications of the practices of research funding bodies? How can Aboriginal communities gain control of research funds?

The Lowitja Institute (www.lowitja.org.au/about-us) has produced resources that consider the ethical issues in the commissioning, assessment and conduct of research with Aboriginal communities, including a practical guide for researchers (Laycock et al., 2011).

CONCLUSION

This chapter has argued that the new public health makes new demands on researchers. In the past public health has relied heavily on epidemiological methods, the limitations of which have been recognised, and the value of a greater repertoire of methods is accepted. These methods include those derived from the social sciences, especially qualitative. A more critical approach to public health also calls for more reflection from researchers so that seemingly 'objective' truths are questioned and the subjective nature of experience recognised. Finally, this chapter considered ethical issues in research.

CRITICAL REFLECTION QUESTIONS

6.1 What are advantages of using both quantitative and qualitative methods in public health research?

6.2 How do the different hallmarks of the research world and the policy world impede research transfer?

6.3 What are the hallmarks of ethically sound Aboriginal health research?

6.4 Visit the NHMRC website (see below) and examine the projects funded in the latest grant outcome notifications. What is the balance of 'health' as opposed to 'medical' projects?

Recommended reading

Crotty (1998) *The foundations of social research: Meaning and perspective in the research process.* Provides a comprehensive overview of epistemological issues to consider in designing research.

Sarantakos (2005) *Social research*, is a clear and concise introduction to all aspects of social research and a very good guide for newcomers to social research.

Useful websites

National Health and Medical Research Council www.nhmrc.gov.au
Lots of useful information about funding schemes for research, ethical and health advice

www.lowitja.org.au
Lowitja Institute Australia's National Institute for Aboriginal and Torres Strait Islander Health Research

7

EPIDEMIOLOGY AND PUBLIC HEALTH

The study of the distribution and determinants of health-related states or events in specified populations, and the application of this study to control of health problems.

Last, 1995, p. 55

KEY CONCEPTS

INTRODUCTION

This chapter provides an overview of epidemiology in order to demonstrate its usefulness to the new public health. The main forms of epidemiological research design and the challenges of applying the methods in unruly field settings are described. Established techniques for measuring health status are described and their validity and reliability assessed.

WHAT IS EPIDEMIOLOGY?

Despite Last's definition given above, epidemiological approaches often focus on disease. Brown (1985) pointed out that the discipline paid far less attention to health and its creation, probably reflecting the origins of epidemiology as a medical speciality primarily concerned with curing disease. Certainly its strength appears to be in understanding and explaining disease. Büttner and Muller (2011) identify the following uses of epidemiology:

1   describing the distribution and extent of disease in populations including measuring disease frequency

2   identifying the causes of diseases (including infectious disease outbreaks and assessing diagnostic tests and screening)

3   assessing the effectiveness and efficacy of interventions to prevent, control and treat disease.

During the 1990s epidemiology became more sophisticated (for instance the use of multi-level models), and much more epidemiological research has been funded and conducted since then. Subspecialties of epidemiology—population and clinical—have become more prominent, and each has further areas of specialty.

POPULATION EPIDEMIOLOGY

Population epidemiology focuses on studies describing and explaining diseases in whole populations, such as the workers in a company, students in a school, or populations defined by geographical boundaries. This can be concerned with studying acute and chronic illness, and communicable and non-communicable disease, and the focus can be on environmental or behavioural factors.

The most often cited example of the value of population epidemiology illustrates its power in determining environmental causes of disease. In the mid-nineteenth century Dr John Snow was able to show, through careful documentation, that there was an association between the number of deaths in part of London and the companies that supplied water to that district. On the basis of his data (see table 7.1) he constructed a theory that maintained that contaminated water could cause cholera.

Population epidemiology has been used to document the links between ill health and many substances, including asbestos, lead and tobacco smoke.

Behavioural epidemiology concentrates on describing the extent of particular health-related behaviours (e.g. smoking, exercise, work habits) in a given population and how these behaviours might relate to disease. This research has been criticised as detracting attention away from the social and structural determinants of health. Schrecker (2012) notes of the authors of an Ontario study based primarily on behavioural risk factors that they 'are remarkably unreflective about their focus on proximate risk factors, ignoring the contextual influences that shape individuals' opportunities to lead healthy lives'.

TABLE 7.1   DEATHS FROM CHOLERA IN DISTRICTS OF LONDON SUPPLIED BY TWO WATER COMPANIES, 8 JULY TO 26 AUGUST 1854

Water supply company	Population 1851	Number of deaths from cholera	Cholera death rate per 1000 population
Southwark	167 654	844	5.0
Lambeth	19 133	18	0.9

Source: Bonita et al., 2006, p. 2.

CLINICAL EPIDEMIOLOGY

Clinical epidemiology applies epidemiological principles and methods to the practice of clinical medicine. Bonita et al. (2006, p. 133) define its central concerns as: definitions of normality and abnormality; accuracy of diagnostic tests; natural history and prognosis of disease; effectiveness of treatment and prevention in clinical practice. Clinical epidemiology has made important contributions by developing techniques to make accurate estimates of important clinical issues. For example, it allows assessments of risk versus the harm of particular treatments, the accuracy of diagnostic or screening tests and estimates of the expected case survival rates in groups of people with particular diagnoses.

Public health has recently become more interested in assessing the effectiveness of treatments and preventive strategies in clinical practice. The rising cost of medical care and increase in possible medical procedures and treatments has led to a growing interest in 'evidence-based' medicine. The Cochrane Collaboration was developed in the 1980s and 1990s in response to the call from the British epidemiologist Archie Cochrane for more systematic assessment of the effectiveness of health care. The Collaboration involves researchers from around the world, including Australia, in preparing, maintaining and disseminating systematic up-to-date reviews of randomised control trials of health care. These reviews are compiled through a process that relates to primary medical care (Silagy, 1993), to the effectiveness of promoting lifestyle change in general practice (Ashenden et al., 1997) and to pregnancy and childbirth (Chalmers et al., 1989). The Collaboration maintains a database of systematic reviews and holds annual colloquiums to bring together those undertaking the reviews (Chalmers, 1993, and see www.cochrane.org).

SOCIAL AND ECO-SOCIAL EPIDEMIOLOGY

Social epidemiology has emerged as a new paradigm for epidemiology. It promises to overcome some of the shortcomings of epidemiology that have been noted above. Fundamentally social epidemiology studies 'the social distribution and social determinants of states of health' (Berkman and Kawachi, 2000, p. 6). It draws on social psychology and sociology and engages more effectively with social science theory than forms of epidemiology. Krieger (2011) provides a very thorough guide to epidemiological theory and the importance of social epidemiology. She proposes the use of an eco-social approach which she described as 'ecologically oriented integrative, multilevel and dynamic epidemiological frameworks, explicitly linking societal and biophysical determinants of disease distribution and health inequities' (Krieger, 2011, p. 203). This framework includes socioeconomic status, social networks, social capital, discrimination, work demands, and sense of control. Social epidemiologists use multi-level analysis to try and disentangle the effects of compositional (that is, the characteristics of people in a particular location) from the contextual effects (the features of the location such as amount of litter or graffiti, condition of housing, availability of shops or leisure facilities). Path analysis is also increasingly used and enables researchers to model predicted relationships and consider the statistical associations. It is particularly useful with complex concepts that are composed of a number of variables such as trust, socioeconomic status or early life experience. They will also often adopt a developmental or life-course perspective that tries to track the ways in which experiences over a lifetime can affect health. Social and eco-social epidemiology are a good fit with the new public health agenda and promise to make an increasingly important contribution to our understanding of how social and economic factors shape the health and disease experiences of populations.

POPULAR EPIDEMIOLOGY

Popular epidemiology has evolved from the environmental justice movement (Novotny, 1994) and involves epidemiologists working with community people in social movements who want to research environmental threat to their health (Brown, 1992). Two examples highlight the type of work popular epidemiology gives rise to. Sebastián and Hurtig (2005) describe their work in the Ecuadorian Amazon where they worked with peasant movements and environment groups to research the impact of the oil industry on people's health. Local organisations set the agenda of the research, were involved in formulating hypotheses, consulted during the study and then took responsibility for dissemination of the findings and lobbying on the basis of them. Potts (2004) describes a similar experience in terms of the breast cancer movement, which has forced a focus on the potential environmental causes of breast cancer rather than individual risk and genetic susceptibility, which have dominated the discourse on breast cancer risk. She sees that the movement has forced attention on the role of factors outside the individual by examining patterns of cancer in relation to potential carcinogenic agents.

Thus popular epidemiology is responding to the criticism of epidemiology as having become divorced from public health practice and policy (Beaglehole and Bonita, 2004) and to the charges that epidemiology is only concerned with individual risk factors.

KEY CONCEPTS AND METHODS IN EPIDEMIOLOGY

Information on health status is crucial to epidemiology. Such data typically come from national institutes of health. The types of data and how to judge their quality are shown in boxes 7.1 and 7.2.

BOX 7.1   TYPES OF HEALTH DATA

- **Administrative data** are collected during the delivery of health services—for example, at the point of care, such as a hospital admission or visit to a general practitioner, or through other processes, such as registration of birth, death or marriage events.
- **Population health surveys** involve the collection of data related to health and disease within a sample of a defined population. Australian examples include the National Health Survey and the National Drug Strategy Household Survey.

- **Health registers** are collections of records containing data about individuals who are typically patients or clients of a health service or health program. They aim for complete coverage of the relevant population and timely data supply, although this can vary between registers. They can be used to measure incidence and prevalence, follow up individuals for further treatment, plan services, monitor survival rates and recall following adverse events, and evaluate the effects of treatments or other interventions. Examples are the Australian Orthopaedic Association National Joint Replacement Registry, Australia and New Zealand Dialysis and Transplant Registry, Australian Childhood Immunisation Register, state and territory cancer registries, and state and territory breast screening and Pap smear registries.
- **Health surveillance** is the ongoing systematic collection, assembly, analysis and interpretation of health data, and the communication of information derived from these data. It can be used to measure the incidence of selected diseases and identify emerging threats. Examples include the National Notifiable Diseases Surveillance System (for selected infectious diseases) and OzFoodNet (for the incidence and outbreaks of foodborne disease).

Source: AIHW, 2012, pp. 21–2.

BOX 7.2   HOW TO JUDGE THE QUALITY OF HEALTH DATA

The Australian Institute of Health and Welfare lists the following dimensions to determine the quality of data:

- institutional environment—the factors that may provide insight on the effectiveness and credibility of the agency producing the statistics
- relevance—how well the statistics or product meets the needs of users
- timeliness—the delay between the reference period (to which the data pertain) and the date at which the data become available
- accuracy—the degree to which the data correctly describe what they are designed to measure
- coherence—the internal consistency of a product, its comparability with other sources of information and over time
- interpretability—the availability of information to help provide insight into the data (for example, information on variables, concepts and classifications used)
- accessibility—the ease by which the data can be obtained by users.

Source: AIHW, 2012, p. 24.

To understand epidemiology, it is essential to appreciate the meanings of, and difference between, incidence and prevalence. 'Incidence' refers to the number of new cases of disease occurring in a defined population over a specified time period. The incidence rate is determined by taking the number of cases (the numerator) over a specified time period and expressing these as a proportion of the total population (the denominator).

'Prevalence' refers to the number of cases of disease that exist in a defined population at a particular point in time. The prevalence rate is determined by taking the cross-sectional count of disease (the numerator) and expressing this as a proportion of the total population at that time. Prevalence rates will depend on the duration and incidence of a disease. A chronic disease from which people suffer for many years will obviously have a higher prevalence than one from which people die relatively quickly.

Epidemiology relies heavily on demographic data that describe the composition of populations, such as the overall size of populations, breakdown according to age and sex, and a host of other variables, including housing tenure, employment status, area of residence, household and family size and female fertility patterns (see part 4).

Epidemiological studies are descriptive, analytical or experimental. The main methods are outlined in table 7.2.

TABLE 7.2   TYPES OF EPIDEMIOLOGICAL STUDIES

Type of epidemiological study	Main methods
Descriptive	Routine data collection (e.g. death certificates)
	Ecological study (study of whole populations and links with disease)
	Descriptive cross-sectional study
Analytical	Cohort study (longitudinal or prospective)
	Case–control study
	Comparative cross-sectional survey (usually random)
Experimental	Randomised controlled trials (RCTs)
	Community trials

Source: Last, 1998; Bonita et al., 2006; Büttner and Muller, 2011.

DESCRIPTIVE STUDIES

Descriptive studies do not attempt to link exposure to any particular agent with a disease effect. They are generally based on routinely collected data relating to mortality and morbidity. In most developed countries, including Australia, the data collection is routine and reliable, but this is not possible in many low-income countries where specific household studies are necessary to gain estimates of mortality rates. It has become increasingly common to display descriptive data in mapped form. Data portrayed in this manner is usually easier to follow for those not accustomed to statistical data.

Descriptive studies involve those based on correlational data or on descriptive studies of individuals, which may be taken from a case series (for example, what happened to a series of people who contracted a particular strain of influenza in a specific region) or from cross-sectional surveys. They provide useful data, but they do not allow the cause of a disease to be ascertained. Correlational data are sometimes, mistakenly, taken to represent causality. But simply showing that a particular disease and a suspected risk factor are both high in a particular population does not necessarily mean the risk factor has caused the disease, which could be explained by other aspects of lifestyle, an environmental factor or work habits.

Routinely collected data should be treated critically like any other data. Mortality statistics, for example, are often treated as the 'true' picture of mortality in any given population, whereas they are based on clinicians' assessments of cause of death, which are always open to judgment and may reflect cultural norms such as strong societal taboos against recording a death as suicide (AIHW et al., 2014a). Bartley's (1985) review of the uncertainty relating to death certification of coronary heart disease demonstrates that the cause of death on a certificate may well be inaccurate. Doctors feel pressured to record a cause even when they are uncertain. Often heart failure results from years of chronic illness, yet only the final heart problem appears on the certificate. Survey data, such as that generated from the Australian Health Survey, are only as accurate as the survey instrument allows.

In ecological studies, the units of analysis are populations or groups of people rather than individuals, which means the link between exposure and effect cannot be made. Associations observed at the group level do not necessarily represent the associations at the individual level. The mistake of imputing to individuals the characteristics of aggregates has been called 'ecological fallacy' (Schwartz, 1994). Much data on inequities in health status, for instance, are based on aggregates. Census areas are sorted according to median income and those with low median income tend to have higher mortality rates. But it is a fallacy to then assume that individuals in the low-income census areas have a higher risk of dying than do individuals in areas with lower mortality rates. Ecological studies are also particularly vulnerable to confounding factors, because an ecological association between two variables in a population may, in fact, be reflecting correlations with other variables that intervene between the original two.

Descriptive studies are often the starting point for more sophisticated epidemiological and other public health research.

ANALYTICAL STUDIES

Cross-sectional studies may be aimed at simple fact finding or occasionally to test a hypothesis (Last, 1998, p. 78). They are useful to establish the prevalence of a disease or health-related behaviour, and are most typically based on random surveys of populations defined by geography (e.g. a local government area) or characteristic (schoolchildren or army personnel). Details of survey methods and potential pitfalls are given in the following chapter.

Australia has conducted a series of cross-sectional health surveys, the most recent being the Australian Health Survey conducted in 2011–13. This collected data on a wide range of conditions suffered by people, as well as demographic information. It is the most comprehensive illness database available on the whole Australian population.

The main weakness of cross-sectional surveys is that they do not include a time dimension, which is crucial in assessing whether an association is causal. A cause must be shown to have come before an effect.

Cohort studies may also be called 'longitudinal' or 'prospective'. People in the study are not defined in terms of having or not having a particular disease, but in terms of exposure to a possible cause of that disease. The population is then followed over time until a specified end point (such as death) when rates for the disease are calculated in relation to the exposure to the potential cause of the disease (see figure 7.1).

FIGURE 7.1   DESIGN OF A COHORT STUDY

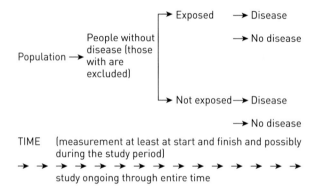

Cohort studies differ in the frequency with which people are assessed as part of the study. Doll and Hill's (1954) cohort study of the smoking habits of 34 440 male British doctors in relation to cancer of the respiratory tract made simple postal enquiries about smoking habits at 6, 15 and 21 years after the initial identification of smokers and non-smokers in 1951. Last (1998) points out that this study highlights two difficulties. First, it proved difficult to track down the doctors for the subsequent postal surveys. Second, many more of the study population gave up smoking than was the case with the general population. Last suggests the doctors' smoking behaviour may have changed because they knew it was under observation—the Hawthorne effect at work. The Hawthorne effect operates when the very fact of a study influences the phenomena being studied.

Cohort studies of disease causation are usually based on large samples, especially when the particular disease is rare. This, together with the time period in which they are conducted, means they require a considerable investment of resources. Costs can be saved by using existing data sets. For example, Doll and Hill, in their study of smoking and doctors, were able to set up a system for obtaining the death certificates of all respondents and cross-checking these with the names of doctors who had been removed from the medical register because of death. Costs can also be reduced by using historical data relating to a cohort. For example, records of exposure of members of the armed services to radioactive fallout at nuclear bomb testing sites have been used to examine the possible causal role of fallout in the development of cancer since the 1950s (Bonita et al., 2006).

Case–control studies look retrospectively at people with a particular disease, comparing them with a control group who are unaffected by the disease (see figure 7.2). These studies are far more economical and quicker than cohort studies. This design is able to determine whether people with the disease have been exposed to a suspected risk factor significantly more often than the controls, and so estimate the relative risk.

FIGURE 7.2   DESIGN OF A CASE–CONTROL STUDY

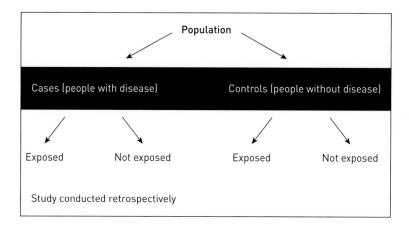

Comparative cross-sectional studies investigate the relationship between a disease (or other health-related characteristic) and other variables of interest in a defined population at one particular time. The presence or absence of disease are determined in each member of the study population or in a representative sample at a particular time (Büttner and Muller, 2011, p. 281). The results of a cross-sectional study allow the observation of an association between the study factor and the outcomes but do not provide evidence of a causal relationship.

EXPERIMENTAL DESIGNS

RANDOMISED CONTROLLED TRIALS

Randomised controlled trials (RCTs) are experiments designed to test new preventive or therapeutic interventions. People in a population are randomly assigned to groups (but not randomly selected from the population), usually called the treatment and control group (see figure 7.3). The treatment group receives the treatment, and then the outcomes for the two groups are compared. People may also be randomly selected from a source population, but most typically RCTs are based on random allocation.

The process of randomisation is designed to ensure that treatment and control groups are comparable at the start of the intervention, and differences at this stage are assumed to result from chance.

FIGURE 7.3   DESIGN OF A RANDOMISED CONTROLLED TRIAL

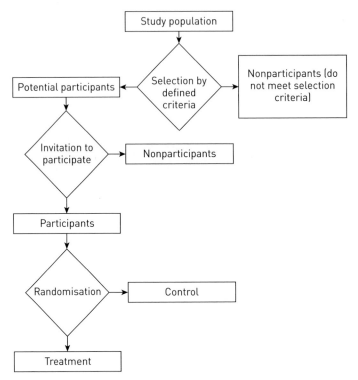

Source: Woodward et al., 2001, p. 877.

Evidence-based practice became increasingly important in the 1990s. An international collaboration—the Cochrane Collaboration—has been established to conduct meta-evaluations of the results from RCTs. Last (1987, p. 94) points out that the RCT has overturned much conventional wisdom in medical practice, supporting this with examples such as studies that have shown that bed rest in patients' own homes is as effective as treatment in a coronary care unit or acute short-stay general hospital for many men who have had an acute myocardial infarction; and that bed rest confers no benefits upon older people who undergo surgery for cataract removal.

The strength of the RCT depends on the ability of researchers to achieve internal validity by establishing statistical control over systematic and random error. Daly and McDonald (1992) point out that this may be difficult to achieve in practice. One of the challenges of conducting an RCT is ensuring true randomisation. Studies have shown that clinicians believing in the effectiveness of a particular technique will undermine the randomisation to ensure particular patients receive the treatment the clinician believes is effective (Keirse, 1988, 1994). This may be overcome by a double-blind experiment where, while people know they are part of an experiment, neither they nor their clinicians know whether they are part of the experimental group or the control group.

The other crucial issue is that the randomisation is of the participants in the trial, not of any particular population. Thus, generalisation from RCTs cannot automatically be made to a wider population than that represented by the people in the trial. It is not unusual for the assumption of randomisation of people to two groups to be a sufficient basis for generalisation to populations. Jelinek (1993) says the major reasons clinicians may not accept the results from RCTs is the lack of knowledge of patients excluded from trials. He adds that it does not matter how well the trial is conducted if the people in it are atypical of usual patients. Feinstein (1983), while acknowledging the many achievements of randomised clinical trials, suggests that there are many clinical settings in which they are not appropriate for logistical or ethical reasons. He points out that trial designers are 'fastidious' about the patients they admit to a trial, whereas clinicians are 'pragmatic', wanting to treat all patients and being generally interested in more complex outcomes than trial designers, who want outcomes that can be easily measured statistically. Public health research generally requires more pragmatism in choice of settings and inclusion of communities, and so RCTs are often of limited usefulness in practical public health work, such as evaluation of community-based initiatives.

RCTs have conventionally been seen as the gold standard for epidemiological research. Christie et al. (1987, p. 77) claim they 'are the most scientifically rigorous method available in epidemiology'. While this claim may be true for determining the effectiveness of a particular therapy (although Feinstein (1983) questions whether this is so), it is not necessarily the case for other purposes, such as evaluation of health promotion in a community setting or of complex community-based coordinated care trials. Community-based interventions have to deal with the complexities of the social world. It is simply not possible to control what happens in these settings to the extent that an RCT depends on for its claims of rigour to be valid. Nutbeam et al. (1993b) report that in the evaluation of the Heartbeat Wales health promotion project, the community selected as a control decided to institute its own heart health program, thus undermining its value as a control. In any case, there simply are no two communities that are identical. The RCT methodology, however, relies on the assumption that the experimental and control groups are identical. Similarly, in the case of community interventions the process of allocating people to groups is unlikely to result in identical comparison groups. In addition, if one treatment or care regime is perceived as superior, it is highly likely that the people in the control group will seek that treatment or care for themselves. Researchers cannot control such factors. It is also likely in most community care trials or health promotion interventions that the randomisation process will not be to compare care or health promotion intervention with no care or no health promotion intervention at all, but rather to compare the value of one form over another. Given that RCTs are not particularly well suited to monitoring subtle effects in comparing treatments, the methodology is usually difficult to use in the evaluation of community-based initiatives. Box 7.3 summarises the conditions that have to be met in order for an RCT to be used in a community setting. It is extremely rare for each of these to be met.

> **BOX 7.3   CONDITIONS THAT HAVE TO BE MET IN A COMMUNITY TRIAL**
>
> - The control community has to be largely the same as the one experiencing the intervention
> - The control community has to have no health promotion intervention of its own
> - The external forces on the control and experimental communities have to be very similar.

The impact of social and economic factors on health status and the importance of participation in health decision-making at a collective and individual level have been stressed as key aspects of the new public health. The track record of RCTs on either of these aspects has not been good. They do not usually consider the effects of social factors such as socioeconomic status, employment status or social support on the people in the trial, which are generally not controlled for, even though a large body of research (see part 4) has shown that they are crucial in determining patterns of ill health.

Randomisation is not easily compatible with the new public health philosophy of encouraging people to participate in decisions about public health and health care. The methodology sits more easily with more paternalistic concepts of decision-making, in which the expert is assumed to know best and decision-making is left to those experts. Oakley (1989) has argued, however, that RCTs can be designed to be more consumer-friendly in health care settings. Büttner and Muller (2011) describe the value of using a multiple baseline design when conducting community trials of disease prevention. A key feature of these designs is that interventions are staggered across the sites over time, providing credible experimental control and minimising the possibility that observed changes occurred by chance. Each community is used as its own control in this design (see figure 7.4). A change in the outcomes measured after the intervention strategies are put into place in one community, together with the absence of change in the others yet to receive the intervention, can allow the deduction that the change resulted from the intervention and not from other factors.

FIGURE 7.4 MULTIPLE BASELINE DESIGN

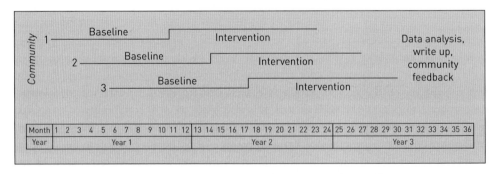

Source: Büttner and Muller, 2011.

Newell (1993) suggests that the usefulness of information from RCTs can be increased if they include qualitative data collection. He describes the importance of qualitative data in an RCT that examined whether frail elderly people should be cared for in a nursing home or in a long-stay ward. The RCT alone could help provide a yes/no answer to the research question, but qualitative interviewing was required to unpack the elements of a nursing home that made it a preferable environment for frail elderly people.

QUALITY AND ERROR IN EPIDEMIOLOGICAL STUDIES

Epidemiological studies are best suited to documenting the links between a particular biomedical, individual, behavioural or social structural risk factor. In isolation, any particular epidemiological study will not be convincing about these relationships, but a number of studies, taken together, will be more revealing about causality. Causal inference in epidemiology depends on accruing sufficient consistent evidence so that there is a high probability that the observed association between 'exposure' and disease reflects an underlying causal relationship (Marley and McMichael, 1991). Thus, the links between lung cancer and tobacco smoking gained credibility as more and more studies pointed to the relationship.

The studies have to ensure that the inference was not due to chance (random error), some kind of bias (systematic error) or uncontrolled confounding. Random error can never be completely eliminated, but can be reduced by careful measurement (Bonita et al., 2006, p. 52). Systematic error results primarily from bias in the selection of participants and from measurement or classification bias. It can be reduced by careful selection and ensuring that measurement is as accurate as possible.

Confounding can be a real problem in epidemiology unless it is controlled, as it may suggest a cause–effect relationship that does not exist. The term describes the situation in which the effects of two processes are not separated (Last, 1998, p. 100). This is when an exposure to a risk occurs in a study population, and is associated with both the disease and the exposure. For example, in a study of the effects of occupational exposure to a risk substance upon respiratory disease, smoking would be a confounding variable unless allowance were made. This could be done by stratifying smokers and non-smokers in the analysis of results. Age can be a confounding variable when the age distribution of two populations being compared is very different. Age standardisation techniques are used to remove the confounding effects of age differences.

Epidemiology, like other natural sciences, does not produce absolute proof, but the best available understanding of the causes of disease, given the aggregate of research available to date. The aim of epidemiological research is to eliminate as many of the sources of error and bias as possible. This is challenging but possible.

CONCLUSION

Moving from an observed epidemiological risk to action is complex and involves processes of advocacy, political action and lobbying. Examples of public health advocacy (sometimes based on epidemiological evidence) are provided in chapter 20.

The limitations of epidemiology were discussed in the last chapter. The need for epidemiology that focuses more on the health of populations and less on the diseases of individuals was recognised. It was also noted that epidemiology should put more emphasis on the social, cultural and economic factors that either create health or cause disease.

Collaboration of epidemiologists with social scientists can improve the quality of public health research and interventions. Robert et al. (1989) provide an example of how epidemiological knowledge of schistosomiasis was combined with a socio-anthropological study to design a health education program that would be culturally and locally appropriate. They show how both forms of knowledge were important. In the field of environmental effects on health, for instance, there has been an increased call for public health workers to give more credence to Indigenous knowledge and perspectives on the effects of environmental hazards. There are an increasing number of examples where those with local knowledge have discerned the link between a particular substance and a disease (Popay and Williams, 1996). Workers, for instance, have identified links between their working environment and disease before studies measured the link (Phillimore and Moffatt, 1994).

Abbott (1990) labelled the local awareness of environmental effects as sensory data, which often come into conflict with that based on epidemiological evidence. He describes how local residents' perceptions of the effects of emissions from local factories in the Port Adelaide area were discounted because scientific air monitoring had not shown a damaging level of pollutants. Tesh (1988) claims that many epidemiological investigations support a strongly individualistic view of the causes of ill health because they focus mainly on measuring individual causes of disease and ill health. So epidemiologists are more likely to research the question 'why do these particular people smoke?' than 'why do large numbers of people continue to smoke?' The first question directs attention to the psychology and physiology of individual people; the second question to the tobacco culture in which everyone lives (the growing of tobacco, the advertising of cigarettes, the social meaning of smoking). Similarly Krieger (2000) asks whether we should see race or racism as the public health issue of concern, a distinction driven by a sociological rather than medical concern. A focus on race tends to focus on non-white people as a 'problem'. A focus on racism, by contract, sees racist attitudes as the cause of health problems.

Epidemiology is a vital tool for the new public health. Developments in epidemiology such as the increasing consideration of social factors and research on the impact of contextual factors make it even more important. Combining epidemiology with social science, especially qualitative methods, makes a powerful evidence basis for public health advocacy.

CRITICAL REFLECTION QUESTIONS

7.1   What is added to the epidemiological perceptive through the introduction of an eco-social perspective?

7.2   What methodological challenges does the eco-social approach raise?

Recommended reading

Krieger (2011) *Epidemiology and the people's health: Theory and context*

Impressive discussion of the scientific, social and political roots of epidemiology. Krieger proposes an epidemiological eco-social approach with which to frame research aimed at reducing health inequities.

Useful website

International Epidemiological Association http://ieaweb.org

8

SURVEY RESEARCH METHODS IN PUBLIC HEALTH

A survey is the systematic collection of people's self-reported information at a particular point in time.

Feuerstein, 1986, p. 65

KEY CONCEPTS

Introduction

Strengths of surveys

Weaknesses of surveys

Planning and conducting surveys

Is the research question amenable to questionnaire or interview survey?

What type of survey to use?

Selecting respondents

How many people should be included in a survey?

Designing a survey instrument

Survey fieldwork

Self-completion questionnaires

Telephone surveys

Face-to-face surveys

Response rates to surveys

Analysis of survey results

Conclusion

INTRODUCTION

Surveys are the most widely used method of data collection in public health research, being used as part of evaluations, epidemiological designs, needs assessments and planning exercises. These surveys may be government initiated, such as the five-yearly Census or specific national health surveys, or conducted within universities or health departments for specific purposes. They can be used to collect data on occupational histories, obtain a snapshot view of a community's health status to use as part of a needs assessment, assess the level of participation in community activity or collect opinion data to inform decision making.

Interview surveys collect data orally. Questionnaire surveys collect data in a written form and are self-administered. We will consider surveys that collect structured or semi-structured data that are primarily destined for use in statistics, but may also include some qualitative data.

STRENGTHS OF SURVEYS

Surveys collect data in which the same variables are measured across units. Units can be individuals, households or health care settings. They produce a standard set of data for each unit that can be analysed using statistics to produce patterns within the data and to examine the relationships between the variables measured (Browne, 2005). Surveys are particularly suited to descriptive data that describes the extent of a phenomenon. The Census is a classic example, collecting factual data on age, marital status, religion, citizenship, country of birth, date of arrival in Australia, Aboriginality, languages spoken at home, educational level, number of live births (for women), income and employment. While some of these data may be sensitive (especially income), none of them are a matter of opinion. Some may be biased by recall (for example, age of leaving school or address five years ago) but the margins for error are relatively small.

Surveys are a relatively cheap means of collecting data, not requiring the detailed, time-consuming data collection and analysis of most of the qualitative methods. They can also, as a consequence, produce relatively quick results as the data are collected in preselected categories. They may also be more acceptable to the participants as they do not take up too much time. This is especially true of questionnaires, which people can complete at their own convenience. They are not dependent on an interviewer finding people at home. On the other hand, interviews mean that people can be encouraged to take part in a survey and allow clarification of the meaning of a question or any aspect of the interview schedule. Interviews can also deal with more complex issues and be longer than questionnaires. They also permit the possibility of recording spontaneous answers. Interviews usually obtain higher response rates, partly because of the personal contact, but also because they are more acceptable to people with low levels of literacy or who do not write English well. In Australia the provision of interviewers with a variety of language skills or interpreters is necessary to ensure that non-English-speaking people are included in surveys.

The easy administration of surveys means that they can be used with large numbers of respondents, which is especially useful when the aim of the research is to generalise to a particular population. For instance, the findings of a needs assessment may be more convincing if data relating to health service use are collected from a representative sample of the population.

WEAKNESSES OF SURVEYS

Surveys evolved within a social science that was modelled on the natural sciences, and were seen as a precise means of measuring social phenomena—the social world's equivalent of the laboratory's experiment. Yet surveys are very different from experiments, which offer researchers total control over the experiment's environment. This control is not possible with surveys. Each person will answer in different circumstances and bring to the survey a different set of assumptions, history and values. In experiments, researchers can precisely control the composition of experimental and control groups and the independent variable in the study, which means they can be more certain about patterns of causality. Surveys simply do not have the same power to establish causality, but they can suggest correlations between variables. Consequently, the descriptive data they produce are most useful.

The types of data produced by surveys have also been criticised by researchers from within social science. Aggregation assumes that meanings are unproblematic and that words have uniform and agreed meanings that are not contingent on their context. Busfield and Paddon's (1977, p. 99) reflection on their use of a survey to study fertility behaviour illustrates this problem. Respondents frequently stated their reason for their particular family size as 'That's all we can afford'. But when they analysed these seemingly similar responses in the context of individual interviews, they found that the phrase's meaning varied considerably: 'For some people it was a comment made only when serious financial difficulties were encountered, for others it was made when no financial problems were in sight either from a desire to maximise material standards of living or to plan for a better future or whatever.'

Aggregation means that a particular variable is rarely considered in the context of the respondent's wider set of ideas and values. This means the pattern and structure of variables are not considered, and also assumes a uniformity in significance of variables that consideration of them in context will demonstrate does not exist.

Oakley (1981) has questioned the ethics and humanity of some traditional survey interviewing practices. Building on her experience of interviewing women about the process of becoming a mother, she recounts the standard advice given in survey methodology textbooks to interviewers who find respondents asking them questions. It was to say 'I guess I haven't thought enough about that to give a good answer right now' or 'It's your opinion we are interested in, not mine'. Oakley (1981) asks how that advice would stand up when the questions asked by pregnant women are 'Which hole does the baby come out of?' or 'Does an epidural ever paralyse women?' Her interviewing style developed to become less objective and detached but rather based on empathy and friendship. She maintains that the women became more willing to discuss issues openly. Oakley's critique of standard interviewing procedure has been widely quoted and supported and, together with other critiques, has led some social researchers to soften their interviewing approaches, and encouraged the use of more in-depth interviewing practices.

The limitations of surveys mean they are most suited to questions with a narrow range of meanings and to the collection of factual data. Busfield and Paddon (1977, p. 110) go so far as to say that when surveys are used for 'explanatory rather than descriptive purposes, then the value of survey data diminishes almost totally'.

The generalisation of findings from surveys is often cited as one of their benefits. Typically, they are based on a randomly selected sample that is taken to be representative of the population from which it is drawn. However, the full benefits of generalisation can only be realised if the sample is indeed representative, and then only to the relevant population. Public health research based on surveys of randomly selected samples of particular populations is often reported as if it can, by virtue of its randomness, be extrapolated to a more general population, but this rests on the assumption that the sampled population is representative of the more general population (Atkins and Jarrett, 1979, p. 97). In practice, error can affect random sampling, especially error in the original sampling frame, non-contact or refusal. Sampling frames may not replicate the real structure of the population because of deaths or people who have moved away, or because the sampling framework does not include all members of a community. For example, the electoral register is a fairly complete list of people in a population but it tends to under-represent young people and people from a non-English-speaking background as they are less likely to be eligible to vote. Most surveys suffer from non-response and refusals, and there is evidence that people are becoming less willing to cooperate (Browne, 2005). These factors all combine to undermine the generalisability of research based on surveys and need to be borne in mind when interpreting and using their results.

Most public health surveys are cross-sectional, based on some form of random sampling and analysed using cross-tabulation. McQueen (1993) criticises this form of survey for being static and not capturing the dynamic aspects of behaviour and attitude, especially in relation to time. He reports that the sum total of errors in a survey (non-response, coverage, sampling, interviewer, respondent, instrument) is rarely taken into account in assessing the value of survey research.

Results from surveys can be powerful but should not be used unquestioningly. The inaccuracies in the research process should be both reported and taken into account when interpreting the meaning of the survey results. Surveys are most powerful when used to collect factual data. Data that relate to social processes and attitudes are much harder to collect from surveys and are likely to be less reliable. Surveys can, however, be combined with other methods that are more suitable for studying social processes and attitudes and are an essential part of good public health practice. For example, surveys have frequently been used to help understand sexual behaviour and practices so as to plan and evaluate HIV prevention campaigns.

PLANNING AND CONDUCTING SURVEYS

Planning surveys involves a series of decisions about whether the survey methodology is appropriate, and what form is most suitable, given the available resources. The first decision to be made is a crucial one that is often not given sufficient attention in public health.

IS THE RESEARCH QUESTION AMENABLE TO QUESTIONNAIRE OR INTERVIEW SURVEY?

Consideration should be given to whether a survey is the best methodology to answer the research question. It takes time to design, implement and analyse. The first step is to check that the information required is not already available. Existing databases and routinely collected data may provide the necessary answers. Census data, for example, provides considerable demographic information.

A survey may be suitable for collecting factual information and straightforward behavioural data. If there are existing scales, which have been used and tested on other populations and which measure the phenomenon of interest, then a survey may be more attractive. For instance, there are a number of scales that measure self-perceived health status (Bowling, 2005), such as the Nottingham Health Profile (Hunt et al., 1986) and the SF36 (Ware et al., 1994) or specific instruments for disease states (Bowling, 1995). More complex issues concerning attitudes and feeling, which need to be viewed holistically, may mean a survey is not the answer. Sometimes it is preferable to combine a survey with other methods of research. For instance, community health needs assessment research on carers was illuminated by survey research reporting the extent of people caring for others in a community (Kalucy and Baum, 1992), whereas the experience of caring was best derived from detailed interviews (McColl, 1985).

Surveys are appealing because they yield a set of statistics that suggests the legitimacy of traditional science. But surveys are usually fairly crude instruments that rarely give more than partial insight to the complexity of public health issues and the interconnected influences on them.

WHAT TYPE OF SURVEY TO USE?

There are three types of survey:

- internet or mailed self-completion questionnaires
- telephone interview surveys
- face-to-face interview surveys.

Each type has specific strengths, discussed at length in research methods textbooks (Sarantakos, 2005; Dillman et al., 2014). Internet and mail surveys are cheaper to conduct but generally cannot be used to collect more complex information. They are usually not acceptable to people with low literacy levels. Telephone surveys are increasingly popular as they are cheaper than face-to-face interviews and may be perceived as less intrusive. Bias can be introduced by not everyone having a telephone, and the absence of complete listing for mobile phones. Face-to-face interviews generally result in a higher response rate, probably because they require less motivation for completion and people are more likely to agree to be interviewed when personally approached. This form of interview also allows the researcher to exert more control, and the interviews can be longer. All the above factors need to be taken into account when deciding what form of survey to use.

SELECTING RESPONDENTS

> If you want to know how a pot of food tastes, take a spoonful. You don't need to eat the whole pot!
>
> *Feuerstein, 1986, p. 69*

It is rare that a survey will include a total population, except in a census or survey of all people involved in a health promotion initiative. Surveys in public health typically involve some form of sampling. When populations are large, surveying the whole population is impractical, expensive and unnecessary. There are two types of sampling—random or probability sampling and non-probability sampling. Surveys of the type we are discussing in this section are almost always based on random sampling.

Simple random sampling relies on identifying a sampling frame—a list of people in the group the research is focused on. This may be a community based on geography, or with a particular characteristic. The accuracy of the sampling frame determines how well the final sample represents the group of interest. Common methods used in Australia are the electoral roll, telephone books or recruitment by door-to-door survey (the most accurate method). The likely bias from the other two is described by Smith et al. (1997).

A variety of methods is available to select the sample from the sampling frame: lottery, which involves putting numbers representing each person in the sampling frame in a hat and picking out the required sample size; using a random numbers table; or, if the information is available, selecting people at random according to their randomly selected birth date. Random samples can be stratified when you want to ensure that your final sample contains sufficient numbers of particular groups of the population.

Another popular method is cluster sampling. Here you randomly select a setting from which you randomly select individuals. For instance, in a survey of attitudes to general practice, you could randomly select general practices from the Yellow Pages and then select your respondents from the patients of the practices in your random sample.

Dillman et al. (2014) list a number of other methods for selecting samples. The likely biases stemming from the choice of sampling method should be noted when findings are reported.

HOW MANY PEOPLE SHOULD BE INCLUDED IN A SURVEY?

The main purpose of a random survey is to provide data that will be representative of the sampled population so that generalisations can be made to the total population. Most public health researchers will refer to a statistician to calculate the required sample size, taking into account the confidence level, population size, expected results and type of analysis to be done. A practical guide to determining sample size in health studies has been published by WHO (Lwanga and Lemeshow, 1991). Statisticians will be able to estimate what size population would be required for your survey if you assume a 60 per cent response rate, 95 per cent level of confidence and a 5 per cent error range. The crucial issue to understand in talking with statisticians is that their assurances are estimates based on probability, not absolute truth, and that by adjusting the size of a sample the error range can alter.

Analysis of survey data according to different subgroups in a total sample (for instance, according to particular income levels) requires ensuring that the subgroups are of sufficient size to allow generalisations. Part of the process of estimating sample size is estimating the non-response rate to the survey—those people who refuse to answer the survey or cannot be contacted for one reason or another. Obviously, the non-response rate should be kept as low as possible. Three mailouts for mailed surveys (which generally have a lower rate than face-to-face or telephone surveys) can increase the response rate considerably (Dillman, 1983). There is increasing concern among social researchers that response rates to surveys are declining. This probably reflects greater concerns with privacy and also competition with market research companies.

DESIGNING A SURVEY INSTRUMENT

Survey instruments have to be carefully designed if they are to be useful. There are many pitfalls for the unwary and few shortcuts that can be taken, no matter how experienced the researcher. New researchers underestimate the care, attention and time needed to produce a useful interview schedule or questionnaire, which should be carefully planned, piloted and revised. A broad guide to designing questions is given here. More detailed guides are given in Sarantakos (2005) and the South Australian Community Health Research Unit (1991).

The design process involves the ordering and content of questions, and, for mailed surveys, the design of the questionnaire. The order of questions is important. Sensitive questions should be left until later in the survey so people are not discouraged from continuing. Layout is important for all surveys, but particularly so for mailed questionnaires. However, a clear and easy-to-follow layout also helps interviewers conduct good interviews. The South Australian Community Health Research Unit (1991, pp. 143–4) recommends the following guidelines for designing self-administered questionnaires:

- Use a plain easy-to-read typeface.
- Leave lots of space between questions so the questionnaire does not look cramped, but has an open, airy look.
- Consider including cartoons that may help keep the respondent interested in the form, especially when the questionnaire is long and time-consuming.
- Do not precode the questions as these may make the questionnaire appear intimidating.

Interviewers need to have as many aids as possible to make the process smooth. In face-to-face interviews, cards detailing the options that people can choose are useful.

In telephone surveys, the questions must not be overly complicated or respondents will find the interview difficult to follow. One means of overcoming this is to mail a copy of the questionnaire to the respondent and then conduct the interview a day or two later. This way the respondents can have the interview schedule in front of them.

Questions asked in public health surveys may be factual, behavioural or attitudinal. Factual questions are generally the simplest, but they still have to be designed very carefully. For example, the question 'How many children do you have?' sounds straightforward but it can be interpreted in a variety of ways: how many children are living in the household, regardless of whether they are adopted, fostered or natural; how many children the person completing the form has ever had, regardless of the age of the children or whether they are living in the household; how many children the person currently defines as children (and this would vary according to when the person defines childhood as finishing). If all the information available is a number in a box, there is no way of knowing how the person interpreted the question, and it would be difficult to make sense of the answers. To provide useful information, the question needs to be more precisely worded, exactly how depending on the purpose of the information. To estimate lifelong female fertility, a question such as that included in the 1996 Census: 'For each female, how many babies has she ever had? (include only live births)' would be necessary. If, however, the question wanted to determine how many children were currently living in the household, it would need to read 'How many children under 16 are living in your household?'

Public health has become increasingly concerned with inequities in health status and the conditions that produce health. It is therefore often important for public health researchers to use surveys to gain a picture of the pattern of inequities within the population with which they are concerned. Measuring social class or socioeconomic status is surprisingly difficult (see discussion in Travers and Richardson, 1993, chapter 1).

The complexities of designing behavioural and attitudinal questions are much greater than is the case with factual questions. People may not accurately recall their behaviour. In nutrition surveys, for instance, it is very difficult for researchers to obtain an accurate picture of people's food intake from a question or interview survey. 'Food diaries' are more effective, even though they require a considerable commitment from respondents. People may not want to disclose particular behaviours. Surveys of safe sex practices may find that the extent of unsafe sex is underestimated as people are reluctant to admit to it. Asking people to report on things retrospectively or prospectively generally produces unreliable data.

Attitudes deal with abstract concepts and are difficult to measure. Psychologists have invested considerable energy in the accurate measurement of attitudes and many scales have been developed and tested. When designing a survey, check whether an existing set of questions that have been tested can be used. This could save much time and effort and provide more accurate and useful data.

A central concern of public health is obtaining measures of people's health status so that the distribution of health in a population can be studied in detail. The discussion in chapter 1 highlighted the complexity of defining and measuring health, but it is now widely accepted that health concerns more aspects of life than simply the absence of disease. Individuals' health status is affected by many social, economic and environmental factors, and many scales have been developed to measure health aspects through survey research. Bowling (2005) reviews many of these, categorising them as:

- measures of functional ability (for example, the Index of Activities of Daily Living, the Quality of Well-being Scale)
- broader measures of health status (for example, the Sickness Impact Profile, the Nottingham Health Profile, the Rand Health Insurance Study Batteries)
- measures of psychological well-being (for example, the General Health Questionnaire, Hospital Anxiety and Depression Scale, the Symptoms of Anxiety and Depression Scale)
- measures of social networks and social support (for example, the Social Support Questionnaire, the Revised UCLA Loneliness Scale, the Family Relationship Index)
- measures of life satisfaction and morale (for example, the Delighted–Terrible Faces Scale, the Self-esteem Scale).

These scales can be used as part of a health needs assessment exercise (especially to compare the health of different populations or subgroups within a population) or to evaluate the effectiveness of a particular clinical or health education intervention. Bowling's (2005) review covers 53 instruments. The researcher needs to decide which, if any, will be appropriate for the particular purpose. Each has its own strengths and weaknesses, which have to be assessed in terms of the aims of particular research. If one is suitable to help answer the research question, the great advantage is that the scale and its questions have been previously validated and proved to be reliable. When using these scales it is important to remember that, however well they have been tested and validated, the data they yield will only ever be an approximation of health, social support, functional ability or whatever other construct they set out to measure.

Questions asked can be either open-ended or closed-ended. Closed-ended questions contain a list of answers from which the respondents are instructed to pick one or more. Types of closed-ended questions are shown in box 8.1.

BOX 8.1   EXAMPLES OF CLOSED-ENDED QUESTIONS

Simple yes or no

- *Have you ever used Anytown Community Health Centre?*   Yes   No

A multiple-choice question where the respondent is asked to choose one answer

- *How often do you have a pap smear test for cancer?*
    - More than once a year
    - Once a year
    - Every two years
    - Every three years
    - Every four years
    - Every five years or less often

- – Never
- – Not sure what a pap smear is

A multiple-choice format that allows multiple responses

- *Which of the following services have you used at the Anytown Community Health Centre in the past 12 months? Please tick one or more boxes*
  - – General practitioner
  - – Podiatrist
  - – Nutritionist
  - – Counselling service
  - – Health education group activity
  - – Community development service

Likert scales, which require respondents to rate the extent to which they agree with a statement on a verbal-numerical scale

- *How safe for your children is the area where you live? Please circle the appropriate number.*

1	2	3	4	5
Very safe	Safe	Neither safe nor unsafe	Unsafe	Very unsafe

Open-ended questions are ones that ask the respondents to reply in their own words. In mailed surveys the quality of the answers will depend on the time the people devote to the questionnaire and their literacy level. An example is: 'What characteristics make a good general practitioner?' Answers could vary from one word such as 'competence' or 'patience' to a few words, 'good listening skills' or 'the ability to empathise with me', to a paragraph detailing the person's experience with GPs. Open-ended questions work best in interviews as the respondents can be encouraged to be more forthcoming if they initially do not have much to say, thus producing more detailed data. Direct quotes from respondents are helpful in bringing statistical data to life.

The aim of question construction should be to make the questions as clear and unambiguous as possible. Common problems are:

- leading questions that bias the respondent in a particular direction—for example: 'Have the cutbacks in health services made it more difficult for people in this community to get access to the hospital?'
- double-barrelled questions that treat two or more separate pieces of information together—for example: 'Does your environmental health office have a procedure for dealing with food safety and air pollution?'
- jargon that may not be familiar to people in the survey—for example: 'Does your local government have an integrated plan for managing environmental sustainability?'
- double negatives that will confuse people—for example: 'Would you rather not use a condom when having sex?'

It is essential that a new questionnaire or interview schedule be piloted. It is useful to do this first with friends and colleagues to iron out any obvious faults, and then with a group as close as possible to the main sample. If many errors are discovered, the survey should be piloted again.

Designing effective and meaningful survey instruments takes time and practice, but it is a skill that will be useful to most public health practitioners and researchers. Knowing how to design a survey means being able to assess the value of others' questionnaires and interview schedules.

SURVEY FIELDWORK

The larger the survey to be conducted, the more complicated the fieldwork. A survey used to evaluate a health education group attended by 50 people could simply be handed out and collected by the health worker or a colleague. If, however, the survey is intended for a random sample of a large population, more thought should be given to its organisation.

SELF-COMPLETION QUESTIONNAIRES

Self-completion questionnaires used to be mailed but are now often internet based using software such as Survey Monkey. Mailed questionnaires should be mailed with a stamped addressed envelope and a covering letter that serves to motivate the respondents to complete the questionnaire. If the survey is to be conducted over the internet then a covering email substitutes for the letter. This should stress the importance of the survey and the respondent's cooperation (even if they do not believe they are qualified to answer all the questions), detail how the respondent may benefit from the research, estimate how long it will take to complete, explain how the respondents were selected, give reassurances about confidentiality, detail the ethical clearance for the research, explain how the research will be reported back, express your appreciation and give a contact person to answer queries. Any factors that might be anticipated to affect responses should be addressed in the covering letter or email.

Dillman et al. (2014) recommend sending a reminder to non-respondents after two to three weeks, and another copy of the questionnaire a few weeks later. This method has yielded response rates of 70–75 per cent in general population samples. However, there is a general trend for declining responses to surveys—see discussion of response rates below.

TELEPHONE SURVEYS

Telephone surveys have increased in popularity because they are cheaper and less intrusive than face-to-face surveys although the increase in telemarketing has seen people become less tolerant of interruptions by phone. McQueen (1993) describes computer-assisted telephone interviews (CATI) in which the interviewer reads the question to the respondent from the screen and records the response straight into the computer. The computer can draw the sample, choose the telephone number and dial the respondent through a self-dial system. This technology offers obvious benefits in streamlining the survey process and reducing costs. McQueen (1993) also suggests that this technique can improve the validity and reliability of surveys for public health purposes by making possible techniques such as continuously collected data.

FACE-TO-FACE SURVEYS

Face-to-face surveys require interviewers who are sufficiently trained to ensure they ask questions in a consistent way. Survey textbooks used to advocate a neutral and 'objective' stance, but it is now accepted that people are more likely to give honest answers and open up to someone with whom they can empathise. If the survey is being done in people's homes, it may help to send out introductory letters a few days earlier or to telephone beforehand to book the interview. Safety protocols need to be introduced for solo interviewers visiting private homes.

RESPONSE RATES TO SURVEYS

The response rate to a survey is an important aspect of assessing the quality of the information from the survey. This is calculated by determining how many people of those eligible to respond to the survey did so. A higher response rate means the survey is more representative of the population it is conducted in. When non-random methods are used to select the survey sample it is obviously not possible to ascertain such an accurate response rate because the characteristics of the whole population are not known. This is often the case when there is no listing of a particular group of people who are being surveyed. For example, it would be hard to find a listing of all population health planners and a survey of them is likely to be non-random.

There is increasing evidence that responses to surveys are dropping (Galea and Tracey, 2007) and this has led to questioning of whether response rate is necessarily the only way to judge study quality and validity. It has been suggested that 'participation rates' may be more appropriate (Morton et al., 2012). These include (Morton et al., 2012):

- a description of the eligible study population and how they were contacted (to ascertain external validity)
- the cooperation rate (the number of completed interviews from those who were able to be contacted)
- the refusal rate (those who refused at some point to complete the interview process after some contact was made).

Dillman et al. (2014, pp. 56–93) contains an excellent chapter on encouraging a higher response rate.

SURVEYS IN ABORIGINAL COMMUNITIES

As noted in the section on ethics in chapter 6, there are now agreed guidelines for conducting research in Aboriginal communities. Donovan and Spark (1997) suggest that using face-to-face interviews is likely to be the most effective method in Aboriginal communities, particularly remote ones. They propose a series of guidelines that have the aim of ensuring that 'interviewing of Aboriginal respondents is done with maximum sensitivity to Aboriginal cultural difference and with minimum discomfort to the respondents' (p. 90). In summary they advise:

* direct questioning is inconsistent with Aboriginal culture
* information gathering is an exchange process for Aboriginal people
* the concept of privacy is important in Aboriginal culture
* use of an appropriate language, as English will not be the first language for most Aboriginal people in remote areas
* concepts of numeracy, intensity and specificity are different in Aboriginal and Western cultures
* concepts of, and attitudes towards, time are different in Aboriginal cultures
* interpersonal interaction styles are different
* appropriate interaction with the Aboriginal community as a whole is crucial
* Aboriginal communities fluctuate considerably, and this needs to be taken into account in sampling.

Miller and Rainow (1997) also stress sensitivity when conducting surveys in Aboriginal communities. They suggest that 'ethical surveys' involve the researchers being prepared to meet immediate needs. If conducting a survey of older people, you should be prepared to collect firewood, or provide a plumber to fix broken toilets in a survey of sanitation. Research budgets should allow for this.

ANALYSIS OF SURVEY RESULTS

The analysis of quantitative data involves setting up a coding guide, coding the collected information, putting it into a computer file and then analysing it. There are numerous statistical packages to assist the analysis of survey information (especially useful is the Epi-Info package produced by the US Centers for Disease Control and Prevention).

Much of the factual data collected can be analysed by using descriptive statistics such as frequencies and percentages (just over half the sample (n = 1012, 52 per cent) had used a community health centre in the past year), and means (on average the women (n = 1102) in the sample made 3.2 visits to their GP each year).

More complex statistical tests require a statistician, but there are many tests to determine whether differences between subgroups in any population are statistically significant and which determine which variables are exerting most influence (Dean, 1993). When interpreting quantitative analysis, bear in mind the limitations of statistical inference (Morrison and Henkel, 1970; Eversley, 1978; Atkins and Jarrett, 1979; Miles and Evans, 1979). Survey data analysis provides correlation data, not causative data.

For instance, you might be able to say that lower levels of household income are correlated with poorer reported health status as measured by the Nottingham Health Profile, which does not mean that low income causes the poor health status. To make this claim you need to develop a theory that explains why the correlation is likely to be causative, drawing on existing theory and other research findings. Researchers often confuse statistical association with substantive importance or causation.

Critics also point out that statistical tests assume random selection of survey respondents. In most public health surveys, non-response reduces the power of statistical tests, and precludes this being true. In the wider debate about the relative value and contribution of quantitative and qualitative research, the power of quantitative research rests on assumptions that are usually not realised in practice.

There is little doubt about the benefits of survey research, especially in relation to factual data, but there are many sources of error. The demographer Eversley (1978) warned against aiming for increasing complexity in statistical modelling as a means of overcoming the limitations of statistical analysis of survey data. He claimed (p. 299):

> The search for purity is, in fact, a quest for scientific sterility. The answers obtained
> from the use of refined models may, in some abstract sense, be truth, but they are
> neither real nor useful and they are probably not even true, if by that we mean that
> they must have some use in helping us to understand a current situation or make
> some future provision.

Dean et al. (1993) discuss the burgeoning of public health data collection made possible by the advent of high-speed computers and survey techniques such as CATI and Survey Monkey. They suggest the value this has brought to our understanding may not be very great because so many data are never thoroughly analysed, and that the computer substitutes (inadequately) for theorising and creative thinking. They acknowledge that statistical modelling techniques for analysing survey data can now examine multiple variables at one time, and so offer better mechanisms for studying interrelationships.

Multivariate analysis enables researchers to assess more than one study factor and allow adjustments for the influence of other study factors (confounders). An overview of multivariable statistical analysis is given in Büttner and Muller (2011, pp. 462–70).

CONCLUSION

Surveys have a valuable role in public health, but they tend to be overused and often stretched beyond their competence. They are most suited to the collection of factual and straightforward behavioural data. Combined with the various qualitative methods, they can be a powerful part of the public health researcher's tool kit.

CRITICAL REFLECTION QUESTIONS

8.1   Why do you think response rates to surveys have dropped consistently over the last 40 years?

8.2   What measures do you think researchers can take to increase response rates to surveys?

8.3   Why do you think surveys are better at describing rather than explaining public health problems?

Recommended reading

For more detailed discussion of sampling techniques see Sarantakos (2005), chapter 7.

Dillman et al. (2014) *Internet, phone, mail and mixed-mode surveys: The tailored design method* An excellent guide to all aspects of survey design and execution.

Useful website

For an example of a program to create web-based surveys, see www.surveymonkey.com

9

QUALITATIVE RESEARCH METHODS

Paradoxically, the 'softer' a research technique, the harder it is to do.

Yin, 1989, p. 26

KEY CONCEPTS

Introduction

What is qualitative research?

Application to public health

Qualitative research methods

Case studies

Participant observation

In-depth interviewing

Focus groups

Document analysis

Common issues of concern

Analysing qualitative data

Conclusion

INTRODUCTION

It is only in the past two decades that the potential of qualitative research methods to public health has been appreciated. Denzin and Lincoln (2011) suggest the way what they term the 'qualitative revolution' has overtaken the social sciences and related professional fields has been nothing short of amazing. The acceptance of qualitative methods has been slower in public health, possibly because public health has long drawn on the same traditions of modernity and science as the biomedical paradigm, which maintains that only the classic experimental design can produce valid results. This method is based on hypothesis testing and is effective in cases that can be easily randomised and controlled in a laboratory setting. Unfortunately, public health research rarely has such opportunities for control. Humans and their communities are typically messy, idiosyncratic, complex and continually changing. Qualitative methods offer considerable strengths in understanding and interpreting this complexity both as a complement to epidemiology and in their own right.

WHAT IS QUALITATIVE RESEARCH?

Most professional disciplines have increasingly adopted qualitative research methods, as they are better suited for coping with complexity and naturalistic settings. Most public health writers now argue for methodological pluralism when advocating the value of qualitative methods (Daly and MacDonald, 1992; Davies and Kelly, 1993; Baum, 1995b; Scott-Samuel, 1995). Daly and MacDonald (1992, p. 6) argue: 'The logical and scientific approach is to choose that study design which is capable of providing the most comprehensive and valid answers in the face of inevitable constraint.'

There has been extensive methodological and epistemological debate about the nature of qualitative research, positions taken including postpositivism, various degrees of relativism, critical theory and interactionism. There are a number of excellent texts that examine the various theories of qualitative research in detail. Starting points for a newcomer to qualitative research might be Miles and Huberman (1994), who see themselves as 'realists' and take a more pragmatic and less theory-driven approach than others; Denzin and Lincoln's edited collection of essays, which assesses and presents 'the major paradigms, histories, strategies and techniques of inquiry and analysis that qualitative researchers now use' (2000, p. ix); and Patton (2015), which provides an excellent guide to the use of qualitative methods applied to evaluation.

Here we discuss the value of this type of research to public health, provide an introduction to some of the most commonly used methods and consider issues of sampling, validity, reliability and analysis; thus providing a framework for more detailed investigation of qualitative methods.

APPLICATION TO PUBLIC HEALTH

Four main applications of qualitative research methods to public health have been defined as (NHMRC, 1996, p. 13):

- to study and explain the economic, political, social and cultural factors that influence health and disease in more depth than is possible through a survey or other quantitative methods. For instance, a survey can tell you how many people participate in community activities in a given community, but interviews are needed to explain why they take part in such activities
- to understand how people interpret health and disease and make sense of their health experiences
- to elaborate causal hypotheses emerging from epidemiological and clinical research. For instance, experimental and quasi-experimental research explains the link between tobacco smoke and lung cancer, but not why people continue to smoke despite evidence about the health effects
- to provide contextual data to improve the validity and cultural specificity of quantitative survey instruments.

QUALITATIVE RESEARCH METHODS

There are three kinds of qualitative data (Patton, 2015):

- in-depth, open-ended data collected from individuals or groups
- direct observation and description of people's activities, behaviours, actions and interactions, including analysis of audio- and video-taped material. Kellehear (1993) extends the definition to include the study of 'material culture' (graffiti, garbage, cemeteries)
- written data, usually excerpts, quotations or entire passages from organisational, clinical or program records, personal diaries, official records or publications and open-ended written responses to questionnaires.

The main methods used to collect these data are:

- case studies
- participant observation
- in-depth interviews
- focus groups.

CASE STUDIES

Case studies are empirical enquiries, using multiple sources of evidence, that investigate contemporary phenomena within their real-life context. The boundaries between the phenomena and their contexts are not obvious (Yin, 2014, pp. 16–17). Case studies are useful when researchers cannot control contexts and want to offer an accurate and detailed view of a particular phenomenon. A 'case' can be an individual, institution or community, and typically will involve more than one quantitative or qualitative method. Case studies can be descriptive or explanatory and may involve testing hypotheses. They were once seen as an inferior methodology that was appropriate for the exploratory stage of research only, but are now seen as more powerful (Stake, 1995; Yin, 2014). A case study is 'a method for learning about a complex instance, based on a comprehensive understanding of that instance obtained by extensive descriptions and analysis of that instance taken as a whole and in its context' (US General Accounting Office 1990, cited in Mertens, 2005, p. 237).

Their main advantages are that they allow study in a natural setting, as well as the complexity of the subject. Usually carried out over a reasonably long period of time, they permit the study of interactions between people. They also allow a researcher to develop methodology as more is discovered about the particular setting.

USES OF CASE STUDIES

Significant insights have been produced in sociology through the use of case studies (Whyte, 1943; Goffman, 1961; Willis, 1977; Williams, 1981).

Case studies have become accepted as evaluation tools, their value now recognised by such development agencies as the World Bank and US AID (Patton, 1990), which previously had preferred large-scale quantitative studies, but in developing countries these approaches involved problems so severe as to call into question their validity and reliability. Patton comments (p. 100): 'Case studies are manageable, and it is more desirable to have a few carefully done case studies with results one can trust than to aim for large, probabilistic and generalisable samples with results that are dubious because of the multitude of technical, logistic and management problems in third world settings.' Evaluators are now likely to use case studies for their inherent strengths, not just because they are more manageable.

The Healthy Cities project is an international initiative that builds on the evaluation of case studies of particular projects. Most of these are descriptive (Ashton, 1992), but others have provided more analysis. The Noarlunga Healthy Cities project evaluation (Baum et al., 1990), for example, presents a number of case studies embedded within the overall study and then, on the basis of the analysis, presents a series of factors that help or hinder the implementation of a community-focused project.

Ritchie (1996) used a case study approach to evaluate a health promotion project introduced by the author to the blast furnace site at an Australian steelworks. Through the evaluation, she determined the meanings the workers attached to health and health promotion, thus helping to explain why health promotion programs affect working-class people less than middle-class people. Ritchie's main methods were in-depth interviews and observation.

A study of best practice in primary health care (Legge et al., 1996) used reviewers to select 25 case studies for detailed analysis. The researchers developed a proforma for analysing each case study that required the identification of outcomes achieved, aspects of process described and the apparent preconditions for best practice. They suggest these case studies or vignettes offer possible benchmarks of best practice for public health practitioners, managers or teachers. They distilled elements of best practice in terms of the outcomes to be achieved, successful strategies of practice and the preconditions. Case studies have also been used to study the extent to which primary health care services are comprehensive, and the contrast between the services highlighted the value of comparative case studies (Baum et al., 2013c).

Case studies were also used in the evaluation of the United Nations HIV and Development Program (Parnell et al., 1996). They provided a deep understanding of the impact of HIV on people's lives, how these people could develop effective responses and how development programs, based on partnerships, could be used to develop capacities. Heymann (2006) in her study of global inequalities at work uses case studies to great effect to illustrate the impact that changes in the global economy are bringing to workers and their families. Her detailed descriptions of the reality of the lives of workers in poor countries are compelling and show the texture of impoverished lives and the stresses and strains brought by a production system focused on profit above all else.

PLANNING, DESIGNING AND ASSESSING CASE STUDIES

Yin (2014) suggests that a case study should be planned according to a protocol that contains:

- a description, justification for the choice and particular characteristics of the case(s) to be investigated, aim of the study and expected outcomes
- documentation of the field procedures and main respondents and how access has been negotiated with relevant people and institutions
- a plan for data analysis—how it will be coded, prepared for analysis, and the process for analysing patterns and drawing more general insights.

Case studies usually collect information from a variety of sources, including in-depth individual or group interviews, interview or self-completion surveys and the collection and analysis of relevant documents. Stake (1995) suggests a checklist with which the quality of a case study proposal can be judged (box 9.1).

BOX 9.1   CHECKLIST FOR RATING A CASE STUDY PROPOSAL

COMMUNICATION

- **Clarity**: Does the proposal read well?
- **Integrity**: Do its pieces fit together?
- **Attractiveness**: Does it pique the reader's interest?

CONTEXT

- **The case**: Is the case adequately defined?
- **The issues**: Are major research questions identified?
- **Data resources**: Are sufficient data resources identified?

METHOD

- **Case selection**: Is the selection plan reasonable?
- **Data gathering:** Are data gathering activities outlined?
- **Validation**: Is the need and opportunity for triangulation indicated?

PRACTICALITY

- **Access**: Are arrangements for start-up anticipated?
- **Confidentiality**: Is there sensitivity to protection of people?
- **Cost**: Are time and resource estimates reasonable?

Source: Stake, 1995, p. 54.

WHAT MAKES A GOOD CASE STUDY?

Factors making for an exemplary case study include (Yin, 2014 pp. 200–06):

- *Significance*, because the case or cases are unusual and of general public interest, or the underlying issues are important in theoretical, policy or practical terms. It is important for a researcher to detail the contribution that would be made by the successful case study.
- *Completeness*, which involves ensuring that the case has clear boundaries and includes all relevant evidence. It must be obvious that all critical evidence was given full attention and the case study continued until the researcher was happy that all relevant evidence had been collected and analysed.
- *Consideration of alternative perspectives*, which implies that a researcher must seek alternative culture views, different theories and consult a variety of people about the interpretations being made from the data. Yin (2014, p. 204) suggests that a critical listener offering alternative interpretations is a useful way of canvassing alternative perspectives. Participatory research approaches build this process into the interpretation of case studies.
- *Sufficient evidence*, so that a reader can make an independent judgment about the quality of the analysis. The case study should include sufficient evidence to support its conclusions, but should not be so weighed down with evidence that the sheer volume bores the reader.
- *An engaging manner*, so that people are keen to read the case study. Usually case studies are presented in written form, but videos may also be used. Enthusiastic researchers usually communicate results effectively. An advantage of case studies is that they generally tell a complete story and so can easily be made engaging.

Case studies probably have more potential application in public health than has been realised, being useful for providing a complete view of a particular community or people. They can be an effective means of presenting information to policy makers and politicians, who may relate to a case study and its story better than to a set of statistics. Combining statistics with case studies may be a particularly effective way of bringing issues to the attention of politicians.

PARTICIPANT OBSERVATION

Participant observation focuses on the meanings of human existence as seen from the standpoint of insiders. It seeks to uncover, make accessible and reveal the meanings people use to make sense out of their daily lives (Jorgensen, 1989). This methodology, dating from the late nineteenth century, has its roots in anthropology. Bogdan and Taylor (1975, p. 15) define it as research 'that involves social interaction between the researcher and informants in the milieu of the latter, during which data are systematically and unobtrusively collected'. Observation is non-interventionist, the observers merely following the flow of events (Adler and Adler, 1994). Participant observation in public health is typically used in conjunction with other research methods—it can be a powerful way to validate interview data.

Participant observation has not been extensively used in public health research, although its use has increased in the past decade (de Laine, 1997). Two participant observation 'classics' are, however, in the health field—*Asylums* (Goffman, 1961) and *Boys in White* (Becker et al., 1961), a study of medical education. Reading these studies would be a good way of becoming familiar with the methodology. Participant observation has been used to study the process of doctors becoming family doctors (Bogdewic, 1992) and as part of a needs assessment in South Australia (Traynor, 1989). This study resulted from my presenting census-derived data to a community group. I pointed out the 'black spots' where there were numerous indicators of social disadvantage, but was somewhat taken aback by some of the group's angry response. They felt their area was again being stigmatised and that the statistics missed the area's positive features. My reflection on their anger led our research team to design a study that would gain a more detailed picture of the community. The subsequent study was based on a researcher spending six months getting to know the community in detail through observation and interviewing. Typical of his field notes was:

> 8 September 1989, 4.00 p.m. Weather sunny. I cycle through the street named by many respondents as the 'worst' street in Christies Downs. It's a warm sunny late afternoon. The housing is single storied, joined housing. There are open front gardens. Lawns are nearly all well kept and flower beds are generally neat. An older woman is tending her plants. She is carefully trailing some climbers up her front wall. Further down, two young men stand talking in a front garden. Four or five stand around talking by another front door. All hold stubbies. All look at me as I go past. I am afraid to hold their gaze for too long. Down on the corner, in another garden, a small child rides happily up and down the drive on a toy train that makes the sound of a whistle as she moves ... A bright orange panel van, with wide wheels, slides around the corner and parks. Another two young men wearing black t-shirts and faded jeans climb out and walk over to a neighbouring house. They too are carrying stubbies. There is a lot going on.

The community people were right—there were positive features of the community that had not shown up in the statistical picture. Together, the researcher's observation and interviews offered some insight into the complexity of the community, and so demonstrated the dangers of drawing conclusions on the basis of only one type of data.

The method was also used in a study of perceptions of occupational risk (Holmes and Gifford, 1997), when observation was collected from a work setting in the painting industry and the resulting data analysed to highlight differences in the way in which employers and employees viewed risk. The researchers chose this method because it 'is well suited to eliciting subjective and collective views of risk in the context of everyday work' and permits this to be done with minimal disruption to people (p. 13).

THE PROCESS OF PARTICIPANT OBSERVATION

The basic process of doing participant observation involves (based on Jorgensen, 1989):

- defining your research question
- selecting and entering a setting
- participating in the life of the setting, maintaining and sustaining relationships with the people who are part of your setting
- observing, gathering and documenting information
- analysis.

Obviously the research question should be amenable to investigation through participant observation, which is a time-consuming process. This is likely to be because the research question requires 'thick description' and a detailed understanding of the social environment and its social interaction. Denzin (1989, p. 83), who coined this term, defines it as follows:

> A thick description does more than record what a person is doing. It goes beyond mere fact and surface appearances. It presents details, context, emotions and the webs of social relationships that join persons to one another. Thick description evokes emotionality and self-feeling. It inserts history into experience. It establishes the significance of an experience, or the sequence of events, for the person or persons in question. In thick descriptions, the voices, feelings, actions and meanings of interacting individuals are heard.

Participant observation is also suited to studying deviant behaviours, about which people may be reluctant to be interviewed, including sexual behaviours, drug use and mental illness. A key question is whether assuming the role of participant observer is ethical. Humphrey's (1970) research on the nature of homosexual sex in public toilets has been criticised because he was observing and reporting on 'illegal acts' and did so in a covert manner. Researchers also have to remember that assuming the role of participant observer involves assuming responsibilities and possible lifestyle changes.

Researchers have a choice of position to adopt when undertaking participant observation (Gold, 1958)—complete observer, observer as participant, participant as observer and complete participant—each having particular advantages and disadvantages. Kellehear (1993) provides a thorough guide to the observer-only category. The researchers can be known to be such by the people they are observing (overt) or not known (covert). The Canadian study of nutritional inequities described by Travers (1996) is an example of overt participant observation. The researcher negotiated entry to the community drop-in parents' centre and then discussed the research with the women who came to the centre. Travers was there 70 per cent of the time it was open, and describes her role thus (p. 545):

> the researcher helped serve meals in the soup kitchen, unpacked food which had arrived from the food bank, helped pack grocery bags of donated food for program participants, ran errands and answered telephones. The researcher also participated in the 'life' of the centre, helping to prepare and eating noon meals with the staff

and volunteers, taking coffee breaks and checking the newspaper for sales, talking and/or gossiping and/or asking questions.

Covert research is controversial and unlikely to be approved by an ethics committee. The dividing line between participant and observer is not always clear—if public health workers undertake structured observation as part of their routine work, should this be disclosed to their clients and co-workers?

It is important to gain the trust of people so they become reliable informants. Bogdewic (1992) recommends the following:

- *Be unobtrusive.* If the goal is to fit in, then don't dress or behave in ways that will make you stand out.
- *Be honest.* Be open about the purposes of your research and assure people that you will not reveal their identities and will treat all information confidentially.
- *Be unassuming.* Don't try to impress with your knowledge, and play down your expertise.
- *Be a reflective listener.* This will help you learn the use of language in your particular setting, and to gain a deeper understanding of it.
- *Be self-revealing.* A willingness to be open will lead to a more trusting relationship.

Trust is crucial if people are to reveal insights and details of their lives that they would not through interviews or other methods. Building trust can be a complicated process, however, and there is no reason to assume that people will be sympathetic to the aims of the researcher. Peberdy (1993) describes her relative failure to collect information from a Papua New Guinea village because her original research question about the relationship between Western and indigenous medicine did not concern the Tolai women. She recommends that participant observation should only be attempted when the researcher has sufficient knowledge of a community, culture and language to be able to identify what may be of interest to the community.

There is a delicate balance between building trust in a community and understanding the issues of relevance to it while still being able to act as a critical observer. Traynor (1989, p. 10) reflects on this issue in relation to the needs assessment described earlier:

> The researcher involved with this project came from a white, male, British middle-class background. As a recent migrant he may have had some insight into the British migrant's view of Christie Downs—and a large number of residents were from this ethnic group. As a middle class individual he may have found it difficult to understand the experience of living in a largely working class culture. However the 'newness' of the encounter with Australian suburban life was felt to be of some advantage as it precluded too deeply-rooted a set of presuppositions.

Data are usually collected in the form of field notes, which are written up as soon as possible after the observation. Notes will typically include description of events that occurred (who, what, when, where, how), notes of theoretical importance that try to derive meaning from the descriptive observation and on methodological issues that may impinge on the credibility of the research. Researchers are typically reflecting continually on the meaning of their observations during fieldwork, and beginning to analyse by generating categories and ideas.

Atkinson and Hammersley (1994) describe how ethnography and participant observation have been variously interpreted by different disciplines. Ashworth (1995) suggests that the theoretical orientation of researchers is crucial, and that some advocates of the method (Spradley, 1980; Jorgensen, 1989) are neo-positivists who strive to be as objective as possible in their observations and discuss how their interactions may affect this objectivity as they defend the technique against criticisms from conventional science. Others come from a postmodern perspective (Atkinson and Hammersley, 1994) in which the observer aims to produce a text that is viewed as the joint product of the researched and researcher. Ashworth believes this postmodern perspective to be dismissive of a humanistic perspective that emphasises the importance of the relationships established. He sees participant observation as a technique that entails conscious social engagement. As a public health researcher, you need to consider these various positions towards participant observation, but you are advised not to lose sight of the key aim of achieving better understanding of your research problem.

Another crucial issue for participant observation is the apparent contradiction in its title. The problems of subjectivity and excessive reliance on the observations of (usually) one researcher have hampered the wide acceptance of the technique (Adler and Adler, 1994). Combining observation with other methods does, however, overcome such objections.

IN-DEPTH INTERVIEWING

Interviews can vary from a quick interview in a busy shopping centre through to a number of sessions over many hours with the same person. They can involve only fixed choice questions (the Census) or be entirely open-ended with only broad topic areas to guide the conversation. Structured surveys (whether face-to-face or self-completed) have already been described. Now we consider interviews that are based on semi- or unstructured interviews that produce mainly textual data. In essence these are discussions to collect information for subsequent analysis. Their advantage is that they usually provide richer, more complex data than tick-in-a-box questionnaires.

Feminist researchers have emphasised the importance of in-depth interviews as a way of gaining a perspective on women's experience in a way that more structured forms of research cannot (Olesen, 1994). Through detailed interviews, Hunt et al. (1989) talked with women who did not comply with doctors' orders, and reported that they were not difficult, but had real reasons for not following advice, that made sense in the context of their lives.

Oakley (1981, p. 49) points out that there is 'no intimacy without reciprocity' in interviewing, so a sympathetic position is more than just ethics—it is also likely to lead to a better quality of data. Fontana and Frey (1994) discuss the issue of gendered interviews, pointing to past paternalistic bias towards women in anthropological and sociological fieldwork. They point out that 'objective' interviews may be biased as rapport and trust are not established in the process, yielding an inaccurate picture.

Factors influencing the decision of the style of interview to use are:

- *Resources available to the researcher.* Long, in-depth interviews are costly to conduct and transcribe, and limited by the volume of data they generate. It is necessary to decide on a smaller number of detailed interviews or a larger number of less detailed ones.
- *Tolerance of the interview group.* Some groups may be more willing to take part in detailed interviews than others. Medical practitioners are generally unwilling, but older people are often more willing to spend time being interviewed.
- *Topic of the research.* Some research topics demand longer, more detailed interviews than others. Research into people's fertility behaviour, for instance, requires an in-depth detailed interview to build up a rapport with the people being interviewed.

There are numerous uses for in-depth interviews in public health. They can be used to explore meanings attached to diseases (Davison et al., 1992), to detail people's understandings of health (Cornwell, 1984), to elicit different key players' perspectives on a particular program (McGuiness and Wadsworth, 1992) or to gain an understanding of why particular factors affect health. An example of the latter is van Eyk's (1996) study of the interaction of isolation and loneliness and their impact on the health of Spanish-speaking women living in Australia.

IN-DEPTH INTERVIEW PROCESS

In-depth interviewing is a skilled process that can only really be carried out by people who are familiar with the research purposes and aims, usually the researchers who are also responsible for most of the analysis and writing. Public health research (except PhD research) is often done in teams, and senior members of the team may find that they have limited time for in-depth interviews, so the task falls to a junior. Whenever possible, however, all members of a research team involving in-depth interviews should do at least some of the interviewing. It is rarely possible, as it is with more structured forms of interviewing, to use casual interviewers to conduct in-depth interviews. You need to consider gender, culture, language and social class when planning the interviews. Is the quality of the interview, and so the accuracy and depth of the data, likely to be improved if a woman interviews a woman or a man a man? To interview people who do not speak English fluently, will you use an interpreter or employ an interviewer who speaks their language? Would Aboriginal people living traditional lives feel more comfortable being interviewed by an Aboriginal person? How effectively can a middle-class woman interview a blue-collar worker? These are the types of questions to consider, and there are no rules to determine the answer. The important thing is that you consider all these factors and explain the rationale for your decisions when you report your research.

Textbooks dealing with in-depth interviewing contain a considerable amount of advice for researchers (Seidman, 2005; Minichiello et al., 2008). The key stages are contacting and explaining the research to potential respondents; establishing rapport and empathy; and ensuring that appropriate information is collected. Glesne and Peshkin

(1992) conceptualise interviewing as the process of making words fly, and suggest some attributes that may contribute to successful interviews:

- *anticipation*, which means being prepared to explain the purpose of the research, reflecting on each interview in order to improve the next
- *establishing rapport* by showing a genuine interest in what the interviewee is saying and encouraging discussion of the central issues. This will be helped by being warm and caring
- *taking a naive position*, meaning that you keep an open mind and search for meaning from your respondent, rather than assume you know what they mean. Glesne and Peshkin note that 'Casting yourself as a learner correspondingly casts the respondent as teacher' (1992, p. 81). They suggest this is both respectful and assists in developing rapport. Of course, you should never do anything to make an interviewee feel ignorant
- *being analytic*, meaning that interviewing is not just data gathering but also an analytical act that begins the process of understanding and meaning. Being analytical assists in appropriate prompts for more information and new avenues to explore. Glesne and Peshkin note that, while a good interview may be like a conversation, it should also be more because the aim is to obtain good, accurate data
- *paradoxically bilateral*—dominant but also submissive. Glesne and Peshkin point out that, despite calls from some researchers for power balances between researched and researcher to be reduced, researchers still generally hold power. They also say that hierarchical relationships are not inevitably devoid of mutual warmth and caring. Researchers are dependent on the willingness of their respondents to participate, and so detailed interviewing involves a delicate balancing act
- *patiently probing*. In-depth interviewing requires considerable patience and probing if you are to understand the topic being researched.

In-depth interviews require people to be open and honest about their lives, habits and behaviours, and so interviews can be quite intrusive. People also have to give up time, possibly two or three hours.

Data from the interview can be handwritten or recorded. Accurate recording by hand is hard work as the interviewer also has to concentrate on maintaining eye contact and responding appropriately. In most circumstances, recording is the preferred option, so long as respondents agree. The disadvantage is the cost of transcribing, which usually takes three to four times the interviewing time.

In-depth interviews are a powerful way of getting detailed pictures of how people experience and explain their world, which can be crucial in public health for understanding why people behave as they do and how structural factors come to impact on their health.

FOCUS GROUPS

Focus groups involve open-ended interviews with between five and ten people (usually a homogeneous group) on a particular focused issue for up to two hours. The methodology originated as a way of gaining accurate information about consumer product preference

(Merton et al. 1956). Participants are asked to reflect on the interviewer's questions. Brown, et al. (1989, p. 40) comment: 'Groups are not just a convenient way to accumulate the individual knowledge of their members. They give rise synergistically to insights and solutions that would not come about without them.' Davidson et al. (2006) used 14 focus group discussions in Scotland and the north of England to explore how group participants view health inequality and how they theorise its impact on health. The study enabled the researchers to provide a detailed contrast of the views of people from different socioeconomic backgrounds.

The focus group method was pioneered in market research (Krueger and Casey, 2009) to test products and create new marketing strategies. This method is often used to develop questionnaires for use with randomly selected samples. The reliance on focus groups by commercial companies suggests they yield useful and accurate information. They can be used to both supplement and validate quantitative and other qualitative techniques (e.g. developing survey questionnaires or obtaining participant interpretation of results from earlier studies) or as a self-contained means of data collection. They may also be one of a range of methods in a larger research project, when the focus-group results can be triangulated with results of other data-collection methods.

Focus groups are now commonly used in health promotion needs assessment (de Koning and Martin, 1996), and in both exploratory and theory-building research within public health. Cortie et al. (1996) used four focus groups (stratified according to low and high socioeconomic status and extent of physical activity) to determine the factors in a local community that determine people's decision to exercise or not.

STRENGTHS AND WEAKNESSES OF FOCUS GROUPS

The advantages of focus groups are that they:

- are an economical and efficient method of collecting qualitative data, as they save on interviewer time and travel. Information can be gathered from up to 10 people in one hour, instead of from one person
- yield lively interaction between participants, leading to discussion and debate that may not occur in a one-to-one interview
- can be exploratory and open-ended, allowing participants to formulate their opinions within the group in a way that would not be possible in individual interviews (Morgan, 1988, p. 28)
- allow more control over the agenda of the discussion than in individual interviews
- are particularly useful when researchers do not know much about the issue they are beginning to research, as they allow an open-ended format in which topics of interest can be explored.

The weaknesses of focus groups may be that:

- a form of 'group think' operates, discouraging participants from expressing opinions that are at odds with the majority of the group. 'Devil's advocates' may overcome this tendency (MacDougall and Baum, 1997)

- focus groups may not tap into emotions. For some people, an individual interview is better, although others may find that a supportive group encourages emotional sharing (Krueger and Casey, 2009)
- dominant individuals can influence results so skilful moderation to minimise this risk is vital
- the researcher's control over the data collected is less than that possible in one-to-one interviews
- high-quality recording equipment and transcription are required if direct quotations are to be taken.

PLANNING AND CONDUCTING FOCUS GROUPS

The number and size of focus groups will often be dictated by research goals as well as available resources. Krueger and Casey (2009) recommend using groups of between 5 and 10, but say size can be as few as 4 to as many as 12. Smaller groups tend to make a high demand on each participant, while larger ones may inhibit some members, but sometimes researchers cannot control the size of their groups. A large group may not prevent the collection of useful information, but will do little for the researcher's stress levels! Sometimes people not turning up can result in smaller groups. It is advisable to plan for between a 10 and 20 per cent no-show rate, depending on the group from which the focus group is drawn. The group must be small enough for everyone to have an opportunity to provide insights and yet large enough to give a range of perceptions (Krueger and Casey, 2009).

There is no easy formula for determining the number of focus groups that should be run, but the extent of heterogeneity in the group being researched is relevant. The more heterogeneous, the more groups will be needed. A trite answer is as many as necessary to obtain an answer to the research question, which cannot always be predicted in advance. It may be necessary to keep open the possibility of conducting more groups than originally planned if new themes continue to occur in successive groups.

You will also need to consider how you might want to divide your population. For instance, if gender is likely to be a significant variable, you may wish to have separate groups for men and women. Or, in an evaluation, you may collect more 'honest' data if you have a focus group for each category of stakeholder. In the case of a health promotion program these might be the funding body, the managers of the service auspicing the health promotion program, those actually running it and the community people involved in it. Each group is likely to have valid but differing perspectives, and be more likely to express their attitude and opinions frankly in a group of peers.

The topic of your research will usually determine how easy it is to recruit people to your groups. Generally, the more sensitive the topic is, the more trouble you will experience with recruitment. Brown (1995), in her research on the medical power of attorneys, believed her difficulty with recruitment was because of society's attitudes towards death and dying. It is likely to be difficult for research on any deviant behaviour. One way of overcoming recruitment difficulties is to conduct the research in partnership with any community, consumer or advocacy groups who are affiliated with the people you want to encourage to attend your group.

Market research companies typically pay participants in their groups, and this is beginning to happen in some social research. It raises the ethical question of the point at which the payment becomes an unfair inducement to take part in the research. Some ethics committees are uncomfortable with the practice. In terms of validity, payment may encourage people to take part primarily to earn the money, possibly even claiming membership of a group in order to qualify. Providing money for expenses (travel and child care) may often be necessary, especially for people on low incomes.

RUNNING A FOCUS GROUP

How you run your focus group is crucial to obtaining higher quality information. It is important to ensure that you have a skilled facilitator with a knowledge of group dynamics, and the ability to ensure no one person dominates the group discussions and that interesting 'leads' are followed up. Most groups within the health sciences are directed by a facilitator, but Morgan (1988) claims that his favoured approach is non-directive, using 'self-managed groups'. He argues that this approach is well suited to exploratory research and that 'if the goal is to learn something new from participants, then it is best to let them speak for themselves' (p. 49). A more highly facilitated approach is important when there is a strong agenda or a specific research question. Morgan found that focus group members were able to handle typical group problems, such as getting irrelevant discussion back on track, avoiding 'dry' periods, controlling dominant participants and engaging reticent group members. He mentions these as potential problems before handing over to the group, after which they will generally handle them. The role of the facilitator is to set up the group with instructions and then move to the side, intervening only if it goes badly off track. In these cases the initial topic is kept broad (Morgan and Spanish, 1985).

A suitable venue must be found before the event, where participants will feel comfortable and relaxed. The location must also be suitable for achieving high-quality recordings. It is preferable to have one researcher facilitating the group and another taking notes, which can be useful in sorting out who said what when the group interview is transcribed.

Focus groups work best when participants feel comfortable, respected and free to give their opinions without being judged (Krueger and Casey, 2009). It is important to be able to establish trust so that people feel able to self-disclose.

The data from a focus group are most easily recorded on audio tapes or digital media, and you need to know how your recording equipment operates. A well-run group will be wasted if you have only a blank tape at the end of the day. Depending on your research it may be sufficient to have a co-researcher take notes during the focus group, merely supplementing them with a recording. If you are interested in a more detailed content analysis, it may be necessary to have your recording transcribed verbatim.

Focus group information, like other textual data, may be analysed according to its content and themes drawn out.

DOCUMENT ANALYSIS

Document analysis is a valuable method in qualitative research that provides a 'particularly rich source of information about many organizations and programs' (Patton, 2015, p. 176), and is often used in combination with other methods (Bowen, 2009). It has been used to assess uptake of research evidence in health policies (Flitcroft et al., 2011; Rosella et al., 2013) and the extent to which health policies address equity issues (Keleher, 2013), and is emerging as a method to assess uptake of evidence on social determinants of health and health inequities (Pinto et al., 2012; Borrell et al., 2013; Fisher et al., 2014a).

COMMON ISSUES OF CONCERN

SAMPLING

The rationale for sampling in qualitative studies is quite different from that in quantitative studies, in which samples tend to be purposive rather than random (Kuzel, 1992; Miles and Huberman 1994), aiming to select cases that will provide rich data and enable detailed study (Patton, 2015). It is often not desirable to select a sample at the outset of a study because 'Initial choice of informants leads you to similar and different ones; observing one class of events invites comparison with another; and understanding one key relationship in the setting reveals facets to be studied in others' (Miles and Huberman, 1994, p. 27).

Although qualitative research does not aim for statistical representativeness, researchers want their theories to be meaningful to a wider population. Consequently, it is usually important to explore many aspects of the topic of interest. Depth rather than spread provides meaning.

Samples are usually theory-driven, either starting from a theory that is being tested or growing progressively (as with grounded theory) (Miles and Huberman, 1994, p. 27). A study considering why people use a particular health service may opt to select people according to their use of the service as this could be a key variable. Qualitative sampling usually has the following features (Kuzel, 1992, p. 41):

- the sample design, although decided at the outset, is flexible and can evolve as the study develops
- people or cases to be included are selected serially
- the sample can be adjusted as the theory develops, the aim being to thoroughly explore the topic of interest and to consider as many different angles and perspectives as possible. Glaser and Strauss (1967) refer to this as 'theoretical sampling'
- sampling continues until little new information is being gained
- sampling includes a search for 'negative cases' (for example, not neglecting those people who choose not to use a health service).

A common technique in qualitative research is 'snowball sampling', which involves contacting a few members of the group that is the focus of the research, and asking them to help find others. This technique is particularly useful for reaching groups that are not readily identifiable, such as illicit drug users. For many groups of people there will be no readily accessible list and researchers may have to advertise in order to make contact with them (for example, voluntarily childless people).

There are no closely defined rules for sample size in quality enquiry. It depends on what you want to know, the purpose of the enquiry, what is at stake, what will be useful, what will have credibility, and what can be done with available time and resources (Patton, 2015, p. 311). However, as a rule of thumb, six to eight data sources or sampling units will often be sufficient for a homogeneous sample, while 12 to 20 might be needed when looking for disconfirming evidence or trying to achieve maximum variation (Lincoln and Guba, 1985; Patton, 1990; Kuzel, 1992, p. 41).

Qualitative researchers may be asked to specify their sample size in advance because of the requirements of a research funding body (particularly those with members who are used to statistically driven research). In these cases the crucial feature is the logic underlying the sampling strategy, as this is the main criterion used to judge the strength of the sampling method. A proposal should be able to describe the sampling strategy clearly, state the selection criteria and provide an approximation of the sample size (NHMRC et al., 2007 [updated 2014]).

ASSESSING THE QUALITY OF QUALITATIVE RESEARCH

Validity, reliability and generalisability are concepts that enable the value of positivist research to be judged. Three types of validity are defined: face, which is concerned with whether the methods assess what they set out to do; internal, which refers to the rigour of the methods used; and external, which refers to the extent to which the results can be generalised beyond the selected sample. Reliability refers to research consistency. Generalisability refers to the extent to which the research findings can be applied to other settings and still have some meaning.

There is uncertainty in the literature about the extent to which these concepts can be applied to qualitative research. Kirke and Miller (1986) use the terms and argue that they can be adapted to qualitative research methods, but others suggest alternative terms. Lincoln and Guba (1985) suggest using the terms 'credibility', 'transferability', 'dependability' and 'confirmability'.

CREDIBILITY

Patton (2015) expresses preference for the term 'credibility' in terms of assessing qualitative research. He suggests it can be determined by reference to:

- the rigour of the techniques and methods used
- the credibility of the researcher, which is 'dependent on training, experience, track record, status and presentation of self' (p. 653; also Miles and Huberman, 1994, p. 38)
- the philosophical orientation and assumptions that underpinned the study.

Techniques that are particularly important to the credibility of qualitative research are the search for negative cases and the efforts to use triangulation. Denzin (1978) originally defined four types of triangulation:

- *data source triangulation*—the use of a variety of data sources
- *researcher triangulation*—the use of several different researchers or evaluators
- *theory triangulation*—the use of multiple theoretical perspectives to interpret the same set of data
- *methodological triangulation*—the use of multiple methods to study a particular problem.

To these Janesick (1994) adds a fifth type, which is relevant to public health:

- *interdisciplinary triangulation*—having researchers from a variety of professions. This can be extended to include non-researchers who have an interest in the findings of research, such as residents in the case of a community needs assessment.

The analysis of negative cases is an important way of obtaining credibility, involving the researchers seeking out instances and cases that do not fit the broader pattern. Patton (2015, p. 656) comments: 'Dealing openly with the complexities and dilemmas posed by negative cases is both intellectually honest and politically strategic.'

TRANSFERABILITY

Transferability refers to the ability of qualitative researchers to extend their findings to other settings. Lincoln and Guba (1985) suggest that research reports should provide sufficient detail (about methods, parameter, setting) for other researchers and users to make judgments about transferability.

DEPENDABILITY/RELIABILITY

The features dependability and reliability refer to whether the research is likely to be consistent over time and across researchers and methods. It is an issue that should be considered at all stages of the research. The reliability of data depends on skills in interviewing, observing and recording. Analysis procedures, such as having two people assess the data for themes and codes, are likely to increase the reliability of interpretation.

CONFIRMABILITY

Confirmability refers to the need to confirm research results with a source outside the research team, such as the research participants. Confirmability shifts the focus from the objectivity of the researcher to the data. De Laine (1997, p. 279), building on the work of Lincoln and Guba (1985) and Sandelowski (1986), suggests that an 'audit trail' can be used to assess credibility by tracking through the raw data, sampling decisions, methods of recording, analysis, coding, construction of themes, and personal notes. A successful audit trail would enable an outsider to follow the logic of each stage and understand how and why the researchers had drawn the conclusions they had.

There are, therefore, distinct ways of assessing qualitative research, but it is still common for the validity and reliability of public health qualitative research to be called into question by positivist scientists who consider the method to be subjective, and so invalid and unreliable. The issue of subjectivity and objectivity in science has been extensively debated by science philosophers, and today there are few who still maintain that positivist science is as objective as it was once thought to be, and there is broad acceptance that qualitative data can make a crucial contribution to aspects of public health. Patton (2015, p. 58) takes a pragmatic position that appears also to make sense for public health when he suggests we avoid the words 'objective' and 'subjective' and 'stay out of futile debate about subjectivity versus objectivity'.

ANALYSING QUALITATIVE DATA

> This involves reducing the volume of raw information, sifting trivia from significant data, identifying significant patterns, distinguishing signal from noise, and constructing a framework for communicating the essence of what the data reveal.
>
> *Patton, 2015, p. 630*

Analysis imposes meaning and interpretation on mainly textual data, which are usually unstructured and unwieldy. As with other aspects of qualitative research, there are several schools of thought about the preferred methods of data analysis, which will be heavily shaped by the theoretical framework within which a study is conducted. The crucial point is that rigour, duration and procedures will be very different according to the study's purpose. Three of the significant theoretical traditions that have given rise to particular forms of data analysis are: interpretivism, social anthropology and collaborative research (Miles and Huberman, 1994, p. 8). Interpretivism emphasises the meaning and 'essence' of the data, and works from the premise that researchers' perceptions and understanding are affected by, and affect, the process of research. Included within this general tradition are ethnomethodologists, phenomenologists, deconstructionists and discourse analysts. Social anthropology, including ethnography, focuses more on the accuracy of description and less on the conceptual or theoretical meaning of the observations. There is an interest in 'discovering' the underlying patterns or rules in a situation, which often begins with a theory that is tested during analysis. Within this broad tradition come researchers in life history, grounded theory and much applied qualitative research and evaluation in education and health. Finally, there is collaborative research, in which analysis is either through reflective (questioning) or dialectic (opposing) enquiry with the research participants. The analysis happens over a period of time, interspersed with further rounds of data collection.

Despite these different traditions, there are some processes common to most forms of qualitative inquiry. Qualitative data are primarily textual and typically comprise a mix of field or observational notes and transcripts from interviews. Miles and Huberman (1994) define analytical procedures that are used across different research types:

- coding observation or interview data
- noting reflections or other remarks in the margin

- sorting and shifting through these materials to identify similar phrases, relationships between variables, patterns, themes, distinct differences between subgroups and common sequences
- isolating these patterns and processes, commonalities and differences and taking them out to the field in the next wave of data collection
- gradually elaborating a small set of generalisations that cover the consistencies discerned in the database
- confronting those generalisations with a formalised body of knowledge in the form of constructs or theories.

Most public health research using qualitative methods will be applied in a relatively short time frame in order to produce information on which action or decisions can be based. It is easy, especially for a new researcher, to be overawed by the growing number of books on qualitative data analysis. Ritchie and Spencer (1994) address the needs of applied qualitative researchers directly, describing a framework that has been extensively used by the UK Social and Community Planning Research (SCPR; now National Centre for Social Research, NatCen). The approach involves a five-stage process of shifting, charting and sorting material according to key issues and themes. These stages are briefly described here.

FAMILIARISATION

The researcher becomes familiar with the range and diversity of data collected. This is especially important for a researcher who will be writing up the data but has not been directly involved in its collection. A selection of the data should be read and (if recorded) listened to. Key ideas and themes should be listed.

IDENTIFYING A THEMATIC FRAMEWORK

From the researcher's initial notes, themes are identified to form the basis of a thematic framework within which the material can be shifted and sorted. This will draw upon the original ideas that formulated the research, issues raised by the respondents themselves and analytical themes that occur when reading the data. The process involves logical and intuitive thinking. The data are gradually sorted into indices that contain major subject headings and categories.

INDEXING

This is the process whereby the thematic framework is applied to the data. Each transcript or set of field notes is coded by reference to the index, usually using a computer package. A commonly used package is NVivo (Bazeley and Jackson, 2013, provide an excellent guide to using the package, and the value of it is given in box 9.2). The system of annotating the text means the process is visible and accessible to others, and can be checked by two or more researchers.

BOX 9.2   USING COMPUTER PACKAGES TO ANALYSE QUALITATIVE DATA

There are five main ways a computer package can assist your analysis:

- Managing data: managing and keeping track of all data, including raw data files from interviews, survey and focus groups or field observations, images, diagrams, audio, video, web pages, policy documents, and ideas recorded as memos.
- Managing ideas: organising and providing rapid access to the conceptual and theoretical knowledge you generate.
- Querying data: asking simple or complex questions of the data allows the program to retrieve all relevant information from your database. Queries can be saved and so become part of an ongoing equiry process.
- Visualising data: showing the content of cases, ideas, concepts or timelines and representing these relationships visually.
- Reporting the data: including information about the original data source, the ideas and knowledge developed from them, and the processes by which these outcomes were reached.

Source: Bazeley and Jackson, 2013, p. 3.

CHARTING

Charting refers to the process of taking the data from its original context and rearranging it according to the appropriate thematic reference. Data are usually analysed by considering each theme across all respondents. Ritchie and Spencer (1994) recommend using charts to do this work, but computer packages can be used to extract all the data relating to a particular theme, which is then available for further analysis.

MAPPING AND INTERPRETATION

Here the researcher pulls together key characteristics of the data and so makes sense of the study as a whole. The process involves reviewing the themes, comparing accounts and experiences, searching for patterns and connections and seeking explanations for them within the data. The researcher has to weigh up the importance of issues, looking for structure within the data rather than a multiplicity of evidence. The process also requires 'leaps of intuition and imagination' (Ritchie and Spencer, 1994, p. 186).

Qualitative data analysis involves two distinct processes: description and interpretation. Patton (2015, p. 534) warns against rushing into interpretation before the work of analysis has been done. He observes that the rigour of qualitative analysis depends on 'thick description' (Geertz, 1973; Denzin, 1989), and that it is important that data are presented so that others reading the results have enough description to be able to draw their own conclusions.

PARTICIPATION IN RESEARCH

> Community-based participatory research holds immense potential for addressing
> challenging health and social problems, while helping bring about conditions in
> which communities can recognize and build on their strengths and become full
> partners in gaining and creating knowledge and mobilizing for change.
>
> *Minkler and Wallerstein, 2003, p. 20*

Many new public health researchers have argued that, whenever possible, research
should be participative (Wadsworth, 1984; Feuerstein, 1986; Baum, 1988; Bruce et al.,
1995; de Koning and Martin, 1996; Minkler and Wallerstein, 2003). This reflects partly
a desire to reduce professional dominance in public health and partly the increasing
recognition that lay people can offer a form of expertise not necessarily held by
professional researchers. Participants in all types of research have an acknowledged
right to receive information about the research, formally agree to their participation and
withdraw at any time. But the new public health is calling for a more meaningful form
of participation, including the defining of the research agendas, selection of methods,
conducting the research and interpreting the findings. Minkler and Wallerstein (2003)
have defined a practice of community-based participatory research and described its
growth and increasing acceptability in the past years. Their edited volumes contain many
examples of this form of research.

Traditional models of social science and biomedical research involved what
Wadsworth (1984) called 'data raids' in which researchers swooped down from their
ivory towers, collected data, returned to their towers and never communicated the
results of the raid to the subjects. A more participative form of research could involve
ordinary people in defining research questions, determining methods and deciding how
research findings could be reported back to communities in a readily comprehensible
way. Action research adds the dimension of using research as a mechanism to achieve
positive change in people's lives. A participatory approach is essential when conducting
Aboriginal research, given the history of the use of research as part of the processes
of colonisation and inhumane treatment of Indigenous peoples. The Lowitja Institute
(www.lowitja.org.au) provides many resources to assist researchers in conducting ethical
research with Indigenous peoples.

Research participation has tended to be associated with qualitative research methods,
but this is not inevitable. Oakley (1989) argues that quantitative methodologies, such as
randomised control trials, can be emancipatory in practice and could be designed to
involve research participants more than at present. There is certainly a strong case for
more community involvement in discussions about public health topics, content, method
and ethics of quantitative research.

PARTICIPATORY ACTION RESEARCH

In recent years there has also been growing interest in participatory research, which is
not related to any particular method. It represents a different understanding of research
in which the privileged position of the researcher is challenged. Participatory approaches

have drawn on the work of the adult educator Paulo Freire (1972) and place as much emphasis on the process as on the product of research.

The term 'action research' is often used interchangeably with participatory research; but sometimes the two are combined into 'participatory action research'. In medical and public health circles the idea of action research is innovative and, along with calls for more participative styles of research, has received more attention in recent years. A history of action research (McTaggart, 1991) demonstrated its origins in the work of educationists, particularly the North American Kurt Lewin (1946). Since the 1940s it has become an accepted and respected research tradition in educational and management research. Cornwall (1996, p. 94) notes that participatory research 'aims to substitute a cyclical on-going process of research, reflection and action for the conventional, linear model of research, recommendation, implementation and evaluation'. Kemmis and McTaggart (1988, p. 11) describe an action research spiral that is based on the processes of planning, acting, observing and reflecting (see figure 10.4, chapter 10). A definition of participatory action research is offered in box 9.3.

BOX 9.3   WHAT IS PARTICIPATORY ACTION RESEARCH?

Participatory action research (PAR) seeks to understand and improve the world by changing it. At its heart is collective, self-reflective enquiry that researchers and participants undertake, so they can understand and improve upon the practices in which they participate and the situations in which they find themselves. The reflective process is directly linked to action, influenced by understanding of history, culture and local context and embedded in social relationships. The process of PAR should be empowering and lead to people having increased control over their lives.

PAR pays careful attention to power relationships, advocating for power to be deliberately shared between the researcher and the researched: blurring the line between them until the researched become the researchers. The researched cease to be objects and become partners in the whole research process: including selecting the research topic, data collection and analysis and deciding what action should happen as a result of the research findings.

PAR draws on the paradigms of critical theory and constructivism and may use a range of qualitative and quantitative methods.

Source: Baum et al., 2006, adapted from Minkler and Wallerstein, 2003, and Grbich, 1999.

A group of people going through a participatory action research process would start by developing a plan of action (developed from a process of critical reflection) with the intention of improving what is already happening. The plan would be implemented and the process observed within the context in which it occurs. Reflections on these observations would determine the next plan of action, and so on through a succession of cycles. This means project actors are involved in the research process. This process calls for the researchers and the project actors to engage in a reflective spiral. Wadsworth (1991) recommended the use of an action evaluation research process (see figure 9.1), which sees evaluation as a spiral process of planning, fieldwork, analysis, reflection and then spiralling up to planning again. At first consideration the action process may seem like the process that any reflective practitioner might go through. Kemmis and McTaggart (1988, p. 10) define the difference thus:

> to do action research is to plan, act, observe and reflect more carefully, more systematically, and more rigorously than one usually does in everyday life; and to use the relationship between these moments in the process as a source of both improvement and knowledge. The action researcher will carry out the four activities collaboratively, involving others affected by the action in the action research process.

FIGURE 9.1   PARTICIPATORY ACTION RESEARCH CYCLE

Source: Created by Simon Kneebone for Fran Baum.

Participatory action research is about more than the generation of knowledge—it is a process of 'education and development of consciousness and of mobilisation' (Gaventa, 1988, p. 19). Kennedy (1995) describes the participative evaluation methods used in the Drumchapel Healthy Cities project and concludes that the process of involvement in the evaluation had an empowering effect on the local community members as they understood the project in a more detailed way and could appreciate the perspectives of the different interest groups. Some of the most exciting experimentation with participatory research has been in developing countries where research has been used as one part of development projects (see for example Kroeger and Franken, 1981; Smith et al., 1993). De Koning and Martin (1996) quote experiences from developing and developed countries in the use of participatory research. In development projects in poor countries there is a potential for participatory research to raise people's consciousness and encourage them to become involved in actions that could liberate them from poverty and despair. Bloem et al. (1996) describe how a non-government organisation in Bangladesh recognised the limitations of many development approaches and sought alternatives. From this began the People's Participatory Planning (PPP) process in 1990. These authors' description of the process stresses that the PPP is not a universal panacea and, without continual critical appraisal, will cease to be effective. Overall, however, they conclude that 'the continual process of reflection and action can enable people to change and transform from one level of functioning to another' (1996, p. 150). Similar conclusions were reached by Howard-Grabman (1996) in her analysis of participatory action research designed to address maternal and neonatal health problems in rural Bolivia. The involvement of men and women in the project led to the development of new reproductive health practices that have seen maternal, perinatal and neonatal mortality decline in the communities that were part of the project. At the heart of participatory action research is the concept that research and evaluation must be flexible and responsive to shifting circumstances and understandings, which has been recognised in the evaluation of community development in Australia, where the developing and changing nature of initiatives demands very flexible research methods (Baum, 1992).

Participatory research is based on the belief that objective truth is a problematic concept and that there are multiple ideas of truth. Exploring the impact of values becomes central to participatory research. Wadsworth (1991) suggests that one way in which to incorporate the various perspectives in a research or evaluation project is to form a critical reference group, comprising those people who are meant to be served by the services or actions being planned, provided or evaluated. She suggests (Wadsworth, 1991, p. 11) that researchers must have a profound respect for the critical reference group they are working with, accept the legitimacy of their viewpoint and have a 'sharply felt dissatisfaction' with the conditions that impinge on those people. They must also be prepared to adopt a collaborative problem-solving style of research that aims to change and improve (not just study) those conditions. The evaluation of a consumer perspective of an acute psychiatric hospital is an example of how such a critical reference group has been used as part of a participatory action research project (McGuiness and Wadsworth, 1992).

Participatory research does not imply that researchers disavow or downplay their specialist knowledge. The key issue is how that knowledge is used in relation to participants in the research process, who in traditional social science and experimental research had been referred to as 'subjects'. This term immediately placed them apart from their normal social roles and objectified them only in terms of their usefulness to the researchers. Reason (1994, p. 328) argued that in participatory research the relationship between the researcher and the participant is viewed critically, explaining: 'A key notion here is dialogue, because it is through dialogue that the subject–object relationship of traditional science gives way to a subject–subject one, in which the academic knowledge of formally educated people works in a dialectic tension with the popular knowledge of the people to produce a more profound understanding of the situation.'

The literature on participatory research tends to eulogise the potential contribution of research participants. There is a sense in which it is, ironically, seen as superior to other forms of knowledge, but, in fact, not all statements made by research participants should be taken as truth. Just as researcher perspectives are subject to critical reflection, so are those of participants. Reason (1994, p. 333) suggests the term 'critical subjectivity' to indicate that participatory research involves a rigorous process of reflection to arrive at new forms of interpretation and knowledge.

The practice of action research in public health presents dilemmas. Boutilier et al. (1997) describe a number of these and argue that the perspectives of academic researchers, practitioners (managerial and frontline) and community members differ and that, to be effective, action research requires considerable negotiation and reflection on practice. They suggest (p. 76) that key issues in the practice of what they describe as 'community reflective action research' are those to do with whose knowledge is valued, who owns the research, why the research is being done and the importance of recognising the complementary skills of community members, researchers and practitioners. Community participation and control of research processes may lead to some conflicts with the demands of scientific rigour (Allison and Rootman, 1996), as participants are generally not experts in research or primarily interested in the validity of research. This tension should be addressed so that evaluations are scientifically acceptable and ethically participatory.

The processes of negotiation in participatory action research make it a long and complicated process, which often conflicts with the needs of funding bodies. It is not easy to reconcile these conflicting demands.

CONCLUSION

This chapter has shown the considerable value of qualitative research to public health. Fortunately this form of research has been more frequently used in the past decade and its value in explaining many of the patterns that epidemiology describes is being valued. It offers a range of methods that should be part of the toolkit of all public health research endeavours.

CRITICAL REFLECTION QUESTION

9.1   Think of a public health issue that you are currently involved in and consider how qualitative research could help you understand it better. What methods would be most helpful?

Recommended reading

Bazeley and Jackson (2013) *Qualitative data analysis with NVivo*. Helpful guide to the nuts and bolts of using NVivo to analyse qualitative and mixed methods data.

Denzin and Lincoln (2011) *Handbook of qualitative research* contains 42 chapters on all aspects of qualitative research, describing the theories behind qualitative research, the practicalities of doing it and ways of analysing and reporting its findings. Most authors adopt a constructivist perspective.

Krueger and Casey (2009) *Focus groups: A practical guide for applied research* is an excellent introduction to running focus groups.

Patton (2015) *Qualitative research and evaluation methods* is an excellent guide to methods.

Useful website

The Lowitja Institute, www.lowitja.org.au, provides many resources to assist with the conduct of ethically sound and participatory research with Indigenous peoples.

10

PLANNING AND EVALUATION OF COMMUNITY-BASED HEALTH PROMOTION

Suit the action to the word
The word to the action.

William Shakespeare, Hamlet, *Act 3, Scene 2, lines 20–1*

KEY CONCEPTS

Introduction
Planning for community-based public health projects
Tools for needs assessment
Setting priorities and ongoing planning
Evaluation of complex public health initiatives
Objectives and outcomes
Ensuring a reflective approach
Methods for community-based evaluation
Validity of evaluation
Conclusion

INTRODUCTION

Increasingly public health involves community-based initiatives that focus on social, policy, organisational and individual change. These initiatives often focus on particular settings and pose particular challenges for planning and evaluation. They are very rarely amenable to evaluation using conventional medical techniques such as randomised controlled trials. They are typically long-term developmental activities that seek to change the ways in which organisations work, and to put health and the environment on the top of their agendas. These projects may be complex and involve multiple activities with evolving and changing objectives. Consequently the planning and evaluation has to be similarly complex. This chapter discusses methods for planning and evaluating community development projects (see chapters 20 and 21 for details of these projects) and complex community-based initiatives such as the UK Health Action Zones, Healthy Cities projects, Local Agenda 21 projects (see chapter 23 for details of these projects) and other settings projects such as Healthy Schools and Healthy Workplaces.

PLANNING FOR COMMUNITY-BASED PUBLIC HEALTH PROJECTS

The broadness of the new public health agenda means that decisions about which health issues to tackle are crucial. Practitioners have to choose between many competing priorities, and there are a variety of techniques to determine needs. Beyond this, it is necessary to set priorities and plan accordingly.

Most projects start with an assessment of needs, which has variously been called a community diagnosis, a needs assessment, a situational analysis or a rapid appraisal. Without such planning work, public health will be reactive and respond to threats to health as they arise rather than plan for and create health. There are two key aspects to assessing needs and planning: the principles underlying the exercise, and the methods used to collect information.

KEY PRINCIPLES

Community-based public health projects, including healthy settings, are based on a social and environmental understanding of health, and any planning work should also use this framework. This means that the needs assessments will be typically wide-ranging and involve a number of government sectors, community groups and private industries. The process of amassing and interpreting the data, however, may be a useful means of initiating and consolidating the work of an intersectoral committee. In the past, 'health' needs assessments have usually been 'disease' needs assessments, with little focus on those aspects of the physical and social environment that help keep people healthy. They have tended to focus on health service use and on documenting individuals' morbidity profiles. A shift in mindset from disease to health is crucial in developing proactive new public health projects such as Healthy Cities.

In this model, data would be collected on the physical environment of the city or community and details of the social supports and community structures. Hancock (1994) has proposed a model for integrated Healthy Cities planning (see figure 10.1)

FIGURE 10.1 TOWARDS HEALTHY AND SUSTAINABLE LIVING

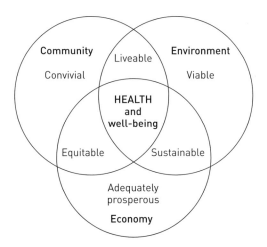

Source: Hancock, 1994, p. 248.

that integrates the concepts of conviviality, viability, adequate prosperity, liveability, sustainability and equity. Chapman and Davey (1997) describe the process used by a number of Queensland local governments to produce municipal public health plans within this framework. They report that it was used successfully as a tool to explore local quality of life issues within a socio-ecological framework, and to keep a positive focus on health needs assessment.

VISIONS, STRENGTHS AND LATERAL THINKING

Encouraging a sense of vision in planning for new public health initiatives is important. Vision is important to establish the overriding goal of the initiative and to keep people focused and committed to it. Vision can also be used to encourage people to think outside the square—a great skill for the new public health.

One way of encouraging people to move from their accustomed thinking is to visualise what their city or community organisation might look like were it to become health promoting. In Liverpool (the location of one of the first Healthy Cities projects in Europe) Ashton (1988) wrote an inspirational vision of a healthy Liverpool, which stressed the strengths of the community as well as its problems. Here is a typical section:

> motor cars were one of the easiest things to do something about. They were now banned from the city centre except for taxis; bicycles had become very popular and in many quarters communal bicycles had become commonplace. The tram line had been rebuilt in 1993 to link up the historic site of Liverpool, the pastoral areas of the inner city and the farm communes which had become a feature of the large band of fringe land between town and country which had become progressively blurred
>
> *Ashton, 1988, p. 31*

Other communities followed suit. Toronto developed a vision workshop method that drew on guided imagery to encourage people to relax and imagine what their community would look, feel and sound like were it to become ideally healthy. This method was also used in Noarlunga in South Australia, involving more than 300 community group members. The visions produced by the disparate groups were surprisingly similar (Baum et al., 2006). The vision method is now used widely in local government. One example is Redlands in Queensland, which produced a community plan on the basis of extensive consultation with its community and noted that in '2030, the Redlands will be a well-designed, vibrant city of mainland and island communities, each with distinctive character, heritage and lifestyles. Our shared values will shape the way we care for each other and how we protect the land, seas and waters where we choose to be' (Redland City Council, 2010, p. 6).

The vision technique was also used in the planning of the WHO Healthy Islands initiative in the Western Pacific Region. The vision stated was that Pacific islands shall be places where:

> Children are nurtured in body and mind
> Environments invite learning and leisure
> People work with age and dignity

Ecological balance is a source of pride

The ocean which sustains us is protected

WHO Regional Office for the Western Pacific, 2002

ACKNOWLEDGING AND BUILDING ON STRENGTHS

An important factor in planning and needs assessment is to ensure that there is not a sole focus on problems. For this reason WHO (1996, p. 7) recommends that planning for a Healthy City vision should start with the question: 'Why is this city or community a fine place to live?' The strengths of a community should be documented and capacities that can be developed sought. A problem-driven needs assessment can be disempowering to communities, especially poor ones where low self-esteem could be reinforced, and where the potential for a negative portrayal is greatest. Weeramanthri (1996) discusses the importance of being sensitive to community perspectives when primary health care practitioners work with Aboriginal communities. He is critical of the problem focus of much needs assessment research, saying it can be both disempowering and disrespectful of local perspectives on health issues. He recommends that practitioners ask 'What do people know?' and 'What do people value?' These questions are relevant to needs assessments in other settings, as they recognise that perspectives and solutions from within a community have a far greater chance of informing effective plans than those based on external perspectives.

SORTING OUT PROBLEMS AND SOLUTIONS

Health professionals have a tendency to define need in terms of their own solutions. Widespread community back pain might be interpreted as a need for physiotherapists, orthopaedic surgeons, ergonomists, massage therapists or yoga teachers, depending on who defines the need. Some social workers may define lack of parenting skills as a problem for low-income communities, while others may see styles of parenting as a reflection of people's poverty and their consequent limited options for creative parenting. Often people jump to solutions without analysing the problems to be addressed fully.

IMPORTANCE OF PROCESS AND INVOLVEMENT

The usefulness and effectiveness of planning will be greatly increased if it involves a broad section of the community and encourages learning among the organisations and individuals involved.

The challenge of making community involvement in public health a real rather than a token process is discussed at length in chapter 18. Research faces the same issues. Researchers need to apply the same community-developed principles as health promoters. Useful tips for public health needs-assessment researchers were summarised thus (South Australian Community Health Research Unit, 1991, p. 55):

- take enough time—consultation can't be hurried
- develop good listening skills—open your ears and be receptive to other people's perspectives even if you don't agree with them

- try to consult those who are not vocal members of the community
- think of creative ways of consulting the community
- ensure that findings from the needs assessment are reported back in an appropriate way to the community, using such methods as the media, plain language reports and community forums
- tap into existing community groups.

The obvious benefits of involving community members in the planning exercise are that they know their community well and probably have perspectives that differ from those of paid, professional workers. Also, understanding how local people see issues and what they see as crucial is essential to establishing priorities that will have community support.

TOOLS FOR NEEDS ASSESSMENT

There is no set formula for carrying out a needs assessment. The particular mix of methods will reflect the resources available, the scope of the planned project, the size of the community and the skills of the project staff. In many ways, assessing needs is like doing a jigsaw, as it involves fitting together different bits of information to produce a complete picture of the issues. The needs assessment should consider the strengths of a community as well as the problems it faces.

EXISTING DATA

One of the first steps in deciding which methods to use is to audit what information is routinely available from sources such as the Australian Bureau of Statistics, social health atlases and databases, existing reports produced by other agencies and community groups and routine service data. This will help identify gaps in the knowledge and determine what additional information will best fill these gaps.

HEALTH AND ENVIRONMENT INDICATORS

Indicators can be used to:

- provide information about a setting
- monitor changes in a setting
- determine issues of most concern
- inform decision making.

COMBINE QUALITATIVE AND QUANTITATIVE DATA

Indicators can be based on quantitative or qualitative information. The arguments in favour of using both forms of data in public health have already been detailed. The International Council for Local Environmental Initiatives (ICLEI) (1996, p. 3) has commented that 'the key to achieving accurate baseline data is to link participant assessment (made by the local community) and expert technical assessments.' The advantages they specify are:

- Priorities for action may be negotiated.
- There may be more understanding of the systemic nature of issues.
- Community involvement may foster detailed dialogue among community groups and between the community and technical experts.
- Sole reliance upon the assessment of experts is prevented.
- Heightened understanding of the technical aspects and complexities of problems enable stakeholders and residents to define effective options for action.
- Innovative informal approaches used by residents or local communities to solve problems may offer practical solutions that can be upgraded and applied more widely to address issues on a larger scale.

Abbott (1990) recast 'qualitative' and 'quantitative' as 'sensory' (based on perceptions) and 'standards' (based on expert definition) data. The data produced are different but complementary. Table 10.1 shows how each type defines problems, producing different forms of evidence and analysis.

TABLE 10.1   DIFFERENCES BETWEEN 'SENSORY DATA' AND 'STANDARDS' APPROACHES: AIR POLLUTION

Sensory data	Standards approach
Definition of the problem	
Nasty, smelly, dust-laden air	What are the levels of particular elements in the air?
Makes for bad quality of life	What is the risk of these leading to disease?
Form of evidence	
Experiential	Statistical
Based on sensory data	Based on numerical indicators of both what is in the air and its effects
Grounded on how it feels	Abstracted
Form of analysis	
Holistic and largely qualitative	Reductionist and largely quantitative
Emphasises different aspects of living in the area that make up the whole picture	Emphasises limited particular quantifiable components of the problem

Source: adapted from Abbott, 1990, p. 5.

COVERAGE OF INDICATORS

The search for a perfect set of indicators for Healthy Cities projects has been somewhat like the search for the Holy Grail—the indicators have remained elusive. In the international Healthy Cities movement there has been debate about the value of local versus international indicators for the projects. Some WHO publications have recommended the development of a set of indicators that could be used internationally. Werna and Harpham (1996) cite a number of these including Draper et al. (1993). Other views suggest that indicators should be locally derived, as those that are appropriate for one community may not be for another. The European project started off with ambitious plans for common indicators that could be used throughout Europe. As the project progressed, the need for indicators came to be seen as less pressing and it was evident that agreement on a detailed common set was unlikely. In Australia there was initial interest in developing indicators, but in the event indicator development happened at an individual project level, but was guided by a broad framework (Baum and Brown, 1989). The Commission on Social Determinants of Health (2008) stressed that data are vital to taking a social determinants approach and that they should be disaggregated according to basic indicators of equity. It recommended ensuring that routine monitoring systems for health equity and the social determinants of health are in place locally, nationally and internationally (CSDH, 2008, p. 179), and including a minimum health equity monitoring framework (see box 10.1)

BOX 10.1   TOWARDS A COMPREHENSIVE NATIONAL HEALTH EQUITY SURVEILLANCE FRAMEWORK

HEALTH INEQUITIES

Include information on:

- health outcomes stratified by
  - sex
  - at least two socioeconomic stratifiers (education, income/wealth, occupational class)
  - ethnic group/race/indigeneity
  - other contextually relevant social stratifiers
  - place of residence (rural/urban and province or other relevant geographical unit)
- the distribution of the population across the subgroups
- a summary measure of relative health inequity: measures include the rate ratio, the relative index of inequality, the relative version of the population attributable risk, and the concentration index
- a summary measure of absolute health inequity: measures include the rate difference, the slope index of inequality, and the population attributable risk.

HEALTH OUTCOMES

- mortality (all cause, cause specific, age specific)
- early child development
- mental health
- morbidity and disability

- self-assessed physical and mental health
- cause-specific outcomes.

DETERMINANTS, WHERE APPLICABLE INCLUDING STRATIFIED DATA

- daily living conditions
- health behaviours
  - smoking
  - alcohol
  - physical activity
  - diet and nutrition

- physical and social environment
  - water and sanitation
  - housing conditions
  - infrastructure, transport and urban design
  - air quality
  - social capital

- working conditions
  - material working hazards
  - stress

- health care
  - coverage
  - health care system infrastructure

- social protection
  - coverage
  - generosity

STRUCTURAL DRIVERS OF HEALTH INEQUITY

- gender
  - norms and values
  - economic participation
  - sexual and reproductive health

- social inequities
  - social exclusion
  - income and wealth distribution
  - education

- socio-political context
  - civil rights
  - employment conditions
  - governance and public spending priorities
  - macroeconomic conditions.

CONSEQUENCES OF ILL-HEALTH

* economic consequences
* social consequences.

Source: CSDH, 2008, p. 182.

The work of the Commission on Social Determinants of Health led to WHO's URBAN Heart project, which has produced a framework for local governments to collect data on equity and enable comparison between neighbourhoods, and to pinpoint areas where action is most needed (WHO Kobe Centre, 2015b).

Table 10.2 provides a summary of the key areas in which indicators may be collected in a Healthy Cities or similar project. For each subcategory a combination of sensory and standards data will normally be applicable.

TABLE 10.2 BROAD INDICATORS FOR HEALTHY CITIES, LOCAL AGENDA 21 OR SIMILAR INITIATIVES

Topics	Type of information (mix of standards and sensory; disaggregated where possible to enable equity analysis)	Data source/method
People		
Demography—population make-up and epidemiology	Total population, age distribution, birth and fertility rate, death rates, household types, income and employment profiles, ethnic profile, languages spoken. Main causes of morbidity and mortality. Illness risk factors (smoking, nutrition, drink-driving, immunisation status)	Australian Bureau of Statistics (ABS), Department of Social Services, federal and state health departments
Perceptions of area	Residents' views about desirability, safety, nature of area. Attitudes and beliefs concerning health and illness and available services. Perceptions of key health problems	Population surveys, participant observation, focus groups, interviews, document analysis including media (e.g. newspapers, especially letters to the editor in locally circulated press)
Support networks	Contact between residents, informal caring, methods of information dispersal, social interaction, community 'hub(s)' (or lack of): what is the impetus or driving force of this interaction?	Mainly participant observation, some in-depth interviews, limited use for surveys

(continued)

TABLE 10.2 BROAD INDICATORS FOR HEALTHY CITIES, LOCAL AGENDA 21 OR
SIMILAR INITIATIVES (*CONTINUED*)

Topics	Type of information (mix of standards and sensory; disaggregated where possible to enable equity analysis)	Data source/method
People		
Community norms, values and traditions, history	A feel for local beliefs and variations in these. Awareness of local ethnic groups: their attitudes, values, concerns. Presence or absence of significant museums, festivals, traditional rituals. Religious expressions and involvement in health-related concerns. Gender relations: their expression in domestic and wider social life. Print and electronic media: state and local	Mainly participant observation, some in-depth interviews, local history documents
Crime	Homicide rates, domestic violence, house break-ins, perceived safety of the city	Police, Attorneys-General, family and community services
Locality and infrastructure		
Housing and planning	Overview of type and suitability of housing. Adequacy of planning and provision of services. Housing needs of different groups (e.g. young people, and those with disabilities). Private and public ownership; rental market, numbers of homeless	Department responsible for environment and planning. Housing Trust/Commission. Plus qualitative methods for perceptions. Local government, housing surveys, ABS, public housing bodies' libraries. Residents' associations/ action groups, cooperatives
Transport	Level of vehicle ownership. Adequacy of public transport provision for bicycle tracks— perceived gaps	ABS, bicycle clubs, automobile associations, transport and highway departments, mortality data
Water, sewerage, energy sources	Availability and type of supply. Drinking water quality	State government information services
Organisations and services		
State and local government	Inventory of local services (with focus on health and welfare), and professional and non-professional perceptions of gaps. Extent of cooperation or conflict between agencies	From information services in local and state government

TABLE 10.2 BROAD INDICATORS FOR HEALTHY CITIES, LOCAL AGENDA 21 OR
SIMILAR INITIATIVES (*CONTINUED*)

Topics	Type of information (mix of standards and sensory; disaggregated where possible to enable equity analysis)	Data source/method
Organisations and services		
Non-government and community groups	Inventory of these, including self-help groups and lobbying groups	Local government, community information services and associations.
Intersectoral groups	Social planning committees, community forums	State and local government, community information services and associations
Business and economic	Main businesses and trades, (including business in the home). Occupational health provision, unions. Occupational illnesses and accidents	Chambers of commerce, retail traders, business directories, local press, unions, occupational health section of state government
Administration and power	Analysis of administrative structure and politics (federal, state and local government; lobby groups). Perceptions of power holders and others. Analysis of extent of communication between different sectors and levels of government. Assessment of ability of the community to influence decision making. Where is power centred? Balance/imbalance of health-related expenditure. Clinical/medical health versus social/community health	Formal documents, council minutes, Parliament records (e.g. Hansard), in-depth interviews; analysis of business and organisations as above. Labour market, housing market, class and status considerations
Natural environment		
Climate, geography, environmental health	Description of topography, location, rainfall, temperature ranges, etc. Air and water quality/pollution, percentage of green space	Year books, meteorological offices, departments that deal with the natural environment and monitor air and water quality, and soil pollution
State of the physical environment	Coastal, river, canal pollution; solid waste disposal facilities; community perceptions	Environmental protection agencies, surveys

SURVEYS

Surveys can produce useful information, but are relatively expensive to do properly and may not always produce the expected information. Often appropriate information will be available elsewhere and at little cost. They are most useful for providing descriptive data on a community and assessing the extent of its issues and opinions.

Needs assessments in Australia have used surveys to good effect, especially when the information from them has been combined with qualitative data to provide a detailed picture. For instance, a community-wide survey showed that young people were heavy drinkers of spirits (Gallus, 1989). The research team were bemused but then discovered by talking with young people that spirits provide the quickest and cheapest way to 'get off your face'. Further exploration found that they saw this as relief from boredom.

QUALITATIVE DATA

The various methods for collecting qualitative data can be applied to needs assessment. Qualitative accounts of a community's or city's health can help bring a report to life and make it more meaningful than simple statistics. Hancock and Duhl (1986) advise that unless 'data are turned into stories that can be understood by all, they are not effective in any process of change, either political or administrative'.

Increasingly, the importance of community stories is being recognised. Most indigenous cultures value storytelling as a means of passing information between generations and expressing the meaning of daily life. Their importance to public health has been described thus: 'Stories contain elements of uncertainty and ambiguity and can be a way of forging agreement or a way of dividing people. Stories help people to imagine the future as well as connecting them to the past. Stories, therefore, encapsulate possibilities for change and can heal both the teller and the listener' (Weeramanthri, 1996, p. 9).

Stories have great potential as a way of engaging people in a needs assessment process. There is no better way of bringing home the reality of unmet needs than stories about a community's or individual's plight. Combined with relevant statistics and survey data, a full picture can be obtained. In terms of bringing about change, stories are likely to be effective because politicians relate far better to stories than other forms of information. Consequently, stories can become an important part of a subsequent advocacy campaign.

RAPID APPRAISAL

Rapid appraisal is a means of doing needs assessment quickly and economically. The method was developed in developing countries but is applicable to most settings. A guide to conducting rapid appraisal is available in Ong (1996). A description of the technique's development and a thoughtful critique of its benefits and limitations are available in Manderson and Aaby (1992). Two of the main benefits are its relative cheapness and its ability to provide information rapidly so that it is of maximum use when the assessment is needed by decision-makers.

Conducting a needs assessment requires a consultative process that includes all the key players in a community, and it should bring together different forms of data from a variety of sources. At the end of the day, unless the information is perceived as relevant and recent, it is unlikely to lead to action or change. Most Healthy Cities projects start with fairly intensive data-gathering exercises. It is then necessary to have some ongoing way of feeding information about the changing situation of the community or city into ongoing planning.

SETTING PRIORITIES AND ONGOING PLANNING

WHICH PRIORITIES?

Most new public health projects based in communities or organisations will never be able to tackle all the issues they identify—hence the need to establish priorities. It is rare that the process of establishing these will be based purely on the information collected. Other considerations are: the particular interests of the project staff and members of the steering or management group; political interest or lack of interest in particular topics; issues considered important by the community; and local policies relating to health and previous work in the area. Werna et al. (1998) stress that priority setting within a Healthy Cities project is a complex process and inevitably involves value judgments. They recommend (p. 71) that 'priority should normally go to those solutions that can reduce the health burden the most, at the lowest cost and with the highest chance of success, given the local circumstances and opportunities for exploiting non-financial resources such as human capital, existing infrastructure and community involvement'.

Priorities that gain the commitment and passion of local people and people who work in their communities are most likely to win the support and commitment necessary for success.

APPROPRIATE PLANNING FRAMEWORKS

The developmental and long-term nature of new public health initiatives such as Healthy Cities or a Healthy Schools project means both a structure for ongoing strategic thinking (rather than one-off strategic planning exercises) is required as well as a commitment to good project planning for particular projects within the broader framework.

The broader framework aims to equip communities and health promoters with the ability to respond to locally defined illness problems and health issues and to establish priorities for action. It is based on the belief that it is more important to establish structures that encourage local people to work in partnership with professionals from a number of different sectors than it is to spend time establishing specific goals and targets. Local communities will establish their own, appropriate goals that will inevitably change over time. Planning frameworks also need to be capable of coping with the complexity of change. Duhl (1992, p. 17) put the problem of assuming simple causal relationships nicely when discussing his experience with Healthy Cities projects: 'Linear change is a rare phenomenon. It occurs only when time is short, goals clear and scale small. Urban issues are instead complex, unclear, confusing and ever changing. Change is full of ambiguous goals on multiple time lines. In fact, control of the intervention process is close to impossible.'

Detailed project planning ensures goals and aims are achieved by 'designing feasible means, managing workloads, making the best use of everyone's talents and establishing the basis for good decision making' (Dwyer et al., 2013, p. 99). There is a huge literature on planning for specific projects (such as a safe workplace project for local businesses that is developed within a Healthy Cities project) and Dwyer et al. (2013) provide a useful summary of the literature. The basic planning sequence of rationale, goal, objectives, strategies, timelines, resources and evaluation is based on a rational approach that uses aids such as Gantt charts to develop logical sequences to assist achievement of objectives. However, it is important to accept that plans rarely go exactly as planned and, as Dwyer et al. (2004b, p. 134) say, 'project management is a set of methods but also an art'. This is not least because of the complex interests involved.

COMPLEX INTEREST GROUPS

The new public health is characterised by complex interest groups who are unlikely to have a single voice. Communities are never entirely homogeneous. In Australia, communities comprise people from diverse cultures, ethnic groups, age groups, gender groups, political orientations and values. Planning to meet the public health needs of these people will inevitably be complex. Intersectoral action, while likely to be rewarding, will make planning and priority-setting more complex, simply because more interests and agendas have to be juggled. Throw in the professional and organisational jealousies that are inevitable and it becomes clear why planning for the new public health has to capture the complexity of reality and remain dynamic enough to cope with the rapid change that is so characteristic of the late twentieth and early twenty-first centuries. This means, of course, that planning cannot be entirely rational.

MUDDLING THROUGH

Basing his opinion on extensive experience with new public health initiatives, Hancock (1992, p. 25) recommends that:

> we should be very suspicious of master plans that will, supposedly, bring us to our goal in 10 or 20 years. In an era of unprecedented change, it is very difficult, if not impossible, to be able to forecast what the world will be like … So instead of trying to develop a master plan for attaining a Healthy City that will in all probability join all the other master plans on the bookshelf, we need a process of 'goal- (or vision-) directed muddling through'.

This muddling through should be guided by an overall vision that provides direction, but does not stifle innovation and creativity. Hancock (1992) recommends the use of a regular process of environmental scanning to discern the current major issues, threats and opportunities. He suggests that this process be kept relatively simple so that people are not swamped by detail.

CITY AND MUNICIPAL HEALTH PLANS AND STRATEGIC PLANS FOR LOCAL AGENDA 21

Many local governments (often through Local Agenda 21 projects) and healthy cities or communities projects produce some form of city health plan to use as their key planning document for future action.

Most of these plans do not specify detailed and exact targets that could stifle initiative. Tsouros (1996, p. 9) comments that in Europe the projects have proved to be 'vehicles for strategic growth and powerful tools to deal with change, uncertainty and the building of alliances'. He adds that the project has not been implemented as a closed system but 'shaped locally through the commitment, persistence and creativity of cities; its diversity gives it strength.' Another important feature is that plans and priorities need to build on and exploit the existing strengths of a community.

Box 10.2 provides an example of the planning the Kiama Council in New South Wales has done, which has involved the local community in innovative ways and drawn directly on the new public health philosophy. Ideally, the preparation of a city or municipal plan will generate awareness of health and environmental problems by city authorities and non-government agencies and communities, and lead to the provision of resources to tackle the problems. The process of producing a plan is crucial because it determines the extent of commitment to implementing it. The process followed by the Glasgow Healthy Cities project is similar to that used in other cities. Black (1996, p. 94) explained:

> The production of a plan could conceivably be undertaken by a single person locked in a small room. The document that would be produced while it may be excellent would stand little chance of being implemented. It would be seen as an imposition, as not reflecting the needs of practitioners as they hadn't been consulted and people would be unwilling to implement it as they would have no investment in it …

Black goes on to say that the process of working jointly is essential for the success of a city health plan. Necessary changes of size and complexity to make any city healthy are only possible through joint work.

BOX 10.2   KIAMA COUNCIL HEALTH PLAN 2011–17

Kiama Council is a member of the Global Alliance for Healthy Cities and has used the values and principles of the Healthy Cities movement to develop its Health Plan 2011–17. It explicitly uses the new public health as a reference point and reproduces the Hancock diagram (figure 10.1, p. 229) to frame the plan.

The plan was based on widespread community consultation including mailouts, surveys, kiosks at community events, touch-screen surveying at council facilities, media campaigns, focus groups and strategy development workshops. Its vision and mission are expressed as:

Vision

A municipality working together for a healthy, sustainable and caring community.

Mission

Kiama Council will work to create a municipality that has a healthy, vibrant lifestyle, beautiful environment and harmonious, connected and resilient community.

The plan contains four sections:

Strengthening community
Supportive social environments for health
Healthy economy
Healthy environment.

In each area, a strategy, time frame and performance indicator are listed.

Source: Kiama Municipal Council, 2011.

EVALUATION OF COMPLEX PUBLIC HEALTH INITIATIVES

Evaluation assists sense-making about policies and programs through the conduct of systematic enquiry that describes and explains the policies' and programs' operations, effects, justifications, and social implications. The ultimate goal of evaluation is social betterment, to which evaluation can contribute by assisting democratic institutions to better select, oversee, improve, and make sense of social programs and policies. (Mark et al., 2000, p. 9). There has been increasing attention to the need for evaluation to cope better with the complexity of social and community settings and how these affect public health programs. Evaluation designs are being constructed that are better able to do this (see Patton, 2011).

OBJECTIVES AND OUTCOMES

The objectives of community development and healthy settings initiatives can only partly be specified in advance, as they depend on definition by the community and may take some time to evolve. They are also likely to change as the context, setting and people involved change. These shifting objectives make it more difficult for evaluators, but have to be considered if they are to do their job of describing and assessing the progress of the initiative. Rather than viewing the shifting objectives as a difficulty, the evaluator should incorporate regular reassessment of objectives into the evaluation design. It is also common that government goals will force a change on a project, such as in the case of the UK Health Action Zones. In this case, shifting central government priorities meant these projects were never implemented as initially planned. Only a flexible evaluation is able to capture the impact of such policy changes.

CRUCIAL ROLE OF DIFFERENT PERSPECTIVES AND VALUES

Defining and measuring outcomes in community development and healthy settings projects means recognising that the choice of outcome measure will depend on the perspective adopted. In evaluating community development and healthy settings projects, different groups may not even agree on what the outcome of a project should be, let alone whether or not it has been achieved (Hunt, 1987). Funding bodies, community members, community health workers and community health managers may all have different perceptions of what the crucial outcomes should be. The evaluation process needs to specify these differing interests in advance and ensure that the evaluation pays attention to each. Costongs and Springett (1997) summarise the importance of these issues to healthy settings projects when they say 'It is important to consider the values, aspirations and motivations of the people involved, which cannot be quantified' (p. 348).

Obviously, each group's values become crucial in this process and should be assessed in the course of an evaluation. A range of perspectives therefore need to be incorporated into an evaluation of such an initiative. Evidence suggests that evaluation is most likely to be used if stakeholders are involved in and committed to the evaluation process (Patton, 2011; Robson, 2000).

DETERMINING IF THE INITIATIVE IS READY FOR EVALUATION

The nature of community development and healthy settings initiatives means that evaluation cannot happen realistically in less than five years. A sustained effort is required to develop a group of people, and for them to define their objectives, take action, learn from their mistakes and successes, and establish sufficient confidence in their ability to effect change. Funders often want evaluation results in a much shorter time frame. Process information on the establishment of an initiative is all that is likely to be available in the short term.

PARTNERS IN EVALUATION

Evaluation of community development and healthy settings projects is as much about partnerships and community participation as the projects themselves. The aims and objectives of the evaluation process need to be negotiated with those who are participating in the project (Springett, 2003). Wadsworth (1991) uses the term 'critical reference group' to refer to those people for whom the particular project was designed. She believes it is crucial to ensure they are primary partners in the process and that their values, aspirations and concerns should guide the evaluation. Robson (2000, p. 25) points to some of the problems that should be avoided when staff from an initiative are involved in evaluation. These points are applicable to both community and professional evaluation partners:

- Develop a realistic, feasible and adequately resourced evaluation plan so that it is possible to deliver on promises about what will happen and when it will happen.
- Allow for adequate release time from normal duties to cover the time needed for the evaluation. Remember that the time required is usually underestimated.
- Avoid really problematic settings (e.g. where the organisation is under severe strain).
- Work hard on utilisation as frustration will occur if there is no action taken on the findings of an evaluation.

Participatory evaluation is more likely to produce findings that can be used to improve the project. It enables communities to build their own evaluation skills and helps establish collective responsibility for the project activities. Feuerstein (1986) provides an easy-to-read guide to participatory evaluation that is accessible to people who are new to the concept of research and evaluation.

MEASURING CHANGE AT A COMMUNITY LEVEL

It is recognised that, while the rationale for community development and healthy settings programs is the promotion of health, the immediate objectives relate more to empowerment and the creation of conditions likely to promote health (Baum, 1995b; Legge et al., 1996). Sheills and Hawe (1996) argue that evaluation of community development should ensure that it measures change at the community and individual levels. They suggest that communities are more than the sum of the individuals within them and that this is a challenge to conventional research techniques that focus on individuals and the aggregate of individual change.

One example of an attempt to measure change at a community level is the work by Bjãrås et al. (1991) on participation. They suggest a mechanism for measuring participation that can be used to illustrate change over time in a project's operation. They suggest that the extent of participation can be measured by assessing the extent of participation in these activities:

- needs assessment
- leadership
- organisational focus and operation
- style of resource mobilisation
- management and decision-making processes.

A score is allocated to each indicator to note the extent of participation, ranging from narrow (1) to wide (5). These scores can be mapped as shown in figure 10.2. The technique can be used to assess change over time or to compare the different players' perceptions of the extent of participation. The debate that may ensue between players about their differing perceptions of participation could itself become part of the participatory process.

FIGURE 10.2   MEASURING PARTICIPATION

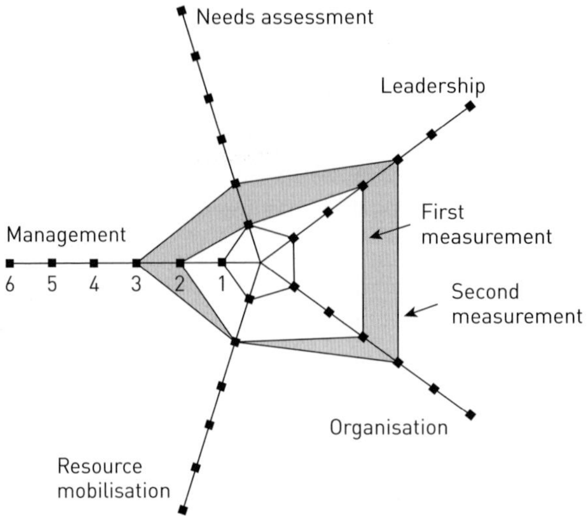

ATTRIBUTING CAUSALITY

Determining the impact of a community development or healthy settings project on health status and the quality of the social and physical environment is difficult. It is relatively straightforward to produce a set of indicators, but making inferences about the causes of any changes in the indicators monitored is far more hazardous. In order to attribute any change to a particular intervention, it is necessary to be able to show that these factors were causally related, as shown in figure 10.3.

The 'gold standard' for epidemiology in dealing with the issue of attribution is the randomised controlled trial. But community projects can rarely, if ever, use a control as no two communities are identical. Even if such communities were found, it is not possible to stop all initiatives in the control community. Nutbeam et al. (1993b) describe how the control city being used to evaluate the Heartbeat Wales health promotion project decided to start its own heart health project, thereby undermining its usefulness as a control community. Similarly, with a healthy settings project it is likely that any community chosen as a control would develop a project of some description that would have similarities to the aims of the setting of interest. Also, the precision of most community indicators will often not be great enough to compare two similar communities and monitor change over several years.

The Commission on Social Determinants of Health (see box 3.3) has recognised that the issue of attributing causation from complex social and health interventions is crucial. In its final report (2008, p. 179) it noted 'Generating evidence on what works to reduce health inequities is a complex process. Randomized controlled trials are often not practically and/or ethically feasible. Moreover, evidence on the social determinants of health can be context dependent.' The type of 'proof' required from much health promotion is more

FIGURE 10.3   COMMUNITY-BASED HEALTH PROMOTION EVALUATION: OUTCOMES
AND ATTRIBUTION

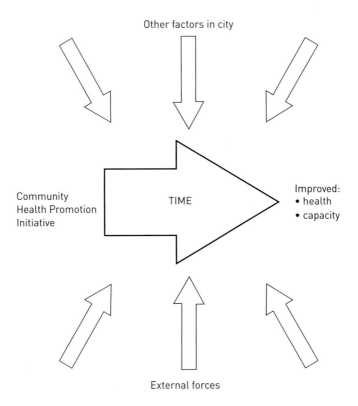

akin to that expected in a court of law than from a tightly controlled experiment (Kelly et al., 2006). Consequently a clear articulation of the logic underpinning it is important. Causality is not established through statistical tests of correlations but by a 'burden of evidence' that supports logically coherent chains of relations that emerge through the contrasting and comparing of findings from many forms of evidence (Baum et al., 2014b). The quality of argument and plausibility of the claims and attributions become vital.

An increasingly popular means of framing evaluation of complex initiatives is to develop a program logic model to scope and describe the intervention including its context. The program logic is based on a theory of change. Connell and Kubisch (1998) define this as a 'systematic and cumulative study of the links between activities, outcomes and contexts of the initiative'. It involves a process of making explicit links between the original problem or context with which the initiative began and the activities planned to address the problem, and the intermediate and longer term outcomes planned. Judge and Bauld (2001, p. 25), who used this approach to evaluate the UK Health Action Zones, comment that if a theory of change is articulated early in the life of an initiative and stakeholders can agree to it, this 'helps to reduce problems associated with causal attribution of impact'. Figure 10.4 shows a program logic model developed at the start of research evaluating the South Australian Health in All Policies (HiAP) initiative (Baum et al., 2014b).

FIGURE 10.4   PROGRAM LOGIC FOR THE EVALUATION OF THE SOUTH AUSTRALIAN HEALTH IN ALL POLICIES INITIATIVE

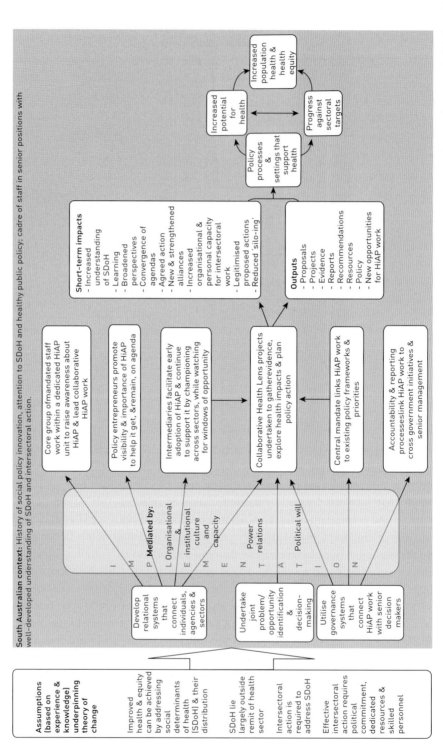

South Australian context: History of social policy innovation, attention to SDoH and healthy public policy; cadre of staff in senior positions with well-developed understanding of SDoH and intersectoral action.

Source: Baum et al., 2014b, p. 136.

Figure 10. 4 highlights the importance of the particular context of South Australia and the impact this has on implementation. It also sets out the assumptions that underpin HiAP—for example, that population health is largely determined by social factors and processes, which are mainly affected by policies and actions outside the health sector. It then shows the main features of the implementation and the ways in which these link to short-term impacts, which can be argued to ultimately link to improved population health. The form of evidence for the program logic does not rely on random assignment but rather on the burden of proof (i.e. the strength of the argument linking HiAP's implementation to short-term impacts and then longer term health gains).

A program logic approach can also deal with anticipated long-term changes beyond the life of the initiative or evaluation, as the burden of evidence approach enables extrapolation on the basis of established evidence to likely future change in population health and equity. The use of the logical theory of change or program logic can enable evaluators to ascribe the likelihood of change being brought about by an intervention if the predicted preconditions for change are documented as resulting from the intervention. For instance, a Health in All Policies program might seek to influence urban planning processes so that they are more likely to incorporate active transport such as walking and cycling, and to reduce the number of outlets selling high-fat and high-sugar food. While it would be very difficult to make a direct causal link between the work of the HiAP project and improved health, reference to broader evidence on the health benefits of lower fat and lower sugar food and increasing exercise could lead to argument that the changes in urban planning were very likely to have a beneficial impact on health.

ENSURING A REFLECTIVE APPROACH

The dynamic nature of community-based health promotion initiatives means that frequent reflection on the methods used is vital. This reflection enables evaluators to be responsive to changing policy and community circumstances and adjust methods to adapt to the changes. This also enables project actors to be involved in the evaluative process. This process calls for the evaluators and the project actors to engage in a reflective spiral. Wadsworth (1991) recommended the use of an action evaluation research process (see figure 9.1), which sees evaluation as a spiral process of planning, fieldwork, analysis, reflection and then spiralling up to planning again. The spiral analogy is particularly suited to community development as it begins to capture its dynamic nature, and incorporates the shifting and changing nature of this work.

METHODS FOR COMMUNITY-BASED EVALUATION

Chapters 6–8 described the wide variety of methods available for public health research, ranging from randomised controlled trials (RCTs) to participant observation. RCTs offer very little for community development and healthy settings evaluation, as the naturalistic setting and evolutionary nature of these approaches mean that a control community is impractical. Even in health promotion programs that are largely based on behaviour change goals, control communities are not as attractive a methodological solution as they might appear. While communities may often be similar, they are never going to be identical, so their power to determine patterns of causality is reduced. The activity in a community development and healthy settings initiative is essentially human centred and about the interaction and relationships between people, and between people and organisations and institutions. Evaluation methods have to be powerful enough to capture the subtleties and nuances of relationships, to capture the dynamics and tease out successes and failures and the likely reasons for them. Suitable methods are focus group and individual interviews, questionnaire surveys, participant observation, journals kept by project participants, analysis of community and organisational networks, records from official documents (including minutes and policies) and analysis of media coverage. These are discussed in detail in chapters 8 and 9.

VALIDITY OF EVALUATION

The use of a number of evaluation methods enables triangulation of the different data sources, thus increasing their validity. A further important means of validation is checking the data with key participants in the particular initiative. This process is typically used to reflect on, and make changes to, the way of working. It enables people to play a role in the evaluation and will increase ownership of the evaluation findings.

Validity is increased through the process of critical thinking. To be critical does not mean attacking the initiative, but reflecting on the meaning of the data in a way that questions taken-for-granted assumptions.

CONCLUSION

This chapter has described the challenge of evaluating community-based health promotion and shown that effective evaluation has to embrace the complexity of such practice and adopt a suite of research methods that are able to shed light on the effectiveness of the initiative. The evaluation also has to be flexible enough to accommodate policy and community changes, and encourage and respond to reflective practice.

CRITICAL REFLECTION QUESTIONS

10.1   Why is it so important that a plan for a healthy community is based on extensive community consultation?

10.2   What makes an action research approach appropriate for the evaluation of community-based health promotion?

Recommended reading

Minkler and Wallerstein (eds) (2003) *Community-based participatory research for health* is a comprehensive resource on the theory and practice of community-based research.

Patton (2015) *Qualitative research and evaluation methods* demonstrates the value of qualitative research methods to evaluation. Patton writes in an engaging way to stress that evaluation problems should dictate methods rather than the other way around. Design, fieldwork, analysis, interpretation and methodologies are covered.

Useful website

WHO's Urban Health Equity Assessment and Response Tool (Urban HEART) www.who.int/kobe_centre/measuring/urbanheart/en is an example of a user-friendly guide for national and local governments to identify health inequities and plan actions to reduce them.

PART 4

HEALTH INEQUITIES: PROFILES, PATTERNS AND EXPLANATIONS

Central to the new public health is a concern to reduce health inequities that result from unfair social and economic arrangements and processes. The following three chapters provide an overview of patterns of health, illness and mortality globally and in Australia, a detailed consideration of the range of health inequities in Australia and then finally, a consideration of why these inequities exist.

Chapter 11 describes changing health and illness profiles in Australia and globally. The sources of data on health and illness are described, data relating to causes of mortality and morbidity presented and the changes over time discussed. Chapter 12 describes the social patterning of health, illness and mortality according to social and economic status, occupation, Aboriginality, migrant status, gender and area of residence. The considerable inequities that characterise patterns of health are highlighted. The reasons underlying the inequities are discussed in chapter 13 and a variety of explanations are explored.

Part 4 is written on the assumption that equity is desirable and achievable, and that the quest to achieve it should form one of the basic objectives of social and economic policy. The Commission on Social Determinants of Health established as a feasible goal closing health gaps in a generation and in Australia a similar national goal has been established through the Closing the Gap policy in relation to Indigenous and non-Indigenous Australians.

11

CHANGING HEALTH AND ILLNESS PROFILES IN THE TWENTY-FIRST CENTURY: GLOBAL AND AUSTRALIAN PERSPECTIVES

We reaffirm that health inequities within and between countries are politically, socially and economically unacceptable, as well as unfair and largely avoidable, and that the promotion of health equity is essential to sustainable development and to a better quality of life and well-being for all, which in turn can contribute to peace and security.

Rio Political Declaration on Social Determinants of Health, WHO, 2011

KEY CONCEPTS

Introduction

Data sources

Life expectancy

Social determinants of health

Cause of death

Deaths from violence and injury

Resurgence of infectious diseases

Chronic disease

Disability

Conclusion

INTRODUCTION

This chapter presents an overview of the main causes of death internationally and contrasts the patterns according to the economic development level of countries. It also provides a more detailed picture of the patterns of health and illness in Australia. The picture painted is of a world in which health is unevenly distributed and where more wealth does not necessarily translate to more health.

DATA SOURCES

INTERNATIONAL

WHO produces an annual World Health Report that contains data on each country. The World Bank also has an extensive database on health and other statistics. Although data quality for rich countries is much better than that for poorer countries, even in rich countries there is no generally accepted measure of well-being, and self-perceived measures of health are becoming more accepted as valid. Generally, however, the health status of a population is commonly assessed by mortality rates. Morbidity data are also used but are generally harder to obtain.

AUSTRALIAN

The Australian Bureau of Statistics (ABS) concentrates on demographic and economic data, but also produces a range of health-related statistical information. It conducts National Health Surveys of the population at regular intervals, as well as more specific surveys such as consumption of selected foods, use of private health insurance, child health screening, immunisation status, women's health, lifestyle and health prevalence of various chronic conditions and population norms for the SF36 health and well-being questionnaire.

The Australian Institute of Health and Welfare conducts specific purpose surveys and collates and analyses ABS and health service use data. It publishes a report every two years—*Australia's Health*—detailing these, and listing other sources of health information and statistics.

A Social Health Atlas of Australia (Public Health Information Development Unit, 2014) contains a considerable amount of data related to illness and social status, often in map form. The main purpose of the Atlas is to describe the patterns of distribution of socioeconomic disadvantage and health status at a local level, integrating information on health, education, housing, welfare and other measures of social status.

Data collected by state and territory health authorities include hospital in-patient separation data, notifiable diseases information and (in some cases) cancer and other disease register data. Some states also collect behavioural risk factor data and conduct regular surveillance on a range of health-related matters.

LIFE EXPECTANCY

> There is something desperately wrong with our health care systems and with our societies when, in this amazingly rich world, there is still so much ill health and premature death. What is at least as wrong is that there are such enormous differences between the rich and the poor.
>
> *Mooney, 2012, p. 3*

The difference in health status between the poorest and richest countries is vast. Table 11.1 compares life expectancy at birth, the infant mortality rate and the gross national income for selected countries in three country income categories: low, middle and high. This table shows that among high-income countries, Australia performs well, with the second highest life expectancy after Japan. The table also highlights the middle-income countries that have high life expectancies. China, Cuba, Sri Lanka and Costa Rica stand out in this regard.

TABLE 11.1   INTERNATIONAL COMPARISON OF HEALTH INDICATORS, SELECTED COUNTRIES, 2012 AND 2013

Country	Life expectancy at birth, 2012			Infant mortality rate per 1000 live births, 2013	Gross national income per capita, Atlas method (current US$), 2013
	Total, males and females	Males	Females		
Low income	**62**	**60**	**63**	**53**	**664**
Kenya	61	59	63	48	930
Nepal	68	67	69	32	730
Sierra Leone	45	45	46	107	680
Zimbabwe	58	57	59	55	820
Middle income	**70**	**68**	**72**	**33**	**4,721**
Brazil	74	70	77	12	11,690
China	75	74	77	11	6,560
Costa Rica	80	78	82	8	9,550
Cuba	79	77	81	5	5,890
Indonesia	71	69	73	25	3,580
Jamaica	73	71	76	14	5,220
Mexico	77	75	80	13	9,940
Sri Lanka	74	71	77	8	3,170
Thailand	74	71	78	11	5,370
High income	**79**	**76**	**82**	**5**	**39,312**
Australia	82	80	84	3	65,520
Japan	83	80	86	2	46,140
Norway	81	80	84	2	102,610
USA	79	76	81	6	53,670

Source: World Bank, 2014.

As in all industrialised countries, life expectancy increased dramatically in Australia during the twentieth century (table 11.2). In the 1890s, life expectancy at birth was 54.8 years for women and 51.1 years for men; by 2010–12 the figures were 84.3 and 79.9 respectively.

TABLE 11.2   COMPLETE EXPECTATION OF LIFE IN YEARS[a], AUSTRALIA, 1881–91 TO 2010–12

Birth cohort	Females	Males
1881–91	50.8	47.2
1891–1900	54.8	51.1
1901–10	58.8	55.2
1920–22	63.3	59.2
1932–34	67.1	63.5
1946–48	70.6	66.1
1953–55	72.8	67.1
1960–62	74.2	67.9
1965–67	74.2	67.6
1970–72	74.5	67.8
1975–77	76.6	69.6
1980–82	78.3	71.2
1985–87	79.2	72.7
1990–92	80.4	74.3
1993–95	80.8	75.0
1994–96	81.1	75.2
1995–97	81.3	75.6
1996–98	81.5	75.9
1997–99	81.8	76.2
1998–2000	82.0	76.6
1999–2001	82.4	77.03
2000–02	82.6	77.4
2001–03	82.8	77.8
2002–04	83.0	78.1
2003–05	83.3	78.5
2004–06	83.5	78.7
2005–07	83.7	79.0
2006–08	83.7	79.2

(continued)

TABLE 11.2 COMPLETE EXPECTATION OF LIFE IN YEARS[a], AUSTRALIA, 1881–91 TO 2010–12 (*CONTINUED*)

Birth cohort	Females	Males
2007–09	83.9	79.3
2008–10	84.0	79.5
2009–11	84.2	79.8
2010–12	84.3	79.9

a Average number of additional years a person of a given age and sex might expect to live if the age-specific death rates of the given period continued throughout the lifetime

Source: ABS, 2014b, Life expectancy data cube, Tables 6.2 and 6.6.

The general trend around the world through the twentieth century was for life expectancy to increase. In the 1990s and in the early twenty-first century this trend has been reversed in Africa, where the HIV epidemic has resulted in a reduction in life expectancy in a number of countries. Table 11.3 provides time-series data for nine sub-Saharan African countries and this shows that there has been very little health gain (only Senegal and Uganda had significant improvements over the period) and, for a period during the first impact of the HIV epidemic, life expectancy fell dramatically in many countries. There has been recovery in the last decade but Africa still has less health improvement than the rest of the world.

TABLE 11.3 LIFE EXPECTANCY (LE) AND UNDER-FIVE MORTALITY RATE (U-5 MR), SELECTED AFRICAN COUNTRIES, 1993, 1999, 2004, 2012

	1993		1999a,b		2004b		2012	
	U-5 MR	LE	U-5 MR	LE	U-5 MR	LE	U-5 MR	LE
Botswana	58	61	98	39	116	40	53	62
Chad	209	48	175	49	200	46	150	51
Ghana	170	56	114	55	112	57	72	62
Kenya	74	59	100	48	120	51	73	61
Nigeria	191	53	172	48	197	46	124	54
Senegal	145	49	130	55	137	55	60	64
Sierra Leone	249	43	312	34	283	39	182	46
South Africa	70	63	76	49	67	48	45	59
Uganda	185	43	159	42	138	49	69	57

Notes
a. Rates provided are an average of male and female rates reported and rounded to whole numbers
b. Ranges are provided in World Health Reports 2000, 2004

Source: WHO, 1995b, 2000, 2006, 2014j.

In 2012, life expectancy at birth for both sexes globally was 70 years, ranging from 62 years in low-income countries to 79 years in high-income countries, giving a ratio of 1.3 between the two income groups. Women live longer than men all around the world. The gap in life expectancy between the sexes was 5 years in 1990 and remained the same in 2012. The gap is much larger in high-income countries (more than 6 years) than in low-income countries (around 3 years). Since 1990, life expectancy at birth has increased globally by 6 years, but during the 1990s life expectancies stagnated in Europe, and in Africa they even decreased. For Europe, the phenomenon is due mainly to adverse mortality trends in the former Soviet Union countries. The decrease in life expectancy in Africa has been caused by HIV, but the increasing availability of antiretroviral therapy has reduced the spread of the epidemic, and the mortality due to HIV has been decreasing since about 2005, allowing life expectancy at birth to increase again. Average life expectancy at birth in Africa was 50 years in 2000, whereas it was 58 years in 2012 (WHO, 2015d). In 2012, the global population aged 60 years can expect to live another 20 years on average—2 years longer than in 1990. Life expectancy at age 60 in high-income countries (23 years) is 6 years longer than that in low-income and lower-middle income countries (17 years). Life expectancies at age 60 were longer and the increases larger in high-income countries (WHO, 2015d).

INFANT AND CHILD MORTALITY

Globally, infant mortality rates (the number of deaths in children aged up to 12 months per 1000 live births) range from 53 per 1000 live births in low-income countries to 5 per 1000 live births in high-income countries (table 11.1). Table 11.1 shows that there are considerable differences in infant mortality rates between high-, middle- and low-income countries, and also especially within the low-income grouping. Cuba's rate is on a par with the average for high-income countries at 5 per 1000 live births, while, at the other extreme, Sierra Leone has 107 infant deaths per 1000 live births. The richest

Many children in India, like these children in Mumbai, live on the streets with considerable threats to their health and well-being. (Piers Benatar)

country in the world, the USA, has an infant mortality rate of 6 per 1000 live births, which is three times that of Japan (2 infant deaths per 1000 live births).

Globally, under-five mortality has decreased by 47 per cent, from an estimated rate of 90 deaths per 1000 live births in 1990 to 48 deaths per 1000 live births in 2012. About 17 000 fewer children died every day in 2012 than in 1990, the baseline year for measuring progress.

While progress has been made, it is unequally distributed. At the regional level, under-five mortality rates declined by more than 60 per cent between 1990 and 2012 in three WHO regions: the Americas, Europe and the Western Pacific. However, under-five mortality rates have not decreased in the WHO African Region, so health inequity has increased (WHO, 2015e).

In 2013, the Australian infant mortality rate was 4.1 per 1000 live births (ABS, 2014b) (figure 11.1). In the last 100 years Australia's infant mortality has decreased by 95 per cent. In the past 20 years (since 1990) the rate has declined by 50 per cent, from 8.2 per 1000 live births to 4.1 per 1000 live births. Glover et al. (2006) show evidence of inequities in health by comparing the infant mortality rate in areas of socioeconomic disadvantage in the city of Adelaide, showing them to be higher than in better-off areas, and by showing that rural areas in South Australia have higher rates than the city (5.1 per 1000 live births in rural areas compared with 4.5 per 1000 live births in urban areas). The Northern Territory has the highest infant mortality rate of 7.2 per 1000 live births, and Victoria the lowest at 3.3 per 1000 live births (ABS, 2014b).

FIGURE 11.1   INFANT MORTALITY RATES IN AUSTRALIA PER 1000 LIVE BIRTHS, 1901–2010

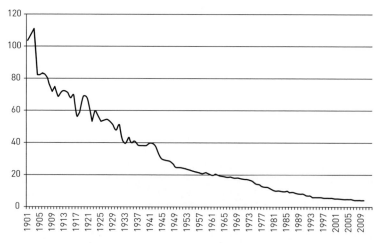

Source: ABS, 2014b, Deaths data cube, Table 5.4.

A lack of high-quality data makes it difficult to provide true national figures for Aboriginal infant mortality. Data from South Australia, Western Australia and the Northern Territory (the only jurisdictions for which reasonably accurate data over time are available) suggest a steady decrease in infant mortality rates from 1972, when the Northern Territory (the area with the highest infant mortality) recorded a figure of 83.4 infant deaths per 1000 live births, or five times the Australian average (ABS, 1999, p. 75).

Over the period 1991 to 2003, infant mortality for Indigenous infants decreased by 44 per cent. In the period 1991–2001, the Aboriginal and Torres Strait Islander infant mortality rate was three times higher than that of other Australian infants and had declined a little to 2.8 times higher in 2002–04. By 2008–12, the Australian Institute of Health and Welfare was reporting that the death rate for Indigenous infants was two times higher than the rate for non-Indigenous infants (6 per 1000 live births compared with 4 per 1000 live births) (AIHW, 2014c).

SOCIAL DETERMINANTS OF HEALTH

A key argument in this book is that health is determined by social and economic factors. Table 11.4 demonstrates that the pattern of some of these determinants follows that of the mortality patterns shown in table 11.1. Thus people in poor countries have lower incomes, are less likely to enrol in secondary schools or use the internet, and have limited access to drinkable water and sanitation.

TABLE 11.4   SELECTED SOCIAL DETERMINANTS OF HEALTH, COMPARISON BY DEVELOPMENT LEVEL OF COUNTRIES

Determinant of health	Heavily indebted poor countries	Low-income countries	Middle-income countries	High-income countries (OECD)
Gross national income per capita (Atlas method, US$) (2013)	816.2	708.8	4751.0	44490.4
Malnutrition prevalence, weight for age (percentage of children under 5), 2013	No data	21.4	15.8	No data
Percentage of secondary school enrolment (2012)	30.3	37.0	66.0	90.7
Percentage of population with access to improved water source (2012)	62.8	68.7	90.3	99.6
Percentage of urban population with access to improved sanitation facilities (2012)	37.8	46.0	75.5	99.9
Internet users per 100 people (2013)	7.2	7.1	32.7	81.2
Access to electricity (percentage of population) (2010)	26.0	31.0	87.0	100.0

Source: World Bank, 2014.

Accessing the internet in public libraries is the only option for some people. (Fernando M. Gonçalves)

CAUSE OF DEATH

There are distinct differences in the patterns of mortality between countries according to their wealth, as shown in table 11.5, which details the 10 leading causes of death by broad country income groups. Coronary heart disease and stroke feature in the top 10 killers in all three income groups (although accounting for a higher proportion in richer countries).

TABLE 11.5   TEN LEADING CAUSES OF DEATH BY COUNTRIES ACCORDING TO BROAD INCOME GROUP, 2012 (DEATHS PER 100,000 POPULATION)

High-income countries		Upper-middle-income countries		Lower-middle-income countries		Low-income countries	
Ischaemic heart disease	158	Stroke	126	Ischaemic heart disease	95	Lower respiratory infections	91
Stroke	95	Ischaemic heart disease	107	Stroke	78	HIV	65
Trachea, bronchus, lung cancers	49	Chronic obstructive pulmonary disease	50	Lower respiratory infections	53	Diarrhoeal diseases	53
Alzheimer's disease and other dementias	42	Trachea, bronchus, lung cancers	31	Chronic obstructive pulmonary disease	52	Stroke	52

High-income countries		Upper-middle-income countries		Lower-middle-income countries		Low-income countries	
Chronic obstructive pulmonary disease	31	Diabetes mellitus	23	Diarrhoeal diseases	37	Ischaemic heart disease	39
Lower respiratory infections	31	Lower respiratory infections	23	Preterm birth complications	28	Malaria	65
Colon, rectum cancers	27	Road injury	21	HIV	23	Preterm birth complications	33
Diabetes mellitus	20	Hypertensive heart disease	20	Diabetes mellitus	22	Tuberculosis	31
Hypertensive heart disease	20	Liver cancer	18	Tuberculosis	21	Birth asphyxia and birth trauma	29
Breast cancer	16	Stomach cancer	17	Cirrhosis of the liver	19	Protein energy malnutrition	27

Source: WHO, 2014i.

Infectious and perinatal conditions are much more evident as causes of death in low-income countries whereas cancers and other chronic diseases are more prevalent in high-income countries. This pattern is further evidenced in figure 11.2, which shows that

FIGURE 11.2 PROJECTED DEATHS BY MAJOR CAUSE AND WORLD BANK INCOME GROUP, ALL AGES, 2030

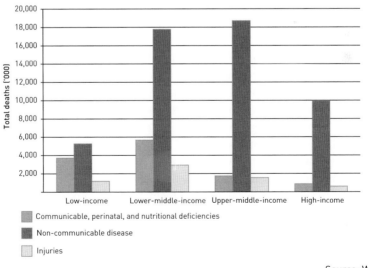

Source: WHO, 2013a.

the most striking differences are that deaths resulting from communicable (infectious and parasitic) diseases are much more common in low-income countries and chronic diseases are more common in middle- and high-income countries.

Communicable diseases are responsible for more than 40 per cent of deaths in low-income countries, but for 1 per cent in richer countries. Although in wealthy countries people are most likely to die of non-communicable diseases, in low-income countries they are most likely to die from infectious diseases. More chronic disease occurs in low- and middle-income countries (80 per cent of total).

Three trends in diseases are of particular note internationally: deaths from injuries and violence, the emergence of new infectious disease from the late twentieth century to the present, and the growth in the prevalence of chronic disease in both rich and poor countries. Each of these will be examined in more detail.

DEATHS FROM VIOLENCE AND INJURY

In 2012 an estimated 1.4 million people worldwide lost their lives to intentional injury (table 11.6). Just over half of these deaths were suicide, nearly one-third were homicides and just over 10 per cent were casualties of armed conflict and other collective violence.

Rates of violent death vary according to country income levels (table 11.6). People in low-income countries are more than three times more likely to die as a result of interpersonal violence, while those in high-income countries are more likely than other income groups to kill themselves. Figure 11.3 shows that the rates for homicides and suicides vary greatly between the regions of the world, with homicides being high in the African and American regions, and suicide being higher in Europe and the Western Pacific region.

ARMED CONFLICTS

War had a dramatic impact on population in the twentieth century. In World War I (1914–18) 26 million were killed and in World War II (1939–45) 53.5 million were killed (Renner, 1999). The Global Burden of Armed Violence Report (Geneva Declaration Secretariat, 2011) reports approximately 55 000 deaths per year from armed conflict or terrorism, and 21 000 from legal interventions. Toole and Waldman (1997) report that approximately 130 armed conflicts occurred worldwide between 1980 and 1996, and that more than five million deaths were caused by civil conflicts between 1975 and 1989. They note that this toll has increased significantly since the end of the Cold War and that in 1993 alone there were 47 active conflicts, of which 43 were internal wars. Increasingly, these conflicts target civilians and result in high casualty rates, widespread human rights abuses, forced migration and, in some cases, total collapse of governance, as in Rwanda in 1994. They lead to numerous public health problems, including population displacement, food shortages, the rapid spread of communicable disease and collapsed

TABLE 11.6 ESTIMATED GLOBAL VIOLENCE-RELATED DEATHS, 2012

	High-income countries		Upper-middle-income countries		Lower-middle-income countries		Low-income countries		Total	
	Deaths	Rate per 100 000	Deaths	Rate per 100 000	Deaths	Rate per 100 000	Deaths	Rate per 100 000	Deaths	Rate per 100 000
Intentional injuries	242 691	18.7	399 556	16.4	587 920	23.4	197 782	23.4	1 427 949	20.2
Self-harm	197 201	15.2	191 576	7.9	332 770	13.3	82 353	9.7	803 900	11.4
Interpersonal violence	43 158	3.3	190 791	7.9	181 289	7.2	89 349	10.6	504 587	7.1
Collective violence and legal intervention	2 332	0.2	17 190	0.7	73 861	2.9	26 080	3.1	119 463	1.7
Population (000s)	1 293 593		2 429 453		2 506 068		846 348		7 075 456	

Source: WHO, 2014c.

FIGURE 11.3   HOMICIDE AND SUICIDE RATES BY WHO REGION, 2012

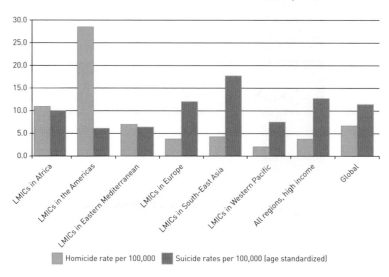

LMICs = low- and middle-income countries

Source: WHO, 2014d, p. 8; WHO, 2014f, p. 17.

basic health services (Toole and Waldman, 1997). Effective public health intervention can make a significant difference to mortality rates following armed conflict.

Since World War II, armed conflicts have taken a much heavier toll on the populations of low-income countries than on industrialised countries. War is continuing to be a feature of life in the twenty-first century, with civil war and attack from a US-led alliance taking a significant toll in Iraq and Afghanistan and the continuing struggle between Israel and Palestine causing the loss of significant numbers of people. Murray et al. (2002) discuss the difficulty of assessing with any accuracy the number of deaths from armed conflict. For instance, estimates of death in Rwanda in the 1994 civil war vary from 500 000 to a million, and in the Bosnia conflict of 1993–95 from 35 000 to 250 000. The number of deaths resulting from the US and allied invasion of Iraq is uncertain but an article in the Lancet estimated the figure to be 600 000 in 2006 (Burnham et al., 2006), while the website www.iraqbodycount.net estimated the figure at around 68 000 in April 2007. Armed conflict also causes significant disability, especially from land mine injuries.

Children and young people comprise a significant proportion of the population in many of the countries that suffer most from armed conflict or political emergencies. Some die as a result of conflict, others are forced into military services, and others have to take on new family responsibilities when members of their families are killed or injured. Children's lives are often severely disrupted by conflict, for example when they are faced with living in refugee camps in very difficult conditions (Zwi et al., 2006).

RESURGENCE OF INFECTIOUS DISEASES

> The Ebola outbreak points to the need for urgent change in three main areas: to rebuild and strengthen national and international emergency preparedness and response, to address the way new medical products are brought to market, and to strengthen the way WHO operates during emergencies … As events since the start of this century have shown, outbreaks rarely have only local or regional consequences in our highly interconnected and interdependent world … Health systems also need adequate numbers of well-trained health care workers and these people need to be appropriately paid. This is one of the biggest lessons the world learned last year. Well-functioning health systems are not a luxury. Well-functioning health systems are the cushion that keeps sudden shocks from reverberating throughout the fabric that holds societies together, ripping them apart. As we learned control depends on community engagement and community leadership at every stage.
>
> *Dr Margaret Chan, Director-General, WHO, 2015*

This quote from Dr Margaret Chan (Director-General, WHO) highlights the importance of infectious diseases to the international public health agenda. Infectious disease outbreaks cause fear, insecurity and even panic. This was clearly seen in the Ebola outbreak in West Africa in 2014–15. Chan reminds us that a strong, effective WHO is vital to support well-functioning national health systems and community engagement. The spread of new and old infectious diseases is described at length by Garrett (1994). Her thesis is that changing social and environmental conditions around the world have fostered the spread of infectious disease—a problem that has been noted by public health authorities, globally and within Australia. The concerns they are raising include the increasing prominence of pathogens resistant to current antimicrobial drugs, an inadequate and in some cases diminishing capacity to prevent or respond to epidemics, and changes in the social and physical environment that have the potential to enhance the spread of communicable diseases.

The resurgence of infectious diseases around the world has been attributed in part to the increased volume of international travel. Lederberg (1996), a Nobel Prize–winning geneticist who specialised in the evolution of microbes, warns that we face a global cauldron with mass movements of people between continents (one million per day cross national borders) creating ideal situations for microbes to evolve and pose a major threat to humans. Lyons et al. (1995, p. 25) argue that internationalisation creates opportunities for communicable diseases to flourish throughout the world: 'The creation of a global village with the virtual free movement of people will create a massive monoculture, a plump mass of unprotected human protoplasm, primed for viral and microbe invasion.'

Global infectious disease trends include increasing mortality in developed countries, including the USA, whose death rate from infectious disease increased by more than 50 per cent between 1980 and 1992 (Pinner et al., 1996). Conditions such as tuberculosis and pneumonia, especially antibiotic-resistant strains, which were predicted to decline in developed countries, are actually increasing in the twenty-first century. New human pathogens such as HIV and Ebola virus have focused lay and professional attention on the issue of infectious diseases. These have been categorised into four groups (Jarlais et al., 1996, p. 130):

- infectious agents that are new as human pathogens (e.g. HIV, Ebola)
- 'old' human pathogens that are spreading beyond their traditional geographic areas
- new strains of 'old' pathogens, including drug-resistant strains (e.g. malaria)
- pathogens re-emerging as public health problems after decades of declining incidence (e.g. tuberculosis).

Aside from HIV, which is covered below, other newly infectious diseases are severe acute respiratory syndrome (SARS) and bird flu—these have so far caused few deaths but have created major concerns about the potential for such viruses to do so. The 2003 SARS epidemic spread to 30 countries and resulted in 8422 cases and 916 deaths (WHO, 2003a). This epidemic illustrated the speed with which infection can be transmitted via air travellers. Its economic cost was estimated to be massive. Chan (2007) says it cost the Asian economies $30 billion in less than 4 months. In 2003 the Wall Street Journal estimated the cost (in terms of lost business as well as direct costs) to be US$10 billion (Saywell et al., 2003). A study by the US National Center for Infectious Diseases in 1999 estimated that the economic impact of an influenza pandemic in the USA would range from US$71.3 billion to US$166.5 billion (Meltzer et al., 1999). The bird flu has caused few deaths—by March 2015 there were 440 deaths from 826 cases worldwide (see www.who.int/influenza/human_animal_interface/H5N1_cumulative_table_archives/en). The Ebola outbreak in West Africa was estimated by WHO to have caused more than 10 000 deaths by April 2015 (WHO, 2015a).

Drug-resistant pneumococcal infections have been recorded in diverse sites across the USA from the early 1990s, and represent a problem for public health workers, clinicians and laboratories (Simberkoff, 1994;). The development and spread of multiple-resistant bacterial pathogens is believed to be linked to the use of antimicrobial agents in veterinary medicine, animal husbandry, agriculture and aquaculture and has come at a time when the development of new antimicrobial drugs has slowed. Some researchers have called for wider surveillance and wiser, more restricted use of these antibiotics to slow the development of resistant strains (Tenover and Hughes, 1996). Tuberculosis (TB) typifies an 'old' infectious disease pathogen re-emerging as a public health concern from the 1990s. In 1993 WHO declared the incidence of TB to be a global emergency and continues to see it as such, and while the majority of cases are in developing countries, increases are happening elsewhere, notably in the USA. The emergence of multi-drug resistant strains and the interaction with HIV are major concerns. WHO and the US Centers for Disease Control and Prevention (CDC) reported that 2 per cent of TB cultures performed at 25 supranational reference laboratories fulfilled the criteria for extensive drug-resistant tuberculosis (XDR-TB) and it was concluded that XDR-TB was present in all regions of the world (UNAIDS, 2006). Africa is particularly vulnerable because of the prevalence of HIV. For example, at the beginning of 2005 XDR-TB was detected in KwaZulu-Natal, South Africa, and this has highlighted the lethal combination of HIV and TB in South Africa, where an estimated 60 per cent of TB patients are also HIV-positive. TB drug resistance arises mainly because of inadequate TB control, failure to follow drug regimens, poor quality drugs or inadequate drug supply. In Australia there was a steady decline in TB during the twentieth century, but especially from 1950, when a national campaign against the disease began. While Australia has maintained a rate of 5 to 6 cases of TB per 100 000 population since the mid-1980s, there has been a steady increase in incidence over the past decade. In 2011, Australia's overseas-born population continued to represent the majority of TB notifications (88 per cent) with a notification rate of 20.2 per 100 000 (Bareja et al., 2014).

Australia may be fortunate in escaping two of the socio-environmental factors believed to be promoting the resurgence in infectious diseases globally. 'Megacities' (more than 10 million population) have increased in number from 20 in 2003 to 28 in 2014 (see figure 15.1). In 2010, for the first time in human history, more than half the world's population was living in urban areas, and WHO has predicted that by 2050 that figure will rise to 70 per cent (WHO, 2010b). Large concentrations of people in relatively small areas enable infection to spread rapidly. The rapid growth of these megacities means that their public health infrastructure, including water supply, sanitation, food safety and vaccination programs, are usually inadequate (Friel et al., 2011). In some developed countries, on the other hand, and particularly in the older industrialised cities of North America, the decay of inner city areas has also left disadvantaged black and Hispanic neighbourhoods with inadequate housing and poor infrastructure services. It has been argued that this 'hollowing-out' or 'doughnut' effect leaves a pool of people at the geographic centre of the city who are most vulnerable to outbreaks of infectious diseases, and ideally located for that infection to spread rapidly to the remaining population (Wallace and Wallace, 1993).

Infectious diseases have not had a significant impact on mortality rates in Australia, and while there are causes for concern, these fade into insignificance compared with the disaster being experienced in sub-Saharan Africa (see the section below). Deaths from AIDS in Australia have remained low. Outbreaks of emerging infectious diseases have included haemolytic uraemic syndrome, equine morbillivirus and food contamination by *Salmonella* enteritidis. The incidence of Ross River virus appears to be increasing, with long-term debilitating symptoms that can persist for up to 12 months. Dengue fever was reported in northern Australia in 1992 for the first time since World War II (Plant, 1995), and outbreaks in far northern Queensland in 1997–98 and in the early 2000s have caused further concern. McMichael (2001, 2005) noted that malaria was present in northern Australia until recent decades and may return as a consequence of climate change. He also foreshadows an increased prevalence of other mosquito-borne diseases, including Murray Valley encephalitis, dengue and Ross River virus, as a likely consequence of increased temperature and rainfall in Australia that could result from global climate changes.

HIV PANDEMIC

At the end of 2012, an estimated 35.3 million people globally were living with HIV (UNAIDS, 2013). Table 11.7 shows the spread of HIV according to different regions of the world. Sub-Saharan Africa accounts for the largest number of existing and new infections. The developed world accounts for only a very small proportion. In many parts of the developing world, the majority of new infections occurred in young adults, with young women especially vulnerable. About a third of those currently living with HIV are aged 15–24. Most of them do not know they carry the virus. AIDS is having a devastating effect in sub-Saharan Africa, where it is now the leading cause of death.

TABLE 11.7 GLOBAL REGIONS HIV AND AIDS STATISTICS, 2012

	Adults and children living with HIV	Adults and children newly infected with HIV	Adult prevalence rate (%)	Adult and child deaths due to AIDS	Number of HIV-positive adults who are women
Sub-Saharan Africa	25.0 million [23.5 million–26.6 million]	1.6 million [1.4 million–1.8 million]	4.7 [4.4–5.0]	1.2 million [1.1 million–1.3 million]	12.9 million [12 million–13.7 million]
North Africa and Middle East	260 000 [200 000–380 000]	32 000 [22 000–47 000]	0.1 [0.1–0.2]	17 000 [12 000–26 000]	100 000 [74 000–150 000]
South and South-East Asia	3.9 million [2.9 million–5.2 million]	270 000 [160 000–440 000]	0.3 [0.2–0.4]	220 000 [150 000–310 000]	1.4 million [1.1 million–1.8 million]
East Asia	880 000 [650 000–1.2 million]	81 000 [34 000–160 000]	0.1 [0.1–0.1]	41 000 [25 000–64 000]	250 000 [180 000–350 000]
Latin America	1.5 million [1.2 million–1.9 million]	86 000 [57 000–150 000]	0.4 [0.3–0.5]	52 000 [35 000–75 000]	430 000 [350 000–550 000]
Caribbean	250 000 [220 000–280 000]	12 000 [9400–14 000]	1.0 [0.9–1.1]	11 000 [9400–14 000]	120 000 [110 000–140 000]
Eastern Europe and central Asia	1.3 million [1.0 million –1.7 million]	130 000 [89 000–190 000]	0.7 [0.6–0.9]	91 000 [66 000–120 000]	430 000 [340 000–550 000]
Western and central Europe	860 000 [800 000 –930 000]	29 000 [25 000–35 000]	0.2 [0.2–0.2]	7600 [6900–8300]	210 000 [190 000–230 000]
North America	1.3 million [980 000 –1.9 million]	48 000 [15 000–100 000]	0.5 [0.4–0.8]	20 000 [16 000–27 000]	260 000 [190 000–380 000]

	Adults and children living with HIV	Adults and children newly infected with HIV	Adult prevalence rate (%)	Adult and child deaths due to AIDS	Number of HIV-positive adults who are women
Oceania	51 000 [43 000–59 000]	2100 [1500–2700]	0.2 [0.2–0.3]	1200 [<1000–1800]	18 000 [15 000–21 000]
TOTAL	35.3 million [32.2 million–38.8 million]	2.3 million [1.9 million–2.7 million]	0.8 [0.7–0.9]	1.6 million [1.4 million–1.9 million]	16.1 million [14.8 million–17.4 million]

Note: The ranges around the estimates define the boundaries within which the actual numbers lie.

Source: UNAIDS, 2013.

Box 11.1 highlights the social and economic impact of the pandemic in Africa. Tackling this pandemic is the biggest public health issue the world is facing in the early twenty-first century. The pandemic itself reflects existing global inequities. The poorest continent is suffering most dramatically. Developed countries have an infrastructure with which to respond to the epidemic. While the accessibility of treatment has increased in Africa in the past few years, weak health service infrastructure means most people living with HIV in Africa receive much less than optimum care. (See box 20.3 for an account of the highly effective Australian response to the epidemic in the 1980s.)

BOX 11.1 SOCIAL AND ECONOMIC IMPACT OF HIV IN SUB-SAHARAN AFRICA

The poverty that is endemic in sub-Saharan Africa has made that country fertile ground for the spread of HIV. People who struggle to survive on extremely low incomes (typically less than US$2 per day) are less likely to be concerned about infection from a virus that kills in a few years' time. In 2012, 1.2 million Africans died of AIDS, down from 2.1 million in 2006 thanks to the widespread use of antiretroviral drugs. Many of these were young adults who should have been entering the most productive time of their lives. Instead, children are left as orphans, and grandparents find themselves with grandchildren to take care of and no adult children to care for them. Health service infrastructure is extremely weak in most of Africa and there is no home-based care available. What care there is comes from already overburdened family members. Studies in Rwanda have shown that households with a person living with HIV spend, on average, 20 times more on health care annually than a household without a person living with HIV. Girls are often removed from school to care for sick family members, despite the importance of girls' education. In rich countries people living with HIV have access to cocktails of antiretroviral drugs, but in Africa the availability is much less. However, the provision of antiretroviral therapy has expanded dramatically in sub-Saharan Africa so that by 2013 an estimated 11 million people were receiving treatment, a ten-fold increase since December 2003 (see www.who.int/hiv/data/art_2003_2015.png?ua=1).

The loss of productive adults has had a serious effect on African economies and economic development because so many in the population are sick. For example, a survey of 15 firms in Ethiopia has shown that, over a 5-year period, 53 per cent of all illnesses among staff were AIDS-related. Millions of farm workers have also died from AIDS-related causes, which has affected food security.

The poor suffer most from HIV and AIDS. In Botswana adult HIV prevalence is more than 35 per cent, and one-quarter of households can expect to lose an income-earner within the next 10 years. Per capita household income for the poorest quarter of households is expected to fall by 13 per cent, while every income earner in this category can expect to take on four more dependants as a result of HIV.

The epidemic is also affecting the teaching, health care and other professional groups. In Zambia, teacher deaths caused by AIDS are equivalent to about half of the total number of new teachers trained in the country each year. Children orphaned by AIDS are much more likely to drop out of school.

Despite the depth and reach of the epidemic, in 2015 the picture is looking less gloomy. UNAIDS states that with sufficient scaling up of responses the epidemic could be beaten by 2030. The benefits of doing so are estimated as 28 million HIV infections averted between 2015 and 2030, and 21 million AIDS-related deaths averted between 2015 and 2030. The economic return on fast-tracked investment is expected to be 15 times, averting US$24 billion of additional costs for HIV treatment.

Source: UNAIDS, 2006, 2014.

CHRONIC DISEASE

Total deaths from non-communicable diseases are projected to rise to 52 million in 2030. The rapidly growing magnitude of such diseases is driven in part by population ageing, the negative impact of urbanization and the globalization of trade and marketing.

United Nations Secretary-General Ban Ki-moon, 2011

Estimates suggest that by 2030, 51.7 million people will die of chronic disease worldwide (see figure 11.2) and the majority will live in low- and middle-income countries. Chronic diseases are the leading cause of illness, disability and death in Australia, accounting for 90 per cent of all deaths in 2011 (AIHW, 2014a). Chronic diseases such as cardiovascular disease, cancer, diabetes, asthma, arthritis and musculoskeletal conditions are all areas of national health priority in Australia. This is typical of rich countries where the significance of the chronic disease burden has been recognised for some decades and has led to chronic diseases being viewed as diseases of affluence, which has also created the inaccurate view that chronic disease is not such a problem for low-income countries. Yet the projected deaths by 2030 from chronic disease suggest some 80 per cent will occur in middle- and low-income countries—this is much higher than the projected deaths from infectious diseases (including HIV, tuberculosis and malaria), maternal and perinatal conditions, and nutritional deficiencies combined.

Chronic disease incurs many costs for society. Traditionally, much of the work associated with the care and management of chronic conditions has fallen to the family, with wives, mothers and daughters usually prominent as carers. Chronic illness frequently has an impact on a person's quality of life and may mean they are unable to work, so lose income and fall into poverty. In many low- and middle-income countries the cost of treatment for chronic disease means patients bearing out-of-pocket payments and so contributing to family poverty.

In Australia and other rich countries, social and economic changes have seen more women in the workforce and seen adult children move long distances from their parents to find employment, making family care less accessible. Professional and institutionally driven illness care services tend to be fragmented, but chronic care often requires close coordination from a wide range of services. Improving the care of older people and people with chronic disease is a top policy priority in all countries.

Prevention of chronic disease is also vital. Strong et al. (2005) estimate that many deaths from chronic disease could be averted. They point to the successes in reducing death rates from heart disease (up to 70 per cent) in the past three decades in Australia, Canada, Japan, the UK and the USA. However, Wilcox (2014) has noted that Australia invests less in prevention than other comparable countries. OECD data suggest that Australia is at the 'low end' in spending on prevention and public health services relative to many comparable countries. In 2011 the OECD reported that Australian spending on prevention and public health as a share of total recurrent health spending was 2.0 per cent, lower than New Zealand (6.4 per cent), Finland (6.1 per cent), Canada (5.9 per cent), Sweden (3.9 per cent), the United States (3.1 per cent) and Japan (2.9 per cent). While the immediate risk factors for chronic diseases are associated with lifestyle (tobacco use, unhealthy diets and physical inactivity), these lifestyle choices reflect a range of broader environmental, social and economic pressures. The United Nations summit on the prevention of non-communicable diseases, held in 2011, noted the importance of cross-sectoral action to reduce non-communicable disease and the impact of patterns of trade on fuelling the epidemic (United Nations, 2012). The importance of developing prevention strategies that address these broader factors is the focus of parts 6 and 7 of this book.

DISABILITY

The first ever World Report on Disability (WHO and World Bank, 2011) suggests that more than a billion people in the world experience disability, and 110–190 million of these experience very significant difficulties. This corresponds to about 15 per cent of the world's population and is higher than previous WHO estimates, which date from the 1970s and suggested a figure of around 10 per cent. The prevalence of disability is growing due to population ageing and the global increase in chronic health conditions. People with disabilities generally have poorer health, lower education achievements, fewer economic opportunities and higher rates of poverty than people without disabilities.

This is largely due to the lack of services available to them and the many obstacles they face in their everyday lives.

Just under one in five Australians (4.2 million people) reported having a disability in 2012. Of these, 1.4 million people needed help with basic daily activities of self-care, mobility and communication (ABS, 2013c). In 2012, one in five older Australians (20 per cent, or nearly 663 000 people) had severe or profound core activity limitation, meaning that they sometimes or always needed assistance with at least one core activity (self-care, mobility or communication), with higher rates among women than men (ABS, 2013c).

This is a much higher prevalence than the world rate, and is almost certainly accounted for by the fact that many more forms of disability are counted in Australia than in many other countries. There is some evidence that as life expectancy increases so does disability. AIHW (2013b) analysis indicates that, between 1998 and 2012, between 37 per cent (females) and 54 per cent (males) of the gains in life expectancy were years free of disability, and between 76 per cent (males) and 89 per cent (females) of the gains were years without severe or profound core activity limitation. Overall, this leaves an increasing rate of disability.

Disabilities include intellectual or physical and include a wide range of conditions that result in different impacts on daily living. Typical effects are difficulties with transport, meal preparation, housework, health care, communication and self care. Overwhelmingly care and support is provided by family members (for Australia see AIHW, 2006, p. 49). This can cause a huge burden for the carers, for instance the parents of children with disability or the spouse of an older person. Few countries provide adequate social support and in most countries of the world families have to bear the full burden of caring for relatives with disability. Werner (1997) provides an empowering resource guide for people with disabilities.

In rich countries the last 20 years have seen some major changes in policy approaches to disability. Advocacy groups have asserted that disability is not an illness and should not be treated as such. Medical dominance of the disability sphere has diminished accordingly, and there has been a movement away from institutional care to home- and community-based living. In Australia this move is being greatly supported by the introduction of a Disability Insurance Scheme from 2014. As populations age there will be a greater need for home-based care to support people with disabilities and to support their carers.

CONCLUSION

Data presented in this chapter demonstrate that health is an extremely unequally divided resource globally. Rich countries like Australia have longer life expectancy and less disease burden and many more resources to devote to the lower disease burden. In Africa, life expectancy is declining in a significant number of countries. Emerging infectious diseases, injuries and deaths from violence, and the growing chronic disease burden will be crucial public health issues for the twenty-first century. The challenge for the global community is to ensure that health is evenly divided so that the gaps in life expectancy between rich countries and others are reduced.

CRITICAL REFLECTION QUESTIONS

11.1 How have the patterns of disease changed globally and in your country in the past 30 years?

11.2 Choose one cause of death that is reported as being on the rise in this chapter—how would you prevent it?

Recommended reading

The WHO annual World Health Reports contain comprehensive data on life expectancy, mortality rates and other health statistics.

AIHW's *Australia's health* is published every 2 years with detailed statistics on Australia's health status and some data on health inequities.

Useful websites

WHO Statistics www.who.int/gho/publications/world_health_statistics/en

World Bank database http://databank.worldbank.org/data/home.aspx

Australian Institute of Health and Welfare www.aihw.gov.au

These websites contain a wealth of useful information on health status.

12

PATTERNS OF HEALTH INEQUITIES IN AUSTRALIA

Inequities in health ... arise because of the circumstances in which people grow, live, work and age, and the systems put in place to deal with illness. The conditions in which people live and die are, in turn, shaped by political, social and economic forces.

Commission on Social Determinants of Health, 2008, Introduction

KEY CONCEPTS

Introduction
Key factors in health inequalities in Australia
Effects of socioeconomic status
Poverty, socioeconomic status and health
Socioeconomic status
Increasing inequities
Unemployment and health
Occupational illness and injury
Indigenous peoples
Refugees, migrants and health
Gender and health
Suicide
Gender and morbidity
Location and health
Rural and remote Australia
Conclusion

INTRODUCTION

Health is a product of people's everyday experience, and is therefore unequal. This chapter uses health inequities within Australia as a means to explore the patterns. The following chapter considers the more contentious issue of why these differences exist.

There are many ways of looking at patterns and trends in health and illness in Australia. Historically there has been a paucity of data to explain why health and illness are socially stratified, but this is changing and there are a growing number of papers and reports that provide a picture of the social distribution of health and illness. From this it is clear that doing well in terms of employment, income, education and the associated material resources is good for health. This chapter presents much of these data according to:

- social and economic factors
- employment
- ethnicity
- gender
- age
- location.

KEY FACTORS IN HEALTH INEQUALITIES IN AUSTRALIA

Box 12.1 provides an overview of the key health inequalities in Australia, showing that while Australia has a very high overall life expectancy this is not shared equally across the population. Each of these areas of inequality are expanded on below.

BOX 12.1 KEY HEALTH INEQUALITIES, AUSTRALIA, 2009–11

Gender	Males have a mortality rate that is 1.5 times higher than the rate for females. There would have been 71 400 fewer male deaths over the 2009–11 period if males had the same mortality rate as females.
Location	People living in remote and very remote areas had mortality rates that were 1.4 times higher than the rates for people living in major cities, and also higher rates of death due to diabetes and land transport accidents.
Socioeconomic status	People living in the lowest socioeconomic status areas had a mortality rate that was 1.3 times higher than the rate among people living in the highest socioeconomic status areas, and higher rates of death due to diabetes and chronic obstructive pulmonary disease.
Country of birth	Overseas-born Australian residents on average had lower mortality rates than Australian-born residents. Asian-born Australian residents had a mortality rate that was 36 per cent lower than the rate for Australian-born residents.

Source: AIHW, 2014d, p. 1.

EFFECTS OF SOCIOECONOMIC STATUS

QUALITY OF EVIDENCE

Data collections in Australia do not usually collect measures of socioeconomic status for individuals that can be linked to health and illness data. One of the most useful resources for examining the social patterning of health and illness is *A Social Health*

Atlas of Australia (Public Health Information Development Unit, 2014). This publication provides data for a wide range of variables, many of them mapped to show geographical distribution. As well as the main volume for Australia as a whole, similar volumes for each of the states and territories are available. The Australian Institute of Health and Welfare publishes the *Australia's Health* report every two years, which contains useful information on health inequities.

In Australia the evidence relating to socioeconomic status and health has to be sought out and fitted together, but it does not form a cohesive pattern as the data have been collected for different purposes using varying assumptions. Turrell et al. (1994) provided a useful guide to the problems of measuring social class in health research, identifying occupation, education, income and geographic area. While acknowledging that each of these measures has methodological and theoretical problems, they agree that the size and consistency of direction of the relationship of health status to class (however measured) do indicate a real relationship. However, they caution:

- it is not entirely clear what is actually being measured. The measures are almost certainly multi- rather than unidimensional, yet the complexities of links between different measures of class are only partially understood
- in Australia, most studies of the relationship between health and social class are based on secondary analysis from data sets collected by agencies such as the Australian Bureau of Statistics or the National Heart Foundation. The nature and extent of the social data thus collected are often not sufficiently detailed or sensitive to provide thorough measures of social class.

Other commentators have pointed out that static demographic variables such as income, occupation and education do not adequately capture the psychosocial dynamics that may be vital to social class experience, and hence to health behaviour and health status (Sen, 1992; Wilkinson and Pickett, 2009). Most epidemiological work on class and health does not discuss theory, despite there being much complex sociological literature on class. However, it has been noted that 'arguing about class is like going for a swim in a country dam. As soon as you put your foot in, you are up to your neck in mud. There is a terrifying confusion of terms and ideas [in which] unwary sociologists have sunk without trace' (Connell and Irving, 1991, p. 81).

There is increasing literature that attempts to understand the complex relationships between variables relating to class, race, ethnicity, gender and education, while recognising the difficulties of health measurement (see chapter 1) and phenomena associated with class, such as poverty.

These difficulties of measuring and understanding social class should not lead to the conclusion that such measures are futile. The social patterning of health and illness is fundamental to the new public health and the challenge is to explain the patterns while refining and improving existing measures. This chapter concentrates on explaining the social patterning of health and the following chapter on explaining why they exist.

POVERTY, SOCIOECONOMIC STATUS AND HEALTH

Poverty has a strong influence on health as shown elsewhere in this chapter. Australian studies of poverty are closely aligned to the British tradition, dating from the great poverty surveys of the turn of the century, in which researchers Booth and Rowntree set out to collect the 'facts' on poverty, using income as the main measure (Travers and Richardson, 1993). Rowntree's work sought to establish a measure of absolute poverty. The notion of the undeserving poor was widespread at the time, and the concept of a technical poverty line below which even the most efficient household managers could not be adequately fed and housed was appealing. Rising living standards in the twentieth century meant that absolute poverty proved to be a less useful concept in developed countries, and the notion of relative poverty was introduced. Townsend's (1979, p. 31) work on poverty demonstrates this approach well: 'Individuals, families and groups in the population can be said to be in poverty when they lack the resources to obtain the types of diet, participate in the activities and have the living conditions and amenities which are customary, or are at least widely encouraged or approved in the society to which they belong.' In Australia the most commonly used measure of poverty is the Henderson Poverty Line, which was developed in the early 1970s by Ronald Henderson while undertaking the Australian Government Inquiry into Poverty. It estimates how much money individuals and families of different sizes need to cover essential living costs. It represents a very basic living standard. Charities advocating for the poor such as the Brotherhood of St Laurence and the Smith Family, or the peak body Australian Council of Social Service, have commissioned or reported research on poverty from university researchers and this has resulted in new ways of measuring poverty. The Brotherhood of St Laurence (2003) provides a definition that considers the social context that poor people live in to be both:

- the lack of access to an adequate material standard of living (in terms of food, shelter, clothing and health) resulting primarily, but not only, from inadequate income
- the lack of opportunity to participate fully in society (for example through employment, education, recreation and social relationships).

ACOSS (2014) uses 50 per cent of median household income (half of the 'middle' income for all households) to identify people in poverty. This poverty line, also used by the OECD, equates to a very austere living standard: a disposable income of less than $400 per week for a single adult (higher for larger households to take account of their greater costs). Figure 12.1 shows an estimate of the percentage living below the poverty line for selected groups. Overall 13.9 per cent of Australians live below the poverty line. The proportion of people in poverty was higher than in 2010, an increase of 0.9 per cent, from 13 per cent in 2010. Poverty is about one-third higher in Australia than the OECD average level (11 per cent).

FIGURE 12.1   PERCENTAGE OF SELECTED GROUPS BELOW THE POVERTY LINE
(EARNING LESS THAN 50 PER CENT OF THE MEDIAN INCOME)

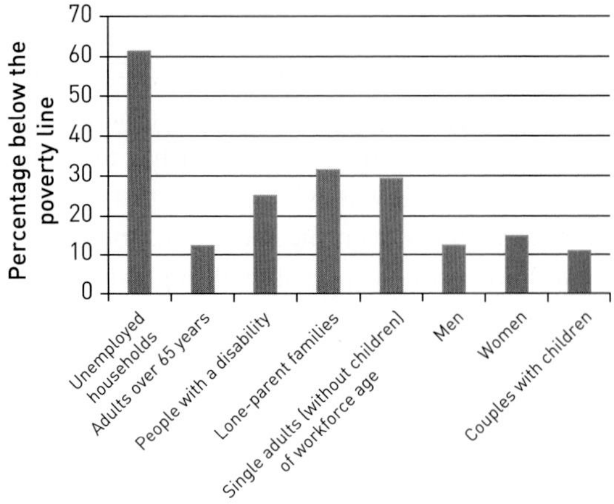

Source: Australian Council of Social Service, 2014, pp. 13–14, Table 4.

SOCIOECONOMIC STATUS

A large body of Australian and international literature (see reviews in Glover et al., 1999; Turrell et al., 1999; Draper et al., 2004; Mackenbach, 2005; Crombie et al., 2005; Hofrichter, 2003; Marmot et al., 2010) shows that there is a consistent trend for people in more disadvantaged circumstances to suffer worse health and die earlier than those in better circumstances. Some selective data from Australia are provided below. Differences are not simply between the worst off in society and the rest, but operate as a gradient across all age groups, gender and countries where data exist, no matter how socioeconomic disadvantage is measured (Wilkinson and Marmot, 2003; CSDH, 2008; Marmot et al., 2010).

In Australia people who are poorer or socioeconomically disadvantaged in other ways generally live shorter lives and suffer more illness and reduced quality of life than those who are well-off.

Mortality rates were 1.3 times higher for the lowest socioeconomic status areas than the highest socioeconomic status areas for males, 1.2 times higher for females and 1.3 times higher for males and females combined (AIHW, 2014d).

Figure 12.2 compares age-standardised death rates in different socioeconomic groups by age group. Relative inequality by socioeconomic status was greatest among people aged 25–44, with 2.1 times as many deaths in the lowest socioeconomic status areas compared to the highest. As age increased, relative inequalities by socioeconomic status were less pronounced, and for people aged 85 and older, there was no difference in the overall mortality rates between the lowest and highest SES areas, showing that death is a great equaliser!

FIGURE 12.2 COMPARISON OF AGE-STANDARDISED MORTALITY RATES ACROSS SOCIOECONOMIC GROUPS, BY AGE GROUP, 2009–11

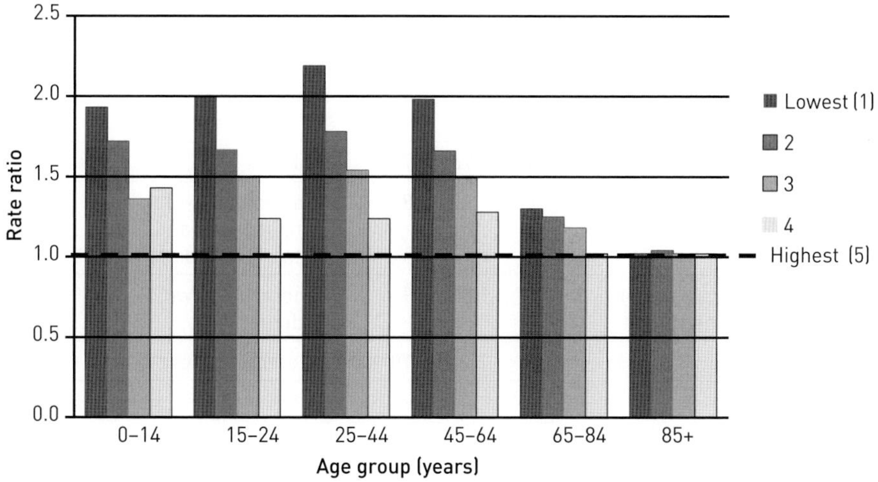

Note: Rate ratio is the age-standardised rate for each socioeconomic status group divided by the age-standardised rate for the highest socioeconomic status group.

Source: AIHW, 2014d, p. 11, Figure 3.4.

Box 12.2 shows the considerable difference in death rates between the most and the least socioeconomically disadvantaged groups for the top ten causes of death.

BOX 12.2   LEADING CAUSES OF DEATH AND SOCIOECONOMIC DISADVANTAGE IN AUSTRALIA, 2009–11

The death rate for Australian males and females living in areas of most disadvantage compared to those living in areas of least disadvantage is shown in the following table. Coronary heart disease was the leading cause of death among all socioeconomic groups, and rates increased with greater disadvantage. The largest relative inequalities in rates for both genders were for diabetes (1.8 and 1.9 times higher for males and females, respectively, in the most disadvantaged areas).

	Males			Females	
Rank	Cause of death	Rate ratio	Rank	Cause of death	Rate ratio
1	Coronary heart diseases	1.41	1	Coronary heart diseases	1.29
2	Lung cancer	1.67	2	Cerebrovascular diseases	1.06
3	Cerebrovascular diseases	1.14	3	Dementia and Alzheimer's disease	0.90

	Males			Females	
4	Chronic obstructive pulmonary disease	1.78	4	Lung cancer	1.48
5	Prostate cancer	1.09	5	Chronic obstructive pulmonary disease	1.65
6	Dementia and Alzheimer's disease	0.96	6	Breast cancer	1.00
7	Diabetes	1.76	7	Diabetes	1.90
8	Colorectal cancer	1.22	8	Heart failure and complications and ill-defined heart disease	1.33
9	Cancer, unknown, ill-defined	1.53	9	Cancer, unknown, ill-defined	1.43
10	Suicide	1.56	10	Colorectal cancer	1.13

Source: AIHW, 2014d.

The most recent Australian Health Survey (2011–12; ABS, 2013a) shows that 23 per cent of those in the most disadvantaged areas smoke compared to 9.9 per cent of those in the least disadvantaged areas. Both factors are risks for major causes of death including cardiovascular and respiratory diseases and cancers. The survey data also showed that, compared with women living in areas of least disadvantage (47.7 per cent), more women living in areas of most disadvantage were overweight or obese (63.8 per cent). For men, overweight and obesity rates were similar for those living in areas of most (69.0 per cent) and least (68.6 per cent) disadvantage (ABS, 2013a).

Indigenous Australians are far more likely than non-Indigenous Australians to be living in a position of socioeconomic disadvantage (Osborne et al., 2013). In 2012–13 the unemployment rate for Indigenous Australians was 20.9 per cent compared to 4.2 per cent in 2011–12 for the non-Indigenous population (Productivity Commission, 2014). The median real gross personal income for Indigenous Australians in 2011–13 was $430 per week, compared with $787 for non-Indigenous Australians (Productivity Commission, 2014). The impact of these data on the lives of Indigenous people is reflected in their poorer health status.

INCREASING INEQUITIES

Economic inequalities are increasing around the world (Piketty, 2014; Oxfam, 2014b, 2015) and these appear to be translating to increased health inequities (Whitehead and Dahlgren, 2006; Wilkinson and Pickett, 2009). Australian data illustrate this. Since the 1970s the trend in Australia has been for the wealth of Australian families to be increasingly concentrated at the upper end of the income distribution (Baekgaard, 1998; Schneider, 2004; Richardson and Denniss, 2014). Richardson and Denniss reported in 2014 those in the bottom quintile (1.73 million households) have less wealth than the ten wealthiest families in Australia as reported by the *Business Review Weekly*'s Rich 200 list. There has been a sharp fall in the real value of wage and salary income for households in the bottom half of the income distribution, and senior executive pay is now 150 times greater than average weekly earnings (Richardson and Denniss, 2014). There has been growing acceptance that inequities are increasing despite some earlier debate (see, for example, Harding, 2005) and ACOSS (2014) is campaigning to reduce this trend.

Australia is not unique in the increasing inequity. Figure 12.3 shows how income inequity has increased in Anglo-Saxon countries, and Piketty (2014) shows this is greater than in European countries. He also shows that wealth concentration is increasing with the richest 1 per cent gaining a greater proportion than in the 1970s. The Nobel prize-winning economist Stiglitz (2012) has documented how the USA has become the most unequal advanced industrial country and the ways in which the recovery from the Global Financial Crisis has only benefited the top 1 per cent. For many citizens, standards of living have been eroded as gross domestic product has grown. Stiglitz suggests that the current levels of inequity are unsustainable in terms of creating a cohesive and well-functioning society.

FIGURE 12.3 INCOME INEQUALITY IN ANGLO-SAXON COUNTRIES, 1910–2010

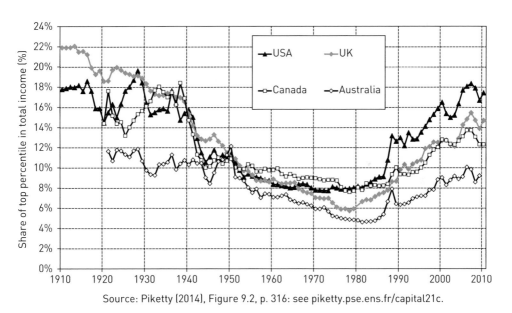

Source: Piketty (2014), Figure 9.2, p. 316: see piketty.pse.ens.fr/capital21c.

Affluent shoppers in Mexico City step over a homeless man. Inequities are increasing in most cities around the world. (Mark Henley, Panos)

UNEMPLOYMENT AND HEALTH

In the 2011–12 Australian Health Survey (ABS, 2012a) few differences were reported between people who were employed and people who were unemployed in terms of long-term conditions, except for 'mental and behavioural problems' where 12.4 per cent of employed people and 26.3 per cent of unemployed people reported having mental and behavioural problems. This represents an increase from the 2004–05 survey, where the figures were 8.2 per cent and 17.3 per cent, respectively. Unemployed people also reported more asthma (14.4 per cent compared with 9.6 per cent of employed people). Unemployed people were also more likely to be current smokers (31.1 per cent compared with 16.8 per cent of employed people) but were less likely to report a risky or high alcohol intake, or be overweight or obese (53.3 per cent compared with 63.1 per cent of employed people). There are no good national data on death rates and unemployment.

OCCUPATIONAL ILLNESS AND INJURY

According to the most recent figures available, mortality rates from all causes of death combined among males working in both manual and non-manual occupations declined markedly during the period 1966 to 2001 (de Looper and Magnus, 2005). In 1998–2000, male blue-collar workers recorded an all-cause mortality rate of 234 deaths per 100 000 people, whereas males employed in managerial, administrative and professional occupations recorded a rate of 115 deaths per 100 000 people—a difference of 104 per cent (Draper et al., 2004). For males in manual occupations, overall mortality declined from 450 deaths per 100 000 population in 1966, to 250 deaths per 100 000 population in 2001, a decline of 44 per cent. For males in non-manual occupations, mortality declined from 390 deaths per 100 000 population in 1966 to 160 deaths per 100 000 in 2001, a decline of 59 per cent (de Looper and Magnus, 2005, p. 5). Consequently, inequality between occupational groups appears to have widened in this period.

Safe Work Australia has the responsibility of providing data on the distribution of occupational illness and injury over occupational and industrial groupings. In 2013 there were 191 worker fatalities in Australia, and 92 per cent of the fatalities involved male workers (Safe Work Australia, 2014). The highest work-related fatality rates were in the agriculture, forestry and fishing industry (15.11 deaths per 100 000) followed by transport, postal and warehousing (7.76 deaths per 100 000); electricity, gas and water (3.35 deaths per 100 000); arts and recreation services (3.34 deaths per 100 000); mining (2.98 deaths per 100 000); and construction (1.85 deaths per 100 000). Each of these industries primarily employs men.

Occupational illness and injury is obviously a crucial public health issue because of the pain and suffering it causes. It also has a significant economic cost for employers, workers and the community as a whole. It has been estimated that the cost of this burden is $60.6 billion (4.8 per cent of gross domestic product) for the 2008–09 financial year (Safe Work Australia, 2012b, p. 3) but this does not include the cost of pain, suffering and early death to injured or ill workers, which was measured in previous reports but has been discontinued. There is a national plan to reduce occupational disease and injury—the Australian Work Health and Safety Strategy. An overview of the strategy is provided in Figure 12.4.

FIGURE 12.4   AUSTRALIAN WORK HEALTH AND SAFETY STRATEGY 2012–2022

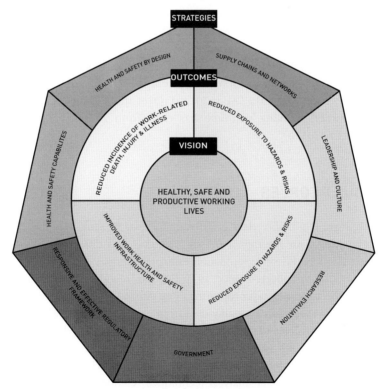

Source: Adapted from Safe Work Australia, 2012a, p. 3.

Occupational health is a neglected area of public health enquiry. There is considerable scope to extend documentation and understanding about the impact of work on health. This is particularly important at a time when the nature of work is changing in ways that are likely to be detrimental to health. There has been a fall in the number of permanent full-time jobs, and a rise in casual, contract and part-time jobs (Forster, 2000). People are much more likely than they were in the past to change workplaces, and so gain less social support from work. Australian households have become increasingly divided between the 'work rich', where two people are in paid employment, and the 'work poor', where there are no people in paid employment, or they have insufficient employment. Concentrated disadvantage is common, where the 'inherent disadvantage of having a low income is compounded by living in poor areas, which tends to produce a self-perpetuating cycle of unemployment, social marginalisation and stigmatisation' (Forster, 2000).

The pressure to work longer hours appears to be increasing in most countries. In Australia workers appear to be working longer and less standard hours than a decade ago (Edgar, 2005). The pressures of work and its impact on home and community life are increasingly the subject of sociological analysis (see, for example, Edgar, 2005; Pocock et al., 2012). Between 1982 and 1994, average hours worked by full-time workers increased from 42 to 45 hours per week (ABS, 2003, p. 119). Since 2000, average hours worked by full-time workers have dropped back to around 39 hours per week in 2012 (ABS, 2013b). In July 2010, full-time men worked 41.0 hours compared with 35.8 hours for full-time women (ABS, 2010a). Between 1982 and 2000, the number of full-time workers working more than 49 hours per week rose from 20.4 per cent to 25.5 per cent (ABS, 2001), then fell back to 20.7 per cent in 2012 (ABS, 2013b). This reduces the time available for family and community work, a situation aggravated by the fact that households have to rely on two external incomes in order to survive. The proportion of women in the workforce has risen steadily from 29 per cent in 1966, 52 per cent in 1993, 54.5 per cent in 2000 (ABS, 2001) and 58.9 per cent in 2012 (ABS, 2013b). Over the same period, the proportion of employed women working part-time has risen from 28 per cent (ABS, 2001) to 45.7 per cent (ABS, 2013b).

INDIGENOUS PEOPLES

Analysing Aboriginal mortality data is difficult for two reasons (Hunter, 1993, p. 76). First, the definition of an Aboriginal person changes, and different ways of collecting data have led to widely different population estimates. Census data relating to Aboriginal people have only been collected since 1971. Second, records of Aboriginal deaths are subject to error because of under-reporting. The Australian Health Ministers' Advisory Council (2006) report that only mortality data from New South Wales, Northern Territory, Western Australia, Queensland and South Australia are regarded as reliable. This is because these jurisdictions validate data against other sources. Data estimations are improving due to the efforts of the Australian Institute of Health and Welfare, which has invested in improving data analysis techniques relating to Indigenous health data (see AIHW, 2014c for details of methods). A very useful website is available to gain access to data on Indigenous health status: HealthInfoNet (www.healthinfonet.ecu.edu.au).

In 2011, Aboriginal and Torres Strait Islander peoples made up 3 per cent of the Australian population and had a much younger age structure than the rest of the population (ABS, 2014e). Aboriginal and Torres Strait Islander people are also more likely to live outside major cities and have a higher population growth rate than the rest of the population. In 2010–12, the estimated life expectancy at birth for Aboriginal and Torres Strait Islander males was 69.1 years, and 73.7 years for females. This was 10.6 and 9.5 years lower than the life expectancy of non-Indigenous males and females, respectively (AIHW, 2014c). After adjusting for differences in age structure, Indigenous death rates were 1.6 times higher than non-Indigenous death rates (AIHW, 2014c). Four groups of chronic conditions account for about two-thirds of the gap in mortality between Indigenous and non-Indigenous Australians: circulatory diseases (24 per cent of the gap); endocrine, metabolic and nutritional disorders (21 per cent); cancer (12 per cent); and respiratory diseases (12 per cent). Mortality rates for Indigenous Australians declined by 9 per cent between 2001 and 2012 (in the five jurisdictions with adequate data over this period). While there were improvements in mortality from cancer in the non-Indigenous population between 2001 and 2012, this did not occur in the Indigenous population, leading to a significant increase in the mortality gap due to cancer for both males and females.

The Indigenous death rate for infants between 2008 and 2012 (6 per 1000 live births) was higher than the rate for non-Indigenous infants (4 per 1000 live births) (AIHW, 2014c). Table 12.1 summarises data on morbidity relating to diseases that disproportionately affect Indigenous Australians. This table highlights the heavier burden of disease in Aboriginal and Torres Strait Islander peoples compared with non-Indigenous Australians and explains the lower life expectancy.

TABLE 12.1 DISEASES THAT DISPROPORTIONATELY AFFECT INDIGENOUS PEOPLES IN AUSTRALIA

Disease/problem	Relevant research findings
Diabetes mellitus	In 2012–13, 3 times as many Aboriginal and Torres Strait Islander people than other Australians reported having diabetes, and 3–5 times as many in all age groups from 25 years and over (ABS, 2014a). Over the period 2006–10, 6.7 times as many Indigenous Australians died from diabetes than non-Indigenous Australians (AIHW, 2013a).
Circulatory system	Circulatory diseases are the largest single cause of death for both males and females, accounting for 26% of all Indigenous deaths in 2006–10 (AIHW, 2013a). Over the period 2006–10, Indigenous Australians died from ischaemic heart disease and cerebrovascular disease at 1.8 and 1.5 times (respectively) the rate of non-Indigenous Australians (AIHW, 2013a).

(continued)

TABLE 12.1   DISEASES THAT DISPROPORTIONATELY AFFECT INDIGENOUS PEOPLES IN AUSTRALIA (*CONTINUED*)

Disease/problem	Relevant research findings
Respiratory disorders	Most marked differences between Indigenous and non-Indigenous Australians occur with infective respiratory diseases, but levels of chronic respiratory disease are also higher for Aboriginal and Torres Strait Islander people, as measured by mortality and hospital separation rates. Over the period 2008–10, Indigenous Australians were hospitalised for respiratory diseases at 2.7 times the rate of non-Indigenous Australians (AIHW, 2013a). Respiratory illnesses are the fourth largest cause of Aboriginal hospital admissions (AIHW, 2013a). In part this can be attributed to higher levels of tobacco smoking.
Ear disease	In the National Aboriginal and Torres Strait Islander Health Survey in 2012–13, 7% of Indigenous children aged 0–14 years were reported as having ear or hearing problems compared to 4% of non-Indigenous Australian children (ABS, 2014a). Studies of active chronic otitis media in Indigenous children show the reported prevalence ranges from 10.5% to 30.3% (Jervis-Bardy et al., 2014, p. S25).
Disability	In non-remote areas in 2008, Indigenous people aged 15 years or over had higher rates of disability than non-Indigenous Australians (rate ratios of between 1.2 and 1.6) (AIHW, 2013a). Indigenous people were also 2.2 times more likely to have a profound or severe core activity limitation (AIHW, 2013a).
Eye disorders	Blindness rates among Indigenous people are 6 times those in the general Australian population (AIHW, 2013a). The 2008 National Indigenous Eye Health Survey showed the leading causes of blindness for Indigenous people were cataract and refractive error.
Dental health	The mean number of decayed, missing and filled teeth per child (aged 5–10 years) was much higher for Aboriginal and Torres Strait Islander children in New South Wales, South Australia, Tasmania and the Northern Territory (AIHW, 2013a). The figure varied with age but was more than twice as high in many age groups. Indigenous children also had higher levels of untreated decay than non-Indigenous Australian children (AIHW, 2013a).
Specific communicable diseases	There is an exceptionally high incidence of invasive diseases caused by *Haemophilus influenzae* type b (Hib) in Aboriginal children. In 2004–05, Indigenous children were 7.5 times more likely to be notified with invasive Hib than non-Indigenous Australians (Wang et al., 2008). In 2010, the overall rate of tuberculosis in the Indigenous population (7.5 per 100 000 population) was 11 times the rate reported in the Australian-born non-Indigenous population (0.7 per 100 000 population) (Bareja et al., 2014). Sexually transmissible infections have higher notification rates for Aboriginal people than for non-Aboriginal people. Among Indigenous children aged 0–14 years, chlamydia notification rates were 28 (for males) and 16 (for females) times as high; syphilis notification rates were

Disease/problem	Relevant research findings
	18 (males) and 44 (females) times as high; and gonorrhoea notification rates were 112 (males) and 122 (females) times as high over the period 2009–11 (AIHW, 2013a). Over the period 2008–10, 2% of HIV notifications and 1% of AIDS notifications in Australia were in Indigenous people, but HIV notification rates for Indigenous females were twice those for non-Indigenous females (AIHW, 2013a). Aboriginal and Torres Strait Islander peoples were also around 4 and 3 times as likely to contract hepatitis B and C, respectively, as non-Indigenous Australians (AIHW, 2013a). Disparity in hepatitis B surface antigen prevalence between Indigenous and non-Indigenous people has decreased since 2000 (when the government implemented a universal infant and adolescent hepatitis B vaccination program). However, hepatitis prevalence is 4 times higher among Indigenous people (Graham et al., 2013).
Cancer	The incidence rates of cancer for Indigenous people are slightly lower than those for non-Indigenous people, but death rates are higher (Cramb et al., 2012; AIHW, 2013a), and the gap between Indigenous and non-Indigenous has increased between 2001 and 2012. Aboriginal and Torres Strait Islander peoples have higher rates of lung and cervical cancer than non-Indigenous Australians but lower rates of bowel cancer, breast cancer and melanoma (AIHW, 2013a).
Mental illness	There are high levels of unmet need in relation to mental health. Aboriginal people suffer mental health problems at a very high rate, rates of self-harm and suicide are higher, and substance abuse, domestic violence, child abuse and disadvantage are contributing additional problems. In 2012–13, Aboriginal and Torres Strait Islander peoples over 18 were almost 3 times as likely as non-Indigenous Australians to have experienced high or very high levels of psychological distress (ABS, 2014a). Over the period 2008–10, Indigenous males and females were hospitalised for mental health-related conditions at 2.2 and 1.5 times the rate of non-Indigenous Australian males and females (AIHW, 2013a). Mortality rates for mental health-related conditions were 7.5 and 6.7 times higher for Indigenous Australians in the 33–44 and 45–54 year age groups (AIHW, 2013a). Trauma and grief are often overwhelming problems (Swan and Raphael, 1995). Alcohol dependence is a frequent comorbidity. The Western Australian Aboriginal Child Health Survey, which uses well-validated measures, found that 24% of Aboriginal children aged 4–17 years were assessed as being at high risk of clinically significant emotional or behavioural difficulties compared with 15% of all children (Zubrick et al., 2005). The effect on mental health of the now discredited policy of removing Aboriginal children from their parents at a young age has recently become a focus for concern. The 1997 Wilson Report on the Stolen Generations, *Bringing Them Home*, focused attention on the lasting trauma and other effects of this abhorrent practice on a whole generation of Aboriginal people.

(*continued*)

TABLE 12.1   DISEASES THAT DISPROPORTIONATELY AFFECT INDIGENOUS PEOPLES
IN AUSTRALIA (*CONTINUED*)

Disease/problem	Relevant research findings
Renal disease	In 2012–13, Aboriginal and Torres Strait Islander peoples were almost 4 times more likely than non-Indigenous Australians to have kidney disease (ABS, 2014a) and for the period 2008–10 the incidence of end-stage renal disease was 7 times higher for Indigenous people. Hoy et al. (1997) have speculated that the rates of kidney disease in some remote communities may be among the highest in the world.
Substance abuse	In 2012–13, 2 in 5 Indigenous people over 15 were current daily smokers, and 2.6 times more likely than non-Indigenous Australians aged over 15 to be daily smokers (ABS, 2014a). Compared to the general population, Indigenous people are more likely to abstain completely from drinking alcohol, but those who do drink are more likely to consume more than 4 standard drinks on a single occasion (ABS, 2014a). Indigenous women aged 35 years or over were significantly more likely than other Australian women to exceed the threshold for single-occasion alcohol risk. At an individual level, substance abuse may be both an effect and cause of further despair (Swan and Raphael, 1995), and is frequently linked to crime and incarceration, which in turn have their own adverse impacts on physical and mental health. At a community level, alcohol has been acknowledged by Aboriginal communities as having had 'a major and generally damaging impact on Aboriginal traditional life, family structure, health and capacity for self determination' (Hunter, 1993).
Violence and injuries	For the period 2006–10, deaths due to external causes, such as accidents, intentional self-harm (suicide) and assault accounted for 15% of all Indigenous deaths, compared with 6% of all deaths among non-Indigenous Australians (AIHW, 2013a). Indigenous Australians died from violence at over 9 times and from suicide at 2 times the rate of non-Indigenous Australians (AIHW, 2013a). Intentional self-harm was the leading external cause of death for Indigenous males for the 2006–10 period (AIHW, 2013a). Indigenous people in the 15–24 and 25–34 year age groups died from intentional self-harm (suicide) at 5 and 3 times the rate of other Australians (AIHW, 2013a).

The excess mortality and failure to reduce the difference between Aboriginal and Torres Strait Islander peoples and the rest of the Australian population to any significant degree is widely seen as an Australian failure (see figure 12.5) but there is cause for hope. While the gap in life expectancy is considerably larger than in other countries where indigenous peoples share a similar history of recent European colonisation (such as the USA, Canada and New Zealand), it is beginning to close. This reflects a national Close the Gap campaign and a government response of a Closing the Gap initiative, which is described in box 22.2.

FIGURE 12.5   PEAK AUSTRALIAN ORGANISATIONS EXPRESS CONCERN AND URGE ACTION ON ABORIGINAL HEALTH STATUS

Indigenous children are dying at almost three times the rate of non-Indigenous children

A CALL FOR HEALTH EQUALITY FOR ABORIGINAL AND TORRES STRAIT ISLANDER PEOPLES

Dear Prime Minister, State Premiers and Territory Chief Ministers, parliamentarians and Australian public,

We, the undersigned, are deeply concerned that Aboriginal and Torres Strait Islander peoples have not shared in the health gains enjoyed by other Australians in the last 100 years. It is a national scandal that Indigenous Australians live 17 years less than other Australians. Indigenous Australians continue to needlessly suffer and die early, not from a lack of solutions or government commitments, but from a lack of political will and action.

We call on all Australian Governments to commit to a plan of action to achieve health equality for Indigenous peoples within twenty-five years.

This commitment must receive bipartisan support from federal, state and territory parliaments as well as all sections of Australian society.

Indigenous Australians die from preventable diseases such as rheumatic heart disease, eradicated among the rest of the Australian population and they have lower access to primary health care and health infrastructure that the rest of Australia takes for granted.

This is not acceptable. We need to intensify our efforts and treat the Indigenous health crisis as a national priority.

There are already national commitments and policies in place to address Indigenous health inequality – what is missing are appropriately funded programs that target the most vulnerable. There are many stories of Indigenous success and high achievement that exist, which we can celebrate and learn from.

The signatories to this letter are committed to working in close and active collaboration with Indigenous peoples, communities and governments to achieve health equality within a generation. We commit ourselves to being engaged in identifying necessary actions and finding solutions.

At minimum, achieving health equality will require:

---} measures to ensure equal access for Indigenous peoples to primary health care and health infrastructure

---} increased support for developing the Indigenous health workforce

---} a commitment to support and nurture Indigenous community controlled health services

---} a focus on improving the accessibility of mainstream health services for Indigenous peoples

---} an urgent focus on early childhood development, maternal health, chronic illness and diseases

---} supporting the building blocks of good health, such as awareness and availability of nutrition, physical activity, fresh food, healthy lifestyles, adequate housing and the other social determinants of health.

It is inconceivable that a country as wealthy as Australia cannot solve a health crisis affecting less than 3% of its population.

Rapid improvements can be achieved in the health of Indigenous peoples by comprehensive, targeted and well resourced government action, through partnership with Indigenous peoples.

We call on the support of the people of Australia to help stop this needless suffering.

Yours respectfully,

Source: http://www.aida.org.au/wp-content/uploads/2015/03/AIDA_Open_Letter.pdf

REFUGEES, MIGRANTS AND HEALTH

> Today immigrants appear as threatening outsiders, knocking at the gates, or crashing gates, or sneaking through the gates into societies richer than those from which the immigrants came. The immigrant-receiving societies behave as though they were not parties to the process of immigration. But in fact they are partners. International migrations stand at the intersection of a number of economic and geopolitical processes that link the countries involved; they are not simply the outcome of individuals in search of better opportunities.
>
> *Sassen, 1999, p. 1*

The number of refugees and asylum seekers is growing and raises new public health concerns. The number of people worldwide recognised by the United Nations High Commissioner for Refugees (UNHCR) as being of concern has risen from just under 15 million in 1990 to almost 42.9 million in 2013 (UNHCR, 2014). Of these, 16.7 million were classified as refugees; the highest number since 2001 (UNHRC, 2014). When people flee their own country and seek sanctuary in a second state, they apply for 'asylum', or the right to be recognised as bona fide refugees with the legal protection and material assistance that that status implies. In 2014 there were 1.2 million asylum seekers worldwide. In addition 23.9 million people were internally displaced, with large numbers in the Syrian Arab Republic, Sudan, the Democratic Republic of the Congo, Somalia and Colombia. Labonté et al. (2011) describe migrants in three groups: asylum seekers and refugees, trafficked persons and undocumented migrants. Their characteristics are shown in table 12.2, which highlights the many health vulnerabilities they face. Globalisation has made travel and communication easier and the numbers of people seeking asylum have increased dramatically. Yet while the World Trade Organization stresses the importance of trade liberalisation, there have been no such arguments for liberalising the control of movement of people. In fact the trend has been in the opposite direction. Mares (2001, p. 187) suggests why this is the case:

> Frontiers, immigration checkpoints and visas form barriers between the wealthier countries and the poorer ones. They are the fortifications that protect privilege and excess, the castle walls behind which global riches are stockpiled for the enjoyment of the few. Removing those barriers would be a revolutionary step towards social justice.

TABLE 12.2 KEY HEALTH ISSUES (POSTMIGRATION) OF VULNERABLE MIGRANTS LIVING IN HOST COUNTRIES (TYPICALLY IN HIGH-INCOME COUNTRIES)

Category	Definition	Key health issues
Asylum seekers/ refugees	A refugee is a person living outside of their country of nationality or habitual residence, who has a well-founded fear of persecution because of their race, religion, nationality, membership in	Because asylum seekers and refugees tend to come from impoverished, often conflict-ridden areas, they may come with prior untreated conditions including infectious and parasitic diseases (e.g. tuberculosis, hepatitis A and hepatitis B, HIV, benign tertian malaria).

Category	Definition	Key health issues
	a particular social group, or political opinion, and is unable to return to their country for fear of persecution. An asylum seeker is a person who is seeking protection as a refugee claimant.	These populations are also likely to experience psychological distress because of suffering traumatic events (e.g. torture) and are at risk of posttraumatic stress disorder, depression, anxiety and other mental health disorders. Stress induced by the migration, isolation, poor social support and racism in their host country may exacerbate their distress and has been linked to type 2 diabetes. Women are particularly vulnerable because of their often limited education and history of gender-based violence.
Trafficked persons	Any person who is recruited, transported or harbored by means of any form of coercion, abduction, fraud or deception for the purpose of exploitation. There are two main categories: (a) forced labor and (b) sex trafficking. Women, adolescent girls, and children are the primary victims.	Trafficked persons (especially women) may be exposed to a range of physical, psychological and sexual abuse. They are often confined and isolated from others and face hazards related to forced labor. Trafficked victims tend to face multiple health problems, including HIV and other sexually transmissible infections, physical injuries, fatigue, psychological problems (including posttraumatic stress disorder), depression, memory loss, and inadequate access to health care.
Undocumented migrants	Persons migrating without necessary documents or permits, usually for employment as laborers. This condition may arise because of entering a country illegally or by entering a country legally and not respecting the permitted time and limits of their visas.	Undocumented migrants tend to be economic migrants from poor countries. They tend to live in poor environments and work in hazardous jobs with unsafe conditions and are generally underpaid (less than minimum wage) without social benefits. Owing to their limited income and their fear of detection by authorities, undocumented migrants are likely to delay or to not seek health care when faced with illness.

Source: Labonté et al., 2011, p. 272, Table 2.

Most refugees and asylum seekers take significant risks with their lives and health when they flee their country of origin. Some may have been victims of political terrorism in their own country and have suffered torture before escaping. Many are also separated from their families and friends. Once they are accepted in a host country, the consequences of torture and years of living as a refugee take their toll. Refugee camps pose massive public health problems in terms of the need for clean water, sanitation and constant vigilance for outbreak of infectious diseases.

Australia, along with some other developed states, has taken a tough stance on asylum seekers. During the 2001 Australian federal election the issue of asylum seekers became central to the Howard government's campaign. This has resulted in subsequent hard-line policies from both major parties that prevented 'boat people' who arrive on Australian territory from coming to Australia to be processed. Instead, those admitted to Australia are detained in centres, most of which were placed in remote desert areas such as Woomera in South Australia (closed in 2003 as a result of widespread protest) and Port Hedland in Western Australia (closed in 2004 due to a decline in illegal arrivals). By 2014 the Australian Government used offshore detention centres in Papua New Guinea, Christmas Island and Nauru. The Abbott Coalition government took the hard line a step further by refusing to settle people judged to be genuine refugees and instead arranging deals with Papua New Guinea and Cambodia to receive the refugees. A majority of Australian people support these hard-line policies but there is a substantial minority who have continued active protests. The disquiet in the Australian public led to the government reducing the number of child refugees held in detention, although all are held there during assessment.

For public health, an important question is: what impact does an increase in intolerance have on a population? It is hard to answer this question, but one might speculate that a less tolerant society is also less accepting of difference. People's fear of the unknown may be exacerbated and so they become less likely to reach out to newcomers to the society. Overall the net effect of the discussion about asylum seekers may be to make Australia a less welcoming and inclusive society. This is likely to have a negative impact on our collective health. For recently arrived migrant groups, the effects may be more tangible. They may be subject to taunting and abuse about their cultural practices, or subject to more direct attack, such as the attacks on mosques and Muslim people in the wake of the Martin Place siege in Sydney in December 2014. There was, however, also a social media network that was supportive of Muslims through the #illridewithyou campaign.

The extent of immigration from low-income countries to high-income countries is becoming more contentious internationally, especially in Europe where an increasing number of right-wing parties are campaigning against further immigration and the 'Islaminisation' of European society.

Migrants, asylum seekers and refugees pose significant challenges to the global community. Effective mechanisms for governing these people in a fair and just manner are essential. Governments should respect and abide by these mechanisms.

Refugees are a special category of migrants who may have particular types of health problems. Some refugees have been the victims of torture and, as a result, may often experience mental health problems (UNHCR, 1995). Australia's policy of detaining asylum seekers in detention centres is likely to have considerable implications for the mental health of these migrants.

Australia's population has increased considerably as a result of immigration over the past 60 years. In 2009–11, Australian-born residents made up 73 per cent of the Australian population, followed by Australian residents born in Asia (including North Africa and the Middle East) (10 per cent); UK and Ireland (6 per cent); 'other Europe' (northwest,

Manus Island Detention Centre, Papua New Guinea. (Vlad Sokhin, Panos)

southern and Eastern Europe) (5.1 per cent); and all other countries (including the Americas, sub-Saharan Africa and Oceania) (5.5 per cent) (AIHW, 2014a, p. 13). On average, overseas-born residents have lower death rates than Australian-born residents. For example, Asian-born Australian residents have a 36 per cent lower death rate (that is, a rate ratio of 0.64 for males and females). Resident males born in northwest, southern and Eastern Europe have a 15 per cent lower rate and females born in those countries a 24 per cent lower rate. The 'healthy migrant' effect reflects two main factors. First, those who opt to move country are likely to be healthier and have less existing sickness and disability. Second, the government selection process uses health status as one of the criteria for excluding potential migrants.

GENDER AND HEALTH

MORTALITY

Gender has a powerful impact on mortality. Table 11.2 shows that life expectancy in Australia, as in other developed countries, has consistently been higher for women than for men over the past 100 years. In the first decade of the twentieth century women's life expectancy at birth was 3.7 years higher than that for men. In the early 1980s this differential had increased to 7.1 years, but in 1996–98 there was a drop to 5.6 years' difference and in 2010–12 the difference had declined again to 4.4 years. Table 12.3 shows that the difference in mortality rates increases with age. In 2009–11 the mortality rate for males was 1.5 times higher than the rate for females, after taking differences in the age structure of the two populations into account. The greatest relative inequality was for people aged 15–24, with males in this age group dying at more than twice the rate of females (AIHW, 2014d).

TABLE 12.3   DIFFERENCES IN AGE-SPECIFIC DEATH RATES BY SEX, AUSTRALIA, 2013

Age group (years)	Age-specific death rate[a]: females	Age-specific death rate[a]: males	Difference
0	3.3	3.7	0.4
1–4	0.2	0.2	0
5–9	0.1	0.1	0
10–14	0.1	0.1	0
15–19	0.2	0.4	0.2
20–24	0.2	0.6	0.4
25–29	0.3	0.7	0.4
30–34	0.4	0.8	0.4
35–39	0.6	1.1	0.5
40–44	0.9	1.5	0.6
45–49	1.4	2.3	0.9
50–54	2.0	3.3	1.3
55–59	3.1	5.0	1.9
60–64	4.7	7.6	2.9
65–69	7.2	12.5	5.3
70–74	12.0	20.0	8.0
75–79	21.8	34.3	12.5
80–84	43.2	63.2	20.0
85 and over	123.6	144.1	20.5

a Deaths per 1000 population of the same age and sex

Source: ABS, 2014d, Table 2.9.

There are distinct differences in causes of mortality between men and women (table 12.4). One of the factors that accounted for the increasing difference between male and female death rates is the dramatic decrease in maternal mortality since the early twentieth century. In 1910 the proportion of maternal deaths relating to pregnancy was 11.5 per cent of deaths for women aged 15–49 years. Between 1999 and 2008, pregnancy and childbirth was the underlying cause of death for 100 females (ABS, 2010b). Since 1964–66, maternal deaths have decreased by nearly two-thirds (AIHW et al., 2014b). In 2006–10, there were 39 direct and 57 indirect maternal deaths, with an average of 8 direct and 11 indirect deaths per year. The reasons for the decline in difference in more recent years reflects in part increased smoking rates and labour-force participation rates for women and decreased smoking rates for men.

TABLE 12.4   LEADING CAUSES OF DEATH BY SEX, ALL AGES, AUSTRALIA, 2011

	Males			Females		
Rank	Cause of death	Number of deaths	Percentage of all deaths	Cause of death	Number of deaths	Percentage of all deaths
1	Coronary heart disease	11,733	15.6	Coronary heart disease	9,780	13.7
2	Lung cancer	4,959	6.6	Cerebrovascular diseases	6,824	9.5
3	Cerebrovascular diseases	4,427	5.9	Dementia and Alzheimer's disease	6,596	9.2
4	Prostate cancer	3,294	4.4	Lung cancer	3,155	4.4
5	Chronic obstructive pulmonary disease	3,278	4.4	Breast cancer	2,914	4.1
6	Dementia and Alzheimer's disease	3,268	4.3	Chronic obstructive pulmonary disease	2,600	3.6
7	Colorectal cancer	2,248	3.0	Diabetes	2,031	2.8
8	Diabetes	2,178	2.9	Heart failure and complications and ill-defined heart diseases	2,024	2.8
9	Cancer, unknown, ill-defined	1,920	2.6	Colorectal cancer	1,839	2.6
10	Suicide	1,726	2.3	Cancer, unknown, ill-defined	1,801	2.5
11	Heart failure and complications and ill-defined heart diseases	1,464	1.9	Influenza and pneumonia	1,356	1.9
12	Pancreatic cancer	1,218	1.6	Kidney failure	1,247	1.7
13	Kidney failure	1,208	1.6	Hypertensive diseases	1,201	1.7
14	Influenza and pneumonia	1,136	1.5	Pancreatic cancer	1,198	1.7

(continued)

TABLE 12.4 LEADING CAUSES OF DEATH BY SEX, ALL AGES, AUSTRALIA, 2011
(*CONTINUED*)

	Males			Females		
Rank	Cause of death	Number of deaths	Per-centage of all deaths	Cause of death	Number of deaths	Per-centage of all deaths
15	Cirrhosis of the liver	1,087	1.4	Cardiac arrhythmias	1,019	1.4
16	Melanoma	1,071	1.4	Accidental falls	950	1.3
17	Land transport accidents	1,003	1.3	Ovarian cancer	903	1.3
18	Liver cancer	980	1.3	Diseases of the musculoskeletal system and connective tissue	790	1.1
19	Leukaemia	933	1.2	Septicaemia	779	1.1
20	Oesophageal cancer	903	1.2	Non-rheumatic valve disorders	773	1.1
	Total (20 leading causes)	**50,034**	**66.4**	Total (20 leading causes)	**49,780**	**69.5**
	All deaths	75,330	100	All deaths	71,602	100

Source: adapted from AIHW, 2014, pp. 72–3, Figure 3.2.

Ischaemic heart disease and cerebrovascular disease accounted for a little under a quarter of deaths of Australian men and women in 2011, reflecting the pattern of the past three decades. However, heart disease declined from the main cause in the early 1970s and cancer deaths increased for women (remaining static for men) to become the main cause. Standardised death rates for ischaemic heart disease reached a peak for men and women in 1968. However the biggest relative inequalities were for coronary heart disease among people aged 25–44 and 45–64, where the rates for males were 4.9 and 4.2 times as high, respectively, as those for females (AIHW, 2014d).

The pattern of cancer deaths differs for men and women. Men are more likely to die of cancer than women and the current risk of dying of cancer by age 75 years was 1 in 9 for men and 1 in 13 for women (AIHW, n.d.). For men and women, lung cancer was the most common cause of death. But while male deaths from lung cancer have declined in recent years, female deaths have increased. However, men are still far more likely to die of lung cancer and their death rate from this cause is nearly twice that of women (45 per 100 000 compared with 24 per 100 000 for women in 2010). For women, the second most common cause of death from cancer is breast cancer, which accounted for 4.1 per cent of deaths in 2011 (table 12.4). For men, prostate cancer is the second most common cause of death from cancer (4.4 per cent of deaths in 2011—see table 12.4).

Deaths from accidents, poisoning and violence are the most frequent cause of death for men and women between the ages of 15 and 34 years of age, although men are more likely to die of these causes than women. In 2012, large differences between men and women were seen in the death rates for 25–34 year olds from intentional self-harm (21.0 per 100 000 for men; 5.7 per 100 000 for women), car accident (8.8 per 100 000 for men; 3.6 per 100 000 for women), accidental poisoning (6.3 per 100 000 for men; 2.1 per 100 000 for women), or assault (2.7 per 100 000 for men; 1.2 per 100 000 for women) (ABS, 2014c). Deaths from road traffic accidents have declined over the past 30 years, most significantly for men, for whom the rate has always been higher (table 12.5). Between 2000 and 2014, the death rate for men almost halved (table 12.5). The public health actions that led to this decline are analysed in chapter 24 on healthy public policy.

TABLE 12.5 DEATHS FROM MOTOR VEHICLE ACCIDENTS: AGE-SPECIFIC DEATH RATES[a] IN SELECTED YEARS, AUSTRALIA, 1940–2014

Females									
Age group (years)	1940	1950	1960	1970	1980	1990	2000	2010	2014
0–14	6.3	5.5	7.1	8.3	7.1	3.7	2.4	0.9	1.1
15–24	8.7	9.8	17.4	24.3	19.4	14.4	9.5	5.2	3.7
25–34	4.3	3.8	5.8	11.8	9.5	7.1	5.5	3.8	2.7
35–44	4.7	5.1	8.6	12.1	8.5	4.6	4.3	2.8	2.3
45–54	6.7	6.4	14.0	17.2	10.8	6.4	4	2.8	1.7
55–64	14.5	7.4	20.9	21.5	14.0	8.4	4.7	2.9	3.1
65–74	20.3	10.8	30.6	30.3	19.1	12.9	7.9	4.4	3.9
75 and over	30.1	24.1	33.6	40.3	30.2	17	10.2	6.7	6.7
All ages	8.2	7.0	12.9	16.7	12.6	8.1	5.4	3.3	2.8
Males									
Age group (years)	1940	1950	1960	1970	1980	1990	2000	2010	2014
0–14	12.2	10.4	9.1	12.4	10.6	6.1	3.4	1.7	1.3
15–24	51.0	74.2	69.1	95.4	79.1	39.3	27.6	16.1	9.9
25–34	34.9	37.6	43.9	44.3	37.0	26.4	19.2	11.2	8.8
35–44	29.4	31.6	33.7	36.4	26.4	13.2	13.8	9.4	8.3

(continued)

TABLE 12.5   DEATHS FROM MOTOR VEHICLE ACCIDENTS: AGE-SPECIFIC DEATH RATES[a]
IN SELECTED YEARS, AUSTRALIA, 1940–2014 *(CONTINUED)*

	Males								
Age group (years)	1940	1950	1960	1970	1980	1990	2000	2010	2014
45–54	30.9	31.6	34.5	41.1	23.2	13.4	10.4	9.4	7.1
55–64	37.2	34.1	47.5	48.1	28.2	13.9	8.8	7.9	6.9
65–74	46.1	53.5	64.2	53.4	39.0	16.9	11.8	7.7	7.6
75 and over	74.0	77.4	95.1	102.0	56.5	36.8	24.4	11.9	10.6
All ages	32.5	35.5	36.8	44.7	34.8	19.2	13.7	9	7

a Deaths per 100 000 of the respective female and male population of the same age.

Source: Compiled by Professor James Harrison (Research Centre for Injury Studies, Flinders University)
2015, based on Australian Demographic Bulletins (1940–60); ABS causes of death data (1970–90); and
the Bureau of Infrastructure, Transport and Regional Economics Australian Road Deaths Database (1990
and later years).

These deaths also have a disproportionate impact on younger people, especially males.
In 2007 transport accidents accounted for 35 per cent of all deaths in the 12–24 year old
age group (AIHW, 2011d). Young males accounted for three-quarters of road transport
accident deaths in the 12–24 year old age group, with death rates almost three times
higher among males as females (13 and 5 per 100 000, respectively) (AIHW, 2011d, p. 32).

SUICIDE

Suicide is a significant cause of death in Australia; it was responsible for 1.7 per cent of
deaths in 2012 (ABS, 2014c). In 2010–11, 1755 men and 527 women committed suicide,
which was 20 per cent of all injury deaths (AIHW et al., 2014a). Suicide deaths make
up more than 20 per cent of deaths from all causes in each 5-year age group for males
between 20 to 34 years. The human costs of suicide are huge, and campaigns in recent
years have been successful in bringing about a decline in rates. It has a very distinct
gender pattern. The male suicide rate was higher than the female rate across all age
groups. In 2010–11 male rates were highest for those aged 35–44 years and 80 years and
over. Male rates were three to five times higher than female rates except at ages below
25–29 (AIHW et al., 2014a).

Figure 12.6 demonstrates the varying rates over the course of the twentieth century
for men and women. Rates were high in the 1930s, probably reflecting reactions to the
economic depression of that period. They dropped significantly during World War II and
then rose again in the early to mid-1960s. Rates for women in their middle years were
high in this period, possibly reflecting the availability of tranquillisers and the social
position of women in this period.

FIGURE 12.6 DEATHS BY SUICIDE IN AUSTRALIA: MALES, FEMALES AND TOTAL, ALL AGES, 1921–2010 (RATES PER 100 000)

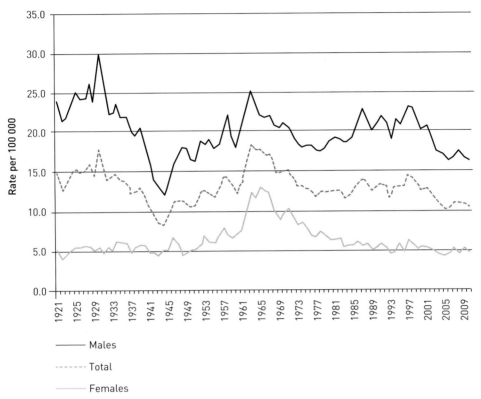

——— Males

------- Total

——— Females

Source: AIHW et al., 2014a, pp. 122–3, Table A4.1.

In the past 30 years young male suicide rates have been a chief cause for concern. The high and rising rate of suicide among men aged 34 years and under made suicide a top focus for public health action since the mid-1990s (Department of Health and Aged Care, 1998). The rate for men aged 15–24 years rose threefold during the 30 years to 1990 (ABS, 2000). For men overall the rate has declined since the peak in 1997 of 23.6 deaths per 100 000 to 10.1 in 2011. This drop has mainly been driven by the drop in the suicide rate of younger and middle aged men, especially the 15- to 24-year-old males, where the rate has halved from over 27 deaths per 100 000 in the early 1990s to 12 in 2012 (ABS, 2014c). Over the course of the past century the number of suicide deaths in men 34 and under increased steadily (apart from declines during World War II and for the 20- to 24-year-old group during the Vietnam War) until the late 1990s from which period the number has declined except for the 30- to 34-year-olds where the number has flattened but remained high (figure 12.7). For the period from 2007–08 to 2010–11, suicide rates for Indigenous males and females were around twice as high as the corresponding rates for other Australian males and females. Rates for females hospitalised as a result of intentional self-harm were at least 40 per cent higher than male rates over the period from 1999–2000 to 2011–12, with female cases outnumbering male cases most markedly in the teenage years (AIHW, 2014a, p. vii).

FIGURE 12.7 MALE DEATHS BY SUICIDE IN AUSTRALIA, 1921–25 TO 2006–10, SELECTED AGE GROUPS (DEATHS PER 100 000)

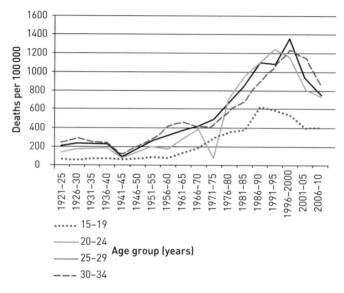

Source: ABS, 2000, p. 21; ABS, 2008, Table 4.1; ABS, 2014c, Table 11.1.

GENDER AND MORBIDITY

Despite women having longer life expectancy than men, they are more likely to report illness, both for recent and long-term conditions, as demonstrated by data from the 2011–12 Australian Health Survey (www.abs.gov.au/australianhealthsurvey).

The Australian Health Survey 2011–12 asked respondents to assess their health status by rating it from excellent to poor. Men and women assessed their health very similarly with slightly more women (55.7 per cent) reporting excellent or very good health than men (54.5 per cent) (table 12.6). Women were more likely to report arthritis (17.7 per cent compared to 11.8 per cent); osteoporosis (5.3 per cent compared to 1.2 per cent); conditions of the circulatory system (18.1 per cent compared to 15.6 per cent). Men were slightly more likely to report diabetes (5.1 per cent compared to 4.2 per cent). Differences were strongly evident in levels of psychological distress with 12.7 per cent of women reporting high or very high levels compared with 8.8 per cent of men.

TABLE 12.6 SELF-REPORTED RISK FACTORS AND GENDER, AUSTRALIA, 2011–12

Risk factors	Female (%)	Male (%)
Health self-assessed as excellent or very good	55.7	54.5
More than two standard drinks per day on average	10.1	29.1
Current daily smoker	14.1	18.3
High level of exercise	8.2	15.7
Sedentary	37.8	32.9
Overweight or obese[a]	55.7	69.7
Does not eat fruit	5.0	7.5

a Based on body mass index for persons whose height and weight was measured

Source: ABS, 2012c.

In 2011–12, women were less likely to consume alcohol at a risky level, be a current smoker, or be overweight or obese, and more likely to eat fruit, while men were more likely to undertake a high level of exercise and less likely to be sedentary (table 12.6). The reasons for these patterns of morbidity are explored further in chapter 13.

LOCATION AND HEALTH

Location has a profound effect on health, as can be seen even within a city or region, where different locations may reflect differing levels of socioeconomic status and related health status, or in the differences between metropolitan and rural lives.

A Social Health Atlas of Australia (Public Health Information Development Unit, 2014) illustrates how health and related variables vary according to location. Table 12.7 shows the comparison between suburbs in four Australian capital cities. This comparison indicates that health disadvantage correlates with socioeconomic disadvantage. The lower socioeconomic areas have higher proportions of one-parent families, low income, low-skilled workers, unemployment, homes without cars, and lower rates of internet use. Australian cities are increasingly divided by socioeconomic status, with the outer suburban areas containing a high proportion of disadvantaged people compared with affluent areas nearer the city centre.

RURAL AND REMOTE AUSTRALIA

In the past two decades there has been a strong policy focus in Australia on the 30 per cent of Australians who live in rural or remote areas. The Australian Institute of Health and Welfare (2014d) reports on the comparison of residents of metropolitan and

TABLE 12.7 COMPARISON OF THE LEAST AND MOST AFFLUENT AREAS IN FOUR AUSTRALIAN CAPITAL CITIES ON SELECTED VARIABLES

Statistical Local Area	Fairfield – East (Sydney)	Woollahra (Sydney)	Inala/ Richlands (Brisbane)	Ascot/ Hamilton (Brisbane)	Playford – Elizabeth (Adelaide)	Unley – East (Adelaide)	Armadale (Perth)	Cottesloe (Perth)
Jobless families (%)	33.5	9.6	19.1	7.2	48.6	5.3	15.3	6.0
Low-income families (%)	23.3	2.6	16.9	3.5	28.8	3.7	12.2	1.6
Single-parent families (%)	29.1	14.3	26.5	17.4	51.2	15.9	25.0	15.2
Low-skilled workers (%)	17.3	1.4	16.1	5.0	22.5	4.7	11.0	2.9
Unemployed (%)	11.1	1.5	14.3	4.6	21.6	3.4	7.2	1.4
Families in government-rented housing (%)	8.1	0.4	10.8	2.7	21.5	2.8	3.2	1.2
Homes without cars (%)	15.6	15.0	9.4	9.0	20.7	10.0	5.4	4.9
No internet connection (%)	29.0	10.6	20.4	13.4	36.0	17.7	18.8	10.2
Disability Support Pension (%)	8.2	1.3	7.1	4.2	16.5	3.6	4.6	1.0
Death of males, 0–74 years (IISDR)	111	53	118	63	176	100	107	61
Death of females, 0–74 years (IISDR)	97	63	112	78	178	97	103	51
Infant deaths (average annual IDR)	4.0	2.4	3.5	na	2.4	na	3.8	0.0
Deaths from cancer (ISDR)	93	66	114	80	152	83	102	80
IRSD	805	1107	927	1082	748	1064	1057	1116

IDR = infant death rate per 1000 live births, IRSD = Index of Relative Social Disadvantage (lower index equals more disadvantaged), ISDR = indirectly age-standardised death ratio

Sources: PHIDU (May 2014 Release); ABS Basic Community Profiles, 2011 Census (for percentage of low-skilled workers).

non-metropolitan areas (box 12.3). The mortality rate for all people living in remote and very remote areas was 1.4 times higher than the rate among people living in major cities. The rate of potentially avoidable deaths also increased as remoteness increased. The increase in mortality rates holds for both males and females. The mortality rate among females living in remote areas was 1.3 times as high as the rate among females in major cities, and in very remote areas it was 1.6 times as high. For males, the rate ratios were 1.2 and 1.4, respectively. The AIHW (2014d) calculated that if people living in regional and remote areas had the same mortality rates as people living in major cities, there would have been nearly 20 000 fewer deaths in regional and remote areas between 2009 and 2011. There would be a 38 per cent reduction in the age-standardised mortality rate for females living in very remote areas, and a 30 per cent reduction in the rate for males. In 2011, for each of the ten leading causes of death, rates were higher for people living outside major cities. People in remote and very remote areas fared the worse. Coronary heart disease rates were between 1.2 and 1.5 times higher in regional and remote areas as in major cities. In remote and very remote areas, the rate of dying due to a land transport accident was more than 4 times higher than in major cities. For deaths due to diabetes, rates were between 2.5 and 4 times as high and, for suicide, between 1.8 and 2.2 times as high.

BOX 12.3 FACTORS CONTRIBUTING TO POORER HEALTH IN RURAL AREAS

Compared with people who live in major cities, people who live outside major cities are more likely to:

- be smokers
- drink alcohol in hazardous quantities
- be overweight or obese
- be physically inactive
- have lower levels of education
- have poorer access to work
- have less access to specialist medical services
- work in physically risky occupations
- have lower socioeconomic status.

Source: based on AIHW, 2014d.

These broad groupings, of course, conceal more specific locational factors such as those between inner and outer metropolitan areas or between country towns and more remote rural locations. The picture is also complicated by the presence of a much higher proportion of Indigenous peoples in non-metropolitan areas, making it hard to estimate what proportion of the higher disease burden reflects Indigenous health issues rather than rural or remoteness. Nonetheless, they provide a guide to locational differences.

From a position of being relatively affluent as a result of its strong agricultural base, rural Australia has undergone dramatic change in the past 30 years. Increasing world competition, falling commodity prices, a decrease in the profitability of traditional primary industries and widespread drought has resulted in rural recession, causing significant social and economic dislocation. For much of rural and remote Australia the decade to the early twenty-first century has seen the withdrawal of services and a declining population with a consequent threat to the viability of rural communities. For some areas the mining boom and the 'sea change' phenomenon has seen a reversal of this decline.

CONCLUSION

This chapter has reviewed the evidence relating to the social patterning of death and disease in Australia. It has demonstrated the stark contrast between different groups in the population according to socioeconomic status, occupation, ethnicity and race, gender and location. The reasons for these differences are explored in the following chapter.

CRITICAL REFLECTION QUESTIONS

12.1 In what ways does socioeconomic status affect health equity outcomes?

12.2 Why do you think more men than women commit suicide whereas more women than men attempt suicide?

12.3 Why do you think women live longer than men? Why do you think the gender gap in mortality is reducing in Australia?

Recommended reading

AIHW (2014a): *Australia's health*, published every 2 years, provides a comprehensive overview of the health status of Australians.

AIHW (2014d): *Mortality inequalities in Australia* provides an excellent overview of inequalities by gender, remoteness and socioeconomic status.

Australian Council of Social Service (2014) *Poverty in Australia 2014*.

Useful websites

Australian Institute of Health and Welfare, www.aihw.gov.au, contains details of the wide range of statistics that AIHW collects.

Public Health Information Development Unit, www.adelaide.edu.au/phidu, contains lots of data on health and determinants of health.

13

THE SOCIAL DETERMINANTS OF HEALTH INEQUITY

This unequal distribution of health ... is the result of a toxic combination of poor social policies and programmes, unfair economic arrangements and bad politics.

Commission on Social Determinants of Health, 2008, p. 1

KEY CONCEPTS

Introduction

Explaining socioeconomic status inequities in health status

Artefact explanations

Theories of natural or social selection

Cultural/behavioural versus materialist or structuralist explanations

Social capital, support and cohesion and health inequities

Gender and health

Inequities: the case of Aboriginal health

Conclusion

INTRODUCTION

Striving to achieve equity in health status is a crucial part of the new public health. In the past 200 years there has been a doubling of the human lifespan and the increase in life expectancy is continuing in most countries (Williams, 2004). However, major inequities in health between groups within populations still exist in Australia (Draper et al., 2004; Glover et al., 2006; AIHW, 2014d) and other countries. Mackenbach (2005) and Crombie et al. (2005) provide evidence for Europe, and Evans et al. (2012) summarise evidence for the USA. Growing policy attention on health inequities and what to do about them was intensified when the Commission on Social Determinants of Health reported in 2008. Its work has been followed by national reports (e.g. UK and Brazil) and reports in the WHO European region (Marmot et al., 2013a), which show that health inequities are increasing in some countries, and the Americas (Marmot, et al., 2013b). The growing interest reflects the fact that economic inequalities are increasing both between and within countries (Stiglitz, 2012) and wealth concentration is increasing considerably (Piketty, 2014) and may also reflect disenchantment with the neo-liberal public policies reviewed in chapter 5. A vital distinction is between health inequalities and health inequities (box 13.1).

BOX 13.1   THE DIFFERENCE BETWEEN EQUALITY AND EQUITY

Much of the literature on health differentials uses the terms 'equity' and 'equality' interchangeably, but their different meanings have implications for policy action: equality is concerned with sameness; equity with fairness. Policies are unlikely to be able to make people the same, but they can ensure fair treatment. The Commission on Social Determinants of Health defined health inequity as 'Where systematic differences in health are judged to be avoidable by reasonable action they are, quite simply, unfair. It is this that we label health inequity' (CSDH, 2008, Executive Summary).

We consider the social patterning of health and disease, looking at four main theories relating to socioeconomic disadvantage before considering the explanations for gender differences in health. We conclude with a case study of Indigenous health inequities (the greatest extreme evident in Australia) that demonstrates the complexities involved in understanding inequities.

EXPLAINING SOCIOECONOMIC STATUS INEQUITIES IN HEALTH STATUS

Inequities in health status appear to be universal across cultures and persistent. They relate to socioeconomic status, gender and ethnicity. Reviews of the research in the area (Turrell et al., 1999, p. 88; Whitehead and Dahlgren, 2006; CSDH, 2008; Evans et al., 2012) concluded that the international evidence based on socioeconomic status and health is consistent, and the relationship:

- has been observed in numerous countries
- has persisted over long periods of time
- exists for virtually all measures of health and health habits
- is evident irrespective of how socioeconomic status is measured
- is evident for almost all health outcomes, irrespective of the measure of health that is used
- is evident for all age groups
- is evident for both men and women
- occurs in a gradient whereby those at the top have better health than those in the middle, who in turn have better health than those at the bottom.

The Commission on Social Determinants of Health has produced an explanatory framework for health inequities that sees health and its distribution resulting from social context, socioeconomic position that leads to distinct patterns of stratification, and a range of intermediary determinants of health that create differential exposure and vulnerabilities (see figure 13.1). This model is designed to be relevant globally and so is broad and conceptual. Figure 13.1 develops the Commission on Social Determinants of Health model by adding other determinants including those from the physical environment that are mediated by social and economic factors.

FIGURE 13.1   KEY DETERMINANTS OF HEALTH AND HEALTH INEQUITIES

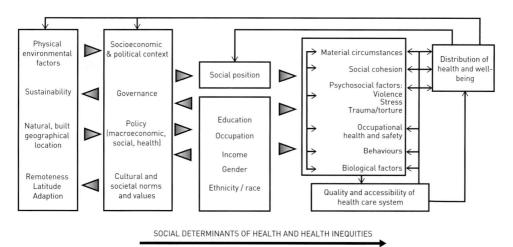

SOCIAL DETERMINANTS OF HEALTH AND HEALTH INEQUITIES

Source: Adapted from CSDH, 2008, p. 43 and AIHW, 2014a, p. 5.

The UK has the most consistent tradition of documenting and analysing inequities. The UK Black Report (Townsend et al., 1992) and the subsequent Acheson Report (1998) are the most comprehensive considerations of health inequities from one country. The framework used to explain inequities in these reports is a useful one for our discussion. Townsend et al. (1992) put forward four possible explanations for variations in health. They are:

* artefact explanations
* theories of natural or social selection
* cultural/behavioural explanations
* materialist or structuralist explanations.

ARTEFACT EXPLANATIONS

This approach is based on the notion that the ways in which social class has been measured in the UK (by the Registrar-General's social class classification of occupations) may be unreliable and artificially inflating the size and importance of observed health differences. The argument (Illsley, 1986; Bloor et al., 1987) hangs on the claim that the classification and nature of occupations have changed so much in recent decades that any comparison with earlier decades is meaningless.

Whitehead (1992) reviews multiple sources of evidence that contradicts this artefact explanation. She quotes a series of UK studies from the 1980s and 1990s that correct some of the methodological problems of time trends and occupational class, none of which alters the pattern of links to social circumstances in any significant way. She also points to a series of studies that have used other measures of social circumstance (income, housing tenure, household possessions or education) and observes a similar pattern of inequities. Other studies have considered retired people and women in different social circumstances and found the same direction of inequities. Finally, results from major longitudinal studies using alternative measures to occupational class have provided additional evidence of a social gradient in mortality (Marmot et al., 1984; Goldblatt, 1990). Whitehead concludes, on the basis of the recent evidence, that the studies reported in the Black Report may be underestimating the extent of health inequities rather than overestimating them. Graham (2000, p. 14) reports that statistical inaccuracies are insufficient to account for the consistency and scale of the association between SES and health.

THEORIES OF NATURAL OR SOCIAL SELECTION

These theories suggest that inequities arise as a result of social mobility—healthier people rise to higher social classes. Those in poor health are less likely to be socially mobile. A similar argument is also made in relation to employment: unhealthy people are more likely to become unemployed and good health 'makes it easier for people to enter, remain and gain promotion in the labour market, and thus to live and work in low-risk environments' (Graham, 2007, p. 108).

The social selection theory received support from the work of Illsley (1986), which demonstrated that taller women tended to marry into a social class higher than their fathers' more often than shorter women. The infant mortality rates and the birthweight of their babies were better than their shorter peers, who remained in their fathers' social class. Fogelman et al. (1989) have collected longitudinal data on health differences at ages 7 and 23 that confirm health to be associated with social mobility. They do note, however, that the effect is not sufficiently large to explain all self-reported health differences. Whitehead (1992) reviews the evidence relating to social selection and concludes that, while there may be some effect at younger ages, it is likely to account for only a small proportion of the mortality differentials between social groups. Power et al. (1996) and Graham (2007) conclude that, while health selection contributes to the socioeconomic gradient, its contribution is modest. Evans et al. (2012) review the evidence on children, socioeconomic status and health status and conclude that socioeconomic conditions are critical in determining life expectancy and health status in adults, suggesting that selection is not an adequate explanation.

A similar debate is evident regarding intelligence and life expectancy. While some studies suggest intelligence predicts mortality (Deary, 2008), the relationship needs unpicking. Marmot and Kivimäki (2009) note that the relationship disappears when education and income across the lifecourse are in the same model, suggesting that it is not intelligence per se that is linked directly to health but the conditions of adult life to which intelligence predisposes. Cognitive function in childhood (which correlates highly with cognitive function in adulthood) is influenced by biological and social conditions in early life. Thus the link is not direct.

Migrants to Australia tend to have lower mortality rates than Australian-born people. Powles and Gifford (1990) suggest this may be a result of the selective effects of migration. As an example, they quote the Levkadian study of immigrant health, which compared migrating and non-migrating siblings on a range of variables. Migrating brothers differed only modestly from non-migrating brothers, but migrating sisters appeared to be relatively advantaged: they were two centimetres taller and much more likely to be literate, implying that selection is a contributor to the mortality advantage of female immigrants.

CULTURAL/BEHAVIOURAL VERSUS MATERIALIST OR STRUCTURALIST EXPLANATIONS

Cultural/behavioural explanations focus on differences in how the various social groups make lifestyle choices. They maintain that people in less well-off groups typically adopt lifestyles that are likely to be damaging to their health. The materialist or structural explanation focuses on the material conditions under which people live, maintaining that health inequalities stem from the less affluent social groups being the victims of unhealthy environments. They have less income for healthy food, engage in more dangerous occupations, have worse housing, more risk of unemployment and fewer resources with which to cushion themselves from illness. These two sets of explanation are often seen as opposing.

Each explanation reflects the different philosophical positions of individualism and collectivism. The behavioural explanations see the cause of greater burdens of illness lying within the individual. The structural stresses the impact of the collective on individuals. Obviously, behavioural approaches see the agency of the individual as crucial while the structural approaches stress the whole structures of society. Evans and Stoddart (1994, p. 43) point out that the emphasis on individual risk factors and disease has tended to maintain existing institutions and ways of thinking about health. These emphases sit comfortably with the underlying individualism of society (Tesh, 1988), and so make the cultural and individual explanations the most readily adopted (see the discussion on individualism in chapter 4).

Behaviour reflects social context and the social consequences of people's different material (stress, self-esteem and quality of social relations) circumstances (Whitehead, 1992; Evans and Stoddart, 1994; Blaxter, 2010). This explanation is evident in the model from the Commission on Social Determinants of Health (figure 13.1), which puts little emphasis on behaviours other than as a reflection of patterns of social stratification and differential exposure to health risks. Most reviews of the evidence now focus on a range of structural factors including income, housing, employment, extent of social support and characteristics of the localities in which people live (see, for example, Marmot and Wilkinson, 1999; Turrell et al., 1999; collection in Graham, 2000; Marmot, 2004, 2006; review of research literature by Whitehead and Dahlgren, 2006; CSDH, 2008). Good health appears to rely on a contribution from material and social factors. The complex evidence on the aetiology of differences in health status is reviewed in detail by looking at the following topics: behaviour, material disadvantage, social networks and support.

BEHAVIOUR: INDIVIDUAL CHOICE OR SOCIAL CONSTRAINT?

We have already seen that behaviours linked to health (smoking, alcohol use, nutrition and exercise) differ between social groups. Evans and Stoddart (1994) conclude that the social grouping of behaviours supports the arguments in favour of structural influences on health. They conclude (p. 50): 'the well-defined clustering of smoking and non-smoking behaviour within the population suggests that such behaviour is also a form of "host" (the smoker) response to a social environment that does or does not promote smoking.'

A population-based Finnish study of the association between measures of socioeconomic status, health behaviours and psychosocial characteristics among 2674 men showed that many adult behaviours and psychosocial dispositions detrimental to health are consistently related to poor childhood conditions, low levels of education and blue-collar employment (Lynch et al., 1997). These data indicate that behaviour-relevant health is powerfully shaped by childhood experiences.

Research on smoking among low-income women concluded that they are aware of the risks associated with smoking, but continue because it is one of the few activities undertaken totally for themselves and that provides some relief from the day-to-day grind of making ends meet (Graham, 1987).

Status and class position appear to have powerful, if subtle, effects on people's ability to control and change their behaviour. Higher senses of personal efficacy typically associated with higher social position encourage beliefs about one's ability to break addictions and make positive changes to lifestyle.

Whitehead (1992) concluded that lifestyle differences could account for some of the health differentials between various groups. The Whitehall study of British civil servants (Marmot et al., 1984) considered coronary heart disease rates and found the disease to be strongly associated with civil service grade. When data were controlled for age, smoking, systolic blood pressure, plasma cholesterol, height and blood sugar, the risk associated with employment grade was reduced by less than 25 per cent.

Similar findings have come from the Californian Alameda County Study (Berkman and Breslow, 1983), which looked explicitly at whether harmful behaviour patterns could account for the increased risk of death among the poorest income group. The poorest group compared to the richest one had double the risk of death over an 18-year period. Yet, even when the data were adjusted to take account of 13 known risk factors, including smoking, drinking, exercise and race, there was still a substantial gradient of risk associated with income. The conclusion was that behaviour factors were not the major factors related to the increased risk of death, but rather the general living conditions and environment of the poor.

Blaxter (1990, p. 223) concluded that only people in more favourable circumstances were likely to either damage or improve their health by changing their behaviour in relation to activities such as smoking and exercise. This finding certainly makes sense if you consider health in extreme circumstances such as the poor in sub-Saharan Africa where it would make little sense to expect personal behaviours to affect lifestyle when other circumstances are so bad. In most settings more risky health behaviour and lower health literacy is associated with socioeconomic disadvantage. Thus, in Australia, 55 per cent of the population in the highest socioeconomic group had at least an adequate level of health literacy compared with 26 per cent of those in the lowest group (AIHW, 2012, p. 184). Those with higher levels of risky dietary behaviours, higher risk of smoking and lower use of prevention and screening services include socially disadvantaged people, people with disabilities and people living in rural areas (AIHW, 2012). These social patterns of risk cannot be considered without looking at the social and economic factors that drive the difference and these include factors such as lack of access to education, the costs of healthy food and the costs associated with disease prevention activities (Baum and Fisher, 2014).

Material, behavioural and psychosocial risk factors cluster together. People in lower socioeconomic groups are likely to suffer from all three and this fact has led researchers to call for a greater focus on a 'lifecourse perspective', which maintains that health inequalities are the outcome of cumulative differential exposure to each of these types of risks (Lynch and Davey Smith, 2005).

MATERIALIST OR STRUCTURALIST EXPLANATIONS

This set of explanations draws on both social epidemiology and social science evidence and considers both the impact of economic and social structures (including class) on individuals and also on comparisons between countries with differing levels of inequality. This understanding underpins the report of the Commission on Social Determinants of Health.

POVERTY AND INDIVIDUAL INCOMES

An increasing body of evidence has linked material resources to health status. Reviews of the evidence on a broader range of material deprivation and health can be found globally (Graham, 2007; CSDH, 2008), in Australia (Turrell et al., 1999; Glover et al., 2006), the UK (Townsend et al., 1992; Acheson, 1998; Graham, 2000; Crombie et al., 2005), Europe (Mackenbach, 2005) and the USA (Hofrichter, 2003; Evans et al., 2012). People with lower incomes report more illness and die earlier in Australia (AIHW, 2014d), the USA (Davey Smith et al., 1996), Canada (Wolfson et al., 1993) and the UK (Marmot et al., 2010). Low income often means people do not have access to those factors that have a direct effect on health, including housing; stable, rewarding, safe employment; nutritious food; and educational opportunity. But it is not only those with lowest incomes that suffer health inequalities, as an increasing body of literature indicates that the relationship of health to socioeconomic status is linear rather than threshold (Marmot, 2004). In other words, it is not only the absolute poor whose health is worse than that of more prosperous people, but people who are relatively less well-off than others also suffer worse health (Adler et al., 1993; Evans, et al., 1994b; Davey Smith et al., 1996; Wilkinson, 1996, 2005).

There is little doubt from this literature that relative and absolute poverty are health hazards. The stark contrasts between industrialised and developing countries are powerful evidence of the impact of absolute poverty on health.

Social science literature on class inequalities that underpin health inequalities has been summarised by Graham (2007). She notes that there are a variety of ways in which class disadvantage is transmitted in high-income countries, including through unfavourable childhood environments. Recent developments in neurobiology have enabled mapping of brain structures and emerging evidence suggests that socioeconomic environments affect brain development (Brito and Noble, 2014). The sociologist Bourdieu (1986) has shown that class advantage stems not just from economic advantage but also from other 'capitals' including social (for example, the 'old boys' network'), cultural (including education), and symbolic middle- and upper-class aesthetic taste, which can inhibit mobility. These factors help explain the gradient in health because they affect status and power (examined below).

INCOME INEQUALITY: COUNTRIES COMPARED

Wilkinson and Pickett (2009) argue that societies in which income is more equally divided also have longer life expectancy and better outcomes on a range of measures (see figure 13.2). While some of the extent and rationale for this relationship has been disputed (Lynch et al., 2000), there is broad agreement that in many circumstances it holds true. Davey Smith et al. (1996) summarised the growing literature on the relationship between inequality and health, noting that studies have related income inequality to infant mortality, adult mortality from several broad causes of death, life expectancy, height and morbidity.

FIGURE 13.2   HEALTH AND SOCIAL PROBLEMS ARE WORSE IN MORE UNEQUAL COUNTRIES

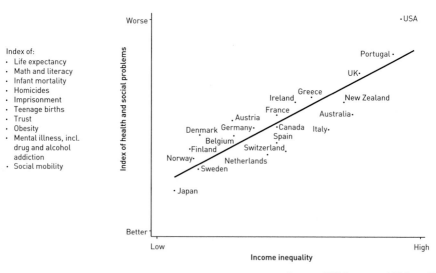

Source: Wilkinson and Pickett, 2009.

Davey Smith et al. (1996, p. 988) conclude that 'those countries now experiencing the largest increases in income inequality are precisely those that have systematically under-invested in human resources for many years'. This argument receives support from those low-income countries (including China, Sri Lanka, Kerala State in India, Costa Rica and Thailand) that achieve relatively good population health status—consistently those that have invested in social infrastructures available to all the population.

The USA is one of the countries that has invested less in social goods, despite being one of the richest countries in the world. Werner and Sanders (1997) detail the situation whereby the USA has the highest real gross domestic product in the world, ranks world first in total health spending, yet has health indices worse than other rich countries and lags behind some countries with much lower gross national products. Box 13.2 discusses the example of countries that have achieved relative high health status without dramatic increases in wealth, indicating that health is about more than level of economic development (Stiglitz, 2012). Some groups within US society have particularly poor health outcomes, especially the African–American, Native American and Hispanic populations (Lillie-Blanton et al., 1996). A stark example is the fact that, in 1990, an African–American man in Harlem was less likely to reach age 65 than a man in Bangladesh (McCord and Freeman, 1990). While mortality rates for this group have dropped they are still nearly two and a half times higher than for the US white population (Geronimus et al., 2011). Explanations for the link between income distribution and population health outcomes have been made by exploring what it is about more egalitarian societies that may make them healthier. These societies appear to be more socially cohesive, more supportive and less conflictual than societies where income differences are larger (Wilkinson, 1999, 2005; Wilkinson and Pickett, 2009), all factors that are discussed below.

BOX 13.2   ACHIEVING HEALTH WITHOUT WEALTH

Some poorer countries (including Cuba, China, Costa Rica, Sri Lanka, Kerala State and Tamil Nadu in southern India, Thailand and more recently Bangladesh) achieved significant health gains without major increases in per capita income, gross national product or institutional health care expenditures. This point is well illustrated by comparing some of these countries with the USA on key indicators:

Indicator	USA	Costa Rica	Sri Lanka	Thailand	Cuba
Life expectancy at birth (years)	78.7	79.7	74.1	74.2	79.1
Infant mortality rate (infant deaths per 1000 live births)	5.9	8.4	8.2	11.3	5
Happy Planet Index	37.3 (ranked 105th)	64.0 (ranked 1st)	49.4 (ranked 35th)	53.5 (ranked 20th)	56.2 (ranked 12th)

(continued)

Indicator	USA	Costa Rica	Sri Lanka	Thailand	Cuba
Gross national income (per capita US$)	53 470	9550	3170	5340	5890
Health expenditure (per capita, purchasing power parity, constant 2005 international $)	8895	1311	189	385	405

Source: Abdallah et al., 2012; World Bank, 2014.

This 'health without wealth' phenomenon appears to be based on:

- greater general social equity
- accessible primary health care services
- decreased income inequities
- improved status for women (especially literacy levels)
- availability of family planning
- lower birth rates
- land reform
- adequate physical infrastructure (water, electricity, transport).

These countries have given top priority to meeting the basic needs of their populations rather than following the growth-at-all-costs model, which hopes for some 'trickle-down effect'. Consequently, the approach of meeting basic needs seems to be the more effective means of promoting population health (Werner and Sanders, 1997, p. 117). These case studies also show that economic growth does not guarantee good health, as it depends how the fruits of economic development are invested.

The high health status of these countries supports the hypothesis that hierarchies can be detrimental to equitable health outcomes. We should note that the imposition of neo-liberal policies on these countries may be putting the health advantage at risk.

Source: Halstead et al., 1985; Balabanova et al., 2011.

Wilkinson and Pickett's (2009) work indicates that inequities in health status are most likely to be reduced in societies that implement redistributive social policies that make an investment in social infrastructures, such as education, affordable housing, welfare support and employment options. The evidence also underlines the necessity for intersectoral health action, as it is only through action in all government portfolio areas that inequities will be addressed.

WEALTH AS A PUBLIC HEALTH EQUITY PROBLEM

Ill fares the land, to hastening ills a prey

Where wealth accumulates, and men decay

Oliver Goldsmith, The Deserted Village, 1770

Inequities in health are underpinned by gross inequities in wealth. Wealth is defined as things people own and use to (1) produce goods and services, and (2) enjoy directly without consuming them in the process. Examples are land, natural resources and shares (Stretton, 2000, pp. 41–2). While the epidemiology of wealth is not as well documented as that for income, all indications are that wealth is extremely unequally distributed between countries and within countries, and that this underpins the distribution of health. Table 13.1 shows that over a third of the world's wealth (official exchange rate basis) is concentrated in North America where the population of that region is only 6.1 per cent of the total, while Africa, with more than 10 per cent of the world's population, has only 1.1 per cent of the wealth. This distribution has not changed much over the last two decades.

TABLE 13.1   GLOBAL WEALTH ESTIMATES BY REGION, MID-2012

Region	Share of adult population (%)	Wealth per adult (US$)	Share of world wealth (%)
Africa	11.7	4 470	1.1
Asia–Pacific	23.7	46 693	22.8
China	21.5	20 452	9.1
Europe	12.7	119 056	31.1
India	16.4	4 250	1.4
Latin America and Caribbean	8.4	22 533	3.9
North America	5.7	258 802	30.6
World	100.0	48 501	100.0

Source: Credit Suisse, 2012, p. 83.

If the conventional public health problem of poverty is recast as one of wealth, the options for achieving equity are broadened. Through this new lens the increasing worldwide concentration of wealth (Piketty, 2014) is a threat to health equity. Understanding the growing concentration of wealth requires a global perspective, as wealth is increasingly held by a global elite operating through transnational corporations, primarily outside the influence of national governments (Korten, 1995, 2006; Saul, 1997). We saw in box 5.4 that wealth distribution is strikingly inequitable. Almost half of the world's wealth is now owned by just 1 per cent of the population. Senior executives of transnational corporations are receiving higher and higher salary packages. For instance, *Business Week* reported that the average large-company CEO received compensation totalling $8.1 million in 2003, up 9.1 per cent from the previous year. While this dropped by 2013 following the Global Financial Crisis, the average was $10.5 million. The gap in pay between average workers and large-company CEOs was 257:1 in 2013. In 1982, it was just 42:1.

Schneider's (2004) analysis indicates that wealth distribution has been persistently unequal. The figures he reports differ somewhat according to method of calculation but in all the countries for which he has data—USA, Sweden, Australia, Canada,

New Zealand, France, West Germany and Belgium—the top 10 per cent of people held at least 50 per cent of the wealth over the course of the twentieth century and the percentage was often higher. A general pattern was that wealth inequality appeared to decline in the first 70 years of the twentieth century, but started to increase from the 1980s, and more so in the 1990s. Piketty (2014) analysed data on the very rich (top 1 per cent) and found that before World War I this 1 per cent received around a fifth of total income in both Britain and the USA (see figure 13.3). By 1950 that share had been cut by more than half. But since 1980 the 1 per cent has seen its income share increase again to the extent that in the USA it is the same as a century ago. Piketty warns that if current wealth concentration trends continue, we are heading to a hyperinegalitarian society. Put together with Wilkinson and Pickett's data on the negative health outcomes from less equal societies, this trend does not bode well for health and health equity.

FIGURE 13.3   INCREASING WEALTH INEQUALITY

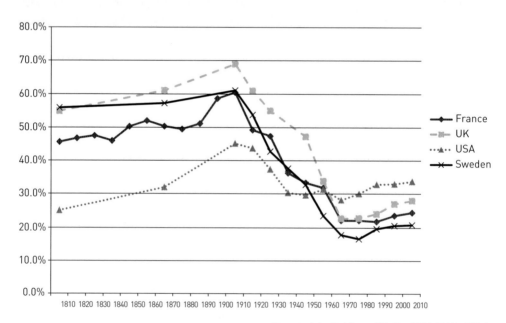

Source: Adapted from Piketty, 2014, Table S10.1.

USE OF HEALTH SERVICES

The notion that social inequalities in health might be due to uneven access to medical care has largely been discounted as a major contributing factor (Marmot et al., 1995), even though access to health care is an important social determinant of health. The UK Black Report and the work of McKeown (1979) and Szreter (1988) (see chapter 2) concluded that medical care has a limited role in improving life expectancy. Mackenbach (1996) does suggest, however, that medical care made a larger (though not the major) contribution to extending health expectancy in the second half of the twentieth century. In the Netherlands and the USA, for example, more effective health care has been estimated to have added five years to life expectancy at birth. Medical care is important not just because of its contribution to life expectancy but also because of its role in reducing morbidity and disability, and relieving pain and suffering. Access to care might be more significant in determining how well people cope with disability and chronic conditions. The comparison of health expenditure between the USA and Costa Rica (see box 13.2)—which shows that Costa Rica achieves one year greater life expectancy than the USA but spends less than 15 per cent of the amount that the USA spends on health services per capita—demonstrates that good health is about much more than absolute health services spending. It also concerns the nature of health services and the extent to which they emphasise disease prevention and health promotion rather than acute, expensive hospital care.

HOUSING

Adequate housing has been recognised for centuries as a fundamental requirement for health (Shaw, 2004). Adequate housing includes reasonable quality materials, facilities and infrastructure; habitability; affordability; accessibility; legal security or tenure; viable location; and cultural suitability (Shaw, 2004). Poor housing was one of the key issues driving the public health revolution of the Industrial Revolution period in nineteenth-century Europe. The sight of slums in rapidly growing cities is a familiar image of that period and is now reproduced in many fast-growing cities in low- and middle-income countries. Friel et al. (2011) have noted that the restructuring of cities through the processes of globalisation has led to benefits for some, but this has occurred through the rapid, often unplanned, urbanisation that has outstripped the ability of governments to build essential infrastructure and services and provide basic needs for living. They note that this has contributed to a growing gap between rich and poor in terms of adequate urban housing, employment opportunities, transportation, levels of pollution, and sanitary conditions (these issues are further considered in chapter 15). The efforts of public health reformers in the nineteenth century industrial revolution focused on health hardware issues, including housing. The reports of Edwin Chadwick provide graphic descriptions of the conditions in which people were living in the industrial Midlands and the north of England. In low-income countries slums are once again a major health risk (CSDH, 2008). In non-slum areas housing still affects health, for example through homelessness and its associated health consequences; through its inadequacy, such as dampness (which can lead to respiratory illness) or safety (which can lead to injury); and through its location and design, which can lead to problems such as isolation and lack of ready access to goods and services. Housing tenure has also been shown to be associated with mental and physical health status, even after controlling for age, sex, income and self-esteem (MacIntyre et al., 2000), which suggests that tenure affects people's life experience and is more than a marker of social class. One billion people live in urban slums and face major health risks as a result of their inadequate housing. This may be through dampness, overcrowding, indoor pollution from fires for heating and cooking or a range of other injury risks (Friel et al., 2011). Many of the world's population in urban areas also suffer from insecurity of tenure.

Homeless young people. (Fernando M Gonçalves)

Homelessness appears to have a particularly strong impact on health (Darnton-Hill et al., 1990). A study in Philadelphia found that the age-adjusted total mortality of homeless persons aged between 15 and 74 years in the 1980s was four times the rate of the general population (Hibbs et al., 1994). A UK study found that among rough sleepers in London, Bristol and Manchester in 1995–96, death rates were 3.6 to 5.6 times the rates in the general population and that average life expectancy was only 42 years (Webster, 1997, p. 444). In Australia, the Homeless Children Report (Human Rights and Equal Opportunity Commission, 1989) by Brian Burdekin detailed the devastating impact homelessness has on young people's health. It estimated that between 20000 and 25000 young people were homeless in Australia. Despite such reports, the problem continues. Those who experience long-term homelessness often suffer from disabilities, substance abuse, and medical and psychiatric comorbidity (Burns et al., 2009). Homelessness has also been shown to be an independent risk factor for mortality in individuals who are already in poor health and socioeconomically disadvantaged (Morrison, 2009). Homelessness may be the result, as well as the cause, of illness. Being poor and without a home means lacking the basic requirements to maintain health. Being homeless also makes it very difficult to hold down a job and lead a healthy lifestyle (Department of Families, Housing, Community Services and Indigenous Affairs, 2008). In low-income countries, being homeless means being completely destitute. Millions of people live on the street and, while there is hardly any research on their health status, it will obviously suffer from the living conditions.

Housing quality in Australia is generally less of a problem than in regions with older housing stock, such as Europe. Design and location are likely to be more significant issues. There are some capital city housing estates where the concentration of low-income people and a range of indicators of disadvantage (single parent, high unemployment, low school retention) coincide and these are likely to also be areas of poor health. Indigenous peoples suffer the most inadequate housing in Australia. Bailie (2007) describes how this has an impact on infectious diseases in particular through factors such as inadequate water supplies, overcrowding and inadequate waste-collection services.

Housing in rural South Australia. (Fernando M Gonçalves)

EMPLOYMENT

Employment has a significant impact on health status, either because it is insecure and/or unsafe or not available. Employment plays a significantly different role in people's lives as they juggle work and family responsibilities.

Blaxter (1990, pp. 66–9) found that self-reported health status for both men and women was related to employment status. Those in skilled and professional jobs were more likely to report good or excellent health than those in semi- or unskilled employment. The most comprehensive study of the effects of job status has been the UK Whitehall study of different grades of civil servants (Marmot et al., 1984). This found a threefold difference in mortality rates between the highest and lowest grades.

The nature of employment variously affects health. Some work is physically dangerous, and mining dangerous substances, such as uranium and asbestos, has been particularly dangerous. The rate of death, serious injury and illness in the Australian mining industry is over 2.4 times the national rate for all wage and salary earners (Foley, 1997). There are gross inequalities between countries in terms of standards of health and safety in the workplace. In many poor countries regulations are few and enforcement non-existent. In these countries working conditions can be appalling, especially in industrial sweatshop settings. Work is a significant and often undocumented cause of illness and death. Over the course of the past 50 years work for most of the world's population has changed dramatically (Heymann, 2006). People are more likely to work in factories, agribusiness or formalised service sectors where employers control hours and location of work. Heymann (2006) describes the impact of the changes to global working conditions in terms of their effects on families. She points to the increasing number of children left alone because all adults in their family work. The poignant picture she paints of the desperation of poor working families (based on a decade of global research) and the magnitude of the crisis in work and family life clearly demonstrates the ways in which work contributes to health inequities. The processes of urbanisation mean people have moved away from family and community support. Globalisation means that companies can readily move their operations to wherever labour is cheapest and its conditions least regulated. This has meant the loss of millions of jobs (especially in manufacturing but increasingly in white collar jobs such as information technology) in industrialised countries—for example in car manufacturing plants in Birmingham in the UK, Detroit in the USA and Adelaide in Australia. With the ever-present threat of job loss, workers are forced to accept lower wages and fewer benefits. The Employment Conditions Knowledge Network (EMCONET) of the Commission on Social Determinants of Health developed a heuristic model of micro-pathways that link employment and working conditions to health inequalities, combining physical, chemical/biological, ergonomic and psychosocial factors. This model shows that while risk factors may be associated with different health outcomes, the production of health inequalities is shaped by a limited number of social mechanisms, namely exploitation, domination and discrimination. Social class, gender and ethnicity were identified as key axes of inequality that help explain why some workers, their families and communities are exposed to multiple risks (Benach et al., 2010).

The psychological demands of various jobs can also have a differential effect. Precarious work has been associated with poor mental health (Ferrie et al., 2008; Kim et al., 2012). Powles and Salzberg (1989, p. 152) note: 'the problem with many jobs may not be so much that they lack intrinsic rewards as that they are positively stultifying'. They go on to quote Karasek, who classified jobs on a two-dimensional grid according to whether they are high or low on demands and autonomy (ability to make decisions). Karasek's empirical work with US and Swedish workforces shows it to be the combination of high demand and low autonomy that produces most subjective distress. Self-reported depression, exhaustion, job dissatisfaction, life dissatisfaction and days off work all peak at the high demand – low autonomy corner of the job distribution. Powles and Salzberg (1989, p. 163) suggest that job stresses may reinforce behaviours, such as smoking, that are bad for health. They remind us that for many people, work may be a negative influence in their lives, 'something which occupies time but which fails to engage the spirit'. Workplace gender inequality mean women are more likely to be employed in casual work, sometimes out of limited choices to enable their caring responsibilities (Pocock et al., 2012; Tausig, 2013) and so experience different health and well-being consequences from exposure to more work-related stressors such as low decision-making latitude, low wages and unsociable hours.

UNEMPLOYMENT

That unemployed people suffer worse health than those in employment is beyond dispute. For example a UK longitudinal study (Moser et al., 1984) found higher mortality and suicide rates among unemployed men. Two systematic reviews indicate that unemployment has an adverse impact on health, particularly mental health (McLean et al., 2005; Modrek et al., 2013). What is disputed in the literature is the direction of the causal relationship: does unemployment lead to poor health or do people with poor health become unemployed? The 1958 British birth cohort study (see summary in Bartley et al., 1999, pp. 84–5) provides strong evidence that longer term unemployment causes deterioration in mental health in those who were previously healthy. Unemployment appears to affect health through the poverty it brings, and the fact that it is a stressful life event in which people lose status and social contact and a reason to exist; it may also lead to health-damaging behaviours such as smoking and drug use. Unemployment is also likely to be one experience among many for people who accumulate disadvantage through the course of their lives. Not surprisingly, the economic crisis from 2007 significantly affected people's health, especially their mental health. A WHO report (WHO Regional Office for Europe, 2011) reviewed the evidence and found substantial research suggesting that people who experience unemployment, impoverishment and family disruptions have a significantly greater risk of mental health problems such as depression, alcohol use disorders and suicide, than their unaffected counterparts. Men are especially at increased risk of mental health problems and death due to suicide or alcohol use during times of economic adversity. Unemployment contributes to depression and suicide, and young unemployed people have a higher risk of developing mental health problems than young people who remain employed.

NUTRITION AND FOOD CHOICES

Poverty and low income have a significant effect on nutrition and food choices, affecting health in many and varied ways, including effects on growth, links with specific disease, such as coronary heart disease and cancer, and through general resistance to infection. Hunger is a major health issue in all low-income countries. Malnutrition is a major cause of death for children in these countries. In more affluent countries hunger per se is much less common but food insecurity is common among low-income people and leads to worse health status (Tarasuk, 2004). Studies from the UK have suggested that providing a healthy diet may be beyond the means of many women on low incomes (Graham, 1984; Cole-Hamilton and Lang, 1986). An Australian study has shown that a basket of food costs more in low socioeconomic suburban areas and rural and remote areas than in higher socioeconomic suburban areas (Meedeniya et al., 2000). A study by the National Children's Homes (Whitehead, 1992, p. 331) found that one in five low-income parents surveyed said they had gone hungry in the previous month because of a lack of money. One in 10 children under five had gone without food in the previous month and two-thirds of the children and over half the parents had poor diets. This study also calculated that a 'healthy' diet cost 17 per cent more than an 'unhealthy' one. Travers (1996) conducted a detailed qualitative study of the food and nutrition practices of socially deprived Nova Scotia families. She found that among five families, the mothers were well aware of nutritional requirements and did, in fact, take these into account when planning the family's diet. Their planning and preparation of meals was restricted by severe material constraints, but their ability to accommodate individual taste preferences, nutrition and family members' schedules was a testimony to their skills and knowledge.

A systematic review of studies concluded that a healthy diet is more expensive than an unhealthy one (Rao et al., 2013). Food considered 'bad' for health (that which is high in fat and sugar) is cheaper to provide than healthier foods like fresh vegetables and fish. Egger and Swinburn (2010) have put forward the hypothesis that the increase in high-fat and high-sugar foods has created an obesogenic environment that is fuelling the growth in obesity worldwide. Changes to the ways food is marketed and sold have brought about three key changes to the food systems: domestic markets have been opened to international food trade, which has led to the local expansion of transnational food corporations (TFCs), which heavily market unhealthy food. Friel et al. (2014) note these three changes affect population diets; raise concerns about obesity and non-communicable diseases by altering the availability, nutritional quality, price and promotion of foods in different locations; and are likely to significantly affect people living on low incomes. Rising levels of overweight and obesity and the resultant chronic disease burden in the past decade around the world have heightened concerns about the need for policies to encourage healthy eating. These are explored further in chapter 24.

EDUCATIONAL OPPORTUNITY FROM EARLY CHILDHOOD TO ADULTHOOD

Education is important for health for three reasons: 1) early childhood development has a powerful impact on later health and well-being, 2) educational qualifications play a considerable role in determining employment opportunities, and 3) education increases knowledge, which may in turn improve health.

In terms of early childhood development there is a significant body of evidence (mainly from rich countries) that poverty has an impact on the physical, social/emotional, and language and cognitive development of children (Maggi et al., 2005). Poverty expresses itself through family, neighbourhood and community life, as well as through broader political and social structures that shape poor people's lives. As with other inequities there is a gradient of how well children flourish that largely follows the economic circumstances of families. The reduction of poverty and health inequities will require the fostering of environments for babies and children that are stimulating, supportive and nurturing. Such environments will benefit children regardless of geography, ethnicity, language or societal circumstances. Evidence from the Early Childhood Knowledge network of the Commission on Social Determinants of Health indicates that what children experience during the early years sets a critical foundation for their entire lifecourse and will influence their learning ability, school success, economic participation, social citizenry and health (Irwin et al., 2007). There is huge variation in the quality of early childhood experience, influenced by the level of development of countries and the economic and social resources available to families.

Formal educational opportunities are socially structured at all levels. Connell (1994) notes that poverty is an effective barrier to access in the school system, where opportunities are distributed according to wealth. Graham (2007) notes that young people from advantaged backgrounds are the most likely to benefit from higher education and gain better jobs, so that privilege is passed down the generations. In low- and middle-income countries, children will be pressured to leave school and contribute to the family income instead, and girls are less likely to be kept in school (CSDH, 2008).

SOCIAL CAPITAL, SUPPORT AND COHESION AND HEALTH INEQUITIES

The evidence (see below) that social support, social capital and social cohesion is beneficial to the health of both individuals and communities and that social isolation leads to high incidence of disease is now considerable. These factors (like most others that affect health) are structured according to the social and economic resources available to communities and individuals. Bourdieu's (1986) theory on social capital is most useful for understanding the ways in which social capital can reinforce an individual's position of privilege. He sees that networks act as a resource to provide people with access to other benefits such as jobs, educational opportunities or helpful legal or financial advice: literally the 'old boys' network' in action. Thus people in better-off economic and class positions are more likely to have useful networks that will assist their further advancement. He also sees that cultural capital (such as education) reinforces social capital.

In public health social capital has been used as shorthand for a measure of the level of trust, positive social networks and extent of cooperative relationships (through, for instance, voluntary associations and resident action groups) that exist in a society and the resources that flow from these. In the past decade public health researchers have become interested in social capital and health. While social capital is theoretically and methodologically complex the research does enable some conclusions to be drawn about the ways in which the social aspects of life affect health and health equity (Baum and Ziersch, 2003). The public health research has examined the role of social networks and the role of social cohesion in producing health.

Epidemiological research on the psychosocial risk factors for poor health has identified some that appear to be mediated through an individual's social circumstances (Berkman and Syme, 1979; Schoenbach et al., 1985; Ross and Huber, 1985; House et al., 1988; Kawachi et al., 1996):

- isolation
- lack of social support
- poor social networks
- levels of civic engagement
- low self-esteem
- high self-blame
- low perceived power.

There is robust prospective evidence that having strong social support is protective of health (Berkman and Syme, 1979; House et al., 1982; Kaplan et al., 1988; Berkman and Glass, 2000). Social support is the major psychosocial risk factor on which there is considerable agreement and research (see Stansfeld, 1999, for a review) and has a longer history than the social cohesion and health research. It is present if a network of people are able, emotionally and materially, to support one another. There are two slightly different dimensions to the concept: the actual number of persons—family, friends, colleagues, coworkers—with whom one meets regularly, and the quality of support offered by persons in the networks. Durkheim's classic study analysed suicide rates in various regions and interpreted a low rate as a healthy indicator of effective social bonds and a high rate as indicating ineffective social bonds and pathology (Cheek et al., 1996). Since then an increasing body of evidence has linked social factors to health outcomes. Table 13.2 summarises the findings of five prospective studies on social network support, indicating that social support is an independent risk factor for mortality. A US study found the socially isolated to be 6.59 times more likely not to survive a stroke than those with lots of social ties, 3.22 times more likely to commit suicide and 1.59 times less likely to survive coronary heart disease (Kawachi et al., 1996).

TABLE 13.2   AGE-ADJUSTED RELATIVE RISKS FOR MORTALITY, LOW VERSUS HIGH SOCIAL NETWORK SCORES

Study	Total number	Length of study (years)	Relative risk—males	Relative risk—females
Alameda County[1]	4 775	9	2.44	2.81
Tecumseh Study[2]	2 754	12	3.87	1.97
Gothenburg Study[3]	4 989	9	4.00	–
East Finland Study[4]	13 301	5	2.63	1.92
Evans County[4]	2 059	13		
Whites			1.5	1.3
Blacks			1.3	1.1

Notes
1. Controlled for self-reports of physical health, socioeconomic status, smoking, alcohol consumption, physical activity, obesity, race, life satisfaction and use of preventive health services.
2. Controlled for biomedically assessed blood pressure, cholesterol, respiratory function, ECG, and self-reports of behavioural risk factors.
3. Controlled for blood pressure, cholesterol, and self-reported smoking, alcohol and health status; study of men only.
4. Controlled for biomedical and self-reports of behavioural risk factors.

Sources: Berkman and Syme, 1979, pp. 186–204; House et al., 1982; Welin et al., 1985, pp. 915–18; Schoenbach et al., 1985, p. 585; House et al., 1988, pp. 540–5.

Studies of disease levels (morbidity), although weaker than mortality studies, also support the hypothesis that strong social networks provide a buffer against disease, especially coronary heart disease (Berkman, 1984). The Israeli Ischaemic Heart Disease study of 10 000 men suggested that psychosocial problems (particularly family ones) and support from a spouse were important predictors of angina pectoris (Medalie and Gouldbourt, 1976). Another study found that male myocardial infarction survivors were at a three to four times greater risk of death if they scored high on measures of life stress and isolation. There is some evidence that recovery from cancer is assisted by social support (Ell, 1996).

Lower levels of social support are more frequent among poorer and less educated people (Ruberman et al., 1984; Berkman, 1986; Auslander, 1988; Blaxter, 1990). Blaxter's UK survey, for example, found that people in lower socioeconomic groups experience more stress and greater isolation. Other studies (Eisenberg, 1979; Revicki and Mitchell, 1990) have found that having a close and intimate confidant can reduce depression at times of stress. They also suggest the quality of social support is more important to well-being than quantity. Parenting on low incomes was found to be associated with stress and depression among women (Brown and Harris, 1978). A British study suggested that poor women with a history of low birthweight babies (and their babies) benefited from formal social support (Oakley, 1985), but that it could not compensate for the women's multiple economic and social disadvantages.

Social cohesion also appears to be related to health. Kawachi (2007) describes the 1940s research of Shaw and McKay that maintained that certain structural characteristics of urban neighbourhoods (for example chronic poverty and high population turnover) do not allow the development of secure social attachment to a community. This in turn results in the lowered ability of communities to control problems such as crime and vandalism. Sampson et al. (1997) have argued that in poor communities informal social control is lower and there are less sanctions on unsocial behaviours. This in turn leads to a less desirable environment, which will compound the existing poverty. They also argue that collective efficacy is often lower in poor communities and so they are less able to organise and lobby to protect their own health and that of their environment and respond to trauma and threats. Public health researchers have examined the factors that make for more cohesive communities. Wilkinson (1996, 2005) argues that there is a link between high levels of income inequality in a society and lower levels of social cohesion. He further maintains that better social cohesion promotes better health by reducing the adverse psychosocial consequences of larger socioeconomic gaps such as feelings of inferiority, social exclusion, envy and shame.

Social capital (as measured by levels of trust, networks and reciprocity) is associated with better population health outcomes (see Kawachi et al., 1997; Islam et al., 2006; Kawachi, 2007) and mental health (Almedom, 2005; De Silva, 2006). The reviews found a fairly consistent association between social capital (measured in various ways) and health. Interestingly, however, the relationships seemed to operate differently between relatively egalitarian countries and relatively unequal countries. Social cohesion appeared to exert more influence on health in the unequal societies. A likely explanation is that egalitarian societies provide better universal health, welfare and education systems and so the quality of the neighbourhood becomes less important to health outcomes. In other words, in unequal societies, local social capital can act as a safety net.

Responses to building social capital and cohesion and reducing social isolation have concerned social exclusion and inclusion. These are discussed in chapter 21.

NEGATIVE EFFECTS OF SOCIAL SUPPORT AND SOCIAL CAPITAL

Kunitz (2001) reviews the literature on social support and health and notes that social relations are not always supportive. Much depends on the structure, functioning and effectiveness of networks. Networks are more likely to be unsupportive when poverty, unemployment, insecurity and inadequate infrastructure of formal organisations are prevalent. He notes that (2001, p. 167), 'Under such conditions, people have little choice in those upon whom they must depend, and the consequences of enforced dependence on kinsmen may be quite mixed, for they may be oppressive as well as supportive'. This provides a good example of how social determinants of health interact to provide health outcomes in complex ways. Some evidence suggests that stressful and negative interactions can have a negative impact on health (Franks and Campbell, 1992). Sudden increases in the number of an individual's social relationships may produce demands for reciprocal support that exceed their ability to meet them and so result in anxiety (Broadhead et al., 1983). Involvement in community groups can lead to conflict and have an adverse impact on health (Ziersch and Baum, 2004). Generally people with less power, control and resources will have less ability to cope with stressful social relationships and so stand a greater chance of being damaged by them. It is also clear that when groups are highly bonded and trusting they may also be exclusionary and suspicious of other groups who are different. This may have an adverse impact on newcomers such as migrants, or groups excluded because of, for instance, racism, such as Indigenous people.

SOCIAL HIERARCHY, STRESS AND ILLNESS

We have noted earlier that Wilkinson and Pickett (2009) and figure 13.2 argue that among rich countries more egalitarian societies have better population health and outcomes on a range of measures of social well-being, and explain this by the social effects of less distinct hierarchies. Certainly Evans (1994) suggests that the differences in health according to socioeconomic status may have something to do with the experience of hierarchy per se. He draws on studies of animal behaviour and recent findings from immunology to suggest that stress (induced by various causes but including being in an inferior position to others and lack of social support) can act to suppress immune systems and so make people more vulnerable to disease. He and colleagues (Evans et al., 1994b) also examine the evidence on genetics, biological pathways and the health of different groups in the population. They claim that subtle biological effects of the social and cultural environment affect groups of people and that the relationships between social and physical factors are 'extraordinarily subtle and complex' (p. 182). In a summary of their review of the literature, Evans et al. (1994b, p. 184) comment:

> There is a chain that runs from the behaviour of cells and molecules, to the health of populations, and back again, a chain in which the past and present social environments of individuals, and their perceptions of those environments, constitutes a key set of links. No one would pretend that the chain is fully understood, or is likely to be for a considerable time to come. But the research evidence currently available no longer permits anyone to deny its existence.

The immune system has received particularly close attention, but Evans et al. indicate that it is only one component of the network of physiological systems responding to (and in turn influencing) the electrical and chemical output of the nervous system as it responds to perceptions of the external world. These biological responses are shown in figure 13.4. Stress is one biological mechanism that explains how social determinants get under the skin.

FIGURE 13.4   STRESS AND HEALTH INEQUITY: BIOLOGICAL RESPONSES

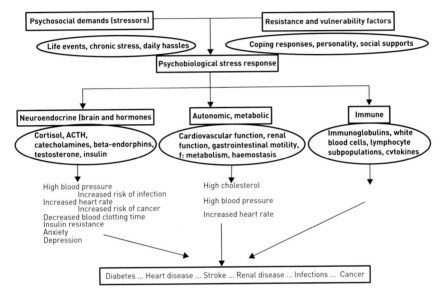

Source: Based on Marmot and Wilkinson, 1999.

There is an increasing amount of research that links low self-esteem, unhappiness, low perceived power and high levels of self-blame to people in the most disadvantaged socioeconomic circumstances (Harding, 1987). Lerner (1986) adapted the theory of 'learned helplessness' to persons with little objective power. He hypothesised that people living in poor socioeconomic conditions tend to internalise their own powerlessness, create a psychological barrier and 'begin to accept aspects of their world that are self-destructive to their own health and well-being, thinking that these are unalterable features of what they take to be reality' (Lerner, 1986). This internalising process leads to isolation and further removal from active group participation.

Lerner also maintains that self-blame and internalised anger are associated with increased health risks and poorer health. He designed a clinical trial to reduce self-blame and internalised anger through the provision of social support to blue-collar workers experiencing occupation stress. The experimental group demonstrated statistically significant improvements on all measures used. The groups took place under union sponsorship, which may have been an important factor. Many stresses are embedded in the structure of work and so actions to change these structures usually require organised political effort.

The literature on health inequalities indicates that their existence has to be accounted for by more than the operation of absolute poverty and its effects. Marmot et al. (1995) review the evidence from the Whitehall studies and make the point that as each grade has worse health and higher mortality than the grade above, there must be factors working across the whole of society. Executive grade civil servants are not poor by any standards, yet they have higher mortality rates than administrators who occupy higher grades. Evans (1994) hypothesises that the difference may be explained by the stresses of inferior position, and uses biological models to support his hypothesis. He suggests lower ranking civil servants may experience learned helplessness. Marmot et al. (1995) believe that the inequities between civil servants may result from job strain, low social support and low control. They point out that high psychological demands and low control at work are linked to cardiovascular and other diseases. Marmot (2004) has developed this thesis and argued that the 'status syndrome' exerts a powerful influence on health and that it is vital that we improve opportunities for control and engagement for all people.

LOCATION AND HEALTH

Location has a powerful impact on health (see table 12.7 in previous chapter). Debate continues on the relative contribution of compositional factors (that is, the makeup of the households in a particular location) and contextual factors (that is, the differences reflecting the characteristics of the location itself—factors such as amenities, pollution levels and social environment). Developments in multilevel statistical techniques have enabled researchers to untangle these complex factors. Turrell et al. (2007), using a multilevel analysis, showed that in Australia, area level and individual level socioeconomic factors make an independent contribution to the probability of premature mortality. Research to date suggests that both compositional and contextual factors are important in determining health outcome and that poorer people may in part have poorer health because they live in places that are health-damaging (Macintyre and Ellaway, 2000; Kawachi and Berkman, 2003; Stock and Ellaway, 2013). The features of areas that contribute to poorer health are material hazards like environmental pollution, traffic volume and rates of road accidents, and the nature of resources such as shops, recreational facilities, public transport and primary health care services. The impact of location on health is clearly demonstrated in remote Indigenous communities in Australia, where basic public health facilities such as clean water and sewerage are lacking, health care access is inadequate, housing is unsuitable and food supply is less healthy than in cities.

INTERACTION BETWEEN FACTORS

All the factors relating to socioeconomic status reviewed above do not operate independently but rather interact with one another and combine to result in inequitable health outcomes. This interaction is shown in relation to mental health in figure 13.5.

FIGURE 13.5   HOW SOCIAL DETERMINANTS INTERACT TO AFFECT MENTAL HEALTH

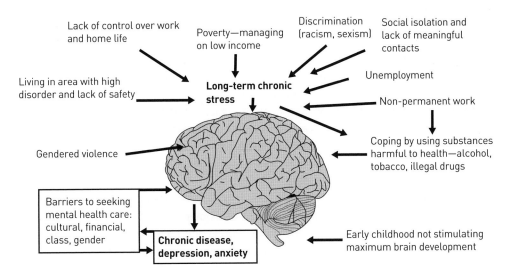

GENDER AND HEALTH

Research on socioeconomic inequalities shows that people living in poorer circumstances are likely to die earlier than their richer counterparts. Women in industrialised countries are more likely to be living in poverty than men. In OECD countries women live longer than men, although the gap is reducing. However, despite their greater longevity women are more likely to report and be treated for illness than are men. They also make more use of health services.

How can we explain this? The reasons are complex and we are far from understanding them. There was intensive research attention on explaining these reasons in the 1970s and 1980s when the life expectancy gap was greatest, especially in the USA (Waldron, 1976; Gove and Hughes, 1979; Verbrugge, 1979; Walsh et al., 1995). Macintyre (1986) reviews British research on social patterning and health and comments on the fact that social variables (occupational class, gender, marital status, age, ethnicity and area of residence) are generally treated as separate research strands. Gender differences in health have to be seen through a lens that incorporates socioeconomic differences. For both men and women socioeconomic circumstances across the lifecourse shape their life chances and their material and psychosocial environments. Rogers et al. (2010) examined US data and found that while co-variates can explain some of the difference (especially men's higher smoking rates) there is still a 62 per cent difference in mortality.

It is worth noting that in developing countries men have a longer life expectancy than women (Sen, 2001b). This is partly due to higher maternal mortality and to the social and economic position of women in many low-income countries. Sen (2001b) sees this as reflecting seven types of inequality, including social preference for boys over girls, discrimination against girls and women in employment and education, unequal ownership of property, and unequal distribution of resources within families so that, for instance, boys and men receive better and more food. This underlines the fact that understanding gender differences in health requires a complex framework of understanding how patterns differ across time, cultures, cohorts and geographical locations. Moss (2002) offers such a framework, which demonstrates the complexities (see table 13.3). Her framework permits consideration of the geopolitical environment (for example policies, legal rights and economic factors); cultural norms and sanctions (for example discrimination and socio-demographic characteristics); social roles; health-related mediating factors (for example access to social capital and networks, health services availability and psychosocial factors); and actual health outcomes (mortality and morbidity) that affect women's health.

In developed countries much of the research on socioeconomic status and health has concentrated on the position of men, and that of women has been secondary. Analysis of households with married couples by occupation is often based on men's occupations. Yet single women are analysed according to their own occupations, making meaningful comparisons between women with different marital status difficult.

It is true that women report more sickness than men but the mortality differentials are considerable. Three main explanations have been put forward:

- biological
- gender-based variations in the reporting of illness
- social.

BIOLOGICAL DIFFERENCES

The biological arguments to explain gender differences in male and female morbidity and mortality rates suggest that there are innate physiological differences in the constitutional resistance to disease. This is supported by the fact that female foetuses have better survival rates than male foetuses (Hart, 1991). Kane (1991) reviews the relevant literature and concludes that the supporting evidence is limited with the exception of susceptibility to infectious disease among infants. Female babies do appear to be less vulnerable, possibly because of the protective effect of immunoregulatory genes carried on the X chromosomes. Whitehead (1992, p. 334) observes that the sex differential in mortality is reversed in less developed countries and that the morbidity differential is not fixed and appears to vary, suggesting that social and cultural factors do play a powerful role.

Women's excess morbidity may partly be explained by biological reasons associated with childbirth, menstruation and the menopause. However, Verbrugge (1985) found that, even when all aspects of reproductive health are controlled, women have acute illness rates roughly 20–30 per cent higher than men's, apart from injury. Table 13.3 shows that there are many other factors than biological ones that affect women's health status.

TABLE 13.3   A COMPREHENSIVE FRAMEWORK OF FACTORS INFLUENCING WOMEN'S HEALTH

Geopolitical environment	Culture, norms, sanctions	Women's roles in reproduction and production	Health-related mediators	Health outcomes
Geography Policy and services:	**Discrimination:**	**Household:**	**Social capital/social networks/ support:**	Chronic disease
Transportation	Ethnic	Structure		Infectious disease
Welfare	Gender	Division of labour	Friendship	Disability
Employment	Age	Ownership/ property	Family	Functioning
Health care	**Socio-demographic characteristics:**	Support/caretaking	Work mates	Mortality
Child care		Equality of access to household resources, e.g.:	Other ties	Mental health/illness
Legal rights:	Age		**Psychosocial:**	
Women's	Gender	• Wages	Stress	
Health	Ethnicity	• Other income	Mood	
Human	Birthplace	• Land	Coping	
Employment	Education	• Other assets	Spirituality	
Organisations:	Marital status	Community roles	**Health services:**	
Banks	Language	Labour market role	Availability/use	
Credit co-ops		**Workplace:**	**Behaviours:**	
Political parties		Sector	Sexual	
Advocacy		Formal	Substance use	
Unions		Home/ market-based	Physical activity	
Economic:		Hierarchies, control	Diet	
Policy			Contraception	
Extent of inequality		Authority, discretion	Breastfeeding	
		Sex	Smoking	
		Segregation/ discrimination	Drinking	
			Violence	

VARIATIONS IN THE REPORTING OF ILLNESS

Some commentators have argued that men have lower rates of illness because they are less likely than women to perceive symptoms, articulate them and then seek professional help (Mechanic, 1976; Verbrugge, 1977), possibly because of societal expectations for male behaviour. They are expected to be in control of their emotions, not show pain, be self-sufficient and not appear weak. Recent years have seen discussion of 'new men', who are more in touch with their feelings and who would be happier to seek help. If this is a sustainable and real trend, it might contribute to reducing male mortality rates in the future.

SOCIAL DIFFERENCES

Socially determined behaviour and the fact that men typically work in high-risk industries could account for higher male mortality rates. Waldron (1991) estimates that these factors together account for about 5.1 per cent of excess male mortality in the USA. Men account for 82 per cent of the homicide deaths globally, of which there were six million between 2000 and 2013 (WHO, 2014d). The rate for young men (15–29 years old) is particularly high at 18.2 per 100 000 compared with 3.2 per 100 000 for women. Astbury (2002) argues that the social and economic position of women, the roles they typically play and the power imbalance with men results in worse mental health status. A US study that measured women's level of depression and their status between state jurisdictions found that levels of depression were highest where employment and earnings, reproductive rights and economic autonomy were lowest (Chen et al., 2005).

Social norms such as these may be changing in contemporary Australia but they have shaped the lives of many Australian men and may account for some of the differences in mortality between women and men. Lopez (1983) estimated that approximately 90 per cent of the male excess mortality at ages 15–44 years in Australia resulted from motor vehicle and other accidents and suicide. Rogers et al. (2010) calculate that in the US smoking accounts for 22 per cent of the sex difference in mortality between men and women, and that the reduction in male smoking has made a significant contribution to the narrowing gender differences in mortality.

Some commentators (Nathanson, 1975; Clarke, 1983) argue that women's higher morbidity reflects the consequences of a largely patriarchal society on women. Saltman (1991, p. 129) reports that women are more likely to experience depression, anxiety, and sleeping and emotional problems than men. This could result from the roles and responsibilities women perform. A UK study (Brown and Harris, 1978) found the highest rates of undiagnosed depression among working-class women who were full-time (and often housebound) carers. Suicide for middle-aged women was higher in Australia when most women in that age group were not in the workforce and were far more likely to be full-time carers experiencing social isolation. This lends support to the argument that social roles are important in explaining the gender differences.

Kane (1991, p. 35) speculates on how social change could have contributed to different exercise patterns among men and women, observing that manual work for men is much less common and car ownership more common than in the past, decreasing the amount of male exercise. On the other hand, women still do most housework, which although less physically demanding than in the past, involves a healthy amount of exercise. There are many other ways in which different male and female roles may affect health. These have not been systematically researched and so the pattern of causality is largely a matter of speculation.

Women are increasingly carrying a double burden of workforce participation and being the main child and home carer. It is unclear how this double burden compares with the strains of being a full-time isolated housewife. Employment offers financial benefits, companionship and may contribute to self-esteem but the double work load inside and outside home may be detrimental to health (Arber et al., 1985; Pocock et al., 2012). Pocock et al. (2012) present the double burden women face as a 'timebomb'.

A further clue to these complex relationships comes from a longitudinal study (Moser et al., 1990) that found women in non-manual jobs had a lower mortality rate than those in manual jobs. Part-timers had lower mortality than full-timers among non-manual workers, but not among manual female workers. Differences in mortality rate according to social class were most pronounced for women not in the workforce. Among them women married to manual workers had a death rate one and a half times that of those married to men in non-manual occupations. A Scandinavian study (Haavio-Mannila, 1986) found that in countries where many women work outside the home, rates of illness and hospitalisation for women were lower than those for men. In countries where women do not tend to work outside the home, women's morbidity tends to be higher than that for men. Walsh et al. (1995, p. 137) review the literature on multiple roles and point out that 'studies have implicitly or explicitly tested a "scarcity hypothesis" (more roles produce more demands on scarce personal resources) versus an "expansion hypothesis" (more roles produce expanded horizons and greater actualisation)'. The literature does not clearly support either hypothesis.

Specific aspects of women's social circumstances can have an impact on health. Women are also more likely to be caring for a sick or disabled person than are men (Bulmer, 1987, p. 24), which may take its toll on the physical and emotional health of the carer. Violence in the home is a significant, but largely unrecorded, cause of physical injury and emotional distress among women (WHO, 2002). A WHO report states, 'acts of violence against women are not isolated events but rather form a pattern of behaviour that violates the rights of women and girls, limits their participation in society, and damages their health and well-being' (WHO et al., 2013a, p. 1). While domestic violence has become more visible in recent years, routine data collections do not capture the extent of morbidity it causes. However, WHO (WHO et al., 2013a) has produced estimates that indicate that, overall, 35 per cent of women worldwide have experienced physical or sexual intimate partner violence or non-partner sexual violence.

The evidence from longitudinal research linking social ties to longevity (see table 13.2) combined with evidence that women generally have more and stronger social ties than men may account for women's mortality advantage. Other evidence reviewed by Walsh et al. (1995) urges some caution in that caring and being close to people is not without its costs and, as noted above, participation in community activities is not necessarily beneficial for health.

The differences in mortality between men and women have narrowed in industrialised countries in the past decade, reflecting aspects of the recent rapid change in gender roles and relationships. However, there is no clear understanding of how these changes impact on health.

INEQUITIES: THE CASE OF ABORIGINAL HEALTH

Social patterns of health illness are obviously complex, as this section demonstrates. The interrelationship of many factors in bringing about poor health is no better illustrated than by Aboriginal people's health. The following constellation of disadvantages are experienced by Aboriginal people (House of Representatives Standing Committee on

Family and Community Affairs, 2000; Aboriginal and Torres Strait Islander Social Justice Commissioner, 2005).

EDUCATION

A far higher proportion of Aboriginal than non-Aboriginal Australians have never attended school and a lower proportion participate in education after the age of 15. Although educational attainment has been increasing among Indigenous Australians in recent years, in 2008, 37 per cent of Indigenous adults aged 18 years, and 22 per cent of people aged 15 years and over, had achieved a minimum of Year 12 or a vocational qualification, and 5 per cent of adults had completed a bachelor degree (ABS, 2012b). Between 2006 and 2011, the rate of Indigenous Australians achieving Year 12 or equivalent increased by 6.5 percentage points, from 47.4 per cent to 53.9 per cent (COAG Reform Council, 2013). While Aboriginal and Torres Strait Islander students are still less likely than non-Indigenous students to complete their final years of schooling (49 per cent compared with 81 per cent in 2011), the gap between the two groups has narrowed. For Year 10, the difference between apparent retention rates for Aboriginal and Torres Strait Islander and non-Indigenous students decreased by 10 percentage points between 2001 and 2011. Differences in the Year 12 apparent retention rate decreased by 7 percentage points over the same period (ABS, 2012b). Groome (1995) reports that Aboriginal adolescents are typically identified by education systems as having low levels of achievement and retention and high levels of failure, absenteeism and behaviour problems. Aboriginal students become alienated by a school system that is usually unsympathetic to their needs. They are likely to meet indifference and harassment, leading to disillusionment and frustration with school, and ultimately dropping out. Educational institutions may often not be culturally appropriate for Indigenous peoples and care needs to be taken to ensure that participation in mainstream education is not accepted uncritically as automatically leading to improved health for Indigenous peoples (Dunbar and Scrimgeour, 2007). The improved education outcomes suggest the education system may be responding to some of these criticisms.

EMPLOYMENT STATUS

In 2011, the employment rate for Indigenous workers aged 15 to 64 was 44.1 per cent, compared with 71.4 per cent for non-Indigenous Australians. Indigenous Australians were more likely than non-Indigenous Australians to be unemployed (that is, looking for work) in 2011—9.2 per cent compared with 4.2 per cent for non-Indigenous Australians—and they were out of the labour force at a greater rate. Employment rates also deteriorated for Indigenous Australians between 2006 and 2011, more so than for non-Indigenous Australians (Karmel et al., 2014).

ECONOMIC STATUS

Aboriginal people are more likely to have low incomes and derive their incomes from social security benefits than other Australians (see chapter 12).

HOUSING

In 2006, Indigenous Australians comprised 9 per cent of the homeless population in Australia, despite only comprising 2.5 per cent of the Australian population overall (AIHW, 2011b). In remote areas, poor-quality housing and inadequate health hardware can create serious health risks and adversely affect physical health through problems such as infectious and parasitic disease, eye and ear infections, skin conditions, and respiratory tract infections (Garvey, 2008).

According to data from the 2011 Census, higher proportions of Indigenous households than non-Indigenous households are in housing stress (Biddle, 2012). This is particularly the case for non-remote regions outside of capital cities, which have the highest rates of Indigenous households in housing stress (for example, the south-eastern region and central and north coasts of New South Wales) (Biddle, 2012). Large capital cities also have rates of Indigenous households in housing stress that are significantly higher than non-Indigenous households (Biddle 2012).

LACK OF APPROPRIATE ENVIRONMENTAL INFRASTRUCTURE

The conditions under which many Aboriginal Australians live are more equivalent to those of a developing country than the rest of Australia. Typical problems are contaminated drinking and washing water, poor sanitation, unsafe housing and lack of nutritious food choices (Osborne et al., 2013). The appalling state of the living conditions of Aboriginal people in remote areas was highlighted by the comments from the president of the Australian Medical Association following a visit to the Kimberley region of northwest Australia in 1996:

> In some of these communities, a 'home' for up to 20 people consisted of a tin shed without a toilet or washing facilities. The toddlers' playground consisted only of a car body strewn with broken glass. Could any of us honestly hope to sustain our motivation and self-esteem if forced to live in a community with appalling physical conditions, where alcohol abuse was rife and with unemployment running close to 100%? One community I visited still had raw sewage on the ground when we arrived and had recently been through a two-month period with no drinking-water supply (Woollard, 1996, p. 7).

Even when infrastructure is provided, the outcome is often disappointing, due largely to lack of funding, equipment, education and training. A further problem is the inappropriateness of the technology to the setting in which rural and remote Aboriginal people live, with infrastructure often being imposed on communities without sufficient participation and devolution of decision-making power, which means communities feel no sense of ownership (Moss, 1994).

CRIMINAL JUSTICE SYSTEM

Imprisonment rates of Indigenous Australians are approximately 12 times the rate of the non-Indigenous population, and Indigenous Australians comprise 40 per cent of people imprisoned for assault offences (Australian Institute of Criminology, 2013).

Despite comprising less than 3 per cent of the Australian population, one-quarter of the prison population in Australia is Indigenous (AIHW, 2011c; Senate Select Committee on Regional and Remote Indigenous Communities, 2010). Indigenous Australians are also overrepresented in juvenile detention, and Indigenous 10–17 year olds are 24 times more likely to be in detention than non-Indigenous young people (Australian Institute of Criminology, 2013). Data from the 2008 National Aboriginal and Torres Strait Islander Social Survey show that, of Indigenous Australians aged 15 and over, 15 per cent have been arrested in the previous 5 years, 3.2 per cent have been incarcerated in the previous 5 years, 17.5 per cent have used legal services in the previous 12 months, and 23.2 per cent report being the victim of physical or threatened violence in the past 12 months (ABS, 2009).

APPROPRIATE HEALTH SERVICE PROVISION

The history of health service provision for Aboriginal people reflects the colonialism and racism that has characterised their experience since white colonisation. For instance, public hospitals provided segregated accommodation for Aboriginal people until the 1960s, and until the 1970s the provision of health services for Aboriginal people was pitiful (Saggers and Gray, 1991). There is also evidence of discrimination in health services. A study based on data from the National Morbidity Database for hospital separations over 1997 and 1998 reported that Aboriginal and Torres Strait Islander patients with cardiovascular disease were significantly less likely to undergo major procedures, such as angiography: at a rate of about half of that of non-Indigenous patients. There were also significant differences in the rates of bypass surgery or angioplasty between the two groups (Cunningham, 2002). It is estimated that in 2004, Aboriginal and Torres Strait Islander peoples enjoyed 40 per cent of the per capita access of the non-Indigenous population to primary health care provided by general practitioners.

Morgan et al. (1997) argue that the nature of Aboriginal philosophical thinking has been greatly misunderstood by the Western health care system and that this has contributed to poor health outcome experiences by Aboriginal people. They illustrate their argument by discussing Aboriginal notions of:

- *Importance of identity*: This is intimately and holistically connected with kinship, ritual, and spiritual relationships and responsibility, which are inseparable from each other and the land.
- *Preference for contextual, concrete knowledge*: This means that the objective science that shapes Western health services may be alien to Aboriginal people.
- *Significance of 'shame'*: Shame is described as 'a powerful emotion resulting from the loss of the extended self' that 'profoundly affects Aboriginal health and health care outcomes' (Morgan et al., 1997, p. 598). It can result from separating a person from the group. Individual recognition separates a person from their extended identity and ultimately from life.

Morgan et al. (1997) conclude that health services will only be appropriate and effective for Aboriginal people when these philosophical differences are recognised and incorporated in health care provision. Some progress towards this was made with the development of a Cultural Respect Framework for Aboriginal and Torres Strait Islander Health to guide policy and service delivery in mainstream health services. This document aims to ensure cultural safety and to legitimise traditional healing practices. It also aims to embed cultural respect at the 'corporate, organisational and care delivery levels' of the health system (Australian Health Ministers' Advisory Council, 2004, p. 13).

Demonstration in support of the Close the Gap campaign, steps of Parliament House, Adelaide. (Fernando M Gonçalves)

One of the most significant positive developments has been the growth of community-controlled Aboriginal medical and health services throughout Australia (see box 18.3). These services have a national organisation—the National Aboriginal Community Controlled Health Organisation (NACCHO). Aboriginal health workers have also played an instrumental role in extending health services to Indigenous people, especially in rural and remote areas (Tsey, 1996).

Dwyer et al.'s (2004a, p. 33) review of primary health care provision to Indigenous peoples concluded that:

> The available evidence of health impact in Indigenous populations, and the known effective interventions of primary health care, indicates that the impact of effective primary health care is seen in:
>
> * reduced prevalence and incidence of communicable diseases that are susceptible to immunisation programs;
> * reduced complications of chronic disease through effective chronic disease management programs;
> * improved maternal and child health outcomes (such as birth weight) through the implementation of culturally appropriate antenatal and early childhood programs; and
> * reduction in social and environmental risks through effective local public health advocacy, such as changes to liquor licensing regulations.

They noted that a focus on primary health care interventions that addressed chronic disease could be expected to have a significant impact on Aboriginal and Torres Strait Islander peoples' life expectancy if they were funded at a much higher level. Money spent on primary health care would also be likely to reduce Indigenous people's use of clinical and hospital services over the longer term.

IMPOSED WELFARISM

Imposed welfarism has been an entrenched response of white Australia to the Aboriginal 'problem', which has structured contemporary Aboriginal life, especially in remote areas. 'One need only visit towns in remote Australia on successive weeks to see the difference between "pension week" and "slack week". The social impact is unavoidable, reflected in the periodicity of resources and behaviours such as drinking and its social consequences' (Hunter, 1993, p. 261). Every major report on Aboriginal health status has defined the importance of community consultation, community control and self-determination to improving Aboriginal health (most recently, see Osborne et al., 2013).

Perhaps the most extreme form of imposed welfarism, however, was the practice of forcibly removing Aboriginal children from their families to bring them up either in white foster homes or institutions. The national inquiry into the separation of Aboriginal and Torres Strait Islander children from their families (Human Rights and Equal Opportunity Commission, 1997) saw the effect of the forced removal of children from their families as having caused major cultural, psychological and social damage to the individuals removed and their communities. The practice resulted in the breakdown of culture links with their Country, and the deprivation and destruction of family relationships. While this practice officially ceased in the 1960s and 1970s, Indigenous children are still taken into care by state government welfare departments at an alarmingly high rate (see box 13.3 for discussion of health consequences for the Stolen Generations).

BOX 13.3   SOME HEALTH CONSEQUENCES FOR THE STOLEN GENERATIONS

In Western Australia, 'in the 1970s ... one-third of Aboriginal people who spent part of their childhood in a children's home or foster home were subsequently imprisoned either for juvenile or adult offences'.

In New South Wales and Victoria, analysis of legal service clients seeking assistance on criminal charges 'showed that 90–95 per cent of clients had been in placement, either in foster homes, institutionalised or adopted, with the majority having been in the care of white families or authorities. There is strong argument that fewer children would have ended up in the criminal justice system had they been brought up in the Aboriginal community.'

In Victoria, 54 per cent of all health service clients surveyed had a psychiatric disorder and, of these, 50 per cent had been separated from their parents and more than 25 per cent had been brought up outside their communities in foster homes or institutions. A disturbing number of Indigenous separated children have fallen victim to violence, abuse, accidents, addictions and suicide. Such children, many of whom are now adults, have a legacy of 'broken families, broken culture and broken spirit'.

Childhood removal ensured that these children were denied their identity, as a matter of policy, as well as access to other information and knowledge, such as stories, songs, dances, how to obtain bush tucker and many other components of their heritage, which is so important to Indigenous people. These children were actively discouraged from speaking their own language in favour of learning English and refused access to their families, communities and lands. The ability to transmit culture, traditions and practices between generations was systematically broken down as a consequence of removal and institutionalisation.

Source: Aboriginal and Torres Strait Islander Commission (ATSIC), 1997, in its submission to the Human Rights and Equal Opportunities Commission Stolen Generation inquiry.

Noel Pearson, one of the most prominent Indigenous leaders in Australia, has been very critical of the imposed welfarism. The Cape York Institute, of which he is director, notes:

> Far from participating in a real economy, people in Cape York have been almost completely dependent on passive welfare for over three decades. By removing the incentive to work, passive welfare delivery has embedded dependency, effectively sapping people of motivation and eroding personal responsibility (Cape York Institute, 2007).

Other commentators have noted that the prominence of welfare dependency in political debates about Indigenous health has the effect of seeing the welfare as the root cause of Indigenous inequality and so obscures the role of social, political and economic inequality in Indigenous health inequality (Walter and Mooney, 2007).

The past decade has seen significant changes to the system of Indigenous governance. In 2005 the Howard Coalition government abolished the Aboriginal and Torres Strait Islander Commission, which had directly run service provision, and moved control back to mainstream services. In addition the federal government has changed the basis of welfare payments and instituted policies based on 'mutual responsibility', which echo back to an earlier paternalistic period (I.P.S. Anderson, 2006) and certainly continue the tradition of welfarism. This trend was intensified by the Northern Territory Emergency Response (NTER), which was introduced as the Howard government's response to the *Ampe Akelyernemane Meke Mekarle 'Little Children are Sacred'* report (Wild and Anderson, 2007), with the overarching aim of protecting children and making communities safer. The *'Little Children are Sacred'* report documented, as had many previous reports, evidence regarding the disproportionate levels of violence and abuse, including child abuse and neglect, in Indigenous communities in the Northern Territory. The report also advocated for genuine consultation and partnership with Indigenous communities to design any initiatives to address these issues. The NTER is described in box 13.4. It has been highly controversial.

BOX 13.4 THE NORTHERN TERRITORY EMERGENCY RESPONSE

The Northern Territory Emergency Response (NTER) set out to address a range of social and economic factors that are associated with child abuse, neglect and poor health, including poverty and economic disadvantage, mental illness, substance abuse, overcrowding, poor housing and unsafe communities (Higgins and Scott, 2012). It operated in 'prescribed' Indigenous communities (including townships and town camps) across the Northern Territory. Activities that were initially involved included:

- comprehensive health checks for Indigenous children under 16 years
- increased policing levels
- widespread alcohol restrictions
- banning possession of pornography
- introduction of compulsory income management for welfare recipients
- provision of school meals
- suspension of the *Racial Discrimination Act* with respect to the overall NTER legislation
- allowing the Australian Government to take leases over parcels of Indigenous-owned land, whereby ownership remained with Indigenous land owners, and the government paid rent to land owners for 5-year leases
- improvements to essential infrastructure to clean up and repair communities.

The NTER was controversial for a number of reasons, including the speed with which it was introduced and the lack of consultation and involvement on the part of Indigenous communities. An independent review commissioned by the newly elected Australian Government in 2008 reported that Indigenous Australians felt humiliated and shamed by the introduction of the NTER, and many believed that the NTER measures were imposed on the basis of race, and felt a sense of being blamed for abuse, violence and neglect (Osborne et al., 2013). The Rudd Labor government elected in 2007 retained the NTER, with a stronger focus on 'Closing the Gap', including the introduction of the Closing the Gap strategy in the Northern Territory National Partnership Agreement (NTNPA), and most recently, the 'Stronger Futures in the Northern Territory' package of services and programs, announced in 2012 (Department of Families, Housing, Community Services and Indigenous Affairs, 2013). The legislative changes made to the NTER in 2010 are summarised in the policy statement: 'Landmark reform to the welfare system, reinstatement of the *Racial Discrimination Act* and strengthening of the Northern Territory Emergency Response' (Department of Families, Housing, Community Services and Indigenous Affairs, 2009). The NTNPA was structured around programs and initiatives in the following areas: income management and welfare reform; law and order; family support; early childhood; improving child and family health; enhancing education; remote service delivery; 'resetting the relationship' between Indigenous people and government; housing and land reform; and other activities (Department of Families, Housing, Community Services and Indigenous Affairs, 2011). The *Racial Discrimination Act* was reinstated. Income management was retained in a substantially altered way (see Bray et al., 2012, for a full description of these changes). Other aspects of the NTER that were retained in modified ways included alcohol restrictions (which communities were able to tailor to their own circumstances) and alcohol management plans (Department of Families, Housing, Community Services and Indigenous Affairs, 2011). Osborne et al. review the evidence on the effects of the NTER and find they were mixed. Some progress was made in terms of improving basic services, infrastructure, safety in the communities, increased policing, increased availability of food and health checks. But the evaluation identifies shortfalls in the health, education, police and governance workforces as a problem (Department of Families, Housing, Community Services and Indigenous Affairs, 2011). There is also low employment in NTER communities, and further economic development was identified as essential. The rates of alcohol-related offences and violent crime remain high. One of the most controversial aspects of the NTER was the income management. Welfare recipients reported problems with using their 'basics card' (a card supplied by Centrelink for individuals whose payments are managed, to spend their money on approved goods), and support for income management was reported as divided in communities. A health impact assessment of the NTER (Australian Indigenous Doctors' Association and Centre for Health Equity Training, Research and Evaluation, 2010) was critical of its overall impact and concluded that compulsory income management is likely to have significant negative long-term implications for health, social and emotional well-being. Altman (2010) argues that income management is a paternalistic measure which is not consistent with basic human rights and smacks of the welfare abuses of the past.

LOSS OF CONTROL OF CULTURE AND LAND

The traditional culture and lifestyle of Aboriginal Australia were split asunder by the British invasion and subsequent colonisation of the continent. There are many stories of how Aboriginal people were treated by their white invaders. Here is one example, related to disease control. Hunter (1993, p. 61) quotes E.L. Grant Watson, who was commenting in 1946 on the treatment of Aboriginal people suspected of having syphilis:

> The method of collecting the patients was not either humane or scientific. A man unqualified except by ruthlessness and daring, helped by one or two kindred spirits, toured the countryside, raided the native camps and there, by brute force, 'examined' the natives. Any that were obviously diseased or were suspected of disease were seized upon. These, since their hands were so small as to slip through any pair of handcuffs, were chained together by their necks, and were marched through the bush, in the further search for syphilitics.

Hall (1990) suggests the term 'diaspore' to describe the experience of colonised peoples. Inevitably, they have a diverse and heterogeneous experience that means they have to construct identity from their past histories before and after colonisation. This means identity formation is difficult. Aboriginal people struggle with their knowledge of the injustice of dispossession and the value of their traditional culture in the face of white indifference or outright racism.

Indigenous Australians legally own 15.9 per cent of the total Australian landmass and of that 98.6 per cent is located in remote or very remote locations (Burgess and Morrison, 2007, p. 178). The land rights movement has been fiercely opposed and the fight for those rights a slow process. The High Court decision in the Mabo case in June 1992 destroyed the concept of terra nullius (that prior to European colonisation there was no ownership of the land) as the basis for determining legal property rights. It was therefore possible for native peoples to have title to the land dating from pre-white settlement—unless that title had been since extinguished by other forms of land use and legal property acquisition. That was followed in December 1996 by a further decision of the Court in the case of the Wik Peoples versus the Queensland Government. This case found that native title was not extinguished by pastoral leases granted by the Crown to graziers. In 1997, however, the Commonwealth Government made clear its intentions to legislate against the effects of the Wik judgment. Given that some public health issues for Aboriginal people can be addressed only through action by Aboriginal communities themselves, and that the concept of links to the land is central to community integrity, particularly for traditional rural Indigenous people, the issue of land rights has immediate and practical, as well as symbolic, importance in the struggle to advance Aboriginal health.

Australia's dominant European culture has not appreciated the deep significance of land and place for Indigenous Australians (Burgess and Morrison, 2007) over two centuries of dispossession of land, language and culture. The history of Indigenous peoples in Australia paints a picture of people denied control over their lives, culture and land. Marmot (2004) describes a body of research that stresses how important a sense of control is to health. Consequently the lack of control experienced by Indigenous peoples since the white invasion suggests that the process of colonisation has in and of itself had a negative impact on health. It provides a strong health rationale (in addition to a social justice argument) for effective community governance and the community control of services.

RACISM

Racism affects societies as a whole through institutional and structural racism and acts as a collective stressor. Racism is hard to define precisely but can be thought of as an oppression, which along with its opposite, privilege, has a direct effect on the distribution and operation of power in society (Paradies, 2007). It also affects individuals' mental and physical health. For example, a review of 53 studies in the USA found a decline in mental health status as racism increased. Eight out of 11 studies found links between the elevated prevalence of high blood pressure in Afro-Americans and racism (Williams et al., 2003). An Australian study found that exposure to racism was associated with worse mental health (Ziersch et al., 2011a) and that racist assaults are an aspect of everyday life for Australian Aboriginal people, which has an impact on mental health (see box 13.5).

Don't Let Racism Win. (Fernando M Gonçalves)

**BOX 13.5   RACISM AND MENTAL HEALTH IN ADELAIDE:
               ABORIGINAL PERSPECTIVES**

It plays like on your brain, on your mind. People's minds, and stress. It gets people down yeah. Makes people drink you know. Drinkin' and smokin' to escape reality.

(Anthony, 45 years old)

Metaphorically if you get slapped down for so long you have. If you're ignored and you're pushed to one side then that affects your health. That affects your mental [health], your physical [health], you end up smoking, you drink, you become frustrated, you get violent.
    And it not only affects you, it affects people around you.

(John, 52 years old)

Makes you sick. It's hard to deal with it. Yeah it's stressful. But you get over it.

(Carol, 53 years old)

Every time I have a racial discrimination experience, it's a shameful experience and it is humiliating and you've got to keep your cool.

(Diane, 56 years old)

Source: Ziersch et al., 2011b.

Aboriginal people are seen negatively by many Australians. These negative images go back to the colonial view of Aboriginal people as inferior beings needing to be 'civilised and Christianised'. More recently, descriptions of Aboriginal people as deficit, victims, at risk, of inferior intelligence have all added to the negative view (Groome, 1995). The racism underlying white Australia's attitude to Indigenous Australians was seen at the founding of the nation. The Australian Constitution of 1901 section 127 stated: 'In reckoning the number of people of the Commonwealth or of a State or other part of the Commonwealth, Aboriginal natives shall not be counted.' Thus racism towards Indigenous people is deeply engrained in Australia, as in other Western countries with similar colonial histories, and is characterised by a complex web of indifference, distortion and harassment. This web shapes the relationship between Aboriginal and non-Aboriginal Australians and means that most white Australians are largely ignorant of, and indifferent to, Aboriginal people (Groome, 1995). While the process of reconciliation established by the Keating Labor government in the early 1990s (including the formation of a Council for Aboriginal Reconciliation) has gone some way to reducing racism, racism remains a serious problem. The Human Rights and Equal Opportunity Commission's (1997) report on the long-term effects of the policy and practice of removing Aboriginal and Torres Strait Islander children from their families recommends a series of reparations for the practice, which it described as a 'gross violation of human rights'. The reparations were recommended to consist of:

- acknowledgment and apology
- guarantees against repetition
- measures of restitution
- measures of rehabilitation
- monetary compensation.

The formal apology was issued by prime minister Kevin Rudd in February 2008 and was a historic step in acknowledging the past wrongs (see extracts of the speech in box 13.6). This apology was viewed as having profound symbolic meaning in enabling healing and facilitating respectful relationships between Indigenous and non-Indigenous Australians (Reconciliation Australia, 2010).

BOX 13.6   EXTRACTS FROM KEVIN RUDD'S SPEECH, FEBRUARY 2008

That today we honour the Indigenous peoples of this land, the oldest continuing cultures in human history.

We reflect on their past mistreatment.

We reflect in particular on the mistreatment of those who were Stolen Generations—this blemished chapter in our nation's history.

The time has now come for the nation to turn a new page in Australia's history by righting the wrongs of the past and so moving forward with confidence to the future.

We apologise for the laws and policies of successive parliaments and governments that have inflicted profound grief, suffering and loss on these our fellow Australians.

> We apologise especially for the removal of Aboriginal and Torres Strait Islander children from their families, their communities and their country.
>
> For the pain, suffering and hurt of these Stolen Generations, their descendants and for their families left behind, we say sorry.
>
> To the mothers and the fathers, the brothers and the sisters, for the breaking up of families and communities, we say sorry.
>
> And for the indignity and degradation thus inflicted on a proud people and a proud culture, we say sorry.
>
> We the Parliament of Australia respectfully request that this apology be received in the spirit in which it is offered as part of the healing of the nation.
>
> *Source: Rudd, 2008.*

The history and politics of racism are reflected in the appalling Aboriginal health statistics that we have reviewed earlier. After a concerted campaign by civil society and professional associations including the Australian Medical Association (2007), the Australian Government adopted a Closing the Gap policy and strategy that aims to close the gap in life expectancy by 2040 (COAG, 2008a).

CONCLUSION

This chapter has shown that explanations for health inequities should deal with the complexity of the determinants of health and reflect consideration of all the factors that have an impact on health (biological, genetic, behavioural, social, economic and political). Material circumstances are a significant cause and the role of behaviour in health status is not as significant as many believe. Increasing evidence suggests that inequities in health status in populations reflect the broader socioeconomic inequities. Hierarchy appears to be bad for population health, and equity good for it. Consideration of gender and health and of Indigenous health illustrated the complexities of explaining the health status of any particular group.

CRITICAL REFLECTION QUESTIONS

13.1   Why does having a relatively low income limit your options for leading a healthy lifestyle?

13.2   Which social determinants have the strongest impact on your life?

13.3   Why do you think women live longer than men?

13.4   How do you account for the 11-year gap in life expectancy between Indigenous and non-Indigenous Australians?

Recommended reading

AIHW (2014) *Australia's health*, published every two years, provides a comprehensive overview of the health status of Australians.

Carson et al. (2007) *Social determinants of Indigenous health* provides a guide to social determinants such as racism, poverty and class, social capital, education, employment and welfare, land, housing and policy processes.

Commission on Social Determinants of Health (2008) *Closing the gap in a generation: Health equity through action on the social determinants of health* This is a seminal report on the social determinants of health and has been widely quoted and used to inform policies around the world.

Graham (2007) *Unequal lives: Health and socioeconomic inequalities.* An important book that makes the link between epidemiological perspectives on health inequalities and social science literature on social inequalities. Helpful chapters on how health and social inequalities are measured and conceptualised.

Townsend et al. (1992) *Inequalities in health* draws together the UK's 'Black Report' and its sequel 'The Health Divide'. Together they provide a comprehensive account of the evidence on inequalities in health status in the UK and thoughtful discussion on the likely causes. Classic and essential reading for new public health students.

Useful websites

The Equality Trust works to improve the quality of life in the UK by reducing economic inequality and its website contains a lot of information about health inequities: www.equalitytrust.org.uk

The World Health Organization Social Determinants of Health site contains useful information on global responses to health inequities: www.who.int/social_determinants/en

PART 5

UNHEALTHY ENVIRONMENTS: GLOBAL AND AUSTRALIAN PERSPECTIVES

---

the inextricable links between people and their environment constitutes the basis for a socio-ecological approach to health ... the conservation of natural resources throughout the world should be emphasised as a global responsibility ... the protection of the natural and the built environments and the conservation of natural resources must be addressed in any health promotion strategy.

Ottawa Charter for Health Promotion, World Health Organization, 1986

Will humans survive the global threats of environmental disaster? This is the most significant public health question for the twenty-first century. The effects of human development are well known: industrialisation, population growth, misuse and/or overuse of natural resources, damage to the stability and productivity of the ecosystem. These factors threaten the planet on which we live, and therefore the survival of the human species. Further, the pace at which the detrimental changes are happening appears to be speeding up. Growing technology, unprecedented world population growth and ever-faster economic growth make more and more demands on the increasingly fragile natural system on our planet, threatening to overwhelm our ability to survive (Brown, 2011). With each edition of this book the evidence that human survival is threatened in the face of human-induced global warming has become stronger and is being voiced by a growing number of scientists. These are the crucial public health issues of the coming decades.

Public health workers have a responsibility to educate people and governments about the nature and extent of the social and environmental hazards to human life and health, and on the measures required for their control. Brown (2011) points out that this will not be an easy task because doing so will, inevitably, incur the opposition of powerful vested interests.

Our current situation on the earth has been likened to cowboys living in spaceships with no appreciation of the limits to expansion. Korten (1995, p. 26) says the consequences of living like this are twofold:

- Life-support systems are overburdened, resulting in their breakdown and a decrease in the level of human activity they can ultimately sustain.
- There is intense competition between the more powerful and weaker members of the crew for the shrinking pool of life support.

Korten sees that ultimately this will lead to an erosion of the legitimacy of government, increasing social tension and then social breakdown and violence. The public health movement began to wake up to the implications of the ecological crisis in the 1990s and this has continued into the twenty-first century. Overall there is a shift in public health that reflects a growing general concern about the environment and WHO is providing more leadership on the topic (see, for example, WHO 2014g, 2014k). From the early 1960s, and particularly since the publication of Rachel Carson's (1963) *Silent Spring*, there has been a steady increase in environmental awareness. Environmental lobby groups have multiplied, governments have taken on green policies to varying degrees and internationally the world is awakening (somewhat like a giant from slumber!) to the crisis that it is facing. Most significant in the international arena have been the United Nations Conferences since the one on Environment and Development held in Rio de Janeiro in 1992 and the Kyoto Protocol in 1997. Time will tell if the UN Paris Climate Conference (late 2015) becomes seen as the turning point in effective global responses to halt and ultimately reverse climate change.

Public health is intrinsically linked to issues of environmental sustainability, and separating them makes no sense. It is quite clear in the first decade of the twenty-first century that deteriorating environmental conditions are a major contributory factor to poor health and quality of life in many settings. Prüss-Üstün and Corvalán (2006), in a WHO report, conclude that environmental quality is directly responsible for 24 per cent of the global burden of disease, and that 23 per cent of all deaths can be attributed to environmental factors. Among children 0–14 years of age the proportion of deaths attributed to the environment is as high as 36 per cent, with diarrhoeal disease, respiratory infection and unintentional injuries heading the list. Recent WHO reports (2014g, 2014k) also show that other diseases, such as malaria, vector-borne diseases, chronic respiratory diseases, and perinatal and childhood infections, are strongly influenced by adverse environmental conditions and that climate change is having a growing effect on mortality.

Environmental issues are complex, with the science sometimes disputed and uncertain. When the complexity and uncertainty threatens to overwhelm us, it is tempting to carry on with a more straightforward public health practice that focuses on measurable and controllable issues. To do so, however, would be akin to rearranging the deckchairs on the *Titanic* as the ship disappears below the ocean.

This part of the book describes the dimensions of a growing environmental crisis and its likely impact on human health. It contains two chapters. Chapter 14 deals with the physical threats to health and considers why, despite an increasing body of evidence, adequate responses to the environmental crisis are slow in coming. Direct threats to health are discussed in detail: climate change, energy use, chemicals, rapid urbanisation, and population and consumption growth. Chapter 15 considers rapid urbanisation and the health consequences of transport, the question of population growth and rising consumption, and the resurgence of infectious disease in the face of globalisation. Both chapters focus on issues that are global and consider some of the local consequences. Solutions to the problems described in this part of the book are considered in parts 6 and 7.

14

GLOBAL PHYSICAL THREATS TO THE ENVIRONMENT AND PUBLIC HEALTH

Human influence on the climate system is clear, and recent anthropogenic emissions of greenhouse gases are the highest in history. Recent climate changes have had widespread impacts on human and natural systems ... Until mid-century, projected climate change will impact human health mainly by exacerbating health problems that already exist (very high confidence). Throughout the 21st century, climate change is expected to lead to increases in ill-health in many regions and especially in developing countries with low income, as compared to a baseline without climate change (high confidence).

Intergovernmental Panel on Climate Change, 2014, pp. 2, 15

KEY CONCEPTS

INTRODUCTION

This chapter looks at the physical threats to the environment and their impact on human health, concluding with a fundamental question: why, given the increasing certainty of the environmental problems we face, are the efforts to reverse the existing damage and prevent future damage so half-hearted? Steffen et al. (2011) argue that the impact of human activity on the global environment is now so great it rivals that of major natural forces, and so have coined the term 'the Anthropocene' for the current era. Population growth, the energy intensiveness of lifestyles in many countries and the misuse of non-renewable resources are greatly affecting our planet and its capacity to sustain human life. While the details of the changes and the precise nature of the strain on the planet are disputed, there is increasing consensus that business as usual will result in ecological disaster. There are many ways in which the strain on our ecosystem will affect human health. The evidence in each area is far from certain. The exact nature of causality is difficult to determine and most outcomes are produced by a number of different influences. The challenges for public health are numerous. Perhaps the most crucial initially is understanding the magnitude and complexity of the threats.

CLIMATE AND ATMOSPHERIC CHANGE

> Warming of the climate system is unequivocal, and since the 1950s, many of the observed changes are unprecedented over decades to millennia. The atmosphere and ocean have warmed, the amounts of snow and ice have diminished, sea level has risen, and the concentrations of greenhouse gases have increased.
>
> *Intergovernmental Panel on Climate Change, 2014, p. 2*

The scientific evidence on climate change is complex and difficult for a non-expert to understand, but there is now a strong acceptance that the earth's climate is changing more rapidly than might be expected and that this change is likely to have resulted from human activity. The somewhat conservative Intergovernmental Panel on Climate Change (IPCC) has become increasingly firm in its predictions about climate change. Its fifth report, released in 2013–14, unequivocally confirms that warming is happening, that it results from human activities and that limiting climate change requires substantial and sustained reduction in greenhouse gas emissions. Box 14.1 provides some of the evidence that supports the IPCC's conclusions that the earth's atmosphere is warming. Tim Flannery's (2005) impassioned appeal for action to be taken to prevent the drastic changes in the atmosphere also includes a relatively simplified description of the ways in which even a small increase in the amount of carbon dioxide in the atmosphere will have an impact on the earth's climate. He uses James Lovelock's Gaia framework to describe the dynamic and systems nature of the atmosphere, which he maintains is a more useful analytical framework than a more conventional scientifically reductionist one. He produces evidence to show that, as a result of climate change, the polar ice caps are retreating, species are becoming extinct and the atmosphere of the earth does appear to be warming significantly. His chilling conclusion is 'If humans pursue a business-as-usual course for the first half of this century, I believe that collapse of civilisation due to climate change becomes inevitable' (Flannery, 2005, p. 209).

BOX 14.1   EVIDENCE IN SUPPORT OF GLOBAL WARMING

There appears to be an increase in the average global temperatures. The 2013–14 Fifth Assessment Report of the Intergovernmental Panel on Climate Change (produced by around 600 authors from 39 countries) concludes that warming of the climate system is unequivocally happening and that it is more than 90 per cent likely due to human activity. The IPCC's 2014 Synthesis Report predicted that without additional efforts to reduce greenhouse gas emissions, global mean temperatures are expected to rise by between 3.7 and 4.8 °C (IPCC, 2014, p. 40).

Ocean warming dominates the increase in energy stored in the climate system, accounting for more than 90 per cent of the energy accumulated between 1971 and 2010. It is virtually certain that the upper ocean (0–700 metres) warmed from 1971 to 2010. Over the last two decades, the Greenland and Antarctic ice sheets have been losing mass, glaciers have continued to shrink almost worldwide, and Arctic sea ice and Northern Hemisphere spring snow cover have continued to decrease in extent (IPCC, 2014, p. 9).

The rate of sea level rise since the mid-nineteenth century has been larger than the mean rate during the previous two millennia. Atmospheric concentrations of greenhouse gases have increased to levels unprecedented in the last 800 000 years. Carbon dioxide concentrations are up by 40 per cent since pre-industrial times, directly contributing to global warming and leading to ocean acidification. Both rising ocean temperatures and ocean acidification have led to measurable changes in the species composition of certain marine plankton in the North Atlantic, including a northward extension of more than 10 degrees latitude of warm-water species and a decrease in the number of cold-water species (Beaugrand et al., 2002) with effects likely to cascade through the marine food chain (Hays et al., 2005).

THE CRUCIAL EFFECTS OF BURNING FOSSIL FUELS

The underlying cause of global warming appears to be fossil fuel burning. Currently about 10 billion tons of carbon are released into the air annually (Le Quéré et al., 2013). Oil and coal play a central place in our civilisation. Prugh et al. (2005, p. 100) note 'Oil saturates virtually every aspect of modern life, and the well-being of every individual, community and nation on the planet is linked to our oil-based energy culture'. If the earth's climate were to return to equilibrium over the next few centuries, carbon emissions would have to be reduced to the rate at which the oceans and forests can absorb them—one or two billion tons a year or at least 80 per cent below the rate of the mid-1990s (Flavin, 1996). People in the industrial world account for only 21 per cent of the world's population, yet consume 75 per cent of its energy and are most responsible for the build-up of greenhouse gases (see table 14.1). Some developing countries have caught up, however. By 2009, China had become the world's leading emitter of carbon dioxide. The McKinsey Global Institute reported on the growth of the Chinese middle class and projected that by 2025, 520 million will fall into a middle-class bracket (Li, 2010). It projects that urban households in China will come to make up one of the largest consumer markets in the world. The implications for climate change are profound and far reaching.

TABLE 14.1   SHARE OF WORLD TOTAL CARBON DIOXIDE EMISSIONS, SELECTED COUNTRIES, 2000 AND 2009

Country	Percentage share of world production of carbon dioxide		Carbon dioxide emissions per capita	Percentage change in emissions, 2000 to 2009
	2000	2009		
Australia	1.5	1.37	19.64	+17.2
China	12.1	21.09	5.83	+170.6
India	4.7	5.27	1.38	+59.7
Japan	5.2	3.61	8.64	−8.6
Russia	6.2	5.17	11.23	+1
UK	2.5	1.71	8.35	−7.2
USA	24.4	17.84	17.67	−7.5
World	100	100	4.49	+27.9

Source: US Energy Information Administration (EIA) (n.d.).

Global warming results from the emission of several greenhouse gases, most notably carbon dioxide (CO_2), chlorofluorocarbons (CFCs), methane (CH_4) and nitrogen dioxide (NO_2). According to the IPCC, global average temperature is due to increase by 1.4 to 7.8 °C between 1990 and 2100 with the most likely increase around 2.0 °C. Average sea level is projected to rise by 0.26 to 0.82 metres. A continued decrease in snow cover and sea ice and a more widespread retreat of glaciers and ice caps are predicted. In 2012 the IPCC reported that more frequent and severe extreme weather events such as droughts, floods, storms and heatwaves were likely in the future warmer world (IPCC, 2012). These are, of course, only predictions, but the prudent response is to assume we are facing a change in the global climate that will be greater than any change over the last 10 000 years at least.

In early 2007 the IPCC's fourth assessment report was produced by around 600 authors from 40 countries, and reviewed by over 620 experts and governments. Before being accepted, the summary for policymakers was reviewed line-by-line by representatives from 113 governments. It concluded that: 'Warming of the climate system is unequivocal …' (IPCC, 2007, p. 5) and 'Most of the observed increase in global average temperatures since the mid-twentieth century is very likely (>90 per cent) due to the observed increase in anthropogenic greenhouse gas concentrations' (p. 10).

Lord Rees, the president of the UK Royal Society, said:

> This report makes it clear, more convincingly than ever before, that human actions are writ large on the changes we are seeing, and will see, to our climate. The IPCC strongly emphasises that substantial climate change is inevitable, and we will have to adapt to this. This should compel all of us—world leaders, businesses and individuals—towards action rather than the paralysis of fear. We need both to reduce our emissions of greenhouse gases and to prepare for the impacts of climate change. Those who would claim otherwise can no longer use science as a basis for their argument (BBC, 2007).

Also in 2007 the UK Treasury released the report of the Stern Review on the economics of climate change (Stern, 2007). The conclusion of the report was simple and unequivocal: climate change could damage global gross domestic product by up to 20 per cent if left unchecked, but curbing it would cost about 1 per cent of global gross domestic product.

The global financial crisis of 2009 saw a pause in the growth of greenhouse gas emissions in developing countries that has been sustained in some, but overwhelmed by the increase in emissions in low- and middle-income countries, and particularly in China. However, while global emissions continue to rise each year, the *rate* of increase has fallen from 2.7 per cent per year for the previous decade to 1.1 per cent in 2012 (Olivier et al., 2013). In 2009, 192 governments convened for the United Nations climate summit in Copenhagen with high expectations of a new global agreement, but the summit only produced a non-binding accord of general intent to reduce carbon emissions. By 2012 Arctic sea ice had reached a minimum extent of 3.41 million square kilometres (1.32 million square miles), a record for the lowest summer cover since satellite measurements began in 1979. By 2013 the Mauna Loa Observatory in Hawaii reported that the daily mean concentration of carbon dioxide in the atmosphere had surpassed 400 parts per million (ppm) for the first time since measurements began in 1958. In the same year the IPCC's fifth assessment report confirmed that scientists were 95 per cent certain that humans are the 'dominant cause' of global warming since the 1950s. The report showed that emissions of greenhouse gases were tracking at the rate of the worst scenario model identified in earlier reports. Extrapolating that projection gives a rise of around 4 °C by 2100 and carries on to exceed pre-industrial levels by 10–12 °C. Woodward (2014) notes that while such longer term predictions are subject to enormous uncertainties, they take us beyond 'the limits of human adaption'.

EFFECTS OF CLIMATE CHANGE ON HUMAN HEALTH

These changes to the global climate have the potential to affect many aspects of human life in ways that are complex and involve interactions between many systems. The effects will depend on several factors, including the rate of change in the environment, the sensitivity of the biosphere and the degrees to which humans can respond to the changes (WHO, 2014g). The cumulative effect of these possible consequences on human health could significantly stretch public health resources, especially at a time when, around the world, public health infrastructures are being reduced rather than strengthened. The IPCC concluded that the effects of climate change are expected to be greatest in low- and middle-income countries in terms of loss of life and relative effects on investment and the economy, and saw multiple direct and indirect connections between climate and health (table 14.2).

TABLE 14.2   MAIN EFFECTS OF GLOBAL CLIMATE CHANGE AND ATMOSPHERIC CHANGE ON POPULATION HEALTH

Direct	Direct injury risks and follow-on outbreaks of infectious diseases resulting from extreme weather events
	Mass displacement and disruption of livelihoods in low-lying coastal areas and small island states due to storm surges and sea level rise
	Lack of nutrition and mental stress resulting from extreme weather events
	Excessive heat exposures resulting in heat stroke, which may lead to deaths
	Heat exhaustion that reduces work productivity and heat stress that interferes with daily household activities
Indirect (mediated through other environmental systems such as rising air pollution and social mediated effects such as occupational heat stress)	Malnutrition and undernutrition due to failing agriculture
	Spread of vector-borne diseases and other infectious diseases, mental health and other problems caused by forced migration from affected homes and workplaces
	Changes in access to clean drinking water (particularly in conditions of crowding and poverty) can cause diarrhoeal diseases and other water-related diseases, including cholera
	Mental illness and conflict-prone tensions caused by forced migration from affected homes and workplaces
	Potentially increased risk of violent conflict associated with resource scarcity and population movements

Source: Kjellstrom and McMichael, 2013; Watts et al., 2014.

DIRECT EFFECTS OF CLIMATE CHANGE ON HUMAN HEALTH

RESPIRATORY ILLNESS

Climate may affect the respiratory tract in three ways: seasonal effects, direct effects of specific weather conditions (such as thunderstorms and cold fronts) and the combined effects of weather conditions and other environmental or topographical factors (WHO, 2014g). Predicting changes is difficult (WHO, 2014g). Increased temperatures in winter could result in declines in bronchitis and pneumonia, but this could be counteracted by possible increased summertime asthma and hay fever.

HEALTH EFFECTS OF THERMAL EXTREMES

Thermal extremes can make life difficult for people. Heatwaves, in particular, can have a severe effect on the very old and very young (McMichael, 1993; WHO, 2014g). The effects of heat are particularly acute for inner city residents in large cities who have little relief from the heat-intensifying artificial environment. Poor people will generally suffer most as they are less likely to have air-conditioning or other means of maintaining a cool environment and are likely to live and work in warmer neighbourhoods and in buildings that are poorly ventilated and absorb heat (Friel, 2014). Poor neighbourhoods with weak infrastructure, inadequate housing and unplanned developments with little green space are likely to be more exposed to high temperatures compared to more affluent neighbourhoods. However, global warming could also reduce mortality from hypothermia or infectious diseases such as influenza (McMichael, 1993, p. 149).

Australia's Commonwealth and Industrial Research Organisation (CSIRO) has found the national impact and cost of recent extreme events has been significant. The Victorian bushfires in early February 2009 killed 173 people and more than one million animals. They destroyed more than 2000 homes, burnt about 430 000 hectares, and cost about AU$4.4 billion. The south-east Australian heat wave in late January 2009 resulted in 374 more deaths in Victoria than would be expected (CSIRO, 2012). In the spring of 2013 the Bureau of Meteorology reported the warmest September (BOM, 2013). In October the worst fires in New South Wales for 50 years destroyed 248 houses and caused damage estimated at over AU$94 million (Owens, 2013).

OZONE LAYER DEPLETION EFFECTS

Chlorofluorocarbons (CFCs) have caused a thinning of the ozone layer, permitting more ultraviolet-B (UVB) radiation to enter the earth's atmosphere. Implementation of the 1987 Montreal Protocol has successfully reduced the emission of CFCs, but because they remain in the atmosphere so long, the ozone layer will not fully repair itself until at least the middle of the twenty-first century (Sivasakthivel and Reddy, 2011). However, there are concerns that growth in the space industry and an increased number of rocket launches could introduce trace-gas radicals with ozone depletion effects that could exceed those of CFCs by 2050 (Ross et al., 2009). Meanwhile, UVB is expected to increase the incidence of eye damage and skin cancers, especially among fair-skinned people. The NHMRC

(1989) has estimated that each 1 per cent increase in UVB will cause, on average, a 1.5 per cent increase in the incidence of skin cancer and cataracts in Australia. There is some evidence that an increase in UVB may depress the body's immune system, leading to an increased rate of diseases such as HIV (McMichael, 1993). Increased exposure to UVB is also likely to have wide-ranging effects for terrestrial and aquatic biota. For example, increases in UVB penetrating below the ocean's surface could deplete the phytoplankton population, which is the basis of the aquatic food chain (McMichael, 1993).

NATURAL DISASTERS

In 1995 the IPCC predicted a rise in natural disasters resulting from climate change, especially storms, flooding and drought. In January 2011, the European Environment Agency (EEA) published a report that found that the frequency and damages from disasters had increased in Europe between 1998 and 2009. There is clear evidence that insurance companies, reeling under the ever-increasing costs of natural disasters, have accepted, without doubt, that human-induced global warming is the cause of the increase in the number of severe natural disasters. Since the 1970s insurance losses have risen at an annual rate of around 10 per cent, reaching $100 billion by 1999 (Flannery, 2005). Munich Re, a major international re-insurer, compiled a 30-year database of natural disasters and found a steady increase in weather and climate related disasters, with 2011 delivering a record $308 billion worth of losses (Raloff, 2012). While scientists disagree about the contribution of climate change to these disasters, there is little doubt they are occurring more frequently. The US National Climatic Data Centre found that the incidence of billion-dollar-loss events (adjusted for consumer price index) more than trebled between the early 1980s and 2011. Total costs for extreme weather events between 1980 and 2011 exceeded $881 billion, including droughts, floods, bushfires, tropical storms, hailstorms, ice storms and hurricanes, but excluding the indirect costs of losses in productivity, revenue from tourism and local industries (Smith and Katz, 2013).

Changes in climate are, in part, being blamed on the El Niño–Southern Oscillation. This refers to sudden changes in sea temperatures in the Pacific Ocean that seem to cause large disturbances in atmospheric circulation over Australia and the eastern Pacific every two to seven years (State of the Environment Advisory Council, 1996, pp. 2–8), reducing rainfall. The effects of El Niño are now thought to be worldwide, but it is unclear whether they represent random fluctuation in climate or a significant deteriorating trend (State of the Environment Advisory Council, 1996).

Drought in Africa and Australia has also been linked to climate change (McMichael, 1993, 2005; Lowe, 2005; Weston, 2014). Again there is a complex chain of causality that includes forest clearance, soil erosion, civil war and an inadequate social and political infrastructure, but it is likely that climate change has made some contribution. Drought was prevalent in East Africa in 2005–06 where rains have been low for six years. Millions throughout Sudan, Eritrea, Ethiopia, Somalia and Kenya were in need of food aid as a result. The severe Australian drought of 2006–07 was also seen to result from global warming.

In Australia, bushfires are more likely in drought conditions. The New South Wales bushfires in the summers of 1994 and 2001–02 came after a number of years of unusually extended drought, while the New South Wales fires of 2013 have been attributed to an exceptionally warm and dry spring. Dramatic climatic variation has been a feature of Australia, being noted in the nineteenth century when, for instance, a series of unusually wet seasons encouraged South Australian farmers into normally non-viable areas, with disastrous results. This variation makes it difficult to estimate how much environmental damage is the result of climate change brought about by human activity.

Black Saturday, Kinglake West, Victoria, 2009. (Fiona Hamilton)

Forest fires in Indonesia in late 1997 led to smoke haze over much of southern South-East Asia, causing respiratory problems and, possibly, the crash of an aircraft. The effects of El Niño extended the dry season, causing drought in the region. This meant rains came late and so the fires could not be easily controlled. In Papua New Guinea the drought resulted in starvation for many in 1997–98.

Natural disasters not only bring considerable physical damage to people and their homes and livelihood, but also result in long-lasting psychological damage. Post-traumatic stress syndrome is now a well-recognised phenomenon and studies of disaster survivors are common (Van der Kolk et al., 1996). The prevalence of posttraumatic stress disorder, major depressive disorders and substance abuse disorders have been found to be significantly elevated in survivors of natural disasters (Goldman and Galea, 2014).

INDIRECT EFFECTS OF CLIMATE CHANGE ON HUMAN HEALTH

COMMUNICABLE DISEASES

Both vector- and water-borne diseases may increase in the face of climate change. Vectors are the agents that transfer disease to humans. Mosquitoes act as vectors for malaria, rats and fleas for the plague, and snails for schistosomiasis, for example. It has been suggested that insects, microbes, parasitic worms and flukes may thrive in an era of global warming (McMichael, 1993, p. 152; McMichael, 2005).

Increases in malaria have been connected with climate change (Bouma et al., 1994; Loevinsohn, 1994). McMichael (1993) notes that slight changes in climate can alter the viability and geographical distribution of vectors. The malarial mosquito can only survive where the winter mean temperature is above approximately 15 °C. Temperatures between 20 °C and 30 °C and humidity of at least 60 per cent are optimal for the mosquito to survive long enough to acquire and transmit infection. A number of variables that affect vectors—breeding rates, maturation, location of breeding sites and habitats, time between feeding cycles—are sensitive to factors such as temperature, humidity and rainfall. Changes in these may expand the geographical distribution of parasites and vectors, affect their behaviour and increase the rate of transmission of diseases (Platt, 1996). The risk of this happening is particularly great in areas next to current endemic areas and among people with no immunity. Platt (1996) quotes a study from the Netherlands that predicts that a temperature rise of 3 °C by 2100 would double the epidemic potential of mosquitoes in tropical regions and increase it tenfold in temperate regions. The study suggests that more than a million people could die each year as a result of the 'impact of a human-induced climate change on malaria transmission' (Platt, 1996, p. 122). Reported malaria cases in the US reached a 40-year high of 1925 in 2011 (US Centers for Disease Control and Prevention, 2014b).

Vector-borne diseases, including viruses, may be more prevalent in the face of global warming. Arbovirus infections such as Ross River virus, Murray Valley encephalitis and dengue fever may extend further south with increased temperature and rainfall in Australia (NHMRC, 1991). The incidence of dengue in a 2007 outbreak in Taiwan exceeded the total for the previous four years and was attributed in part to a warm autumn with high rainfall associated with a typhoon (Hseih and Chen, 2009). Climatic disturbances, such as floods, storms or earthquakes, may also increase the transmission of vector-borne disease. Some experts connect the outbreak of plague in Surat, India, in September 1994 to the flooding of the local river after an earthquake a year earlier. The food sent in for the survivors encouraged rats to flourish, allowing the plague bacterium (in the fleas that infest the rats' fur) to extend its range. The flood forced people to leave their homes and seek refuge on higher ground, along with the flea-infested rats. The combination of weather patterns and resultant environmental damage, together with the local living conditions—shanty towns, squalid living conditions, excess food and inadequate health care and public health surveillance—resulted in a plague outbreak (Platt, 1996). This was controlled but received worldwide media coverage, and may have gone some way to shaking complacency about diseases that were believed to be under control.

Global warming may also lead to an increase in water-borne diarrhoeal diseases, such as cholera and dysentery, especially in poor countries that lack sufficient clean water supplies and have poor sanitation. There is some suggestion that cholera may be associated with changes in the climate; it occurs seasonally when the temperature, sunlight, nutrient levels and acidity are adequate. In Bangladesh, outbreaks of cholera are linked with plankton blooms, heavy rains and warm ocean temperature associated with the El Niño effect (Epstein et al., 1993).

Finally, while the scientific literature in this area is far from conclusive, the indications are that climate change is contributing to an increase in communicable diseases (Garrett, 1994; McMichael, 2005).

SEA LEVEL RISE

Sea levels have already risen by 20–40 centimetres during the past century. The IPCC estimates that by 2100 the rise could be up to 88 centimetres, flooding many deltas and rendering some cities uninhabitable (Dunn and Flavin, 2002). Developing countries will be particularly vulnerable to these rises. Much of Bangladesh and a number of Pacific Island states, and many of the world's large, low-lying cities would probably disappear. Rising seas would cause salt water to encroach upon freshwater estuarine and tidal areas (McMichael, 1993; McMichael et al., 2003), damaging the wetlands whose health is important to the viability of many coastal areas.

As before, the precise effects of global warming are unknown. The IPCC predictions have been based on a linear rise in line with temperature change relating primarily to thermal expansion and the increased melting of glaciers. However the IPCC also acknowledges that the rate at which the Greenland and Antarctic ice shelves break down is also crucial (IPCC, 2013). Hansen and Sato (2012) argue that current data suggest an exponential rate of melt is possible, resulting in a 1-metre rise by 2045 and 5 metres by 2057. In 2013 *The Guardian* (Goldenberg, 2013) reported that Native Alaskan villages are under imminent threat from sea level rise. A report by the US Army Corps of Engineers predicts that the highest point in the village could be underwater by 2017. Alaska is reported as warming twice as fast as the rest of the USA in the past 60 years.

Flooding in low-lying countries like Bangladesh is likely to become more severe and make the existence of millions of Bangladeshis even more marginal. (Trygve Bolstad, Panos)

FOOD SECURITY

The links between food security and climate change have to be seen in terms of other factors that are reducing food security, including desertification, deforestation, waterlogging and salinisation of land. Research on the effects of climate change on agricultural areas suggests they will be uneven: some areas may become drier while others will get more rain. Potential beneficiaries are North America and Russia, where grain belts may receive more rain. The IPCC predicted adverse consequences for food security for south and South-East Asia, tropical Latin America and sub-Saharan Africa (Flavin, 1996) and by 2014 reported impacts on global food production and availability, food prices and access. They failed to repeat the 2007 report suggestion that climate change could be good for crops in higher latitudes, noting predominantly negative impacts on global crop yields. The 2014 report noted that 75 per cent of the available studies predict yield declines of up to 50 per cent by the 2030s with impacts on major crops including wheat, maize and rice. Oxfam (2014a) quotes the IPCC as concluding that it is 'very likely' that climate impacts will result in estimated food price increases from between 3 per cent and 84 per cent by 2050. The fifth of the world's population that suffers from malnutrition will become more vulnerable. Climate change may also allow new types and combinations of food parasites to emerge (McMichael, 2005).

SUMMARY: CLIMATE CHANGE AND HUMAN HEALTH

There is an overwhelming research consensus that the world's climate is changing as a result of human activity. Scientists note that climates do not change in a linear manner and the system of feedbacks within the earth's atmosphere mean that some seemingly small change could lead to catastrophic consequences. Positive feedback loops such as the release of methane from the tundra permafrost could dramatically increase the greenhouse effect and lead to very dramatic warming (Nisbet, 1991; Lowe, 2005; Smith and Maddern, 2011). Damage to population health may also be mediated via the social inequity resulting from the direct effects of climate change. Thus poorer people with fewer resources or access to what will be diminishing overall amounts of water, food, housing and health care will be further disadvantaged with more risk from armed conflicts and less access to options such as migration. Inequity will increase and population health will decline (Bowles et al., 2014).

DECLINING AIR AND WATER QUALITY

Cities are polluted to varying degrees, according to local circumstances such as housing, forms of transportation, level of industrialisation, water supply, sanitation and removal of refuse. Environmentally related communicable diseases are the most pressing problem in developing countries. In both developing and developed countries, however, environmental pollution poses a significant risk to health. The problem is far from new. The severe London smogs of the 1950s led to a series of Clean Air Acts that resulted in improved air quality. Developed countries have been better able to afford and implement the imposition of stricter controls than poor countries experiencing rapid industrialisation.

Every corner of the planet is affected by chemical and other forms of pollution that affect the soils, air, groundwater and animals, including people. WHO (2014b) estimated that around 3.7 million premature deaths were attributable to ambient (outdoor) air pollution in 2012, with about 88 per cent of these occurring in middle- and low-income countries. Air pollution in many cities, especially in the rapidly growing urban areas in developing countries, has reached severe levels, damaging the human respiratory system in a variety of ways. Older and younger people, smokers and those with chronic respiratory diseases are the most vulnerable. Industrial processes, power production and personal transport in urban areas are the main causes, but the effects differ from city to city and have varying impacts on different groups in the population. For example, tropical cities are more likely to suffer from photochemical smog, while children are more susceptible to lead pollution than adults (WHO, 1993).

The first 15 years of the twenty-first century have seen an overall rise in the level of global ambient particulate emissions, but small reductions for other key pollutant emissions including sulfur dioxide and nitrogen dioxide (Cofala et al., 2012). However, these aggregates mask reductions in emissions in developed countries combined with dramatic increases in India and particularly China. Outdoor air pollution was estimated to have contributed to 1.2 million premature deaths in China in 2010, or nearly 40 per cent of the world total (Wong, 2013). A 2014 analysis of WHO's (2014a) urban air quality database covering 1600 cities across 91 countries found that only 12 per cent of those residents had air that complied with WHO air quality guidelines. While air quality has improved in many developed countries in the past decades, in most of the monitored cities, where trend data was available, air quality was declining.

Air pollution in Beijing. (Economist)

THE HEALTH EFFECTS OF AIR POLLUTION

Determining the health effects of air pollution is methodologically difficult because it is made up of a cocktail of pollutants that varies in concentration and does not have a standard effect (Abramson and Voigt, 1991, p. 551; Utell et al., 1994, p. 159). Exact health effects depend on exposure level and the characteristics of the individuals exposed, particularly smoking behaviour and their susceptibility to allergies. The cocktail of pollutants combines with sunlight to produce a photochemical smog. Despite the difficulties of determining precise effects, there is increasing evidence that this smog is damaging to human health.

A review of US research on air pollution and health (Romm and Ervin, 1996) paints a worrying picture of the link. A 16-year, six-city study (Dockery et al., 1993) tracked the health of over 8000 individuals and showed a nearly linear relationship between particle concentrations in the air and increased mortality rates. The risk of early death in high-level areas was 26 per cent higher than in areas with the lowest levels of pollution, even after controlling for other risk factors such as smoking and occupation. The risk of cardiopulmonary disease was found to be 37 per cent higher.

Further evidence came from an American Cancer Society and Harvard Medical School study involving more than 550 000 people living in 151 cities (Romm and Ervin, 1996, pp. 392–3). Over seven years there was a 17 per cent increase in mortality rates in areas with higher concentrations of fine particles relative to areas with lower concentrations, and a 15 per cent increase for sulfate aerosols. The risk of death from cardiopulmonary disease was 31 per cent higher in the most polluted cities. Cancer has been associated with leaded petrol, particularly when combined with other carcinogenic substances in vehicle exhaust fumes (Hillman, 1991). McMichael (2001) reports that several long-term follow-up studies of populations exposed at different levels of air pollution implicate fine particulates in raising death rates, especially from heart and respiratory diseases. A WHO (WHO Regional Office for Europe, 2013) review of the evidence on the health impacts of air pollution recommended that the WHO air quality guidelines (last updated in 2005) be further tightened because in some cases, ill effects from fine particles, ozone and nitrogen dioxide occur at concentrations lower than those specified as safe limits.

Indoor air pollution is a significant cause of ill-health in developing countries. In tropical countries, health problems can arise from the burning of fossil fuels in houses with poor ventilation. This is particularly a problem in low-income rural and urban slum settings. In colder climates, where people spend a lot of time indoors and buildings are designed to retain heat, pollutants trapped inside homes and workplaces can reach dangerous levels (WHO, 1993). Poorer people may try to save on heating costs by restricting ventilation, further reducing indoor air quality.

Exhaust fumes from cars are a key contributor to pollution. In urban areas, the most common emission is carbon monoxide (CO). The range of health effects associated with vehicle pollution is shown in table 14.3. There is Australian evidence that there has been a concurrent rise in mortality from asthma and the output of motor vehicle emissions (Woodward et al., 1995, p. 402). Levels in urban areas can be sufficient to cause headache, lassitude and dizziness in normal people, as the gas interferes with the ability of the blood to carry oxygen. The effects can be serious for all groups sensitive to lower oxygen levels, particularly pregnant women, young children and older people (Hunt, 1989; Hillman, 1991). Internal combustion produces photochemical smog, causing eye irritation and plant damage and also heightens the effects of sulfur dioxide and nitrogen oxides (Hunt, 1989). Nitrogen can also contribute to the acidification of water and air. Air pollution from cars is adding to the problems of the rapidly industrialising cities in east Asia, south Asia and Latin America. For example, the number of cars in New Delhi doubled in the 10 years to 2014, while levels of airborne coarse particulates have increased by 75 per cent since 2007 (Koutsoukis, 2014). Increased car densities in cities and longer commute times mean increased exposures. Urban commuting is an increasingly important factor in determining exposure to ambient air pollutants (WHO Regional Office for Europe, 2013).

TABLE 14.3   POTENTIAL HEALTH EFFECTS OF VEHICLE POLLUTION

Pollutant	Source	Health effect
Nitrogen dioxide (NO_2)	One of the nitrogen oxides emitted in vehicle exhaust.	May exacerbate asthma and possibly increase susceptibility to infections.
Sulfur dioxide (SO_2)	Some SO_2 is emitted by diesel engines.	May provoke wheezing and exacerbate asthma. It is also associated with chronic bronchitis.
Particulates (PM_{10}), total suspended particulates, black smoke	Includes a wide range of solid and liquid particulates in air. Those less than 10 micrometres in diameter (PM_{10}) penetrate the lung fairly efficiently and are most hazardous to health.	Associated with a wide range of respiratory symptoms. Long-term exposure is associated with an increased risk of death from heart and lung disease. Particulates can carry carcinogenic material into the lungs.
Acid aerosols	Airborne acid formed from common pollutants including sulfur and nitrogen oxides.	May exacerbate asthma and increase susceptibility to respiratory infection. May reduce lung function in those with asthma.
Carbon monoxide (CO)	Mainly from petrol car exhaust.	Lethal at high doses. At low doses can impair concentration and neuro-behavioural function. Increases the likelihood of exercise-related heart pain in people with coronary heart disease. May present a risk to the foetus.
Ozone (O_3)	Secondary pollutant produced from nitrogen oxides and volatile organic compounds in the air.	Irritates the eyes and air passages. Increases the sensitivity of the airways to allergic triggers in people with asthma. May increase susceptibility to infection.
Lead	Additive present in leaded petrol to help the engine run smoothly.	Impairs the normal intellectual development and learning ability of children.

Pollutant	Source	Health effect
Volatile organic compounds (VOCs)	A group of chemicals emitted from the evaporation of solvents and the distribution of petrol fuel. Also present in vehicle exhaust.	Benzene has given most cause for concern in the group of chemicals. It is a carcinogen that can cause leukaemia at higher doses than are present in the normal environment.
Polycyclic aromatic hydrocarbons (PAHs)	Produced by incomplete combustion of fuel. PAHs become attached to particulates.	Include a complex range of chemicals, some of which are carcinogens. It is likely that exposure to PAHs in traffic exhaust poses a low cancer risk to the general population.
Asbestos	May be present in brake pads and cloth linings, especially in heavy duty vehicles. Asbestos fibres and dust are released into the atmosphere when vehicles brake.	Asbestos can cause lung cancer and mesothelioma (cancer of the lining of the chest cavity). The consequences of the low levels of exposure from braking vehicles are not known.

Source: NSW Health Department 1995, cited in ABS, 2009, p. 111.

Polycyclic aromatic hydrocarbons (PAHs) are a group of chemicals that are formed during the incomplete burning of coal, oil, gas, wood, garbage, or other organic substances, such as tobacco and chargrilled meat. PAHs enter the environment mostly as releases to air from volcanoes, forest fires, residential wood burning, and exhaust from automobiles and trucks. They can also enter surface water through discharges from industrial plants and wastewater treatment plants, and they can be released to soils at hazardous waste sites if they escape from storage containers. There is some evidence they are carcinogenic and can cause other health effects (for details see http://envirocancer.cornell.edu/Bibliography/General/bib.pah.cfm#sources).

Persistent organic pollutants (POPs) are another group of chemicals that are increasingly being recognised for their adverse effects on health (see box 14.2).

PERSISTENT ORGANIC POLLUTANTS (POPS)

Persistent organic pollutants (POPs) are prime pollutants of soil and water that have received growing attention in recent years. They are carbon-based chemical compounds and mixtures that include industrial chemicals like PCBs (polychlorinated biphenyls), pesticides like DDT and unwanted waste by-products of industry like dioxins (Stott, 2000). They are usually fat-soluble and tend to accumulate in the fatty tissue of animals and then concentrate many millions of times as they move up the food chain. They have been implicated in the disruption of the workings of the reproductive system, the immune system and the neurological system. While it is very difficult to study the effects of POPs on human health (McMichael, 2001), they have been associated with declining sperm counts, and seen to have carcinogenic properties, possibly accounting for the unexplained, widespread increases in non-Hodgkin's lymphoma, brain cancer, kidney cancer and multiple myeloma, infertility, genetic defects among sons, and heart disease (McGinn, 2000). Many POPs have an effect on foetal development, inducing low-birthweight babies. Some are immunosuppressive agents similar to, but not as devastating as, HIV (Stott, 2000). Increasingly the manufacture and use of POPs is banned as their multiple impacts on human health become better understood. Phasing out POPs will require fundamental changes in regulations, business, agriculture and society at large. By 2014 the Stockholm Convention on POPs had identified and listed 23 POPs to be phased out (Stendahl, 2014).

THE HEALTH EFFECTS OF WATER POLLUTION

Coastal ecosystems are deteriorating because of contamination from inadequately controlled industrial, agricultural and domestic waste disposal. One example is the increasing incidence and intensity of marine algal blooms, which have been reported from coastlines around the globe. Health impacts come from ingestion of toxic shellfish and finfish, skin contact and aerosol inhalation. An Australian prospective study indicates that people using water that contains higher concentrations of cyanobacteria (blue–green algae) report more health effects (diarrhoea, vomiting, flu-like symptoms, skin rashes, mouth ulcers, fevers, ear and eye irritation) (Pilotto et al., 1997). Algal blooms have also been associated with the die-back of seagrasses, which in turn contributes to coastal erosion. Harmful algal bloom events increased in both frequency and severity over the last decades of the twentieth century (Glibert et al., 2005).

Ensuring an adequate supply of clean drinking water is a central public health issue, and was one achievement of the first public health revolution of the nineteenth century. One of the Millennium Development Goals is to reduce by half the number of people without sustainable access to safe water by 2015. Diarrhoea is still a major killer in poor countries and WHO has estimated that 88 per cent of all cases globally were attributable to water, sanitation and hygiene (Prüss-Üstün and Corvalán, 2006, p. 34). Twenty-five million people in developing countries die annually from pathogens and pollution in contaminated drinking water (Platt, 1996). Human pathogens that thrive in water can cause hepatitis A, salmonella, cholera, typhoid and dysentery. Some pathogens are spread by drinking contaminated water or eating contaminated fish and shellfish, and others by swimming or bathing in contaminated water. While there have been some major successes in providing safe drinking water to hundreds of millions of people in recent years, rapid population growth means that an estimated 768 million people did not use an improved source (protected from faecal contamination) for drinking water in 2011, and 185 million relied on surface water to meet their daily drinking water needs. By the end of 2011, 2.5 billion people lacked access to an improved sanitation facility that hygienically separated human excreta from human contact (WHO and UNICEF, 2013b).

Australia is the driest of the world's inhabited continents and also has the most variable rainfall and stream flow in the world. Because of this variability, it has a very high per capita water storage capacity. Only about 5 per cent of reticulated water is used in the kitchen for drinking and cooking. Water in the Murray–Darling Basin has been over-allocated to irrigation, putting these aquatic environments under extreme stress. There, and in southwestern Western Australia, rising water tables and salinity threaten the viability of agriculture. Sediments from erosion, pesticide residues and high phosphorus levels, which can result in toxic algal blooms, all threaten aquatic environments.

BOX 14.2   WATER WARS: THE APPROACHING RISKS

Water resources are finite. Growing population levels with accompanying growth in water-intensive agriculture and industry are coming at a time when climate change is reducing rainfall in some existing dry regions. The resulting conflicts of interest at national, corporate and local levels may lead to social, economic and even military friction. In Iraq and Syria, rivers, canals, dams and sewerage and desalination plants are military targets. Control of water supplies means control of the cities that depend on them. Moves by China to dam upstream sections of the Mekong River threaten major impacts on communities downstream in Burma, Thailand, Laos, Cambodia and Vietnam. Existing tensions between India and Pakistan may be exacerbated as both rely on the Indus River for irrigation and hydro-electricity.

Intra-national water conflict and conflicts between local communities and global corporations are likely to be more common than those between nation states. The conflict between Australian states over management of the threatened Murray River is one example, with any effective action to stop unsustainable over-extraction for irrigation stymied by intransigent state governments for many years. Corporate activities such as Coca-Cola's excessive extraction of groundwater at its Mehdiganj bottling plant in India have also triggered vigorous and persistent protests, as have other water-hungry industries such as pulp mills and agribusiness projects such as large palm oil plantations. Large scale corporate agricultural production consumes 70 per cent of the world's fresh water. Community groups facing these threats argue that access to clean water is a human right and that water should be regarded as a resource we hold in common.

Source: Adapted from Balch (2014)

WATER SUPPLY

Water availability is becoming a key issue for the future. This has obvious implications for public health as a supply of clean, drinkable water is one of the fundamental requirements for health. 'Water stress' has become a commonplace expression as well as a widespread phenomenon, of which lack of clean water is a substantial part (Middleton and O'Keefe, 2003, p. 50). Three factors have increased the stress on water: population growth; the fact that one-third of the world's population lives in areas suffering water stress; and climate change, which has brought increased drought to many arid or semi-arid regions. It has been predicted that five billion of the world's 9.7 billion people (52 per cent) will live in water-stressed areas by 2050. Further, an area is deemed 'overly exploited' when water demand exceeds water supply. By this measure, an increase of 1.0 to 1.3 billion people living in overly exploited water-stressed areas is expected by 2050 (Schlosser et al., 2014).

The United Nations (2014) estimated that around 700 million people in 43 countries suffered from water scarcity in 2014. However, by 2025, 1.8 billion people are predicted to be living in countries or regions with absolute water scarcity, and two-thirds of the world's population could be living under water-stressed conditions.

Brown (2013) has argued that unsustainable extraction from major aquifers and rivers has lead to a 'peak water' scenario, such that expansion of irrigation farming has stopped while fresh water reserves diminish at an unparalleled rate. Desalination is offered as a solution but has significant environmental threats.

Protest at the building of a desalination plant in Adelaide. (Fernando M. Gonçalves)

Freshwater toxic algal blooms in rivers, lakes and reservoirs have the potential to bring about acute shortages of potable water for human consumption, as well as threatening irrigated food crops.

NUCLEAR POWER

In the past decade the possibility of the expansion of the nuclear power industry re-entered the policy agenda in many countries around the world as a response to the threat of global warming. This idea had been largely discredited in the wake of some of the accidents that have occurred, most notably at Three Mile Island in the USA and Chernobyl in Ukraine. The willingness to once again consider nuclear power was in response to the growing evidence on global warming and the fact that nuclear power does not add to carbon dioxide greenhouse gases. The case against nuclear power was weakened when the iconic environmentalist James Lovelock (2004) came out in favour of the energy source as being the only 'green' solution that would save the world from what he saw as the bigger threat of global warming. A coalition of six non-government organisations in Australia (including the Public Health Association of Australia and the Medical Association for the Prevention of War) commissioned a report (Green, 2005) to examine Lovelock's claim. It pointed out the very considerable risks of nuclear power including potentially catastrophic accidents, routine releases of gases and liquids from nuclear plants, the intractable problem of dealing with nuclear wastes, the huge costs of decommissioning power reactors and storing the waste and the heightened risk of terrorism and sabotage. It also pointed out that even a doubling of global nuclear power by 2050 would only reduce greenhouse gases by 5 per cent.

The case for nuclear power suffered a major setback in March 2011 when the Fukushima Daiichi plant in Japan was damaged by a tsunami, resulting in a meltdown of three of the plant's six reactors and a major release of radioactive material. Around 300 000 people were evacuated from the region. A massive amount of water was contaminated, and some was released into the sea. Japan shut down its remaining nuclear plants and in 2012 prime minister Noda announced plans to make the country nuclear-free by

2030. However, the following year, a new government was arguing for a renewed role for nuclear power, despite an inquiry revealing major failures in planning and safety measures at Fukushima. As a result of the Fukushima disaster Germany decided to phase out its nuclear power program.

Go Nuclear Free. (Fernando M Gonçalves)

THE HEALTH IMPACT OF NUCLEAR POLLUTION

One of the most potent threats to environmental health comes from the possibility of an accident at a nuclear power plant, such as the 1986 accident at Chernobyl, Ukraine, which had a huge health impact on the surrounding population. More than 90 000 people living within a 30-kilometre radius were evacuated and permanently rehoused. They received a single dose of radiation equivalent to at least 250 times the International Commission on Radiological Protection's recommended maximum annual dose. In addition, the accident sent radioactive pollution across Europe and Russia.

Because of the different ways in which people can be exposed to radiation and the different parts of the body affected by each radionuclide, there is a wide range of possible health effects. Delayed effects occur as different contaminants move through, and are absorbed by, different levels of the food chain. While the amount of radioactive material released at Fukushima was estimated at around 10–30 per cent of that from Chernobyl, much was made up of contaminated water released into the adjacent Sea of Japan, a major source of seafood (von Hippel, 2011). The psychosocial effects of evacuation and lack of information are also significant. Children around Chernobyl proved particularly vulnerable, with thyroid cancer rates elevated by a factor of 40 in Belarus between 1986 and 1994. Other malignancies have latency periods of up to 25 years (WHO, 1995a). A 2013 report found that thyroid cancer rates in children in Fukushima Prefecture had increased seven-fold from 2007 (Nose and Oiwa, 2013). Katz (2008) has raised concerns that the industrial and military interests of the nuclear industry may have obscured the real risks of continued exposure to radionuclides, which can accumulate in human tissue over decades. She cites the apparent suppression of the ongoing health impacts of Chernobyl and notes the requirement for WHO to consult with the International Atomic Energy Agency before initiating any program in the area of nuclear power.

There is controversy over the effects of living near a nuclear installation. The World Bank (1993, p. 96) reports that individuals are only exposed to a tiny amount of extra ionising radiation from safely operating nuclear power stations or other installations (such as medical and dental systems). However, there seems to be no threshold dose for ionising radiation below which exposure causes no increased risk of malignancy. In the USA (where the overall annual cancer incidence is about 22 000 cases) exposures at the allowable limit of 0.02 to 0.026 rem per year are estimated to be responsible for 20–60 additional cases of cancer per year in the 7.7 million people occupationally exposed to radiation (Last, 1998, p. 185). Evidence from Sellafield in the UK suggests that exposure of fathers to ionising radiation during their employment at a nuclear power plant is associated with the subsequent development of leukaemia in their children (Gardner et al., 1990). Several countries have investigated links between certain cancers and exposure to radon in houses, electromagnetic fields created by high-voltage cables (World Bank, 1993), and (non-ionising) electromagnetic radiation from televisions, mobile phones and other transmissions (Hocking et al., 1996). The results are inconclusive.

LOSS OF BIODIVERSITY

There is increasing recognition of the importance of biodiversity to human health (Chivian, 2003; State of the Environment Advisory Council, 1996; State of the Environment 2011 Committee, 2011). In all natural environments—including wetlands, saltmarshes, mangroves, bushland and inland creeks—the destruction of habitat is causing loss of biodiversity at an alarming rate (see box 14.3). Biodiversity refers to the variety of all types of life (plants, animals, micro-organisms) and the ecosystems of which they are a part.

BOX 14.3   GLOBAL BIODIVERSITY CONTINUES TO DECLINE

Despite a UN-sponsored global convention and a 2010 strategic plan for protecting and enhancing biodiversity, the 2014 *Global Biodiversity Outlook 4* (Convention on Biological Diversity, 2014, p. 10) report found that little progress had been made towards the 2020 targets, and indeed that, in key areas, biodiversity continued to decline. The report noted the vital role that achieving key biodiversity targets would make to meeting sustainable development goals, including reducing hunger and poverty, improving human health and ensuring a sustainable supply of food and clean water. Of 55 target elements, only five are on track for 2020. Global rates of deforestation have slowed, but are still at unsustainable levels. The report concluded that despite the convention and strategic plan, 'The average risk of extinction for birds, mammals, amphibians and corals shows no sign of decreasing.'

Both Chivian (2003) and the State of the Environment Report (State of the Environment Advisory Council, 1996, pp. 4–5) consider biodiversity at three levels: ecosystem, species and genetic. They stress that each type is of crucial importance to human health for the following reasons:

- Biodiversity is essential to healthy functioning ecosystems, controlling pest plants, animals and disease, for pollinating crops and for providing food, clothing and many kinds of raw material.
- Ensuring the survival of species and preserving biodiversity for future generations is an ethical value of importance.
- Ecosystem disturbance to biodiversity can increase the spread of infectious disease among humans.
- Biodiversity will help preserve places of beauty, tranquillity and isolation. These features are important to human health and well-being. Preserving biodiversity will also help maintain the culture of indigenous people.
- Biodiversity is important to the human spirit. Being in natural surroundings gives us hope and comfort. Nature's beauty has inspired painters, writers and musicians and enhances emotional well-being for many people.
- Biodiversity has economic value by preserving areas for ecotourism and providing food stocks.
- Preserving biodiversity means preserving plants that may have pharmaceutical properties.
- Biodiversity is important to ensuring sufficient world food production, and monocultures threaten global food security.

The 2011 State of the Environment Report confirmed that little further had been done to halt the decline in biodiversity and that data on long-term trends on biodiversity are limited, making it difficult to monitor. The report also noted that, while each state and territory has goals, they are not matched by practical plans or sufficient resources (State of the Environment 2011 Committee, 2011). One of the causes of declining biodiversity worldwide is the invasion of exotic flora and fauna, threatening the native ecology. The pathways by which it is occurring are deeply enmeshed in the basic trade and travel patterns of the world (Bright, 1996). Container traffic is efficiently moving goods and exotics. Ballast water is responsible for carrying many creatures all over the world. In Tasmania, for example, a Japanese starfish has infected much of Hobart's harbour area. Air traffic is also rapidly expanding and opening the way for new bioinvasions.

The world's forests are a particularly important reservoir of biological wealth. They harbour more than half of all species on Earth and provide a range of other protective functions including flood control and climate regulation (French, 2000). Yet half the forests that once covered Earth have been lost and nearly 14 million hectares of tropical forests are felled each year.

Agricultural practices also encourage the depletion of native biodiversity. Cattle in Australia have destroyed the habitat of many native species, and tree plantations have often involved the prior destruction of native forests and the ecosystems they support. Aquaculture opens up the possibility of introduced diseases that can have a dramatic impact on native species. Global warming appears to be affecting coral reefs around the world. 'Global climate change, by itself or acting synergistically with other environmental changes secondary to human activity, could well become the factor most responsible for species extinctions over the next 100 years' (Chivian, 2003, p. 14).

Bioinvasions can directly threaten human health and well-being. A number of examples are provided by Bright (1996). A large fish from the Nile was introduced to Lake Victoria (Africa) to improve fishing, but it led to the mass extinction of native fish, destroying an important food source for 30 million people. Native forests are being destroyed to make room for introduced species, some of which directly cause disease. The Asian tiger mosquito may have been a factor in the 1986 yellow fever epidemic in Rio de Janeiro that affected about a million people. Deforestation can also lead directly to disturbances of the forest floor, providing depressions that catch and hold water and create new sites for the development of mosquitoes (Chivian, 2003).

Overall, the importance of biodiversity reminds us that that people are an integral part of nature and must learn to live in balance with its other species and within its ecosystems. If we fail to do this then the prospects for human health and even survival are bleak.

CONSUMERISM, NEO-LIBERAL GLOBALISATION AND THE ENVIRONMENT

It is necessary to see global warming as one of a suite of problems arising from the system of capitalist political economy, a system that is now globalised (Weston, 2014, p. 3). The world is experiencing a period of unprecedented social and economic change as well as the environmental crisis. Globalisation of economies and social and environmental issues have been key features of the last two decades. A United Nations report prepared for the World Summit for Social Development noted that social institutions are seen as obstacles to economic progress and are being dismantled: 'This has happened at every level. At the international level, social organisations have been overtaken by transnational corporations and international financial institutions. At a national level, many state institutions have been eroded or eliminated. And at a local level, the imperatives of market forces and globalisation have been undermining communities and families' (United Nations Research Institute for Social Development, 1995).

The impacts of neo-liberal globalisation on health were discussed in chapter 5, showing its adverse effect on health. Consumerism, driven by an aggressive advertising industry worldwide, is causing people to consume more and more. Most significantly, this consumerism heightens the massive inequities between rich and poor countries. Barbara Kingsolver (1998) sums up the immorality and astounding nature of these differences when the African family in her novel, *The Poisonwood Bible*, go into a US supermarket. Adah, the aunt of the half-American, half-African boy who has lived all his life in Africa until now, notes:

> When I go with them to the grocery, they are boggled and frightened and secretly scornful, I think. Of course they are. I remember how it was at first: dazzling warehouses buzzing with light, where entire shelves boast nothing but hair spray, tooth-whitening cream, and foot powders …

'What is that, Aunt Adah? And that?' their Pascal asks in his wide-eyed way, pointing through the aisles: a pink jar of cream for removing hair, a can of fragrance to spray on the carpet, stacks of lidded containers the same size as the jars we throw away each day.

'They're things a person doesn't really need.'

'But, Aunt Adah, how can there be so many kinds of things a person doesn't need?'

I can think of no honorable answer. Why must some of us deliberate between brands of toothpaste, while others deliberate between damp dirt and bone dust to quiet the fire of an empty stomach lining? There is nothing about the United States I can really explain to this child of another world (Kingsolver, 1998, p. 498).

That we have an economic system that revolves around encouraging excessive consumption despite the fact that millions live on less than US$1 per day is quite literally amazing. This economic system manages to insulate rich people from facing the fact that the lifestyles we lead depend on the overuse of the natural environment and exploitation of poor countries.

As the discussion in chapter 5 showed, it is in the interests of transnational corporations to encourage the growth of consumerism by expanding the markets available. This is happening rapidly with expanding urban middle classes in India and China that are projected to develop massive and growing appetites for consumer goods over the next 20 years (see, for example, Li, 2010). However, while this is good for the corporations who see these markets as translating into their profits, the implications for the environment are potentially devastating. The question is: how will the relentless quest for profits from expanding consumer markets be curtailed? By regulation and controls, the introduction of which would almost certainly be resisted? By persuasion, which seems to have little hope of success in the face of the relentless search for profits and returns to shareholders? Western societies and the newly emerging middle classes in poor countries will have to be weaned off overconsumption if our planet is to survive. Some may argue that it is up to individuals to change their behaviour. But just as behaviour change strategies have proved largely ineffective in modifying people's health-threatening lifestyles, so are they likely to be unsuccessful in terms of making environmental changes. Policies and collective decisions to change the choices people face are most likely to succeed. Thus people will use private cars less if the costs of doing so are high and if there are affordable public transport alternatives. However, for much more fundamental change, humanity has to move away from capitalist modes of production and profit seeking to systems of economics that are most benign to the planet. Such systems are discussed in chapter 16.

These issues are complex and overwhelming, but they are central to the quest of the new public health to improve health. If we are unable to find the answers, then the future for our health is bleak. What is most hopeful is that the environmental health problems we face are largely a result of social and economic arrangements. This means they are, therefore, open to change and adaptation. Ensuring this happens should be a central task for public health in the twenty-first century, and some of the emerging solutions are canvassed in part 6 of this book.

GLOBAL EFFORTS TO ADDRESS CLIMATE CHANGE

The United Nations has led efforts to address environmental problems starting with the UN Conference on the Urban Environment in Stockholm in 1972 and then the World Commission on Environment and Development (WCED) in the 1980s, which led to the Brundtland Report (1987) that promoted the notion that economic growth was sustainable. In the wake of this report, the IPCC was established and the Rio Earth Summit established the UN Framework Convention on Climate Change, which set voluntary limits on greenhouse gas emissions, and the UN Convention on Biological Diversity was established in 1992 (followed by a Strategic Plan for Biodiversity 2011–2020).

In December 1997 the UN International Conference on Climate Change met in Kyoto, Japan, with the aim of developing a global consensus on reducing greenhouse gas emissions. The USA signed but has not ratified the protocol because it does not put any constraints on developing nations, and Australia signed the protocol but did not ratify until 2007 under the Rudd Labor government. Many commentators have noted that the Bush administration in the USA received large campaign donations from petrochemical companies and that its opposition to the Kyoto Protocol may reflect its indebtedness to these companies. Australia was notable at Kyoto for its opposition to uniform global reduction targets for all nations and successfully arguing that Australia should increase its emissions. This is despite Australia being a high per capita emitter of greenhouse gases, with large fossil fuel subsidies and poor standards for appliance energy use. While there are debates about the value of the Kyoto Protocol because its target of reducing carbon dioxide emissions by 5.2 per cent will have a marginal impact at best, it is the only international treaty in existence created to combat climate change and so is significant from that point alone (Flannery, 2005).

A protestor dressed as US President George Bush holding a placard condemning earlier US refusal to sign the Kyoto Protocol to combat climate change. Many protestors attended this demonstration in Brussels against the visit of President Bush. (Tuen Voeten, Panos)

The major climate talks in the past few years have been the Lima Climate Change Conference in December 2014, which was preparatory to a new global agreement to be concluded in Paris in late 2015 that aims to keep warming to less than 2 °C (for updates, see http://unfccc.int/meetings/items/6240.php). If the agreement at Lima is enacted then for the first time all countries would be committed to cutting their greenhouse gas emissions. A major focus will be on national action plans, and a Green Climate Fund has been established to assist low-income countries with climate adaptation. The USA and China also negotiated a bipartite deal to reduce carbon pollution in December 2014.

WHY DON'T WE TAKE ACTION?

> When we get our story wrong, we get our future wrong. We are in terminal crisis because we have our defining story badly wrong.
>
> *Korten, 2015, p. 1*

Concern about the environment is receiving an increasing amount of attention, yet action to protect and promote the health of the environment is slow. Public health's reluctance to recognise and incorporate the implications of environmental crisis has mirrored the broader myopia on this topic. In 1992, three-quarters of all Australian adults, or 8.6 million people aged 18 years or older, stated they were concerned about environmental problems. By 2004, the proportion of people concerned had declined to 57 per cent (ABS, 2006), before rebounding by 2008, possibly associated with a long drought. However after that, up to 2012, concern faded once more.

Yet the problems have become worse since this time. The evidence on the impacts of climate change and other environmental threats to our health continues to amass. Yet despite serious efforts by the United Nations, no binding global treaty has been reached and many still deny the evidence that climate change is real. Increasingly the links between the global economic system, the ecological crisis and growing inequities are hard to mask. Extremes in weather patterns over the past decade (e.g. Hurricane Katrina in the USA, and an increase in cyclonic activity in Queensland) may contribute to breaking the complacency.

In the UK, the Cameron Coalition government has shown some commitment to action on climate change following the publication of the Stern Review. In Australia the Gillard Labor government introduced a modest carbon tax but in 2014 the Abbott Coalition government repealed it. The Abbott government is generally seen as a climate change–sceptical government and is reducing funding to climate science, climate change mitigation measures and renewable energy. The US Obama government has taken some action. China is moving fast to adopt renewable energy but its use of carbon is still increasing. Many experts believe we have only a few decades in which to take action, yet action is occurring very slowly. Why is this so?

NO APPRECIATION OF ECOLOGICAL DEPENDENCE

> Economos is the rules and regulations for running the domain, while eco-logos is the reason for it all, the underlying principle, the spirit. Normally, the logos should determine the nomos, but in the late 20th century, this is not the case.
>
> *Susan George, political economist, cited in Suzuki, 2010*

Boyden (1996, p. 55) sees the reasons for the failure to take action as reflecting complex assumptions in the basic culture and organisation of our societies. He points out that vested interests in the corporate sector are likely to obstruct changes that reduce consumption and pollution.

The dominant culture globally has lost sight of the fact that people are living organisms that arose out of, are part of, and are totally dependent on the processes of life and the living system of the biosphere. Boyden (1996, p. 55) believes that our culture has to 'embrace, at its very heart, a profound understanding of, and interest in, nature and the human place in the natural world'. He suggests that this should be an aspect of culture shared by people throughout the world. Imbuing the dominant culture with such an understanding would profoundly affect decision making; significant change could then take place.

LACK OF VALUE ATTACHED TO THE ENVIRONMENT

Another obstacle to action is the relationship between economics and the environment. Our global system of economics has no way of valuing the environment unless it is developed and exploited. Petrochemical companies, for instance, have a vested interest in a continuing growth in the use of cars and the sale of fossil fuels. Thus, a forest becomes valuable not because it sustains our ecosystem, absorbs carbon dioxide and produces oxygen, but because it can be felled and converted to wood chips. Areas of great natural beauty have no economic value other than tourism, despite the benefits they offer of rest, relaxation and recharging of the mind and spirit.

We saw in chapter 4 that the hegemony of neo-classical economics has been challenged by feminist and green economists, but to date our economic system rewards destruction rather than protection of the environment. Conventional economists argue that economic growth will improve the position of the poor in all countries. Recent experience has suggested this is not the case—the gap between the rich and poor is widening in many countries around the world. A belief in the desirability of continued economic growth remains a powerful barrier to tackling the problem of environmental degradation, but a growing number of economists are criticising this (Robertson, 1989; Daly and Cobb, 1990; Jacobs, 1991; Korten, 1995; Stretton, 2000). Alternative economic systems that are not based on the need for economic growth are discussed in chapter 21.

An acknowledgment of the traditional relationships that indigenous peoples had with the land and environment is seen as one means of valuing the environment (see box 14.4).

BOX 14.4 INDIGENOUS PROTECTION OF THE ENVIRONMENT

The Wisdom of the Elders (Suzuki and Knudtson, 1992) describes the traditional ways in which indigenous peoples of the earth protected and promoted the health of their environment. Their approach was based on intuition rather than science—an example of the sensory approach. Suzuki and Knudtson (1992) argue that this intuition offers many lessons for dealing with the current environmental crisis. They maintain that, unlike Western science, which offers a reductionist view of ecology, indigenous cultures acknowledge its complexity. These cultures also have a very deep understanding of the need to balance human needs with those of the natural environment. Suzuki and Knudtson offer an example: the Red Kangaroo Dreaming of the Aranda peoples of Central Australia. Krantjirinja is revered as an original ancestor of the Aranda people. Krantji (the spring through which the ancestor first made his appearance on earth) is sacred. There are taboos on killing kangaroos in the area that is also the best habitat for their survival. As a biologist subsequently noted, the legends appear to be based on sound ecological principles, and so helped preserve the kangaroo population.

Traditional Aranda beliefs about the sacred site called Krantji appear to represent a remarkable fusion of ecological and spiritual knowledge. They encode genuine ecological truths about the population dynamics and dietary preferences of local red kangaroos. At the same time, unlike sterile scientific findings, they contain a moral code mandating irrevocable human responsibility to honour and nurture those precious, life-sustaining animal populations in perpetuity (Suzuki and Knudtson, 1992, p. 166).

POWER OF INDUSTRY LOBBYISTS

The cartoon below shows the way in which renewable energy may not be popular with large energy corporations because of its potential impact on their profits.

It is clear that this desire to control markets leads to strong political lobbying from the fossil fuel and forestry industries to protect their commercial interests, which are likely to be most affected by any move to reduce greenhouse gas emissions or to curtail logging. The impact of this on the US Bush administration was widely recognised. Hamilton (2007) notes the considerable power that members of the fossil fuel lobby (who he names the 'greenhouse mafia') had over the Howard Coalition government through very close relations between the relevant bureaucracy and industry lobby groups. This resulted in the rejection and undermining of the Kyoto Protocol under the Howard government. Middleton and O'Keefe (2003) report that Bush did his utmost to discredit the findings of the IPCC when he asked the US National Academy of Sciences to review them. Gelbspan (1997) has argued that the fossil fuel industry used questionable science and private funding to build evidence against climate change in much the same way as the tobacco industry has attempted to influence research into smoking and health.

The mining industry lobbied heavily against the carbon and mining taxes introduced by the Gillard Labor government and contributed significantly to the defeat of Labor in 2013 and the subsequent repeal of the taxes. While there are also 'green' lobbyists and political parties, their resources to mount effective campaigns to change popular opinion are limited.

COMPLACENCY AND LACK OF TIME

Most of the environmental threats are happening over long time scales. McMichael (1993, p. 295) noted that 'those processes, which are dramatic when measured against geological time scales, are trends, not events—and they therefore register only faintly on our alarm system'. Flannery (2005, p. 237) suggests that the failure of the US and Australian governments to ratify the Kyoto Protocol initially, reflects the fact that they both come from frontier mentalities in which there were no limits to growth. Others suggest the failure to ratify reflected the power of the oil and coal industries' lobby groups to influence government policy. Complacency is also encouraged by the belief that a technological 'fix' will be found to solve problems.

The public health challenge is to respond to long-term and gradual danger. It seems the threat will have to be perceived as extreme if concern for the environment is to take priority over the need for economic growth. Public health activists know that it is far easier to win a battle when something happens to bring a health issue to public attention. The Tasmanian Port Arthur massacre, in which 35 people were murdered by a single gunman in April 1996, provides an example. After the tragedy the anti-gun lobby achieved far greater change in gun ownership legislation than would have been possible had it not occurred.

Pocock et al. (2012) note in their study of work and life pressures that while many people would like to take actions that would make their lives more sustainable (recycling, growing their own food, using public transport, cycling and walking), their work and caring commitments do not allow them to do this. They note that time is at the heart of adapting to climate change, yet people are generally time-poor: 'Work–life configurations that leave substantial proportions of working Australians time-pressured, fatigued and lacking sleep are a substantial barrier to more sustainable lifestyles' (p. 207).

ENVIRONMENTAL JUSTICE

It is also universally true that poor people live in the worst deteriorating and health-damaging environments (Watts et al., 2014). WHO (2014g) notes that the health impacts of climate change will be felt most by those in low- and middle- income countries (mainly because of the underlying social determinants of vulnerability). They will also disproportionately affect vulnerable groups within each country including the poor, children, older people and those with pre-existing medical conditions. Poor people also have less resources with which to challenge and prevent the sources of pollution of their local environment. Poorer countries are likely to suffer more, and be protected less, from environmental hazards than richer countries, and industrialised countries can afford cleaner environments than developing countries. Newly industrialising nations are often attractive to multinational companies because they do not have strict levels of control over environmental considerations. A World Bank official has even suggested that poor countries should be used to dump toxic wastes because the mortality and morbidity is less expensive (Pearce, 1992; Beder, 2000). This is despite the fact that the industrialised world is responsible for most of the environmental degradation.

Developing countries bear the brunt of many of the degrading activities (Sachs, 1996). Agriculture, forestry and mining in developing countries (often for the benefit of industrialised nations) degrade the environment for the export industry. Local people see few of the benefits, as these go to overseas companies or elite groups within the developing countries. The populations of poorer countries are much less able to oppose polluting practices.

Climate change is likely to affect different parts of the world to different extents. Some Pacific Island nations face drastic consequences as sea levels rise because their land will be submerged. On the African continent people cannot easily move across boundaries to agriculturally more fortunate regions, and the climate of Africa is complex and diverse—small changes in weather patterns can produce large changes in the viability of particular areas. Droughts have been prevalent in Africa in recent years. A weakening of the monsoon in the north equatorial belt across Africa could accelerate the ecological decline, and Nisbet (1991) predicts this may lead to biological and economic collapse in some nations. He says of most countries in Africa that in the third decade of the twenty-first century (p. 138):

> the prospects are of rapid reduction in per capita incomes and episodic famines. Millions or even tens of millions of people may die or try to migrate, emigrate or burst the locked doors of the West. The land will be in collapse, its ecosystems devastated and many of its animals and plant species eradicated. There will be strife or revolution in many countries. In short the Malthusian correctives—war, disease, famine—will be in full power, the continent a surging sea of misery.

The emergence of the Ebola epidemic in 2014 suggests his predictions may arrive earlier than anticipated. The People's Health Movement (2014) describes the ways in which exploitation of natural resources and deforestation in West Africa have forced populations to move, and says it is likely that these movements brought humans into contact with the Ebola virus. Weston (2014) analyses the impact of climate change on Africa and paints a very bleak picture of the way in which colonial exploitation is continuing, resulting in groups such as small-scale peasant farmers (for centuries a source of food security) losing their livelihoods in the face of climate change. It is estimated that the loss of healthy life years as a result of climate change will be 500 times greater for Africans than Europeans (Costello et al., 2009). The populations of low-income countries, especially African countries, are, and will continue to be, very vulnerable to a variety of forms of environmental degradation that will affect their physical environment and eventually their health. Inevitably, this type of scenario will affect the rest of the world, because a world of stark inequalities would be ethically extremely uncomfortable and it is likely that demands from those suffering the worst excesses of environmental degradation would require other parts of the world to become fortresses.

Orr (2013) argues that the increasingly obvious and undeniable threat of climate change may force governments to become increasingly centralised and authoritarian. Orr cites Klein as calling for a more robust form of democracy as a balance to an increasingly powerful state (see box 16.5).

Within rich countries, it is the poorer populations who suffer most from environmental impacts on health. Brulle and Pellow (2006) point out that environmental pollution appears to have a major impact in the creation of health inequities. They observe that producing evidence on environmental effects on health is difficult because of the problems of designing robust studies and attracting research funds for this topic. However, they cite evidence from California that race was a strong predictor of the location of hazardous waste facilities and also in explaining cancer risk distribution, even after controlling for socioeconomic status and other demographic factors. They note that there are debates about whether class or race is responsible for the injustice in the USA but, either way, the increased recognition of the injustice has led to an environmental justice movement.

FEMINISM AND ENVIRONMENTAL JUSTICE

A further perspective on environmental justice has been offered by the Indian feminist and scientist Vandana Shiva (1988). She examines the position of women in relation to development as it is driven by Western scientific thought, and concludes that it has been particularly detrimental to women in developing countries, who have been violated and marginalised. She maintains that women in 'ecological societies of forest dwellers and peasants' have played a key role in maintaining the sustainability and ecological diversity of these societies. The feminine principle is central to them. Western development processes totally ignored the value of indigenous wisdom and this has had dramatic consequences: 'What local people had conserved through history, Western experts and knowledge destroyed in a few decades, a few years even' (Shiva, 1988, p. 26). She maintains that a return to environmental justice for women in developing countries will only be achieved when indigenous knowledge is valued and their societal values more widely adopted. Central among these is the importance that women in these societies accord nature as the 'very basis and matrix of economic life through its function in life support and livelihood' (p. 224). She sees indigenous women as experts in survival and as offering the key to the planet's ecological survival: 'The intellectual heritage for ecological survival lies with those who are experts in survival. They have the knowledge and experience to extricate us from the ecological cul-de-sac that the Western masculinist mind has manoeuvred us into' (Shiva, 1988, p. 224). In this case Shiva sees restoring environmental justice as benefiting the whole earth.

THE PRECAUTIONARY PRINCIPLE

Around the world, whenever there is conflict over an environmental threat, the onus is on the community to prove that an environmental hazard is dangerous rather than on the industry or developer to prove that it is safe. The 'standards' approach only monitors actual effects on human health. On the other hand, the precautionary principle holds that once there is reasonable evidence, but still some level of uncertainty, that a particular practice might be harmful, it is advisable to take preventive or ameliorative action. In practice, implementation of this principle depends on operational 'reasonableness', especially when powerful groups have vested interests in not acknowledging the possibility of harm. The basis of the precautionary principle is 'better safe than sorry'. Advocates of the approach argue that the burden of proof should be on the proponent of an activity. They should bear the burden of assessing its safety and of showing that it is both necessary and the least harmful alternative. They also argue that decisions affecting public and environmental health should be fully participatory.

A counter-argument to the precautionary principle is that all activity involves some (public health) risk and that the fundamental question is in deciding (and in who decides) what is a socially acceptable level of risk. The question for public health is whether community concerns about potential environmental effects can be taken as warnings and so accorded more credence than at present.

CONCLUSION

This chapter has reviewed the considerable evidence that demonstrates that environmental factors have a very significant impact on human health. These impacts are likely to continue through this century and, unless action is taken to mitigate them urgently, human survival will be under threat. The chapter also discussed the threat posed to environmental sustainability by our existing system of political economy. It is obvious that there are major tensions between the type of economic development that is so highly valued in the world and the protection of the natural environment. At every turn in this debate, these two factors come into conflict. This debate is highly political and once again illustrates a central point of this book—that public health is an inherently political activity. Part 6 will return to our examination of the protection of the natural environment and the importance to human health of doing this, and examine how the conflicts between economic development and the environment can be tackled.

CRITICAL REFLECTION QUESTIONS

14.1   What factors combine to mean that low-income countries will suffer most from the health effects of climate change?

14.2   Do you agree with the central argument of this chapter that there is no room for climate scepticism and that it is urgent we take action to protect human health and the health of the planet? If not, why not?

Recommended reading

State of the World reports from the Worldwatch Institute report on progress towards a sustainable world. Each edition contains a variety of chapters on different environmental problems and so provides an assessment of current evidence and prospects for the future. Highly recommended as a digest to current thinking. *State of the World 2014: Governing for sustainability* is a clear-eyed yet ultimately optimistic assessment of citizens' ability to govern for sustainability.

BBC has provided an animated guide to how the greenhouse effect works, which is available at: http://news.bbc.co.uk/2/shared/spl/hi/sci_nat/04/climate_change/html/greenhouse.stm

Weston, D. (2014) *The political economy of global warming: The terminal crisis* An assessment of the ways in which the current capitalism economic paradigm is driving climate change.

Useful websites

WHO has a web section devoted to climate change and human health www.who.int/globalchange/environment/en

For examples of activists' sites, see World Wide Fund for Nature at www.panda.org/climate, and Friends of the Earth at www.foe.co.uk/campaigns/climate

The Intergovernmental Panel on Climate Change (IPCC) website contains details of its latest reports, which provide a clear and up-to-date view of the current state of scientific knowledge relevant to climate change www.ipcc.ch

15

URBANISATION, POPULATION, COMMUNITIES AND ENVIRONMENTS: GLOBAL TRENDS

He let his mind drift as he stared at the city, half slum, half paradise. How could a place be so ugly and violent, yet beautiful at the same time?

Chris Abani, quoted in Davis, 2006, p. 20

KEY CONCEPTS

Introduction

Urbanisation

Violence and crime

Living conditions

Crowding and health

High density: a health hazard?

High density and social disorder

High density and environmental sustainability

Slums

Affluent suburbia: dream or nightmare?

Social impact of urban life: from community to anomie?

Social capital declining?

Transport in urban areas

Population, consumption and equity

Conclusion

INTRODUCTION

This chapter considers two key trends that are putting strain on our ecological systems: urbanisation and population growth. The shift of populations form rural to urban areas has been one of the defining features of the world since the 1950s. This chapter will consider the impact of urbanisation in rich and poor countries in terms of both physical and social aspects of the urban environment. A consideration of the impact of cars as the dominant form of urban transport continues the theme of factors contributing to climate change considered in the previous chapter. Privatised cars are seen as a threat to human health. The chapter then considers the debate about the extent to which population growth threatens health and argues that overconsumption poses more of a threat.

URBANISATION

Cities create both problems and opportunities. Throughout recorded history people have been drawn to cities to experience the excitement, the variety of people and the wide range of social and employment opportunities they can offer. Yet cities also create problems and challenge our ingenuity to the limit. In terms of health, cities can promote and create health. Urban density and economies of scale can provide services and resources that would not be possible in more dispersed populations (WHO, 1993). Experiences of cities are remarkably varied—from very rich people living in downtown Manhattan, harbourside in Sydney, London's West End or Shanghai, to slum dwellers in Mumbai, Rio de Janeiro, Nairobi or Manila. Some urban communities have achieved high levels of health and well-being, but health risks appear to be increasing for most urban dwellers, especially in low- and middle-income countries.

At the beginning of the nineteenth century only 5 per cent of the world's population were living in urban areas.

FIGURE 15.1 GLOBAL URBANISATION AND LEVEL OF DEVELOPMENT

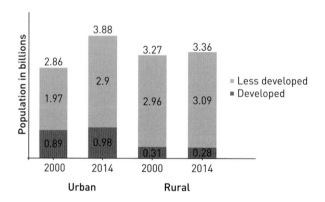

Source: WHO Kobe Centre, 2015a.

In 2010, for the first time in human history, more than half the world's population was living in urban areas, with WHO predicting that by 2050 that figure would rise to 70 per cent (WHO, 2010b).

In the first 14 years of the twenty-first century, a billion more people were added to urban areas (WHO Kobe Centre, 2015a).

The United Nations (UN Department of Economic and Social Affairs, 2014) listed 20 'mega-cities' with a population in excess of 10 million in 2003. By 2014, that figure had risen to 28 with a total population of 453 million (table 15.1). Most of these mega-cities are in developing countries. This urbanisation represents one of the great mass migrations of history.

TABLE 15.1   CITIES WITH A POPULATION OF 10 MILLION OR MORE, 2014

City	Population (million)
Tokyo, Japan	37.8
Delhi, India	25.0
Shanghai, China	23.0
Mexico City, Mexico	20.8
São Paulo, Brazil	20.8
Mumbai, India	20.7
Osaka, Japan	20.1
Beijing, China	19.5
New York, USA	18.6
Cairo, Egypt	18.4
Dhaka, Bangladesh	17.0
Karachi, Pakistan	16.1
Buenos Aires, Argentina	15.0
Kolkata, India	14.8
Istanbul, Turkey	14.0
Chongqing, China	12.9
Rio de Janeiro, Brazil	12.8
Manila, Philippines	12.8
Lagos, Nigeria	12.6
Los Angeles, USA	12.3
Moscow, Russian Federation	12.1
Guangzhou, China	11.8
Kinshasa, Democratic Republic of the Congo	11.1
Tianjin, China	10.9
Paris, France	10.8
Shenzhen, China	10.7
London, UK	10.2
Jakarta, Indonesia	10.2

Source: United Nations Department of Economic and Social Affairs, 2014, p. 26, Table II.

While cities may provide their residents with access to services and employment, rapid urbanisation can also produce damaging combinations of health determinants. WHO and UN-HABITAT (2010) have described a triple threat from infectious diseases (such as HIV, tuberculosis and pneumonia, spread via crowded unsanitary conditions), non-communicable diseases (from poor diets, pollution and lack of exercise) and injuries (including road traffic accidents and violence). The problems created by this rapid urbanisation are summarised in box 15.1.

BOX 15.1   PROBLEMS ASSOCIATED WITH RAPID URBANISATION

- Rapid urbanisation happening in a context of deregulation and liberalisation
- Increased population density, overcrowding and congestion
- Transport and mobility problems and pollution
- Increasing biological, chemical and physical pollution of air, water and land from industrialisation, transportation, energy production, and commercial and domestic waste
- Increasing pressure from climate change
- Large populations in squatter settlements and shantytowns, often occupying urban land subject to landslides, floods and other hazards. These people have come to form an underclass without full citizen rights
- Inadequate sanitation, sewerage and solid waste disposal
- Inadequate provision of clean water
- Increasing number of people living in extreme poverty (especially women and children) and consequently at high risk of violence and sexual and other forms of exploitation. This leads to increasing inequities between different groups within cities
- Social isolation and anomie, possible decline in social capital
- Increasing violence and crime and rates of imprisonment of the poor and marginalised
- Unemployment, especially of young people, and lack of job opportunities
- Inadequate social services
- Inadequate healthy food supplies
- Lack of empowerment and unequal distribution in power, prestige and resources

Source: WHO, 1993, 2010b; Global Research Network on Urban Health Equity, 2010.

Although the relative severity and exact nature of these problems differ from city to city and between developing and developed countries, there are global problems of urbanisation and industrialisation being faced by nearly all cities and countries in the world for the first time in history. Developed countries have far superior resources with which to cope with the problems of rapid urbanisation. Certainly the poorer the city, the less resources available to deal with problems.

In all large cities there are marginal groups or underclasses, who live in extreme poverty. They are either unemployed or underpaid in the informal economy, lacking social organisation and experiencing inadequate nutrition, hygiene and housing.

They are also open to exploitation (WHO and UN-HABITAT, 2010). Sassen (2014, p. 211) sees the emergence of underclasses as part of a pattern of exclusions in which 'the move from Keynesianism to the global era of privatisation, regulations and open borders for some, entailed a switch from dynamics that brought people in to dynamics that push people out'. She sees that cities are becoming the spaces of exclusion in which marginalised people are abandoned, whether this is whole cities (as in Africa) or parts of cities (as in the USA or the suburban fringes of Australian cities). Thus while absolute poverty is prevalent in the lowest income countries, there are also increasing inequities in developed countries (see chapters 11 and 12). For example, the age-adjusted black male death rate for inner metropolitan areas in the USA in 2010 was 9.2 per 1000, compared to 7.1 for white males in the fringing suburbs in the same cities (US Centers for Disease Control and Prevention, 2013). Such statistics have led commentators to speak of a Fourth World to be found within the cities of developed countries. Most cities are characterised by spatial distributions of poverty and the disadvantaged, while many developing cities have the rich elite living well away from the absolute poverty of squatter dwellings on the city outskirts. In the USA residential segregation is driven by racial/ethnic difference more than by class and evidence indicates that this segregation results from institutional racism, which dictates access to housing markets (Acevedo-Garcia and Lochner, 2003). In the USA segregated minorities are concentrated in central cities, which are typically the oldest, most dilapidated and most socioeconomically deprived part of the metropolitan area (Acevedo-Garcia and Lochner, 2003, p. 267). Kawachi and Berkman (2003, p. 9) review evidence of the effect of neighbourhood characteristics on mortality and conclude that most of the evidence indicates 'a moderate (statistically significant relative risk between 1.1 and 1.8) association between neighbourhood environment and health, controlling for individual socioeconomic and other characteristics'. In this chapter we look at various aspects of the impact of place on health in rich and poor countries.

Many aspects of the problems being faced in the large urban areas are featured in other parts of the book. Four aspects warrant particular attention here: violence and crime, living conditions (including housing), social stresses, and transport.

VIOLENCE AND CRIME

WHO (Krug et al., 2002) reports that problems of crime and violence have become increasingly serious in all cities, particularly in Africa and the Americas. People in low-income countries are over three times more likely to die as a result of interpersonal violence (see details in chapter 11). The variation between homicide rates in different countries was shown in chapter 12.

Chasin (1997) notes the links between violence and economic deprivation in the USA, and provides and develops the concept of 'structural violence', which results from a system in which class and gender inequities are entrenched and increasing. The Commission on Social Determinants of Health (2008) noted that while most responses to violence tend to focus on 'downstream factors' the underlying causes are deeply rooted in social and economic structures. An unpublished working paper on violence and injury written for the Commission on Social Determinants of Health concluded 'Compressed disadvantage, systematic exclusion from social participation, associated with severe deprivation and the experience of economic and other forms of inequality contextualise interpersonal violence (in particular, perhaps, with rapid urbanisation and exacerbated poverty, creating new pathways between exclusion, identity and public violence)'. The crucial issue for the new public health is how our communities, and especially the rapidly growing urban areas can be encouraged to develop so that they encourage low-violence environments in which people have a sense of personal safety. Unfortunately in many parts of the world the opposite seems to be happening. The Commission's final report called for newer approaches to violence prevention, including regulatory control (such as limiting access to alcohol and improved gun control), conflict transformation, environmental design and community-based approaches to social capital.

Gender violence is pervasive throughout the world (Fischbach and Herbert, 1997; WHO, 2013a)—from rape to dowry-related deaths. It is systematically underreported, but undoubtedly leads to serious psychological problems for victims and perpetrators. The invisibility of women in the privacy of households means that the exact level of violence is unknown but as the topic of intimate partner violence receives more research attention its extent and impact (especially on children) receive more recognition. The level of gender violence reflects underlying social and cultural expectations that have to be challenged if the violence is to be reduced.

In recent years there have been a number of mass murders in which a lone gunman runs amok, killing many people, not always in urban areas (for example the Virginia Tech massacre of 33 people in April 2007, and the 2012 Sandy Creek Elementary School massacre, where 20 children and 6 adults were killed). The profile of the gunman is often a single male who appears to have led an isolated life with few meaningful social contacts. The apparently random killing of strangers increases people's fear of violence. The difference in gun laws between the USA and Australia is credited with the lower rates of gun deaths in Australia (see chapter 24 for further discussion).

Crime rates, especially violent crime, are contributing to a change in the spatial form of cities (WHO, 1996). Richer people in all countries increasingly live, work, shop and take their leisure in fortified enclaves with sophisticated security systems. Shopping malls, office complexes and leisure activities are being moved to the outskirts of cities because of the high level of crime and violence in inner city areas. Some city centres are now only inhabited by the poor, who have few choices. These trends are less evident in Australia but community concern about violent crime is high, although it has been suggested that the perception of risk is greater than the reality in Australia. In Australia crime rates tend to be higher in outer suburban areas where socioeconomic disadvantage is concentrated. The fear, however, can mean that people are scared to venture out of their homes and so, ironically, make communities less safe as fewer people mean less surveillance by the community. Crime levels can be affected by and affect the extent to which people are trusting of their communities. Issues of crime, trust, social inclusion and safety are important determinants of a community's health, as are the ways in which communities respond to crime. Sassen (2014) points to the growth of incarceration as a response. She notes that 1 in 100 US citizens were in jail in 2013 (p. 67) and that this number had steadily increased over the last decade, and that there is a powerful lobby from corporations to encourage this trend.

LIVING CONDITIONS

>You can kill a man with a tenement as easily as you can kill a man with an axe.
>
>*American social reformer and journalist Jacob Riis, quoted in Ross, 1991, p. 37*

People in developed countries take the supply of safe drinking water and the efficient and safe disposal of waste water and sewerage for granted. Living conditions in industrialised countries are far superior to those in the developing world, where the problems faced are similar to those tackled in the industrialised cities in the nineteenth century. The fact that the living standards achieved in industrialised countries rested, in part at least, on the fruits of the colonial era adds to the moral argument for ensuring improved living conditions in all countries. Industrialised countries consume far more resources than non-industrialised countries, and the challenge for public health is to contribute to a world in which resources are shared more equally and living standards for the world's poorest citizens are significantly improved.

In developed countries in both rural and urban areas there are pockets of housing where conditions are inadequate to support healthy living. Remote Indigenous communities in Australia lack adequate water and sanitation, power supply, access to a variety of nutritious food and appropriate housing. In cities poor suburbs have poor-quality housing stock, and inadequate facilities such as shops, transport, parks, schools and community space.

Environmental conditions helping to spread communicable diseases include insufficient and unsafe water supplies, poor sanitation, inadequate disposal of solid wastes, inadequate drainage of surface water, inadequate housing and overcrowding (WHO and UN-HABITAT, 2010). Developing countries today do not have the resources that were available in the rapidly growing nineteenth-century industrial cities because they were able to expand quickly with resources coming from their colonies. They were rich cities with the capacity to provide an urban infrastructure for public health. Many of today's fast-growing urban areas do not have the resources to provide supportive infrastructures for health. The disposal of sewerage and waste water is a problem for all cities. The most vulnerable populations are the informal communities living on the edge of cities in developing countries, who typically have no housing, no safe water supply or provision for safe disposal of sewerage. Infectious diseases are rife in these circumstances and are a major cause of mortality, especially of children. At a global level large inequities exist in under-five mortality rates, with children from poorest urban families roughly twice as likely to die as those from richest urban families (WHO and UN-HABITAT, 2010).

Assessing poverty in poor urban environments and slums is difficult. It is most conventionally done by including people earning below one or two US dollars a day as poor. Mitlin and Satterthwaite (2004) point out that this misses out so many dimensions of the experience of being poor in an urban area. Box 15.2 provides a more nuanced list of the aspects of poverty that make life so difficult for people in impoverished urban environments. Sassen (2014, p. 147) notes that whereas poor used to mean having a plot of land that did not produce much, those living in slums now own nothing but their bodies.

BOX 15.2   DIFFERENT ASPECTS OF POVERTY

- Inadequate and often unstable income (and, thus, inadequate consumption of necessities, including food and, often, safe and sufficient water; frequent problems of indebtedness, with debt repayments significantly reducing income available for necessities)
- Inadequate, unstable or risky asset base (non-material and material, including educational attainment and housing) for individuals, households or communities
- Poor quality and often insecure, hazardous and overcrowded housing
- Inadequate provision of 'public' infrastructure (for example piped water, sanitation, drainage, roads and footpaths), which increases the health burden and often the work burden
- Inadequate provision of basic services such as day care, schools, vocational training, healthcare, emergency services, public transport, communications and law enforcement
- Limited or no safety net to ensure that basic consumption can be maintained when income falls or to ensure access to housing, healthcare and other necessities when these no longer can be paid for
- Inadequate protection of poorer groups' rights through the operation of the law, including laws, regulations and procedures regarding civil and political rights; occupational health and safety; pollution control; environmental health; protection from violence and other crimes; and protection from discrimination and exploitation
- Poorer groups' voicelessness and powerlessness within political systems and bureaucratic structures, leading to little or no possibility of receiving entitlements to goods and services; of organising, making demands and getting a fair response; and of receiving support for developing their own initiatives. In addition, there is no means of ensuring accountability from aid agencies, non-government organisations (NGOs), public agencies and private utilities and of being able to participate in defining and implementing their urban poverty programs
- Low-income groups may also be particularly seriously affected by high or rising prices for necessities (such as food, water, rent, transport, access to toilets and school fees).

Source: Mitlin and Satterthwaite, 2004, p. 15.

Africa's largest slum, Kibera, Kenya, where women spend hours each day fetching water.
(Fran Baum)

Housing is an essential element of a safe living environment in urban areas. There is a vast literature exploring the issues of housing and health, highlighting that not only is the provision of shelter essential but the nature of it is a vital contributor to health (GRNUHE, 2010). The World Health Organization (1989, p. viii) noted:

> Housing is intimately related to health. The structure, location, facilities and uses of human shelter have a strong impact on the state of physical, mental and social well-being. Poor housing conditions and uses may provide weak defences against death, disease and injury or even increase vulnerability to them. Adequate and appropriate housing conditions, on the other hand, not only protect people against health hazards but also help to promote robust physical health, economic productivity, psychological well-being and social vigour.

The link between housing and health is well established and the impacts range from the need for shelter to the importance of secure housing to mental health (Kingsley, 2003; Thomson et al., 2003). There are aspects of urban living about which the health impacts are less clear (for example, the impact of crowding and urban density) and these are considered next. Both areas demonstrate the complexity of social determinants of health.

CROWDING AND HEALTH

Crowding is a relative concept. Despite numerous psychological studies, it has not been possible to determine at what density abnormal behaviour occurs. Housing size varies from country to country, and the dangers of overcrowding can be used to justify very different housing sizes. Data from 2009 on average household size and house area suggest that the average resident of Hong Kong had around 15 square metres of floor space, compared to 89 square metres for the average Australian. For China (urban only), the figure was 20 square metres, for the UK, 33 square metres and for Germany, 55 square metres. Australians had the largest average house size in the world at 214 square metres (James, 2009). There is no doubt that in many slum areas in poor cities, overcrowding is a major issue made much worse because of very poor quality housing. Yet the relationship between residential density and health is a complex one, as demonstrated by the long life expectancy of the citizens of Hong Kong. Newman and Hogan (1981) point out that questions of equity have to be brought into play if one country justifies the provision of far greater space to each person than others. Perceptions of overcrowding appear to be a culturally determined concept, rather than determined by density. The perception is intimately related to privacy as Lang (1987, p. 147) explains:

> Too much privacy leads to feelings of social isolation, and too little privacy leads to subjective feelings of crowding ... Crowding is stressful because it limits personal autonomy and expression and breaks down communication patterns. It must be distinguished from population density ... Crowding is associated with a feeling of lack of control over the environment ... Crowded conditions lead to negative behaviours because they are related to social overload ... density, on the other hand, does not seem to be causally linked to such behaviours.

Newman and Hogan (1981, p. 283) report that the 'stultifying effect of high-rise living on children finds almost universal support'. They note that it is important to draw a distinction between 'high density' and 'high rise'. The two do not need to go together and cities can be low-rise and dense. Even though low-density suburban environments are considered by many people to be safe, optimum conditions for children, research by Lynch (1977) and Berg and Medrich (1980) suggest that such environments do little for a child's imagination. It would appear that what is needed for a satisfying urban community is a high density of street-level interaction, involving corner shops and small businesses, as well as residents and children at play. It is this type of environment that the new urbanists seek in their quest to create human-scale 'main street' developments that encourage interaction and safety (see details in the section on healthy neighbourhood design in chapter 17).

HIGH DENSITY: A HEALTH HAZARD?

Epidemiological perspectives on density were most prominent in the nineteenth-century public health revolution, when overcrowding and high-density living were seen as the enemies. Dr Duncan, a Liverpool (UK) general practitioner, carried out a housing survey in the 1830s, to discover that a third of the population lived in the cellars of back-to-back houses with earth floors, no ventilation or sanitation, and as many as 16 people to a room (Ashton and Seymour, 1988, p. 15). Such living was seen to be an ideal breeding ground for infectious disease, and similar concerns continued throughout the nineteenth century.

The desire of reformers to eradicate poor housing and infectious disease led directly to ideas like garden cities and the Bourneville (UK) ideal settlement (Sarkissian and Heine, 1978). The suburb came to be associated with health, light, sunshine and the good life, compared to the horror of the disease-ridden, overcrowded city centre. The slum was viewed as dense, dirty, unnatural, disorderly and disease-ridden; the suburb viewed as open, clean, natural, orderly and healthy (Davison, 1994, p. 100). There is little doubt that public health considerations were prominent in the early justification of suburban development, as is shown by a quote from the Report of a Royal Commission on Housing of the People of the Metropolis, which sat in Melbourne in 1917. Above all the arguments for legislation for a minimum size for suburban allotments was sanitation and health:

> In a general view, it is regarded as insanitary, and otherwise undesirable practice, for two or more families to occupy at the same time a dwelling house of ordinary design and size, when evils due to overcrowding are to be looked for. So it is agreed amongst sanitarians that similar evils, on a larger scale, are to be expected where dwellings are built on allotments having dimensions so limited as to have insufficient space for entrance of sunlight and fresh air around and into the house, or for privacy, or for adequate yard space, clothes drying ground, play areas for young children, or for fire breaks for the spread of fire from house to house, to say nothing of possible advantage presented by such open spaces in reducing risk from supposed aerial convection of infection.
>
> *Report of the Royal Commission, 1917, pp. 25–6, quoted in Davison,*
> *1994, pp. 108–9*

The epidemiological evidence at the time of the Industrial Revolution led to the assumption that high-density living was a health hazard. A more likely explanation is that the lack of the most basic public health measures, such as sewerage, solid waste collection, water treatment and control of air pollution, was the real culprit. Today extremely high-density living in countries such as Singapore, Hong Kong and Japan suggests it can be compatible with health, as these countries have achieved long average life expectancies.

One of the issues associated with high density is the extent to which a city environment provides green space. Green space is generally seen as desirable although, as Galea and Vlahov (2005) found in their review of literature in this area, empirical data evaluating the relationship between green space and health remains limited and the relationship is a complex one with many confounding variables (Lee and Maheswaran, 2011). Recent work has shown that living in areas with walkable green space is associated with greater likelihood of physical activity, higher functional status, lower cardiovascular disease risk and longevity among older people independent of personal characteristics (Lee and Maheswaran, 2011).

The density of populations within cities makes them particularly vulnerable to man-made or natural disasters. Terrorist attacks such as those on 11 September in 2001 in New York and 7 July 2005 in London show the impact these attacks can have on concentrated urban populations. The New Orleans Hurricane Katrina disaster of August 2005 demonstrated the way in which natural disasters have a greater impact on the poor of a city than on the rich.

HIGH DENSITY AND SOCIAL DISORDER

From as early as 1903, the sociologist Georg Simmel claimed an association between high urban density and social disorder (Press and Smith, 1980, pp. 19–30), but his claims were based on casual observation rather than systematic studies. Although a little more cautious than others about 'the mechanisms underlying these phenomena', Wirth wrote in 1938: 'Personal disorganisation, mental breakdown, suicide, delinquency, crime, corruption and disorder might be expected … to be more prevalent in the urban than in the rural community' (Press and Smith, 1980, p. 47).

Association does not necessarily imply cause and, when ethnicity, poverty, education and other factors are considered, Craig (1989) concluded that urban density itself bears little relation to social pathology. Research in the USA and UK has concluded that crime and vandalism are more common in urban designs featuring anonymity, lack of surveillance and availability of alternative escape routes (Jacobs, 1961; Newman, 1972; Coleman, 1985). Galea and Vlahov (2005) note that a substantial body of research has established a relation between stress and social strain and mental and physical health and that recent research is suggesting an association between urban neighbourhood contexts and adverse health behaviours.

HIGH DENSITY AND ENVIRONMENTAL SUSTAINABILITY

Well-managed high-density urban environments have the potential to minimise their ecological footprint (the resources required to sustain their inhabitants). Owen (2004) has argued that Manhattan is the 'greenest' city in the USA in terms of the per capita consumption of resources by its residents. Apartment dwellers do not run cars, they spend less on heating and cooling, ride bicycles and don't have lawns to put chemicals on. While they require a major infrastructure in terms of water supply, power sanitation, public transport, food distribution and so on, this infrastructure caters for millions, so per capita costs are kept low.

Informal housing in Khayelitsha, Cape Town, South Africa. (Trish Struthers)

Thus there is much debate about how the urban structures we live in affect our health. This discussion will be taken further through the exploration of two types of urban environments—slums in poor countries and suburban living in rich countries. Both types of urban environments pose health problems albeit of a different magnitude and type.

SLUMS

The health impacts of living in informal settlements have been well documented and are summarised by Unger and Riley (2007) and GRNUHE (2010) as:

- lack of basic services, especially water and sanitation
- substandard housing or illegal and inadequate building structures
- overcrowding and high-density living
- unhealthy living conditions and hazardous locations
- insecure tenure or informal settlements
- poverty and social exclusion
- minimum settlement size.

The Commission on Social Determinants of Health (2008, p. 60) notes that 43 per cent of the urban population in developing regions live in slums and that in the least developed countries that figure is 78 per cent. Slums are areas of extremely high ecological stress, which has implications for physical and mental health status. For instance, they are breeding grounds for infectious diseases. In Nairobi, where 60 per cent of the city's population live in slums, child mortality is 2.5 times greater than in other areas of the city, while in Manila, the rate of tuberculosis infection in slum areas is twice the national average (CSDH, 2008, p. 60). Slums are also very emotionally stressful environments to live in. Yet despite the high disease burden, health service access for slum dwellers is poor. People living in these communities have little chance of obtaining a more adequate house with space, security and services because they are typically extremely poor and could not afford the rent. Insecure tenure goes hand in hand for most people in informal settlements, meaning that fear of eviction is a constant worry.

The Commission on Social Determinants of Health (2008) notes that slums represent a failure of urban governance. Municipal governments are unable to cope with the exponential population growth by expanding public provision of adequate shelter, basic infrastructures and services and provision of gainful employment. So most slum areas simply do not have the provision of public services to meet basic needs for health and well-being.

AFFLUENT SUBURBIA: DREAM OR NIGHTMARE?

Most people in rich countries live in urban areas (in Australia the figure is just under 80 per cent) and within these areas in suburbs. Suburbs of course, compared to the slum dwellings experienced by most urban dwellers, are health promoting environments.

However, suburbia has been criticised quite extensively, particularly in terms of its impact on mental, communal and spiritual health. The thrust of this criticism is summarised by Alexander (1967, p. 88):

> But autonomy and withdrawal and the pathological belief in individual families as self-sufficient units can be seen most vividly in the physical patterns of suburban tract development. This is Durkheim's dust heap in the flesh. The house stands alone: a collection of isolated, disconnected islands. There is no communal land and no signs of any functional connection between different houses.

More recently the design of suburbia in North America and Australia has been blamed for encouraging sedentary lifestyles and lack of social contact as a consequence of their physical design, which caters to the needs of cars and does not encourage casual social interaction between neighbours. Some critics have referred to outer suburban areas as urban wastelands in which there is little opportunity for the development of a sense of community and where personal crimes are encouraged by the absence of community surveillance. Australian outer suburban areas usually lack vitality, intimacy and neighbourliness compared with areas nearer city centres. They contain tracts of new housing with little mixing between shops, businesses or community facilities, and represent a significant public health issue in Australia. Isolation and loneliness can pose health issues. This is especially the case for older people and migrants (Colson, 1986). Van Eyk (1996) describes in detail the extreme isolation of older Spanish-speaking migrants in Adelaide, resulting from a complex mix of class, gender, migration and language. Australian communities lacked the warmth and closeness of those they had left. This quote contrasts life in El Salvador with Australia:

> We had a completely different life there than here where people live more separately in their own house and they don't talk to each other. There the atmosphere in which I was living in the neighbourhood was like a festival, you know, all the time you can hear music in the street from the houses and people come out and they talk to each other and children play. So it is completely different from here. I had a wonderful life in that time. Here it is very quiet, very quiet.
>
> *Van Eyk, 1996, p. 74*

The desirability of different urban forms has been hotly debated. Compared to Asian and European cities, North American and Australian cities often appear empty and devoid of life and excitement and lacking in social capital. Apart from central business districts, Australian cities are suburban, but their density has increased in the last decade even as the average house size has increased. In North America city centres often contain poor and marginalised people whereas more affluent people live in suburbs, increasingly in gated communities that have high security and low social contact.

Suburbs have also been criticised as being heavy carbon users and highly dependent on fossil fuels (Newman et al., 2009). Rising oil prices as peak oil is reached will make outer suburban areas less attractive and more isolated. In Australian and US cities the people living on the urban fringes are on lower incomes and will not be able to afford the costs of transport to city centre amenities and jobs. This is likely to increase social and eventually health inequities.

FEMINISTS AND SUBURBS

Feminists have been critical of cities and other communities as they cater for the needs of male car owners to the detriment of women, particularly older women and those with young children (Saegert, 1985; Harman, 1988; Fincher, 1990). They argue that urban forms have been based on masculine assumptions about domestic life in suburbs and work in cities, and that the interests of capital accumulation have taken precedence over the areas of consumption, reproduction and daily life (Huxley, 1994). In this view, suburbs have created ghettos of isolation in which women and children find it difficult to establish contact with others (Harman, 1988). More recent feminist critiques have argued the need for policies that go beyond local and place-based strategies, to address the influences of institutions and organisations operating at regional and national levels (Fincher, 2007). Mitchell (2004) notes that a retreat from multiculturalism may be part of a wider pressure for assimilation and a reduction in diversity in urban planning, making urban areas less amenable to the increasingly diverse populations.

BENEFITS OF SUBURBAN LIFE

Of course, compared to urban slums with their extremely unhealthy environments, suburbs in rich countries offer their residents a physically safe environment. While Australian and US urban densities have increased in the last decade, many suburbs do offer their residents private garden space, which is used for socialising, growing produce, and providing play space for children. Home ownership also provides secure tenure and so promotes control.

The suburb is also seen as an attractive expression of the strong rural tradition in English-speaking culture. This was reflected in the garden city concept that influenced planning in Australian cities. Much of this tradition of town planning was inspired by reactions to slum conditions in the newly industrialising cities of the nineteenth century and by a strong desire to promote public health through urban planning. The irony, then, in the twenty-first century, is that the reliance on the private car and so sedentary and privatised lifestyles is creating a new set of public health issues because of the tendency of these suburbs (together with changes in work patterns) to promote lack of exercise and withdrawal from community (see collection in Dixon and Broom, 2007) as discussed in the section below.

URBAN CONSOLIDATION AND EQUITY

Urban consolidation and denser cities have been the target of much public policy in the past few decades. Concerns have been raised that these policies may benefit rich more than poor people. Troy (1996) argued some two decades ago that suburbs with land have given lower-income people access to the benefits of space, clean air and hazard-free environments that were hitherto unavailable to them. He argued that trends to urban consolidation are deepening divisions between rich and poor. Rich people will always be able to buy the space and environments they desire, but low-income people's choices will be limited to smaller and less desirable homes in consolidated areas.

Similar concerns have been expressed by a community movement in Sydney—Save our Suburbs (www.sos.org.au/new_sustain.html)—which argues that consolidation favours developers and, despite seeming to be more ecologically sound, in fact is not.

Urban development policies struggle continually with the tensions between development interests, whose aim is to maximise profits, and the need to create equitable cities. Cities that are characterised by areas with less employment, recreational and cultural opportunities and access to health and welfare services will underpin growing inequities in health status (de Snyder et al., 2011). The challenge for the future is to create cities that are both sustainable and equitable in the face of growing population pressures, economic crises and climate change.

SOCIAL IMPACT OF URBAN LIFE: FROM COMMUNITY TO ANOMIE?

Modern urban society has been characterised as isolating and anomic. Cities and suburbs, in particular, have been accused of encouraging isolation and not providing a sense of community, in contrast with earlier communities where people had more intense and meaningful ties. The appeal of strong, well-connected communities is common to many modern social movements, including the new public health. Utopian socialism and the modern green movement also feature visions of close, cooperative and supportive communities (Pepper, 1996).

THE LOSS OF COMMUNITY?

In Europe, visions of community often stem from the ideal of a village life in which people know each other well, have entwined lives and offer support to one another. In most cities around the world the appeal of a lost romantic past of community is strong, reflecting, in part, the traditions of European and other immigrants. All in all, there is a sense that modern life does not provide the opportunities for intimacy and support that were available to past generations. This feeling might come from European migrants to the USA, Canada or Australia or from a slum dweller in a large Asian city who has left their close-knit rural community. Such feelings are also experienced by many indigenous peoples who have been dispossessed from their traditional lands. This experience has been summarised by the German philosopher Ferdinand Toennies, who attempted to make sense of social change between the period before the Industrial Revolution, when social relationships were small-scale, personal and particular, and nowadays, when they tend to be large-scale, impersonal and more universal. Toennies labelled the pre-industrial, close-knit communities as Gemeinschaft and the communities of the industrial age Gesellschaft.

How far does the shift described by Toennies reflect reality? A contrary view might hold that life in pre-Industrial-Revolution society was nasty, brutish and short, characterised by high infant mortality, widespread infectious disease and poor-quality housing and other infrastructure. This is certainly the picture for many people living in poor urban or rural environments around the world. The appeal to a golden age of close communities may be a romanticisation, and small close communities can also be limiting and repressive in their demand for conformity. Perhaps what is important is that the idea of such communities has a powerful appeal, and there is certainly plenty of evidence that social ties and connections can be good for your health. Much community development work aims, among other things, to create closer communities in which people know each other and can work together towards collective goals. Such collective action can reinforce self-esteem and people's sense of belonging. The rapidity of globalisation has increased the trend towards Gesellschaft, possibly leading to a rise in nationalistic feelings and an increase in xenophobia. This is a reminder that close-knit communities are not inevitably benign. They may foster and promote attitudes and behaviours that are damaging to the health of other groups and, ultimately, to themselves. Certainly the rapid urbanisation that is happening across the globe is seeing the loss of the close ties of community but it is also leading to challenges to traditional systems of inequity such as caste.

SOCIAL CAPITAL DECLINING?

An important concern about life in urban areas is that the stocks of social capital appear to be declining. 'Social capital' is the term used for 'the processes between people which establish networks, norms, social trust and facilitate coordination and cooperation for mutual benefit' (Cox, 1996). Generally communities high in these characteristics are seen to be more functional because people are able to get along better and achieve more for the collective good.

The two theorists on social capital whose work has been used in public health are Bourdieu (1986) and Putnam (1993a, 1993b, 1996, 2000). Bourdieu (1986) presents social capital as a resource that assists people in getting on in life. Like other forms of capital he sees that its distribution is uneven, so that people who are economically more advantaged also have access to more social capital. Bourdieu's definition of social capital places much more emphasis than Putnam does on the power dimensions of society and the way these shape social and other interactions. Putnam stresses a more consensus view of society and does not link social capital to individual economic advancement as Bourdieu does. Despite increasing agreement that social capital is a valuable term there is debate about its definition and no agreed way in which to measure it. The complexity of the constructs behind social capital (trust, the ways in which people interact and their level of cooperation) are not easy to measure in any reductionist way. Nonetheless, these constructs are important to life in all settings, rich or poor, including cities and slums, and are an important way of understanding the ways in which cities can detract from and enhance health.

Throughout the literature on social capital, the existence of trust in relationships emerges as the key factor in determining the extent to which a community or society can be seen to have a high level of social capital. Together, reciprocity and trust characterise societies in which people are able to cooperate effectively to achieve common civic goals. These societies are those that provide their citizens with multiple opportunities to interact and network through groups, associations and societies.

Social capital takes on particular importance in a world in which cultures, peoples and customs are increasingly mixing. Globalisation is bringing with it a significant emphasis on difference and all cities around the world are learning to govern and manage cities in which populations are heterogeneous rather than homogeneous. The urban sprawl associated with the growth of suburbs in many North American and Australian cities has been associated with a trend towards greater social stratification (partly because houses of a similar price level tend to have been clustered together) and less social capital (Frumkin, 2002). The lower social capital has been seen to reflect a decline in trust, less engagement in civic life (because of the isolating nature of suburban life) and the fact people have less time because of the time devoted to commuting in sprawling cities.

There is some evidence that social and civic trust are declining around the world. Levels of trust differ considerably, with the highest levels in Scandinavia and most of the low-trust countries being those with low income and unstable states (Roser, 2014). In the USA, researchers (Twenge et al., 2014) found that as income inequality and poverty rose, public trust declined, indicating that socioeconomic factors may play an important role in driving this downward trend in public trust. Cross-nationally, trust is highest in countries with a more equal distribution of income (Wilkinson and Pickett, 2009). In Australia, while trust declined between the early 1980s and mid-1990s (Hughes et al., 2000), surveys conducted in 2006 and 2010 (ABS, 2013d) found that the level of generalised trust was unchanged, with 54.1 per cent agreeing or strongly agreeing that most people could be trusted. Many social philosophers (for example, Hobbes and de Tocqueville) have seen trust as central to social order, and therefore declining trust is commonly seen to signify a troubled society. Hughes, Bellamy and Black note that trust of people from different cultures and backgrounds is of vital importance to multicultural societies. They comment that if multiculturalism is to continue to be a success 'trust needs to be built in such a way that people from all cultures and backgrounds feel accepted and included in Australian society'. Riots in the Sydney suburb of Cronulla in December 2005, which stemmed from tension between local Muslim young people and Anglo-Australians, was seen at the time as an indication of a growing intolerance and decline in trust between the groups. By contrast, the response to the Sydney Martin Place siege of December 2014 with campaigns such as #illridewithyou was seen as a community determined to maintain trust. The health effects of declining trust are likely to be significant, even though they are hard to measure or prove directly. Trust levels are likely to be harder to maintain in rapidly growing urban areas where many new residents are arriving from different places. This will be especially the case in informal settlements.

Increasing social capital cannot be expected to solve problems that are essentially those of poverty and deprivation (Portes and Landolt, 1996). Communities are likely to benefit from participation and mutual trust, but these may well rely on sufficient material opportunities, such as jobs, good affordable housing and clean safe environments. It is likely that richer communities have more opportunities and potential to create and maintain social capital than poor ones. Nonetheless, high levels of social capital may be an important coping mechanism for some poor communities, as shown by British studies of working-class terraced housing in east London where women had close and effective networks (Young and Willmott, 1957). However, evidence from slum areas of the burgeoning cities in poor countries suggests that the exposure to high rates of crime and violence and the consequent constant fear for one's own safety, creates high levels of mistrust and low social capital, which increase as long as the crime and fear prevails (World Bank, 2011). The challenge for public health is to determine those aspects of the urban environment that might be more likely to be supportive of social health and the health benefits it may confer and determine what investments might promote these.

Putnam's work has been criticised for assuming an overly homogeneous view of societies in terms of levels of social capital. Arneil (2006) notes that much of Putnam's work is blind to gender and class and does not consider the ways in which power is mediated through local communities. The challenge in the ever-growing urban areas around the world is how to create societies that have increasing rather than decreasing social capital and that do reduce inequities that result from gender and class difference. In a world of growing inequities this will prove difficult, and part 6 will argue that greater equity is very likely to be a prerequisite for increasing the extent of components of social capital such as trust, solidarity, social and civic networks across different groups, and reciprocal behaviour.

TRANSPORT IN URBAN AREAS

Heavy traffic and pollution from carbon-fuelled vehicles are major problems for cities around the world. The health problems caused by transport include respiratory diseases from vehicle emissions, road accidents, stresses associated with extreme traffic and the social dislocation caused by car-dominated cities. Most of the rapidly growing cities in Asia and South America face extreme problems from traffic congestion and traffic flow. For example, the number of vehicles in Jakarta has grown by around 10 per cent per year for 6 years, reaching 4.1 million cars and 11.9 million motorcycles by 2013. Airborne lead levels increased more than tenfold over the same period. Around 5.4 million daily commuters struggle with traffic congestion estimated to lose Jakarta US$3 billion a year (Rukmana, 2014; Maulia, 2014). The *China Daily* (29 September 2014) reported that despite anti-congestion strategies, Beijing's annual bill for traffic congestion comes to US$11.3 billion, and average daily congestion time is almost two hours, up 25 minutes since 2012. A study conducted by the Chinese Academy of Environmental Planning blamed air pollution for 411 000 premature deaths in 2003 (Watts, 2005b).

The needs of individual car users have shaped the form of most cities in industrialised countries, especially in the USA, Canada and Australia. The car has allowed cities to sprawl in a manner that has made walking and cycling much less feasible. North American and Australian cities are the least dense in the world (figure 15.2) and provide much more road space than cities in other parts of the world (figure 15.3).

FIGURE 15.2   URBAN DENSITIES, 1990

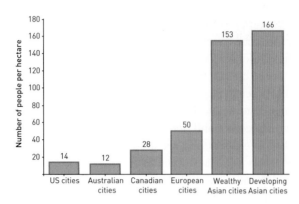

Source: Kenworthy et al., 1999.

FIGURE 15.3   LENGTH OF ROAD PER PERSON, 1990

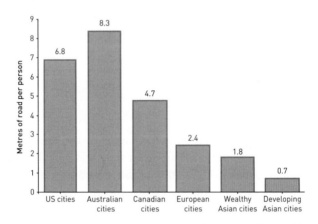

Source: Kenworthy et al., 1999.

This also means our cities are very energy inefficient. An example of urban planning driven by the needs of private cars is urban shopping malls. These are generally not accessible by foot and are surrounded by car parking space. Australia is one of the most car-dependent countries in the world. Newman et al. (1992) have contrasted North American and Australian cities with those of Asia and Europe. The former have high levels of car ownership and use and very low density. By contrast, European and Asian cities are denser and less reliant on cars. Newman et al. (1992) describe how cities in general have changed from being basically a series of urban villages to transit cities, with transit routes radiating out from the cities, and finally to automobile cities characterised by low density, decentralisation and high dependence on cars. The form of Australian and other automobile cities has been criticised as being damaging to the environment and human health. Table 15.2 demonstrates the high public health costs of the car. An updated analysis of selected cities in the USA, Canada, Australia and Europe showed that only very modest progress had been made towards more sustainable transport, but that some cities (Berlin stood out) have done much more than others (Kenworthy and Inbakaran, 2011).

TABLE 15.2   ENVIRONMENTAL AND HEALTH PROBLEMS ASSOCIATED WITH
AUTOMOBILE DEPENDENCE

Environmental	Economic	Social and health
Urban sprawl	Congestion costs	Loss of street
High greenhouse gas contribution (in production as well as use)	High infrastructure cost	Dissection and loss of community
	Loss of productive rural land	
Greater storm water problems—pollution of surface water and groundwater by surface run-offs from impervious surfaces (roads, car parks etc.)	Loss of urban land to bitumen roads and parking	Loss of community safety
		Isolation in remote suburbs
	Long travel to work times	High lead levels
	High mortality of young productive men	Respiratory illness
		Inequities for those without cars (very young, very old, infirm, disabled, impoverished)
Traffic noise and vibration	Costs of road accidents (health services, rehabilitation)	
Local photochemical smog and global CO_2 (greenhouse and acid rain) and CFCs	Oil vulnerability	Stress caused by traffic and driving and road rage
	High transport costs for car users	
Solid waste (abandoned soil tips and rubble from road works; vehicles withdrawn from service; waste oil; tyres)		Corruption, inequity and pollution of local food sources in developing countries exploited for oil
		Traffic accidents: mortality, morbidity
		Lack of exercise leading to obesity
		War over oil reserves

Sources: Adapted and extended from Newman, 1993, p. 5; United Nations Environment Programme, 1993; Newman et al., 2009.

ENVIRONMENTAL AND ECONOMIC PROBLEMS OF AUTOMOBILE DEPENDENCE

The environmental problems caused by cars include traffic noise, air pollution, visual intrusion and disruption of neighbourhoods by roads. The effects of physical pollution from cars were described earlier (table 14.3). Noise pollution is also significant, as it can reduce concentration and exacerbate psychiatric disorders. If the traffic is continuous it can interrupt REM sleep and if intermittent it may induce lighter and so less restorative sleep (Hillman, 1991). Transport is a major contributor of greenhouse gases. Road transport emissions contribute significantly to the greenhouse effect, particularly carbon dioxide (CO_2). Globally, the transport sector contributes approximately 15 per cent of greenhouse gas emissions. Global carbon dioxide emissions increased by 45 per cent between 1990 and 2007, with the bulk coming from road transport (OECD and ITF, 2010). The road transport sector in Australia was responsible for 26.25 per cent of Australia's greenhouse gas emissions in 2010. While emissions from cars were static, emissions from light and especially heavy commercial vehicles were growing (Charting Transport, 2012).

The need to provide space for cars means high urban development costs. Cities sprawl to accommodate the space needed for cars, necessitating more new roads, sewers, schools, community centres, public transport services and so on. In cities where most people use cars, public transport runs with high deficits and diminishing services. Giles-Corti et al. (2012) call for a transition to higher population density levels for Australian cities in order to reduce carbon emissions, preserve arable land for food security, encourage walking, cycling and public transport use, reduce obesity and build social capital.

Being dependent on cars means price increases for oil have a significant inflationary effect on transport costs, and therefore on most goods and services. High levels of car ownership mean that much of the city's land has to be handed over to the car for roads and parking, and as Newman et al. (1992, p. 8) say: 'the more space devoted to cars, parking lots and freeways, the more sprawling and energy wasteful the city.' When communities are bisected by roadways there is less likely to be interaction between residents, which reduces opportunities for the accumulation of local social capital. These spaces are usually not human-friendly and can create an alienating environment. A psychiatrist who has done a considerable amount of research on the impact of the urban environment on mental health commented on the likely impact of car-dominated cities:

> mental health is unlikely to be promoted by incomprehensible urban sprawl, severed by dangerous motorways and full of monotonous blocks with unwelcoming spaces between them. There are also the likely ill-effects of population dispersal, such as the time and energy wasted by millions of people every day in commuting, together with the stress that must come from this frustrating activity. As a consequence, city centres are deserted at night and weekends and suburbs empty during the weekdays, causing further undesirable social and psychological effects (Freeman, 1992, pp. 25–6).

Car dependency also breeds social inequities. For instance, a third of the US population is too old, too young or too poor to drive. In developing countries only the very rich can afford a car, yet cars are coming to dominate the form and development of cities in low- and middle-income countries and so make the city less useable for pedal bikes and pedestrians, cheap and environmentally friendly forms of transport (O'Meara, 1999).

SOCIAL AND HEALTH PROBLEMS OF AUTOMOBILE DEPENDENCE

> every private car should carry a government health warning because of the … enormous impact of the car on disease, death, disability, quality of life, integrity of the environment, social intercourse and social inequalities, and from the huge public cost to the social purse.
>
> *Hunt, 1989, p. 101*

RESPIRATORY DISORDERS

Cars make a significant contribution to air pollution, leading to adverse health effects, such as asthma and bronchitis. We saw in chapter 14 that WHO (2014b) estimated that around 3.7 million premature deaths were attributable to ambient (outdoor) air pollution in 2012, with about 88 per cent of these occurring in middle- and low-income countries. Those suffering from existing respiratory conditions, the very young and the very old are particularly affected (Cohen et al., 2014). The health effects of car exhaust gases, especially those of yellow-brown summertime photochemical smog, have been well documented (McMichael, 1993, p. 286). As car ownership increases, so does the impact of air pollution.

CAR USE AND STRESS

The traffic associated with cars has been associated with increased stress. Noise from traffic is a source of stress (Cohen et al., 2014). Negotiating heavy traffic and finding parking are frustrating, especially in those cities where traffic is particularly dense and heavy, such as London, New York and Bangkok. Many of the world's most automobile-dominated cities (for example, Houston, Phoenix, Perth, Adelaide) have so much space devoted to the car that moving about in cars may be less frustrating than in some of the denser cities. However, traffic is a major issue in Los Angeles, the world's most car-dependent city. Freeway congestion and gridlock are chronic and seem to be getting worse, the 'peak hour' lasting for most of the day (Newman et al., 1992). The sensory and social isolation of driving a car can encourage antisocial behaviour in drivers (see box 15.3). It has been suggested by Frumkin (2002) that the increasing traffic volume and distance travelled associated with urban sprawl is encouraging aggressive behaviour from drivers. He quotes a number of surveys that point to high levels of frustration on the roads of sprawling cities and hypothesises that when these angry drivers arrive at work or home they will also exhibit angry behaviour and so concludes that transport use associated with urban sprawl is bad for health. Cohen et al. (2014) also point out that regular exercise reduces psychological stress and, as car use reduces levels of exercise, this may increase overall population stress levels.

BOX 15.3   PEDESTRIAN ROAD RAGE

Hunt (1989, p. 109) contrasts a driver's behaviour with that of people on foot and imagines a pedestrian behaving in the following manner:

- Getting close behind another pedestrian and maintaining the same speed while flashing a torch at them
- On being overtaken by another pedestrian, speeding up and then immediately pulling up in front of her
- On rainy days, throwing large quantities of water over people standing at the kerbside

- Speeding up when about to be passed by another pedestrian—whereupon the other also speeds up, and so on until both are proceeding along the High Street at breakneck speed, endangering other pedestrians
- Walking along the street making obscene gestures, swearing at other pedestrians merely for being there and criticising the way they walk.

Road rage is now recognised as a hazard of urban living. Aspects of driver behaviour have attracted the attention of public health and other reformers—in particular speeding and drink driving. In Australia, the road toll has declined significantly since the 1970s, largely as a result of policies such as seat-belt legislation and random breath-testing. Those public policies that have resulted in a significant reduction in the road accident death rate are reviewed in more detail in chapter 24. Young men are by far the most likely to be killed and do seem to exhibit the most antisocial behaviour on roads. No evidence is available, but it is likely that the most automobile-dominated societies also encourage the most extreme antisocial behaviour on the roads.

ROAD ACCIDENTS

Road accidents are one of the great public health hazards of the car. The Global Burden of Disease Study 2010, published by the *Lancet*, reported road traffic crashes as the number one killer of young people, accounting for nearly a third of the world injury burden—a total of 76 million disability-adjusted life years in 2010, up from 57 million in 1990 (Make Roads Safe, 2012). A Global Burden of Disease report (2015) found that the rate of deaths from transport injuries increased significantly from 1990 to 2013, largely due to increased use of motor vehicles in the developing world. WHO (2013c) estimates that about 1.24 million people die each year as a result of road traffic crashes, with 91 per cent of deaths occurring in low- and middle-income countries, which have only half the world's vehicles. The cost has been estimated at US$518 billion globally. Rapid urbanisation of cities in low- and middle-income countries with poor urban planning, little safety legislation (such as mandatory seat-belt or helmet wearing, drink-driving testing or safety checks on cars) has resulted in the death rate increasing. In rich countries the introduction of road safety measures has resulted in a decline in the death rate in the past 20 years (see table 12.5 in chapter 12).

TRANSPORT EXCLUSION: WOMEN, CHILDREN, OLDER AND POOR PEOPLE

Cities that encourage almost total dependence on cars for travel disadvantage those without access to them. Women, children, older people and people with disabilities are the least likely to have access to a car. High densities of traffic and complex intersections that call for rapid decisions and high levels of driving skill may effectively preclude older people from driving in larger cities. Car dependence often makes poor people poorer as lower income groups are housed in fringe suburbs where car transport is essential but expensive. Ironically, people often move to these suburbs because housing costs are cheapest, but then find they have high transport costs. If a family with two parents and children cannot afford two cars, the car will often be used by the breadwinner, usually the man, for commuting to work, while the woman runs a high risk of becoming isolated. In a pamphlet entitled 'Winning Back the Cities', Newman et al. (1992) argue that car domination in the suburbs has a detrimental effect on social interaction, as people retreat from streets dominated by cars.

TRANSPORT AND INEQUITIES

Transport arrangements and options contribute significantly to health inequities. Above, we have seen how urban planning options may fuel the obesity epidemic. In many cities poorer people are most likely to live in neighbourhoods affected by traffic air and noise pollution, and also to live close to major traffic infrastructure that creates community severance (Cohen et al., 2014). Poorer people are also more likely to be dependent on walking as a mode of transport and yet have to do so in cities and suburbs that are designed primarily with cars in mind. Cities with accessible public transport are likely to be the most equitable. People with disabilities are most greatly disadvantaged by transport systems and special efforts have to be made to increase accessibility for them.

CAR DOMINANCE OF CITIES

The other areas that appear to suffer because of the car are inner cities. Central cities and regional centres have become functional and sterile corporate or commercial centres lacking in human appeal and increasingly dangerous outside business and shopping hours. Many cities around the world are at risk of losing their unique character as they turn more space over to parking areas and freeways. Local governments often provide subsidies to private cars (see box 15.4). Developing countries have had far more sustainable forms of transport than most developed cities, but how many of them will manage to maintain this?

BOX 15.4   LOCAL GOVERNMENT SUBSIDISING PRIVATE CARS

A study by the International Council for Local Environmental Initiatives (ICLEI) highlights hidden subsidies by municipalities of motorised private transport (MPT). In the framework of a study funded by the German Federal Environment Agency (UBA), ICLEI-Europe's Cities for Climate Protection (CCP) team looked for obvious as well as hidden sources of income and expenditure in the budgets of three major German cities—Bremen, Dresden and Stuttgart.

The findings demonstrate that German municipalities pay a significant amount of money towards MPT, which far exceeds the income from that source. Working with a very conservative calculation, the subsidies amount to over 84 million Euros in Stuttgart, 56 million Euros in Dresden, and over 60 million Euros in Bremen. When a projection for all 82 million German citizens was made, the result was more than 10 billion Euros in subsidies towards MPT during the year 2000 (local expenses only, excluding central and state governments).

> The highest figures originate from the maintenance and upkeep of roads, city drainage, the cleaning and lighting of streets, and the building of parking lots. The fire brigade, the police, economic development, parks and recreation departments also represent large sources of expenditure that strain a municipality's budget. Sources of income through MPT include fines, tolls and parking fees.
>
> Most municipalities would not be able to answer the question whether road traffic pays for itself. There is a lack of local budget transparency because income and expenditure are typically not shown relative to MPT. These are often tucked away in other subgroups in the budget. Municipalities need to assess MPT income and expenses in order to determine if they are in fact subsidising MPT and, if so, should this continue?
>
> *Source: Erdmenger and Führ, 2005.*

Australian and US cities sprawl, and so are car dominated. Suburbs were only made possible with the mass ownership of cars. North America and Australia have given their cities over to the needs of the car far more than other parts of the world. Some of the rapidly developing Asian cities such as Bangkok and Beijing are following the car dominance of developed economies.

LACK OF EXERCISE

Cities that encourage car use above other forms of transport also encourage sedentary lifestyles and overweight. Physical activity levels have declined as cars are increasingly used for trips that previously involved walking (Cohen et al., 2014). Hinde (2007) notes that the reliance on cars in urban environments has made a considerable contribution to the creation of 'obesogenic environments'. This is because urban planning in many developed countries is done to accommodate cars, which are used in preference to walking, and as 'motorised shopping trolleys' they encourage the consumption of mass-produced and pre-prepared products that increase the intake of energy. Brisk walking and cycling are ideal methods of exercise for most people, and if people walk to public transport or their destination and ride bicycles, then exercise is built into their everyday life; they don't have to find special time to keep fit. One of the reasons people on lower incomes are more likely to be obese in Australia might, in part, reflect the fact that low-income people are more likely to live in car-dependent outer suburbs. Car-dominated environments also discourage people from walking and cycling because of the risk of accidents and the lowered quality of the environment, discriminating particularly against the very young and the elderly. The British Medical Association has concluded that the health benefits of cycling—reduction in coronary heart disease, obesity and hypertension—outweigh the risks of accidents by around 20 to 1 (O'Meara, 1999). A rapid review of the evidence on the use of public transport and exercise (Rissel et al., 2012) found that 30 per cent of public transport users met all their recommended levels of physical activity just from their transport walking, and public transport users were 3.5 times more likely to be sufficiently active compared with car drivers. MacDougall (2007) notes that people's willingness to exercise is influenced by the perceived safety of suburbs and the extent to which there are local facilities such as shops, cafes and post offices to walk to.

POPULATION, CONSUMPTION AND EQUITY

We have enough for everyone's need but not everyone's greed.

Gandhi

MALTHUS REVISITED

The world's population in 2012 was estimated at 6 916 183 000 (United Nations Department of Economic and Social Affairs, 2013). By 2015, the estimate was 7 285 229 972 and growing at a rate of 1.14 per cent per year. The recent population growth of humans is unprecedented (see figure 15.4). In 1900, the world's population was estimated to be 1.6 billion, in 1950 it was about 2.5 billion, by 2000 it was more than six billion and growing by approximately a billion people every 12 or 13 years (Dimick, 2014). Analysis of UN 2012 population data (Gerland et al., 2014) reported an 80 per cent probability that world population will increase to between 9.6 billion and 12.3 billion in 2100. An increasing number of authors now maintain that the global population growth and consumption patterns are unsustainable (Robertson, 1989; Daly and Cobb, 1990; Goodland et al., 1992; Brown, 1996; Campbell et al., 2007). McMichael and Butler (2011) state that the present population growth will increase the destruction of farmland and forest, the contamination of air and water, the disruption of climate and the extinction of species. For instance, the increased carbon dioxide emissions of rapidly industrialising developing countries (such as China, India and Indonesia) largely reflect population growth.

FIGURE 15.4 WORLD POPULATION TRENDS, 1950–2030

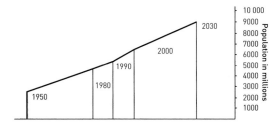

Source: United Nations Department of Economic and Social Affairs (UN DESA), The World Population Prospects, http://esa.un.org/unpd/wpp/index.htm.

In the 1960s and 1970s, population growth was arguably seen as the most crucial environmental issue. Calls for zero population growth abounded and the Club of Rome's Limits to Growth was a bible for many environmental activists. In the 1980s the topic went off the agenda to some extent. Other issues threatening the environment were seen as more pressing—greenhouse gases, the preservation of wilderness areas, campaigns against carbon-fuelled vehicles, the threat of nuclear war. Even the United Nations Earth Summit, held in Rio de Janeiro in 1992, did not pay much attention to population. Yet the historical unprecedented rate of population growth continues to put increasing pressure on global environmental and social viability. Meadows et al. (2004) in the third update on the Club of Rome's 1972 book *Limits to Growth* posit two possible futures: either a transition to sustainability or to 'let nature force the decision through lack of food, energy or materials, or through an increasingly unhealthy environment'.

DEVELOPMENT REDUCES DEATHS MORE THAN BIRTHS

On the face of it, the solution to the world's growing population seems obvious: encourage those countries with high birth rates to introduce contraceptive technology that reduces the fertility of the women. In this view the problem is purely technical and a matter of mobilising family planning clinics around the world to bring about the desired outcome. Of course, the problem is far more complex. It is in developing countries that the population is growing very rapidly. Industrialised countries experienced a demographic transition between the late eighteenth century and the mid-twentieth century, resulting in an eventual parallel decline in death and birth rates after a period in which birth rates exceeded death rates and, for a period of about 50 years, an increasing population. Demographers associate the transition with improvements in living standards. The final phase of the transition, characterised by low birth and death rates, involves high educational levels for women and effective contraception.

Demographers predicted that the transition to low birth and death rates would be experienced by all countries as they developed, but this has not happened in Africa, where death rates have declined in response to the success of some primary health care measures such as oral dehydration for diarrhoea, eradication of smallpox, improvement in sanitation and, in some periods over the past 40 years, economic growth. Meanwhile, birth rates have remained high. HIV/AIDS brought yet another change to the pattern in the 1990s and the failure to respond to the epidemic has meant a dramatic falling life expectancy in many sub-Saharan African countries, as noted in chapter 12.

McMichael and Butler (2011) note that international data show, in general, a positive correlation of a population's health with level of affluence and size of per-person footprint. Yet, beyond a modest threshold, larger footprints afford negligible health gain and may impair health (such as by increasing obesity). They argue that dramatic change is necessary to ensure that future growth in population combined with intensified economic activity does not result in an unsustainable planet that will not support human life.

GENDER EQUITY IS CRUCIAL

Moves towards gender equity are a key to effective population policy. Engelman et al. (2002, p. 143) maintain: 'As long as girls and women are envisioned as less able than boys and men to navigate human experience and decide for themselves how to live, population policy will always be imperfect.' This means that girls need to be free of sexual violence and free to make choices about their sexuality, and that they need to receive education. Women need access to contraception, integrated reproductive health care, access to educational and employment opportunities and freedom from violence. In both developed and developing countries women's work is undervalued. Women typically work longer hours than men—nurturing children, caring for older people, maintaining homes, farming, and hauling wood and water home from distant sources. Box 15.5 lists the priorities for population and gender equity for discussion at the Johannesburg World Summit, 2002.

BOX 15.5   WORLD SUMMIT PRIORITIES ON POPULATION AND GENDER EQUITY

- Meet the goals of the 1994 International Conference on Population and Development, including funding universal access to reproductive health care and closing the gender gap in education
- Respond aggressively to the global HIV/AIDS pandemic, stressing prevention of further infection as well as treatment of those already infected
- Change laws and work for social change to ensure that women enjoy equal protection and equal rights
- Increase female participation in all levels of politics
- Correct gender myopia in all levels of private and public planning including international lending, natural resources policy and globalisation
- Guarantee equal access to economic opportunities for women and men
- Enact and enforce strong laws to protect women from all gender-based violence
- Involve men in reproductive health services and discussions and educate them about the importance of gender equity
- Ensure that young people have better access to reproductive health care choices and to education on sexuality and the changing roles of men and women.

Source: Engelman et al., 2002, p. 148.

Sen (2001b) points out that reducing fertility is not only important because of its consequences for economic prosperity but also because of the impact that high fertility has in diminishing the freedoms of people, especially young women, to live satisfying lives. Reducing high fertility, in Sen's view, will result from the promotion of female literacy, work opportunities and free and open discussion about family size and fertility.

EQUITY, CONSUMPTION AND POPULATION—INSEPARABLE AT THE GLOBAL LEVEL

The international debates about how to solve the world's population problem often come unstuck because of the differences in perspective between developed countries (stable populations) and developing countries (high population growth).

Korten (1995, p. 33) highlights this: 'We have endured far too many debates in which the representatives of rich countries condemn the population growth of the poor and refuse to discuss overconsumption and inequality, and the representatives of poor countries condemn overconsumption and inequality and refuse to discuss population growth.'

This underlines the fact that equity, consumption and population have to be seen as interconnected. It makes no sense to try to solve one issue in isolation. People living in industrialised countries tread far more heavily on the earth than those in developing countries. They are responsible for most of the hazardous waste created by the mining and smelting of aluminium and iron ores, the clearing of forests for paper, air pollution, greenhouse gases from burning fossil fuels, and severe soil erosion caused by grazing animals for meat (Sachs, 1996, p. 144).

The two studies described in box 15.6 compare the resource use by people in rich countries with poor countries and so put the population 'problem' in perspective. As poor countries develop through economic growth (especially as India and China have in the past decade) then the overall ecological footprint of people on the planet will increase to a point that is not sustainable.

BOX 15.6   IS POPULATION GROWTH OR OVERCONSUMPTION THE LARGER PROBLEM?

INEQUITABLE LAND USE

Four to six hectares of land are required to maintain the consumption of the average person in a high-income country (Rees and Wackernagel, 1994). Yet in 1990, the total available ecologically productive land area in the world was only an estimated 1.7 hectares per capita. Industrial countries make up the deficit by using their own natural resource stocks and by engaging in trade that allows them to expropriate the resources of lower-income countries.

What would equitable and sustainable consumption be like?

A study by O'Brien et al. (2014) produced a vision of a resource-efficient Europe. They suggested that compared to 2007 there would need to be a 70 per cent reduction in mineral extraction, a 45 per cent reduction in cropland to halt the loss of biodiversity and increase forests, and a 90 per cent reduction in carbon use compared to 2010. Clearly current consumption levels in rich countries are unsustainable and reaching sustainability will require significant changes in policies and social priorities.

EQUITY THROUGH GREATER CONSUMPTION IS IMPOSSIBLE IN MANY COUNTRIES

The major problem for the world is that it cannot solve the problem of inequity by increasing the consumption available to the 80 per cent of the world's population in developing countries. Jared Diamond (2008) observes that the per capita rate of consumption for the estimated one billion people living in the developed world is around 32 times that for the other (now) six billion. If those billions in the developing world consumed at the same rate as those in the USA, Europe, Japan and Australia, the global consumption level would equate to that of a developed world population of 72 billion. He concludes:

> Yet we often promise developing countries that if they will only adopt good policies—for example, institute honest government and a free-market economy— they, too, will be able to enjoy a first-world lifestyle. This promise is impossible, a cruel hoax: we are having difficulty supporting a first-world lifestyle even now for only one billion people.

Poor countries have a valid argument that the problem of overpopulation is as much one of overconsumption. Two Indian feminists also maintain that if the population problem is viewed in isolation, women in developing countries will be further disadvantaged (Shiva and Shiva, 1995). They note that the UN Cairo conference on population did not view the issue of population in conjunction with development and the need for the emancipation of women. Rights were viewed narrowly as reproductive rights and the environmental problem was seen in terms of population growth in developing countries to the exclusion of the broader problems of globalisation, debt and structural adjustment. The importance of linking population control to development and general security has been made forcefully by Seabrook (1995, p. 10):

> the only known pathway to limiting the birth-rate is clear and simple; not by ever more ingenious forms of contraception, not by 'educating' the poor so that they will produce fewer children, not even by rhetoric about 'giving' women control over their own fertility; but by the provision of an assured and adequate social security to all people. The existence of a level of subsistence and health care below which no human being will be allowed to fall is the surest way of confounding the apocalyptic forecasts of population disaster.

He supports this by saying that the poor try to capitalise on their only resource—the hope of a new generation. Children can be a shield against total destitution. Poor people in developing countries have large families mainly because it is to their benefit. Children provide labour, economic benefit and the hope of a secure old age. All nations that have stable populations have universal education, good health care, social security for old age and women who have equal rights with men.

CONCLUSION

Part 5 reviewed the social and economic threats to health, including urbanisation, population growth, and overconsumption. The ways in which the environment of planet Earth is being increasingly stretched so that it is less supportive of human health were reviewed. Growing certainty about climate change in particular is having unpredictable and increasingly evident impacts on health. Urbanisation is increasing in the twenty-first century. For poor countries this means more slum dwellers, who lack access to the basic requirements for good health: clean water, sanitation and safe housing. In rich countries suburban living offers both threats and benefits to health. Benefits stem from the high standard of housing and provision of many services. Threats come from social isolation and car-dominated cities, which do not encourage interaction or exercise.

Population growth and overconsumption also pose threats to health. Gender equity is the key to reducing population growth and it was argued that means have to be found to reduce the ecological footprint of people living in rich countries and increase the living standard of the world's poor so that they have access to education, health care and social security.

Thus this part sets the scene for part 6, which considers solutions to these problems.

CRITICAL REFLECTION QUESTIONS

15.1   What factors help make high-density cities healthier?

15.2   What forms of transport would you recommend to city planners if the main aim was to ensure transport options promote health?

15.3   To what extent do you think cars are a public health hazard?

15.4   Is the biggest threat to the world population growth or overconsumption?

Recommended reading

Davis (2006) *Planet of slums* describes the worldwide growth of slums and the ways in which the practices of the World Bank and International Monetary Fund perpetuate their existence and poverty.

Global Research Network on Urban Health Equity (2010). *Improving urban health equity through action on the social and environmental determinants of health: Final report of the GRNUHE* provides an overview of the problems of urban health and a range of solutions.

Lowe (2012) *Bigger or better? Australia's population debate* explores the complex and often controversial issues about population for Australia, providing a comprehensive, accessible and considered account.

WHO and UN-HABITAT, (2010) *Hidden cities: Unmasking and overcoming health inequities in urban settings*. Excellent guide to the way in which urban settings contribute to health inequities.

Useful website

www.who.int/kobe_centre/about/en

WHO Centre for Health Development (WHO Kobe Centre) was established to promote and disseminate urban health research and is a useful source of information on urban health.

PART 6

CREATING HEALTHY AND EQUITABLE SOCIETIES AND ENVIRONMENTS

Where there is no vision, the people perish.

Proverbs 29:18

The central aim of the new public health is to achieve healthy individuals, communities, societies and environments, as well as equity between different groups. This part of the book is concerned with the measures needed to make this happen. You will find it less full of references and facts. It concentrates more on ideas and solutions. Of course these reflect my ideas about the world and how it should be structured to encourage health, equity and sustainability. I am far from sure about my solutions but know that we have to dream about ways in which we can organise our societies and communities differently. This part of the book is not concerned with detailed strategies of how the imagined dreamings could be achieved. Part 7 deals with strategies. Here we will be concerned with what could be. I hope that as you read this section it will encourage you to reflect on your dreams for a better society. Good public health practice rests on having a vision for a better society. If you do not know where you want to go it is extremely unlikely you will ever get there! So please feel free to let your imagination blossom as you read the coming pages. Consider these ideas and think about why you agree or disagree with them. Most of all, generate ideas and imaginings for creating societies in which happiness, health, equity and sustainability are maximised.

When we think about society we often tend to take much of our existing structures for granted. Chapter 5 explored the hegemony of neo-liberal economic ideology over the past three decades, which it is hard to see beyond. Yet if the adverse health effects of this form of economics are to be avoided then alternatives need to be imagined and articulated. I would like to see a world in which governments strive to create equity, and use government powers to protect people's rights and dignity and promote the health of individuals and the environment. Of course it would be profoundly democratic and would not allow the needs of the markets to rule all aspects of society. This part of the book, then, sets out an agenda for achieving more equitable, sustainable and healthy communities. It is divided into two chapters. Chapter 16 argues that the adoption of a more people-centred economic system is fundamental to achieving the goals of a healthy society and suggests what needs to change to achieve this, including in the organisation of the economy, ways of measuring progress and types of work. Chapter 17 considers how the physical infrastructures we live in—city environments, housing, neighbourhood design, the natural environment and our use of non-renewable resources—need to be adapted so that they are more sustainable to support health and well-being. Both chapters give central consideration to healthy and sustainable changes that can be made in ways that are also equitable.

16

HEALTHY ECONOMIC POLICIES

We need an economy for the twenty-first century, one that is in sync with the earth and its natural support systems, not one that is destroying them.

Lester Brown, 2011, p. 183

KEY CONCEPTS

Introduction

Challenging economic growth

Beyond GDP AND GNP: indicators of well-being

Polluter-pays principle

Retreat from consumerism

Healthier economic options: Keynes, post-carbon and low growth

Controlling the transnational corporations

From global to local

Local action to resist globalisation

Fair taxation, income and wealth distribution

Trade justice

An economy that encourages healthy work

Conclusion

INTRODUCTION

The strategies to achieve health and equity discussed in the following chapters of this part would be much easier to achieve if the global economic system put as much emphasis on promoting the health and well-being of people and the environment as it does on economic growth. Hence the reforms discussed in this chapter are fundamental to the new public health. The present global economic system has been highlighted as a threat to public and environmental health in chapters 5 and 15, and economists and others have promoted alternative economic systems with the potential to be more benign to human health and the physical environment. Introducing these alternative economic systems may prove to be the most important development of human health and sustainability. In the past decades neo-classical economics has assumed an unparalleled centrality in government and public service policy making (see chapter 5). Five main areas in which the current system needs to change have been identified:

- the need to challenge the assumption of economic growth
- the need for new indicators of economic and social development
- the need to restrict the domination of the international economic system by multinational corporations and for the development of smaller scale locally controlled economic systems
- the need for a fair taxation and income system within countries
- the need to create fair terms of global trade and especially fair trading conditions for poor countries.

CHALLENGING ECONOMIC GROWTH

> Just as Copernicus had to formulate a new astronomical worldview after several decades of celestial observations and mathematical calculations, we too must formulate a new economic worldview based on several decades of environmental observations and analyses.
>
> *Lester Brown, 2011, p. 15*

The need for economic growth has become paramount throughout the world. Neo-liberal economics was founded on growth as the means by which wealth is produced, and wealth production is assumed to be good for well-being. From a health perspective it seems to be beneficial, as wealthier countries are mostly healthier. The links between wealth, well-being and health, however, are being increasingly questioned on both environmental and social grounds. For a thorough analysis of economics as a discipline and the models that exist in addition to dominant neo-liberal economics see Stretton (2000). Anderson and Draper (1991a) argue for a new health economics, pointing out that one might be forgiven for thinking that a discipline calling itself 'health economics' might be of great help in addressing the overall health impact of the economy. In fact, that discipline is almost entirely concerned with the micro-economics of treating illnesses. New health economics would address such questions as adding to national accounting a consideration of how different forms of production affect health, demonstrating that not all production is healthy.

Economic growth and the resultant wealth production are only possible because of the exploitation of non-renewable resources, especially fossil fuels. By 2011 humans were using 135 per cent of the resources that can be sustainably generated in one year (Ecological Footprint, 2011). Yet neo-liberal economics does not account for the impact of resource extraction and use on the environment. Unquestioned growth and wealth production are only possible in an economic system that does not count these external costs. Systems that consider social and environmental impacts may lead to very different assumptions about the value of economic growth. For instance, in box 13.2 we saw countries that have achieved health without wealth, challenging the assumptions of conventional economic theories. We now look at other challenges to these theories and see how they are more health promoting.

For many people the possibility of our society shifting away from the consumption-oriented system seems extraordinarily unlikely. Therefore, it is important to be visionary about the potential for change. Here is a view, written by a new public health activist imagining a mayor in the year 2020 reviewing how her community has shifted its view of economics and growth:

> While our per capita income may be less than it was 25 years ago, we are immeasurably wealthier. In part, that is because we no longer measure our wealth in mere economic or material terms. We no longer use the grossly misleading concept of GNP [Gross National Product], but instead have adopted, both locally and nationally, the measure of NHB (net human benefit) which subtracts harmful effects from beneficial ones. Moreover we use a battery of indicators—reflecting environment, economy, health and equity—that allow us to more accurately reflect our true wealth. Also, the gradual adoption of a set of values more consistent with sustainability has resulted in conservation rather than consumption, frugality rather than waste, becoming the measure of success.
>
> *Hancock, 1994, p. 253*

What might lead to this shift? A number of green economists and others have addressed this question. Schumacher's (1973) *Small is Beautiful* has been very influential in the new economics movement. He argued for a decentralised economic system and advocated alternative technologies, especially for developing countries. Another influential work is that of Daly and Cobb (1990), who advocate a shift from the current economy to one based on small-scale decentralised communitarian capitalism. Their model is based on values and takes into account the needs of present and future generations. This, they argue, will be more sustainable. Their argument builds on earlier work by Daly in which he promotes a 'steady-state economy'. Such an economy would be one in which the rate of use of renewable resources does not exceed the rate of regeneration; the rate of use of non-renewable resources does not exceed the rate at which sustainable renewable substitutes are developed; and the rate of pollution emission does not exceed the capacity of the environment to assimilate it (Daly, 1992). It is obvious that we are far from this situation at the moment (Monbiot, 2014). An increasing number of commentators consider that the pursuit of economic growth is not a healthy option and certainly not one that will guarantee health and happiness. Eckersley (2005) reviews the literature (including that on happiness) and considers that the consumerism and obsession with economic growth as the central focus on public policy is a major threat to our collective health and happiness. On the basis of his review he notes: 'My sense is that if we remove growth—becoming even richer, regardless of where and how—as the centrepiece of our worldview, things would fall into place, the tensions would be resolved, a sense of coherence and balance would be restored' (Eckersley, 2005, p. 250). Harvey (2014) diagnoses the problem as being one of the capitalist system itself. He identifies 17 contradictions of capitalism, in particular its need for 'endless compound growth', its need to exploit nature to its limits and the resulting, inevitable human alienation as fatal flaws. He sees hope in a revolutionary humanism but predicts that the transition may be violent.

BEYOND GDP AND GNP: INDICATORS OF WELL-BEING

The environment is an essential foundation of economic activity and can be considered to be part of the 'capital' from which income is derived (Jacobs, 1991). Yet the state of this environmental 'capital' does not feature in national accounts. There have been many criticisms of the gross national product (GNP) as a measure of the health of an economy (see box 16.1). GNP measures the annual national revenue of firms and industries, production being valued at the price people pay for it. If GNP goes up (after taking off the effects of price inflation), it is seen as 'economic growth' (Anderson, 1991; Anderson and Draper, 1991a, 1991b). From a public health perspective this is not good for health. Consider these examples. If there is an increase in road accidents, there will be a greater need for crash repairers, so production is up; if more baby formula is manufactured so that fewer babies are breastfed, then this contributes to economic growth; after a bushfire there is need for reconstruction, which is good for economic growth; if a Tasmanian forest is logged for wood chips it contributes to growth, but if it remains pristine wilderness it is accorded no economic value; and if more people become addicted to minor tranquillisers, the pharmaceutical industry will increase production and contribute to economic growth.

BOX 16.1   CRITIQUE OF GROSS NATIONAL PRODUCT AND GROSS DOMESTIC PRODUCT

> Gross National Product measures neither the health of our children, the quality of their education, nor the joy of their play. It measures neither the beauty of our poetry, nor the strength of our marriages. It is indifferent to the decency of our factories and the safety of our streets alike. It measures neither our wisdom nor our learning, neither our wit nor our courage, neither our compassion nor our devotion to our country. It measures everything in short, except that which makes life worth living, and it can tell us everything about our country except those things that make us proud to be part of it.
>
> *Robert Kennedy quoted in Roddick, 2001, p. 257*

Jackson (2009) notes there is a huge literature critiquing the value of GDP as a well-being measure. It does not account for non-market services (household labour) or externalities (pollution), and it counts negative expenditures (war). Also, as GDP has increased, there has been little change in reported measures of life satisfaction.

A further criticism of conventional accounting is that it does not cost in 'externalities', which occur when economic activity affects people or the environment external to that activity (Daly and Cobb, 1990; Jacobs, 1991; Hamilton and Denniss, 2005). An example would be the assessment of the economic worth of a steel production plant in terms of the steel it produces and exports, not in terms of the costs incurred by pollution on human health and the environment. Green economics argues for a system of decision-making that counts externalities and does not have growth as its overriding aim. This thinking is very compatible with the aims of the new public health.

The GNP measure does not allow for any calculation of the distribution of wealth and income. Economic growth has not alleviated poverty. In most developed countries, inequities have remained steady or increased. An analysis of long-term trends (UNDP, 2000, p. 82) shows the ratio between the richest and poorest countries has increased significantly in the past two centuries as follows: 3 to 1 in 1820; 11 to 1 in 1913; 35 to 1 in 1950; 44 to 1 in 1973; 72 to 1 in 1992. The United Nations Development Programme (2010) noted that the distance between the richest and poorest countries has widened to a gulf. The richest country today (Liechtenstein) is three times richer than the richest country in 1970. The poorest country today (Zimbabwe) is about 25 per cent poorer than the poorest country in 1970 (also Zimbabwe). We saw in chapter 13 that within countries the gap between the very rich and poor has increased rapidly in the past 20 years. It is noticeable that there is increasing policy concern about this and statements from politicians about the need to reduce this gap (see for example Leigh, 2013) with a recognition that 'too much inequality strains the social fabric, threatening to cleave us one from another' (Leigh, 2013, p. 149).

Globally, between 1960 and 2013, the world total economic output has increased sixfold (World Bank, 2014) yet about one billion people continue to live in poverty and economic welfare, as estimated by the Genuine Progress Indicator (GPI), which has actually decreased since 1978 (Kubiszewski, et al., 2013). We saw in chapter 5 that the benefits of the increase in economic output are being increasingly concentrated among the top 1 per cent in society and the bottom 50 per cent are stagnating or going backwards (Piketty, 2014). Economic growth as measured by GNP is not benefiting most people.

Not only does the GNP measure view positively many things that detract from health, but it also undervalues activities that are not part of the formal system of production. Waring (1988) shows that conventional economics does not count much of the work done by women, for instance. Housework, childcare done in the home, emotional caring and unpaid caring for sick people all remain uncounted by the GNP (Pocock et al., 2012). The *New Internationalist* described what it called the 'social economy' as 'All non-market activities. This includes subsistence farming, housework, parenting, volunteer labour, home health care and DIY as well as barter or skill exchanges. In northern economies the informal economy is estimated to be one-and-a-half times the size of the visible market economy' (Anon., 1996, p. 11). None of this activity is counted in conventional measures of economic progress. Similarly, the resources of nature, which form the basis of all economic activity, are accorded no value at all, despite the fact that a healthy and sustainable ecosystem is essential to the survival of the official market economy.

The critics of conventional economics often argue for an alternative form of national accounting. Daly and Cobb (1990) propose the Index of Sustainable Economic Welfare. In the UK the New Economics Foundation (NEF) has given rise to much discussion about the limitations of the GNP and suggested alternative measures and approaches that protect health and the environment. The NEF's key slogan is 'economics as if people and the planet matter'. Their work includes critiques of the banking system in the wake of the Global Financial Crisis of 2009. In 2006 they launched a new global measure of progress, the 'Happy Planet Index' (HPI); a summary of this index is given in box 16.2. The Kingdom of Bhutan has also developed the concept of 'gross national happiness' as a set of guiding principles underpinning their modernisation process. The concept is strongly rooted in Buddhist principles and promotes valuing social and environmental factors above the search for economic development. Bhutan has sponsored international seminars on the concept of gross national happiness (Ura and Galay, 2004). A UK report (CentreForum Commission, 2014) has called for the 'pursuit of happiness' to be established 'as a clear and measurable goal of government'. A Genuine Progress Indicator has also been developed, which Kubiszewski et al. (2013) say is not perfect but is a far better approximation of economic welfare than GDP. In Latin America there has been much emphasis on valuing 'mother earth'. In 2010 a Universal Declaration of the Rights of Mother Earth was adopted in Cochabamba, Bolivia. The declaration sees that mother earth is 'an indivisible, living community of interrelated and interdependent beings with a common destiny' which 'is the source of life, nourishment and learning and provides everything we need to live well'. It sees that the capitalist economic system has 'caused great destruction, degradation and disruption of Mother Earth, putting life as we know it today at risk through phenomena such as climate change' (Global Alliance for the Rights of Nature, 2014).

BOX 16.2   HAPPY PLANET INDEX

The Happy Planet Index (HPI) is a league table ranking the nations of the world according to their performance on three criteria that are designed to summarise national performance in delivering long and happy human lives without overstretching natural resources. The three criteria used in calculating the HPI are average life expectancy, life satisfaction and ecological footprint. This is expressed in a simplified way by the formula:

HPI = experienced well-being × life expectancy ecological footprint

Developers of the index, the New Economics Foundation, argue that it represents the efficiency with which countries convert the earth's finite resources into well-being experienced by their citizens. They see it as a far more useful index for guiding public policy and governments than gross domestic product (GDP). GDP, the key headline indicator for government policy in the vast majority of countries, leads to perverse results—for example, by counting disasters as positive activity because of spending generated. In addition, GDP counts resource consumption but does not consider its sustainability.

Applying the HPI brings some surprising results. With a range of 0–100, the highest scoring country, Costa Rica, gets only 64.0. The lowest, Chad, gets just 24.7. No country scores well on all three indicators, although the developers of the index believe that an HPI of 83.5 is a reasonable national target. The results show clearly that there is no necessary relationship between long and happy lives and high levels of resource consumption.

SELECTED COUNTRIES FROM THE HAPPY PLANET INDEX

Rank	Country	Experienced well-being	Life expectancy	Ecological footprint	HPI
1	Costa Rica	7.3	79.3	2.5	64.0
2	Vietnam	5.8	75.2	1.4	60.4
12	Cuba	5.4	79.1	1.9	56.2
14	Indonesia	5.5	69.4	1.1	55.5
16	Pakistan	5.3	65.4	0.8	54.1
30	Palestine	4.8	72.8	1.4	51.2
46	Germany	6.7	80.4	4.6	47.2
60	China	4.7	73.5	2.1	44.7
76	Australia	7.4	81.9	6.7	42.0
105	USA	7.2	78.5	7.2	37.3
115	Zimbabwe	4.8	51.4	1.2	35.3
122	Russia	5.5	68.8	4.4	34.5
130	United Arab Emirates	7.2	76.5	8.9	31.8
150	Chad	3.7	49.6	1.9	24.7

Source: Abdallah et al., 2012.

POLLUTER-PAYS PRINCIPLE

A further key concern of those promoting an alternative or new economics is to develop systems that prevent manufacturers and others from externalising their environmental costs. These costs impose on people without compensating them—when a chemical firm pollutes a river, for example. Policies that encourage the polluter to pay the costs associated with the pollution it creates would discourage the activities that create it (Anderson and Draper, 1991b). As an example, Brown (2011) notes that if the full costs of producing petrol are considered (climate change, treatment of respiratory illness, oil spills and the US military presence in the Middle East to ensure access to oil) then the cost increases threefold.

Robertson (1989, p. 140) argues that the externalisation of environmental costs is most likely to happen in a system of centralised technologies and industries. His argument is supported by Korten's (1995, 2000) work on multinational companies, which demonstrates that these companies will shift their operations to those countries where it is easiest to externalise these costs. The move to systematically internalising costs is most likely to happen with small-scale, decentralised, conserving technologies and industries, owned and controlled by the people who use them and have to live with their impact (Robertson, 1989). This argument is supported by comparing nuclear power with renewable energy. Nuclear power stations appear to offer a cheap form of electricity, but the calculations would look very different if the full costs of research and development and ultimate decommissioning were included, compared to renewable, localised green energy supplies. Large agricultural concerns have benefited from the externalisation of costs. The effects on land of intensive farming, the pollution of rivers and estuaries by agricultural fertilisers, and the cost of treating the illness caused by chemical spraying are not returned to the farmers.

Freudenberg (2014) argues that it is vital to develop new strategies to reduce negative externalities. He lists possible strategies as litigation by governments to recover the costs of corporate-induced harm, taxes on unhealthy and high-carbon products, and eliminating subsidies and tax breaks for unhealthy and high-carbon products.

As the perceivable aspects of climate change have become more evident it is being suggested that personal carbon allowances could become a carbon control mechanism. This might mean that people have an annual carbon ration stored on a swipe card from which credit would be deducted when they consume something that contributes to global warming. Once they have used their allowance they would either have to stop using carbon or pay for extra credits (Hinscliff, 2006). Such a system would encourage corporations to seek low-carbon options.

RETREAT FROM CONSUMERISM

'Make do and mend' was the catch-cry of the generation that had experienced World War II, during which little was wasted. The postwar generations, raised in periods of affluence and optimistic growth, became accustomed to more profligate lifestyles. The booming consumer society of the 1950s and 1960s encouraged a 'disposable' way of life, offering little encouragement for reusing and recycling, and plenty for excessive and sustained consumerism. A society geared to consumerism is unlikely to be sustainable. It requires the manufacture of goods with little functional value, increases the need to travel to shop and increases the demand for costly waste disposal facilities. Durability is designed out of consumer goods and obsolescence is designed in (Jackson, 2009).

The neo-liberal economics system that has dominated public policy since the 1980s is primarily geared to a market that supplies goods and services and encourages demand for these through advertising. Things not mediated through the market are neither literally nor figuratively counted in this system. Coombs (1990, p. 2) noted this when he commented that social well-being is inevitably influenced by more than the flow of goods and services through the marketplace, and that people's physical and social environments, experiences (sunsets and bushwalking, for instance) and the 'quality and richness of the cultural experience and personal relationships' all have a crucial impact on well-being. Yet the dominant economic theory has no mechanisms to protect and support these. Daly and Cobb (1990, pp. 161–75) point out that conventional economics ignores the fact that humans are social beings, living in communities from which they gain much personal satisfaction and well-being. A retreat from consumerism may well involve a re-emphasis of the value of human interaction, trust and networks to health and well-being. There is certainly a case that the competition for improved consumer goods (a bigger house, a larger plasma TV set, the latest mobile phone, for example) is psychologically unhealthy because it induces status anxiety and the fear of being left behind in the consumer marketplace. (Jackson, 2009; Wilkinson and Pickett, 2009).

Since the 1970s there have been challenges to the increasingly consumption-focused society. Recycling schemes for paper, glass and some plastics have been established by many local governments in Australia. Excessive packaging is being challenged. But little has been done to challenge consumption. Economic growth is uniformly seen as a good thing by mainstream society; consumption fuels economic growth. Consequently, curbing patterns of consumption will require a dramatic change in societal views, especially as a huge advertising industry is directed at increasing all forms of consumption.

What role does public health have in challenging the consumption agenda? Increasingly public health academics and activists are seeing the practices of corporations in advertising as unhealthy and unsustainable, and are arguing for legislative control of the marketing of unhealthy products and for generally greater control of the practices of transnational corporations in the interests of health and sustainability (Wiist, 2010; Freudenberg, 2014). Aspects of advertising have been highlighted by public health practitioners as potentially damaging to health. An increasing number (Zuppa et al., 2003; Mehta, 2007) argue that the advertising of food on children's television encourages children to adopt diets high in fat and sugar. The promotion of a thin body shape as desirable for young women has been named as one of the contributors to a rise in anorexia nervosa. Some health and welfare agencies do warn consumers about overspending in periods such as the Christmas holidays, but this is to help them avoid debt in a society where credit is issued with ease, and focuses on the individual's behaviour rather than on social pressures to consume.

A Canadian group, Adbusters, has set about challenging consumer society (Crockford, 1996) by parodying the advertising strategies of major companies and, when possible, putting out its alternative advertisements in the mainstream media. These set out to persuade people that they are being manipulated by advertising, and that excessive consumption is a threat to the health of the environment. They also organise a Global Buy Nothing Day and a Buy Nothing Xmas campaign (see photo on p. 428).

The voluntary simplicity movement aims to reduce dependence on paid employment by assessing lifestyles in terms of how many things and services people really need to

Adbusters 'Buy Nothing Xmas' campaign poster

lead satisfying lives. The Natural Step (a movement that started in Sweden) begins with a scientific assessment of the nature and limits of the ecosystem and then mobilises individuals, households, government bodies, business enterprises and professional groups to help bring human activity back into balance with these limits. Two Australian books, *Growth Fetish* and *Affluenza* (Hamilton, 2003; Hamilton and Denniss, 2005), argue that economic growth does not guarantee happiness. They pose the question of whether society will ever be satisfied with the level of economic development or whether 'the relentless emphasis on economic growth and higher incomes simply makes us feel more dissatisfied?' (Hamilton and Denniss, 2005, p. 4). Box 16.3 highlights the often-false promise of economic growth with an individual story that suggests why economic growth in and of itself might not lead to happiness.

BOX 16.3   ECONOMIC DEVELOPMENT: THE PATH TO HAPPINESS?

An American investment banker was at the pier of a small coastal Mexican village when a small boat with just one fisherman docked. Inside the small boat were several large yellowfin tuna. The American complimented the Mexican on the quality of his fish and asked how long it took to catch them.

The Mexican replied, 'Only a little while.'

The American then asked why didn't he stay out longer and catch more fish.

The Mexican said he had enough to support his family's immediate needs.

The American then asked, 'But what do you do with the rest of your time?'

The Mexican fisherman said, 'I sleep late, fish a little, play with my children, take siesta with my wife Maria, stroll into the village each evening where I sip wine and play guitar with my amigos. I have a full and busy life.'

The American scoffed, 'I am a Harvard MBA and could help you. You should spend more time fishing and with the proceeds buy a bigger boat. With the proceeds from the bigger boat you could buy several boats, and eventually you would have a fleet of fishing boats. Instead of selling your catch to a middleman you would sell directly to the processor, eventually opening your own cannery. You would control the product, processing and distribution. You would need to leave this small coastal fishing village and move to Mexico City, then LA and eventually NYC where you will run your expanding enterprise.'

The Mexican fisherman asked, 'But how long will this all take?'

To which the American replied, 'From 15 to 20 years.'

'But what then?'

The American laughed and said, 'That's the best part. When the time is right you would sell your company stock to the public and become very rich. You would make millions.'

'Millions ... then what?'

The American said, 'Then you would retire. Move to a small coastal fishing village where you would sleep late, fish a little, play with your kids, take siesta with your wife, stroll to the village in the evenings where you could sip wine and play guitar with your amigos.'

Much of the consumerist push is linked to competing for status with friends and neighbours. This competition inevitably makes people dissatisfied. Thus Hamilton and Denniss (2005) quote the results of a Newspoll that asked people whether they agreed or disagreed with the following statement: 'You cannot afford to buy everything you really need.' Sixty-two per cent of Australians—nearly two-thirds—believe they cannot afford to buy everything they really need. Hamilton comments

> when we consider that Australia is one of the world's richest countries, and that Australians today have incomes three times higher than in 1950, it is remarkable that such a high proportion feel that their incomes are inadequate. It is even more remarkable that among the richest 20 per cent of households—the richest people in one of the world's richest countries—almost half (46 per cent) say that they cannot afford to buy everything they really need.

Turning this consumerism around will require a considerable change in the way goods, especially luxury goods, are marketed and a change in the ways in which people gain satisfaction from life.

HEALTHIER ECONOMIC OPTIONS: KEYNES, POST-CARBON AND LOW GROWTH

Keynesian economics, which dates from the Great Depression, has been the main challenge to neo-classical economics. John Maynard Keynes' General Theory of Employment, Interest and Money gained popularity through policies such as the New Deal in the USA and the post-World War II welfare state in the UK and Australia. Galbraith (1994) isolates two fundamental points that distinguish Keynesian from neo-classical economic thinking. First, Keynes maintained that depression was neither a temporary thing nor just a self-correcting manifestation of the business cycle, but might itself become the equilibrium. Classical economic theory could lead to a downward spiral in which wages were reduced and worker income and spending lowered, resulting in less sales and more unemployment. Keynes pointed to the importance of aggregate demand, which was the effect of any economic development or public action on the larger flow of purchasing power. Galbraith (1994) defines the second of the Keynesian fundamental points as the need to supplement aggregate demand or purchasing power by breaking the unemployment equilibrium to increase output and employment. He advocates doing this through government borrowing and spending to increase aggregate demand and employment. In other words, government debt could be good for an ailing economy.

Until the early 1970s in Australia, as elsewhere in the West, Keynesian thought underpinned most government policy making. Galbraith (1994, p. 113) pointed out that Keynes' theories worked best in times of recession. In times of growth government expenditure should be cut back as private demand was so high. This was hard to achieve, and burgeoning demand led to increasing inflation. Carroll and Manne (1992) see the failure of Keynesian economics to deal with inflation as leading to a resurgence of neo-classical economic theory. However, it has been clear that government spending in the wake of the Global Financial Crisis has been important in maintaining employment and jobs. Thus in Australia, the Rudd Labor government, in the USA, the Obama government, and some European governments invested in infrastructure and other projects. Despite this, it has been essentially business as usual in economic terms following the Global Financial Crisis. Monbiot (2014) argues that a much more radical direction needs to be taken that deals with the environmental crisis as being much more serious than the financial crisis. Proponents of neo-classical economics argue that economic growth is ultimately beneficial to human health and well-being. They support this argument by pointing to the tremendous gains in life expectancy experienced in Western industrialised countries since the nineteenth century. However, a thorough analysis of the impact of rapid economic growth on British populations in the mid-nineteenth century (Szreter, 1995) provides ample evidence that unfettered economic growth brings with it 'the four Ds': disruption, deprivation, death and disease. Through careful analysis of records of economic growth and demographic statistics, Szreter demonstrates that living conditions and health only improved when economic growth was combined with state market intervention and collective organisation.

Before there was adequate intervention in Britain (approximately 1830–60), economic growth did not bring rising living standards. This lesson is fundamental for contemporary industrialising societies and for all who maintain that economic growth alone is sufficient to promote health. Neo-liberal economics argues that economic development leads to a trickle-down effect but the evidence suggests it is how the profits of economic growth are used by societies that is important to well-being—are they invested in public goods or used for private profit? The rapid growth in economic inequalities in the past two decades indicate that the latter is the case and suggest that increased revenue raising is vital in order to fund public health and other health-producing infrastructure (see section on taxation later in this chapter). Add to this the reality of the ecological limits to growth and there is an urgent need for governments to make ecological sustainability the key aim of economic policy. They are not separate concerns but are deeply intertwined.

A new economic system that would result in all countries scoring higher on the Happy Planet Index needs to be developed. Box 16.4 outlines the principles that the Worldwatch Institute has determined need to underpin such an ecological economy. Brown (2011) lists some of the measures that would make practical contributions towards an ecological economy:

- moving to efficient energy sources (wind, solar, geothermal, LEDs bulbs, electric cars)
- building energy-efficient buildings
- designing cities for people, not cars, with emphasis on public and active transport (see chapter 17)
- developing high-speed intercity rail links
- changing the 'throwaway economy' to one based on recycling (e.g. recycling waste, building materials and household appliances; eliminating bottled water)
- restoring the economy's natural support systems (protecting remaining forest, reducing paper use, protecting the earth's soil, reforestation, reverse desertification, restoring fisheries)
- supporting local sustainable food supplies that are low carbon
- eradicating poverty, stablising populations and rescuing failing states (noting for the first time in history that we have resources to do this).

BOX 16.4   PRINCIPLES FOR ECOLOGICAL ECONOMICS

1   Our material economy is embedded in society, which is embedded in our ecological life-support system, and we cannot understand or manage our economy without understanding the whole interconnected system.

2   Growth and development are not always linked, and true development must be defined in terms of the improvement of sustainable human well-being, not merely improvement in material consumption.

3   A balance of four basic types of assets is necessary for sustainable human well-being: built, human, social, and natural capital.

4   Growth in material consumption is ultimately unsustainable because of fundamental planetary boundaries, and such growth is or eventually becomes counterproductive (uneconomic) in that it has negative effects on well-being and on social and natural capital.

Source: Costanza et al., 2013, p. 160.

CONTROLLING THE TRANSNATIONAL CORPORATIONS

The growth and power of multinational companies are seen by an increasing number of commentators as threatening the sustainability of the environment and human health and well-being (Daly and Cobb, 1990; Korten, 1995, 2000; Freudenberg, 2014). These companies are seen to be lacking in social and environmental awareness. They put company profits above all other considerations; exploit workers, especially those in poor countries; feel no allegiance to local communities; encourage unproductive currency speculation; and realise profits that are increasing rapidly and irresponsibly. A healthy economy would rest on companies that operate very differently. The earlier discussion of globalisation (chapter 5) demonstrated that our current global trading system is organised to serve the needs of these companies more than it is to encourage the health and well-being of people and the environment. How can the corporations be tamed?

The UN Johannesburg Declaration on Sustainable Development (2002) (www. johannesburgsummit.org) states that 'we agree that there is a need for private sector corporations to enforce corporate accountability. This should take place within a transparent and stable regulatory environment'. Similarly, the World Commission on the Social Dimension of Globalization (2004) noted that good corporate governance is essential to both market economies and democracy, but only devoted two short paragraphs to this and made no recommendations concerning the strengthening of corporate responsibility. Both documents are unclear how corporate accountability will be ensured or what the nature of the regulatory environment will be.

There are some companies that do try to operate in a more responsible manner and their corporate social responsibility is to be encouraged. The Body Shop is one example. An increasing number of international non-government organisations are lobbying transnational corporations (TNCs) about the detrimental effects of their activities and advocating protests and actions against them. One such example is the campaign run by Greenpeace against Esso (ExxonMobil in the USA). Their website (www.greenpeace.org/usa/news/it-s-time-to-bury-the-fossil-f/chronology-of-our-campaign) describes the company as 'one of the worst global polluters'. They maintain that Esso has been consistently undermining the accepted scientific consensus on climate change, deliberately misleading the public and policy-makers about the economic implications of tackling global warming and funding a range of climate-change sceptics to argue against the need for change. Corporate Accountability is a non-government organisation devoted to stopping corporate abuses. Their current campaigns include one against corporate control of water resources, and the 'corporate hall of shame'. Other campaigns have been run against Nike, the athletics shoe manufacturer, on the grounds that they exploit workers in poor countries such as Indonesia; against Nestlé for their marketing of bottle-feeding over breastfeeding in poor countries; and against tobacco companies for their continued production and marketing of tobacco products.

Such is the power and wealth of the TNCs that it is hard to imagine what actions will bring them back to a situation where they show corporate responsibility (see figure 5.2 for size of TNCs). Korten argues that while particular campaigns against corporations have some effect they are limited in what they do to establish an alternative economic system. He suggests six items that would bring such a system about. Achieving these items will require legislation, programs of direct action and political mobilisation strategies. Korten's agenda is wide-ranging and radical but appears to offer the sort of changes necessary to restore health to our economic system. The six agenda items are:

1. *Restore political democracy* by reforming the system of political campaign finance. The aim of this is to remove TNCs from the political process and so restore trust in democratic systems.
2. *End the legal fiction of corporate personhood.* Korten claims that the legal fiction that the corporation is a natural person under the law is the means by which corporations have acquired rights to act in the way that they do. He says this measure would place strict limits on corporate privileges and facilitate the conduct of business in the public interest.
3. *Establish an international agreement regulating international corporations and finance.* Currently international agreements under the World Trade Organization such as GATS and GATT (see chapter 5 for details) maintain an international trading regime that permits TNCs to operate with very few controls on their activities and to appeal to the WTO if there are restrictions that impede their profit-taking. This item would use international agreements to control the TNCs and hold them accountable to the public good.
4. *Eliminate corporate welfare.* TNCs receive considerable direct public subsidies and tax breaks. They also externalise a range of costs like pollution (as discussed earlier in this chapter), worker health and safety, and dangerous and defective products. Steps to internalise the costs of their operation would include eliminating direct public subsidies and tax breaks, charging environmental-use fees for the full cost of natural resource extraction and the release of pollutants into the environment and estimating and charging for other indirect subsidies. These include the costs to society when corporations fail to provide a living wage adequate to support a family, health insurance, pension contributions and safe working conditions for their workers. It would also be possible to recover the costs associated with harmful and defective products such as cigarettes and unsafe cars.

5. *Restore money's role as a medium of exchange.* Korten (2000) argues that financial speculation should be eliminated and money should be restored as a medium of exchange. He proposes a series of mechanisms by which this reform could be achieved. The aim is to prevent unproductive financial speculation, which only serves to make short-term profits for the speculators.

6. *Advance economic democracy.* Public policy should be proactive in promoting human-scale, stakeholder-owned enterprises to displace the subsidised TNCs. Korten suggests that the TNCs could be broken down into smaller firms that are linked to their local communities and controlled by people with a stake in the community.

Korten argues that while his proposals are radical and would require a massive shift in power, they are possible, especially as there is 'evidence of deep concern among thoughtful corporate leaders, bankers and even economists that they may be sitting atop an increasingly unstable system on the brink of collapse' (Korten, 2000, p. 201). It is certainly easier to imagine the new public health agenda being achieved in a world governed by Korten's agenda than in the current world of domination by a handful of unaccountable corporations that put profit above health at every turn. His views have been recently reinforced by Klein (2011), who argues for dramatically increased democracy (box 16.5).

BOX 16.5   TAMING THE CORPORATIONS

Naomi Klein writes that saving ourselves from rampant capitalism and the consequent climate change:

> requires that we break every rule in the free-market playbook and that we do so with great urgency. We will need to rebuild the public sphere, reverse privatizations, relocalize large parts of economies, scale back overconsumption, bring back long-term planning, heavily regulate and tax corporations, maybe even nationalize some of them, cut military spending and recognize our debts to the global South. Of course, none of this has a hope in hell of happening unless it is accompanied by a massive, broad-based effort to radically reduce the influence that corporations have over the political process. That means, at a minimum, publicly funded elections and stripping corporations of their status as "people" under the law.

Source: Klein, 2011

The United Nations has developed Guiding Principles on Business and Human Rights following the appointment of a Special Representative on the issue of human rights and transnational corporations, who was charged with identifying and clarifying standards of corporate responsibility and accountability regarding human rights, including the role of states. However, these principles are non-binding and 90 countries, trade unions and non-government organisations have called for a binding treaty (Branson, 2014).

FROM GLOBAL TO LOCAL

As dissatisfaction with the power and control exercised by transnational corporations grows, there are an increasing number of alternative visions of economic systems published. Daly and Cobb's (1990) vision of steady state economy rests on a shift away from TNCs to national and decentralised economic activity. They fear that a world economy dominated by large TNCs will be to the detriment of workers (in terms of their working conditions and wages) and to the environment. Companies with no allegiance to particular communities or nations will not feel responsible for environmental conditions, particularly in the longer term. With respect to both the environment and workers' rights they believe that competitive free trade at the international level will 'come to rest only at the lowest common denominator' (p. 235). Korten (1995, 2000, 2006) argues that modern corporations are designed to concentrate economic power and to maximise profits for those who invest in them. Free-trade agreements like NAFTA and GATT, he maintains, are designed to guarantee the rights of global corporations to move goods and investments wherever they wish with a minimum of public interference or accountability. National governments give these corporations tax concessions and even direct subsidies, and compete to be the most attractive.

Once again the solution to the ever-increasing global concentration of economic power and wealth is seen to be an increase in community ownership and control of productive enterprises:

> We need to break up large concentrations of economic power, re-establish the connection between investment returns and productive activity and root the ownership of capital in people and communities engaged primarily in local production to meet local needs. We need a vision of a global system of localised economies that reduce the scale of economic activity and link economic decisions to their consequences (Korten, 1996, p. 13).

Korten (2015) has further argued for locally owned, community-oriented 'living enterprises' in which success is measured by the impact on people. These sentiments are compatible with those of the new public health's emphasis on community empowerment and control. Frustration with public health community-development strategies centres on the fact that they rarely give economic power to communities. An economic agenda could usefully be added to public health's advocacy for community empowerment. The power of multinational companies and stock markets currently seems so great that alternatives appear to be well-intentioned but pie-in-the-sky rhetoric. Nonetheless there are some long-standing examples of alternative economic models that demonstrate great democracy and cooperative principles (see Mondragon example in box 16.6).

BOX 16.6   MONDRAGON COOPERATIVE: AN EXAMPLE OF INDUSTRIAL DEMOCRACY

Industrial democracy has a history going back to the Owenite and anarcho-syndicalist communes of the nineteenth century. The term also covers a broad range of models including share ownership, participatory decision making and profit sharing. Various combinations and degrees of these features mean that worker involvement can range from tokenistic involvement in marginal management issues through to worker ownership and control. The concept faded from prominence after the Spanish Civil War and during the Thatcherite period of the 1980s, but has seen a recent revival.

Wilkinson (2005, pp. 305–10) argues that an extension of industrial democracy has the potential to reduce social inequality, and to build healthier and more fraternal societies. He sees the spread of industrial democracy as a particularly valuable strategy as it works within the dominant market model and can establish institutions and cultures that are relatively immune to the macro-level policy initiatives of regressive governments. His crucial point is that 'the spread of such forms of economic democracy ... provide mechanisms through which employees can decide on the magnitude of differentials in earnings, as well as what happens to profits' (p. 305).

The Mondragon Cooperative Corporation (MCC), based in the Basque Country of Spain, embodies many of the features that Wilkinson admires, and some of the tensions and challenges of a socially based corporation in a neo-liberal commercial world. From its beginnings in 1956 as a small co-op manufacturing heaters and cookers, the organisation went on to establish a credit union, other manufacturing ventures (using credit union capital), a transport cooperative and a retail chain. These merged in the 1980s to form MCC. There has always been a strong focus on education, and in the 1990s the University of Mondragon was opened. Ten years later it had 4000 students on three campuses. The MCC is now the largest corporation in the Basque region and the seventh largest in Spain, and is believed to be the world's largest worker cooperative. Workers have an extremely strong relationship with the corporation, preferring to take pay cuts when a cooperative gets into financial difficulties. If a cooperative must fold, workers are offered positions in other group companies.

The MCC's growth has created tensions with its socialist ideals. Workers must complete a probation period and then buy a share (at around the cost of one year's base salary) in order to gain full membership. Rapid growth has meant full memberships have not kept pace and in 2004 less than half the 70 000 workforce were full members. It has 38 industrial plants abroad and there is criticism that these workers (largely in Latin America) do not have the same membership rights. The principle of minimising the gap between managers and workers has been compromised to better compete for high-level professionals, and in recent years some cooperatives have withdrawn from the MCC to try to restore some of the more personal and fraternal features of the cooperative model. The MCC did suffer during the Global Financial Crisis but has largely recovered (although its whitegoods company entered bankruptcy in November 2013) and is credited with helping keep jobless levels in the Basque region to less than half the national average (Mathews, 2012).

Source: Mondragon Corporation (n.d.).

Maleny in Queensland shows how a community can establish an economy based on local control and cooperation (Tilden, 1996). Since 1979, the community has established more than a dozen cooperative ventures, including a community credit union, a food cooperative, a land settlement venture, Wastebusters (a recycling venture) and a publishing company. They have also introduced a Local Exchange and Trading System (LETS), which is a computer-based barter system. People can accumulate credit for goods or services they provide. LETS functions like a local currency, except that neither credits nor debits accrue interest. A similar cooperative ethos is behind the slow movement (see, for example, www.slowmovement.com). This movement stresses local food production, the development of local economies and greater connection between people. In regard to local economies the Slow Movement website says 'What is needed is a "new" economy—a green economy, which focuses on local production for local use. Where production cannot meet local needs a source is found from the next level out ie from as close as possible to the need. Only when necessary would we source goods and services from outside our local area or state or country.'

A further example comes from the Bendigo Bank, Australia, which was established because other banks were not seen to serve the needs of communities. Bank branch closure had become commonplace and foreign ownership was a concern. What was perceived as the anti-community behaviour of large banks was seen as the downside of globalisation. The Bendigo Bank developed the 'Community Banking' concept, (www.bendigobank.com.au/community/community-banking/about-community-bank). The bank provides communities with the opportunity of increasing control over their capital and ensuring that more money stays in the district for local investment by opening a franchised branch of the bank. The community banks also invest their profits in local community projects.

Systems such as LETS and local cooperatives rest to a significant degree on trust between people, and a task for public health researchers is to determine those conditions in which community trust most easily flourishes.

LOCAL ACTION TO RESIST GLOBALISATION

If the global trend towards increased corporate concentration and rampant consumerism is to be halted and reversed then an important part of the change is likely to stem from popular action and protest. Globally there are numerous examples of local action to resist economic globalisation (see chapter 22).

These local sites of resistance to the negative effects of globalisation may well prove to be one of the most effective means of providing an alternative vision of what society can be like if they are based on the needs of people in their local communities rather than on the needs for large TNCs to make a profit.

FAIR TAXATION, INCOME AND WEALTH DISTRIBUTION

> Taxation is not a technical matter. It is pre-eminently a political and philosophical issue, perhaps the most important of all political issues.
>
> Piketty, 2014, p. 493

Taxation cuts to the heart of creating a society based on the principles of solidarity and fairness. It is also vital to creating a society that has the ability to control and regulate the private sector and its profit-seeking motives, which so often win ahead of questions of public health and environment. Frank Crean, Treasurer in the Whitlam government, was fond of paraphrasing the American jurist Oliver Wendell Holmes: 'With taxes we buy civilisation' (http://evatt.org.au/news/what-chance-fair-and-decent-country.html). People are rarely happy to pay taxes and the media tends to reinforce this picture with screaming headlines about any taxation increases. However, recent polls have shown that people in Australia, for example, may be prepared to pay more taxes if that can guarantee them better public services (Wilson and Breusch, 2004).

Representatives of the charity Christian Aid wearing masks and straitjackets take part in a Trade Justice demonstration in London outside the House of Parliament, condemning the World Trade Organization (WTO), World Bank, International Monetary Fund (IMF) and the European Union's (EU) trade agreements. (Penny Tweedie)

Rob Moodie, former Chief Executive of VicHealth, sees the paying of local taxes as good for health promotion. In the introduction to the VicHealth Letter he noted:

> I really appreciate paying my rates. This may sound a little odd but not when you consider the huge demands and responsibilities of local councils. Whether it's parklands, street lighting, planning for new buildings, community festivals or strengthening links between different groups in the community, all these elements impact on our health and well-being and they are all the responsibility of councils (Moodie, 2006, p. 3).

Thus taxation can be seen as a useful form of public health regulation. Reynolds (2011, pp. 175–6) notes it can achieve the following purposes:

- provide general revenues for governments
- contribute to the costs of particular problems associated with the consumption or use of a commodity
- emphasise that activities that place a stress on 'public goods' (e.g. production of carbon) ought to have price signals attached to them (carbon tax)
- provide revenue for a particular purpose (by hypothecating or dedicating a proportion of the receipts for health promotion)
- encourage a change in consumption or behaviour, for public health or other purposes (e.g. sugar tax).

A healthy global society would find means of preventing TNCs avoiding taxation and playing one country off against another by implementing a system of effective global regulation of taxation on transnational corporations. These taxation regimes would ensure that the TNCs were contributing to externalities such as damage to the environment, costs of injury to workers that are borne by the state, and costs of education and training. Such taxes are likely to require globally ratified treaties so that the TNCs cannot play off one country against another. There are increasingly loud calls for fairer taxation regimes. The World Commission on the Social Dimension of Globalization (2004) saw that taxation could be a powerful tool by which to make globalisation fairer. Its report suggested a Tobin tax on financial transactions and taxes on the use of global resources, especially through taxes such as one on carbon. They also recommended the exploration of means of establishing a framework for global taxation, the revenue from which could be used to make the world fairer. More recently the G20 meeting in Sydney in 2014 included on its agenda ways of instituting measures to ensure corporations pay their fair share of taxes. Piketty (2014), in his book that details the growth of inequality in the twentieth century, argues that a progressive global tax on capital coupled with a very high level of international financial transparency is required 'to avoid an endless inegalitarian spiral and to control the worrisome dynamics of global capital concentration' (Piketty, 2014, p. 515). Revenue is vital to development and this is most pressing for poor countries who lack the public infrastructures that richer countries have established (see box 16.7 for ways of increasing revenues available to states).

BOX 16.7   OPTIONS TO INCREASE GOVERNMENT REVENUE EXIST EVEN IN THE POOREST COUNTRIES

Each of these options is supported by policy statements of the United Nations and international financial institutions:

- **Increasing tax revenues** through other tax sources—e.g. corporate profits, financial activities, natural resource extraction, personal income, property, imports or exports—or by strengthening the efficiency of tax collection methods and overall compliance, including fighting tax evasion.
- **Restructuring debt**: For those countries at high debt distress, restructuring existing debt may be possible and justifiable if the legitimacy of the debt is questionable (e.g. nationalised private sector debts) and/or the opportunity cost in terms of worsening growth and living standards is high. There is ample experience of governments restructuring debt, but in recent times creditors have managed to minimise 'haircuts', a popular term that refers to investor losses as a result of debt restructuring.
- **Domestic borrowing**: Many developing countries have underdeveloped domestic bond markets and could tap into them for development purposes.
- **Using fiscal and central bank foreign exchange reserves**: This includes drawing down fiscal savings and other state revenues stored in special funds, such as sovereign wealth funds, and/or using excess foreign exchange reserves in the central bank for domestic and regional development; for instance, a country like Timor-Leste, where the share of people living in poverty increased from 36 per cent to 50 per cent between 2001 and 2007, has an estimated US$6.3 billion stored in a sovereign wealth fund invested overseas.
- **Adopting a more accommodating macroeconomic framework**: This entails allowing for higher budget deficit paths and higher levels of inflation without jeopardising macroeconomic stability (e.g. quantitative easing in the USA).
- **Curtailing illicit financial flows (IFFs)** could also free up additional resources for economic and social investments. IFFs involve capital that is illegally earned, transferred or utilised and includes, among other things, traded goods that are mispriced to avoid higher tariffs, wealth funnelled to offshore accounts to evade income taxes and unreported movements of cash. In 2009, it is estimated that US$1.3 trillion in IFFs moved out of developing countries, mostly through trade mispricing, with nearly two-thirds ending up in developed countries; this amounts to more than ten times the total aid received by developing countries.

Source: Ortiz and Cummins, 2013.

A healthy taxation system would be progressive—in other words it would tax people on high incomes more than those on low incomes, or those with more capital more than those with less. In most OECD countries the trend has been towards the reverse on the basis that less taxation acts as an incentive for people. Ironically many of these countries had more progressive taxation systems at a time when economic prosperity was widespread. Thus Stretton (2005, p. 250) notes that in Australia under the conservative Menzies government in the 1950s the marginal tax rate on the top incomes was kept at 66 per cent for six years and then at 60 per cent for the rest of its term in government. Since that time this rate has been reduced by successive governments (Labor and Liberal). Taxes as a percentage of GDP have been declining in Australia since 2000 (McAuley, 2013). There is no evidence to support the argument that taxes impede economic growth (McAuley, 2013). Increasing taxes is generally seen as political suicide, yet in Australia, civil society has begun to campaign for tax increases (see box 16.8).

BOX 16.8   SOUTH AUSTRALIAN COUNCIL OF SOCIAL SERVICE 2014 STATE ELECTION CAMPAIGN: 'WITHOUT TAXES, VITAL SERVICES DISAPPEAR'

The South Australian Council of Social Service (SACOSS) ran this campaign in the lead-up to the South Australian state election in March 2014 to highlight the problem of South Australia's declining tax base and falling revenues. SACOSS saw this impacting on the ability of governments (of any persuasion) to fund the infrastructure, policies and programs that are needed to support vulnerable and disadvantaged people. They argued that unless a fair and sustainable revenue base was secured, services would continue to be cut.

Source: South Australian Council of Social Service (SACOSS), n.d.

The Australian taxation system disproportionately favours the well-off, who are likely to have higher superannuation balances, own more expensive houses and hold private health insurance (McAuley, 2013). Goods and services taxation systems are regressive in that they are flat rates and so favour the better-off because the tax is a smaller proportion of the income of the rich than it is of the poor. Progressive taxation (such as income tax that is progressively higher as income increases) is more equitable because people who earn more pay more tax. A challenge for low- and middle-income countries is to ensure the development of a fair public financing system that provides the state with resources for the range of health promoting services that constitute a healthy society (including education, health and transport), and which favours the poor rather than the rich.

The argument against higher taxation is that people do not want to pay higher taxes. Yet polling in Australia indicates that the proportion willing to pay more taxes in exchange for better public services is increasing. About three-quarters of respondents would prefer any budget surplus to be spent on improving services rather than on tax cuts according to a May 2004 AC Neilsen Age public opinion poll (Wilson and Breusch, 2004). Similar poll results are evident in the UK where it was reported (Adam and Wintour, 2006) that 63 per cent approved of a green tax to discourage behaviour that harms the environment. The same poll found that when asked which two areas should be priorities for the government, 28 per cent highlighted action to tackle climate change and 16 per cent wanted the economy to grow faster. The signal from those aged 18–24 was clearer: 35 per cent picked climate change and 9 per cent the economy.

A healthy global economy would also ensure a much fairer distribution of income and wealth than the current one (chapters 5 and 13). Within many countries there are also extremes of wealth and poverty and the general direction is for the magnitude of inequities to be increasing rather than decreasing. Poverty reduction clearly needs to be targeted globally but so does income and wealth redistribution. In the USA a group called Responsible Wealth (http://faireconomy.org/responsible_wealth) is a 'network of business leaders, investors, and inheritors in the richest 5% of wealth and/or income in the US who believe that growing inequality is not in their best interest, nor in the best interest of society' (United for a Fair Economy, 2011). Taxation is the major public policy tool that can ensure a fairer distribution of incomes and wealth.

TRADE JUSTICE

A healthy economic system will depend on the evolution of a global system of fair trade. The problems of unfair trade and its health impacts were described in chapter 5. A just trade system would mean:

- millions of small farmers and their families around the world would be able to afford a decent income
- currently low- and middle-income countries would be able to move closer to the standard of living in currently rich countries
- poor and middle-income countries could collect taxation and use this to invest in health-promoting state infrastructure
- a crucial step towards global social justice would be achieved.

Small moves are being made towards greater trade justice through the fair trade movement but there have been criticisms of this movement as being marginal to the more structural changes required (see box 16.9 for details of argument).

BOX 16.9   FAIR TRADE OR TRADE JUSTICE?

The Fair Trade movement has evolved in response to the constraints and distortions facing producers who have been economically disadvantaged or marginalised by the conventional trading system. The core principles of the movement (Fair Trade International, 2009) are to enshrine market access and greater empowerment for marginalised producers and help them receive more through shorter supply chains; endorse more sustainable and equitable trading relationships; and institute an improved social contract between producers and consumers. However, some critics have argued that the movement is a misguided attempt to compensate for global market failure, and so provides an escape valve for pressures that might otherwise more powerfully challenge the underlying inequities. The Fair Trade movement's focus on labelling and voluntary consumer choice to purchase labelled products avoids the hard challenges of changing national and multilateral trade policies. Advocates of 'trade justice' may argue that 'fair trade' is more akin to a charity than a social justice movement.

Boris (2005) has argued that immediate trade policy changes would have a much larger impact on disadvantaged producers' lives than 'fair trade' campaigns. 'Trade justice' is a campaign by non-governmental organisations for changes to the rules of world trade so that poor people can work their own way out of poverty. Monbiot (2003) points out that the World Trade Organization has the potential to act in the interests of trade justice—if the dominance of rich countries in its decisions can be broken. Free trade, as advocated by the first world nations, serves to perpetuate their competitive advantages against developing country producers who need some degree of protection. A sliding scale of trade privileges that permits the very poorest nations to fully protect their infant industries could redress this imbalance. As they become richer, they would be forced to gradually drop these protections. A world trading environment that addressed social justice, health and equity concerns would ensure:

- A World Trade Organization that is not dominated by the interests of rich countries and transnational corporations, but which acts in the interests of fair trade.
- *Abolition of agricultural subsidies and dumping*—this would particularly apply to practices of developed nations against poorer countries.
- *Limited protection* on a sliding scale for infant industries in developing countries—as they become more prosperous, protection is diminished.
- *Poor countries have free use of rich countries' intellectual property*—this would apply within their own borders and in trade with other poor countries.
- *Transparency and accountability*—transparent management and commercial relations would help to ensure that deals are made fairly and respectfully with trading partners.
- *Payment of a fair price*—a fair price in the regional or local context is one that has been agreed through dialogue and participation. It covers not only the costs of production but enables production that is socially just and environmentally sound. It provides fair pay to the producers and takes into account the principle of equal pay for equal work by women and men. Fair traders ensure prompt payment to their partners and,

whenever possible, help producers with access to pre-harvest or pre-production financing.

- *Gender equity*—fair trade means that women's work is properly valued and rewarded. Women are always paid for their contribution to the production process and are empowered in their organisations.
- *Working conditions*—fair trade means a safe and healthy working environment for producers. The participation of children (if any) does not adversely affect their well-being, security, educational requirements and need for play and conforms to the UN Convention on the Rights of the Child as well as the law and norms in the local context.
- *Environment*—fair trade actively encourages better environmental practices and the application of responsible methods of production.

AN ECONOMY THAT ENCOURAGES HEALTHY WORK

Work patterns and conditions are changing rapidly in Australia and overseas. The postwar pattern of full-year, full-time employment with the almost certain prospect of lifetime tenure that was common in many rich regions (including Australia, the USA and Europe) is being replaced by more casual, part-time, contract and insecure work opportunities and multiple job holding (Pocock et al., 2012). In the era of globalisation companies are able to shift work to countries offering the lower production costs and least regulations (Heymann, 2006). Working conditions are being eroded by legislative changes, the process of globalisation and the reorganisation and downsizing that have characterised managerial reforms of the past decades. Diderichsen (2002, p. 59) notes that around the world, capital is pressing 'for fewer standardized secure employment relationships and more just-in-time jobs with precarious temporary contracts, greater wage differentiation and increased self-employment'. Jackson (2009) argues that an important aspect of changing the social logic of our current materialistic society is to reduce working time so that opportunities for work are spread more evenly (and so helps reduce inequities), and that less work would reduce outputs, which would be good for the environment.

CREATING WORK AND LIFE BALANCE

An increasing concern among people in developed countries is the trend towards longer working hours and both adults in a family working, which leaves less time for other aspects of life such as family, caring for children and older people, civic engagement, exercise and other leisure. Pocock (2003) notes that in Australia as many as one-fifth of families are 'downsizing' and accepting less money by changing the amount they consume. Generally the choice to downsize is more available to better-off families.

While rich country workers are facing worsening working conditions, the situation is much worse in the world's poor countries. In those settings the conditions are often appalling and this appears to be particularly the case in the fast-developing and industrialising countries of the world such as China and India. Many industrial jobs are now shifting from richer countries, where higher wages and tighter industrial laws make manufacturing more expensive, to poor countries, where wages are a fraction of the cost and legislation governing the rights of workers is more or less non-existent.

If the quest of international capital for cheap and flexible labour continues to drive government agendas then the prospect for workers around the world is not bright. There appears to be a need for an international trade union movement to protect the rights of workers and for some innovative and lateral thinking about forms of employment that can be conducive to social, family and community life (Heymann, 2006). Part of the package that emerges would need to include good provision of childcare and leave for parents to ensure that the important work of raising children is adequately protected from the demands of work. These rights need to be guaranteed globally for all families. International conventions and trade union action and campaigning appear to be the most promising way of ensuring that adequate and family-friendly workplaces become policy goals in all settings.

Crucial goals for healthy public policy in the twenty-first century are to protect working conditions and redistribute and create more work opportunities. The role and distribution of work and the balance between work and family will be crucial to reducing health inequities and promoting population health. There are clear conflicts between the desire of companies for cheap labour that is at its beck and call, and the need of families for a decent living that leaves time for family, community and civic life. It will be crucial to public health to find a global means of resolving this conflict and protecting the health of workers and their families.

CONCLUSION

This chapter argues for a global economic system that is fairer. Currently the global economic system is stacked in favour of rich countries and transnational companies. The sorts of changes outlined in this chapter are likely to be crucial in this period of rapid globalisation if we are to create a world in which the opportunities of development and globalisation are shared more equally. If fairer systems are not instituted then the result is likely to be more environmental deterioration and growing inequities between and within countries, with the resultant lack of harmony, unrest and increased motivation for terrorism.

CRITICAL REFLECTION QUESTIONS

16.1   Why is continued economic growth unhealthy?

16.2   What changes need to be made to our economic systems to make them more healthy and sustainable?

16.3   What are the arguments in favour of using indicators of progress other than economic ones such as gross national product?

16.4   What changes would you make to the taxation system in your country to make it more likely to promote health and equity?

Recommended reading

Brown (2011) *World on the edge: How to prevent environmental and economic collapse* Examines how to solve the most pressing environmental issues

Jackson (2009) *Prosperity without growth: Economics for a finite planet* argues against continued economic growth and that increasing consumption may even impede human happiness.

Korten (2015) *Change the story, change the future: A living economy for a living earth* Provides an account of an alternative economy system designed to respect people and the environment

Useful websites

The New Economics Foundation (www.neweconomics.org) contains a wealth of information about alternative ways to organise economics and society that put people and the planet first.

The Australia Institute (www.tai.org.au) is a progressive think tank whose website contains many helpful reports on how economic and social systems can become more sustainable, healthy and equitable.

The work of David Korten, who has published a number of books that imagine an alternative economic system, is presented at http://livingeconomiesforum.org

In May 2015 Radio National Big Ideas hosted a panel discussion on fair taxation and health. The panellists included Fran Baum, and you can hear the program at: www.abc.net.au/radionational/programs/bigideas/fair-taxation-and-health/6448918

17

SUSTAINABLE INFRASTRUCTURES FOR HEALTH, WELL-BEING AND EQUITY

We are all engaged in the creation of our future. The future is not somewhere we are going, but something we are creating. We take decisions every day that make some futures more probable and others less probable. It should be a goal to make our future a sustainable one. This will involve some big changes.

Lowe, 2005, pp. 220–1

KEY CONCEPTS

INTRODUCTION

Ecological sustainability is at the heart of the aspirations of a public health for the twenty-first century. The environmental stresses and burdens we are collectively placing on the earth offer compelling evidence that the physical support systems for human life are already threatened and will continue to be so in the coming decades. Climate change has moved to the top of public concerns and is on political agendas around the world. Crafting an ecological public health is an absolute priority for public health practitioners. To ensure sustainability, changes will have to be made to the ways our cities and communities operate. Changes are required to make cities less polluted, more energy efficient, less carbon-burning dependent, more human-scale with social space and trees, and less wasteful with more emphasis on recycling and reducing refuse and more self-sufficiency in food production. If these changes were to happen they would promote health and make cities healthier places to live.

The concept of sustainability first came to prominence in *Our Common Future* (the Brundtland report) (World Commission on Environment and Development, 1987), whose emphasis on sustainable development was criticised because it linked the concept with economic growth in both industrial and developing countries, and because of the report's belief that economic growth and diversification 'will help developing countries mitigate the strains on the rural environment, raise productivity and consumption standards and allow nations to move beyond dependence on one or two primary products for the export earnings' (World Commission on Environment and Development, 1987, p. 89). This tension remains central to environmental debates.

Environmental issues are now global. Many environmental resources are shared globally: oceans, forests, genetic diversity, climate, the ozone layer (Sachs, 1996; Lowe, 2005), which means that the solutions to many environmental problems need to be arrived at internationally. Environmental problems do not respect national borders, and so international organisations such as the United Nations have a central role in coordinating global responses to environmental problems.

THE GLOBAL FRAMEWORK

The environmental events that play out in local communities around the world are shaped by international action on environmental issues. Concern over environmental action has been gathering pace since the original publication of Carson's *The Silent Spring* in 1962 (Carson, 1994). The United Nations has led a series of initiatives to improve environmental sustainability, including the following highlights:

- United Nations Conference on the Human Environment 1972, which ended with the declaration of 26 common principles relating to the environment.
- World Commission on Environment and Development (WCED), chaired by Gro Harlem Brundtland. Its report, *Our Common Future*, was published in 1987.
- United Nations Conference on Environment and Development, held in Rio de Janeiro in June 1992, variously referred to as the UNCED, Earth Summit or Rio Summit. This event really thrust the environment into the centre of politics. It produced a series of important and aspirational documents. Agenda 21 was the framework for an ambitious program for sustainable development and Local Agenda 21 was targeted at local government to encourage local action on sustainability. Two agreements were reached: the UN Framework Convention on Climate Change (UNFCCC) and the Convention on Biological Diversity (CBD).
- The Kyoto Protocol, which resulted from the UNFCCC. The success of the protocol 'lay not in any serious effect it might have on the environment, that could only be minimal, but on the fact that it was reached at all' (Middleton and O'Keefe, 2003, p. 9).

- World Summit on Sustainable Development held in August 2002 in Johannesburg, where two documents were agreed upon—the Declaration on Sustainable Development and the Plan of Implementation. The crucially important thing about these documents is that while they are very rhetorical they do establish the principle that environmental issues are closely tied with issues of health, politics, economics and poverty.
- The UN climate summit in Copenhagen in 2009, attended by 192 governments with high expectations of a new global agreement. However, the summit only produced a non-binding accord of general intent to reduce carbon emissions.
- The United Nations Conference on Sustainable Development in 2012, also called Rio+20, which resolved to develop sustainable development goals to replace the Millennium Development Goals and so converge the development and climate change goals of the UN.
- UN Climate Summits in New York and Lima in 2014, in preparation for concluding a Climate Agreement in Paris in 2015. At Lima the initial draft was watered down so that there was no binding agreement to provide plans for curbing greenhouse gas emissions, known as 'Intended Nationally Determined Contributions' (INDCs). Lima also confirmed a goal for developed nations to mobilise $100 billion a year, in public and private funds, in climate aid for developing nations by 2020.
- UN drafted the Sustainable Development Goals (see box 17.1), to be adopted in 2015.

UN Sustainable Development Goals for Post-2015 Agenda. (from http://sustainabledevelopment.un.org/post2015)

Although these UN summits and declarations have been criticised for being too consensual and accommodating the needs of transnational corporations (Middleton and O'Keefe, 2003) they do provide a vision and a framework for establishing what should happen. Certainly the Local Agenda 21 that came from Rio has established a framework within which many local governments and communities are able to take action. Consequently the strengthening of the United Nations so that it is able to provide a counterpoint to the lobbying power of industry is going to be important to the future of a sustainable environment. Nonetheless a key goal for achieving a world in which environmental sustainability is a reality should be strong international organisations that can negotiate international consensus and mediate conflicts.

The international frameworks are essential to ensuring global cooperation and agreement on environmental protection and restoration. The actual work of creating sustainable environments will happen in countries, cities and local communities. There are already numerous initiatives and plans underway and it is the creativity and commitment of those involved in these that provides great hope for the future.

The Agenda 21 statement was adopted by the United Nations Conference on Environment and Development held in Rio de Janeiro in 1992. It is a work program agreed upon by 179 states and based on principles specified in the accompanying Declaration on Environment and Development. Newman and Kenworthy (1999) have summarised the four key principles behind the Brundtland Report and the Rio Declaration thus:

1   The elimination of poverty, especially in developing countries, is necessary not just on human grounds but as an environmental issue. This principle recognises that unless there is social and economic development for poor countries then the 'commons' will continue to be degraded as more forest is cleared, more soil overgrazed, more fisheries destroyed.

2   The developed world must reduce its consumption of resources and production of wastes. This recognises that the main consumers of natural resources are people in rich countries (the average North American or Australian consumes natural resources at a rate 50 times that of the average Indian). Reducing this consumption will require significant changes to the way cities are run and to lifestyles of people in rich countries.

3   Global cooperation on environmental issues is no longer a soft option. So many environmental issues have to be tackled globally (see chapters 14 and 15), such as hazardous waste, greenhouse gases, particularly CFCs, and the loss of biodiversity, that this cooperation will be essential to ensure sustainability.

4   Change towards sustainability can occur only with community-based approaches that take local cultures seriously. International action through the UN and other bodies is essential but changes will only happen when local communities determine how to resolve their economic and environmental conflicts in ways that create simultaneous improvement in both.

Some ecocentrics have been very critical of the Rio Declaration approach, although others have welcomed the incorporation of localism and the calls for negotiation and participation among all stakeholders. Pepper (1996, p. 105) summarised the critics of the UN Declaration and Agenda 21 thus:

- poverty is not the root cause of environmental degradation but American-style wealth
- overpopulation is caused, not cured, by modernisation because it destroys the traditional balance between people and their environment
- the 'open international economic system' of the Declaration will extinguish cultural and ecological diversity
- the problem of externalisation of pollution within conventional economics will not be solved by pricing the environment but instead by reversing the enclosure of the commons, so there is nowhere to 'externalise' to
- the calls for more 'global management' will in practice mean Western cultural imperialism and, in any case, global agreements cannot be verified and enforced

- the attitude that transfers of Western technology to developing countries are urgent reflects a Western 'scientific imperialist arrogance' that assumes ignorance and laziness characterise people in those countries.

These criticisms are also relevant to the current round of climate talks in which the high-carbon emitting rich nations say they are reluctant to take action until middle- and low-income countries do, and in which the current organisation of the economy around increasing profits through increasing consumption and high carbon is not seriously questioned in the talks. What has changed dramatically since the 1990s (as we saw in chapter 14) is that some previously low-income countries are developing rapidly (most significantly China and India) and are increasing their carbon use considerably. Thus it is crucial that the international solutions proposed put these countries at the centre of negotiations. Significantly, the USA and China made an agreement on tackling climate change that suggests both countries are serious about taking action. This contrasts with Australia's continued failure to act and the Abbott government's reversal of the previous Labor government's carbon tax.

An important part of achieving the big changes that Ian Lowe says are necessary in order to achieve a sustainable future (see quote at start of chapter) is establishing goals for sustainable development. The Worldwatch Institute has produced annual reports since 1984 that have documented the increasing urgency of the need to develop sustainable goals. The 2013 report addresses the questions of 'is sustainability possible?'. Table 17.1 provides a summary of the recommendations that are advocated in that report to achieve sustainability. You will see that these are entirely compatible with the new public health approach. These include the changes to economic measures and growth targets that were discussed in chapter 16. The report noted, 'We desperately need—and are running out of time—to learn how to shift direction toward safety for ourselves, our descendants, and the other species that are our only known companions in the universe' (Engelman, 2013, p. 5).

TABLE 17.1   CHANGES REQUIRED TO HUMAN ACTIVITY TO MAXIMISE CHANCES OF A SUSTAINABLE FUTURE

Type	Goals
Stopping unsustainable practices	Use concept of ecological footprint to respect planetary boundaries and reconnect to the biosphere by reducing global warming and decline in biodiversity
	Use sustainable sources of fresh water
	Prevent ecological collapse by making fisheries sustainable
	Promote use of renewable energy
	Conserve non-renewable resources
	Make the world more equal by sharing resources
	Encourage environmental protest and activist movements

(continued)

TABLE 17.1   CHANGES REQUIRED TO HUMAN ACTIVITY TO MAXIMISE CHANCES OF A SUSTAINABLE FUTURE (*CONTINUED*)

Type	Goals
Moving to true sustainability	An economy-in-society-in-nature—replacing gross domestic product as a measure of progress (see chapter 16 for details)
	Embrace Indigenous wisdom concerning the importance of nature and land to human spirit
	Transform corporations and rules that govern them so externalities are factored into costs
	End the fossil fuel era
	Create energy-efficient urban areas (see box 17.2)
	Make agriculture low carbon
	Protect the sanctity of native foods
	Develop new narratives to support sustainability

Source: Drawn from The Worldwatch Institute, 2013.

The UN has devised a series of Sustainable Development Goals (see box 17.1), which from 2015 will replace the Millennium Development Goals. If these goals are achieved then the sustainability of the earth and human life on it will be much more likely. Ivanova (2014) notes that the Millennium Development Goals illustrate the power of global goals to provide meaning, purpose, and guidance, which can lead to political attention and action. She says they offer a structure to focus advocacy, spur motivation, and target investment, and have improved the ability of countries to meet many of the targets. Partly as a result of the Millennium Development Goals, extreme poverty has been reduced and more people have access to clean water.

BOX 17.1   THE UNITED NATIONS DRAFT SUSTAINABLE DEVELOPMENT GOALS

Goal 1	End poverty in all its forms everywhere
Goal 2	End hunger, achieve food security and improved nutrition and promote sustainable agriculture
Goal 3	Ensure healthy lives and promote well-being for all at all ages
Goal 4	Ensure inclusive and equitable quality education and promote lifelong learning opportunities for all
Goal 5	Achieve gender equality and empower all women and girls
Goal 6	Ensure availability and sustainable management of water and sanitation for all
Goal 7	Ensure access to affordable, reliable, sustainable and modern energy for all
Goal 8	Promote sustained, inclusive and sustainable economic growth, full and productive employment and decent work for all

Goal 9	Build resilient infrastructure, promote inclusive and sustainable industrialization and foster innovation
Goal 10	Reduce inequality within and among countries
Goal 11	Make cities and human settlements inclusive, safe, resilient and sustainable
Goal 12	Ensure sustainable consumption and production patterns
Goal 13	Take urgent action to combat climate change and its impacts
Goal 14	Conserve and sustainably use the oceans, seas and marine resources for sustainable development
Goal 15	Protect, restore and promote sustainable use of terrestrial ecosystems, sustainably manage forests, combat desertification, and halt and reverse land degradation and halt biodiversity loss
Goal 16	Promote peaceful and inclusive societies for sustainable development, provide access to justice for all and build effective, accountable and inclusive institutions at all levels
Goal 17	Strengthen the means of implementation and revitalize the global partnership for sustainable development

Source: *United Nations Department of Economic and Social Affairs, n.d.,*
http://sustainabledevelopment.un.org/focussdgs.html.

SUSTAINABLE DEVELOPMENT: OXYMORON OR SALVATION?

The word 'sustainability' has been popular and so widely used that its meaning can become confused. Engelman (2013, p. 3) says that we live in an 'age of *sustainababble*, a cacophonous profusion of uses of the word *sustainable* to mean anything from environmentally better to cool'. Retailers use it to persuade consumers to buy their green products (often dubbed 'greenwashing'); politicians find it useful to persuade voters; and academics from many different disciplines and theoretical perspectives use the term in varying ways. The following definition, however, captures the spirit of many others and combines cultural and physical considerations and has stood the test of time:

> Sustainability is a relationship between dynamic human economic systems and larger dynamic, but normally slower changing, ecological systems, in which human life can continue indefinitely, human individuals can flourish and human cultures can develop; but in which effects of human activities remain within bounds, so as not to destroy the diversity, complexity and function of the ecological life support systems.
>
> *Costanza et al., 1991, pp. 8–9*

Other definitions add the importance of taking into account the needs of future generations in current decision making and avoiding living as if there is no tomorrow. This argues the ethical imperative of intergenerational equity. Critics of the concept of sustainable development maintain that it unrealistically offers the best of both worlds. They argue that if economic growth meant growth along historic patterns of economic activity 'then this is clearly inconsistent with ecological sustainability' (Hare et al., 1991 [1990]). There has been much discussion in the literature about the feasibility of the Brundtland notion of sustainability (Lyons et al., 1995), concluding that any degree of further economic development is likely to have a detrimental impact on the global environmental situation. Increasingly the idea of a 'triple bottom line' is promoted whereby economic, environmental and social goals are set in a way that one does not threaten the other. Industry is particularly fond of this conceptualisation. The notion, though, often seems to be a means of having your cake and eating it too. McMichael (2005, p. 134), the foremost global public health expert working on environmental sustainability issues, stresses that the idea of the triple bottom line is a means not an end. For him the crucial end point is the long-term optimisation of human experience of which health and maintaining social cohesion are crucial aspects.

CREATING ECOLOGICALLY SUSTAINABLE AND HEALTHY COMMUNITIES

Ecological sustainability is at the heart of the aspirations of a public health for the twenty-first century. The environmental stresses and burdens we are collectively placing on the earth offer compelling evidence that the physical support systems for human life are already threatened and will continue to be so in the coming decades. Communities that place priority on sustainability (for instance by recycling, being more self-sufficient in food production and energy use) and that are designed to provide a healthy and satisfying life seem much more feasible in rich countries where some communities are well-advanced towards these goals while the burgeoning slums (where over one billion people live) are much further away. Optimism comes from the fact that the world does have the resources to provide healthy and sustainable lifestyles to all the world's population. It is a matter of how we distribute the available resources. This is a political and moral question that will require the shift in the ethics of public decision-making suggested in table 17.2.

TABLE 17.2   A NEW ETHIC FOR SUSTAINABILITY AND EQUITY

Current ethic	Ethic of sustainable place
Individualism, selfishness	Interdependence, community
Shortsightedness, present-oriented ethic	Farsightedness, future-oriented ethic
Greed, commodity based	Altruism
Material, consumption based	Non-material, community based
Arrogance	Humility, caution
Anthropocentrism	Kinship

Source: Beatley and Manning, 1997, p. 195.

CHARACTERISTICS OF HEALTHY AND SUSTAINABLE CITIES AND COMMUNITIES

The nature of these characteristics have been given much thought by urban and social planners, sociologists, ecologists, architects, development economists, residents of cities and other communities including slums and, in the past decades, people working on locally and regionally based projects such as healthy cities, neighbourhood renewal or slum upgrading. There is general agreement that cities and communities have to move to being more sustainable in their use of resources, especially non-renewable resources.

Urban planners have been particularly influential in designing cities since the nineteenth century, some working on idealistic views of towns and cities in which people could live most healthily and happily. Adelaide's Colonel Light was an example of a nineteenth-century planner with such vision. He set out to plan Adelaide so as to minimise disease and maximise the attractiveness of the urban environment. Visions of garden cities led to the suburban living that widespread car ownership made possible. Even high rise flats, the planning disaster of the 1960s, resulted from visions of better housing than the slums so many people experienced in cities. One of the major challenges for the twenty-first century will be to create liveable cities for all, devoid of slums. Slums first became a political issue in the nineteenth century and were a focus for public health action and have been since then. However, as we move further into the twenty-first century, slums are becoming the norm in most cities in developing countries (Davis, 2006). The sections on healthy neighbourhood (p. 396) and on healthy infrastructure (p. 421) consider what needs to be done about slums.

Other commentators have stressed the importance of designing for equity. Cities should be designed for the benefit of all residents, not just those in a particularly privileged position. Yet the reality is that this does not always happen: 'Cities give physical expression to relations of power in society. The population of cities is very varied; citizens can be rich, poor, young, old, men and women, but these diverse experiences, needs and aspirations are not given equal weight in our cities' (Short, 1989, p. 54). Short argues that children, women, the poor and people with disabilities are disadvantaged by the structure of today's cities, and that healthy cities should pay more attention to their needs. So healthy cities and communities should fulfil the needs of all citizens.

Newman et al. (2009) call for the development of resilient cities to enable them to respond to peak oil and climate change. The characteristics of such cities are shown in box 17.2. These characteristics are explored further in the rest of the chapter.

BOX 17.2   A VISION FOR RESILIENT CITIES

Renewable energy: whole regions and buildings powered by renewable energy

Carbon neutral: every home, neighbourhood and business

Distributed: shift from large centralised power, water and waste systems to small-scale and neighbourhood-based systems

Photosynthetic: harness renewable energy and provide food and fibre locally from urban green infrastructure

> Eco-efficient: move from linear to circular or closed loop, where substantial amounts of energy and material needs are provided from waste streams
>
> Place-based: built on local economy and nurture a unique and special sense of place
>
> Sustainable transport: design cities to use energy sparingly by offering walkable, transit-orientated options for all, supplemented by electric vehicles.
>
> **Source:** *Newman et al., 2009, pp. 55–6.*

TENSIONS IN CREATING HEALTHY CITIES AND COMMUNITIES

What characterises attempts to create healthy and sustainable cities and communities? Three factors stand out:

1. *The complexity of sustainability.* The environmental issues previously discussed interrelate in ways that defy simple or straightforward causal links. Sustainable cities and communities will only evolve with multidisciplinary, lateral thinking and cooperation across sectors.
2. *The constant tension between the desire for economic growth and development and the need to protect and maintain the viability of the physical ecosystem and the social, welfare and health needs of people.* In all cities and communities the battle between these forces is being waged. Sustainability requires a shift to development that is not dominated by short-term economic decision making and which stresses hope rather than fear (Newman et al., 2009).
3. *Effective management and leadership capacity within communities.* The scale of the task facing city government is often frightening and will require considerable skills and capacity building as cities in many poor countries grow rapidly and throw up massive problems (WHO, 1993; Friel et al., 2011). Rural areas face equally challenging situations, often shaped by de-population as people leave for the bright lights and opportunities offered by cities. Leaders will have to juggle the different values and politics in determining acceptable strategies and principles for tackling problems in both urban and rural contexts.

In communities around the world discussions on the influence of values in public health are taking place. They all have a local flavour and reflect particular local political traditions and culture. However, there are also some fundamental questions about the dominance of economics, the importance of local participation and the priority placed on environmental protection that lead to disagreement and sometimes conflict in cities, especially those that are developing rapidly. Cities and other communities need to learn to deal with areas of major disagreement at an operational level. There must be scope for negotiation, mediation and consensus building, and this calls for a new kind of political and professional leadership that takes a facilitating rather than a controlling role. The Commission on Social Determinants of Health (CSDH, 2008) is clear that the world has sufficient resources to alleviate urban poverty and that empowerment of poor people and ensuring their full engagement in the process of making cities healthier is essential. In support of the claim about resources they note that providing the conditions necessary for a decent quality of life for the urban poor in Ahmadabad, India, would cost only US$500 per household and that it costs US$2 per person annually to support community development that results in better access to health and living conditions for the urban poor in Manila, Philippines.

A crucial task for healthy urban governance is to encourage economic development that will provide jobs, economic security and healthy living conditions for families, but doing this in a way that is sustainable and does not put undue stress on the urban environment. The Commission on Social Determinants of Health (2008) stresses that effective and participatory urban governance to ensure this outcome is vital.

The chapter will now consider the ways in which urban neighbourhoods can become healthier, ways of reducing energy use, and how agriculture and rural areas can be made more sustainable.

HEALTHY NEIGHBOURHOOD DESIGN

> Well-planned cities can offer unique opportunities to positively influence people's health. Through decisions about appropriate mixes of land use, strategic density, and the various policies, design and review processes, cities can arrange for health-promoting environments in all neighbourhoods.
>
> *WHO Kobe Centre, 2014*

Convivial neighbourhoods that promote health will encourage interaction, make people feel safe because people are evident in the streets and public spaces, and ensure that the needs of the car do not dominate. The recipe for healthier neighbourhoods has to be a comprehensive one that melds the physical infrastructure requirements for more sustainable cities with the social needs of people for interaction and connectedness.

Around the world there is growing focus on the ways in which urban planning can be used to advance health goals and particularly to reduce the growing burden of non-communicable disease. The Commission on Social Determinants of Health has noted the need to place 'health and health equity at the heart of governance and planning'. The focus is developing urban environments that encourage physical activity, reduce car dependence, increase access to public transport and healthy eating options, reduce obesity risk, and create more convivial cities. Box 17.3 describes the initiatives of New York City to bring these elements into city planning. There is also a Fit City movement that promotes similar design principles to those from New York.

BOX 17.3   NEW YORK CITY *ACTIVE DESIGN GUIDELINES*

The *Active Design Guidelines* are the city's first publication to focus on the role of designers in tackling one of the most urgent health crises of our day—obesity and related diseases, including diabetes. The guidelines are aimed at designers who are responsible for the planning and construction of buildings, streets, and neighbourhoods. The publication seeks to educate designers about opportunities to increase daily physical activity, including measures such as making stairs more visible and providing inviting streetscapes for pedestrians and bicyclists. The guidelines were the result of a collaboration between four departments of the city government—the New York City Departments of Design and Construction, Health and Mental Hygiene, Transportation, and City Planning. In the introduction to the guidelines, the city commissioners say, 'The goal of the guidelines is to make New York City an even greater place to live, by creating an environment that enables all city residents to incorporate healthy activity into their daily lives'. The guidelines recommend that urban planning should:

- develop and maintain mixed land use in city neighbourhoods
- improve access to public transport
- improve access to plazas, parks, open spaces, and recreational facilities, and design these spaces to maximise their active use where appropriate
- improve access to full-service grocery stores and fresh produce
- design accessible, pedestrian-friendly streets with high connectivity, traffic-calming features, landscaping, lighting, benches and water fountains
- facilitate bicycling for recreation and transportation by developing continuous bicycle networks and incorporating infrastructure like safe indoor and outdoor bicycle parking.

They recommend that designers of new buildings should:

- increase stair use among the able-bodied by providing a conveniently located stair for everyday use, posting motivational signage to encourage stair use, and designing visible, appealing and comfortable stairs
- locate building functions to encourage brief bouts of walking to shared spaces such as mail and lunch rooms, and provide appealing, supportive walking routes within buildings
- provide facilities that support exercise such as centrally visible physical activity spaces, showers, locker rooms, secure bicycle storage and drinking fountains
- design building exteriors and massing that contribute to a pedestrian-friendly urban environment and that include maximum variety and transparency, multiple entries, stoops and canopies.

Source: City of New York, 2010.

http://centerforactivedesign.org/guidelines/

Neighbourhoods built according to the New York criteria should be healthier, more sociable and use less carbon. Encouraging such urban development should be part of public health advocacy. This advocacy would be supported by research that documents the effects of urban developments, such as freeways and large shopping centres, on the social aspects of cities, on the health of inhabitants and on the lives of people who are not car owners.

Increasingly concerns about creating environmentally sustainable communities are taking prominence. An inner-city development in Adelaide (see box 17.4) was designed to be both convivial and ecologically sustainable. The benefits of greening urban areas for health are clear (Lee and Maheswaran, 2011). More experiments of this type are needed.

BOX 17.4   CHRISTIE WALK: A MODEL FOR ECOLOGICAL AND SOCIAL SUSTAINABILITY

Christie Walk is a medium-density co-housing development located in downtown Adelaide, South Australia, that combines many ecologically sustainable and community-enhancing features. The development comprises houses and apartments over 2000 square metres and accommodates more than 40 people in 27 households. Ecological criteria have been fundamental to the design, which incorporates townhouses, apartments and straw bale cottages. It includes a five-storey building with 13 apartments and a ground-level community area with a kitchen, dining or meeting room, library, toilet (disabled access) and shared laundry; all set in a creatively landscaped, pedestrian-friendly space.

Christie Walk was completed in 2006 and features:

- pedestrian-friendly spaces
- shared gardens including a roof garden
- local food production in onsite community food garden
- onsite storage of stormwater—water used on gardens and to flush toilets
- passive solar/climate–responsive design: heating, cooling and humidity control using solar orientation, breezes and vegetation
- solar hot water
- power from photovoltaics—panels installed adjacent to roof garden
- recycled, non-toxic materials with low embodied energy
- reduced car dependency due to inner-city context and a resident car-sharing scheme.

The overall design strategy was to use high internal mass within highly insulated envelopes with multiple user-controlled ventilation options and thermal flues. Vegetation and outdoor spaces were included as an integral part of the passive house design approach. Smaller house plan areas were favoured, with quality of space considered more important than mere quantity. Community space includes natural meeting places, and shared and public-access garden areas, which support a strong neighbourhood network. The end result is a highly energy and water efficient community with strong local social capital.

The community group responsible for the development reported a long struggle against the constraints of normal practice in the residential development industry—but many valuable lessons were learnt, which they share with others on guided tours of the site. Christie Walk was a winner of a National Energy Globe Award (Prague 2009), a finalist in the UNAA World Environment Day Awards (2009), and a finalist in the World Habitat Awards 2005, run by the Building and Social Housing Foundation. The development provides a good example of a healthy and sustainable living environment for the twenty-first century.

For more information, see the Urban Ecology Australia website (www.urbanecology.org.au) and the Christie Walk YouTube video (www.youtube.com/watch?v=T4N0XeMadjc).

View of Christie Walk, Adelaide. (Paul Laris)

Achieving healthy urban environments in poor countries has so far proved impossible. Yet the material resources to do so are available if wealth were shared more equally between and within countries. Wealth redistribution could be used to upgrade slum areas and introduce urban governance that ensures that poor people have a strong voice about the deployment of resources. The most important means of bringing healthier neighbourhoods to poor cities is the public provision of basic infrastructure, especially water; sanitation; affordable, dependable and clean household energy supplies; and adequate housing. Public provision of housing would go a long way to ensuring better lives for poor people in cities. It may be that one advantage that poor cities have over suburbanised environments in rich countries is that they are more convivial places. Many slum areas, for instance, do have third places (see box 17.5) in the form of markets and eating places. This is not to romanticise what are unacceptable environments but to point to strengths that should be retained in slum upgrading or relocations of slum dwellers or other poor communities.

HIGH SOCIAL CAPITAL AND CONVIVIAL CITIES, NEIGHBOURHOODS AND COMMUNITIES

> Resilient cities will have strong social capital, which will strengthen their ability to respond to the challenges of rethinking how their city is powered, where and how their resources originate and how they travel ... The necessary technologies will be adopted if we are able to create strong communities of hope that will take on these issues with confidence and strong political commitment.
>
> *Newman et al., 2009, p. 85*

Lindheim and Syme (1983, p. 341) stressed three factors that appear to be important for healthy environments:

- *Building social relationships* that include support, social ties and family relationships. They note that urbanisation and industrialisation have decreased the likelihood that supportive social relationships can exist, and that one of the tragedies of our time is that architectural and planning policy have made it difficult for people to maintain support networks.
- *Minimising hierarchical relationships* as evidence suggests that being lower in a hierarchy is related to poor health status, apparently because individuals have less opportunity for participation and control over their lives, and so have worse living and working conditions and may feel stigmatised, leading to low self-esteem.
- *Encouraging connection to people's cultural heritage and the natural world* because most people relate to these factors. Who has not been in a large city and found respite in a public park? Similarly a sense of history and culture is important to self-esteem and well-being. These links of cultural heritage are under significant stress in the cities of the world that are growing rapidly—often development wins over cultural preservation.

In recent years, the struggle to create cities that are liveable, healthy and people-focused has been conceptualised as involving the creation of social capital. Some commentators have suggested that social capital is as important as other forms of capital (Cox, 1996; Putnam, 1996; Gillies, 1997) and as we saw in chapter 13, higher levels of social capital do appear to be related to better health status. The dislocation and pressures faced by people (and especially poor people) often make it difficult for cities and other communities to be generators of social capital. Traditional community values and strengths are under threat from the pressures of urban life, yet social capital is crucial for health.

Globalisation brings a significant emphasis on difference and all cities around the world are learning to govern and manage cities with populations that are heterogeneous rather than homogeneous. This means social capital takes on particular importance in a world in which cultures, peoples and customs are increasingly mixing. It is crucial that our cities and communities are places where people from different backgrounds, cultures and ages can meet and interact in ways that overcome difference and reduce the potential for conflict and social disorganisation.

Evidence is accumulating (Wilkinson and Pickett, 2009) that living in an equitable community can assist the creation of conviviality and the elusive sense of community. This is hardly surprising. Given the strong evidence on the power of status on relative health outcomes (evidence summarised in Marmot, 2004), communities in which there are very wide and evident differences in wealth will lead to resentment from those worse off. This will display itself as crime, vandalism and other anti-social behaviour. On the other side, better-off people who live in inequitable communities are more likely to surround themselves with security and cut off their communications from the other, threatening parts of society. South Africa and Latin American countries such as Brazil are perfect examples of this. In South Africa the legacy of apartheid—extreme and legally sanctioned division—is a society in which crime is very high and the middle-class population is locked behind barbed wire fences, protected by security companies and security alarms. Brazil is one of the most unequal countries in the world (although it is becoming less so since the introduction of pro-equity public policies) and also has a high crime rate. So to create equity is a crucial aspect of creating healthy and convivial societies.

In cities, neighbourhoods and communities, social policies need to be directed to encouraging people to live varied and fulfilling lives and to interact with people other than their family and workmates. This can happen through the vibrancy that comes from the amenities in cities and larger communities—cinemas, restaurants, theatres, night clubs, art galleries, sports events, street cafes, buskers, clubs, voluntary groups, parties. These facilities provide relaxation and entertainment and perform the vital function of linking people together—social glue, as they have been described. Often these services are provided by the private sector or may be supported by the state in part or fully. These structures are essential to effectively functioning communities. They are vital to a healthy community. Social policy should aim at making access to a range of social and recreational amenities as inclusive as possible. Third places have been identified as important features of welcoming and convivial neighbourhoods (see details in box 17.5).

BOX 17.5   THIRD PLACES: COMMUNITY SITES FOR HEALTH PROMOTION

Countries such as the USA, Canada and Australia are characterised by cities with sprawling suburbs. They are places where people return home from work and spend few of their waking hours. Oldenburg (1997) suggests suburbs need 'third places' that offer a balance to the increasing privatisation of home life. These are informal gathering places to which people can walk. Many suburbs contain few such places and are made up solely of individual family housing. Such third places would allegedly offer the following benefits relevant to promotion of health (Oldenburg, 1997):

- Help to unify neighbourhoods. Where they are absent, people may live in the same vicinity for years without ever getting to know one another. By becoming acquainted, people are likely to find others with similar interests.
- Serve as 'ports of entry' for visitors and newcomers to the neighbourhood, where directions and other information about a community can be easily gained.

New suburbs, on the outskirts of cities, are those most lacking in third places, yet it is there that people most need them.

- Bring people of different ages together. There are few opportunities for young people to mix with adults other than their parents and teachers. Local public venues would provide these and, perhaps, do something to alleviate the gap between young people and older generations. Many older people do get lonely, especially if they live alone. Third places provide accessible, safe venues for meeting people.
- Become the sites of civic and political debate. They could provide a forum for people to debate and discuss ideas, especially those related to local issues such as the need for traffic control, who to vote for in the local elections and the desirability of a mobile phone tower or overhead cables.
- Lead to development of mutual support and help for people in times of crisis.
- Provide entertainment, being somewhere to go that wouldn't cost a lot of money. They should be places where people can develop friendships. Oldenburg calls them 'a very easy form of human association' because no one is in the position of guest or host, and people come and go very easily.

Third places have been eliminated from our suburbs because of rigid zoning regulations. Residential areas are cordoned off from industrial and commercial areas, which, together with the development of large shopping centres that are rarely accessible by foot, means cafés and shops are no longer an integral part of the immediate areas where people live. Oldenburg (1997) says third places operate best as local, independently owned commercial establishments, and will flourish in communities that encourage walking and are not dominated by cars. Third places are natural places of health promotion. They are generators of social capital that could help protect communities from the effects of low social cohesion and the absence of community spirit. An example of their application is their use on the Healthy Alberta Canada website (www.healthyalberta.com/688.htm), where it is noted: 'Third places are the "anchors" or driving forces that help to keep a community connected on social levels. They really help to create a sense of belonging for community members.'

Community development strategies (described in detail in chapter 22) can also be used by government (especially local and regional) to create more cohesive and supportive communities. These strategies should challenge attitudes that undermine cohesion such as racist attitudes and caste and gender-based discrimination. Cities and communities should ideally be places that welcome all groups, genders and ages and afford each equal opportunity.

CREATING SPACE FOR CIVIC DEBATE AND DEMOCRATIC DECISION MAKING

Friel et al. (2011, p. 864), building on the work of the Commission on Social Determinants of Health, note that 'Urban health equity depends vitally on the political empowerment of individuals and groups to represent their needs and interests strongly and effectively and, in so doing, to challenge and change the unfair distribution of material and psychosocial resources.'

Another crucial aspect of a healthy, equitable and sustainable society is the encouragement of civic debate and engagement. An important means of bringing about social change is through concerted action from civil society. This sector plays a role in opposing the actions of transnational corporations and governments when they are not in the interests of community health. There are numerous examples of international and local non-government organisations and community groups that advocate for health and environment causes (see chapter 23). Increasing social media, especially Twitter, has emerged as one of the sites of civic debate. It allows quick responses to new policy and enables activists to communicate rapidly. Governments can help create the spaces for civic debate by funding groups who facilitate such debates and sponsoring events such as the Adelaide and Brisbane Festivals of Ideas at which alternative futures are imagined and debated. The South Australian Government funded a Thinkers in Residence program through which leading thinkers from around the world were brought to South Australia for a period of about three months, expressly with the idea of raising debate and discussion about issues crucial to the state's future such as water management, homelessness, development of science, and health in the twenty-first century. In 2007 Ilona Kickbusch (one of the authors of WHO's Ottawa Charter) was an Adelaide Thinker in Residence and her recommendations led to the South Australian Health in All Policies initiative.

The Adelaide Thinkers in Residence program encourages innovative thinking about ways of creating healthy and sustainable futures. (Adelaide Thinkers in Residence Office)

ENERGY USE

> The old energy economy, fuelled by oil, coal and natural gas, is being replaced with an economy powered by wind, solar and geothermal energy. Despite the global economic crisis, this energy transition is moving at a rapid pace, and on a scale that we could not have imagined even 2 years ago.
>
> *Lester Brown, 2011, p. 117*

Sustainability and prevention of global warming depends on shifting energy systems towards non-polluting and renewable options. Energy policies at federal, state and local levels need to move towards sustainability. Policies relating to energy are not conventionally related to public health, but the consequences for human health of continuing with current high energy consumption suggest that public health should at least advocate and lobby for, and preferably encourage, the adoption of healthy energy options. Rich countries including Australia are the highest energy-using countries in the world, and it will be necessary to shift away from fossil fuels and nuclear options to others less stressful to the biosphere, such as solar, wind, geothermal and tidal power (see box 17.6 for examples). Increased investment in these technologies is urgently needed and many places are moving to do this. Investment in renewables has increased over the last decade. China is an outstanding example—since enacting its Renewable Energy Law and introducing many different measures to move to a low-carbon economy, by 2013 its production of renewables had reached in excess of 200 gigawatts (Newman and Matan, 2013). By contrast Australia's Abbott Coalition government disinvested in renewable energy from 2014. Previously the Renewable Energy Target aimed to supply at least 20 per cent of Australia's electricity supply from renewable sources by 2020, or more than 45 000 gigawatt hours of renewable energy.

BOX 17.6   RENEWABLE ENERGY: SIGNS OF PROGRESS

WIND POWER

- In 2000, world wind electricity-generating capacity was 17 000 megawatts. By 2013, this had increased to 318 105 megawatts. In the last few years China has been the biggest investor. Asia is likely to be the biggest investor for the foreseeable future (Global Wind Energy Council, 2013). China has 28.7 per cent of world capacity, the USA has 19.2 per cent (and falling) and Germany has 10.8 per cent.
- Denmark is the world leader—in 2010, it had 21 per cent of electricity supplied by wind and aims to have 50 per cent by 2025.
- South Australia is the leader in Australia, on track to have 40 per cent of electricity generated by wind by 2015 (on track to around 30 per cent in 2013) (Australian Energy Market Operator, 2013).

SOLAR ENERGY

- Germany is the world leader in solar energy—in 2010, it had capacity of almost 10 000 megawatts.
- The move from small-scale rooftops to utility-scale photovoltaics (PV) and solar thermal plants makes renewable energy more feasible.
- More and more countries are setting solar installation goals, such as Italy, which projects 15 000 megawatts by 2020, and Japan, which projects 28 000 megawatts by 2020.

- There is widespread adoption of rooftop solar heaters in China, with the technology leapfrogging into villages that do not yet have electricity installed. In 2010 it was estimated that there were 1.9 billion square feet of rooftop solar thermal collectors in China, which provide free hot water.
- Hawaii requires that all new single-family homes have rooftop solar water heaters.

GEOTHERMAL ENERGY

- Roughly half the world's installed geothermal-generating capacity is in the USA and the Philippines, and there is great potential for further expansion. The greatest resource is in Indonesia with 40 per cent of potential supply. By 2025, Indonesia aims to produce more than 9000 megawatts of geothermal power, becoming the world's leading geothermal energy producer. This would account for 5 per cent of Indonesia's total energy needs (Ampri, 2012).

HYDROPOWER: DAMS, TIDAL AND WAVE

- South Korea and New Zealand have invested in tidal power, and Scotland has invested in wave power. Wave power could generate 10 000 gigawatts of electricity (double the current world electricity capacity from all sources).

Sources: Brown, 2011; Global Wind Energy Council, 2013.

Starfish Hill Wind Farm, Delamere, South Australia. (Fernando M. Gonçalves)

The potential for a less-polluting energy system is considerable. Manufacturing, transportation and buildings all contribute to the burden of energy use, and suggestions for reducing energy use, especially fossil fuels, have been well documented. The challenge is a political and economic one that involves ensuring that research and development investment are directed towards those options that do not affect environmental and human health. The signs for a change to renewable and less-polluting energy use are a little more encouraging now. Commentators in the debate about policy solutions to climate change are suggesting that renewable energy will benefit at the expense of fossil fuels (Lowe, 2005; Hamilton, 2007; Brown, 2011).

More and more countries are taking action to move to a low-carbon economy, as we saw in the discussion of the UN efforts to conclude a treaty to reduce emissions.

REDUCING FOSSIL FUEL USE

One of the most urgent issues in moving to a low-carbon economy is greatly reducing the extraction of fossil fuels. McGlade and Ekins (2015) calculate that if an average temperature rise above 2 °C is to be avoided then one-third of oil, half the gas and 80 per cent of coal reserves need to remain unused. Renner and Prugh (2014, p. 36) propose a global pact to leave the bulk of the world's proven fossil fuel reserves in the ground. This aim is at direct odds with the economic growth aims of most governments and a powerful fossil fuel industry that makes its profit from extraction of those fuels. Renner and Prugh note that 'leaving the bulk of the world's fossil fuel deposits untouched will require quasi-revolutionary change', which will require a combination of regulation, litigation, shareholder activism, and dogged divestment and civil disobedience campaigns. Examples of civil disobedience against fracking (hydraulic fracturing of rock to access fossil fuel reserves) are provided in chapter 22 through the discussion of the Lock the Gate campaign. Major institutions are beginning to disinvest from fossil fuels. Governments are increasingly intervening to encourage low-carbon use through taxes, regulation, the introduction of carbon markets, and energy-efficiency standards for industrial equipment, buildings, motor vehicles and consumer goods. Binding emissions limits are another option, such as proposed carbon pollution standards for US power plants that would effectively rule out conventional coal units. These government measures have to be made in the face of a very powerful fossil fuel industry which will resist all attempts to curb its profits and to restrict the use of fossil fuels.

One of the pressing needs in order to curtail energy use is restricting the use of private cars.

TAMING THE CAR

The public health problems associated with car dependence have been fully detailed in chapter 15. In a number of ways cars are an enemy of good health. They create pollution, contribute to global warming, remove people from street and community life and have driven the style of urban development in cities across the world. This is evident in cities from Los Angeles to Cairo and Beijing. Rising petrol prices are putting strain on the urban sprawl that typifies many cities in rich countries and peak oil means that petrol-fuelled cars are likely to be phased out. Thus the challenge in the coming decades is to ensure that our cities are less car dependent. This will involve changing urban design and developing public mass transit systems.

THE CAR AND URBAN DESIGN

In Europe and in some cities in North America urban design has been used to reduce the dominance of cars in cities. These designs encourage more sustainable and energy-efficient cities. Traffic calming has been advocated as an approach to urban design that minimises the intrusion of the car on city design and the environment. The idea originated in the Netherlands and has been translated to other cities around the world. The key principles and health benefits of traffic calming are shown in box 17.3. The traffic calming ideas have, however, not permeated urban planning ideology to any significant degree. In Australia during the past decade, Sydney, Adelaide and Melbourne have all engaged in major freeway building projects, while new investment in public transport infrastructure has been small by comparison and shows no signs of being sufficient to switch public preferences from the car to public transport. A similar pattern is evident in the USA, where some cities have made some moves towards taming the car but where the dominant feature of the urban landscape is dominance by the car. Urban consolidation has some currency but there is little sign of the imaginative urban design seen in some of the Canadian and European urban villages. However, across the world there are encouraging developments in some suburbs where the opening of street cafés and a degree of traffic calming have created suburban environments that offer an alternative to those areas dominated by the single family dwelling with no other type of land use.

Inner, more affluent suburbs tend to have better public transport infrastructure and are more dense and mixed in their land use. Outer suburbs are heavily car dependent, with some households using 40 per cent of their income just to travel around to jobs and services (Newman, 2006, p. 4). Newman suggests it is absolutely vital that there is a crash program in public transport infrastructure for middle and outer suburbs. He suggests extension of electric rail lines integrated with local buses to ensure the transport is quick. He sees that 'oil-proofing' cities is vital. This desire will have to have appeal globally, as in countries with fast developing economies such as India and China car ownership is increasing rapidly, so these governments also need to invest in good public transport systems rather than encourage private cars. Cities in these countries also need to protect the low oil dependencies they currently have compared to US and Australian uses, which are the highest in the world.

The challenge for planning in the world's poor cities is to ensure that poor people have easy access to employment and education. Currently poor people often have to travel for many hours each day from slum communities to places of employment (Davis, 2006). A focus of urban planning should be to enable people to have much easier access to employment and education.

CYCLING AND WALKING

One of the most effective ways of taming cars and reducing energy use is to provide people with alternatives. Cycling and walking have been identified and promoted as forms of transport that are energy efficient, clean, inexpensive and environmentally friendly. Cities in Europe and Asia have much higher rates of cycling and walking than in North America and Australia. This does not reflect characteristics of the people so much as urban environments that encourage and promote cycling and walking. The Netherlands and Denmark provide environments that particularly encourage cycling. In Copenhagen, Denmark, 36 per cent of people go to work on a bike compared to just 27 per cent by car (Newman et al., 2009, p. 100). Both walking and cycling also offer direct health benefits in terms of muscular, respiratory and cardiovascular fitness, weight control, stress reduction and improved well-being (Roberts et al., 1995). Asian cities have been far more bike dependent than other cities and they need to institute measures to maintain this on health and environmental grounds.

Newman and Kenworthy (1999) suggest that walking and cycling are also encouraged by an efficient transit system whereby people undertake short trips on foot or by bike and longer cross-city trips by transit. Land use planning that is compact and mixed in character also helps to enhance the role of walking and cycling.

EFFECTIVE PUBLIC TRANSPORT

Sustainability will require a shift away from private cars to public transport. Cities differ considerably in the extent to which they provide this. North American and Australian cities typically provide less public transport than most European or Asian cities (table 17.3).

TABLE 17.3   PROVISION OF PUBLIC TRANSPORT IN CITIES

City	% motorised transportation that is mass transit
Atlanta, San Diego, Denver, Houston, Phoenix, Riyadh	<1
Washington, San Francisco, Chicago	5
Ho Chi Minh City	8
New York	9
Perth, Glasgow, Marseille, Geneva	10
Kuala Lumpur	11
Sydney	12
Toronto	14
Zurich	24
Singapore, Seoul	40
Beijing, Tunis	50
Tokyo, Osaka	60
Dakar, Chennai, Shanghai	70
Hong Kong	73

Source: Newman et al., 2009 pp. 86–7.

The sprawling nature of North American and Australian cities makes it expensive to provide a system that would encourage people to switch from their cars. In European cities the environmental effects of private cars are so great that the environmental and public health consequences are far more evident, and the benefits of using a car far less so than in North America or Australia. Some European cities, therefore, provide excellent models of what an effective public transport system can be like (see box 17.7). China appears to be moving to adopt more mass transit systems with many cities seeing rapid bus transit as a better solution than freeways. Delhi in India is building an electric metro rail which will see all residents within a 15-minute walk of it (Newman and Matan, 2013).

BOX 17.7   INNOVATION IN PUBLIC TRANSPORT SYSTEMS

Zurich in Switzerland has one of the best public transport systems in Europe. It is based on a tram network, which works well because of the following features (Walter, 1996, p. 40):

- large numbers of reserved lanes were built so that trams did not have to contend with cars for most of their journeys
- the city installed a traffic-light system that gives priority to trams. A transmitter in the cab of the tram triggers a sensor in the traffic lights that gives it priority over other traffic. This means trams do not have to wait at lights and makes trips on them much quicker than those by car
- construction of a sophisticated operational centre to control the timekeeping of the network's trams and buses and to provide spare parts in cases of breakdown.

The public transport system was planned and implemented with extensive public consultation. Walter (1996, pp. 40–1) quotes one of the architects of the system as saying that the Swiss political system, with provision for numerous referendums, is a huge advantage when it comes to transport decision making because:

> In countries where there are no referenda it is predominantly men aged between 25 and 60 who make the decisions. This is the population group that uses cars most. In the Swiss system 50 per cent of decision-makers are women.

Zurich citizens also voted for restrictions on cars, rejecting plans to build more car parks in the city centre. Fines of over A$200 for a minor parking violation are accepted.

Perth in Western Australia has invested $2 billion in a 280-kilometre modern electric rail system with 72 stations, which has explicitly been linked to reducing oil-dependency in successive elections (Newman, 2006). In the world's poor countries the challenge is to maintain their low dependency on private transport and encourage the provision of good public transport infrastructure. Encouragingly, China is investing in a network of high-speed trains. It is vital for global health (because of global warming and subsequent climate change, pollution levels and oil depletion) that the growing economies in poor countries are given every incentive to build effective public transport systems and not to encourage private car ownership. Singapore, where traffic control is strict, has given rise to innovative ways of offering car ownership such as car cooperatives.

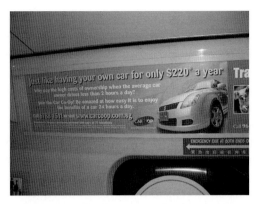

Car sharing in Singapore is a means of using cars so they have a lower ecological impact. (Paul Laris)

EQUITABLE PROVISION OF HEALTHY INFRASTRUCTURE

Resilient cities will also be equitable. Health is dependent on access to affordable infrastructure: housing, transport, clean environment, clean water, sanitation, education, nutritious food and supportive social relations. Ensuring supportive environments for all groups, especially for children, is an important way of reducing inequities (Lynch et al., 1997; Wadsworth, 1997). The most efficient way to reduce the disease burden associated with poor health behaviours and psychosocial characteristics is to improve the socioeconomic conditions that generate them (Lynch et al., 1997, p. 817).

The most pressing need for poor communities is access to clean water and sanitation. Some progress has been made, as documented in a recent WHO and UNICEF (2014b) report. By the end of 2012, 89 per cent of the global population used improved drinking water sources, an increase of 13 percentage points in 22 years, and 64 per cent of the global population used improved sanitation facilities, an increase of 15 percentage points since 1990. Although open defecation is declining across all regions, it is still practised by one billion people. Nine out of 10 people defecating in the open live in rural areas. There are still significant equity gaps as wealthy people universally have higher access to sanitation than the poor. In some countries this gap is narrowing. The gap is increasing, however, in rural areas of countries with low coverage and for marginalised and excluded groups. Around 5.7 million people—90 per cent of these living in sub-Saharan Africa (43 per cent) and Asia (47 per cent)—still use unimproved drinking water sources; 82 per cent live in rural areas. A major challenge in the coming decade is to ensure that water and sanitation is available to all citizens. We have the knowledge, but we need the political will.

The results of the WHO-commissioned analysis point out that improving water supply and sanitation would bring economic benefits. US$1 invested would give an economic return of between US$3 and US$34, depending on the region. Achieving this target would require an estimated additional investment of around US$11.3 billion per year over and above current investments. The benefits would include an average global reduction of diarrhoeal episodes of 10 per cent and a total annual economic benefit of US$84 billion, in addition to the dignity and health benefits.

In poor countries, providing clean water and sanitation and reducing pollution will require significant change in the way that most cities are managed. The growth of slum areas is proceeding rapidly and there needs to be dramatic action to arrest this development and ensure that all people have access to the basic requirements for healthy living in their environment. The improvement of slum areas is only likely to come about when the underlying problem of land price speculation and the need for fair and just city planning are accepted. Davis (2006) describes how far this is from the reality in large cities around the world. Massive commitment from international agencies to making city development more legal, planned and just could go some way to making cities much healthier places for the world's poor. Equity should be a central goal of urban development policies, because of their impact on the health of cities and suburbs. They should aim at achieving cities that are as unsegregated in terms of socioeconomic characteristics as possible. Essential to achieving a voice for poor people in city affairs is a reform of governance so that it is more inclusive and democratic (WHO, 2005). This will require broad-spectrum reform in systems of governance to ensure more transparency and accountability. The Commission on Social Determinants of Health (2008) notes that the characteristics of urban governance that should be strived for are:

- participation
- rule of law
- transparency
- responsiveness
- consensus orientation
- equity, effectiveness and efficiency
- accountability
- strategic vision.

It is clear that most governance in poor countries does not fit these criteria despite the efforts of many United Nations projects such as the United Nations Development Programme's Urban Governance Initiative, Local Agenda 21, Healthy Cities and HABITAT slum upgrading projects. The Commission on Social Determinants of Health's Knowledge Network on Urban Settings (KNUS) details actions that governments can take to upgrade slums in addition to improving governance, including:

- explicitly providing security of tenure
- implementing a low-cost, user-friendly system of land titling
- allowing community contracts to enable small-scale infrastructure work in slums
- reforming building codes to enable incremental building by slum dwellers and facilitating micro-credit to enable this to happen
- encouraging the private sector to provide credit.

Source: WHO Kobe Centre, 2005.

In rich countries such as Australia nearly all people have access to safe and clean water and sanitation, and basic aspects of public health are largely taken for granted by most people. This is not the case for many Indigenous populations. The greatest potential to improve equity in health status in Australia is by improving the environments in which Aboriginal and Torres Strait Islander Australians live. Studies suggest that relatively straightforward public health interventions have the potential to make a considerable difference to the health of Aboriginal people in rural areas (Pholeros et al., 1993; Bailie, 2007). The Close the Gap strategy has included funding for improving infrastructure—for example, in 2010, $51.7 million was made available through the Council of Australian Governments for a strategy to improve water and waste water services in 17 Indigenous communities (Department of the Environment, 2010). A Housing for Health program operating in 71 communities across New South Wales was shown to be effective in an independent assessment of the 10-year program. It found there had been a 40 per cent drop in 'hospital separations' (a measurement of the number of people admitted to hospital) for infectious diseases among people whose houses had undergone repairs—faulty plumbing, drainage and electrical systems repaired, or bathrooms and kitchens rebuilt (Laurie, 2011). Healthabitat (www.healthabitat.com), who ran the scheme, have extensive experience in housing for remote Aboriginal communities and are now extending their work to low-income communities in other countries, including South Africa.

HOUSING

> Housing is necessary material capital for developing human capital.
>
> Stretton, 2005, p. 131

Secure, appropriate housing is an essential element of a healthy lifestyle, and policies that ensure cheap, safe, reasonable-quality housing are important in reducing health inequities. In poor countries the challenge of providing housing is massive because many people do not have secure housing, and the quality of the housing is extremely poor. Africa has many refugee camps where people fleeing conflicts live, and within cities across the developing world people live in shantytowns and slum areas. Within rich countries housing appears to be becoming more of an issue than in the immediate past. In both Canada (Shapcott, 2004) and Australia (Stretton, 2005), for instance, it is noted that housing is less affordable, that renters face increasing costs, and homelessness is more of a problem than in the past. Stretton (2005) concludes that affordable housing that is provided equitably requires state intervention to control the free market. He cites the example of South Australia under the conservative Playford government of the 1940s and 1950s, which successfully used control over land and house prices to provide cheap housing for workers. In most rich countries around the world cutbacks in government expenditure mean less state-owned housing stock, which has always been an important resource for people who could not afford their own home.

Davis (2006) notes that state control strategies are required in contemporary cities in poor countries. The lack of government controls means poor people are subjected to evictions, exploitation, and suffer poor housing and insecurity of tenure. State control is needed to ensure that poor people are protected, and that city land allocations are made fairly and are not unduly generous to those with power and influence.

Appropriate housing for Aboriginal people living in remote areas has been identified as crucial to improving Indigenous health, but it is difficult to provide because of the extreme conditions and remoteness of many Aboriginal communities (Bailie, 2007). A recent response has been the National Partnership Agreement on Remote Indigenous Housing (COAG, 2008c), which is one part of the Close the Gap strategy established by the Australian Government as the major funder of remote Indigenous housing. A review conducted in 2013 (Department of Social Services, 2013) indicated that the program was likely to achieve its targets of reducing overcrowding and homelessness, redressing the shortage of housing and improving its quality by 2018.

Homeless people are particularly vulnerable to illness and premature death, and so policies to prevent people sliding into homelessness are particularly important. Evidence from rich countries suggests that intervention in the first three weeks is particularly effective in ensuring that temporary homelessness does not become entrenched (Webster, 1997). Emergency shelters, support to encourage people back to secure housing and the provision of appropriate health care for homeless people are all policies that should reduce the impact of homelessness on health. The best solution to homelessness is the provision of affordable and appropriate shelter. Housing for all is a crucial part of ensuring health for all and should be considered in any national public health strategies.

PRESERVING AGRICULTURAL LAND AND NATURAL SPACES

Around the world, communities are juggling the tensions between development and preservation of natural environments. Australia, as one of the oldest continents, provides a good example of the challenge of maintaining the delicate balance. The following three key components of the environment have been defined as crucial to sustainable development (State of the Environment Advisory Council, 1996, p. 9):

- biodiversity—the variety of species, populations, habitats and ecosystems
- ecological integrity—the general health and resilience of natural life-support systems, including their ability to assimilate wastes and withstand stresses such as climate change and ozone depletion
- natural capital—the stock of productive soils, fresh water, forests, clean air, ocean and other renewable resources that underpin the survival, health and prosperity of human communities.

Sustainable agriculture that provides sufficient food and the availability of areas of natural beauty are obviously crucial to healthy and sustainable environments. There is considerable effort in Australia to preserve agricultural land from further degradation and to protect areas of pristine beauty, but there is a continual struggle between traditional exploitative practices (usually for immediate profit) and protection of the environment.

Many anecdotal accounts tell of the health-promoting benefits of spending time in areas of natural beauty and wilderness. People have seen the benefits of time in such areas as renewing the spirit, getting in touch with nature and stepping out of the daily urban grind. Consequently, apart from the environmental arguments for the preservation of such areas in terms of protection of biodiversity, there is a strong public health rationale for maintaining these places for recreation. However, keeping a balance between tourism and environmental preservation is often tricky. There are increasing calls to restore previous wild ecosystems and reverse the destruction of the natural world through a process of 'rewilding' (Monbiot, 2013).

Examples of successful measures to protect ecosystems are the LandCare program (see box 17.8); the preservation of the Gordon-below-Franklin river system in Tasmania; and the preservation of the Kakadu National Park in the Northern Territory. Kakadu is co-managed by the National Parks and Wildlife Service and the local Indigenous people, a partnership that has fostered effective nature conservation, a tourist industry that provides the local people with a steady income and the preservation of traditional communities and their cultural legacy (Sachs, 1996, p. 142). Indigenous cultures around the world have strong links with their ancestral lands and provide deep knowledge about living in harmony with nature. Key to protecting the remaining wild areas on the planet is to respect the rights—especially the right to self-determination—of Indigenous peoples worldwide. Thus in 2010 a new Kenyan constitution was passed that recognised the traditions, customs, languages and rights of Kenya's indigenous peoples, the Maasai, and acknowledged the legitimacy of hunter-gatherer, pastoral, and nomadic ways of life (Adamson et al., 2013). In addition, the Maasai have been recognised, for the first time, as important stewards of the land, whose environmental knowledge and practices—including rotational livestock grazing and the fostering of beneficial wildlife habitats—can help build resilience to climate change, improve water conservation and protect biodiversity.

BOX 17.8   LANDCARE: PARTICIPATION FOR SUSTAINABILITY

LandCare was initiated in 1986 by the National Farmers' Federation and the Australian Conservation Foundation, and has been supported as a national program by the Australian Government since 1989. Both its parent organisations were committed to finding common ground in an area often fraught with conflict. LandCare was a policy response to the problem of land degradation through salinisation of agricultural lands, soil erosion, depletion of fertility and loss of vegetation. Federal and state governments developed participatory programs to involve land users, local community groups, schools and interest groups in a wide range of educational, remedial and strategy development activities, leading to the creation of thousands of LandCare groups in rural and urban Australia, linked by an impressive network of regional, state and national conferences, newsletters, training schemes and decision-making bodies. In 2007, LandCare had more than 4000 volunteer community LandCare groups—including BushCare and Urban LandCare, RiverCare, CoastCare and sustainable agriculture groups, all of which are tackling land degradation.

Considine (1994) says the success of LandCare has been its strong participatory structure, which has drawn together key community interests in a constructive manner. The program has enabled farmers to resist criticism that they are responsible for the nation's worst environmental problems. LandCare has drawn strength from its policy centre and the local-level activity—an approach to change that reflects new public health principles well. A 1999 national evaluation (Standing Committee on Agriculture and Resource Management and Agriculture and Resource Management Council of Australia and New Zealand, 1997, pp. 27–8) found that increasingly LandCare was focusing on whole catchments and regional themes. The emphasis was shifting to awareness-raising to embrace the integrated management of soil, water and biological resources, the development of economic instruments to encourage sustainable natural resource management, a stronger voice for LandCare groups in determining the direction of research, and greater involvement of Aboriginal and Torres Strait Islander people and people from culturally and linguistically diverse backgrounds. The program's future is somewhat uncertain after the Abbott government's first Budget, which cut 30 per cent of its funding. Case studies of the valuable work done are available on the LandCare website www.landcareonline.com

THE SUSTAINABILITY OF RURAL AREAS

We have seen in chapter 15 that one of the most significant social trends globally is rapid urbanisation. An important factor driving that trend is the lack of opportunities in rural areas for people to advance the well-being of themselves and their families. Policy responses to this need to consider how rural living can be made more attractive, including in terms of environmentally sound practice. Thus maintaining a heavily water-dependent cotton industry in drought-stricken areas of Australia or encouraging poor farmers to stay in unsustainable industries in India and China would not make policy sense.

Yet many rural areas require a regenerated economy to improve the prospects for health and well-being. The policy challenge is to find how to do this in a way that also ensures that the social and physical environments are sustainable and promote health. Public services and private companies have to be persuaded to support rural infrastructures so that people can, and want to, continue living in the communities. Across many rich countries the picture in rural areas is one of cutbacks to public services and the centralisation of private enterprises such as banks. In poor countries, rural areas have little access to services, and unfair trade compounds the disadvantage of small farmers. Policy mechanisms to make rural living more attractive in all countries will be essential in the coming decades. These mechanisms will also need to ensure that the development is sustainable and low-carbon. Regional Australia is heavily dependent on diesel, which the government subsidises. One example of a move away from diesel is in the Pilbara mining region, which is adopting a sustainability strategy that suggests that the Pilbara could become a demonstration region for how to become diesel-free (Newman et al., 2009). Box 17.9 provides an example of a rural town's desire to reinvent itself by moving from coal to solar power. Rural areas close to cities can become vital to the local and sustainable production of food, which results in lower food miles for the city and reduces its reliance on imported food.

BOX 17.9   PORT AUGUSTA: A SOLAR FUTURE

Port Augusta is a city of 13 257 people in the north of South Australia, 322 kilometres from Adelaide. It has relied on two coal-fired power stations, which are coming to the end of their lives. There is a strong community movement to replace these power stations with Australia's first solar thermal plants or a combined cycle gas plant.

The Repower Port Augusta Alliance (http://repowerportaugusta.org) developed a solid proposal to replace the coal plants with six solar thermal plants and 95 wind turbines. The proposal claims this would secure the 250 jobs at local power stations, create 1300 construction jobs and 225 manufacturing jobs, save five million tonnes of greenhouse gas emissions, improve the health of the local community, and ensure energy security and stable electricity prices. The proposal is supported by most of the community (a poll showed 4053 for solar and 43 for gas), the local council and local business. The campaign states that the proposal 'would enable South Australia to become a world leader in renewable energy, and Port Augusta would become an iconic global hub for baseload solar power generation'.

Source: Repower Port Augusta.

CONCLUSION

Much of the public health work of creating sustainable and healthy communities will challenge the vested interest of powerful groups within society and, consequently, be difficult and demanding. But contributing to the growing movement for a greener, cleaner and more sustainable society will be one of the crucial tasks for public health in the twenty-first century. Most critically, public health advocates have to keep plugging away with the message that we have the collective resources to live sustainable lives and to ensure that everyone in the world has a decent standard of living that is compatible with health for all.

CRITICAL REFLECTION QUESTIONS

17.1   How sustainable is the city you live in?

17.2   How can cities be made more supportive of the ecosystem and human health?

17.3   What public health benefits do effective and well-used public transport systems offer?

Recommended reading

Commission on Social Determinants of Health final report (2008) and website (www.who.int/social_determinants/en) provide many policy ideas for improving the conditions of daily life.

Newman, Beatley and Boyer (2009) *Resilient cities: Responding to peak oil and climate change* provides a guide to how cities can respond to the dual demands of peak oil and climate change, and presents lots of practical solutions.

Newman and Matan (2013) *Green urbanism in Asia: The emerging green tigers* is an excellent guide to the ways in which Asia is moving towards low-carbon economies, with lots of practical case studies of how cities are doing this.

Useful websites

The Worldwatch Institute (www.worldwatch.org) works to accelerate the transition to a sustainable world that meets human needs. It publishes an annual report on different aspects of sustainability annually.

The Transition Network (www.transitionnetwork.org) and Global Ecovillage Network (http://gen.ecovillage.org) both provide practical examples of sustainable living practices.

PART 7

HEALTH PROMOTION STRATEGIES FOR ACHIEVING HEALTHY AND EQUITABLE SOCIETIES

Part 6 set out a vision of communities and societies that are designed to be health creating and to promote equity of access to the resources and services that promote health and more equity in health outcomes. Part 7 is concerned with the approaches used by health promotion and public health to achieve such communities and societies.

The past 30 years have seen a continuous tension between approaches to health promotion and public health that emphasise the agency of individuals and try to change their behaviours directly, and those stemming from the Ottawa Charter for Health Promotion, which pay more attention to the need to create supportive environments and make healthy choices the easy choices. In the 1970s and early 1980s behaviour paradigms reigned supreme but the lack of success, especially in producing equitable outcomes, meant the new public health evolved. Public health practitioners increasingly appreciated the need for a broad social, economic and environmental approach to health promotion, to encourage action across sectors and to involve non-experts in public health decision making. This century has seen the added imperative to think globally about health promotion. Health promotion strategies have varied widely in different countries, at different times and among different professional groups, but all approaches to health promotion reflect the values and beliefs of the promoters. In parts 1 and 2 we saw that individualistic philosophies tend to lead to behavioural strategies that focus on individuals, while others try to change social and economic structures. The questions that a structural perspective gives rise to are shown in table 1.1. There are three main approaches to health promotion: medical/health services, behavioural and socioeconomic. Chapter 18 describes medical and health service approaches to health promotion and discusses the role of the general practitioner, screening and immunisation. Chapter 19 considers the dominance of behavioural paradigms in health promotion, examines the limitations of the paradigm and considers alternative ways of incorporating agency into initiatives. Chapter 20 considers community-wide campaigns designed to change behaviours that are direct risk factors for various diseases. A critique of behavioural approaches to health promotion is offered. Chapter 21 considers the importance of participation in health promotion: it describes community development strategies and highlights a number of contradictions and dilemmas. Chapter 22 describes the landscape of public health advocacy and activism including leadership, strategies and dilemmas. Chapter 23 describes the variety of approaches to health promotion based on settings or organisations, and chapter 24 reviews the ways in which policy and legislation have been used to restrict and control individual behaviour.

Although medical and health services and behavioural interventions are limited in their contribution to achieving health communities and societies, they are reviewed here for three reasons:

- they have been the dominant form of formal health promotion activity
- their limitations need to be acknowledged so that, wherever possible, they can be overcome. This is particularly important as behavioural health promotion is still widely, and often inappropriately, practised in isolation, even though evidence suggests its power derives from being one aspect of a broader strategy that seeks to make healthy choices the easy choices
- medical and behavioural approaches to health promotion do have a place in the portfolio of health promotion approaches, particularly if they can be modified to be part of a broader socio-environmental approach and to empower people to act on their own agency in favour of their health.

Labonté (1992) provides a framework to consider three different approaches to health promotion. These are the medical approach, which tries to return sick people to a disease-free state; the behavioural approach, which promotes healthy lifestyles; and the socio-environmental approach, which is concerned with the totality of health experiences and the factors that help to maintain health, including those connected directly with people (behaviour, self-esteem and genes) and environment (income, housing and employment). The three approaches defined by Labonté differ in terms of how they view health, how they define health problems, the intervention strategies they advocate, their focus and how they determine success. An overview of these differences is provided in table 7A. As we see in chapter 19 health and services systems do not necessarily (although they often do) have to adopt a medical approach. When they are based on comprehensive primary health care they will encompass all three approaches.

The Labonté model does not reject the value of medical and behavioural approaches but sees that they are more powerful when incorporated within the broader framework offered by the socio-environmental model. We have seen that people who experience the worst health status are more likely to have low incomes, live in poverty, have unsafe and stressful jobs and have inadequate housing. This, linked to the fact that they also have fewer social resources and support and have lower self-esteem and perceived power (which may affect their physiological functioning and ability to change behaviour), means that the socio-environmental view of health makes most sense. The dynamic links between these factors mean that health promotion is only likely to be successful when based on an understanding of these complexities.

TABLE 7A   APPROACHES TO PROMOTING HEALTH

	Medical	Behavioural	Socio-environmental
Focus	Individuals with unhealthy lifestyles	Individual's and group's conditions	Communities and living environments
Definition of health	Biomedical, absence of disease and disability	Individual practice of healthy behaviours (e.g. exercise, nutritious food)	Strong personal and community relationships. Feeling of ability to achieve goals and be in control
How problems are defined	Disease categories (and physiological risk factors e.g. cardiovascular disease, HIV/AIDS, diabetes, cancer). Medical definition	Behavioural risk factors (e.g. smoking, poor nutrition, lack of fitness, alcohol abuse, poor coping skills). Expert definition	Socio-environmental risks (e.g. poverty, unsafe or stressful living and working conditions). Psychosocial risks (isolation, lack of social support, low self-esteem). Equity key factor. Community involved in problem definition
Main strategies	Illness care, screening, immunisation, medically-managed behaviour change	Mass media behaviour change campaigns, social marketing, advocacy for policies to control harmful agents (e.g. drink-driving, smoke-free public places)	Encouraging community organisation, action and empowerment. Political action and advocacy
Success criteria	Decrease in morbidity and mortality and decrease in physiological risk factors	Behaviour change, decline in risk factors for disease	Individuals have more control, social networks are stronger, collective action for health evident, decrease in inequities between population groups

18

MEDICAL AND HEALTH SERVICE INTERVENTIONS

Reorientation requires health services to shift from predominantly reacting to individuals requiring treatment for illnesses and injuries, to a position where health services take a broader view of 'health' and incorporate prevention and health promotion as part of their core business.

Johnson and Paton, 2007, p. 8

KEY CONCEPTS

INTRODUCTION

Health care systems are dominated by medicine. Health care approaches to health promotion concentrate on the prevention of disease mainly through primary medical services. However, numerous barriers have been identified to general practitioners (GPs) being more involved in prevention, including the fee-for-service structure, short consultation times, the traditional focus of medicine on curative interventions and GPs' lack of health promotion skills and knowledge, and these have continued over the past three decades (Bauman et al., 1989; Ward et al., 1991; Peckham et al., 2011). There are moves in most settings to encourage GPs to place more emphasis on health promotion. In Australia the General Practice Reform Strategy, for example, has sought to increase the amount of health promotion work being done. Comprehensive primary health care, in which medicine is one of a range of approaches, offers the best care setting in which to practice health promotion.

GENERAL PRACTITIONERS

GPs often have a longer term relationship with their patients than other medical practitioners and this provides a basis for health promotion and disease prevention (Powell-Davies and Fry, 2004). General practitioners are being urged to become more involved in health promotion and disease prevention because approximately 80 per cent of Australians visit a GP each year and every patient is likely to have at least one behavioural risk factor (Bonevski et al., 1996). A review of the international literature concluded that there was 'a great unrealised potential for disease prevention in primary care' (Bonevski et al., 1996, p. 29). Ashenden et al. (1997) examined the effectiveness of lifestyle advice provided by GPs by conducting a meta-evaluation of trials of this activity. The review indicated that while many interventions showed promise in bringing about small changes in behaviour, none appear to produce substantial change. The main barriers to GPs' involvement in health promotion aside from the limited basis for its effectiveness are: structural factors (lack of initial and continuing education and training in health promotion, non-standardised guidelines and low financial incentives); office organisation (lack of time in consultations, lack of support staff and sheer forgetfulness); patient reluctance and competing priorities for the time available in the consultation; low confidence; and frustration from doctors because they do not receive rapid feedback.

GPs may be able to influence patients to:

- change their lifestyle
- undergo screening for the early detection of a range of conditions
- present for health-protecting vaccinations
- manage chronic conditions to improve quality of life.

General practice has put more emphasis on these aspects of its work. The Royal Australian College of General Practitioners publishes a 'Red Book' (RACGP, 2012) and a specific guide for preventive assessment for Aboriginal and Torres Strait Islander peoples (NACCHO and RACGP, 2012). A key framework for understanding the way in which prevention is implemented in primary health care is 'the 5 As' (figure 18.1). The 5 As has been widely adopted internationally in addressing behavioural risk factors (Harris and Lloyd, 2012). This framework relies on the ability of GPs to refer patients to other health professionals for additional assistance and to engage in teamwork to make these referrals timely and appropriate. Australian community health services have been recognised as having good multidisciplinary teamwork (Baum, 2014); however, as we see later in this chapter, they have remained a minority model of health care that has not received sufficient investment for them to become mainstream.

FIGURE 18.1   THE 5 As FOR BEHAVIOURAL RISK FACTORS IN AUSTRALIAN
GENERAL PRACTICE

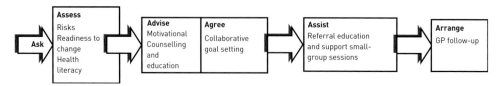

Source: Harris and Lloyd, 2012, p. 4.

In the UK, health promotion and disease prevention through intervention from health professionals has also become more systematic. The National Institute for Health and Care Excellence (www.nice.org.uk) produces 'guidance' on behaviour-change strategies and some limited new public health topics such as transport.

The USA does not have GPs that are a first point of call and responsible for referral to other specialists. Nonetheless their value has been recognised in studies that have shown that primary health care produces better health outcomes for communities (Starfield et al., 2005). However, many approaches to primary health care remain narrow and do not consider the social determinants of health. This is well illustrated by Chaufan (2004), an anthropologist, physician and diabetes educator, who describes the ways in which diabetes in the USA is almost always treated as a medical and behavioural problem. She describes how, while social and environmental factors are acknowledged, these are given little attention, and the problem of high-risk individuals in terms of their genes and lifestyle is stressed. She sees that the policy responses are bounded in terms of the medical responses and exhortations to individuals to make lifestyle changes. Yet little attention is paid to the social and cultural distribution of diabetes—Native Americans, Latinos, African-Americans and lower socioeconomic groups are more likely to have diabetes. These groups are also much more likely not to have private health insurance. Chaufan stresses that campaigns to reduce the prevalence of diabetes and improve its management should be based on a sociological analysis that considers the restraints to improved lifestyle and management. Table 18.1 presents the measures that she suggests would flow from such an analysis.

TABLE 18.1   CONTRASTING SOLUTIONS TO DIABETES

Issue	Medical/behavioural	Structural
Need for healthy food	Advice to individuals about healthy food to prevent or control diabetes.	Restrict advertising of unhealthy foods. Improve food labelling. Make healthy food more widely available and cheaper for groups at risk.

Issue	Medical/behavioural	Structural
Increase exercise	Information about benefits of exercise in preventing and controlling diabetes.	Provide exercise-rich transport options (i.e. improved public transport and facilities for bikes, reduced car use).
		Make neighbourhoods safer and more attractive to walk around.
		Fund exercise programs in schools.
Better management	Provide information to encourage improved 'compliance'.	Ensure health system is accessible and coordinated—universal public health insurance and publicly funded system is best means to ensure an equitable system.
Diabetes educational models	Education aimed at individuals: 'patient empowerment' to self-manage disease and set goals.	Education takes account of and acknowledges constraints on people's lives that make 'compliant' behaviour difficult.
		Health professionals lobby for change in 'landscapes of risk' that make compliance difficult.
Conceptualisation	Risky individuals are the focus.	Risky environments and policies are the focus—aim is to 'make healthy choices the easy choices'.

Source: Based on Chaufan, 2004.

The two areas where medical interventions have made most impact on public health are screening and immunisation.

SCREENING

Screening involves the investigation of individuals to find out whether they are at risk of a particular disease through tests that either seek the existence of a risk factor for the disease or early physiological indications. Risk factors may be physical or behavioural attributes. For disease and risk factor screening to be effective the following conditions need to be met (Naidoo and Wills, 1994, p. 85):

- the disease or risk factor should have a long preclinical phase so that a screening test will not miss its signs
- earlier treatment should improve disease outcomes

- the test must be sensitive so that it will detect all with the disease. People who are told they are clear of a disease or risk factor when they in fact have it are referred to as 'false negatives'
- the test should be specific and detect only those with the disease. People who are told they have a disease when they do not are referred to as 'false positives'
- the test should be acceptable, easy to perform and safe
- it should be cost-effective.

Screening for diseases has become increasingly common in recent decades. In Australia today people are most commonly screened for heart disease and cancer. In the past screening for tuberculosis was prevalent. It is rarely totally hazard-free, and ethical questions can be raised, for instance the impact of a 'positive' test result on a person who does not have the particular risk factor or early signs of the disease. There is evidence that screening could be most effective in the healthiest groups as they are most likely to take advantage of the opportunity to be screened.

SPECIFIC SCREENING TESTS AND THEIR EFFECTS

Heart disease screening involves blood cholesterol levels and hypertension testing. Egger et al. (1990) cite research that suggests an estimated 5 per cent reduction in diastolic blood pressure in the Australian population could result in a decrease of about 30 000 major cardiac events each year. This, of course, relies on people identified as having high blood pressure taking appropriate preventive measures. Most people in high-risk categories are rarely in a position to make changes to their lifestyle, reducing the potential for health gain from screening if strategies rely on individual behaviour change.

Screening for high cholesterol levels is controversial. Petersen and Lupton (1996, pp. 43–5) call into question the strength of the relationship between high cholesterol and heart disease and suggest that the test has gained popularity, in part, because there is a powerful 'healthy foods' and pharmaceuticals industry benefit from products that assist in lowering high cholesterol.

Screening for behavioural risk factors for cardiovascular disease in terms of dietary assessment, stress assessment and lifestyle appraisal has been more prevalent in recent years (Egger et al., 1990). Such screening aims to bring about individual behaviour change, but its success has been very limited.

Mechanisms for cancer screening have become more sophisticated. In Australia, screening for cervical and breast cancers has become the focus of national programs. The national program for the Early Detection of Breast Cancer oversees a network of dedicated, accredited screening and assessment units throughout Australia (Weller, 1997). These mammography tests have led to a substantial increase in the number of diagnosed cases of breast cancer and some decline in mortality. Weller reports that the benefits for women older than 50 appear clear but are still disputed for women aged 40–49. In the 1980s breast self-examination (BSE) was widely promoted as a screening tool but there was no evidence that this strategy was effective in reducing mortality from breast cancer. Systematic screening for cancer of the cervix through the establishment of state-based Pap smear registers, recall systems and various promotion strategies to encourage women to participate in screening is widespread (Weller, 1997).

Screening for prostate cancer is also contentious. Some practitioners advocate the test, but an extensive review of the literature showed that there was insufficient evidence to assess the value of screening asymptomatic men, that widespread screening would require significantly greater resources and that there was a real risk that the early detection and treatment of localised prostate cancer may cause more harm than good (Weller, 1997). The effectiveness of screening has remained uncertain (Basch et al., 2013). Weller also reports uncertainties about the value of population screening for colorectal cancer using the faecal occult blood test. Issues to be considered include the high rate of false positives (about 90 per cent of positive results will not lead to a diagnosis of colorectal cancer) and that the cost per life saved has to be weighed against other health expenditure.

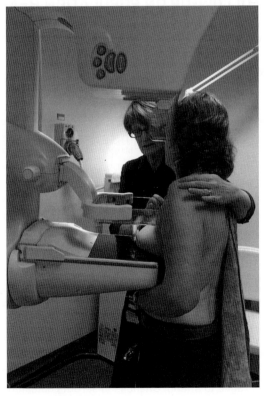

Mammography screening is credited with saving the lives of many women. Medical Imaging, Flinders Medical Centre (Professor Neil Piller)

Recent research on the benefits and harms of screening for cancer has suggested that the two are finely balanced and that evidence should be available to individuals so they can make an informed decision (Barratt et al., 2005). There have also been calls for assessment of the psycho-social impact of screening on people (McCaffery and Barratt, 2004), which is especially important for those who receive false positive results (that is, they are informed of abnormalities when further investigation shows they do not have cancer).

Crucial to the success of screening is increasing uptake (i.e. the proportion of screening invitees in a given year for whom a screening result is recorded). Evidence shows that lower socioeconomic groups and ethnic, black and Aboriginal communities are likely to have lower rates of uptake (Weller and Campbell, 2009). There is some evidence that factors such as translation of key materials, culturally appropriate materials, group health education and lay workers may help uptake. Without these special efforts screening may actually increase inequities, as those who are most well-off are more likely to benefit from early detection if there are barriers to other groups taking advantage of the screening.

EFFECTIVENESS OF SCREENING FOR BEHAVIOURAL RISK FACTORS AND FOLLOW-UP ON POPULATION HEALTH

The available evidence indicates that screening followed by educational intervention has little impact on risk factors. Two programs that illustrate this relative lack of effectiveness are described in box 18.1.

BOX 18.1 EVIDENCE ON EFFECTIVENESS OF SCREENING AND EDUCATION ABOUT CARDIOVASCULAR DISEASE RISK FACTORS

MRFIT

Heart disease has been the focus of screening and intervention to reduce risk factors. The Multiple Risk Factor Intervention Trial (MRFIT) Randomised Control Trial (Winkelstein and Marmot, 1981) resulted from a screening program in 20 communities over a two-year period from 1973 to 1975. Over 370 000 men were examined—12 866 aged between 35 and 57, who did not have evidence of pre-existing clinical heart disease and who were in the upper 10 to 15 per cent at risk of ischaemic heart disease (IHD) mortality because of hypercholesterolaemia, cigarette smoking and high blood pressure, were identified. A six-year random intervention program was planned and those willing to participate were randomised to either normal medical care or a special intervention program.

The intervention program began with a series of 10 intensive group meetings to provide information about risk factors and initiate behavioural modification programs. Partners were encouraged to attend. Changes in serum cholesterol levels and cigarette smoking were attempted only through behavioural techniques. Reduction of diastolic blood pressure was attempted by weight reduction and 'stepped care' drug therapy.

Participants who had not reached their risk factor modification goals at the end of the 10-week program were invited to participate in an extended intervention program consisting of case conferences and individual consultation. A maintenance program was available for when risk factors were reduced. Despite this intensive intervention, the program had little success (Syme, 1996).

RYDE HEART DISEASE PREVENTION PROGRAM

In Australia the Ryde Heart Disease Prevention Program (see Biro et al., 1981) involved a controlled trial to lower risk factors among employees. Screening of 3401 people identified those with increased risk factors (other than blood pressure, for which people were referred to their doctor) and these pepole were divided into an intervention group, who were offered a variety of group education programs, and a control group, who were only sent the results of their screening test. After one year the two groups had similar changes in risk factor levels, and changes in blood pressure favoured the control group.

There seems to be little evidence for the value of advice in relation to alcohol and smoking. The experiences of programs such as MRFIT suggest that the limited interventions possible in a primary medical care session are unlikely to affect individual risk factors to any great degree. The potential for GPs (or any other health professionals) to contribute to population health outcomes is even less supported by the evidence. Rose (1985) points out that, although the medical model of health promotion regards people with a particular disease or risk factor as differing in some categorical way from the rest of the population, they actually represent one end of a continuum. He considered the distribution of risk factors in 32 countries with different levels of economic development and concluded that the proportion of people at high risk in any population is simply the function of the average blood pressure, cholesterol or other risk factor in that society. He found this to be true of a range of conditions. Rose concluded that you could not reduce the proportion of the population at high risk without reducing the whole society's exposure to the risk. In this view, strategies aimed at individuals would not be expected to achieve much change.

Rose and Marmot's (1981) study of British civil servants also questions the value (in relation to coronary heart disease) of focusing much effort on direct attempts to change individual behaviour in order to reduce risk factors. Their work in relation to health inequalities indicated that the main risk factors combined (cholesterol, smoking, blood pressure and others) explained considerably less than half of the difference in mortality between men in different ranks of the civil service (see figure 18.2).

FIGURE 18.2   RELATIVE RISK OF DEATH FROM CORONARY HEART DISEASE ACCORDING
TO EMPLOYMENT GRADE AND PROPORTION OF DIFFERENCES THAT CAN
BE EXPLAINED STATISTICALLY BY VARIOUS FACTORS

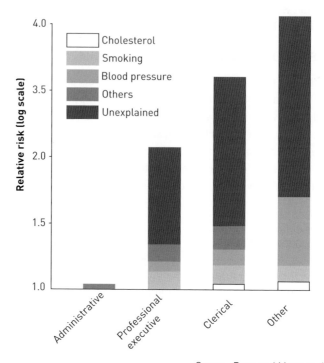

Source: Rose and Marmot, 1981, p. 17.

Wilkinson (1996) observes that strategies focused on individuals leave the underlying societal causes of disease untouched. Inevitably, this means there will continue to be a demand for new services to cope with people identified as being at high risk.

The implications of this are that it would not be sensible to pursue a strategy that solely puts emphasis on individual change. However, within hospitals and general practice, health promotion is still usually seen as being synonymous with individual behavioural change. This reflects the focus of medicine on individuals as opposed to communities and societies, but there are some signs that this is changing. Hospitals are beginning to see health promotion as implying organisational change and having as many implications for staff practices as for patients (Johnson and Paton, 2007).

IMMUNISATION

Ever since Jenner's 1796 discovery of the effects of vaccination with calf-lymph against smallpox infection, immunisation has been an important public health tool. As a public health strategy, immunisation depends on creating a sufficient pool of immunised people to prevent outbreaks from spreading more widely among the population. This is the concept of herd immunity. If immunisation levels in the population are not sufficiently high, it will not be possible to control acute, vaccine-preventable diseases. Worldwide, it has been estimated that immunisation programs prevent approximately 2.5 million deaths each year (Department of Health, 2013a). Many countries provide free vaccines. For instance, in Australia, free vaccines are provided for 16 diseases.

SMALLPOX

On a global level, the eradication of smallpox is the best example of the success of immunisation, and one of the major achievements of public health. Following a campaign initiated by the World Health Organization (WHO) in 1959 and relaunched in an intensified and more coordinated form in 1967, smallpox was eradicated from the planet in 1977. The process was difficult and possible only because the characteristics of smallpox made it responsive to vaccination. Also, the advent of widely available freeze-dried vaccine in 1970 greatly enhanced the effectiveness of vaccination campaigns. However, as Fenner (1984) has pointed out, these biological features of smallpox and its control were necessary, but not sufficient, conditions for success. Several sociopolitical features were also crucial. The success of campaigns at national and regional levels had shown that eradication was achievable, and there were significant costs in treating and containing smallpox that could be avoided altogether. From 1967 onwards, the Intensified Smallpox Eradication Unit of WHO had a strong leadership and was able to attract sufficient funding. From 1969 to 1979, the average cost of the campaign was US$23 million per year. In 1967, smallpox killed some two million people, blinded and/or disfigured 10 to 15 million more and cost nearly US$1.5 billion, so the financial benefits of eradication were obvious (Jezet, 1987).

POLIO

A Global Polio Eradication Initiative (GPEI) (www.polioeradication.org) has been established. Its original goal was eradication of poliomyelitis by 2008, but by 2006 that goal was stated to be achieved 'as soon as possible' (WHO, 2006). By 2014 the campaign aimed to eradicate polio by 2018. In 2014 polio was endemic in just three countries of the world—Nigeria, Afghanistan and Pakistan. India was declared polio-free in 2014 after having no endemic cases for three years. The worldwide number of polio cases had fallen from 350 000 in 1988 to just 1951 in 2005 (WHO, 2006) and to 416 in 2013 (WHO, 2014e). The GPEI estimates that, to date, the effort has saved US$1 billion in terms of polio treatment and rehabilitation costs and, more importantly, that there are five million children walking who would otherwise be paralysed. The key strategies have been enhanced immunisation, certification to ensure standard surveillance including laboratory analysis in all countries, and building a global partnership to mobilise interest and resources. The partners have included WHO, UNICEF, the US Centers for Disease

Control and Prevention, and Rotary International. Despite the massive effort of these partners, in early 2014 the GPEI had received just US$1.83 billion in contributions and is tracking over US$3.13 billion in pledges and projections—even if these funds are fully realised, there will still be a funding gap of US$563 million (WHO et al., 2014a).

IMMUNISATION IN AUSTRALIA

Australia has a long history of immunisation, with smallpox vaccination records going back to the 1850s in New South Wales and South Australia (Woodruff, 1984). Responsibility for providing vaccination services in Australia is shared by local government authorities and private medical practitioners with the policy context set by the federal government. The *Australian Immunisation Handbook* (Department of Health, 2013a) provides a comprehensive clinical guide to vaccination and recommends infants and children be routinely immunised against diphtheria, tetanus, pertussis (whooping cough), poliomyelitis, *Haemophilus influenzae* type B (Hib), pneumococcal disease, rotavirus, measles, mumps, rubella, varicella (chickenpox), meningococcal disease and hepatitis B; and older people should be immunised against pneumococcal infections and influenza. Girls and boys aged 10–15 years are also recommended to receive the human papillomavirus vaccine.

In the mid-1990s Plant warned that vaccination programs in Australia have been unable to establish sufficient herd immunity to prevent outbreaks of diseases such as measles and rubella (Plant, 1995). Since that time vaccination rates have increased and it was calculated that in 2011 at one year of age 91.8 per cent of Australian children were vaccinated and 92.6 per cent by age two (AIHW, 2012). Unlike smallpox, which needed a vaccination level of just 50 per cent to achieve eradication, measles is so highly contagious that 94–97 per cent must be vaccinated to eradicate it (Hawe, 1994). In Australia as a result of a Measles Control Campaign in 1998 and improved coverage as part of the routine childhood vaccination schedule, there has been a large decline in the measles notification rate (AIHW, 2006, p. 116). Measles outbreaks are seen as a useful warning of an inadequately vaccinated population. The key features of an effective campaign are being coordinated across sectors, improvements in service delivery and access, better surveillance, and removal of financial barriers to vaccination (AIHW, 2006). The Australian Government has funded work intended to increase the vaccination coverage in Indigenous communities (Telphia et al., 2006). The AIHW (2012) reported that there are only very small differences in rates of immunisation between advantaged and disadvantaged and Indigenous and non-Indigenous Australians.

Actions undertaken to increase coverage include greater use of opportunistic vaccination and legislation requiring parents to present evidence of vaccination status to schools and childcare centres, so that unvaccinated children can be kept away from school during outbreaks. From 2012 parents had to have their child vaccinated in order to receive family benefits payments. The Australian Government established the Immunise Australia Program, which encourages and provides incentive for GPs and others to vaccinate. Immunisation coverage figures from the Australian Childhood Immunisation Register (ACIR) continue to show increases in the number of fully immunised children.

INDIVIDUAL RISKS AND SOCIAL BENEFITS OF IMMUNISATION

Vaccination offers health, social and economic benefits and can contribute to health equity (Andre et al., 2008). All immunisation involves some risk: the nature, severity and rate of incidence varying with the process in question. It is possible to calculate a reasonably accurate equation of the risks and benefits in a given vaccination program. The Australian Government Department of Health publishes an annual Australian Immunisation Handbook, which provides a comparison of the effects of diseases and the vaccines used to prevent them (available online at www.health.gov.au/internet/immunise/publishing.nsf/Content/Handbook10-home). For the individual parent, the decision to initiate an action with the potential to cause their child harm, and perhaps lacking accurate information on the real risks and benefits of either course of action, may not be so clear-cut. Research suggests that lack of access (transport, unsuitable clinic times, poverty) accounts for about half of people who are undervaccinated, and the other half from lack of acceptance. Of these, only 2 per cent of parents are vaccine refusers. This is much lower than the 'hesitant' group who have some reservations (Willaby, 2014). Anti-vaccination groups exist and an increasingly individualist trend in social values may aid their cause. However, there is also the risk of public health protagonists being seen as unconcerned with the suffering of individuals, or as censoring alternative views. Last (1998, p. 373) makes the point that those who conduct public immunisation programs must ensure that all who consent to have their children immunised are aware of the risks. However, there is also an ethical responsibility to advise the community of the benefits of an adequate level of immunity, and the potential risks of letting that level drop. This implies a role in fostering public education and informed debate, as well as a clinical responsibility.

THE CONTRIBUTION OF THE HEALTH SECTOR TO PROMOTING POPULATION HEALTH AND REDUCING INEQUITY

A review of studies of the impact of medical services on health showed that even the small proportion of deaths that are wholly amenable to medical treatment seems less influenced by differences in medical provision than by socioeconomic factors (Mackenbach et al., 1990; Evans and Stoddart, 1994). However, other evidence does suggest that medicine may have had a greater role in extending life expectancy in the later twentieth century, when most of the gain in life expectancy in industrialised countries occurred in older age groups. Bunker et al. (1994) show that, for a range of causes of death, medical interventions have made an important contribution of about

20 per cent of increased life expectancy this century. Cutler and Meara (2001) suggest that medical intervention in regard to cardiovascular disease has made a significant contribution to reducing death rates among older people. Davey Smith (2003) points out that, as access to medical technology is determined in part by socioeconomic status, then inequities may increase as medical technology comes to have some impact on life expectancy. This means of course that equal access to medical care is an important part of an equitable social policy.

Dahlgren and Whitehead (2006) note that although all European governments make statements about the need for justice, equity and solidarity in their health care systems, in practice the actual experience of low-income households does not reflect these commitments. They note that since the beginning of the 1990s inequities in access to health services and medicines have generally increased in central and eastern, and even in some in western European countries. In many poor countries the imposition of neo-liberal policies has meant that public health service infrastructures have been run down and health service access has grown worse (discussed in chapter 5). One example from the International Monetary Fund is Colombia, which followed the World Bank blueprints very closely, experienced very substantial increases in health care expenditures, had a large percentage of the population remaining uncovered, had high co-payments and no measurable efficiency or medical care quality improvements. As a result, public health care deteriorated and health equity suffered (Homedes and Ugalde, 2005, p. 92).

Although the main determinants of health and health inequities are chiefly outside the health care sector, there is still room for action within the health sector. In Australia the rhetoric of addressing inequities in health status has been accepted by many health departments, but seldom translated into concrete action. There has been a retreat from the language of social justice since the 1980s and that of the market seems to dominate public discourse. Policy statements endorsing action against inequities can be powerful supports for services, which can use them to claim legitimacy for their social justice work.

There may be some frustration among health bureaucrats and service providers that they can do little to reduce inequities, given that the crucial factors are outside the health sector. This is reinforced by the tendency for social factors to be regarded as epidemiologically fixed and unchangeable. Health policy makers need to be reminded that these factors are not inflexible—because they are socially created, they are amenable to change through social and political actions. The task of health departments is to look for opportunities to complement services with actions that may do something to change the factors underlying inequities. Box 18.2 outlines the characteristics of a health equity–oriented health care sector, showing that it requires leadership and willingness to work with other sectors (see chapter 24). Primary health care should be a key feature of a health system designed to promote equity in population health outcomes.

BOX 18.2   CHARACTERISTICS OF A HEALTH EQUITY–ORIENTED HEALTH CARE SECTOR

LEADERSHIP: IMPROVING THE EQUITY PERFORMANCE OF THE HEALTH CARE SYSTEM

1   Focus on comprehensive primary health care
2   Decision-making processes that involve local communities
3   Universally accessible care that is publicly funded, preferably through general taxation; good-quality health services that are free at point of use
4   Planning, including allocation of resources, based on the needs of populations within a social determinants of health framework
5   Policy statements and strategies that are explicit about closing the health equity gap and the need for action on the social determinants of health to achieve this goal in all programs, including those that are disease focused
6   Evidence that the health care system has a systematic approach to increasing its spending on community-based services until a significant proportion of funding is devoted to community-based care, and evidence that it has reformed its financing system so that it rewards keeping people healthy through preventive action rather than throughput of clinical cases

STEWARDSHIP: WORKING WITH OTHER SECTORS TO IMPROVE HEALTH AND HEALTH EQUITY

1   Presence of health sector advocacy program with other sectors regarding the need for action on the social determinants of health and the importance of intersectoral action
2   Development of expertise to establish a health equity surveillance system and to conduct cross-government health equity impact assessments on a regular basis along with private-sector activities
3   Reform of medical and health professional education so that the importance of social determinants is reinforced in theory teaching, clinical training, understanding of population health perspectives, and skill development for interprofessional collaboration
4   Training and education of professionals (including urban and transport planners, teachers, and architects) on the importance of social determinants of health
5   Specified, funded program of research on the impact of social determinants of health and evaluation interventions designed to address them, with a significant proportion of health research funding devoted to studies on social determinants

Source: Baum et al., 2009, p. 1970.

COMPREHENSIVE PRIMARY HEALTH CARE

WHO (1978; 2008) endorses comprehensive primary health care as the basis of good health systems. Primary health care services are particularly effective at working with disempowered, poor communities by using community development strategies in addition to clinical work. They try to change the conditions that create inequities by, for example, working with local environmental action groups or with public housing tenants, providing nutrition education and advice that is sensitive to the constraints imposed by poverty, supporting indigenous people's health action groups and advocacy groups for a variety of people with particular needs, including victims of domestic violence, refugees, outworkers and women from non-English-speaking backgrounds. In Australia it has been noted that community health services have been far more creative and effective at integrating equity considerations into their work than have other parts of the health system (Baum et al., 1992). While GPs have an important role in disease prevention, they are just one part of the primary health care landscape.

In chapter 3 we saw that a comprehensive approach to primary health care includes medical care but also pays significant attention to disease prevention and health promotion. It is also based on multidisciplinary teamwork and uses a range of strategies, from clinical work with individuals, to the provision of health promotion and support groups, and community development work with the broader community. Australia has had a number of good examples of comprehensive primary health care in the past, although these have never received sufficient investment to make them a viable mainstream sector. These services have been greatly eroded in the past decade through a series of reforms that has seen most of the community health services become more like hospital outreach programs (Baum et al., 2013c). In Victoria more comprehensive services have survived and the Aboriginal community controlled sector also embraces the principles of comprehensive primary health care (see box 18.3). Canada retains a vibrant community health sector with a national organisation to support its work (see www.cachc.ca).

BOX 18.3   ABORIGINAL COMMUNITY CONTROLLED HEALTH SERVICES: AN EMPOWERING MODEL

The first community controlled Aboriginal health service was established in Redfern, Sydney, in 1971. Despite initial reluctance by governments to fund such services, by 2014 there were more than 150 Aboriginal community controlled health services (ACCHSs) operating across Australia in all states and territories. A full history is available at www.naccho.org.au/about-us/naccho-history. The National Aboriginal Community Controlled Health Organisation (NACCHO) defines an Aboriginal community controlled health service as an incorporated Aboriginal organisation that is:

- initiated by a local Aboriginal community
- based in a local Aboriginal community

- governed by an Aboriginal body that is elected by the local Aboriginal community
- committed to delivering a holistic and culturally appropriate health service to the community that controls it.

Panaretto et al. (2014) and the Health is Life Inquiry into Indigenous Health (House of Representatives, 2000, p. 38) detail the benefits that a properly resourced community controlled health service can deliver, including:

- providing evidence that they are improving health outcomes for individuals
- significantly improving access—because the local community has ownership and control of the service, and because service delivery is flexible and responsive
- making the full range of primary health care services available in one place, including prevention, early intervention and comprehensive care, with service delivery being integrated and holistic
- providing culturally appropriate care
- overcoming unintentional racism
- delivering value for money, as services can be better targeted because they are based on local knowledge
- providing a major source of education and training for Aboriginal people
- employing Aboriginal people and so developing the Aboriginal workforce
- supporting a pool of knowledge and expertise about Aboriginal health that enables the sector to not only deliver appropriate care but also to advocate effectively for Aboriginal people in health.

It has been noted that the ACCHSs play a significant role in delivering curative services, community development and health promotion to Aboriginal peoples, despite most of them being significantly underresourced. This aspect was summed up by the late Puggy Hunter:

> thing is that we own the bloody thing and it is something that we can't, I can't, explain— about the ownership and the pride that it actually brings' (Hunter et al., 2005, p. 339)

The control that ACCHSs provide to Aboriginal communities is hard to quantify but it is almost certainly important to health. In a submission to the Health is Life Inquiry, Scrimgeour (House of Representatives Standing Committee on Family and Community Affairs, 2000, pp. 41–2) raised a number of issues that affected the operation of community controlled health services. He noted that in towns and cities the services were established to provide an alternative to the mainstream services. In remote locations the community-controlled service is often the only service, so it is not an alternative. He also notes that the quality of non-Aboriginal staff employed in these services is crucial. For community control to be real and effective there needs to be significant capacity building in Aboriginal communities and sufficient resources committed to the services.

A study that compared state-funded primary health care services with the Central Australian Aboriginal Congress found that the services offered by the congress were more comprehensive and better able to address the social determinants of health (Baum et al., 2013c).

Health worker Gladys Wamati at the clinic in Ramingining, Northern Territory.
(Penny Tweedle, Panos)

CONCLUSION

Some medical approaches such as immunisation have achieved some spectacular results, including the global eradication of smallpox, substantial moves towards the eradication of poliomyelitis, and the general reduction of communicable diseases. Medical interventions promise to be most successful as one part of a population approach based on a socio-environmental view of health, and as part of a more general comprehensive primary health care approach. Health systems have an important role in reducing inequities by ensuring equitable access and promoting health in a way that reaches all sections of the community. Comprehensive primary health care offers a strong model for community-based care.

CRITICAL REFLECTION QUESTIONS

18.1   Why did the MRFIT program described in box 18.1 not succeed?

18.2   What are some of the arguments of the anti-vaccine movement? Examine and critique these arguments.

18.3   What factors will improve the equity performance of the health sector?

18.4   Why are Aboriginal community controlled health services likely to be particularly effective forms of primary health care?

Recommended reading

Johnson and Paton (2007) *Health Promotion and Health Services* provides a guide to the processes of change management relevant to all health services.

WHO (2008) *Primary health care: Now More Than Ever* presents clear arguments for the value of primary health care.

Useful websites

Canadian Association of Community Health Centres www.cachc.ca works to expand access to community health services and has an excellent advocacy site.

National Aboriginal Community Controlled Health Organisation www.naccho.org.au represents more than 150 Aboriginal community controlled health services and works to further Aboriginal self-determination.

WHO has a portion of its website devoted to primary health care www.who.int/topics/primary_health_care/en

The World Organization of Family Doctors (WONCA) www.globalfamilydoctor.com website contains lots of information on international family medical practice.

19

CHANGING BEHAVIOUR: THE LIMITS OF BEHAVIOURISM AND SOME ALTERNATIVES

Although the risks and contradictions of life go on being as socially produced as ever, the duty and necessity of coping with them has been delegated to our individual selves.

Zygmunt Bauman, 2007, p. 14

KEY CONCEPTS

Introduction

Social learning theory

Health belief model

Theory of reasoned action

Stages of change model

Health action model

Application of behavioural theories

Second generation of heart health campaigns

Social marketing

Mass media campaigns

Health education through entertainment

Using social media

Criticisms of social marketing

Relational, mindful and positive: other approaches to health promotion for individuals

Conclusion

INTRODUCTION

Ultimately, improving the health of populations may require changes in behaviour. However, one of the crucial premises of the new public health is that behaviours are socially structured and so it follows that changing these behaviours requires changing the structures within which behaviours occur. Despite the acceptance of this within the new public health, behaviourism is still often promoted as a solution to 'lifestyle' diseases while paying little attention to structures. Baum and Fisher (2014) note that this reflects the dominance of neo-liberal ideology, which stresses individualism (see chapter 5) and which has seen lifestyle change programs receive considerable acceptance despite the lack of evidence for their effectiveness. They also see that there is a strong inherent logic to behaviour change strategies. If the problem of smoking is seen as one of people choosing to smoke and obesity as one of people over-eating, then telling them not to do so seems to make sense. Powerful tobacco and food corporations also influence the agenda of the World Health Organization and national governments to persuade them to focus on individual factors rather than systematic issues with the sale and marketing of unhealthy foods. Consequently it is important for new public health advocates to understand the evidence on the limitations of behaviour change in order to argue effectively for more structural change and to design means of changing behaviour that focus on creating supportive environments and which do not focus on individual blame.

THEORIES UNDERPINNING BEHAVIOUR CHANGE APPROACHES

A number of overlapping theories, most of which stem from social psychology, form the basis of behavioural approaches to health promotion. They attempt to explain the influence of different variables on an individual's health behaviour and are concerned with attitudes, beliefs, motivations, values and instincts. Early models of behaviour change were based on the assumption of a relatively stable link between knowledge, attitude and behaviour—if people were given relevant information (that is, too much fat is bad for your health) from a credible source (nutritionist) they would change their attitudes towards their diet and, in turn, their behaviour (reducing fat intake). Experience showed that this was not correct, and so psychologists developed more sophisticated models of behaviour determinants and change. Some of these are described in this chapter. Nutbeam (2006, p. 25) notes that most of these theories have not been rigorously tested when compared with theories in the physical sciences and suggests that they might be more accurately termed 'models'. Their other major limitation is that these models pay scant attention to the social, economic and cultural environments in which people's behaviours occur. This critique is developed in the later stages of the chapter.

SOCIAL LEARNING THEORY

Bandura (1977) was the main proponent of this theory, which argues that most learning occurs by modelling rather than trial and error and that the more positive the consequences of a behaviour change, the more likely people are to engage in it. The theory differentiates between people's beliefs in the outcome (giving up a high fat diet) and their ability to perform the behaviour (self-efficacy). Their behaviour is likely to be strongly influenced by their confidence in their ability to change, and personal behaviour can be learned and unlearned through influences in the family, community, work and the media. Lefebvre and Flora (1988) describe the model as being put into practice in the following way:

- promotion and motivation to interest people in changing a particular behaviour
- skills training to provide people with specific behaviour-changing skills
- the development of support networks so new behaviour can be maintained
- maintenance of behaviour through reinforcement.

The concept of locus of control has also been associated with social learning theory, and in the context of health can be understood in terms of: internal locus of control where people believe they are responsible for their own health, and external locus of control where people see their health as being influenced primarily by outside forces such as other people and chance, fate or luck.

Self-efficacy, which refers to individuals' beliefs about their capacity to perform specific behaviours in particular situations, is crucial to behaviour change, according to Bandura. Stretcher et al. (1986) reviewed programs designed to change behaviour related to smoking, weight control, contraception, alcohol abuse and exercise, and found a constant positive relationship between self-efficacy and health behaviour change and maintenance. They issued a warning that some traditional methods of behaviour change that do not incorporate self-efficacy may diminish rather than enhance efficacy. The ways in which people's social status may affect self-esteem and belief in their ability to control events may be crucial to the concept of self-efficacy, but are rarely considered by behavioural research.

HEALTH BELIEF MODEL

The health belief model (Becker, 1974) was developed specifically to explain health-related behaviour. It is based on the belief that when people consider changing behaviour they do a cost-benefit analysis, which includes:

- the likelihood of the illness or injury happening to them (susceptibility)
- the severity of the illness or injury
- the likely effect of the behaviour change (efficacy)
- whether it will have some personal benefit.

The revised model (Becker and Rosenstock, 1987) added the following two points:

- an assessment of sufficient motivation to make health issues salient or relevant
- the belief that change following a health recommendation will be beneficial to the individual, taking into account the cost involved.

Thus, individuals may be more likely to stop smoking if they are aware of the health consequences and think they are vulnerable to, say, lung cancer. Connected with their risk assessment is their belief in the cessation of smoking benefiting their health and whether it will have any other benefits. ('Kiss a non-smoker—taste the difference' was based on this notion.) However, the individual may decide that the long-term benefits of giving up smoking are not worth the short-term problems of nicotine withdrawal and missing the pleasure of smoking. Outside forces (including the health warnings on cigarette packets) may motivate or maintain behavioural change. The health belief model maintains that 'cues' to behaviour change are important. The health belief model has been most useful when applied to relatively straightforward actions such as encouraging screening and immunisation (Nutbeam, 2006). It has been less effective in long-term, complex and socially determined behaviour changes.

THEORY OF REASONED ACTION

Ajzen and Fishbein's (1980) model maintains that behaviour is governed by intention and that personal attitudinal and social normative factors determine behavioural intentions. Each personal attitude is made up of a belief (for example, too little exercise is bad for you) and people may have a number of conflicting attitudes towards a certain behaviour. The social normative influence on behaviour refers to the individual's perception of what important others will think of their behaving in certain ways. These two major influences combine to form an 'intention' to behave in a particular way and this intention is predictive of the behaviour. So the link between attitude and behaviour is mediated by beliefs and perceptions of normative expectations.

These mediating factors explain why people do not always behave in accordance with their expressed attitudes. For example, a young person may understand the risks associated with becoming a heroin user but interaction with a peer group who use heroin may interfere with a previous intention of not using the drug. So this theory emphasises individuals' motivation to conform with significant others.

STAGES OF CHANGE MODEL

People do not usually change their behaviour suddenly, completely and permanently. Prochaska and DiClemente's (1984) behavioural theory is important as it shows that the changes people make are only part of an ongoing process. This model suggests that people cycle and relapse through five distinct stages:

1. precontemplation with no intention to change behaviour
2. contemplation and making a decision about whether or not to change
3. preparation for changing behaviour in the near future, having experimented with behaviour change in the past
4. action, successfully changing behaviour over a relatively short time
5. maintenance, successfully changing behaviour over a lengthy time.

Few people go through these stages sequentially, typically going backwards and forwards (Prochaska and DiClemente, 1992). Identifying the precontemplative stage is important, as it can remind health workers that change is not likely, and they can focus their attention on other issues, such as minimising the risk associated with a behaviour.

So, for instance, they might suggest the use of a needle-exchange scheme to ensure clean needles (Naidoo and Wills, 1994). For people in the action stage, coping skills may be most important. Prochaska and DiClemente's (1984) work was developed through encouraging people to change addictive behaviours, but it also informed health communication in smoking cessation, dietary habits, mammography, pregnancy and HIV prevention (Maibach and Holtgrave, 1995). This model has been important in encouraging health promoters not to assume that an intervention will be equally applicable to all (Nutbeam, 2006) and to tailor programs to the range of needs in a population, recognising that these may change, and the need to sequence interventions to match different stages of change.

HEALTH ACTION MODEL

The Health Action Model (HAM) (Tones, 1992) posits that the belief, motivation and normative systems all influence the intention to act. Certain facilitating factors (including any necessary knowledge or skills) also need to be present before the action intention is translated into health action. Environmental circumstances must also be favourable if the healthy choice is to be taken. It is this aspect of the HAM that really distinguishes it from others that draw exclusively on social psychology models. Tones (1992, p. 43) says: 'Environmental facilitators include relatively specific factors (such as ready access to condoms). They also include macro-influences (such as poverty) which directly depowers through lack of material resources and indirectly depowers through its alienating effect.'

Intrinsic to the HAM is the two-way interaction between motivation and belief systems. Beliefs about a particular health action (for example, that condom use will reduce the likelihood of HIV infection) will be assessed in the context of an individual's values (for example, the morality of sex outside marriage). Intention to act will depend on the relative strength of these two motivators. An environmental inhibitor could be the cost of condoms or the embarrassment of buying them. The HAM model also considers emotional states, whether instinctive (hunger or fear), acquired (addiction) or derived (anxiety or guilt). Self-esteem is a central factor in the HAM. Tones (1992, p. 42) sees two interrelated factors as important in determining self-esteem:

- the reaction of significant others to the individual
- success in achieving goals that are valued by the individual and the relevant social group.

Beliefs about competence and control are central to self-esteem. High self-esteem is considered to be healthy for the following reasons (Tones, 1992, p. 47):

- it represents a significant feature of mental health (so long as it bears some relation to reality)
- the higher people value themselves, the more likely they are to take care of their health
- people with high self-esteem are less likely to succumb to pressures to conform
- self-esteem is related to the development of better coping skills.

APPLICATION OF BEHAVIOURAL THEORIES

COMMUNITY HEART HEALTH PROGRAMS

Community-wide campaigns to encourage people to adopt healthier lifestyles came into vogue in the 1970s and remained popular through the 1980s. The programs started from the recognition that cardiovascular disease (CVD) was the main cause of death in industrialised countries in the second half of the twentieth century. The community health heart programs focused on improving the health status of entire communities by controlling modifiable risk factors for CVD, including high blood pressure, elevated serum cholesterol, smoking, being overweight and sedentary lifestyle. Most programs are North American, and typical elements are summarised in box 19.1.

BOX 19.1   TYPICAL COMPOSITION OF COMMUNITY HEART HEALTH PROGRAMS

Most community heart health programs include the following range of activities:

- mobilisation of the community (especially community leaders) to contribute time, money and effort to achieving the goals of the program
- social marketing, which usually involves the broadcast and print media
- direct behaviour-change efforts, including the development of cooking skills, weight loss programs and quit smoking programs. Motivation at community level has been built into some programs, including competition between worksites to achieve the most weight loss or smoking cessation (Elder et al., 1993)
- screening to identify asymptomatic people at high risk for heart disease
- environmental intervention to encourage healthier lifestyles have also featured, but they take a definite second place to the behavioural interventions at the heart of the projects. Examples of environmental intervention are the North Karelia work with food producers and distributors to encourage the provision of low-fat products, the provision of bicycle lanes, bike racks, jogging tracks and healthy food choices in restaurants and workplace cafeterias.

EVALUATING COMMUNITY HEART HEALTH PROGRAMS

The community heart health programs put much emphasis on evaluation, which often absorbs half the total budget (Mittlemark et al., 1993). The studies, their strategies and evaluation outcomes are summarised in table 19.1.

TABLE 19.1   OVERVIEW OF SELECTED CLASSIC HEART HEALTH PROGRAMS AND THE STRATEGIES USED

Heart health	Strategies used	Evaluation outcomes (note: methodological limitations of such studies are discussed in text)
North Karelia Project, Finland (Puska et al., 1985). Started 1972. Two counties: one intervention, one reference. Area of high CHD mortality (n = 433 000)	Integrated community-wide approach which included the mass media, leaflets, stickers, the development of a schools program, use of volunteers as lay educators and role models in the community and the production of low-fat foods.	Risk behaviours (e.g. fat consumption and smoking), mean serum cholesterol and blood pressure and death rate from CHD declined more in North Karelia than the rest of Finland (reduced by 24% in North Karelia compared to 12% in the reference area and 11% in the rest of Finland). The results were more positive for men than women.
Stanford Three Community Study (USA) (Farquhar et al., 1977). Three towns: two interventions, one reference (n = 45 000)	One community: intensive mass media campaign. Second community: mass media, screening and face-to-face health education for high-risk people.	Both intervention communities showed an increased knowledge of the risk factors associated with heart disease. Behaviour change was greatest in the community receiving screening and health education.
Stanford Five City Project (Farquhar, 1984). Five cities: two interventions, three reference. Built on the results of study above (n = 350 000)	The program had two basic intervention strategies: a multimedia education program in one community; and a similar program supplemented by an intensive instruction program for high-risk individuals in a second community.	Intervention communities showed significant increase in knowledge in the treatment communities and decreases in cholesterol and blood pressure levels. The changes in risk factors were modest. Morbidity and mortality results not reported yet.
Pawtucket Heart Health Program (Elder, 1986)	Social marketing model, focusing on individual behaviour change. 'Products' to be marketed through 'channels' such as worksites, religious organisations, mass media etc. Creation of supportive social networks at home and in workplaces.	Formative process-focused evaluation noting levels of involvement in community activity. Some evidence that risk factor change was achieved through specific interventions. No morbidity or mortality data.

Heart health	Strategies used	Evaluation outcomes (note: methodological limitations of such studies are discussed in text)
Minnesota Heart Health Program (USA) (Blackburn et al., 1984). Two towns, two cities, two suburbs: paired intervention and reference (n = 356 000)	Mass media campaign, risk factor screening centre with a direct education component. Wide variety of programs aimed at different target groups, schools and community organisations. Education covered smoking, exercise, nutrition, improvement of health professionals' preventive practice. Food labelling protocols. Some emphasis on policy changes.	Comparison of risk factor data shows only minor difference for control and intervention communities. Morbidity and mortality data not reported yet.
New South Wales North Coast Project (Egger et al., 1983). Three towns: two intervention, one reference	Three communities; one control, one media campaigns only, one media plus community-based programs. Intervention aimed to change smoking, dietary fat intake and exercise behaviours over three years.	Smoking prevalence declined in all communities but to a greater degree in the two test towns. Some evidence that the decline was not sustained in the media-only town (Egger et al., 1983). The greatest increase in participation in regular exercise occurred in the town used as a control (Monaem et al., 1985).
Heartbeat Wales (Smith et al., 1994)	Efforts to bring about structural change, e.g. restrictions on smoking in public places, better food labelling, provision of healthy food in shops and restaurants. Mass media.	Only one evaluation results paper published. Based on longitudinal survey data (no control). This shows that smoking prevalence was reduced and there was a trend towards healthier eating but there was little change in exercise or alcohol consumption levels.

These evaluations were complex and failed to produce particularly conclusive findings. Methodological difficulties abounded, and have been summarised by Altman (1986) as follows:

- Most of the evaluations, especially the early ones, concentrated on black box evaluation whereby the focus was on the question 'Did the program work?' rather than on why it worked. This also meant that it was difficult to determine the impact of various aspects of the programs.
- Longitudinal evaluation is necessary to track the progress of projects over time, yet it is extremely difficult to design a study for community-based interventions. Attrition and migration of people into the communities, who are then surveyed without having had much exposure to the intervention, are typical of the problems faced. In the Stanford Three Community Study only 56 per cent of participants completed all three surveys.
- Determining what change can be attributed to the project itself, rather than other influences on population health, is very difficult. Monitoring the specific interventions that influence health outcomes is difficult, let alone trying to determine how these health outcomes came about. The evaluation of the North Karelia Project was criticised for not being able to distinguish whether effects were a result of the program, external forces or existing national trends (Klos and Rosenstock, 1982). The Heartbeat Wales Project found that their control community also received health promotion interventions relevant to heart health (Nutbeam et al., 1993a). There was no way these could be controlled by the evaluators. Although Altman describes some attempts to overcome these methodological problems, he concludes that controlling for extraneous variables over a lengthy period is difficult. Mittlemark et al. (1993, p. 451), after their review of community-based cardiovascular disease prevention projects, concluded that 'ultimately it may be impossible to separate the relative effects of community-based programs, national heart campaigns, mass media and other sources of health improvement information.'
- Very few of the projects conducted in the 1970s and 1980s included any qualitative data, tending instead to depend on physiological outcome data only. No data were collected on the effects of community structure on the programs. Altman (1986) notes that the lack of information about the social system in which the programs were implemented limits understanding of the evaluation as the rules, values and norms of the community are likely to be important in the program's success or otherwise.
- Altman points out that an important aim of many of the large-scale community programs was to assess their generalisability to other communities. Often this was not realistic because of the costs of the program and the fact that particular effects of different parts of the programs were not evaluated, so other communities were not able to select those parts of a program that proved effective. In addition the assumptions underlying the goal of generalisability may not be proven. Altman (1986, p. 485) points out that three of the large US programs (Stanford, Minnesota, Rhode Island) differed because of the nature of the communities they were in: 'In Rhode Island the community is predominantly blue-collar and heavily affected by the national and local economy. In contrast the Minnesota communities consist primarily of white, middle-income people living in relatively stable economic conditions. The Stanford

communities are heterogeneous, composed of agriculture workers, military, tourist industry workers and middle-income people and about a third are Hispanic.'

This makes it clear that the assumption of generalisability cannot be made. An additional problem is that the conditions of the original trials (which normally involve considerable resources) are unlikely to be repeated routinely. Nutbeam et al. (1993a) evaluated two large UK school smoking-education programs based on Minnesota smoking-prevention programs, and a similar Norwegian program. None was found to be successful. The evaluators believe that the original Minnesota smoking-prevention program may have been successful as a result of the experimental classroom conditions under which it was taught, and not easily transferable to normal classroom conditions.

One of the noticeable features of the research on these classic heart health programs is that they typically pay little attention to issues of equity. The evaluations present little data about the differential effects on people in different social circumstances. Obviously such data would be vitally important if the new public health is to ensure that the programs were not simply making the healthy healthier.

RATIONALE FOR COMMUNITY-WIDE PROGRAMS

The striking feature of the community-based large-scale programs is that they use mass media to mobilise and coordinate community resources to promote and support behaviour change. These programs were innovative when they were first developed as health promotion had hitherto been seen only as a clinical activity involving education to individuals, or less commonly, to groups. The rationale for community-wide projects as opposed to individual or group programs was put by Chapman (1985) in his assessment of stop-smoking clinics. He points out that a 5 per cent success rate among 10 000 people is 333 times more efficient than the 30 per cent success rate achieved by groups involving only 50 people, and concludes that clinics make an insignificant contribution to the overall community smoking rate, whereas population-based health promotion programs have the potential to bring about population-wide changes that are far more significant.

THE NORTH KARELIA PROGRAM

The North Karelia program to prevent coronary heart disease (CHD) in a region of central Finland was one of the first community-wide programs based on behavioural change theories, being developed at about the same time as the Stanford Three Community Study. The two research teams developed mutually beneficial scientific exchanges (Puska et al., 1985). The Finnish project's evaluation reported greater changes in CHD risk factors than for the Finnish population as a whole, and in CHD male mortality (Puska et al., 2009). Interestingly, the project's evaluators note that the relative success in North Karelia was not mainly a result of individual increase in health knowledge or changes in health-related knowledge, but rather that 'broad-ranged community organisation—including provision of primary health care services and involvement of various other community organisations—was of central importance. The project was able to disseminate its message through media and opinion leaders so that it created a social atmosphere more favourable to change' (Puska et al., 1985, p. 185). A major conclusion from a retrospective review of the North Karelia project stressed that it had to be flexible and responsive to changing community circumstances (Puska et al., 2009)

It may be significant that the North Karelia Project did not start at the behest of researchers but as a result of a petition signed by community representatives, asking for help in reducing the high morbidity and mortality from ischaemic heart disease in the county. Finland had the highest IHD death rate of all developed countries, and North Karelia the highest in Finland.

LESSONS FROM THE FIRST GENERATIONS OF HEART HEALTH PROGRAMS

The evaluations of the community heart health programs paid little attention to issues of equity, rarely reporting data on the differential impact on different groups within their populations. The first generation of these programs was not easily translated to other settings. They had large budgets and placed limited emphasis on low income, inner city or minority populations (Elder et al., 1993).

Elder et al. (1993) reviewed a range of North American heart health programs and concluded that the following elements helped to achieve their success:

- Community participation in the planning, design and evaluation of interventions helps the community adopt the intervention.
- All aspects of the planning and intervention should be data driven.
- Feedback to the community is essential.
- Primary prevention should be given priority over secondary prevention or treatment.
- Population-wide change should be the main aim.
- Multiple strategies that address multiple risk factors promoted in a range of different ways are more effective than narrowly focused interventions.
- Policy and environmental interventions are often more effective and preferred to direct behaviour change efforts.
- Community capacities to develop, implement and sustain interventions should be a priority.
- State health departments have an important role in facilitating, sustaining and disseminating the efforts from heart health programs.

These messages are very different to those that were being drawn from these programs a decade earlier, when the emphasis was still firmly on individual behaviour modification. By the 1990s the importance of interventions with multiple strategies supported by community organisation and participation was recognised as crucial if any success was to be likely.

SECOND GENERATION OF HEART HEALTH CAMPAIGNS

The lessons from the first generation of heart health campaigns have certainly been taken on board by health promoters. In the early twenty-first century health programs incorporate many of the lessons listed in the previous section. This is best shown by examples. The Canadian Heart Health Initiative operated in nine of the ten Canadian provinces from 1986 to 2006. Over that time 311 projects were instituted with a total expenditure of $36 million, of which 62 per cent was spent on modifiable risk behaviours (Rocan, 2009). The dissemination phase monitored the dissemination of heart health initiatives in the provinces (O'Loughlin et al., 2001). What is striking about the description of the activities in each province is that the focus of the heart health work had shifted from achieving behaviour change in individuals to achieving change in the capacity of provincial health structures to implement heart health strategies. It is also striking that the work was grounded in a '"socio-ecological" approach to health promotion, a perspective that maintains that improvement in the health of populations depends largely on changing environments in ways that promote, extend and sustain health behaviours among individuals.' Thus changing organisational and community environments to enable the creation and sustainability of capacity for health promotion was an underlying theme of each province's activities. This is a very significant departure from the individual behaviour change focus that typified the first generation of heart health campaigns. However, a retrospective assessment of the initiative concluded that it had ended prematurely before achieving its goals and suffered from a number of implementation problems associated with leadership, the problems of multilevel collaboration and lack of institutional support from Health Canada (Rocan, 2009).

Change can also be seen in Australia, where the Heart Foundations have changed from a total focus on individual behaviour as the centre of their health promotion activity to at least acknowledge approaches that consider the ways in which urban environments can support people in undertaking physical exercise. This includes considering the safety of suburbs in terms of lighting, recommending guidelines for urban planning that suggest such measures as designing suburbs that have destinations to walk to and encouraging group walking activities that also include a social element (MacDougall et al., 2002). VicHealth completed similar work until 2006 which considered transport designed to encourage exercise and urban planning to encourage walking, cycling and other exercise options (see www.health.vic.gov.au/localgov/downloads/enviro_ practical_guide.pdf). The Health in All Policies initiative in South Australia has also considered how transport-oriented development can be made health promoting (Baum et al., 2014b). A recent issue of the *Health Promotion Journal of Australia* (Rissel and McCue, 2014) contained a number of articles demonstrating the impact of the built environment and active transport on health.

SOCIAL MARKETING

Social marketing applies marketing techniques to social psychology theories in order to bring about population-wide behaviour change. The most commonly used technique is mass media campaigns (see below), which borrows heavily from traditional marketing and the 'four Ps of marketing'—product, price, place and promotion (Egger et al., 1990). Maibach and Holtgrave (1995) offer the following definition of social marketing:

- a disciplined approach to public health intervention whereby research and management strategies are used to pursue clearly stated objectives in ways that often include mass media
- aimed at well-defined (i.e. segmented) audiences who have been carefully profiled demographically, behaviourally, psycholographically and media-graphically
- offers a set of products and messages that are responsive to consumers' wants and needs, and refined as needed.

The sequences of a social marketing campaign are (Egger et al., 1990, p. 68):

1 defining the target audience
2 developing a concept for intervention
3 developing a message based on the concept
4 testing the message
5 running the message
6 evaluating the message
7 evaluating the outcome.

Recent developments in social marketing have involved the use of marketing data and stage-of-behaviour change models to define specific, relatively homogeneous audiences, and behavioural theories to fit health messages to specific needs. Social marketing is now also able to use social media extensively, with organisations establishing websites and Facebook pages and using Twitter to spread health promotion messages.

MASS MEDIA CAMPAIGNS

The past four decades have witnessed a burgeoning of mass media campaigns aimed at persuading people to change their lifestyle and behaviour to be more health promoting. Mass communication has become a major strategy used by health promoters. Yet the value of mass media campaigns in contributing to behaviour change is much disputed (Naidoo and Wills, 2001, p. 281; Baum, 2011) and only moderate impact is found in most cases (Wakefield et al., 2010). Simple awareness is relatively easy to achieve; to inform or reinforce an attitude is more difficult, and to change behaviour is even more so. The advantages of mass media campaigns are that they can reach large numbers of people, which means they fit with Rose's (1992) famous dictum (see chapter 1) that a whole population strategy will result in the prevention of more disease because a lot of people at slight risk of a disease account for more of the disease than do a few people at greater risk.

Mass media campaigns (along with increased taxation and smoke-free policies) have contributed to more Australians quitting smoking and fewer people taking up the habit, so the prevalence is in decline (Wakefield et al., 2010; Davoren et al., 2014). Australia's first National Tobacco Campaign was launched in 1997, and between 1997 and 2000 six advertisements intended to reduce smoking behaviour involving radio, billboards and a campaign website were launched (Scollo and Winstanley, 2012). Surveys of the campaign's impact showed a decline in smoking prevalence among Australian adults over the period of the campaign: from 23.5 per cent in 1997 to 20.4 per cent in 2000 (Wakefield et al., 2004). Hurley and Matthews (2008) also estimated about 55 000 deaths were prevented, with potential health care savings of $740.6 million. The campaign advertisements have since been adapted in other countries (Scollo and Winstanley, 2012). A recent review of mass media campaigns (Davoren et al., 2014) reported 'negative health effects messages' (p. 127) were most effective on quitting behaviour and in reducing 'socioeconomic disparities in smoking' (p. 127).

Road safety has been an important target of media advertising in Australia. A systematic review of eight studies on the effectiveness of mass media campaigns designed to reduce alcohol-related crashes estimated the average decline in crashes across the studies was 13 per cent (Elder et al., 2004). Over the period 2000 to 2004, the Victorian Government implemented a package of speed-related initiatives, including the 'Wipe Off 5' campaign, which comprised mass media advertising (television, radio and billboard) and aimed to dispel the 'myth that travelling even a few kilometres over the legal limit is safe' (www.wipeoff5.com.au). An evaluation found a 3.8 per cent reduction in casualty crashes associated with the implementation of the speed-related package (D'Elia et al., 2007).

Another example of the effective use of mass media in health promotion is the Anti-Cancer Council of Victoria's Slip! Slop! Slap! program (1980–88) and the subsequent SunSmart program including the recent targeting of pro-tanning attitudes in Australia in the 'Dark Side of Tanning' campaign. These campaigns responded to Australia's high incidence of skin cancer. They were based on a mass media campaign but also used a range of other strategies designed to create a supportive environment for the changes advocated in the mass media to reduce sun exposure (Montague et al., 2001). Other strategies included knowledge dissemination to key groups such as health care workers, teachers, participants in sports, arts and recreational activities and lobbying for structural changes such as the provision of shade by local government, changing outdoor work policies so that workers use sun-protection measures, and encouraging the fashion industry to put less emphasis on tans and support the manufacture of SunSmart projects such as swimsuits that provide whole body coverage. These various measures have resulted in marked reductions in sun exposure, and skin cancer incidence rates are beginning to plateau after years of increase. In younger age groups the incidence of melanoma has declined (Hill and Marks, 2008).

Mass media campaigns are most effective when used as part of comprehensive approaches to improving health behaviours, such as the approach to smoking, which also included substantive public policy measures using supply reduction and taxation (Wakefield et al., 2010). This was also the case in the successful HIV campaign in Australia from the 1980s, which accounts for Australia's success in curtailing the spread of the virus. The National Advisory Committee on AIDS (NACAIDS) developed a range of strategies that involved community groups (especially of sex workers and men who have sex with men). In 1987 they launched a confronting advertising campaign based on images of the Grim Reaper bowling down people with the line 'prevention is the only cure we've got'. This campaign was not effective so much because it persuaded people to change their behaviour but because it reinforced community development strategies and worked to convince policy makers that resources had to be committed to HIV prevention (Sendziuk, 2002).

A joint Australian, State and Territory Government initiative
under the National Partnership Agreement on Preventive Health

'Swap It Don't Stop It' was an Australian Government campaign designed to encourage healthy lifestyles but met with little success and was widely criticised (see Baum 2011).

HEALTH EDUCATION THROUGH ENTERTAINMENT

Public health messages have also been embedded within entertainment. This is predicated on the assumption that entertainment will attract more people than education messages, that people will understand and be receptive to educational messages within entertainment and that the heightened audience size, attention and receptivity can influence cognitive, affective and behavioural outcomes that underlie many public health problems (Maibach and Holtgrave, 1995, p. 228). Typically these techniques will use behavioural modelling. The Johns Hopkins Health Institutions in the USA developed two nationally syndicated health information series designed to pick up on issues in episodes of the popular medical drama series *ER* and *Chicago Hope*. These short programs were screened immediately after the drama episodes and gained national audiences of around four million viewers. The programs were linked to interactive websites and phone advice lines (Langlieb et al., 1999). An evaluation study found that the drama tie-in enhanced the attention and satisfaction of viewers.

Soul City, a South African initiative, has used a soap opera to tackle controversial issues such as condom use and domestic violence, supported by other interventions including community education and print material. The most recent evaluation (Soul City Institute, 2007) found that Soul City had been successful in increasing condom use (21 per cent increase), increasing testing for HIV (5–8 per cent increase), reducing stigma (2–8 per cent reduction) and increasing willingness to care for someone with HIV (19 per cent increase).

South African health promotion body LoveLife produces a series of posters with messages designed to encourage young people to think about their sexual behaviour. (Fran Baum)

USING SOCIAL MEDIA

Social media is now used widely to deliver health promotion messages. The online world is an important part of people's lives. Beyondblue started as a national campaign to reduce the incidence of depression and reduce the stigma associated with it. It has been successful in increasing awareness and access to information (Pirkis et al., 2005) and uses social media effectively (see images). It has also used more traditional advertising such as the beyondblue bus. Although social media has an increasingly wide reach, some groups are digitally excluded for reasons of cost, lack of skills and opportunities to acquire the skills (Baum et al., 2014c).

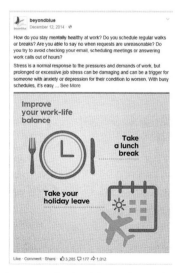

Extract from beyondblue's website promoting healthy work practices.

Beyondblue bus.

SOCIAL MARKETING IN ABORIGINAL COMMUNITIES

Social marketing can be a useful tool if employed within an overall health promotion strategy that is community driven and, therefore, culturally sensitive. A Central Australian Aboriginal community has developed a social marketing approach that is applicable to Aboriginal communities. Maher and Tilton (1994) describe the Central Australian Aboriginal Alcohol Media Strategy or 'Beat the Grog' campaign, claiming it was largely successful because of its community-based approach and use of culturally appropriate materials. The community control of media campaigns can ensure that messages are culturally appropriate and likely to be effective. Other Aboriginal social marketing campaigns are described in box 19.2.

BOX 19.2   EXAMPLES OF ABORIGINAL SOCIAL MARKETING CAMPAIGNS

Source: Aboriginal Health and Medical Research Council, 2015.

'Kick the Habit' is a social marketing campaign designed to reduce the prevalence of smoking in Aboriginal communities in New South Wales. It was developed in close cooperation with three Aboriginal community controlled health organisations and their local communities. The campaign has a positive feel and features well-known local people (Campbell et al., 2014).

Another example is the 'No Germs on Me' handwashing campaign, which aimed to reduce the high rates of infectious diseases among Aboriginal babies and children in the Northern Territory. The aim of the campaign is to motivate men, women and children to regularly wash their hands with soap after going to the toilet, after changing babies' nappies and before touching food (McDonald et al., 2011). The slogan 'No Germs on Me' was well understood by the majority of respondents. Discussions with community members indicated that people appreciated the humorous tone of the campaign and found that they could ask each other 'did ya wash ya hands?' (the campaign catch cry) without being seen as too bossy or rude.

CRITICISMS OF SOCIAL MARKETING

CRITIQUE OF BEHAVIOUR-BASED HEALTH PROMOTION

Behaviour-based health promotion starts from the premise that modification of the lifestyles linked to chronic disease or injury will be beneficial to people's health. Most of the large-scale community projects primarily theorise behavioural problems with some minor recognition of the role of societal factors. Their theory is drawn from social psychology, which is reflected in the primarily individualist focus of the programs. Wallack et al. (1993) have criticised the application of social marketing to public health and convincingly argued that it focuses on individuals and their behaviour, rather than on social and economic structural causes of illness. Mass media is used most effectively in public health when they are accompanied by 'concomitant structural change that provides the opportunity structure for the target audience to act on the recommended message' (Randolph and Viswanath, 2004). Campaigns that help to build community coalitions or influence policy may have more positive long-term effects on health.

This is especially the case when the messages being promoted are competing with powerful counter-messages. For example, messages that advocate for proper nutrition and not smoking face an environment with powerful advertising from food and tobacco companies. In 2012 McDonald's spent almost US$2.7 billion globally (Advertising Age, 2013), and in 2011 in the USA alone the tobacco industry spent $8.4 billion on advertising and promotion (US Centers for Disease Control and Prevention, 2014a). Public health advertising budgets are a tiny proportion of this expenditure.

The individualist focus of most behavioural health promotion creates an undercurrent of victim-blaming, which maintains that individuals are responsible for their own health status, whatever their social and economic circumstances. Yet the available evidence on inequities in health status suggests that behavioural explanations of health inequities would, at best, explain only a small proportion of the difference between groups. The work by Blaxter (2010) suggests that behaviour makes most difference to people's health status when other conditions in their life are favourable. So, if someone is not poor in relation to the society they are living in, if they are not living in absolute poverty, have a reasonably supportive social network, are reasonably free of disease, then behaviour change might make a difference to their health status. For people for whom the reverse is true, behaviour change is most unlikely to be effective. A Scottish study of smoking compared the habits and health knowledge of unemployed and full-time workers and pointed to the effects of disadvantage in determining health behaviour (Lee et al., 1991). They reported, in common with other studies, that unemployed people were more likely to smoke, but that their knowledge of the hazards of smoking was very similar to the employed. The unemployed are certainly a group who usually have very unfavourable economic and social conditions. Lee et al. (1991) concluded that health promotion programs have to recognise the complex interaction of financial and social factors that affect smoking behaviour.

The various models of individual behaviour change conceptualise health behaviour as based on reason and rational choice. The assumption is that once people are provided with sufficient information, provided with support for their decision, then they will change their behaviour. The models assume that people will actively choose their behaviours according to what they believe is good for their health. This, of course, assumes that health is a central consideration in people's decision-making, which, as Blaxter (2010) suggested, is likely only to be true for people in favourable social and economic circumstances, and even then will only be one of a range of factors in decision-making.

The assumption that people will change their behaviour if given sufficient information has also been questioned on the grounds that this implies that knowledge acquisition is a one-way process. New understandings of the gap that may exist between professional and lay understandings of health and associated issues give some clues as to why this one-way process is ineffective. Behavioural approaches to health promotion are based on a linear understanding of knowledge and, generally, have not tangled with more complex understandings and interpretations of people's bodies, health and well-being (Bunton et al., 1995).

One of the most powerful arguments against programs focusing on individual behaviour change has been the lack of evidence on the effectiveness of programs. Syme (1996) discussed the disappointing results of the Multiple Risk Factor Intervention Trial (MRFIT) in the USA. Men in the 10 per cent risk for coronary heart disease group (who would seem to have the greatest motivation to change) succeeded in making only minimal changes in their eating and smoking, despite six years of intensive attempts to persuade them to change. Syme (1996) points out that even if lifestyle programs do meet with some limited success with high-risk people, there will be others who are adopting high-risk behaviours because 'we have done nothing to influence those forces in the society that caused the problem in the first place' (p. 22). The gradual modification of the community-based heart health programs to incorporate more and more community involvement and determination has been an indication of the recognition of the fallacy of the behaviour change models in isolation from a social context. Moving from the position of an advocate for behaviour-based health promotion to a far more sceptical position, Syme has commented: 'The development of prevention programs that focus on places or structural dimensions can influence the lives of more people and for longer periods of time than individually based interventions' (Yen and Syme, 1999, p. 289). There are many examples of how behaviour happens in a social context:

- We eat food in a social context with family and friends and have our food choices determined by food suppliers, advertising and the availability of particular foods. Agriculture, marketing and the social organisation of eating all affect our choices, which are not simply about surviving, but also about social interaction and maintaining agriculture, retail and restaurant industries.
- Smoking also happens in social contexts. We have already seen that smoking represented one of the few pleasures in the life of single mothers in the UK, offering a way of coping. 'Having a fag' helped to create a structure for the day, providing a break and permitting both physical and emotional distance from children in situations where there were very few alternative ways of obtaining release (Graham, 1987). As one of Graham's respondents explained: 'If I was economising, I'd cut down on cigarettes, but I wouldn't give up. I'd stop eating. That sounds terrible, doesn't it? Food just isn't that important to me but having a cigarette is the only thing I do for myself' (Graham, 1987, p. 55). Maté (2012) notes that addictions reflect 'hungry ghosts' and reflect life experiences that often have deep roots in childhood abuse, neglect and lack of loving relationships.
- A study of women's experience of pregnancy also found that smoking reflected far more than an addictive behaviour that might be changed by information about its potentially harmful effects. Oakley's (1985) study found that women who experienced low levels of social support and poor material conditions during pregnancy were particularly likely to smoke. Aboriginal people who were part of the Stolen Generations and forcibly removed from their families are more likely to smoke than those who weren't (Thomas et al., 2008), indicating how smoking may help people to cope with stressful life circumstances.
- Such observations are not novel. In nineteenth-century England, Engels (Reynolds, 1989, p. 398) observed that liquor was 'almost their only source of pleasure for

workers in large industrial towns' and that: 'The working man comes home from his work tired, exhausted, finds his home comfortless, damp, dirty, repulsive; he has urgent need of recreation, he must have something to make work worth his trouble, to make the prospect of the next day endurable.'

- Exercise patterns may reflect the supportiveness of the environment in which people live. For instance, someone in an area where it is perceived to be unsafe to go out alone is less likely to exercise than someone in a safe area. In a non-industrialised society, exercise is far more likely to be an accepted and necessary part of everyday life. The notion of having to encourage people to exercise would be quite absurd.

- Erben et al. (1992) point out that many factors other than health determine our 'body' behaviours. Building on the work of Turner (1984), they point out that body awareness and experience are primarily structures of social communication and interaction. They suggest that the body is used to present an image to the world that is to do with such things as improving work chances, increasing sexual attractiveness and making statements about a person's perception of their place in the world. So, in people's considerations of behaviour and their body, health is likely to only play a small part in the overall picture of their lives.

- Chronic stress has been shown to be damaging to health, especially mental health (Marmot, 2004), and is associated with the conditions that characterise lower socioeconomic status (insecure living and employment conditions, unsafe neighbourhoods, managing life on a low budget) (Fisher and Baum, 2010). Stress levels appear likely to shape the extent to which people are able to make the change to more healthy lifestyles, yet most lifestyle change programs do not deal with people's stress load. Thus mental health and stress are central to how people can change their lives; this has been recognised in other areas but less so in terms of the lifestyle health promotion movement.

RELATIONAL, MINDFUL AND POSITIVE: OTHER APPROACHES TO HEALTH PROMOTION FOR INDIVIDUALS

Perhaps the most promising approaches to changing behaviours in individuals are those that focus on improving mental health but also have flow-on impacts on people's physical health and related behaviours. Advances in the understanding of neurobiology suggest that people's physical health and well-being depend on their mental state. Siegel (2012) notes that humans are rarely rational and that human thinking processes are relational, embodied and embedded in a social context. He says that society shapes synoptic connection. He notes that trust is crucial ('that's how you get good change') and that the brain is reactive when threatened, responding as fight, flee or freeze. The behavioural health promotion movement assumed that people were rational and that when they received information about healthy behaviours this would automatically lead to changed behaviours. In fact, well-being appears to depend on healthy relationships and a sense of awareness and mindfulness about one's own life.

HEALTHY RELATIONSHIPS

Increasing evidence suggests that stressful social relationships are bad for our health. Lund et al. (2014) found from a longitudinal Danish study that stressful social relations are associated with increased mortality risk among middle-aged men and women. Ziersch and Baum (2004) also found that interpersonal conflicts explained why volunteers report more mental health issues than those who do not volunteer. Given this emerging evidence it seems important that public health pays attention to ways of reducing interpersonal conflicts by determining the means by which people can avoid these conflicts in the first place and cope with them better when they do occur. Conflicts can occur in any social settings, including within families, between neighbours and at work. There are a wide range of psychological techniques and training that enable people to communicate better and so avoid conflict or cope better when it does occur.

Box 19.3 provides an example of a Canadian centre that is dedicated to improving the ways people live their lives and conduct relationships, and improving their mental health status.

BOX 19.3   THE HAVEN, GABRIOLA ISLAND, BRITISH COLUMBIA, CANADA

The Haven, 'a centre for transformative learning', notes on its website that 'people leave The Haven with skills to be fully alive, have healthy relationships and communicate effectively'. The Haven offers a range of programs that allow people to reassess their lives, examine their personal goals, determine what brings them happiness and health, and improve their relationships with others. The programs are based on the work of Bennett Wong and Jock McKeen (2013a, 2013b) and emphasise resonance and communication models that position people in relation to others and their own sense of authenticity. The Haven's work explicitly links mind and body health. Bursaries are offered for those who are unable to afford the programs.

Source: www.haven.ca.

Improving relationships has also been one of the aims of the positive psychology movement. This has been informed by the work of Seligman (2012), who stresses the importance of positive psychology. His model is based on the principles of PERMA, which outlines five building blocks of well-being and happiness:

- Positive emotions—feeling good
- Engagement—being completely absorbed in activities
- Relationships—being authentically connected to others
- Meaning—purposeful existence
- Achievement—a sense of accomplishment and success.

His ideas have been used widely in workplaces (including the US Army), schools and communities.

SELF-AWARENESS AND MINDFULNESS

There has been a burgeoning interest in methods of increasing self-awareness and mindfulness (Kabat-Zinn, 1990). These have primarily been promoted as a means of improving mental health (Bohlmeijer et al., 2010) but have also been shown to have benefits for physical health (Grossman et al., 2004). These techniques do not assume a rational link between information and behaviour, but rely on changing how people view themselves and on changing their stress levels.

SOCIAL DETERMINANTS OF HEALTHY RELATIONSHIPS AND MINDFULNESS

The support for healthy relationships and mindfulness-based training is much more available to richer people who can afford to pay for it. This means that the people who already have access to health-promoting social determinants also have the best opportunity to further improve their health. Few people living on an average income or below could afford to enrol in programs that support these approaches unless they receive some subsidy. Some publicly funded programs may incorporate them and be offered on a no-charge basis. Thus some public mental health services provide mindfulness-based training in Australia and the UK. Extension of these programs through public services including health services and schools would be an equitable move, as they would benefit everyone, regardless of their income. Investing in systematic approaches to developing the skills for healthy relationships and self-awareness through mindfulness and other techniques would reap rewards through reduced conflict and so less stress and illness in the longer term. Given the increasing evidence on the links between mental states and physical health, this would also benefit overall population health.

CONCLUSION

This chapter has shown that behaviour is clearly related to the social context in which people live, and that 'to change behaviour it may be necessary to change more than behaviour' (Wilkinson, 1996, p. 64).

Of course, behaviour change is often a prerequisite to changing health status. The key question is whether this is best brought about by programs directed at providing information to individuals or by programs that change the environments in which people make choices about their behaviour or enable people to improve their skills for living a meaningful life. Some of the classic programs based on behaviour change did not simply use methods aimed at individuals, but also sought to change people's decision-making environments and to provide them with a more supportive environment in which to make healthy choices—'making healthy choices the easy choices'. The evaluations of those programs that used a mix of methods were unable to tease out the effects of different parts of their programs, but it seems that community involvement is increasingly emerging as a crucial aspect of programs, and that the best outcomes have been achieved in programs using mixed strategies (Elder et al., 1993). Mass media has been used with mixed success to promote health and again is more effective when combined with a range of other strategies. Social media, including Twitter and Facebook, also offer new tools for health promotion. Increasing the Enlightenment duality between mind and body is being questioned and the importance of strategies that address mental and physical health are favoured. This chapter noted the potential of programs that aim to promote positive mental health through improving relationships and self-awareness, and that these have flow-on benefits for physical health. The rest of part 7 considers the following strategies that are essential complements to behaviour change: community participation and development, public health advocacy, the healthy settings movement and healthy public policy.

CRITICAL REFLECTION QUESTIONS

19.1 Why might health promotion behaviour-change strategies result in inequitable health outcomes?

19.2 Why is it important for health promotion to consider mind and body health?

19.3 What role does social media have to play in health promotion campaigns?

Recommended reading

Bunton et al. (1995) *The sociology of health promotion* provides 16 chapters offering critical analyses of lifestyle health promotion.

Baum and Fisher (2014) Why behavioural health promotion endures despite its failure to reduce health inequities, *Sociology of Health and Illness*, 36, 213–25.

Useful website

VicHealth contains multiple examples of health promotion campaigns that use a range of strategies: www.vichealth.vic.gov.au

20

PARTICIPATION AND HEALTH PROMOTION

The participation of people and people's organisations is essential to the formulation, implementation and evaluation of all health and social policies and programmes ... Participatory democracy in political organisations and civic structures should thrive. There is an urgent need to foster and ensure transparency and accountability.

People's Charter for Health, People's Health Movement, 2000

KEY CONCEPTS

Introduction
Participation in practice
Values and principles for participation
Participation in health
Social media and participation
Lessons from participation in health
Pseudo or real participation?
Types of participation
Participation and power
Who participates? Issues of representation
Citizens or consumers?
The role of professionals in participation
Effective bureaucratic consultations
Conclusion

INTRODUCTION

This chapter examines the importance of democratic participation and the mechanisms that can encourage it in public health. These mechanisms include community development, which has been a crucial way in which people have been involved in public health and which is examined in depth in chapter 21. We start here by considering the role of participation in society in general, and in health services and public health specifically. Lessons arising from participation practice are reviewed in terms of the actual extent of participation, power, issues of representation, the importance of empowerment and the role of professionals.

PARTICIPATION IN PRACTICE

One of the most important threads in the past 200 years or so has been the demand for increased participation in decision-making processes. The suffrage and national liberation movements and European revolutions in the nineteenth century, the industrial democracy, women's rights, civil rights and indigenous people's rights movements in the twentieth century, and the Occupy movement of the early twenty-first century all demanded a wider involvement of people in decision making.

The demand for increased participation has often been a protest against concentration of power. Burgmann (1993) reminds us that as long as there have been entrenched and unequal relations of political power there have been protests:

> From the ancient rebellions of slaves to today's civil rights movements, from the mutinies of press-ganged armed forces to the draft dodgers of the 1960s and 1970s, from the early nineteenth-century women's rights campaigns to contemporary feminism, from the utopian communes of last century to the alternative lifestyle experiments of the 1970s, people have striven to abolish those forms of power by which they felt constrained.
>
> *Burgmann, 1993, p. 1*

Since white settlement, protest has been an important feature of Australian political life, but it was in the 1960s and 1970s that it became most evident. Protest against the Vietnam War, the demand for Aboriginal land rights, the growing women's movement, and resident action groups mobilising against urban development proposals, all included demands for increased participation in public life. The focus on participation in these political and social movements was paralleled in the health field. Globally, the Commission on Social Determinants of Health (2008) has called for increased empowerment as a means of achieving health equity.

VALUES AND PRINCIPLES FOR PARTICIPATION

The International Association for Public Participation (IAP2) developed the following values for public participation based on a two-year international consultation:

1 Public participation is based on the belief that those who are affected by a decision have a right to be involved in the decision-making process.
2 Public participation includes the promise that the public's contribution will influence the decision.
3 Public participation promotes sustainable decisions by recognising and communicating the needs and interests of all participants, including decision makers.
4 Public participation seeks out and facilitates the involvement of those potentially affected by or interested in a decision.
5 Public participation seeks input from participants in designing how they participate.
6 Public participation provides participants with the information they need to participate in a meaningful way.
7 Public participation communicates to participants how their input affected the decision.

These values are useful when supplemented by the following eight key principles (Department of Public Health/SACHRU, 2000, p. 6) that organisations need to bear in mind in order to make participation genuine and effective:

1   Participation means partnerships, which means accepting uncertainty.
2   Bringing about effective participation means organisational change.
3   Community involvement plans need to be aligned with organisational capacity and the capacity of staff has to be developed.
4   Community participation must be supported by the management of an organisation.
5   While the top-down support of management is crucial, the participation must be built from the bottom up.
6   Effective participation is built by using well-developed people skills.
7   The partnerships developed with communities require dialogue and trust, so developing these is crucial.
8   Using a range of strategies and not relying on one is most effective.

PARTICIPATION IN HEALTH

The Whitlam government established the Community Health Program in 1973 with the chief objective being:

> to encourage the provision of high quality, readily accessible, reasonably comprehensive, coordinated and efficient health and welfare services at local, regional, state and national levels. **Such services should be developed in consultation with, and where appropriate, the involvement of, the community to be served.**
>
> *Hospitals and Health Services Commission, 1973, p. 4; emphasis added*

The focus on participation in public policy was a hallmark of the Whitlam government and was also seen in other initiatives, such as the Australian Assistance Plan. Although the original Community Health Program did not achieve its aim of community participation to any significant extent (Owen and Lennie, 1992), it did establish the importance of participation as an element of effective community and public health practice, and set thinking and practice patterns for the 1980s and 1990s. There have been many experiments with participation in health in the 25 years since the launch of the Community Health Program. Examples of the types of participation in health are:

• participation in health services, which can take the form of client feedback and evaluation, membership on boards of management of health services setting priorities, volunteer work, user advocacy, complaints procedures and self-help care. Taking part directly in management is likely to be the most powerful form of participation (see box 20.1). The form of participation varies considerably, but all help to make health services more accountable and responsive to people's needs if the participation is taken seriously by the services. Box 18.3 (chapter 18) provides the example of Aboriginal community controlled health services

- formal consumer bodies such as the Consumers Health Forum (www.chf.org.au), which was formed in 1985, or topic-specific organisations such as Mental Health Carers (www.arafmiaustralia.asn.au)
- participation in bureaucratic process, including responding to the agendas of services and government, commenting on policies and plans, and participating in consultations as requested. The value of participation in bureaucratic processes depends on the extent to which the bureaucracies are prepared to devote resources to the process and to share power
- participation in needs assessment and planning on public and environmental health issues; this may be through a community health centre, a Primary Health Network, a project such as Healthy Cities, or a local government planning process
- participation in pressure groups campaigning on public health issues, including environment concerns such as pollution from local industries, protection of local environments, environmental improvement, and occupational health issues.

BOX 20.1   THE BOARDS OF MANAGEMENT OF COMMUNITY AND WOMEN'S HEALTH CENTRES IN VICTORIA AND SOUTH AUSTRALIA

Community-based management is at the core of community health philosophy, so that health problems can be seen from the perspective of the local community, rather than as technical problems for the health professional to define and treat. In Victoria in the early 1990s there were around 90 community health centres, run by locally elected committees of management (Legge, 1992), and 12 in South Australia. These committees were responsible for the overall management and strategic direction of the community health centres. Legge (1992, p. 97) says there are many instances where the community representatives on these committees have led the way in reordering priorities and recognising problems as social as well as political.

A study of the boards of directors in South Australia (Laris, 1995) found that, overall, they were effective in ensuring an accessible and responsive means of service accountability to the local community. They required considerable support from the funding body and Laris noted that this was rarely forthcoming in sufficient strength. The community health centres themselves, however, often made up for this lack of support and the boards could be very effective mechanisms for local participation. One community health service was reported as having more than 250 people at its public meeting to elect new board members (Laris, 1995, p. 90).

Community boards of management have been targets for change in the search for more 'efficient' health management from the late 1990s. In Victoria, legislation was passed to remove democratically elected boards. Community health centres were merged in both Victoria and South Australia, making it more difficult to link with local communities, but easier for the central health department to communicate with the services. Oke (1997), a community health centre board member, feared the changes would 'threaten the very essence of community health'. In South Australia from 2004 the merged community health centres were absorbed into larger regional health services, which included, and were dominated by, hospital services. Similarly, in Victoria the amalgamations and reduction of the electoral input into boards have led to community health centres focusing more attention on their relationships with the hospitals. In both states the centres have forsaken their earlier role of supporting any popular or political movement for health. The community development end of the health promotion spectrum also appears considerably less prominent in the activities of the centres, which may reflect the lack of community management.

SOCIAL MEDIA AND PARTICIPATION

Communication patterns have changed dramatically in the past decades. New forms of communication via innovations such as social media sites (including Facebook, Twitter, and professional networking sites) have revolutionised communication and opened up new possibilities for citizen participation. These changes have significant implications for the ways in which participatory processes are conducted, as the example in box 20.2 demonstrates. However, while new forms of communication mean there is more potential for citizen participation, the dilemmas concerning power and effective, non-token means of participation raised in the rest of this chapter are just as relevant. Although new forms of communication are increasing, not everyone has equal access to them, meaning that these mechanisms may also exclude groups whose power and representation is already low (Baum et al., 2014c).

BOX 20.2   NEW FORMS OF PARTICIPATORY MECHANISM IN SOUTH AUSTRALIA

In July 2014 the Premier of South Australia, Jay Weatherill (2014), announced that there would be a dramatic reduction in the number of advisory boards and committees in South Australia. In doing so, he said in an email to existing board members:

> Since becoming Premier, I have sought to improve the way Government engages with community, business and citizens in making public policy. The past decade has seen profound technological and social change. Community expectations today favour a direct say, transparency, immediacy and greater use of technology to enable collaboration in public policy development. The boards and committee model, which has served the state well in the past, needs reform to give way to more contemporary approaches that meet these expectations.
>
> Our Building a Stronger South Australia: A Modern Public Service policy outlines a number of ways in which we are seeking to reform the way Government engages, including through more citizens juries, country cabinet, an upgraded YourSAy website and a new GovChat service. My vision is to give a broader range of organisations and individuals more direct access to government advisory and decision-making processes.

Source: Weatherill, 2014.

LESSONS FROM PARTICIPATION IN HEALTH

Key issues have emerged through the practice of participation in health:

- To what extent does participation actually occur? Is it pseudo or real?
- What are the types of participation?
- What is the relationship between participation and power?
- Who participates?
- What is the role of professionals in participation?

PSEUDO OR REAL PARTICIPATION?

To what extent do participatory exercises really involve participation? The idea of a hierarchy of participation has often been used in health to distinguish 'genuine' from pseudo-participation (Arnstein, 1971). Commentators often judge participation to be unhelpful unless it involves the exercise of full citizen control. But giving information and consultation can be useful if it does not masquerade as full participation.

It is crucial that the form of participation and its potential for power sharing are recognised by those seeking partners in a collaborative exercise. So if a bureaucracy intends to consult on a policy, it should be clear what the parameters are. In this way people do not develop unrealistic expectations and are aware of the rules of the exercise. Bureaucrats are often limited in the time they have for consultation and the extent to which they are able to incorporate the diverse views they will gather. Australian government is complex and it is reasonable at times for bureaucrats to consult rather than call for full participation, but this can still be valuable in policy and service development, and should not be overlooked as a valuable strategy of the new public health. There are other powerful players in the health industry (for example pharmaceutical companies) and community interests are always fighting to be heard amidst the many different interests.

There have been examples of attempts by health departments to encourage participation in health planning and issue identification. Victoria established District Health Councils in the 1980s and for a short period these councils were successful in providing a community voice in health decision making. South Australia established four pilot health and social welfare councils in 1988, the aims being to:

- increase community participation in decision making
- increase accountability of the health and welfare systems
- promote community education and awareness
- strengthen local action to promote health and prevent social and health problems.

These councils were able to put forward a community voice on some issues and produced some useful resources (such as a guide for community representatives called 'Not Just a Token Rep'). The councils survived as a program until 1995 when their funding was withdrawn. The program was evaluated in 1991 (Shannon and Worsley, 1991) and has been the focus of other critical assessment (Sanderson and Baum, 1995; Baum et al., 1997). Each concluded that, despite the councils' somewhat contradictory position, they were effective at involving local people on some issues, including mental health, and commenting on proposed changes to Medicare funding.

A major factor that can make participation pseudo is when there is a high turnover of organisational form or people within an organisation. We have seen trust is vital to good participation and this is difficult to establish in the absence of long-term relationships. A recent example has been the Medicare Locals, which were established from 2010 to 2012 and asked to conduct participatory needs assessments. Many resources went into these needs assessments, but the Medicare Locals were cut in 2014 and the links and trust established with local communities were disrupted.

TYPES OF PARTICIPATION

Oakley (1989) discusses the role of participation in health development in developing countries. He distinguishes between participation as a means and participation as an end.

PARTICIPATION AS A MEANS

Participation is the means of achieving a set objective or goal. There is less concern with the act of participation and more with the results. The emphasis is on rapid mobilisation and direct involvement in the task at hand. The participation is abandoned once the task has been completed. An example would be an external agency coming to a community with a predetermined program that required the program implementers to work with the community. Participation would be limited and solely for the purpose of implementing the program.

PARTICIPATION AS AN END

Participation may also be an end in itself. Oakley (1989, p. 11) comments that the process is 'dynamic, unquantifiable and essentially unpredictable'. Participation is not limited to the life of a particular project but is a permanent and intrinsic feature of an organisation or community. The critical elements in the process are to increase people's awareness and develop organisational capacities. Full, engaged participation does not happen immediately. Oakley (1989) indicates that it may begin as marginal participation, in which people have relatively little impact on the activity. This will be especially true when the motivation for the project comes from outside the community. Such participation may lay the basis for substantive participation where people are actively involved in determining priorities and carrying out activities, but the mechanism is still externally controlled.

Structural participation (control by the community) is integral and forms the basis for all activity. Community members play an active and direct role in the initiative and have the power to ensure their opinions are heeded. More affluent communities are likely to have the skills and resources to achieve structural participation in relation to promoting their communities' health to a greater degree than less-resourced communities.

The concept of participation as an end is somewhat problematic in that it can be taken to imply that participation itself is sufficient. However, a situation in which people participate but do not achieve the desired changes is unlikely to be empowering.

TYPOLOGY OF PARTICIPATION

It is important that health promoters recognise what form of participation they are seeking. Structural participation may be a fine ideal but it is not always achievable. Other forms of participation can be useful. What is crucial is that practitioners make a critical examination of the concept of participation they are using and not claim it to be something it is not. Table 20.1 outlines the four main forms of participation used within public health and health promotion in Australia. It offers a useful checklist to determine the form of participation in any intervention.

TABLE 20.1 A CONTINUUM OF PARTICIPATION FOR THE NEW PUBLIC HEALTH

Feature	Consultation	Participation as a means	Substantive participation	Structural participation
Form it takes	Asking for people's opinions and reactions to policies and plans.	Using participation to achieve a defined end.	People are actively involved in determining priorities and implementation but initiative externally controlled.	Participation as an engaged and developmental process in which community control predominates.
Who initiates	Organisations outside the community.	Organisations outside the community.	Initiated by outsiders but may lead to structural participation in time.	Control by the community— initiative may have come from the outside initially but control will have been handed over.
Features	Limited, usually one-off activity controlled by organisation.	Instrumental. Lasts for the life of the initiative. Driven by outsiders. No shift in power. May lead to more developmental participation but this is not initial aim. Scope of activities limited to agenda of those initiating exercise.	Engaged and developmental. Active involvement. Despite this control, still outside the community. Will usually involve a shift in power to the community. Scope initially determined by those introducing initiative but may change over time.	Engaged and developmental. On-going relationship. Driven by community. Potentially empowering to individuals, organisation and community. Scope of activities as broad as the community wishes.

(continued)

TABLE 20.1   A CONTINUUM OF PARTICIPATION FOR THE NEW PUBLIC
HEALTH (*CONTINUED*)

Feature	Consultation	Participation as a means	Substantive participation	Structural participation
Examples	Consultation on policies by federal government. Feedback surveys on quality of services.	Community panels for priority setting in health services.	Self-help groups initiated by community health centre staff. Community heart health programs working with local agencies.	Aboriginal-controlled health services. Victorian District Health Councils. Resident action groups.

PARTICIPATION AND POWER

Demands for more participation inevitably mean some people feel aggrieved that they do not have access to sufficient power to influence events that have a significant impact on their lives. Participation is a complex, dynamic, relational and political process of negotiation in which groups with differing interests and agendas vie with one another for power. In 2003 hundreds of thousands of Australians protested against the invasion of Iraq (see photo) and opinion polls showed that a considerable majority of the population was against the war yet the government joined the invasion despite the popular opinion thus exerting its power.

Protest against the Iraq War, Adelaide, February 2003. (Fernando M. Gonçalves)

Two beliefs appear evident in the literature on participation and health. The first is that involving people in health initiatives improves the quality, relevance and effectiveness of the initiatives. The second is that participation helps overcome community and individual powerlessness and so leads to people being healthier. This issue of power relates to the importance of self-esteem and feelings of control over health outcomes. People gain power by coming together with others, building up networks and relationships and taking collective action. Thinking on social capital suggests that the fabric of civic society is an important determinant of the health of a community. Encouraging participation helps to weave and strengthen this fabric.

Experience of participation in Australia suggests that power is a crucial element in measuring its success. Power was relevant in the case of the boards of management of community health centres when they were abolished by the government (see box 20.1), despite the fact they appear to have been effective because they did control resources and had a say over the philosophical direction of the centres, and in the case of Aboriginal community controlled health services where holding power makes a significant difference to the style of services provided (see box 18.3).

EMPOWERMENT AND HEALTH

> Any serious effort to reduce health inequities will involve political empowerment –
> changing the distribution of power within society and global regions especially in
> favour of disenfranchised groups and nations.
>
> *Commission on Social Determinants of Health, 2008, p. 155*

Empowerment aims to reduce the number of people who are powerless. Parenti (1978) defines powerlessness as 'the inability to get what one wants or needs (social desiderata) and the inability to influence others effectively in ways furthering our own interests'. Concern with empowerment is reflected in literature from a number of relevant disciplines, including radical social and community work (Freire, 1972; McKnight, 1985; Ife and Tesoriero, 2006), health promotion and education (Community Development in Health, 1988; Wallerstein, 1992; Labonté, 1990, 1997; Tones, 1992; Israel et al., 1994; Rissel, 1994; Minkler, 2005) and community psychology (Rappaport, 1987; Zimmerman, 1990; Hawe, 1994). Israel et al. (1994, p. 153) offer the following definition of empowerment, based on a review of literature from a range of disciplines and professions: 'Empowerment, in its most general sense, refers to the ability of people to gain understanding and control over personal, social, economic and political forces in order to take action to improve their life situations.'

They go on to point out that empowerment can operate at the individual, organisational or community level and that it is positive and proactive. Individual empowerment refers to an individual's ability to make decisions or have personal control. Israel et al. (1994, p. 153) see that it combines personal efficacy and competence; a sense of control over life; and the ability to participate to influence institutions and their decisions. Empowered organisations are democratically run and develop processes that enable individuals to increase their control at work and make an input to the design, implementation and control of work processes and outputs. They also influence the wider system of which they are a part. Community empowerment results in a community in which individuals and organisations work together to meet their respective needs. They provide support for each other, deal with conflicts constructively and establish control over the quality of life of the community. Clearly, these three levels of empowerment interact and new public health initiatives should be designed to work simultaneously at each level.

Labonté (1990, p. 3) detailed the elements of personal empowerment:

- improved status, self-esteem and cultural identity
- the ability to reflect critically and solve problems
- the ability to make choices
- increased access to resources
- increased bargaining power
- the legitimation of people's demands by officials
- self-discipline and the ability to work with others.

This recognises that participation empowers when it results in material change in people's situations and when they have increased access to resources. It also recognises the importance of collective action to increasing people's power. A recent outstanding example of increasing power is the introduction of participatory democracy in Porto Alegre in Brazil (Kingsley, 2012). Issues of power imbalances in participatory or consultative exercises are often not taken seriously by bureaucracies, who conduct their business as though everyone were equal. In many low-income countries women's self-help groups have been developed to assist in the process of empowering women, who often lack social or economic power (see photo).

Meeting of women's self-help group, Tamil Nadu, 2006. (Fran Baum)

The earlier discussion of pseudo-participation showed that lip-service is paid to participation, which is often more real in rhetoric than practice. It is probably fair to say that the bigger the stakes, the more chance there is of a merely token participation of marginalised people. Generally those with economic, social and educational resources will be in a better position to participate than those who do not have access to these resources. Public health initiatives that wish to engage communities (especially poor ones) will have to be grounded in an understanding of the ways in which power operates if they are to avoid the traps.

The term 'empowerment' may not always be used in the same way (Mayo and Craig, 1995). National and state health departments and large international organisations such as the World Bank are likely to mean something quite different to a local community health centre or popular social movement. Concepts of power differ considerably. Labonté (1992) pointed out that, given the central importance accorded empowerment in contemporary health promotion, surprisingly little effort is made to understand the concept of power. Political theorists have viewed power in a variety of ways. There have been many attempts to explain and understand the nature of power in society, but no consensus as to which is the preferred meaning of the term. Each of the theories elaborated in table 20.2 has different implications for public health and the strategies to be used.

TABLE 20.2   FOUR MAIN PERSPECTIVES ON POWER

Perspective	View of society	View of power	Empowerment
Pluralist	Competing interests; groups and individuals	Capacity to compete successfully, 'winners and losers'	Teaching individuals or groups how to compete within 'the rules'
Elite	Largely controlled by self-perpetuating elites	Exercised largely by elites through ownership and control of dominant institutions	Join and influence elites, form alliances with elites, confront and seek to change elites
Structural	Stratified according to dominant forms of structural oppression: class, race, gender	Exercised by dominant groups through structures of oppression	Liberation, fundamental structural change. Challenge oppressive structures
Poststructural	Defined through constructed meanings, understandings, language, knowledge accumulation, and control	Exercised through control of discourse, construction of knowledge, etc.	Change the discourse, develop new subjective understanding, liberating education.

Source: Ife, 1995, vp. 59.

STRUCTURALIST THEORIES OF POWER

Structuralists or Marxists see political and economic power as intimately related. Power represents a struggle between the forces of capital (owners of the means of production) and workers (those who actually produce wealth), and empowerment of the poor is very limited under capitalism. While they may make some limited gains in terms of bargaining power or ability to influence the fine-tuning of plans and policies, empowerment is ultimately limited and circumscribed by the wider requirements of the capitalist system to maximise profits. If empowerment moves towards challenging the structural aspects of the political and economic system so that, for instance, a more equitable distribution of resources is achieved, then resistance is likely to increase.

Marx and Weber both viewed power in zero-sum terms. They saw that there was a limited amount to go around and that struggle for control is inevitable. Weber saw that power involves the ability of individuals/groups to realise their will—even against the resistance of others. Power can be asserted through the exercise of force or influence. In this view empowerment will, inevitably, involve the less powerful gaining power from the more powerful, either through negotiation or reconciliation. It must involve the powerful giving something up. Weber saw that some groups have greater status in particular cultures and, therefore, greater power.

The exercise of power also extends to control of the power of ideas. Gramsci (1978) used the concept of hegemony to demonstrate how the prevailing systems of political and economic power are legitimised and protected within capitalist society. An approach to empowerment that involves a belief in this form of hegemony would be concerned with challenging it and working for transformation to a political and economic system in which greater empowerment were possible. Freire's approach to adult education is about enabling people without power to analyse their situation in order to understand the nature of the hegemonic ideas and so be in a position to struggle to transform the existing state of affairs (Mayo and Craig, 1995).

ELITE THEORY

The elite theory of power recognises that all groups and individuals in society do not have equal power and influence over decisions. This theory was first propounded by C. Wright Mills, an American sociologist, who argued that economic and political power in modern industrialised society was becoming more concentrated, so that a power elite has come to control the key institutions in society. Elites are able to reproduce their privilege through institutions such as private schools, clubs and societies with exclusive memberships and professional associations. They are seen to hold more wealth, resources and influential connections than other members of society. This view sees society as hierarchical with a small number at the top controlling the rest of society through the key institutions of society, such as the media, education, policy-making, the senior parts of the state bureaucracy, political parties and the professionals.

PLURALIST THEORIES

Pluralist theorists (stemming from the work of Dahl, 1961) see power being distributed through a variety of institutions and groups in society. In contrast to the more 'elitist' view of the Marxist political theorists, they do not see a concentration of power in the hands of a few. This view allows far more scope to gain power because it is more diffuse and spread within a society. Various groups and individuals within society continuously compete for power. Such groups would include trade unions, churches, pressure groups, resident action groups, professions, media and consumer lobby groups. This view of power basically accepts the status quo and encourages people to be able to engage with the system in a more effective way. From a pluralist perspective, empowerment is concerned with helping people develop skills to engage with the system and win power more effectively. This view of power also fits with Sen's (1999, p. 18) work on the concept of development as lying in the 'capabilities of persons to lead the kind of lives they value—and have reason to value'.

POSTMODERN AND POSTSTRUCTURALIST VIEWS OF POWER

More recent views of power have seen it as inextricably linked with knowledge and woven throughout the fabric of society. Foucault (Cheek et al., 1996, pp. 173–84) sees power operating as a network of relationships throughout society. Rather than seeing a binary system with the powerful on the one hand and the repressed on the other, Foucault sees power operating both horizontally and vertically in society and being deeply enmeshed in social institutions. Such is this enmeshment that power is often invisible. Foucault (1979) defined three main expressions of power: exploitative (the power to control people's economic lives), dominance (the direct power to control people's choice) and hegemony (the power to control people's perceptions so that their actions are controlled by dominance). He believes power can be challenged via the complex 'discourses' that support its maintenance. These discourses are based on claims of superior expertise and knowledge. Because power is diffuse, there are many opportunities for resisting its expression. Kenny (1999, pp. 151–2) points out that two features of Foucault's analysis are particularly significant to community development. First, he argues that power is not an impenetrable, irresistible force and that wherever there is power there is resistance. This offers hope for community development when combined with his other insight, which is that any understanding of power begins from an analysis of the multiplicity of forces and practices at the micro level of everyday life.

UNDERSTANDING POWER

The common theme to all theories of power is that resistance or change to the patterns of power relations must be preceded by analysis, with people understanding their own relationship to power and its expression in their particular context. Without this, the potential for structural participation is almost certainly going to be limited. While many public health practitioners may be frustrated by an abstract discussion on the nature of power in society, these theories are crucial in determining and understanding people's actions. A structuralist approach to public health inevitably means challenging the powerful in society, and seeing public health struggles as conflicts with powerful forces that act to maximise their economic power. An elite perspective may imply joining forces with elite groups in society to persuade them that public health is a cause they should support. A pluralist view will lead to approaches based on compromise, and learning to compete within the established rules. Poststructuralism implies a less clear path of action in which cultural means and discourses are deconstructed in an attempt to understand the multiplicity of perspectives held by different groups within society. This conceptualisation of power as operating through a network of influence implies, however, that participatory processes have the potential to be more powerful than they may appear at first. If people are able to exert power in a variety of subtle ways, through a variety of discourses and networks, they may be able to challenge the hegemony of more economically and socially powerful agents. In order to do this they need to operate in a policy environment that is open to this form of participation.

MEDICAL POWER

Power has particular relevance in the new public health in terms of medical power. Health debates are dominated by a biomedical view of health (see chapter 1). Within health systems medical doctors have a powerful position, documented over many years (Freidson, 1970; Willis, 1983), that depends on their control over diagnosis and treatment of illness, subsequent authority over other health professions and overrepresentation on health boards and policy groups. Germov (2014) notes that the 'medical profession dominates every aspect of health care delivery' and this could be extended to dominance over many aspects of public health. Although there are threats to the power of medicine from deprofessionalisation and corporatisation of health services, it remains powerful. Recent analyses of the politics of health policy (J. Lewis, 2005; Baum et al., 2013b) illustrate the ways in which organised medicine remains highly influential in health policy. This can be seen in Australia by the power of the Australian Medical Association, which is able to directly influence the direction of health debates. Citizen power in the health arena always has to deal with medical power and position itself in relation to it.

WHO PARTICIPATES? ISSUES OF REPRESENTATION

It is often asked to what extent participants are representative of some broader constituency. Community participation initiatives are often dismissed as being unrepresentative. Sometimes this is a convenient response from an organisation or government who does not want to respond to the demands of community groups.

Questions to do with representativeness have also been raised by commentators who are sympathetic to the ideal of democratic participation. Yeatman (1990), in a discussion of the politics of difference, points out that those who participate may not include women, ethnic and cultural minority groups, or inarticulate people. Similarly, children have typically been excluded from participation in community-based health promotion. Power relations can serve to exclude certain groups in the community from participation. These are all reasonable doubts about representation, but they should be viewed in context. While not all members of a community may participate in an activity, the inclusion of some will often ensure a view that is different to that of the usually white, often male, professionals involved. For instance, health services participation by community members can offer a counterpoint to the medical perspective that often dominates. Urban planners will gain a wider perspective on their task if they work in partnership with local people, even if they do not precisely represent the community. Of course, efforts should be made to ensure participation by as broad a range of people as possible.

Abbott (1995, p. 164) argues that there is a tendency in the literature to overstate the problems of representation in community participation and development. He recognises that the notion of community as a coherent entity with a clear identity and a commonality of purpose is a myth. The reality is that most communities are made up of an amalgam of competing interests and factions that often have competing purposes. Successful community development projects recognise this and overcome the problem by focusing on small groups that come together around a common interest.

Possibly the crucial issue is identifying who the participants' constituencies are, rather than asking if they are representative of the total community. The latter is simply not possible, but there can be a network of people within the community whose interests they can represent. In this sense people act as bridges to their community and become leaders because they are respected for their community knowledge and contacts. This concept of a community constituency may also guard against co-option of community initiatives by professional workers. It is also true that, while questions are often asked about the true representativeness of community, rarely are the same questions asked of representatives of government bureaucracies or private industry. These representatives are automatically given legitimacy, though it is likely that their representation is also partial.

Community representatives need to consider how they can represent everyone in their community, and if it is their role to do so. This problem is at the heart of democratic society. It becomes more acute as societies become larger, more complex and more multicultural.

Another common concern that arises in terms of participation is that professionals and others often say people are apathetic and do not want to participate. Often this translates to a desire to 'motivate' people to participate. In fact their reluctance more commonly reflects structures and processes that militate against participation and encourage what Ife (2001) has described as 'a society of passive individual consumerism'. Facilitating participation is the responsibility of bureaucratic organisations rather than the individuals and communities they wish to consult (see box 20.4).

CITIZENS OR CONSUMERS?

In chapter 5 we considered the shift in public policy rhetoric towards using the term 'customer' rather than 'citizen'. Many recent policy statements in relation to health services have used the term 'consumer', and health consumer councils exist. There is plenty of anecdotal evidence that this term is more acceptable to governments than 'citizen'. While this shift in terms can be cast as a rather academic debate with little practical relevance, there are some important distinctions. Consumers are defined by their consumption of a particular service. They are defined by their relation to a marketplace. Thus their interest is seen in terms of the quality and appropriateness of the service. 'Consumer' tends to be a passive construction in which people seek to receive a better service or treatment. Consumers usually need to bargain with the health care system at a time when they are ill or injured and relatively powerless to effect change to systems. A further difficulty with the word 'consumer' is that an industry has developed around their needs in recent years. This means that 'consumer advocates', who are often paid consultants, take on the mantle of speaking for consumers or working alongside them. This highlights the position of 'consumers' as being those who have a primary interest in meeting their own needs in relation to a particular disease. An example of potential conflicts is that pharmaceutical companies sponsor consumer groups and encourage them to advocate for the uptake of drugs they produce and influence policy relating to drug testing, marketing and use. For example, *The Guardian* (Sample, 2013) claimed that pharmaceutical companies in Europe had mobilised patient advocacy groups to campaign against greater transparency in drug trials on the grounds that information might be misinterpreted and cause health scares. The word 'citizen' by contrast implies the rights and responsibilities that are conferred by virtue of citizenship. Citizenships are construed to include civil rights (such as the right to free speech), political rights (the right or, in Australia, the duty to vote) and social rights (which include the right to income support in time of hardship and the right to health care based on need) (Marshall, 1950). Participation in health and community development implies an active citizenship with links to notions of democratic participation. 'Citizens' does not restrict participation in health services to those who are users of the service but also extends to citizens in general who have a legitimate interest in shaping health services and the decisions that they make about resource allocation and types of services and in lobbying to improve environmental and other conditions that may affect their health.

THE ROLE OF PROFESSIONALS IN PARTICIPATION

> When I give food to the poor, they call me a saint. When I ask why the poor have no food, they call me a communist.
>
> *Dom Helder Camara, quoted in preface to Mayo and Craig, 1995*

Many social movements involve the citizens having grievances with the state and setting about, unaided, to establish movements to bring about changes to the status quo. In the new public health movement, the primary push for community participation appears to have come from professionals employed by the state. The form of participation they advocate differs. Legge (1990) points out that health professionals working within a risk factor and disease prevention model tend to see community participation in instrumental terms, or in P. Oakley's (1989) terms, as a means. For them it is a means of encouraging behaviour change or generating community support for a program they have designed. As an example, he quotes the local sporting team sponsored by a quit smoking program aiming to change the climate of opinion in relation to smoking. In the USA much of the community organising associated with heart health community programs appears to be instrumental (Elder et al., 1993). In such programs the health professionals are inviting community people to participate in an endeavour defined by the health professionals, a form of participation that is concerned with neither empowerment nor challenging the broader social and economic conditions that shape health.

Legge (1990) contrasts this instrumental approach with a more developmental one that starts with the concerns of people and encourages professionals to work alongside people in a way that gives them a significant degree of control, but does not manipulate them to achieve their ends to the exclusion of the community priorities. This form of participation has considerable implications for professional practice. First, professionals are required to give up their traditional authority based on professional knowledge and accept the value and contribution of lay knowledge to health promotion. Second, they have to develop the skills of working in partnership with lay people and respecting their priorities for health. Finally, they will find themselves in a contradictory position. They are paid by state authorities, but the strategies that are effective in their work require them, on occasions, to question, advocate and organise against state policies.

Medical dominance within the health field has been well documented by medical sociologists (Freidson, 1970; Doyal, 1979; Willis, 1983). New public health practitioners come from a broader field than medicine, and include social workers, health educators, urban planners, community developers and environmental health officers. To a greater or lesser degree these professionals may establish a position of dominance based on their status and knowledge. They may be reluctant to share appropriate knowledge and skills, and may prefer to work on, rather than with, people. If this sounds rather negative, it should be balanced by the recognition that professionalism brings with it, for most practitioners, competence and a professional code of ethics. Professionals who chose to work in the new public health are often highly motivated by the desire to help people and communities achieve better health, and often have a fair degree of evangelical zeal. To be effective in using strategies based on community participation, they need to develop particularly good skills at working alongside people. Kickbusch (2003), in discussing the shifts required to implement the new public health, noted that health professionals require a new mind-set and professional ethos so that they can follow the Ottawa Charter's (WHO, 1986) strategies of enabling, advocating and mediating. Such a mindset requires a significant shift from the traditional model of professional dominance.

The health promotion winners' and losers' triangles illustrate the shift in patterns of working that might be required (figure 20.1). In the losers' triangle health promoters fall into either the persecutor or rescuer role. The persecutors tend to blame victims and work from a position that people could be healthy if only they would change their habits. Rescuers tend to see people as victims and believe their role is to be their rescuer. The rescuing health promoter will believe they have the solutions and know best. The chances are that this approach will leave communities and their members feeling like victims. Certainly, nothing will be done to challenge their position of relative powerlessness or increase their self-esteem or belief in their ability to initiate and bring about change.

FIGURE 20.1 HEALTH PROMOTION WINNERS' AND LOSERS' TRIANGLES

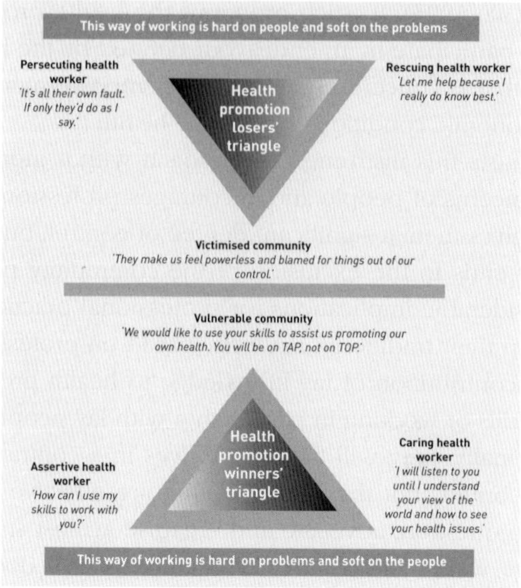

This way of working is hard on people and soft on the problems

Persecuting health worker
'It's all their own fault. If only they'd do as I say.'

Health promotion losers' triangle

Rescuing health worker
'Let me help because I really do know best.'

Victimised community
'They make us feel powerless and blamed for things out of our control.'

Vulnerable community
'We would like to use your skills to assist us promoting our own health. You will be on TAP, not on TOP.'

Health promotion winners' triangle

Assertive health worker
'How can I use my skills to work with you?'

Caring health worker
'I will listen to you until I understand your view of the world and how to see your health issues.'

This way of working is hard on problems and soft on the people

Source: Baum, 1993, p. 37.

By contrast the health promotion winners' triangle puts the community in a controlling position. The health promoters are assertive and caring, offering to use their skills to work with people. The issues selected result from a dialogue between the professionals and community members. In the losers' triangle, problems will tend to be patched rather than solved. The winners' triangle should provide the community with solutions, power, respect, information and control. While it offers a potentially more rewarding way of working, it is, however, at odds with much professional training, which tends to see professionals as rescuers and does little to examine victim-blaming philosophies that are often prevalent. Additionally, most professionals occupy more powerful and privileged positions than the community members with whom they work. Wallerstein and Bernstein (1994) ask whether a relatively privileged group can empower others from this position of dominance (from culture, race, gender, class or status), or whether people have to take power and empower themselves? They go on to say that if the latter is true, then two aspects of the role of health promoters are crucial:

- to serve as a resource and help create favourable conditions and opportunities for people to share in community dialogue and change effort
- to engage in the empowerment process as partners, plunging themselves equally into the learning process. For this role as a partner, health promoters need to ask what they can learn about themselves, their own racism, and how resources are controlled by their own institutions (Wallerstein and Bernstein, 1994, p. 144).

The Australian response to the HIV/AIDS epidemic drew on extensive participation and empowerment approaches and is an excellent example of the successes that can be achieved by doing so (see box 20.3).

BOX 20.3   THE NATIONAL HIV/AIDS STRATEGY 1989–95: PARTICIPATION IN ACTION

The Australian response to the HIV epidemic involved a remarkable and successful alliance of government, community groups and health care services and may be considered one of the most successful examples of participation in Australian health policy history. Government policy support helped build a network of AIDS organisations, including groups of men who have sex with men, people who inject drugs, and sex workers (Altman, 1991). These organisations and the wider alliance were able to stabilise the epidemic by providing peer education, support and treatment services to their own constituencies.

From 1991 to 1995 the total annual number of newly diagnosed cases of HIV infection fell from 1400 to 666 (AIHW, 1996, p. 63). The aims of the National HIV/AIDS Strategy 1989–95 were to eliminate transmission of the virus and to minimise personal and social impact of infection. The key principles of the strategy included (Feacham, 1995, pp. 62–3):

- education and prevention programs to encourage behaviour change to reduce HIV transmission
- encouragement of personal responsibility for avoiding infection and transmission
- elimination of discrimination against people living with HIV and ensuring their human rights are protected
- the cooperation of people living with HIV
- ensuring informed consent is obtained before any person is tested for HIV, guaranteeing confidentiality about the results and providing pre- and post-test counselling
- provision of working conditions that reduce the risk of infection
- encouraging the law to complement and assist education and other public health measures.

The remarkable social and political consensus that enabled the implementation of the HIV/AIDS Strategy, and the better outcomes relative to comparable countries that followed, have earned international recognition (Kaldor, 1996). From the outset, the Australian Government was aware of the necessity of involving people living with HIV in the strategy, and of the risk of excluding them through discrimination or victim-blaming. The then health minister, Dr Blewett, subsequently reflected: 'Authoritarian solutions to this dilemma were risible or ill-thought out. Governments and medical systems simply had to win the confidence of the communities affected. I could not, and still cannot, see any alternative to this kind of a partnership if we were to combat the spread of the disease' (Blewett, 1996).

The key focus for community development approaches was the AIDS councils: bodies in each state that represented the interests of people living with HIV and, particularly, the community of men who have sex with men. The AIDS councils grew out of existing gay movements, galvanised by the threat of the epidemic and the support of Commonwealth funds. According to Altman (1991), federalism, the pre-existence of a gay movement and support from the Australian Labor Party for community health initiatives were the key features that made the AIDS councils into community development success stories. Direct federal funding flowed to the councils, the Haemophilia Foundation and organisations of sex workers and people who inject drugs, enabling them to work with their own respective communities. The resulting activities included support and advocacy, peer education on safe sex, the establishment of needle and syringe programs and a 'safe-house' system for sex workers and their clients. Ironically, this illness that affected some of the most stigmatised groups in the community became notable for the level of polit ical empowerment those groups achieved in dealing with the challenge. As the epi demic progressed, people living with HIV began to emerge as a community in their own right who had certainly been empowered.

EFFECTIVE BUREAUCRATIC CONSULTATIONS

Organisational processes and policies are essential to ensuring practices are compatible with the new public health.

Box 20.4 summarises the features that would maximise a bureaucratic organisation's chances of conducting an effective consultation. Miller (2014) argues that the potential for citizen engagement in public policy making is being eroded despite talk of a 'big society' and an increasing rhetoric of participation. He sees that the dominant neo-liberalism stresses individualism in government rather than solidarity and real meaningful citizen participation.

BOX 20.4   FEATURES OF A BUREAUCRACY THAT PROVIDE CONDITIONS CONDUCIVE TO CONSULTATION

- Official endorsement of consultation at senior levels of the department
- Staff with expertise, experience and skills in consultative practices
- Decentralised and devolved decision-making
- Simple and clear structures and procedures
- Stable functions and continuity of staff
- Economic efficiency and social justice
- Constructive and on-going relationships with communities
- Recognition of the knowledge and experience of communities
- Representative mechanism for diverse communities.

Source: Putland et al., 1997.

TRAINING FOR PARTICIPATION

Most health professionals have not been trained in participation methods. In fact, their training is more likely to have prepared them for a role of professional dominance in which community participation has at best an instrumental role. The World Health Organization funded some research in the late 1990s that looked at ways in which health systems could work more effectively with civil society. The insights from this research were published in a handbook (Laris et al., 2001). There are also numerous guides to community and consumer participation, and a Google search will locate these.

DISSENTERS IN THE SYSTEM

One of the challenges faced by professionals when they work alongside communities to achieve change is that they may find themselves having to oppose actions of the state, even though they are part of the state's work force—the 'state's intellectuals' (Gramsci, 1978). Most governments will be uncomfortable about paying for dissent, so health promoters have to tread a delicate line between their funders and the communities they work with. Petersen (1994) points out that while most health promoters are far from being simply pawns of the state, uncritically implementing what might have the potential to be repressive policies, there are structural constraints on what can be achieved because 'health promotion policy has evolved within a bureaucratic logic that stresses consensual, incremental change rather than radical change' (pp. 216–17). He warns that, in the absence of a theoretically sophisticated approach to community, there is a danger that health promoters who believe they are liberating and empowering people may, in fact, be bringing increased surveillance and regulation but little positive change.

Health promoters are citizens as well as employees of the state. Opportunities for citizen contribution to debates are decreasing as dissenters in systems (whether professionals or citizens) are frozen out of critical debates. Saul (1997) argues convincingly that liberal democracies are increasingly corporatist, and that only highly organised interest groups contribute to public debates. In his view, the voices coming from them are increasingly conformist and so a situation has evolved in which open, creative debate about vital social and economic issues is increasingly rare. Saul extends his argument to say that corporatist society has structured itself to eliminate citizen participation in public affairs. People are encouraged to mind their own business and not be critical, which is dangerous because 'Criticism is perhaps the citizen's primary weapon in the exercise of her legitimacy. That is why, in this corporatist society, conformism, loyalty and silence are so admired and rewarded; why criticism is so punished or marginalised' (Saul, 1997, p. 195). Public health activity at all levels often involves criticising the status quo and arguing for change, which threatens established interests. An open society ready to receive and respond to criticism should be the aim of all public health practitioners, if only because it will make it easier to achieve public health goals. Of course, public health's role as critic and troublemaker will not make for an easy life. But, as Saul concludes, 'a citizen-based democracy is built upon participation, which is the very expression of permanent discomfort' (1997, p. 195).

CONCLUSION

This chapter has examined the complexities of participation in health services and public health in democratic systems. It has shown that effective participation relies heavily on the style of operation and values of health organisations and professionals employed within them. At its heart participation is concerned with the ways in which power is used and distributed. Effective participation should result in redistribution of power so that a broader range of voices is heard and listened to in policy-making processes and service-delivery decisions. Bringing about effective means of participation is central to the new public health and its quest for a more equitable distribution of health.

CRITICAL REFLECTION QUESTIONS

20.1   What is the value of citizen participation in health decision making?

20.2   What factors help ensure that public participation in public policy processes is effective?

20.3   In what ways does power affect participatory processes?

Recommended reading

Germov (2014). *Second opinion: An introduction to health sociology* (5th Edition) provides an excellent guide to issues of power and professional dominance in the health sector.

Miller (2014). Citizen engagement in Australian policy-making, in Miller and Orchard (2014) argues that citizen engagement is being undermined, despite rhetoric to the contrary.

Useful websites

International Association for Public Participation (IAP2) www.iap2.org is an organisation dedicated to improving public participation.

Entering 'participation in health' into Google will produce links to guides on participation produced by health departments and other organisations around the world.

21

COMMUNITY DEVELOPMENT IN HEALTH

Development requires the removal of major sources of unfreedom: poverty as well as tyranny, poor economic opportunities as well as systematic social deprivation, neglect of public facilities as well as intolerance or overactivity of repressive states.

Sen, 1999, p. 3

KEY CONCEPTS

Introduction
What is 'community'?
Community development and social capital
Community development and health services
Community development: ways of working
Dilemmas of community development
Conclusion

INTRODUCTION

The use of community development (or what is called community organising in the USA) strategies to promote health or development was popularised by the South American educator Paulo Freire (1972) through his popular education approaches, and the work of Saul Alinsky (1971), a Chicago community organiser. The popularity of community development has waxed and waned in recent decades. In Australia and in the UK they are most commonly adopted by Labour parties and are often associated with strategies designed to reduce inequities. In the late 1990s community development began to be advocated for under a new policy discourse of social capital, community capacity building or reducing social exclusion. Central to these concepts is the notion of 'community' so this chapter starts with addressing the question of 'what is a community?' Then it proceeds with a discussion of social capital and its links with health. The chapter then considers the ways in which community strategies are used in health promotion, examines community development ways of working and some of the dilemmas associated with this form of practice including issues of accountability and the long-term nature of the strategies.

WHAT IS 'COMMUNITY'?

This question is the old chestnut of community development literature. Much ink has been spilt in attempts to arrive at an agreed definition. Perhaps the most important thing is to recognise that a static definition of community is not helpful. What is meant will differ from place to place, time to time and depend on who is using the term. Undoubtedly the word 'community' is used both symbolically and descriptively. In government statements it often conveys a comfortable and secure image—community care, for instance. This use draws on images of close, caring communities, often invoking the past where problems seemed less pressing and life easier. This notion has had much appeal over the past century and community is often used in a romantic way that is designed to conjure up the possibility of a golden age of close-caring communities.

Critiques of the normative use of 'community' have noted that it can obscure conflicting interests in social and political life and assume consensus that may not exist. In this view it may give the appearance of homogeneity, excluding the leakage of dissent and difference. Feminists have been particularly critical about this. Young (1990), for instance, discusses the politics of difference and notes that while the rhetoric of community may be inclusive, the practice often is not. She suggests that the relationship between different groups in society is far from being cosily comfortable, but 'blotted by racism, sexism, xenophobia, homophobia, suspicion and mockery' (p. 319). She suggests that if a community is to be progressive it must be underpinned by politics of difference that provide for political representation for different groups and that celebrate the distinctive cultures and characteristics.

In another sense 'community' is used more descriptively: 'The people who live in a defined geographic locality, and/or who share a sense of identity or have common concerns' (Fry and Baum, 1992, p. 297). This use of the term is common in public health. Community health refers to services designed to meet the health needs of a defined, usually geographic community. Sometimes, however, they will be directed more specifically. Often a service with a specific geographic responsibility will define particular communities within their area with whom they want to work. These could form the Aboriginal community, Vietnamese community or a particular group of workers such as outworkers in the textile industry. These groups will differ in the extent to which they identify as a community, and individuals within the groups will differ in the extent to which they identify with their particular community. Increasingly people participate in online, virtual communities that transcend geography. These groups share some characteristics of other forms of community and need to be considered as a potential site for community development, although little is known about their structure or operation. Laverack's (2004) definition illustrates that people may belong to a number of communities at any one time and that communities are rarely homogeneous. His four characteristics are:

- a spatial dimension (this could be a street, township, suburb, online)
- non-spatial dimensions (interests, issues, identities) that involve people in groups who are otherwise disparate and heterogeneous (gay men's group, netball club)
- social interactions that are dynamic and bind people into relationships with one another (Public Health Associations, colleagues in a workplace)

- identification of shared needs and concerns that can be achieved through a process of collective action (political party, environmental action group).

'Community' cannot be treated uncritically. It is a word with cultural importance and its meaning is certain to change when used in different contexts by different people. Health promoters should also be wary of assuming that their definition of community is accepted by those they are defining into a community group. This was well-illustrated in the case of Aboriginal people in Australia where it has often been assumed that Aboriginality alone is sufficient to define a community whereas for Aboriginal people the picture of their social organisation is much more complex (Hunter, 1993).

TYPOLOGY OF COMMUNITY INTERVENTION

A classic definition of types of community intervention was first presented by Rothman and colleagues in the early 1970s. This has been updated and adapted by Minkler and Wallerstein (2012) to incorporate more recent thinking about community organising and building, and is shown in figure 21.1. This typology incorporates needs- and strengths-based approaches. The needs-based approach is conceptualised as either more consensual (community development) or conflict based (social action). The newer strengths-based models contrast a capacity-building approach with an empowerment-oriented social action approach. New public health community initiatives usually have a mix of these types of interventions. Community initiatives from within state-funded institutions can rarely adopt 'pure' social action. Even if community members and workers are committed to such an approach, they are generally limited by the constraint of ensuring funding continuity. Often this means adopting a social planning approach that gives the appearance, at least, of rationality.

COMMUNITY DEVELOPMENT AND SOCIAL CAPITAL

One of the important aims of community development is to strengthen communities so that they are better able to support the health and well-being of residents, and enable them to promote and protect health in their community. Two decades of research on social capital suggests that it is important to health, especially mental health (see chapter 15). Community development or organising can be one strategy through which social capital can be strengthened.

WHAT IS SOCIAL CAPITAL?

There is a vast amount of literature on social capital, which has grown particularly rapidly in recent years (Winter, 2000; Halpern, 2005). Common to most definitions is a focus on networks between people that develop trust and then lead to cooperation and beneficial outcomes. Coleman (1988) defines it as coming about through changes in the relations among people that facilitate action, noting that it is less tangible than physical or human capital (skills and knowledge possessed by individuals) because it

FIGURE 21.1 COMMUNITY ORGANISATION AND COMMUNITY-BUILDING TYPOLOGY

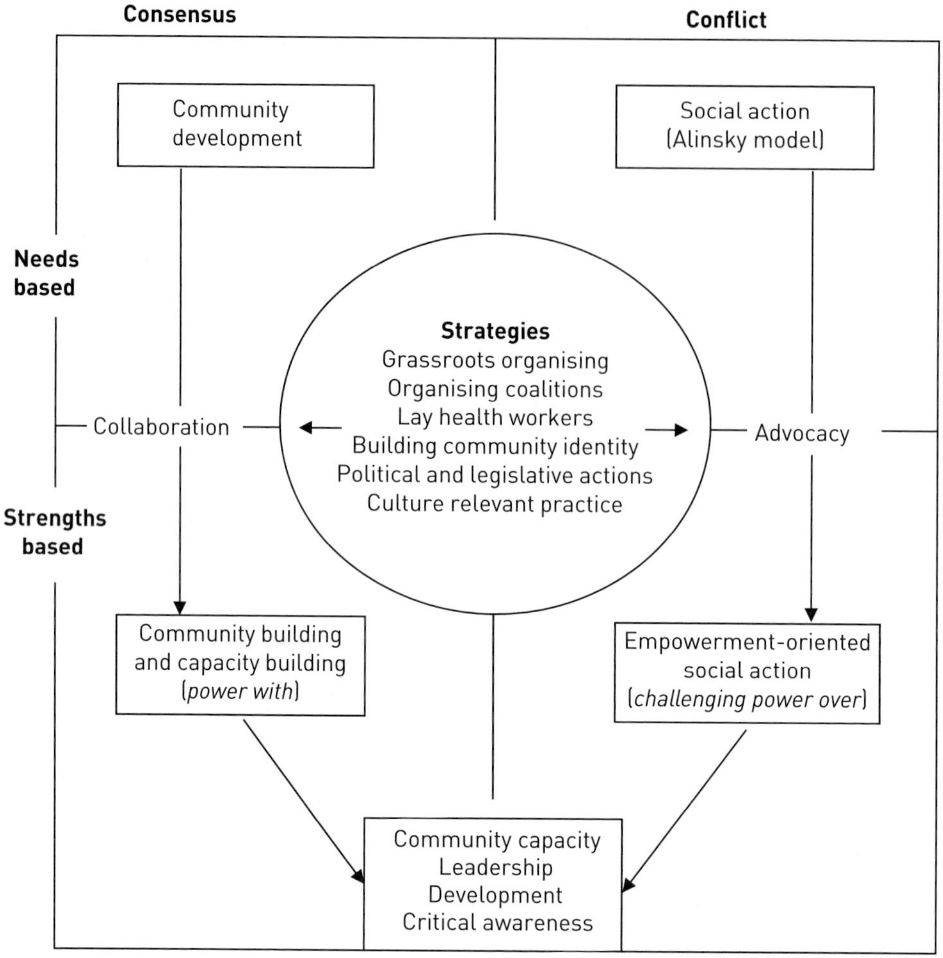

Source: Minkler and Wallerstein, 2012.

exists in the relations among people. This is, of course, why social capital is important to community development as the relationships built through its processes develop networks and trust. Trust is also seen as central to the successful operation of these networks. Beyond this, theoretical definitions of the concept differ, from Putnam (1993a, 1993b, 1996, 2000), whose focus is quite narrow and does not consider issues such as power, to Bourdieu (1986) whose concept is based on the role social capital plays in the reproduction of class relations, especially by mediating economic capital. Bourdieu, in particular, stresses the power and status dimensions of social capital. His view of social capital is particularly relevant to the new public health because he considers social capital as one of the ways in which economic capital is reproduced and consolidated. This happens when the social connections people have lead to gains such as job or investment opportunities—an example is the operation of 'old boys' networks'. Thus it follows that community development designed to increase social capital must focus on the ways in which it can bring benefits to communities and individuals within it. The role of power is central to Bourdieu's theory of social capital and largely ignored by Putnam.

Throughout the literature on social capital, the existence of trust in relationships emerges as the key factor in determining the extent to which a community or society can be seen to have a high level of social capital. The literature on social capital refers to three main types of trust. The first is trust of familiars, which exists 'within established relationships and social networks'. The second is generalised trust, which is extended to strangers 'often on the basis of behaviours or a sense of shared norms'. Third, there is civic or institutional trust, which relates to basic forms of trust in the formal institutions of governance, for example fairness of rules, official procedures, dispute resolutions and resources allocation (Stone and Hughes, 2000). Community development has the potential to contribute to each of these forms of trust.

Together, reciprocity and trust characterise societies in which people are able to cooperate effectively to achieve common civic goals. These societies are those that provide their citizens with multiple opportunities to interact and network through groups, associations and societies. There is significant evidence that both social and civic trust are declining in Australia and in most countries around the world (Hughes et al., 2000).

Increasingly commentators distinguish between three types of social capital: bonding, bridging and linking. Bonding social capital is that between relatively closely knit groups, who are likely to share many characteristics in common. It may be exclusionary and may not act to produce society-wide benefits of cooperation and trust (Baum and Ziersch, 2003). Bridging ties are looser than bonding ties and operate across differences in, say, culture or ethnicity, but not in terms of institutional power and influence. Linking social capital refers to relationships between people and groups that operate across explicit, formal or institutionalised power or authority gradients in society (Szreter, 2002; Szreter and Woolcock, 2004). It is the latter two forms of social capital that community development is likely to contribute to in order to promote health and to reduce health inequity.

SOCIAL CAPITAL, HEALTH AND COMMUNITY DEVELOPMENT

> The vocabulary of social capital formation is an ethical vocabulary: trust, respect, concern, solidarity, dignity. Its practices will lead to a strengthening of moral practice including consensus building and collective decision-making rather than the disempowerment of directives or authoritarian dictums.
>
> *Reid, 1997, p. 6*

It has been noted that as well as consisting of trust and respect, social capital also reflects 'the creation of alliances across difference' (Reid, 1997, p. 5). Where social capital exists or can be created, 'mutual aid societies spring into existence and credit training schemes, social investment funds and other such development initiatives are more effective.' Thus Reid concludes that social capital 'makes possible participatory development and good governance'. This makes the concept particularly important to the new public health and to community development strategies.

Evidence of links between social capital and health has major implications for health promotion and public health policy and practice, suggesting the need for a greater focus in health promotion and public health activity on:

- the levels of social and civic trust and factors that affect the levels of trust
- opportunities for people to come together and establish bridging and bonding networks and trust
- the nature and quality of interactions between people rather than on individual behaviour or risk factors
- increasing collective efficacy, which is the ability of community members to undertake collective action for shared benefit, such as residents collectively lobbying to force the removal of polluting industry from residential areas
- the creation of cities, neighbourhoods and communities that are socially cohesive and supportive (see chapter 17).

Clearly community development is one of the ways in which these points can be made to happen as it focuses on action in whole communities rather than with individuals alone. Putnam (2000, p. 20) notes that a 'well-connected individual in a poorly connected society is not as productive as a well-connected individual in a well-connected society. And even a poorly connected individual may derive some of the spillover benefit from living in a well-connected community'. While community development can rarely challenge the structural basis of economic inequity, such as the debt burden of poor countries or the distribution of wealth, it can make some contribution to building networks and trust that may result in less advantaged people gaining access to the education, employment or capital that will make a difference to their quality of life and eventually their health. An example of the explicit use of social capital in community development is provided in box 21.1.

BOX 21.1   KIDSFIRST: CANADIAN EARLY CHILDHOOD INTERVENTION USING COMMUNITY DEVELOPMENT AND SOCIAL CAPITAL

KidsFirst is a Canadian provincial government-initiated, community-based early childhood intervention program implemented in nine sites with high community needs in Saskatchewan, Canada. It explicitly used the theory of social capital to strengthen community fabric and develop social capital in the community. It built institutional social capital through employing local people (including Cree speakers and elders), encouraging staff to deepen connections with the community, and making the services more user-friendly through measures such as the provision of childcare, food and transportation. The program created bonding relations among the families, linked them to services and bridged them to the broader community. The program required the government's central policy framework and commitment to working with the community to enable them to own and shape the program to the particular characteristics of the communities.

Source: Shan et al., 2014.

Szreter and Woolcock (2004) argue that consideration of the relationship of the state in terms of the initiation and sustaining of networks, trust and social structures is crucial. They show, with illustrations from a case study of nineteenth-century England, that states (local and central) can create and encourage the conditions in which linking social capital can operate. States can do this by ensuring that resources flow from more powerful to less powerful groups. One of the ways by which they can do this is through resourcing and providing empowering institutional support to community development exercises that are designed to increase the resources available to poorer communities and the capacity of the people within those communities.

The ways in which community development can be used to build social capital as a pathway to creating health and well-being are detailed in a manual (Pomagalska et al., 2009).

USES OF COMMUNITY DEVELOPMENT

> A change is brought about because ordinary people do extraordinary things.
>
> *Barack Obama*

Community development in a number of areas, including public health, has grown in both popularity and credibility. This has happened most noticeably in developing countries through the work of development agencies, such as the World Bank and the United Nations Development Program (UNDP), and many non-government organisations (NGOs), which have seen community development as the answer to improving the living conditions of the world's poorest people.

Community development was used in Australia as a strategy throughout the period following World War II. Agricultural projects used it extensively in the 1940s and 1950s (Dixon, 1989), but it was under the Whitlam Labor government (1973–75) that community development flourished. Many of its programs preferred this way of working, thus associating community development in Australia with the Whitlam government, which Australians tend to have either strongly supported or strongly opposed. Public health from the 1980s onwards has shown a considerable interest in community development strategies. There has been much thinking about its practice and theory in relation to public health (Community Development in Health Project, 1988; Baum, 1989; Dixon, 1989; McWaters et al., 1989; Dwyer, 1989; Legge, 1992; Legge et al., 1996; Wass, 2000; Baum et al., 2012). Community development strategies have also been the basis of public health action in the UK including some strategies in the Health Action Zones, Neighbourhood Renewal, and New Deal for Communities initiatives designed to reduce social exclusion and inequities.

The advantages of community development based on popular participation have been summarised by Oakley (1991, pp. 17–18):

- *Efficiency*: participation can increase efficiency as people are more likely to be convinced of the benefits of initiatives they have helped develop. If local people are involved in projects, this reduces the amount of time needed by paid professional staff, and so becomes a cost-effective option. Oakley adds a caution here, however, as he points out that community development can become an excuse for shifting the costs of services and development onto already resource-deprived and poor communities.

- *Effectiveness*: community development can make initiatives more effective by allowing people to have a voice in determining objectives, supporting project administration and making their local knowledge, skills and resources available. Health promotion imposed on people is rarely effective.
- *Self-reliance*: refers to the positive effect on people of participating in community development in health projects. The participation can help to break dependency (which has characterised much health and welfare work in the past) and so promote self-awareness and confidence, helping people examine their problems and be positive about solutions. Community development also involves individual development and increases people's sense of control over issues that affect their lives, helping them to learn how to plan and implement, and equipping them for participation at regional and national levels. This helps create social capital, a hallmark of a healthy community.
- *Coverage*: health promotion has tended to be more successful in reaching those who are already relatively healthy. Community development offers a way of working with people who are the least healthy.
- *Sustainability*: experience from numerous development projects indicates that those who are externally motivated frequently fail to be sustained once the initial level of support is reduced or withdrawn. The chances of sustainability are increased in situations where local people are the main dynamic. Community development can contribute to a momentum of change in an area.

Laverack (2004) suggests that the main point of community development should be to increase the power of communities to be able to take social and political action and suggests five ways in which this might happen (see box 21.2).

BOX 21.2   COMMUNITY EMPOWERMENT: INDIVIDUAL AND COLLECTIVE ACTION

- *Empowering individuals for personal action*—many people in poor communities will not have the skills to take collective action (such as how to work in a group, how to solve conflict, how to write a letter to politicians). These skills form the basis of collective action.
- *Development of small mutual groups*—people come together around issues that they feel are important to their lives. These issues could be about any of the social determinants of health (housing, violence, transport, environmental). A health promoter's job is to work with the group without directing it and to keep the focus on socio-environmental causes of poverty and ill health and avoid focusing on individual problems. A variety of techniques can be used to show the links between individual problems and structural factors.
- *Development of community organisations*—these groups have an established structure, clear leadership and ability to organise their members to mobilise resources.

They might be church, youth or women's groups, farmers' cooperatives, resident action groups.

- *Development of partnerships*—these partnerships may be between groups coming together in a coalition (for example of environmental groups) or between community structures and health services (see example of the World Health Organization's work on health development structures on p. 558) in order to achieve common aims. Effective partnerships are difficult to achieve and some of the features of these are described in the following chapter.

- *Taking social and political action* for the purpose of improving health and redressing inequities. This may include civil protest and other forms of political action. Inevitably gaining power to influence economic, political, social and ideological change will involve community groups in struggles with those in powerful positions. This form of empowerment becomes very difficult for health promoters employed by state bureaucracies and is a central dilemma for community development, as discussed below

Source: Laverack, 2004, pp. 48–54.

COMMUNITY DEVELOPMENT AND HEALTH SERVICES

Comprehensive approaches to public health and primary health care emphasise community development, empowerment and capacity building as the basis of effective strategies (box 21.3). More selective approaches tend to be of limited effectiveness because, while one particular disease may be cured, another comes along to take its place because the underlying structural problems have not been cured.

BOX 21.3   MAIN FEATURES OF COMMUNITY-BASED HEALTH PROMOTION

PRINCIPLES

- Uses a socio-environmental approach to health promotion that encompasses medical, behavioural and community development strategies
- Based on a recognition of the importance of power differentials in determining health outcomes and the abilities of groups to promote their own health
- Recognises the diversity of communities and the particular needs of subgroups within a defined area, in terms of variables such as of gender, ethnicity, class and age
- Concerned with achieving equitable health outcomes
- Is informed and strengthened by the participation of local people in management, program planning, implementation and evaluation.

STYLE OF PRACTICE

- Focuses on the health of the people in a defined geographic area or community of interest
- Community members define the issues on which the health promotion effort focuses
- Development, through a partnership between community members and professionals, of a comprehensive knowledge of local people, their environment and needs
- Uses this knowledge to identify and analyse local health issues, and to develop and implement initiatives
- Rests on models of professional practice that stress partnerships with communities and strives to overcome professional hegemony
- Involves advocacy and the provision of a public voice for the health of the local community
- Main strategies are based on community development practice.

Much of the documented innovative community development in Australian health work has taken place within community health centres, especially in Victoria and South Australia. Unfortunately that style of working appears to have become less prevalent in recent years, during which there appears to have been a greater focus on curative intervention and behavioural interventions in community health services. Box 21.3 defines the main features of community-based health promotion and box 21.4 provides examples of types of community development projects that are run in some Aboriginal health services. These services use community development strategies to connect with people for whom the services are hard to reach. The development approach enables the service to adapt to people's needs and be responsive to different sections of the community they serve.

BOX 21.4   COMMUNITY DEVELOPMENT IN ABORIGINAL HEALTH SERVICES

Some of the best examples of community development in Australia are from Aboriginal community controlled health services. Community development is used by the Central Australian Aboriginal Congress in most of its programs. Examples are the men's health program Injintka, which provides a drop-in service for men where they can have showers (important if they are sleeping rough), talk over problems, hear about other activities they might like to get involved in, join a men's shed program or just hang out. A community health education program also engaged schools to provide a holistic sexual health education. The team employed local Aboriginal staff to educate young people about sexual and reproductive health, including a program of simulated babies for young people to take home.

An Aboriginal Health Team that is part of the state government health service also has used community development strategies to good effect. One example is the regular 'Nunga Lunches' and health camps hosted by the service. The Aboriginal Health Team reported that their lunches and camps were an important way of both reducing social isolation and providing opportunities for the attendees to come and feel comfortable with a range of services relevant to their health:

> With the Nunga Lunches, that is the most important gathering for the community people ... It's an opportunity for us to promote our health professionals in this area as well, and so that the community can put faces to names.

Baum et al., 2012

The camps are designed to have some fun and healthy activities including fishing, walking and sharing meals. They also provide an informal way of talking with people about their health and encouraging them to use services that they might not otherwise have used. The benefits of the camps were described by one of the workers as:

> At camps it's wonderful, there's a lot given out at camps because you've got them, they can't go anywhere, 'You're stuck with us for four days.' We have certain sessions every day, there are maybe three different sessions. You've got your little fun time, it's not a prison so you can go fishing now and when we come back we'll wash up, ... then at three o'clock we're meeting in the boardroom at the camp we're at and we've got someone talking about kidney and alcohol and drugs and everything.

Baum et al., 2014a

The features in box 21.3 are unlikely to be found concentrated in one community health centre. They are benchmarks against which a centre's health promotion activity can be judged, but they demonstrate that community health centres provide an excellent base for community development work to be established.

The centres typically use a range of strategies, including community development. The work of community health practitioners is interrelated so that, while part of it may focus on one-to-one work, this activity is important to other aspects of the centre's work. The individual attending the one-to-one activity may be encouraged (once they have the confidence) to join a group, which may lead to the development of an action campaign on a broader health issue. The work with individuals permits health workers to gain an intimate knowledge of the people they are serving. This information is important to the centres' planning processes. For instance, a counsellor may note that women who are consulting her about relationship problems may be isolated and have no support, so she could respond by forming a support group or finding out what other activities there are in the locality. Community development can be an important means of increasing patient safety (Baum et al., 2012), especially for people who find health services inaccessible for reasons such as perceived lack of cultural safety, language or fear of professional power. Community development strategies enable people to become familiar and comfortable with a health service and so are more likely to use the services when needed. At the social-political activity level the centre might spearhead a campaign on gaining more support for survivors of domestic violence or on behalf of refugees. Inevitably such action becomes difficult for community health centres that are funded by the state. It is, therefore, important for health agencies to develop policies that are supportive of community development work by protecting innovative and politically 'risky' practices and constraining practices that are disempowering. Such policies are described in box 21.5.

BOX 21.5   HEALTH AGENCY POLICIES SUPPORTIVE OF COMMUNITY DEVELOPMENT

They must contain specific social analyses and models that:

1   Locate personal troubles in political systems
2   Recognise community development as concerned with process rather than being static
3   Recognise community development both as a philosophy involving all practitioners and a practice specific to some practitioners
4   Define community development as a practice that supports social action around structural conditions of power/powerlessness
5   Support accountability methods that most embody the ethical stance and social analyses that inform the institution's policies on community development
6   Develop explicit criteria for supporting specific individuals, groups or organisations that are public and publicly defensible.

Source: Boutilier et al., 2000, p. 273.

Existing community groups can also be the basis of community development activities by health services. The World Health Organization (WHO) recognised this through their development of the concept of Health Development Structures (HDSs) that are community groups or organisations that play a role in promoting health (defined in a broad sense). A report on these structures noted 'the majority of HDSs owe their origins to age-old community traditions of mutual support and cooperation and have a long history of community action' (WHO, 1994, p. 70; Baum and Kahssay, 1999). They include social clubs, youth groups, women's groups, mutual aid societies, cooperative societies, some functions of sporting clubs and representative health councils. Most health promotion activity has hitherto concentrated on inviting community people to participate in activities established (and largely controlled) by health agencies and their personnel. WHO funded research that led to the development of guidelines specifying the ways in which health services can develop partnerships with community groups (Laris et al., 2001). The process involves an analysis of existing local community structures to identify the many, often invisible, 'health' roles played by such groups and then demonstrate the way the health service needs to prepare its organisation to form effective partnerships. These groups can be an efficient and effective way for health service personnel to use the knowledge and skills of their local community in planning, service development, fundraising and advocacy work. Underlying this approach is the increasing recognition that a healthy society is one with high levels of civic engagement providing cohesiveness and trust.

COMMUNITY DEVELOPMENT: WAYS OF WORKING

You must be the change you wish to see in the world.

Mahatma Gandhi

Community development practice draws heavily on the work of Freire (1972), who advocated education for liberation using these stages:

- reflection on people's lived reality
- analysis and collective identification of the root cause of that reality
- examination of their implication
- development of a plan of action to bring about change.

In this process professionals and community members should ideally meet as equals and develop a dialogue based on trust. The aim of the process is critical consciousness-raising through critical reflection on the structures that constrain empowerment and the ways in which these structures might be challenged (Labonté and Laverack, 2008). Ife and Tesoriero (2006) discuss the role of professional community development workers and note that social work practice should be driven not only by careful analysis but also by a passion to make the world a better place. The same can be said of community development work within the new public health. The challenge for the new public health activist is to find a means by which passion and rage against injustice can be channelled into a useful practice that results in action that makes a difference. People working for change need to find ways of maintaining their sense of vision, purpose and passion. Ife and Tesoriero suggest that often this can be achieved by drawing inspiration from the struggles of others, such as Aung San Suu Kyi, Nelson Mandela, Xanana Gusmão and Martin Luther King. Community development workers are often also inspired by the lives of the people they work with, who, despite enormous odds, live lives full of dignity and meaning. Obama (2012, p. 31), in discussing his community organising work, said 'organising teaches as nothing else does the beauty and strength of everyday people' and that from their stories 'of dashed hopes and powers of endurance, of ugliness and

Example of a GetUp organised community protest for improved mental health services, which was part of a broader campaign. (Fernando M. Gonçalves)

strife, subtlety and laughter, that organisers can shape a sense of community not only for others, but for themselves'. Learning a respect for the knowledge and determination of such people is an important asset for a community development worker.

Community development workers have to understand how the processes of power affect their work. Kenny (1999, p. 153), drawing on the notion of power as diffuse and as operating locally, notes that the workers 'must always be sensitive to the myriad of manoeuvres, techniques, dispositions, tactics and languages through which power relations are expressed, maintained and altered.'

DILEMMAS OF COMMUNITY DEVELOPMENT

AMELIORATION OR REAL CHANGE?

A crucial issue concerning community development is whether its aim is to bring about real change in people's lives and the structures that constrain them or if it acts to make the conditions poor people are living in a little more bearable. Mayo and Craig (1995) ask whether community development is always used as a tool of democratic transformation or whether, for agencies like the World Bank, it serves as the 'human face of structural adjustment'. It is possible that the World Bank uses terms such as 'empowerment' differently to a progressive NGO, such as Oxfam. But they recognise that, even with problematic questions such as this, community development and participation are still vital. They see these strategies being increasingly advocated in both developed and developing countries in 'the context of increasing poverty, polarisation and social exclusion' (p. 3).

Ife and Tesoriero (2006) discuss many of the dilemmas that will be faced by community development workers and stress that for practice to be useful the workers must be continually reflecting on the contradictions in their work and the fact that their work risks doing little more than making poverty more acceptable. They warn (p. 262) that community development structures and processes 'can easily reinforce the dominant structures of oppression'. They argue that reflective community development should address issues of class, gender, race/ethnicity, age, disability and sexuality. They also acknowledge that doing this is often difficult and the extent to which it can be done will vary between communities and depend on the broader political climate.

CAN LOCAL-LEVEL ACTION EFFECT STRUCTURAL CHANGE?

A further dilemma is the extent to which action at the local level can be effective in bringing about significant, as opposed to superficial, change. Today global influences are evident throughout everyday life: the media is increasingly transnational; transnational corporations control a considerable amount of retailing and manufacturing; communities are more mobile and less geographically bounded, due to travel and migration; and electronic communication and environmental problems do not respect national, let alone local, borders. In this context there is a risk that local action can ultimately be disempowering (Israel et al., 1994). For instance, if a campaign to improve the work conditions of textile outworkers is successful in Australia, but manufacturers decide to shift their operations to a developing country, who has been empowered in the process? If a community group successfully opposes plans for a toxic waste dump, which is then shifted to a community that does not have the capacity to organise, is there actually any shift in power? These issues are tough but important. Wallerstein and Bernstein (1994, p. 144), in an introduction to special editions of *Health Education Quarterly*, note: 'Ultimately, community empowerment strategies must also be linked to the larger society to ensure policy and political solutions that decrease health and socioeconomic inequities and foster healthier places to live.'

It has been claimed (Anderson, 1996, p. 702) that the empowerment movement draws upon the ideology of individualism. It is no accident that the rhetoric of empowerment has gained momentum precisely when governments are moving to cut expenditure on health and social services. Empowerment strategies focus on what people can do to empower themselves and so may deflect attention from social issues. Labonté and Laverack (2008) warn that unless national and international trends are taken into account, the decentralisation of decision-making may shift from victim-blaming of individuals to victimising powerless communities.

Ultimately, shifts in power are unlikely to result from local action but rather from a shift in policy nationally and increasingly internationally. Local action can help bring such shifts about through forcing the hand of policy makers. This nutcracker effect whereby policy action is combined with bottom-up advocacy is vital to progressive change and is described in chapter 24 (see figure 24.2). The importance of advocacy to policy change is detailed in chapter 22.

It is also possible to design community development so that it accounts for the fast-changing political and economic landscape. Wright et al. (2013) describe an arts for social change project that considers the context of the 'geopolitical landscape and "hypercomplexity"' in which it is implemented (see figure 21.2). This big picture informed the strategies of reinventing identities of young people, reinvigorating their communities and reimagining alternative futures.

ACCOUNTABILITY

The ultimate aim of community development in health is to empower people and their communities in such a way that individual and collective health status is improved. The management of community development requires considerable flexibility, as it does not easily fit within rational planning, management and evaluation frameworks. The values base of community development and the relinquishing of at least some control to the community mean that public health organisations have to ensure they develop management and evaluation frameworks that support community development activity.

ACCOUNTABLE TO WHOM?

The experiences of health promoters suggest that community development is often not understood or valued by health funding authorities. The work is often diffuse and it is hard to specify in advance exactly what form the initiative will take. Indeed, if it could

FIGURE 21.2   BIG HART: PERFORMANCE ARTS FOR SOCIAL CHANGE SET IN ITS
COMPLEX SOCIAL-POLITICAL CONTEXT

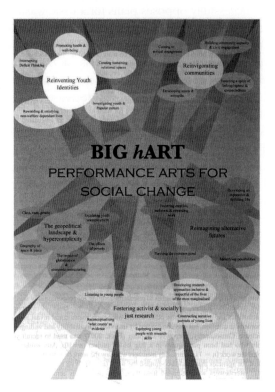

Source: Wright et al., 2013, p. 191.

be specified in advance, it would almost certainly not be community development. The
process of working with communities is invariably messy and a lot less controllable than
traditional health promotion, which selects target groups and delivers pre-set programs,
an approach that fits the rational planning frameworks preferred by most bureaucracies.
Often a health authority may identify the local health problems to be connected with
issues such as child health and parenting, older people, mental health and services for
ethnic minorities. By contrast, the community may identify quite different priorities,
such as developing anti-racist strategies within the health services and the community,
building community health organisations, developing community and user representation
in the health services and raising awareness of health issues. Health workers who work
for the community may risk their employment status, but to meet the expectations of
their employer would mean betraying the trust of the community and so undoing what
is likely to have been a considerable amount of groundwork. Ellis and Walton (2012)
describe the involvement of a local health department in a Healthy Neighbourhoods
project and talk of the importance of the department's staff showing 'cultural humility',
needing to acknowledge institutionalised racism, exposing health inequities, building on
community assets, making a commitment to sweeping change and giving up a degree

of control over outcomes. They note when these factors are in place the payoffs for the communities and the health departments can be substantial.

The trend in many high-income countries is for a withdrawal from community development work as part of health service activities. Where it is likely to survive is in community-controlled services (see box 22.4), where perhaps the benefits are obvious to the community even though direct quantifiable health outcomes cannot be measured.

Activity initiated by community members and without formal links to any health promotion agency is part of public health activity. There are many social movements and community groups that arise from the passion and interest of local people and achieve very healthy outcomes for their local area, although they are not labelled 'health promotion'. An example would be the work of resident action groups to protect the health of their community.

Ife and Tesoriero (2006) see considerable benefit in community development work being independent from the state. They recognise that there is sometimes no alternative to government support, but argue that such sponsorship weakens rather than strengthens the community basis of the initiative. They believe that community development with radical aims is unlikely to happen within a state-sponsored structure and urge communities to look for alternatives. The state-sponsored work generally demands accountability to government-set priorities and styles of working. They see the 'ideal' model of community development as one where any workers are directly accountable to the community they work in. They also note that community development in poor countries is often more in touch with the needs of local people because, of necessity, self-reliance is at the centre of the initiatives.

LONG-TERM AND DEVELOPMENTAL NATURE OF COMMUNITY DEVELOPMENT

A further dilemma of community development, from the perspective of funding bodies, is the developmental and long-term nature of the work. Health promotion projects using community development as their basis are often only funded for a short period. In a disempowered and disadvantaged community it will often only be possible to establish legitimacy, gain trust and work out some priorities for action, and if funding then ceases, very little will have been gained. Much community development work has to be seen as an investment that should take place over a long time scale to yield maximum rewards. This is one reason why community health centres are well positioned to be the centre of community development in health activity. They are already known and trusted. Activities can be built on each other, and the community can develop a sense of ownership—not just of particular initiatives, but of the centre itself. This can help overcome the difficulties of short-term funding.

OUTCOMES AND EVALUATION

Community development is difficult to evaluate using traditional research methods (Baum, 1992, 1998). Action research and qualitative methods are more fruitful, but not always acceptable to funders. This does seem to be changing, however, and the wider acceptance of community development techniques by international funding agencies (including the World Bank) may see this acceptability increase. The particular considerations that need to be considered when evaluating community development were detailed in chapter 10.

Some studies of community development approaches have been able to document outcomes. A community arts program in Northcott, a public housing building in Surry Hills, Sydney, was linked to a reduction in crime in the community (Coggan et al., 2008). In Nepal a controlled trial of women's groups that used an action-learning cycle to identify and address perinatal problems showed a reduction of about one-third in the neonatal mortality rate in the intervention group (Manandhar et al., 2004).

CONCLUSION

This chapter has shown that community development has much to offer long-term strategies to improve health. The approach does this by working with people to empower them to take action to change the conditions that shape their health. Often this means tackling powerful or traditional institutions and so community development can be viewed as a risky strategy for state-funded health departments to use. Despite this, community development is used as part of public health strategies by international, national and local health authorities. It is a core strategy of the new public health.

CRITICAL REFLECTION QUESTIONS

21.1   What is a community?

21.2   How can health services use community development strategies?

21.3   To what extent do health services you know embody the criteria in box 21.5?

21.4   How might community development challenge existing power bases?

Recommended reading

Pomagalska et al. (2009) *Practical social capital: A guide to creating health and wellbeing* provides a guide to using social capital for the purposes of public health community development. Available at http://som.flinders.edu.au/FUSA/SACHRU/Publications/PDF/PSC_A_guide_to_creating_health_and_wellbeing.pdf

Minkler (ed.) (2012). *Community organizing and community building for health and welfare* discusses theoretical and evaluation frameworks for community development and provides practical examples

Useful website

Bank of I.D.E.A.S http://bankofideas.com.au provides lots of examples of the use of community development in practical projects.

22

PUBLIC HEALTH ADVOCACY
AND ACTIVISM

So please, allow this old man in front of you to insist that unless we all become
partisans in renewed local and global battles for social and economic equity in
the spirit of distributive justice, we shall indeed betray the future of our children
and grandchildren.

*Dr Halfdan Mahler, DG Emeritus addressing the 61st World Health
Assembly, May 2008*

KEY CONCEPTS

Introduction
What is public health advocacy and activism?
Who are public health advocates and activists?
What are key advocacy and activism strategies?
Advocacy and activism dilemmas
Conclusion

INTRODUCTION

Public health advocacy and activism involves individuals and organisations who operate
locally, regionally, nationally and globally and work to combat unhealthy policies,
products and practices. The landscape of public health advocacy and activism is laid
out in figure 22.1. Issues may relate directly to health, concern a particular risk factor or
express overarching concerns about structural threats to health. Chapman (2008, p. 29)
points out that advocacy has played a crucial role in translating research into policy
and practice, yet its study and teaching is neglected. He puts this down to a perceived
incompatibility between achieving scientific credibility and the overtly political nature of
advocacy and activism. Yet if researchers want their research to influence public policy,
advocacy is one of the main ways by which this can happen. The strategies used by
advocates and activists are many and varied. Each of these issues is further unpacked
in this chapter.

FIGURE 22.1   THE LANDSCAPE OF PUBLIC HEALTH ADVOCACY AND ACTIVISM

People and organisations

Individuals: Concerned citizens, aggrieved citizens, researchers.

Professional associations: Public health, medical, health promotion, community health.

Consumer driven: Consumer health organisation, treatment action campaigns, healthy food access.

Global movements: Health for all, environmental, anti-neo liberal/capitalism, anti-poverty, pro-equity, peace movement, feminist movement.

AIMS
Improved preconditions for health
Improved health and health equity

Public protest
Non-violent direct action
Civil disobedience
Picketing
Activist art
Petitions

Single issue
Access to health care
Debt reduction
Food access
Tobacco
Responsible drug use
Opposing an unhealthy development

Lobbying politicians and governments
Media advocacy
Social media: Facebook, Twitter
Activist research

Broad overarching concern
Environment
Anti-capitalism/ neo-liberalism
Wealth distribution
Social determinants
Health equity
Anti-poverty
Women's rights

Strategies

Issues

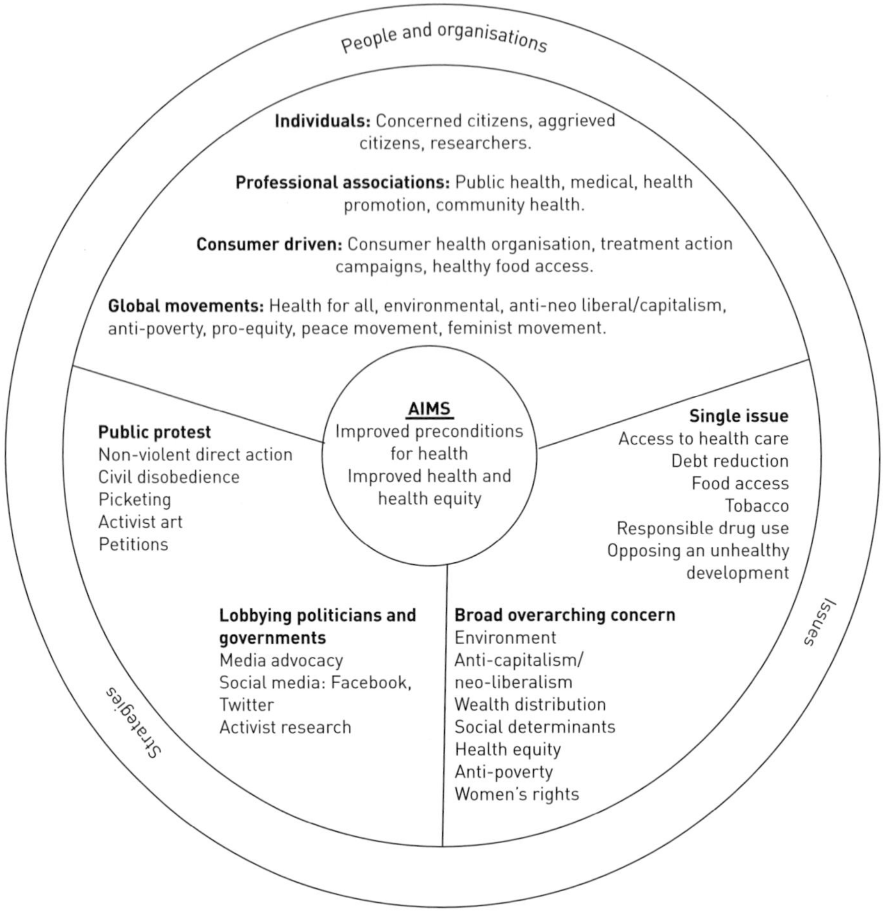

WHAT IS PUBLIC HEALTH ADVOCACY AND ACTIVISM?

One of the most important ways of influencing policy related to health is through advocacy and activism. Chapman (1994, p. 6) notes that public health advocacy is used 'most often to refer to the process of overcoming major structural (as opposed to individual or behavioural) barriers to public health goals'. Advocacy is a powerful tool of the new public health: 'Advocacy is necessary to steer public attention away from disease as a personal problem to health as a social issue, and the mass media are an invaluable tool in this process. Advocacy is a strategy for blending science and politics with a social justice value orientation to make the system work better, particularly for those with least resources' (Wallack et al., 1993, p. 5).

Laverack (2013, p. 1) defines activism as an 'action on behalf of a cause, action that goes beyond what is convention or routine'. Of course what is routine or convention will differ depending on the context so it is hard to pin down exactly what activism is. Box 22.1 provides examples of different types of public health advocacy and activism

---

BOX 22.1   EXAMPLES OF PUBLIC HEALTH ADVOCACY AND ACTIVISM

- The Close the Gap campaign to eliminate the gap in life expectancy between Indigenous and non-Indigenous Australians (see box 22.2).
- A group of brain surgeons who lobby for safer road design with the aim of preventing brain injury.
- Local residents who work together to persuade their local council to reduce the speed limit and implement traffic-calming measures to make their suburb safer for pedestrians.
- The international movement against buying products from the Nestlé company in protest against their marketing of breast milk substitutes in developing countries.
- A group of local residents lobbying against the diversion of a polluted river to their coastal suburb because of the marine pollution and environmental degradation.
- Action from BUGA UP (Billboard Utilising Graffitists Against Unhealthy Promotions) against tobacco products.
- International Physicians Against Nuclear War.
- Action by the People's Health Movement in favour of the 'Right to Health'.
- Action by the Medical Association for the Prevention of War against the devastating health impact of the war in Iraq.
- Corporate Accountability International, who campaign against unhealthy corporate practices

Chapman (2008), one of Australia's foremost public health advocates, reminds us that advocacy is strategy and not an end in itself, and that advocates should always be able to point immediately to the public health objectives they are trying to address. These include (p. 25):

- new law and regulations
- enforcement of existing laws and regulations, including stronger penalties
- more funding for programs
- tax rises or reduction on products to depress or increase demand
- changing clinical or institutional practices
- having other sectors direct energy at a health issue.

WHO ARE PUBLIC HEALTH ADVOCATES AND ACTIVISTS?

There is a high degree of overlap between advocates and activists, and the dividing line between them is blurry. Public health advocacy and activism can be undertaken by:

- individuals in their role as concerned citizens
- organisations
- coalitions of organisations.

Advocacy and activism can happen locally (for example, resident action groups), regionally (for example, state branch of a public health association or an environmental group), nationally (for example, professional associations, consumer associations), globally (for example, People's Health Movement, Oxfam International, Save the Children). Increasingly activism is online. New information technologies have given public health and environmental movements unprecedented ability to make quick responses to events, share information and put pressure on governments and corporations. Avaaz is an online global activist network with 38 million members in 194 countries (www.avaaz.org/en/index.php) and GetUp! is an Australian independent, grassroots, community advocacy organisation online (www.getup.org.au). They use email and other social media to stay in contact with their constituencies on a range of issues relating to health and environmental well-being. The potential of the internet as a global advocacy mechanism was first demonstrated on a major global health policy issue in 1998 by a concerted campaign by non-government organisations (NGOs) around the world against the Multilateral Agreement on Investment (MAI). These groups used websites to publicise the details of the proposed MAI and analyses of its likely impact. The internet was used to maintain communication between NGOs around the world and to spread the latest details of the negotiations. These techniques enabled protest groups to pool their information so that 'they have broken through the wall of secrecy that traditionally surrounds international negotiations, forcing governments to deal with their complaints' (Drohan, 1998).

A Fresh Water Embassy was established in South Australia during the 2009–10 drought to protest at mismanagement of the Murray–Darling River System. (Fernando M. Gonçalves)

Civil society is the heart of advocacy and activism and is defined as 'the associational life that brings people together and allows civic values and skills to develop' and as a public space 'where debate and deliberation allows the negotiation of the common interest. A healthy civil society is one where voluntary associations of people, groups and organisations focused on an identified common good can thrive' (National Council for Voluntary Organisations, n.d.).

Worldwide there are millions of civil society organisations with concerns that relate to health services, the determinants of health and health equity. Foremost among these are the public health and health promotion associations, which bring together professionals in associations that enable them to be more effective lobbyists. The Public Health Association of Australia (PHAA) has a very robust policy development process and this has resulted in a range of policies on diverse issues such as child health, drugs, environmental health, food and health, health care financing, housing, health and international trade, media ownership, public health research funding and women's health. The policies of the association are available at www.phaa.net.au. It lobbies on each of these issues to government. The work of the People's Health Movement has been described in box 5.11. It is a global network of health activists which has campaign circles and country circles in approximately 40 countries. It receives very little funding and runs predominantly on the efforts of volunteers.

WHAT ARE KEY ADVOCACY AND ACTIVISM STRATEGIES?

Successful strategies should set an agenda, frame the issue for public consumption and advocate specific solutions. Wallack et al. (1993) suggest that public health advocates can catalyse public opinion, bolster the public's willingness to support the proposed solution and gain access to key opinion leaders and community decision-makers. Strategies used by advocates are as varied as the imagination of activists and some of the most prominent are listed below. Most campaigns will use a mix of these strategies. Box 22.2 provides details of the range used by the Close the Gap campaign.

BOX 22.2   CLOSE THE GAP CAMPAIGN

One of the highest profile activist campaigns in Australia in recent years has been the Close the Gap campaign, which demands action to close the gap in life expectancy between Indigenous and non-Indigenous Australians. The campaign was mounted in 2006 by a broad-ranging coalition of Indigenous organisations, professional associations (such as the Public Health Association of Australia and the Australian Medical Association) and civil society groups (such as Oxfam and the Fred Hollows Foundation). It has high-profile Olympic gold medal winners Cathy Freeman and Ian Thorpe as patrons and they launched the campaign in April 2007.

Strategies include: lobbying politicians, launching progress reports in Parliament House, public awareness campaign

Successes: establishing a steering committee with an impressive range of support from key national organisations, winning bipartisan support for the Close the Gap Statement of Intent, a Closing the Gap strategy from government with funding and implementation through partnership agreements between the Australian Government and the states and territories, and $4.6 billion of funding.

Co-Chairs of the Close the Gap Steering Group: Ms Kirstie Parker (Co-chair of the National Congress of Australia's First Peoples), Mr Mick Gooda (Aboriginal and Torres Strait Islander Social Justice Commissioner, Australian Human Rights Commission) with Prime Minister Tony Abbott in Parliament House for the launch of the Close the Gap progress report.

Source: Australian Human Rights Commission, 2015.

Lobbying governments and politicians is a key advocacy activity. Emailing and meeting with local politicians or with cabinet ministers is a crucial means of public health advocates and activists putting their points across. In encounters with politicians facts will help the advocacy case but it is also clear that stories are vital too. Marshall Ganz (2009) (who is credited with designing Obama's activist campaign for his successful presidential bid in 2008) notes how powerful stories are to campaigns and in talking with people in power. He notes, 'A story communicates fear, hope, and anxiety, and because we can feel it, we get the moral not just as a concept, but as a teaching of our hearts'. This appeal to emotions is as important as an appeal to facts when working for change. Personal stories were used effectively in the campaign for justice for victims of asbestos-related disease against the company James Hardie, who continued to use the product despite the evidence of its health harms. The Australian Asbestos Network website (www.australianasbestosnetwork.org.au/about) contains a series of stories of victims that bring to life the terrible effects of asbestos on people's lives. Stories of hardship and medical evidence of adverse effects have also been an important aspect of the campaign against the Australian Government policy of holding asylum seeker children in detention. One success of this campaign has been a national inquiry into this practice (www.humanrights.gov.au/our-work/asylum-seekers-and-refugees/national-inquiry-children-immigration-detention-2014).

Protest at Australian Government policy of holding children in detention. (Fernando M. Gonçalves)

This point was also made by Don Hewitt (producer of the USA *60 Minutes* program), (Wallack et al., 1993), who said: 'Acid rain isn't a story, it's a subject. Tell me a story about somebody whose life was ruined by acid rain, or about a community trying to do something about acid rain, but don't tell me about acid rain.' This means public health campaigns will often be more successful if there is someone who can speak with an 'authentic voice' and provide a face to a story.

Dorfman and Krasnow (2014) point to the importance of 'social math' to translate statistics and other data so they become interesting to journalists, meaningful to audiences, and therefore helpful in advancing healthy public policy. Social math includes making data local (e.g. number of obese or overweight people in a local council area compared with local fast food outlets); an ironic comparison (we spend more on locking people up than in keeping them out of jail); breaking statistics down to a time period (number of people who die each hour from smoking-related causes).

Boycotts involve not using, buying or dealing with a person, organisation, country or product as a deliberate protest. This strategy was used effectively in the campaign against Nestlé's promotion of their product as preferable to breast milk (Wallack et al., 1993). The problem of declining breastfeeding in developing countries was initially seen as the failure of the mothers, until the Infant Formula Coalition Action (INFACT) used strategies such as a boycott of Nestlé's products, a CBS documentary and lobbying of WHO to adopt the International Code of Marketing Breast Milk Substitutes. Boycotts were also an effective part of the international protest against apartheid in South Africa. More recently, disinvestment campaigns have been launched as a means of changing unhealthy practices. These involve persuading consumers to withdraw funds from banks and other financial institutions that are engaging in supporting unhealthy or environmentally damaging practices. As an example, GetUp! launched a campaign against Adani, the approved developers of the Carmichael Coal Mine in Queensland, highlighting the company's lack of respect for the environment in previous developments and seeking funds to publicise these facts to potential investors so that Adani would be unable to fund the development because of withdrawal of investment funds (GetUp! Australia, n.d). Chapman (2008) warns that boycotts are more difficult to execute in practice than they appear. In Australia there are also moves to amend consumer laws so that secondary boycotts are illegal (Denholm, 2013).

Public protests take many forms and can include marches of thousands of people, street theatre with an activist message or other staged events. In past years there have been marches against wars (the Iraq war in 2004 was protested by millions of people around the world), the austerity agenda in many countries, the WTO and trade treaties, and meetings of the G8 and G20. Publicity for public protests is greatly aided by the existence of social media and email, as messages can be spread rapidly. Access to health care has been the focus of much advocacy and activist activity (see box 22.3). The Occupy movement has launched many public protests. This movement started as the Occupy Wall Street movement against social and economic inequity and used the slogan 'We are the 99%' in reference to the uneven distribution of wealth.

We are the 99% protest in Berlin, Germany. (Stefan Boness, Panos)

Greenpeace has very effectively used public protests to campaign on environmental issues (see photo). Street theatre can be a very effective way to put across complex messages in a simplified form. An example of its use was by the NSW Nurses and Midwives Association during the July 2014 meeting of the G20 trade ministers in Sydney to highlight the likely impact of the Trans-Pacific Partnership Agreement (NSW Nurses and Midwives' Association, 2014). Adbusters uses cultural jamming to put across messages about the excessive power of corporations and the destructive impact of consumerism on the environment (www.adbusters.org).

Greenpeace protesting in favour of climate treaty, Sydney Opera House, 2009. (David Sewell, Greenpeace International)

BOX 22.3   ADVOCACY FOR UNIVERSAL HEALTH CARE

THE ROLE OF COMMUNITY ORGANISATIONS IN ACHIEVING AN AFFORDABLE HEALTH CARE ACT IN THE USA

In March 2010 US President Barack Obama signed the *Patient Protection and Affordable Care Act* (ACA) into law. It was designed to provide affordable access to health care for more than 30 million US citizens who lacked coverage, and improve coverage for tens of millions more. It would, however, still leave around 20 million citizens uninsured. Part of the reason the ACA could be passed despite powerful corporate forces opposing the measure was because of decades of advocacy from local state and national groups and experience from the failure of the Clinton health reforms. This meant there was a network of advocacy organisations in the USA, including a large and formal organisation, Health Care for America Now. Anderson et al. (2012) identify three premises underpinning the success of the movement:

- engaging the uninsured to protect their own interests and tell their stories, which was a powerful lobbying tool
- engaging other interested stakeholders (unions, faith organisations, multi-issue grass roots community organisations) motivated by justice rather than self-interest
- focusing attention on the more widespread benefits of the reforms beyond extending coverage to the uninsured, including rate increases and denial of coverage to those with pre-existing conditions.

In India Jan Swasthya Abhiyan (JSA), the Indian circle of the People's Health Movement, has held 'People's Health Tribunals', in which they have put state health authorities on trial for not realising the right of access to health care. JSA is also running a campaign against the privatisation of health care in India.

People's Health Movement India holding a People's Health Tribunal in Tamil Nadu. (Jan Swasthya Abhiyan – People's Health Movement, India)

THE DEFENCE OF MEDICARE IN AUSTRALIA

There was strong advocacy for the initial introduction of Medicare, Australia's universal national health insurance scheme, in 1983, and since then any threats to its universality and the introduction of increased co-payments have been met with opposition. In the late 1990s a Friends of Medicare campaign was launched with the campaign slogan 'It works, it's fair, it's Medicare'. It comprised a coalition including the Public Health Association of Australia (PHAA), the Australian Nursing Federation (ANF), the Australian Council of Social Service (ACOSS), the Doctors' Reform Society, The Australian Women's Health Network, and the Health Issues Centre.

Most recently, campaigns have been launched following the 2014 federal Budget, which proposed to introduce a $7 co-payment for general practitioner (GP) visits. The proposal has been met with advocacy from the Australian Medical Association, the ANF, PHAA and ACOSS. Tactics have included public meetings and extensive use of social media. The protests have included demands to retain Medicare and to stop corporatisation of health care. In March 2015 in the wake of the considerable community protest, the Abbott Coalition government dropped the idea of introducing a GP co-payment.

Protest at proposals to cut health funding in the 2014 Budget. (Fernando M. Gonçalves)

Non-violent direct action: This form of activism draws on Greenpeace's actions in the Southern Ocean against Japanese whaling.

A current example of non-violent direct action is the opposition to coal seam gas in the Northern Rivers area of New South Wales, Australia, which has used grassroots organising and has organised blockades of coal seam gas drilling sites. Their website (Gasfield Free Northern Rivers, n.d.) says:

> Late 2012 [and] early 2013 saw almost four months of continuous blockades at Metgasco drill sites near Grafton and Kyogle with hundreds of community members turning out to block drill rigs and determined prote[s]tors sustaining permanent roadside camps in floods and wild weather to maintain a constant presence. People who had never ever protested before put their bodies on the line to defend their environment and community. Indigenous elders stood beside farmers to protect land and water. Massive police numbers, at huge taxpayer expense, were required to facilitate drilling operations.

As a result of this opposition the two companies drilling in the region announced they were suspending their operations indefinitely. A national organisation—Lock the Gate Alliance—links such local campaigns that are using community blockades. Their website says that 'in a David-and-Goliath struggle of farmers against mining giants, everyday citizens against global corporations, our communities are choosing grace under fire and displaying incredible courage, integrity and imagination' (Lock the Gate Alliance, n.d.).

Picketing originated as actions by trade unions against employers when a picket line was formed to protest against employer practices, but it has also been used more generally to draw public attention to a cause. In the Philippines, tobacco shows have been picketed as tobacco companies pay more attention to markets in middle- and low-income countries as rich companies tighten restrictions on tobacco (Teves, 2012).

Civil disobedience

> One has a moral responsibility to disobey unjust laws.
>
> *Martin Luther King Jr.*

Civil disobedience is the active, professed refusal to obey certain laws, demands or commands of a government. This tactic was most famously used by Gandhi during the Indian movement to gain independence from British rule. A contemporary example is the practice of 'tree sitting', in which environmental activists sit in a tree to prevent logging of old-growth forests.

Public art is often used as a form of protest. This can include street theatre, internet games or graffiti in public places. The BUGA UP (Billboard Utilising Graffitists Against Unhealthy Promotions) campaign in the 1980s organised its members to change the message of advertising billboards in a humorous way (for instance, changing 'Welcome to Marlborough Country' to 'Welcome to Cancer Country').

Example of the work of BUGA UP.

Petitions are increasingly used to target single issues and are much easier to organise than in the past because of the internet. Change.org hosts many petitions on a range of social determinants. Victories listed on the website include winning a ban on student guestworkers in the USA who were being exploited and taking away jobs from locals (Change.org, n.d.), and action against homophobia in Australian Rules football (Change.org, n.d). A browse of change.org's site (www.change.org) will highlight many other health-relevant examples.

Media advocacy, including that on social media, has become an increasingly important public health activity because of its society-wide influence. According to Walt (1994, p. 66) mass media serve a number of functions in contemporary society:

- as agents of socialisation by transmitting society's culture, values and norms
- as sources of information
- as propaganda mechanisms, in that they seek to persuade people to support particular policies or buy consumer goods
- as agents of legitimacy for the dominant political and economic institutions.

Dorfman and Krasnow (2014) suggest that media advocacy addresses power gaps between powerful and less powerful groups in society. They see advocacy as a public health strategy that is quite different in focus to the use of mass media to persuade individuals to change their behaviours. They see it as being about 'raising voices in a democratic process' (p. 295). Media advocacy can reshape public agendas so that, instead of public health issues being continually individualised (Tesh, 1988), the underlying structural and environmental issues can be highlighted. The main target is policy decision makers. Media bites are extremely important—these are concise statements that highlight the main points of the campaign (see box 22.4). Social media has offered powerful new strategies for health activists—Facebook is used by many campaigns to publicise and seek supporters for the campaign, and Twitter has proved to be an important way of spreading information very quickly. There is an increasing number of health blogs that provide venues for advocacy (see box 22.5).

BOX 22.4   MEDIA BITES

An analogy that brings a picture to mind: '[Food] marketing is just washing over this country like a tidal wave, and we're trying to give people swimming lessons' [Kelly Brownell, Rudd Center on Food Policy and Obesity, Yale University].

A statement that makes the extent of the public health issue evident: 'it's like two jumbo jets crashing every day with no survivors' [a quote from tobacco advocates illustrating the number of deaths from smoking-related causes every day].

Source: Dorfman and Krasnow, 2014, p. 299.

BOX 22.5   PROGRESSIVE HEALTH BLOGS

Healthy Policies: for a healthier world (www.healthypolicies.com)

This blog site examines the structural determinants of health and considers the upstream political processes that explain how different people have varying levels of resources. The site says:

> Through research, policy analysis and advocacy, Healthy Policies is committed to actively advancing public polices which protect and expand access to resources important for health. We hope to provide a platform for multi-disciplinary action and encourage people

to engage with, and sometimes challenge, discourses surrounding public policy and health.

Healthy Policies, n.d.

Croakey (http://blogs.crikey.com.au/croakey)

The Croakey blog is a forum for debate and discussion about health issues and policy. It is moderated by journalist Melissa Sweet and encourages debate about a wide range of health service and public health issues. For instance, after every Budget it contains contributions that assess the new Budget's likely impact on health.

Using research for activist purposes is increasingly common in civil society. Good recent examples are the report by Oxfam (2015) on growing health inequities, and the *Global Health Watch* produced by the People's Health Movement, which is positioned as an alternative World Health Report (see box 22.6). Professionals who are independent of vested interests play an important role in 'watching' and researching powerful industries such as pharmaceutical companies. Healthy Skepticism provides a very good example of such activity (box 22.6). A study of top public health researchers in Australia (Chapman et al., 2014) found that a willingness and capacity to engage with the mass media was seen as an essential attribute of influential public health researchers. This study also noted, however, that it was rare for the researchers to receive any training that would enable them to engage the media effectively.

BOX 22.6   EXAMPLES OF ACTIVIST RESEARCH

Global Health Watch (www.ghwatch.org)

The *Global Health Watch* is a broad collaboration of public health experts, non-government organisations, civil society activists, community groups, health workers and academics. It was initiated by the People's Health Movement, Global Equity Gauge Alliance and Medact as a platform of resistance to the neo-liberal dominance in health. It is now in its fourth edition (People's Health Movement et al., 2014), which contains 38 chapters that critique neo-liberalism, consider current health systems debates, examine broader determinants of health, 'watch' global organisations and look at 'resistance, actions and change'. The latter section includes chapters on grassroots struggles in Latin America, Europe, India and Australia.

Municipal Services Project (www.municipalservicesproject.org)

An excellent example of activist research is the Municipal Services Project (MSP), which conducts a range of participatory research 'that explores alternatives to the privatization and commercialization of service provision in electricity, health, water and sanitation in Africa, Asia and Latin America'. Its members are academics, labour unions, non-government organisations, social movements and activists from around the globe who are committed to analysing successful alternative service delivery models to understand the conditions required for their sustainability and reproducibility.

Healthy Skepticism: improving health by reducing harm from misleading drug promotion (www.healthyskepticism.org/global)

This organisation invites supporters to subscribe to its regular updates of analyses of drug promotions. It aims to raise awareness about the dangers of misleading drug promotion, to lobby health authorities, to engage constructively with the pharmaceutical industry and to provide support for like-minded individuals and organisations. Its actions have led to products being withdrawn or reformulated on a number of occasions following correspondence with drug companies. It has provided paid services to government bodies, universities and WHO. Its contributions to improving health have been acknowledged in the *Lancet* and the *BMJ*. The approach of Healthy Skepticism is built on the principle that any claims made regarding a medication (or other health product) should be clearly supportable by good-quality evidence. The onus to supply the evidence should be on the person, organisation or company making the claim, rather than on the sceptic to refute an unsubstantiated claim.

ADVOCACY AND ACTIVISM DILEMMAS

Both advocacy and activism pose dilemmas that have to be faced and resolved as part of activism and advocacy work

FOCUS OF ADVOCACY WORK: SINGLE ISSUE OR SYSTEM REFORM?

Single-issue advocacy campaigns are generally easier to pursue than those that focus on broadscale system reform. Thus the campaign against the adverse health effects of tobacco, while facing concerted opposition from the tobacco industry, was able to be very targeted and specific. In rich countries, the campaign had some very real wins—for example, in Australia, daily smoking rates have declined from more than 70 per cent to a rate in mid-2014 of 12.8 per cent (AIHW, 2014e). Jubilee 2000, another single-issue campaign, has focused on debt cancellation.

This single-issue and targeted approach contrasts with more broad-brush campaigns that are concerned with changing whole systems. Aiming for broad system change is more difficult because the goals and strategies are likely to be less clear and more open to debate. The Australian Social Determinants of Health Alliance (www.socialdeterminants. org.au) has as its main goal to have Australian governments commit to implementing the recommendations of the Commission on Social Determinants of Health. This goal presents many dilemmas for advocates, such as where to focus effort (social determinants involve all government departments). The Occupy movement (www.occupytogether.org) also had very broad aims in wanting to reduce inequality and oppose neo-liberalism. The original occupation was forcibly removed but it has inspired a generation of activists, and in the years since the occupation of Wall Street has become an anticorporate protest movement (Wedes, 2013).

CONFLICT OR CONSENSUS?

Activism can be more grounded in either consensus or conflict approaches. Most campaigns will have a mix of both, and judgments have to be made by activists about which approach suits which circumstance. Strategies involving public protest, for example, are more conflictual than those involving direct lobbying of politicians.

BITING THE HAND THAT FEEDS YOU

Civil society and non-government organisations are an important part of the policy-making machine because they provide an independent voice and, as such, are vital to democratic society. Yet their role has been progressively undermined in recent years. Maddison and Hamilton (2007) describe how funding has been withdrawn from non-government organisations who have criticised government policy, and that those that do receive funding are expressly forbidden to play an advocacy role. Such an example is the Public Health Association of Australia, which, in association with other non-government organisations, developed a 'Friends of Medicare' campaign with the slogan 'It works, it's fair, it's Medicare'. The association ran this campaign very vigorously during 1999–2001, much to the annoyance of the Coalition government. One of the results for the association was that its government grant (which it had been receiving since 1987) was removed. This highlights the dilemma for non-government organisations: whether to accept government funding but then find that their ability to speak out is limited, as no government wants to finance an organisation that bites the hand that feeds it.

This is a real dilemma for civil society and professional associations that receive funding from government or whose members rely on government for their employment. On an individual level advocates and activists may be perceived as troublemakers and so suffer personally as a result of their activities.

The Howard Coalition government introduced no-advocacy clauses into contracts with non-government organisations. These were removed by the subsequent Labor governments but are now being reintroduced by the Abbott Coalition government. Maddison and Hamilton (2007, p. 79) argue that criticism of government policy from non-government organisations is an important feedback loop for governments and makes an important contribution to robust and deliberative democracy.

POWERFUL ADVOCATES AGAINST PUBLIC HEALTH

Public health advocates and activists face extremely powerful advocates who are concerned primarily with corporate profits rather than health. A US group, OpenSecrets. org (Center for Responsive Politics, n.d.), estimates that in 2013 the food and beverage industry spent over $30 million, the pharmaceutical industry $226 million, and the oil and gas industry $145 million on lobbying, with total spending on lobbyists (across all interest groups) estimated at $323 billion, using 12 341 lobbyists. Corporations also hire public relations (PR) firms to set up phoney grassroots advocacy groups. PR firm Burson-Marsteller has formed many of these, including the National Smokers' Alliance (NSA), developed for Philip Morris; the World Council for Sustainable Development, an international big business organisation behind hijacking the 1992 Rio Earth Summit; and the Forest Protection Society, which was set up to lobby for the Australian timber industry (Bohme et al., 2005). In Australia it is hard to obtain details of the extent of corporate lobbying, but one example is the Mineral Council of Australia, a coal lobby group that has 39 staff and an income of $58 million. It is estimated that over the two years to early 2014, the coal lobby has spent over $100 million in lobbying (Burton, 2014). This points to the David versus Goliath battle over environmental issues when community groups (such as the Northern Rivers campaign on coal seam gas mentioned previously) rely on volunteer grassroots community action with little or no funding.

A further tactic from corporations that may undermine the work of public health activists is that corporations provide sponsorship to existing consumer groups and this is highly likely to influence the agenda of the group (Freudenberg, 2014). An example would be a drug company funding a consumer group who then argues for that drug to be made available through public health systems.

GETTING COVERAGE IN THE MEDIA

It can be hard for advocates to get their activities reported in the mainstream media. Commentators have also noted the implications of the increasing concentration of media ownership. Chadwick (1998) pointed out that, in Australia, newspaper ownership became much more concentrated over the course of the twentieth century. In 1923 there were 21 separate owners of 26 capital city newspapers. By 1996 there were only four owners of 12 newspapers. Similar concentration of ownership is evident in the electronic media. Chadwick (1998) warned that such concentration has the potential to threaten public health, which does not have the resources to compete with commercial interests. Although the extent of the media influence over society is disputed, there is little doubt that media empires exercise great power over public political debate and determine which issues get aired. Bowman (1997, p. 8) has warned that 'great political and social power exercised by a few media proprietors should be unacceptable to anyone with any instinct for democracy. It's crudely, offensively, anti-democratic.'

Public broadcasters have been one medium for advocates to gain coverage for their issues through investigative journalism. Yet there is considerable pressure to reduce the budgets of these organisations, as seen in the UK, Canada and Australia. Social media is more democratic and enables advocates to bypass the mainstream print and electronic media to spread advocacy messages and organise campaigns.

MANAGING RELATIONSHIPS

Being an advocate or an activist involves being reflective about the different types of relationships involved. These include relationships with the people who are the subject of the activism, and relationships with an activist group. A Canadian organisation, the Inner Activists (www.inneractivist.com), provides a range of courses to help activists develop the emotional and psychological skills 'to be a transforming influence in the world'. Shields (1991, chapter 8) provides an excellent chapter on the challenges of working within an activist group and covers issues such as using conflict constructively, creating safety, encouraging and supporting leadership, making room for fun and humour, and respecting diversity.

PERSONAL PRESSURES OF BEING AN ADVOCATE

Being an advocate is not easy on the personal level. It requires emotional energy and commitment. Activists may be seen as 'troublemakers'. Effective advocates know that they will be subject to criticism from the powerful interests they are opposing—indeed, that criticism is a mark of their effectiveness. On the positive side, activism and advocacy done from within a supportive movement can be an important part of a happy and meaningful life, as is expressed here:

> Every day I get better at knowing that it is not a choice to be an activist; rather, it is the only way to hold on to the better parts of my human self. It is the only way I can live and laugh without guilt.
>
> *Staceyann Chin (Jamaican poet and political activist), 2007, p. 365*

BURNOUT

A real risk for health activists is burning out. People who devote their time to advocacy and being an activist are often doing this on top of work and family commitments. They are often highly committed people who work extremely hard for the cause they believe in. Often this results in burnout where the activism involves too much psychological stress. This may also be exacerbated by the fact that activism often involves conflict, both with the object of the activism but also sometimes with other activists.

This phenomenon of burnout is well recognised. Shields (1991) provides an excellent guide for activists that contains a section on preventing burnout. She also warns that burnout is not just about individuals—groups can also burn out, and they need to be 'burnout-proof'.

CONCLUSION

This chapter has examined the many different ways that public health advocacy can be conducted. It has outlined key strategies and tactics, provided examples of advocacy campaigns and noted some of the dilemmas involved in being an advocate. The following chapter examines the healthy settings approaches to health promotion.

CRITICAL REFLECTION QUESTIONS

22.1 What are the dilemmas involved in being both a health professional and an advocate?

22.2 What factors make for a successful public health advocacy campaign?

Recommended reading

Chapman (2008) *Public health advocacy and tobacco control: Making smoking history*

Laverack (2013) *Health activism: Foundations and strategies*

Useful websites

There are numerous sites with examples of advocacy. Here are a selected few that focus directly on health, and those that address a social determinant of health.

Health

People's Health Movement (www.phmovement.org)

Public Health Association of Australia (www.phaa.net.au) is a professional association that has a very well developed policy development process and advocacy strategy.

Medact (www.medact.org) is a UK-based organisation of 'health professionals for a safer, fairer and better world', which runs many campaigns including those on peace and demilitarisation, health and human rights, economic justice, and climate and ecology.

Determinants of health

Australian Council on Children and the Media (http://childrenandmedia.org.au) campaigns to promote 'healthy choices and stronger voices in children's media'.

Avaaz (www.avaaz.org/en) is an online global activist network with 38 million members in 194 countries.

GetUp! (www.getup.org.au) is an Australian advocacy organisation that campaigns on a wide range of social and economic determinants of health.

23

HEALTHY SETTINGS, CITIES, COMMUNITIES AND ORGANISATIONS: STRATEGIES FOR THE TWENTY-FIRST CENTURY

The key strategic point of the settings approach was to move health promotion away from focussing on individual behaviour and communities at risk to developing a strategy that encompasses a total population within a given setting.

Ilona Kickbusch, 2015

KEY CONCEPTS

Introduction

'Settings' approaches to health promotion

Bringing about change in healthy settings initiatives

Political and policy leadership and commitment is essential

Encouraging action across sectors

Types of partnerships

Detailed examples of healthy settings initiatives

Legislative frameworks that support healthy settings in the workplace

Healthy settings projects in the workplace

Healthy cities and communities

WHO's Healthy Cities program

Healthy Cities in Australia

Healthy Cities: actions for health

Settings with a specific focus: obesity prevention in cities and communities

Sustainability of healthy settings

Critical perspectives on healthy settings approaches

Conclusion

INTRODUCTION

Initiatives around the world are focusing on health promotion strategies that aim to change the social and/or physical environment to promote the health of people and environments. This builds on the thinking of Rose (1992, and see chapter 1) that effective and sustainable public health strategies must lower the risk of the whole population and not just those at the high-risk end of the distribution. These initiatives work across sectors and use strategies to engage communities. Leadership comes from local government, health departments, environment departments, workplaces, schools and community groups. Together they represent a powerful force for change through which to create healthier and more equitable societies. This chapter first looks at the features of successful settings projects. It then considers some crucial tools to assist all settings projects (change management, action across sectors and politics and leadership). Detailed examples of settings initiatives are provided through case studies of workplace health promotion, the WHO Healthy Cities initiative and the use of healthy settings in obesity prevention. The chapter concludes with a consideration of some critical perspectives on settings approaches to health promotion.

'SETTINGS' APPROACHES TO HEALTH PROMOTION

The ideal shape of health promotion in the twenty-first century is that it should be embedded within the operation of organisations such as schools and workplaces and evident in the way local communities and cities plan for the future. This approach involves a shift away from the behaviourally focused health promotion of previous decades. Kickbusch (1996), a key architect of the new public health, has stressed that the healthy settings approach is about asking the question: what creates health in our setting? She stresses that many of the first step solutions are organisational rather than linked directly to health behaviour.

For an organisation or other setting to ask this key question of 'what creates health?' requires a concerted strategy of change in the way the organisation works, relates to the world outside its boundaries and its understanding of the factors that create health and well-being. The settings projects usually reflect the health promotion philosophy expressed in the Ottawa Charter (WHO, 1986) with an emphasis on the achievement of health through an integrated holistic approach. The World Health Organization (WHO, 1998) defines settings for health promotion as 'the place or social context in which people engage in daily activities, in which environmental, organizational and personal factors interact to affect health and wellbeing'. The hallmarks of the settings approach to creating health are shown in box 23.1.

BOX 23.1   HALLMARKS OF A SETTINGS APPROACH TO CREATING HEALTH

- Focus is on the setting and enhancing its ability to create health rather than on changing the behaviour of individuals directly—social, economic and environmental change strategies rather than those based on the psychology of individuals.

- Focus is not only on reducing social, economic and environmental risk conditions but also about creating a healthy setting in a positive and holistic sense.

- Genuine participation by all key stakeholders is encouraged.
- Change in the culture of the setting and organisation is one of the goals, and change management will be a central point of the activity.
- Health creation is the centre point of a planning process that generally includes devising a vision of improved health in the setting and a series of goals and strategies to achieve this vision.
- Health is conceived as being about more than physical safety risks and has a broad socio-environmental conception of health.
- The setting is conceived of as being networked to a variety of other settings and organisations rather than as existing in isolation.
- The creation of more equitable access to goods and services that promote health is a central focus so that more equitable health outcomes can be achieved.

The types of initiatives these can give rise to are shown in box 23.2. These examples demonstrate the way the hallmark features are put into practice.

BOX 23.2   HEALTHY SETTINGS: APPROACHES AND EXAMPLES

HEALTHY SCHOOLS PROJECTS

Key approach

The focus is on the whole of the school environment, including the social, physical and community, and extends beyond health education in the classroom to consider a range of policy issues (for example bullying, nutrition policy for the school canteen) and environmental issues (trees in the school yard, growing vegetables, recycling).

Examples

In Australia the National Healthy School Canteen (NHSC) guidelines and resources provide national guidance and training to help canteen managers across Australia to make healthier food and drink choices for school canteens. The guidelines (Department of Health, 2014b) build on activities of state and territory governments and encourage a nationally consistent approach to promoting healthy food through Australian school canteens. The guidelines include three components: a national food categorisation system for school canteens; training materials for canteen staff; and an evaluation framework.

HEALTHY FOOD MARKETS

Key approach

In developing countries an important aspect of Healthy Cities initiatives has been the development of healthy markets. These are defined as those that promote the safety of the food supply from production to final consumption. Strategies have focused on safe food-handling practices, improving the market premises and training for market vendors. WHO has produced a *Guide to Healthy Food Markets*, which explains the importance of ensuring healthy food supplies and provides guidelines for improving market environments and promoting safe food handling in them.

Example

The Buguruni Healthy Market Strategy in Dar es Salaam, Tanzania, has included a range of local and international donor partners. The project developed an action plan, which was reported by WHO (2006) as achieving the following outcomes:

- improvement in road access
- construction of a solid waste storage bay
- construction of a toilet and hand-washing facilities
- development of a system for the collection and sorting of solid waste for subsequent disposal.

The synergies between these initiatives have contributed significantly to improved hygiene in the markets. The initiative has involved an education program for stallholders and users of the market.

HEALTH PROMOTING HEALTH SERVICES KEY APPROACH

This approach aims to move from an exclusive focus on disease to a mandate to improve and promote health. It involves the health service as a whole and its relationship with the broader community.

Examples

The Women's and Children's Hospital, Adelaide, South Australia, developed an overall organisational change process to encourage the hospital to become more health-promoting, which attracted strong support from senior management (Johnson and Baum, 2001). Specific activities included the development of an advocacy plan to encourage staff, especially senior staff with community authority, to advocate on public health issues such as car safety barriers, encouraging parents to become partners in their children's care, and involving community members in decision-making about hospital resource allocation (Johnson and Paton, 2007).

The European Health Promoting Hospitals (HPH) Initiative began in 1988. The aim of the HPH project is to improve the quality of care by supporting the provision of health promotion, disease prevention and rehabilitation activities in hospitals. Health promotion is considered a core quality dimension of hospital services, as are patient safety and clinical effectiveness. The specific objectives of the Health Promoting Hospitals Network (HPH Network) are:

- to change the culture of hospital care towards interdisciplinary working, transparent decision making and active involvement of patients and partners
- to evaluate health promotion activities in the health care setting and build an evidence base in this area
- to incorporate standards and indicators for health promotion in existing quality management systems at hospital and national levels.

Source: WHO Regional Office for Europe, 2015b.

HEALTH PROMOTING PRISONS

Key approach

The key aims of a healthy prisons approach are:

- building the physical, mental and social health of prisoners (and, where appropriate, staff)
- helping prevent the deterioration of prisoners' health during or because of custody
- helping prisoners adopt healthy behaviours that can be taken back into the community (Baybutt et al., 2007).

Examples

WHO in Europe has a health-in-prison project (www.euro.who.int/prisons), and the UK Department of Health (2002) published a report, *Health Promoting Prisons: A Shared Approach.*

Typical initiatives would be introducing needle-exchange programs, encouraging a non-smoking environment, improving the nutritional quality of the food and working to reduce the bullying within prisons. In June 2014 an expert meeting concluded that all countries should be concerned about the need for better health care in prisons for the benefit both of the health of prisoners and the public health of communities at large. It concluded that prisons are generally unhealthy places and policies are needed to improve this situation (WHO Regional Office for Europe, 2014b).

BRINGING ABOUT CHANGE IN HEALTHY SETTINGS INITIATIVES

> Profound and powerful forces are shaking and remaking our world, and the urgent question of our time is whether we can make change our friend and not our enemy.
>
> *Bill Clinton, inaugural address, 20 January 1993*

Most settings are either based in an organisation (such as a school or workplace) or comprise a series of organisations such as in a Healthy Cities project. Thus healthy settings projects have to be very cognisant of the need to change and adapt the culture of organisations so that they can take on the proactive and positive perspectives health promotion requires. A health-promoting organisation needs to adopt a broader perspective on health, recognising that it has an impact on the health of all its members. A school, for instance, would recognise the impact on the health of teachers, students, parents and the community in which it is located. A hospital would need to develop concern not just for patients and staff but also for the population it serves and the broader communities of interest it relates to.

Healthy settings require change in the orientation and focus of organisations. Managing change has become an industry in its own right and there are numerous manuals, books and courses focused on the issue. This section considers some of the core lessons emerging from this literature and how they might be used to help advance health promotion efforts in settings and organisations. Readers who are interested in the details of how to effect change in workplaces to make them more health promoting should consult Johnson and Paton (2007), which focuses on health promoting health services, and contains much information on practical tools and approaches that is relevant to any healthy setting.

Literature relating to change management is principally aimed at private sector businesses and corporations, whose main aim is making a profit, rather than pursuing social or health objectives. Nevertheless, there are lessons to be learnt from these insights. The shift from an organisation with a limited behavioural view on health promotion to one with a broader perspective (see box 23.1 and apply hallmarks of healthy settings to organisations) will involve significant organisational change. A typology of change strategies (Dunphy and Stace, 1992) suggests that the magnitude of the required change needs to be determined, as it can range from fine-tuning (the organisation might already have many features of a health-promoting organisation) to transformation (where radical shifts in the organisation's core purpose and values base are required). Leadership styles will vary but more radical change may often require more directive and coercive management. Most recent management literature stresses the value and effectiveness of collaborative and consultative management styles. Action learning, action research and participatory action research have all been used in processes of organisational change in the public and private sectors. These methods seek to involve the key players and work in collaborative ways to bring about transformation of various types. Theory, methods and examples of these processes are provided in Cummings and Worley (2004). The basic processes of change management are shown in figure 23.1.

FIGURE 23.1   PHASES OF ORGANISATIONAL CHANGE

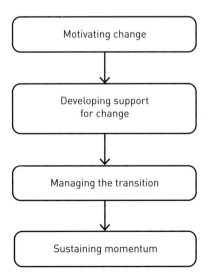

Source: VicHealth, 2011, p. 3.

Recent trends in public sector management have favoured generic managers over those with specialist skills, on the grounds that there is little difference between running a factory and running a public health service, community health centre or hospital.

They are less likely to have specialist understanding of public health or the importance of shifting to a broader health promotion perspective. Alexander (1995) has argued that effective reorientation requires managers with a strong vision and commitment to the ideals of the new public health (which do not necessarily come with health professional training). In fact, some health specialists have a narrow view of health and no sympathy for the need for change. The beliefs and values of senior management are likely to be crucial in determining the effects of change.

Auer et al. (1993) suggest some strategies, based on interviews with managers who have successfully brought about change:

- *Accept that change will also involve conflict*, especially resistance and anger—blocking and even sabotage can be expected. People may be threatened, feeling they have neither the skills nor knowledge to change to the new approach. The new public health is discomforting because its focus on equity and questioning of structures and practices that have long been taken for granted will be resisted by many who are content with the status quo.
- *Discomfort can help the process of change*: managers often gloss over discomfort, hoping it will somehow resolve itself, but recognition of discomfort may release energy for change. For example, speech pathologists at a community health centre may not want to change their way of working as they feel oppressed by the long waiting list. Recognising this may encourage enthusiasm for change. Or a CEO of a hospital may not want to reorientate his or her service to health promotion because they fear the opposition of powerful medical specialists, some of whom may ridicule what they would see as a 'waste' of valuable resources.
- *The need for vision and a plan*: the vision is seen as crucial to determining the future direction of an organisation. Auer et al. (1993, p. 17) comment:

> For all managers, the development of a common vision for change shared by those who must implement it and those most likely to be affected by it was a critical step. Organisational change and development are dependent on the ability of managers to communicate their vision and to effectively involve others in its translation into organisational identity, goals and action.

One of the first steps in reorientation is often the drafting of a vision statement to enshrine an organisation's commitment to health promotion. For organisations with a new public health focus, the vision will often involve a commitment to becoming more responsive to the community or to involving people who have not hitherto been involved. Auer et al. (1993, p. 19) quote the example of Family Planning (New South Wales), which decided to direct its services at women from non-English-speaking backgrounds and young people. Services were regionalised, some existing clinics closed, new service strategies evolved and recruiting practices changed to assist the recruitment of staff from non-English-speaking backgrounds. Many Healthy Cities projects have taken as their first step the establishment of a vision of an ideally healthy community. This has proved a good way of building consensus because while people may differ on the strategies to achieve the vision, what constitutes a healthy community is likely to attract consensus.

- *Clear information*: people feel hostile to change if they believe information is being withheld. As far as possible, clear information about decisions that have been made, what is and is not negotiable and how people will be involved should be provided.
- *Participation and consensus building*: the involvement of staff in decision-making is compatible with the philosophy of the new public health, and will ensure wider understanding of the goals and commitment to the process of change. Involvement can mean a broader group of people actively support the change process. The focus on equity in the new public health means it is crucial to build consensus about the importance of the social and economic determinants of health. People easily see that behaviour influences risk factors but less easily see the impact of upstream causal factors such as income, housing, education opportunity, racism. Behaviours (despite the evidence) seem easy to change directly while action on the social determinants is more complex and generally seen as more politically risky. The strategies for achieving such change in perspective include the importance of focusing on people and encouraging their participation in the running of the organisation; encouraging lateral and innovative thinking; celebrating and recognising achievements; and taking a holistic view of the organisation. Much emphasis is placed on teamwork and creating small, self-managing teams that have a degree of autonomy, especially compared to the situation in the past when hierarchical organisations were prevalent. There are techniques that can be used to develop effective teams, including the use of 'quality circles' (Simnett, 1995, pp. 71–2). These are groups that meet regularly to review and improve work performance and provide support to each other. They often use a facilitator (who is not a line manager of anyone in the group), but are staff- rather than management-led. Staff explore an issue in detail and work out how they could improve their practice.
- *Reorganising or redirecting resources*: this is often crucial to the process, but may require tough decisions and be met with resistance. In public health organisations this process will generally involve shifting resources from curative care to prevention and health promotion. In other organisations, it may require a commitment of new funds to a health promotion process that mainly offers long-term outcomes and few short-term wins. This situation will never be popular with politicians who, with a few visionary exceptions, are focused on short electoral periods and with demonstrating outcomes within that period.
- *Evaluate the process*: encouraging a climate of critical reflection is important so that processes can be realistically measured. Techniques such as the quality circles can be useful here.
- *Personal change*: this process of changing an organisation will often be a time of personal change. Managers may have to develop new forms of leadership and drop those that are unsuccessful. For workers the switch to focusing on prevention and health promotion in their work may require a considerable amount of retraining and support.

The aim of the change is to create a learning organisation that gives scope to question and looks for opportunities to improve existing practice. For instance, the staff in a hospital division attempting to adopt a public health and health promotion perspective might pose the following questions:

- What health promotion and public health goals are we trying to achieve?
- How do we understand health promotion and public health?
- What is our vision for the best possible health promotion work we could do?
- How would we know if we were successful?
- How would outsiders know if we were successful?
- How will the different groups we relate to view quality health promotion work?
- What improvements can we make so that we can meet the needs of these groups while moving closer to our vision of health promotion?
- How can we demonstrate that our improvements are working in practice?

This activity involves a learning cycle in which participants are willing to review their performance and determine ways of improving it. The pace of workplace change makes such flexibility a real benefit to organisations.

POLITICAL AND POLICY LEADERSHIP AND COMMITMENT IS ESSENTIAL

> Our problem is the work is not technical: we already know what the problems are and we know a lot about how to solve them. The problem is to create a political will for action; the challenge to deploy the managerial skill and innovation required to pull together the vast human and other resources that a city possesses in order to bring them to bear on this work.
>
> *Dr J. E. Asvall (then Regional Director of the WHO Regional Office for Europe) at the 1987 Dusseldorf Healthy Cities meeting*

From the start of the European WHO Healthy Cities Project, the crucial role of political and leadership commitment to the success of the projects has been stressed. European meetings of project personnel have often included mayors from project cities, providing them with a chance to discuss the projects and issues from a political perspective. The need for awareness of the political issues is common to all healthy settings projects. Political support is generally essential when introducing a healthy settings project that results in real change. It is certainly true that there have been healthy settings projects that have enabled politicians to launch a new initiative with a fanfare but then players have settled down to business as usual and little is achieved. Duhl (1992, p. 16) notes that often the 'words are hailed, but the process is stymied'. He notes that in some cases Healthy Cities projects may simply become 'old wine in new bottles', primarily because the project participants are not committed to a process of organisational change and development. Projects that stick to trying to change behaviour rather than more structural factors are generally less contentious. Gaining political support from mayors and ministers is vital, and so is having policy actors who are able to take advantage of windows of opportunity when they occur.

Inspirational leadership is crucial to the success of healthy settings initiatives—many of them have flourished because of this. Legge et al. (1996, p. 106) defined such leadership as characteristic of best practice in primary health care. In their view this leadership comprises:

- skills in political insight and an ability to chart a path in confusing territory
- ability to take insights from different places and bring coherence to them in the context of a program in action
- ability to listen to people and reflect back, with added value
- clarity of vision and an ability to depict possibilities as achievable
- confidence and readiness to act (even when full certainty is still not possible)
- ability to inspire others to act even where (and especially where) there is uncertainty about the outcomes
- readiness to examine what happens critically, to take feedback and to learn how to do it differently and better next time.

ENCOURAGING ACTION ACROSS SECTORS

A very common theme in thinking about promoting health and creating sustainability for the environment is the need for action across sectors. This is true for all healthy settings projects. This approach requires organisations to become more outward focused, as shown in figure 23.2. The need for intersectoral action is based on recognition of the complexity of the problems faced by modern societies, including:

- the failure of the market model to protect the more vulnerable members of society
- persistent and widening inequities within countries and between them
- the recognition that health promotion has to incorporate factors relating to both individuals and structures
- the enormity of the environmental problems facing the world
- the need to maintain democratic participation in the process of designing solutions to these serious and challenging problems.

FIGURE 23.2   HEALTHY SETTINGS: SHIFT IN ORGANISATIONAL FORM

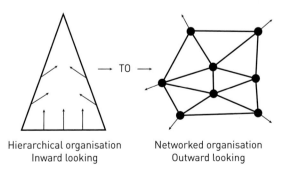

Hierarchical organisation
Inward looking

Networked organisation
Outward looking

No one sector can tackle these fundamental issues and come up with sufficiently innovative and radical solutions on its own. Cooperation and collaboration across sectors become more crucial as health and environment issues grow in complexity. In the UK the need for action across sectors is referred to as 'joined up working' and became a catch cry of the Blair government. Health in All Policies is another approach designed to achieve joined up government and is covered in more detail in chapter 24. Healthy settings are a means to implement this joined up action at a local level.

Healthy settings initiatives are not the only experiments with cross-sectoral models. Carley and Christie (1992) describe similar developments in environment protection. The examples they provide include the local Groundwork Trusts in the UK, the California Growth Management Consensus Project and the Netherlands National Environmental Policy Plan. These intersectoral organisational forms are 'action centred networks'. The necessity for them comes from what Carley and Christie (1992, p. 265) identify as the main institutional constraints on managing sustainable development: 'the fragmented nature of policy-making in key institutions; failure to promote organisational learning; the lack of policy integration in economic management; the massive complexity of environmental problems; the difficulty in balancing "top-down and bottom-up" initiatives in environmental management and planning; and the great turbulence of the world as industrialisation becomes a global condition'. All these factors apply to public health, and the lessons from Healthy Cities and other healthy settings projects suggest there is much potential for learning about and developing more sophisticated approaches to working across sectors, but that numerous problems bedevil existing attempts. There are, however, enough positive examples to indicate the considerable potential.

All the European Healthy Cities projects have established intersectoral committees. The first requirement for participation is that there should be an intersectoral committee, at a political as well as an administrative level. In Australia such committees have been common to the initiatives as described in more detail below. The potential for misunderstanding between people from different professions and organisations working on healthy settings projects is considerable and it is not surprising that there are many blocks to intersectoral collaboration happening effectively. Yet if public health is to be effective in the future, an essential skill for practitioners is to be able to collaborate with people from different sectors who are likely to have different values, ideologies and training. Doing so requires high-level communication and conflict resolution skills, rarely taught in any professional training, but which should become a feature of all future professional training. The skills needed to manage the challenges we are likely to be faced with in the quest for Health For All appear to be flexibility, willingness to question current practice, an entrepreneurial approach to problem-solving and a willingness to take risks and step around bureaucratic blocks. If those things are present, intersectoral action is likely to be fruitful. Ensuring this effectiveness is more important in the twenty-first century, when the pace of social and economic change is continuing unabated. Globalisation (see chapter 5) is affecting all communities around the world by increasing the complexity of life and making it less certain and more multidimensional for most communities. This creates less trusting relationships.

TYPES OF PARTNERSHIPS

VicHealth (2011) suggests that there are four types of partnerships in health promotion:

- networking: is concerned with the exchange of information for mutual benefit. This requires little time or trust.
- coordinating: involves exchanging information and altering activities for a common purpose.
- cooperating: involves exchanging information, altering activities and sharing resources. This will require a significant amount of time, a high level of trust and some sharing of trust.
- collaborating: involves the activities of cooperation and in addition enhancing the capacity of the other partner for mutual benefit and a common purpose. It will often involve sharing resources and giving up some turf.

A really effective healthy settings initiative would fall into the collaborating type. Thus a Healthy Cities project would involve joint activity, having joint goals and a commitment to achieving these.

KEY ELEMENTS FOR SUCCESSFUL PARTNERSHIPS

Box 23.3 provides a checklist developed by VicHealth (2011) to determine the success or otherwise of a health promotion partnership. This would be a good tool for healthy settings projects of all types to use periodically to determine how their partnerships are faring.

BOX 23.3   CHECKLIST FOR PARTNERSHIP IN HEALTH PROMOTION

1 Determining the need for the partnership
   - There is a perceived need for the partnership in terms of areas of common interest and complementary capacity.
   - There is a clear goal for the partnership.
   - There is a shared understanding of, and commitment to, this goal among all potential partners.
   - The partners are willing to share some of their ideas, resources, influence and power to fulfil the goal.
   - The perceived benefits of the partnership outweigh the perceived costs.

2 Choosing partners
   - The partners share common ideologies, interests and approaches.
   - The partners see their core business as partially interdependent.
   - There is a history of good relations between the partners.
   - The coalition brings added prestige to the partners individually as well as collectively.
   - There is enough variety among members to have a comprehensive understanding of the issues being addressed.

3   Making sure partnerships work
- The managers in each organisation (or division) support the partnership.
- Partners have the necessary skills for collaborative action.
- There are strategies to enhance the skills of the partnership through increasing the membership or workforce development.
- The roles, responsibilities and expectations of partners are clearly defined and understood by all other partners.
- The administrative, communication and decision-making structure of the partnership is as simple as possible.

4   Planning collaborative action
- All partners are involved in planning and setting priorities for collaborative action.
- Partners have the task of communicating and promoting the coalition in their own organisations.
- Some staff have roles that cross the traditional boundaries that exist between agencies in the partnership.
- The lines of communication, roles and expectations of partners are clear.
- There is a participatory decision-making system that is accountable, responsive and inclusive.

5   Implementing collaborative action
- Processes that are common across agencies have been standardised (e.g. referral protocols, service standards, data collection and reporting mechanisms).
- There is an investment in the partnership of time, personnel, materials or facilities.
- Collaborative action by staff and reciprocity between agencies is rewarded by management.
- The action is adding value (rather than duplicating services) for the community, clients or the agencies involved in the partnership.
- There are regular opportunities for informal and voluntary contact between staff from the different agencies and other members of the partnership.

6   Minimising the barriers to partnerships
- Differences in organisational priorities, goals and tasks have been addressed.
- There is a core group of skilled and committed (in terms of the partnership) staff that has continued over the life of the partnership.
- There are formal structures for sharing information and resolving demarcation disputes.
- There are informal ways of achieving this.
- There are strategies to ensure alternative views are expressed within the partnership.

7   Reflecting on and continuing the partnership
- There are processes for recognising and celebrating collective achievements and/ or individual contributions.
- The partnership can demonstrate or document the outcomes of its collective work.

- There is a clear need and commitment to continuing the collaboration in the medium term.
- There are resources available from either internal or external sources to continue the partnership.
- There is a way of reviewing the range of partners and bringing in new members or removing some.

Source: VicHealth, 2011.

A further factor that seems to be essential for collaboration is the development of trust. Any organisational arrangements that undermine this will jeopardise effective collaboration. Collaboration between organisations with very different structures will also be difficult. Take, for instance, the two organisational extremes displayed in figure 23.2. Collaboration between the hierarchical and networked organisations will be difficult, as representatives of the networked organisation will have more autonomy to make decisions and take action than their counterparts in the hierarchical organisation. Similarly the networked organisation is designed to be more outward looking and will be more open to collaborative action.

DETAILED EXAMPLES OF HEALTHY SETTINGS INITIATIVES

WORKPLACE HEALTH PROMOTION

Workplace health promotion illustrates that one of the central dilemmas associated with healthy settings projects is that they are limited by the broader frameworks within which they exist. In the case of workplaces these factors will include the legislative framework concerning employment practices, the practices of employers and the state of a national economy, which will dictate the power that employees have in the workplace. Workplace health setting projects will be concerned with the micro issues within a single worksite. Much can be done at this level, as the case examples in this section will demonstrate. But broader protections for workers come from legislation that covers occupational health and safety and the rules concerning the rights that workers have. Historically the rights of workers have resulted from the struggle of social movements, most notably trade unions, which have involved workers in struggles to attain better conditions and safer workplaces. Thus in the past in most industrialised countries workers have fought for rights such as restrictions to the length of the working day, limitations to the grounds on which they can be sacked, and to ensure employers provide safe working conditions. This section will first of all consider the broader legislation framework for workplaces, consider international efforts to improve conditions of work globally and use recent changes to industrial law in Australia as a case example of workers' rights being eroded. Then specific work-based health promotion initiatives in rich and poor countries are considered.

LEGISLATIVE FRAMEWORKS THAT SUPPORT HEALTHY SETTINGS IN THE WORKPLACE

The movement towards global free trade (see part 2 for discussion) has meant that there is pressure for countries to weaken labour laws in the interests of making national economies as competitive as possible. Globally this has led the International Labour Organization (ILO) (2015) to call for 'Decent Work for All'. The main goal of this campaign was 'the promotion of opportunities for women and men to obtain decent and productive work in conditions of freedom, equity, security and human dignity'. There are obviously very different standards of protection and rights while at work across the world. It is also clear that work-related health issues have a significant impact on the health of workers and their families around the world. Healthy settings projects will always have to work within the framework provided to them by national legislation. Economic globalisation means that if a company feels the costs of labour are too high in one setting then they can re-invest in a country where labour regulations are less onerous and wages are lower. Thus many poor countries (for example India, China, and in Latin America) have seen a huge growth in manufacturing in the past few years and the evidence suggests that the new workers have few rights and experience very poor conditions of work (Heymann, 2006). There are huge inequities globally in the conditions of work. Many workers in poor countries experience conditions akin to those of the early days of industrialisation in nineteenth-century Britain.

The ILO reports that approximately two million people worldwide die each year from occupational accidents and work-related diseases. Occupational health and safety is a crucial aspect of a healthy workplace. The potential for preventing death and injury in the workplace was recognised in Australia with the 1985 formation of the National Occupational Health and Safety Commission, replaced in 2009 by Safe Work Australia. One of the most successful aspects of its work (Pearse, 1997) has been the establishment of the National Industrial Chemicals Notification and Assessment Scheme, which has developed far-reaching standards and leads prevention-orientated research. But work in this area has been less effective in reducing the number of workplace diseases and injuries. This is mainly because of difficulties in ensuring that states implement standards and regulations (Pearse, 1997).

The role of Safe Work Australia has been reduced as a result of federal government budget cuts. This, combined with macro-economic theory that stresses deregulation and the loosening of government control on issues such as occupational health and safety, means that the new public health cannot be complacent about gains made by the earlier public health movement in workplace safety. Reliance on self-regulation and market solutions ignores the reality that some companies will put consideration of profits before their workers' health and safety. Aside from the moral argument, there is a strong economic case for ensuring safe and hazard-free environments for workers. The cost of work-related injuries and deaths is borne by the community as a whole through the health and social security system. It is likely that the costs of regulation would be easily recouped through savings to the health and social security budgets.

Workplace restructuring and industrial reform have meant that coping with continual change is the experience of most Australian workers. At the macro-level it would almost certainly be health promoting if the speed of change were slowed, given the emerging evidence on the impact of threatened change on workers' health (Ferrie et al., 1998). In 2005 the Australian Government introduced wide-ranging reforms to the conditions under which workers could be employed. These reforms were fiercely opposed by the Australian Council of Trade Unions (see box 23.4) and this campaign was credited with being a major factor in the Australian Labor Party's electoral victory in the 2007 federal election.

BOX 23.4   YOUR RIGHTS AT WORK: ACTU CAMPAIGN

In 2005 the Australian Council of Trade Unions (ACTU) launched the 'Your Rights at Work' campaign with the chief aim of opposing the reforms to the industrial relations legislation passed by the Howard Coalition government in December 2005. The campaign included television advertisements, mass rallies, national days of action, posters, T-shirts and car stickers.

The ACTU campaign website stated, 'The Howard Government's radical industrial relations reforms will unfairly curtail our rights at work, cut the amount of time Australians can spend with family, and erode job security.' The key concerns about the legislation were stated by the ACTU as:

- *Unfair dismissal*: the new legislation means workers in businesses with fewer than 100 staff lose the right to unfair dismissal protection. This means employees are no longer allowed to seek reinstatement or compensation if they are sacked because of harsh, unreasonable or unjust treatment. People employed by companies with more than 100 staff keep their right to claim unfair dismissal, but employees will not be regarded as unfairly dismissed if employers state their sacking was for 'operational reasons'.
- *Weakened bargaining power*: Individual contracts are favoured under the law rather than collective bargaining. The ACTU says that this will lead to harsher working conditions with fewer holidays, penalty rates, shift work rates and less rights to redundancy pay.
- *Lower pay*: The Australian Industrial Relations Commission will no longer have the right to set minimum wages. Instead this will be done by the newly established Fair Pay Commission, which promises to be more concerned with the competitiveness of the economy than the fairness of wages.
- *Restriction of access to trade unions*: It is more difficult for unions to make workplace visits, more difficult to take industrial action and the penalties for doing so are higher for unions and workers.

The possible impact of these changes on the health of workers stemmed from the fact that they would have less control over their conditions of work, were likely to have to work longer hours for less money and have less rights to take collective action to gain improved working conditions. The campaign continued after the 2007 election to ensure that improvements to rights at work were locked into the new Labor government's workplace laws.

HEALTHY SETTINGS PROJECTS IN THE WORKPLACE

A Queensland study that tracked the development of workplace health promotion (WHP) over recent decades (Chu and Forrester, 1992) showed that the trends in workplace health promotion had mirrored the changes in broader health promotion. In the 1970s WHP focused on providing fitness facilities, usually to executives and often in large corporations.

Unions were critical of the victim-blaming nature of behaviour-based health promotion, preferring to put their efforts into occupational health and safety. More recent developments in WHP have seen more comprehensive strategies that acknowledge the impact of environmental factors, including the impact of shift work on workers' health, the provision of nutritious food in canteens, secure space for bicycles, showers for exercising workers, no-smoking policies, greater employee democracy and more flexible working hours. The Australian Government funded a Healthy Workplace initiative that focuses on making the workplace supportive of healthy lifestyles (Department of Health, 2013b). VicHealth (2014) also has a 'Creating Healthy Workplaces' program, which has funded five large-scale pilot projects in Victorian workplaces to develop and test solutions for promoting good health and preventing chronic disease.

Equity is a key consideration, as behavioural interventions are most likely to be effective for people who have the other aspects of their life going well. Executives are therefore likely to be a good target, with their high incomes, relatively secure employment, higher than average education levels, and excellent material security. Manual workers, by contrast, are likely to experience less favourable environmental conditions and worse health status and will be less able to respond to messages about healthy behaviours.

The best WHP for many workers is likely to be improved basic conditions of work. Blue-collar workers have the worst health status of any socioeconomic group, but little attention has been paid to health promotion for them in the workplace. One exception is Ritchie (1996), who used participatory action research methods to work with 40 male blast furnace workers in an Australian steelworks to identify and work on issues they saw as important to their health. She found that, contrary to her expectations, the creation of a supportive health environment was possible in a blast furnace environment. The men, for instance, found their workmates personally supportive. They felt that occupational health and safety legislation did improve air quality, but her perception differed significantly from the men's. They minimised the risks of their work environment and were positive about the area in which they lived, stressing the natural beauty rather than the pollution from the steelworks.

Ritchie also found that the men did not want to 'take control of their health'. She speculated that the emphasis on control may be a middle-class, professional concept. The men's concerns were: health risk appraisals, measurement of worker stress, monitoring the furnace environment with feedback to the workers, researching the health effects of shiftwork, regulations relating to safety clothing, mud gun fumes, quitting smoking, increasing physical activity for some workers, hearing protection, improving the environment in the control room, quality of the canteen food. Many of these issues were not satisfactorily dealt with, and may have needed organisation-wide commitment. Ritchie's work showed that an action research approach may be effective and that workers will identify a wide range of issues relating to personal health and the structure and organisation of work when thinking about health.

People at work using safety equipment. (Fernando M. Gonçalves)

Organisations should consider equity when devising strategies for a health promoting approach. Chu and Forrester (1992, p. 61) provided the following guidelines for the development of WHP, which remain relevant:

- WHP should be made an integral part of a corporate culture. Programs should be comprehensive, addressing both individual risk factors and the broader environmental and structural issues, and having ongoing and long-term management commitment.
- WHP programs that aim to improve workers' health by addressing individual risk factors should also consider possible improvements to the underlying workplace structure and practices.
- A community development approach emphasising grassroots participation, self-determination and empowerment is an essential strategy to involve workers in the planning, decision-making, organisation and implementation of WHP.
- WHP should be shared by employees and employers, with health professionals acting as facilitating and mediating agents.

HEALTHY CITIES AND COMMUNITIES

There are a burgeoning number of initiatives around the world that use very similar values, principles and processes to the healthy settings approach and are based at a city, municipal or local government level or a sub-area of these. In a nutshell these projects are concerned with integrating economic, social, community and environmental issues, with using participatory processes, encouraging sectors to work together and integrate their activities, building local capacity to act to improve health and well-being in an equitable manner. These projects include Local Agenda 21 (www.gdrc.org/uem/la21/la21.html) and the UN Sustainable Cities Program (http://ww2.unhabitat.org/programmes/sustainablecities). Here we examine the WHO Healthy Cities initiative as one example of a local government–based healthy settings initiative.

WHO's HEALTHY CITIES PROGRAM

WHO's Healthy Cities program was originally an initiative of the WHO Regional Office for Europe in 1986 and was designed to implement the Ottawa Charter at a city level. Since then it has captured the imagination of cities and other communities around the world as a means of tackling complex public health issues at a city, community or regional level. Healthy Cities networks have been established in all six WHO regions. The main goal of the initiative is explained by the European region as:

A healthy city is defined by a process, not an outcome.

- A healthy city is not one that has achieved a particular health status.
- It is conscious of health and striving to improve it. Thus any city can be a healthy city, regardless of its current health status.
- The requirements are: a commitment to health and a process and structure to achieve it.
- A healthy city is one that continually creates and improves its physical and social environments and expands the community resources that enable people to mutually support each other in performing all the functions of life and developing to their maximum potential.
- The Healthy Cities approach recognizes the determinants of health and the need to work in collaboration across public, private, voluntary and community sector organizations. This way of working and thinking includes involving local people in decision-making, requires political commitment and organizational and community development, and recognizes the process to be as important as the outcomes.

Source: WHO Regional Office for Europe, 2015c.

The types of initiatives labelled as 'Healthy Cities' vary between and within countries. The same key ideas are at the heart of each project, so that while the problems and priorities may differ from city to city, the process to be followed is similar. This means that Healthy Cities is both a concept and a project. Common features are evident in the various projects, initiatives and networks:

- A concern with local political and bureaucratic action to promote the health of a local community with an explicit concern for promoting equity and reducing inequities in health status. 'Health' is defined in keeping with current health promotion thinking (see the Ottawa Charter), which recognises the importance of the physical, social, emotional, economic and political environment in shaping the health of populations and individuals. Gaining political support is an important aim of a Healthy Cities project.
- Encouraging a range of local agencies to reassess their practices and policies so that they make a better contribution to promoting health. In the main these agencies have been government ones (including planning, environment, housing, health, welfare). Intersectoral committees usually coordinate the project. Healthy Cities is about organisational change, encouraging government authorities to be more flexible and innovative in their approach. Often encouraging staff to acquire the skills needed for effective intersectoral collaboration is an essential component of a Healthy Cities project.
- Establishment of a core staff and small office (typically two to three staff) to implement and coordinate the Healthy Cities approach, disseminate information and encourage the organisations associated with Healthy Cities to change in a direction consistent with the new public health. This change is unlikely, unless there are change agents working within the organisation.
- Concern with monitoring and assessing the health and health needs of the population. Central to the process is a concern with fostering a visionary approach to the future (Ashton, 1988). The information on local needs and vision for a healthy future is used to devise local action plans for which there are resources, passion, energy and commitment. These plans normally include strategies ranging across the five areas of the Ottawa Charter.
- Fostering of new projects and/or the co-option of existing projects as models of good practice in the new public health.
- Involvement of community members in the Healthy Cities planning and activities in more than a token way. A crucial element of Healthy Cities is ensuring that planning and implementation are done in collaboration with local people, and recognising the particular skills and knowledge they can contribute.

WHO has established *Twenty Steps for Developing a Healthy Cities Project*. These are shown in box 23.5. This framework has been used in many parts of the world. Estimates suggest that there are more than 2000 communities participating worldwide. In some places they are part of a formal project, in others the basis for a more informal network. Healthy Cities has been variously described as a project, network or approach, and one of its strengths is that it is sufficiently broad to accommodate local circumstances and priorities. It offers a framework within which cities and other communities can plan their own public health initiatives.

BOX 23.5   *WHO's TWENTY STEPS FOR DEVELOPING A HEALTHY CITIES PROJECT*

GETTING STARTED

1   Build a support group that includes people with an understanding of the new public health and who have leadership and determination.
2   Understand and explore the concepts behind healthy cities, especially the links between health and the environment. Ensure all the support groups are involved.
3   Know the city by conducting some kind of community needs assessment.
4   Identify potential project partners and if possible obtain some seed monies.
5   Decide on where the project will be located (options include local government, community organisations, or independently).
6   Prepare a sound proposal, which is concise, clear, convincing and pragmatic.
7   Get approval from the relevant authorities, which normally involves seeking the support of powerful political and community groups.

GETTING ORGANISED

8   Appoint a steering committee with clear responsibilities to plan, lead and coordinate the project. Subcommittees for fundraising, personnel and specific projects can be established if necessary.
9   Review, rework and research the project environment to ensure it is feasible and is being implemented in an appropriate way. Check that communication between the relevant organisations is happening effectively.
10   Define project work with a detailed plan, which includes innovative but workable strategies.
11   Set up a project office.
12   Plan strategy and develop a city health plan that provides for the short- and long-term vision of the project.
13   Build capacity in terms of resources and personnel.
14   Establish accountability by putting monitoring and evaluation systems in place. Regular reports should be made available to key people in the city or community.

TAKING ACTION

15   Increase health awareness among politicians, community members and bureaucrats.
16   Advocate strategic planning to ensure that all opportunities are used and plans put into practice.
17   Mobilise intersectoral action so that it is collaborative, not competitive.
18   Encourage community participation from all sections of society. Support local action programs and initiatives for health development.
19   Promote innovation, flexibility and health promotion.
20   Healthy policies make healthy cities and create an urban environment that can promote health.

Source: WHO Regional Office for Europe, 1997.

EUROPEAN PROJECTS

The European project started in 1986. In 2006 there were 56 designated cities in the WHO European Region that were engaged in implementing the Phase IV project. Phase IV (2003–08) has three core themes (healthy ageing, healthy urban planning and health impact assessment) and encourages action to tackle obesity and promote physical activity and active living. The strategic goals for the period 2014–18 are shown in box 23.6. Currently there are national Healthy Cities networks in 29 countries in the WHO European Region, which bring together more than 1300 cities and towns. A common set of accreditation criteria for national networks and their member cities provides a quality standard and a source of international legitimacy for all stakeholders of a national network. The WHO European Office has also produced resources, such as a series of booklets on aspects of project development. A Google search will reveal many different guides. *The Twenty Steps* booklet has been translated into 20 European languages (WHO Regional Office for Europe, 1997) and a set of resources on health impact assessment provides useful and practical tools for cities (see for example WHO Regional Office for Europe, 2005).

BOX 23.6   STRATEGIC GOALS OF THE WHO EUROPEAN HEALTHY CITIES NETWORK PHASE VI, 2014–18

The WHO European Healthy Cities Network has six strategic goals:

1   to promote policies and action for health and sustainable development at the local level and across the WHO European Region, with an emphasis on the determinants of health, people living in poverty and the needs of vulnerable groups;
2   to strengthen the national standing of Healthy Cities in the context of policies for health development, public health and urban regeneration with emphasis on national–local cooperation;
3   to generate policy and practice expertise, good evidence, knowledge and methods that can be used to promote health in all cities in the Region;
4   to promote solidarity, cooperation and working links between European cities and networks and with cities and networks participating in the Healthy Cities movement;
5   to play an active role in advocating for health at the European and global levels through partnerships with other agencies concerned with urban issues and networks of local authorities; and
6   to increase the accessibility of the WHO European Network to all Member States in the European Region.

Source: WHO Regional Office for Europe, 2015d.

The ways in which these goals translate and integrate into local Healthy Cities projects is shown in figure 23.3, which shows the Cardiff (Wales) Healthy Cities model.

FIGURE 23.3   CARDIFF HEALTHY CITY MODEL

Networking is seen as very important, and has enabled cities to bypass national governments and talk directly to each other. Tsouros (1995) reports that exchange visits have been an important part of training for Healthy Cities. Twinning arrangements have been organised between well-established cities and newer ones, particularly those from central and eastern Europe.

AMERICAS

Canada has a Healthy Communities network (see for example the Ontario Healthy Communities Coalition, www.ohcc-ccso.ca/en/chc-csc_network) that includes large cities such as Toronto and far smaller rural communities and provincial initiatives such as the British Columbia and Ontario networks. The British Columbia initiative was launched in 2005 and its aims are stated as:

> BC Healthy Communities (BCHC) is a province-wide not-for-profit organization that facilitates the ongoing development of healthy, thriving and resilient communities. We provide a range of services, programs, events and resources to support communities, local governments and multi-sectoral groups to collaborate around a shared vision for a common purpose. (BC Healthy Communities, 2015)

Its website describes a healthy community as:

> Healthy lifestyles. A vibrant economy. Affordable housing. Protected parks & green space. Accessible community services. Thriving neighbourhoods. Clean air and water. A sustainable environment. Ethnic and cultural diversity. Healthy public policy. Engaged citizens. A healthy community is all of this, and more.

Quebec formed its own network, 'Villes et Villages en santé', in 1988 and now has more than 200 municipal members representing 70 per cent of the population of Quebec. Its website lists diverse projects working for the health of local communities (www.rqvvs.qc.ca).

The Pan American Health Organization notes that the Healthy Cities or Healthy Communities concept in the USA is not easily characterised, as there are multiple interpretations of the movement and Healthy Cities projects are initiated independently, often conforming to the orientations of different funding organisations.

There are more than 200 self-declared Healthy Cities and Communities in the USA and a number of others that are involved in the movement at some level. The movement has not been isolated geographically and is represented by projects at both the state and city level. The first statewide initiatives were in California (Twiss et al., 2000) and Indiana (Rider and Flynn, 1992). Other cities and states currently involved include Boston, Philadelphia, Denver, New Mexico, Maine, Massachusetts, Virginia and New Jersey. Although diverse projects have been implemented within the Healthy Cities framework in the USA, a few common themes have emerged, such as the conservation of resources and environmental health, domestic and youth violence, adolescent services, and job and life skills training.

The Pan American Health Organization has developed the idea of Healthy Municipalities. Each of these different initiatives contains the same core idea of getting the philosophy and practice of the new public health on the local agenda. Increasingly they also incorporate ideas of environmental sustainability. Most of the initiatives have a series of embedded Healthy Settings projects that work under the umbrella of a Healthy District or Healthy City.

ASIA-PACIFIC

The Western Pacific, South-East Asian and African Regions of WHO and the Pan American Health Organization have taken up the idea of Healthy Cities with enthusiasm, and there is an increasing number of cities and communities experimenting with the idea across the world. Examples of regional initiatives include the Healthy Islands project developed by the Western Pacific Region, which has applied the Healthy Cities ideas to the islands in the region. The Western Pacific Region has produced a set of regional implementation guidelines for Healthy Islands. Projects are active in Niue, Samoa, Fiji, Solomon Islands and Papua New Guinea (WHO Regional Office for the Western Pacific, 2002). The WHO South-East Asian Region has adapted the idea of Healthy Cities to Healthy Districts as this is more applicable to developing countries.

The Western Pacific Region of WHO has supported the formation of an Alliance of Healthy Cities that is currently coordinated from Tokyo and has members from across the Western Pacific Region. It supports its members and runs an annual conference at which many examples of Healthy Cities practice are presented (www.alliance-healthycities.com). See box 23.7 for its vision statement.

BOX 23.7   VISION OF THE ALLIANCE FOR HEALTHY CITIES

Building cities and communities of peace where all citizens live in harmony, committed to sustainable development, respectful of diversity, reaching for the highest possible quality of life and equitable distribution of health, by promoting and protecting health in all settings.

Source: Alliance for Healthy Cities, 2007

HEALTHY CITIES IN AUSTRALIA

In Australia the Healthy Cities idea was taken up with gusto in the late 1980s. Dozens of communities around Australia experimented with the concept and some continue to have active and vibrant projects. A three-year pilot Healthy Cities project, under the auspices of the Australian Community Health Association (ACHA) was funded for three pilot cities: Canberra (Australian Capital Territory), Illawarra (New South Wales) and Noarlunga (South Australia).

The objectives of the national project were to test the applicability of the European model to Australia; develop and test models for addressing health issues on an intersectoral basis; encourage the participation of an Aboriginal community; and establish a network of Healthy Cities in addition to the pilot cities.

The 1990s saw a modest extension of the Healthy Cities idea. The Queensland Government funded a state Healthy Cities and Shires network in 1992 and the idea is popular with a number of local government areas there, including Townsville. This network produced a Healthy Cities video and a guide to Healthy Cities, aimed specifically at local government, in 1993 (Low and Rinaudo, 1993). In 1996 the Healthy Cities and Shires network worked with nine local government areas on municipal health plans that use a vision process to bring together issues concerning the local economy, environment and community to plan for the overall health of the area (Chapman and Davey, 1997).

Noarlunga, South Australia, was one of three pilot cities that trialled the Healthy Cities Project in Australia.

Many of ideas behind Healthy Cities have been incorporated into the Victorian Government's Municipal Public Health Planning initiative, which encourages local governments to plan in an integrated way that incorporates social, economic and environmental concerns. This initiative has seen many municipalities develop integrated plans designed to promote the health of the whole community.

Of the originally funded pilot projects, Illawarra continues to thrive; the Canberra project is defunct but its legacy can be seen in many approaches that reflect a new public health philosophy within the city; the Noarlunga (renamed Onkaparinga) initiative, after a period of flourishing in the 1990s and early twenty-first century (see box 23.9), has subsequently lost the active support of the health sector as it withdrew from health promotion towards a tightly prescribed clinical role (Baum et al., 2013c). This meant that two of the sustainability factors shown in box 23.9 no longer existed—the commitment to a social health perspective and inspired leadership from the health sector. Without these factors, the once thriving project has struggled.

HEALTHY CITIES: ACTIONS FOR HEALTH

A description of the Healthy Cities initiatives does not capture their variety and colour. Most projects have many things happening at once. Healthy Cities involves an overall commitment by a municipality to ensuring health considerations are involved in all aspects of the city's or community's decision-making and practices. This is a long-term process necessitating a range of local initiatives. Those described here are each one of a range being implemented under the general Healthy Cities umbrella. A successful project is likely to have activities of each type under way.

Many Healthy Cities projects have initiatives that are tackling specific diseases or risk factors. Examples include the development of a needle-exchange program in the UK city of Liverpool, injury prevention projects, and programs that encourage people to quit smoking. The Illawarra project has a healthy ageing focus (www.healthyillawarra.org.au/healthycities/index.php). Many Healthy Cities projects take action on tobacco smoking. For instance in 2004 Brighton and Hove in the UK used a health impact assessment framework to develop their plans for a smoke-free city and consult widely through the 'big smoke debate' (WHO Regional Office for Europe, 2005). Most are more concerned with bringing about policy change than with attempting to change people's behaviour. The Noarlunga (now Onkaparinga) Healthy Cities project (Baum et al., 2006) at its heyday in the 1990s gave rise to further intersectoral and community initiatives. One dealt with injury (Noarlunga—Towards a Safe Community Initiative) and had a range of injury-prevention measures based in schools, workplaces and homes. The other was the Noarlunga Community Action on Drugs, which brought together people from a wide number of government sectors (including education, police, health, welfare and housing), non-government agencies and community groups to tackle drug issues in the community. This was one of a number of regional intersectoral roundtables in the region (Fisher et al., 2014b).

Other Healthy Cities projects focus on people or organisations. Within the overall framework, the projects often contain initiatives focusing on organisational change in particular settings such as schools, workplaces, markets or hospitals. Belfast Healthy Cities has a number of initiatives including a focus on urban planning and health, health literacy, and empowerment throughout the lifecourse including healthy ageing (Belfast Healthy Cities, 2014). Rio de Janeiro's (Brazil) Healthy Cities Project: Favela do Gato ('Shantytown of the Cat') Housing Scheme project (Rice and Rasmusson, 1992, p. 75) achieved better housing for informal residents. The area grew up on the edge of Rio de Janeiro as rural people came in search of a better life and built houses out of whatever material they could scavenge. With the support of the Group for Community Projects of the University Federal Fluminense, the slum residents negotiated with the national housing authority to gain funding for 71 model houses and a community centre. Individuals were granted finance and the city paid the costs of the land and infrastructure. The residents were able to choose their land and the position of their houses on it. Rio de Janeiro also has an adolescent initiative designed to empower teenagers from slum areas by encouraging them to develop skills, find employment and live healthy lives. The initiative provides courses after school in personal development and capacity building in health, sexuality, citizenship, employment, conflict resolution, leadership, and entrepreneurship. There were over 400 youth social projects running in 2007. The initiative is multi-sectoral and involves most sections of the municipal government.

Environmental initiatives are an important part of many Healthy Cities projects. Many cities in Asia have seen improving the urban environment as vital to their healthy cities, with waterway clean-ups, new cycle paths and increased greening. The Noarlunga project hosted a campaign, led by a local resident, that resulted in cleaning up the Onkaparinga River (Baum et al., 1990, pp. 36–8). This was achieved by bringing together the various government agencies with jurisdiction over the river. Healthy Cities Illawarra works for healthy forms of transport that have a low impact on the environment and promote health through encouraging exercise. Cycling and walking are particularly encouraged. Its newsletters regularly report on local initiatives designed to contribute to greenhouse gas reduction. It also works for healthy urban design and community safety.

Also typical of Healthy Cities initiatives is the engagement in systematic city-wide planning. This has taken various forms in different cities but basically follows a common process, which involves consultation with a wide variety of interests within a city, followed by a plan of action. These city health plans are generally attractive documents designed for use by the local community, but they need to have the support of local agencies if they are to be implemented effectively.

SETTINGS WITH A SPECIFIC FOCUS: OBESITY PREVENTION IN CITIES AND COMMUNITIES

Settings are also used to pursue particular health promotion aims. One of those in the past decade has been the growth of locality-based obesity prevention projects. A national group has been formed to advance these initiatives, the CO-OPS Collaboration, 'enabling best practice to create healthier communities' (www.co-ops.net.au), which links research and practice. In South Australia two initiatives—Eat Well Be Active and OPAL—have been implemented in local government areas and used a range of strategies through schools and community groups to encourage exercise and healthy eating. Both were designed to contribute to the state government's target set in its strategic plan to increase the percentage of the population with a healthy weight. South Australia has also taken a Health in All Policies approach, which has audited all government departments to determine what actions they can take to help the state achieve its healthy weight target (Newman et al., 2014).

The evidence on the effectiveness of settings-based obesity reduction programs is weak and this is partly because of the difficulties of designing strong evaluation methods and the inherent weakness of essential behaviour-based programs, which pay attention to only local factors and not to the broader political economy underpinning the growth in obesity (see for example Freudenberg's (2014) work on the role of the practices of transnational food corporations in bringing about the global increase in obesity). One promising approach is Healthy Together Victoria, which includes strategies for schools, workplaces and local communities (see box. 23.8).

BOX 23.8   HEALTHY TOGETHER VICTORIA

Healthy Together Victoria aims to improve people's health where they live, learn, work and play. It focuses on addressing the underlying causes of poor health in children's settings, workplaces and communities by encouraging healthy eating and physical activity, and reducing smoking and harmful alcohol use. It has four components: healthy eating and living; get healthy together; mobilise your community; and the Healthy Together blog. The 'get healthy together' component is based on local communities and the website lists the active communities. For example, Healthy Together Bendigo is a partnership between the City of Greater Bendigo (a rural city about two hours' drive from Melbourne) and Bendigo Community Health Services that aims to create opportunities for eating healthier and being more active. It works in early childhood services, workplaces, schools and the local community. It also has a focus on reducing smoking in Bendigo.

The advantage of Healthy Together Victoria is that it is a state-wide approach that also has local elements and a community rather than an individual focus.

Source: Victorian Government, 2015.

SUSTAINABILITY OF HEALTHY SETTINGS

Achieving sustainability is very important for healthy settings. Many projects start with a flourish and achieve significant advances, but then prove not to be sustainable. When the novelty wears off or seed funding is withdrawn, projects may fade. The coordinator of the European Healthy Cities project stresses that a Healthy Cities project needs a sustainable commitment to a long-term process that can really ensure that health and environmental considerations are brought into the mainstream of municipal decision-making. This is most likely to happen when health expenditure is seen as an investment and not an expenditure (Tsouros, 1996). The experience from the California network of Healthy Cities and Communities suggests that the work is, by nature, long term. The project staff comment: 'It takes years to build the relationships and corresponding trust that allow community efforts to take root and be fruitful. Too often there is a failure to appreciate how "upstream" this work is, especially when its benefits will not be realised for years or in the terms of political office holders' (Twiss et al., 2000, p. 133).

There is now plenty of evidence that many Healthy Cities initiatives have survived over time. Many of these initiatives have now been active for well over a decade. Sustained evaluation of the factors making for sustainability and the benefits of projects that have remained active for a long period is now essential.

Werna and Harpham (1996) make the point that, while it is possible in developing countries to achieve improvement in environmental and health conditions as a result of international programs, the capacity of local people and institutions must be strengthened if these are to be maintained and sustainable. Box 23.9 shows the nine factors that a narrative review of evaluation, reviews and reports suggested account for the sustainability of the Noarlunga Healthy Cities project from 1987 to 2007.

BOX 23.9   SUSTAINABILITY OF HEALTHY CITIES: EXAMPLE OF NOARLUNGA, SOUTH AUSTRALIA

The following factors were derived from a narrative review (Baum, et al., 2006) of a range of documents relating to the Healthy Cities Noarlunga (HCN) project, including evaluations, government reviews, annual reports and newsletters.

1   *Social health vision*. It was clear that full understanding of the social determinants had been crucial because it was clear that the tendency to revert to medical or behavioural models was strong. The HCN has consistently promoted a social health perspective through vision, training and focus.

2   *Leadership*. The importance of committed, passionate and dogged leadership was stressed continually.

3   *Model adapted to local conditions*. The HCN model was designed to influence and include local and state government agencies and developed as an incorporated non-government organisation.

4   *Juggling competing demands*. The HCN initiative has had to ensure it achieves outcomes in the short term while laying the ground for longer term successes. It has also successfully worked to mediate community perspectives on needs and priorities with those of government planners.

5   *Strongly supported community involvement*. Community people hold a majority of positions on the management committee and their development has been supported (for example being sponsored to attend conferences and training programs).

6   *Recognised as 'neutral gameboard'*. HCN has been useful to policy makers because they have found that the collaborative working arrangements developed as part of

the HCN initiative and its off-shoots—a safety forum and a community action against drugs project—means that they are able to implement new projects smoothly in the Noarlunga area. A good indicator of the success has been the bipartisan support of local politicians.

7 *University links and research focus.* Links with Flinders University led to a focus on evaluation and on training. HCN in partnership with the University has run regular Healthy Cities training courses since 1991 and these have attracted participants from across Australia and overseas.

8 *International links and WHO leadership.* From the outset the HCN project has established international links and received international visitors. These links and visits have acted to reinvigorate the project and provide local legitimacy. The project has also established direct links with projects in other countries including with a community safety project in Bangladesh.

9 *Transition from project to approach.* After the initial pilot period (1987–89) the initiative became more of an approach and way of doing business that can be rapidly transferred from one topic to another.

CRITICAL PERSPECTIVES ON HEALTHY SETTINGS APPROACHES

By their nature healthy settings initiatives focus on local issues. This means, of course, that there is a limit to what they can achieve. If the policies of a national government are unsupportive of health then work at the local level may have some ameliorating impact but will not bring about fundamental changes. Many factors determine health inequities, such as taxation and welfare policy, which are determined nationally and shape what can be achieved through a local or regional healthy settings project.

It is often difficult for evaluators of healthy settings projects (and other innovative initiatives) to report negative results. There is usually strong pressure to present the work in a positive light. This is partly because health promotion is so often a marginal activity and those managing initiatives do not want to risk being seen to have failed for fear it will result in a more general withdrawal of investment from health promotion. Goodwin et al. (2013), in an evaluation of the UK Healthy Towns obesity reduction program, questioned whether there is a real commitment to evidence when there is such a strong political requirement to ensure interventions do not fail.

Healthy settings projects are initiated in the main by bureaucracies and this means that they are unlikely to be supportive of progressive change and may rather be protective of the status quo. The changes they bring are likely to be incremental rather than transformational (Baum, 1993). However, the best healthy settings projects will provide space for input from civil society, who may act to encourage more progressive action. Jones (1995–96) raised concerns about the potential for community collaboration when a health promotion initiative is institutionally based. She comments that the 'tendency to abstract the "setting" from its position within a community' is a major weakness. Her fear is that healthy settings will make individual institutions responsible for their own health, and that this will encourage victim-blaming of institutions. This implies that healthy settings need to consider the social, political, resources and other contextual factors during their implementation.

Questions are raised about the potential for genuine participation in healthy settings, as they are normally introduced by the management of organisations. This means that as students, prisoners, patients or workers (depending on the setting) are unlikely to initiate the project and their participation may be limited (Jones, 1995–96), organisationally imposed projects need to take particular care that they are participatory and empowering in practice as well as in rhetoric.

Not all healthy settings approaches show evidence of a focus on equity. They may concentrate their efforts on health promotion but not consider ways in which they could have a greater impact on poor people in the setting. Some may have a focus on equity by virtue of their position as a disadvantaged area or setting and will have been selected to take part in an initiative on that basis. But it is also possible that settings may be self-nominated and that well-off communities will be in a better position to implement a healthy settings approach and so, albeit unwittingly, act to increase inequities. WHO (2010a) has developed the Urban Health Equity Tool for use by local governments in low- and middle-income countries. It provides a practical manual to help municipalities keep a focus on equity. The use of health equity impact assessments by initiatives such as Belfast Healthy Cities is an example of how equity can be built into settings approaches (Belfast Healthy Cities, 2013). A health equity assessment was used to examine travel plans, planning legislation, regeneration plans and educational outcomes to determine how equity could be a more central consideration.

The Ottawa Charter suggests that settings projects should make a commitment to equity, but this is not always evident. The reasons for this are often political. For instance, under the Thatcher and Major Conservative governments in the UK the use of the term 'equity' in policy statements was simply not permitted, and this also seems to be the case under the Australian Abbott Coalition government. By contrast, equity, including health equity, became a concern of policies developed by the Blair Labour government and the Australian Rudd–Gillard Labor government. Jones (1995–96) expresses the fear that some healthy settings projects have the potential to increase rather than reduce inequities in health. Thus schools in affluent suburban areas are more likely to have the resources and motivation to participate in Healthy Schools programs than their inner city counterparts that are struggling with high levels of truancy, low achievement, a high number of students in poverty and parents with ambiguous attitudes towards the school. Jones (1995–96) points out that health promotion experts are likely to prefer working in settings that hold greater promise of successfully implementing a project and that these are not likely to be those in lower socioeconomic settings.

Within healthy settings projects equity can be handled constructively by determining which settings receive priority for resources on the basis of social justice criteria, thus ensuring that the rhetoric of equity becomes more than that and is used to determine strategies and resource allocation. Obviously a political commitment to equity is essential for this to happen.

CONCLUSION

This chapter has discussed a range of local initiatives that are designed to promote the health of communities by focusing on people and their environments. Healthy settings, healthy cities and like projects are providing templates for new ways of planning, conducting business and working for health. They typically involve local lay and professional people and take a holistic perspective on issues. This combination makes these projects powerful agents for the new public health. Consequently they provide a good focus for investment in health and environmental promotion by governments wishing to take action at a local level to complement national health promotion programs and the development of healthy public policy, which is the focus of the following chapter.

CRITICAL REFLECTION QUESTIONS

23.1   What are the benefits of a healthy settings approach compared to health promotion strategies that target individuals only?

23.2   What are the hallmarks of a Healthy Cities initiative?

23.3   To what extent can a healthy settings approach achieve equitable health outcomes?

23.4   What are the constraints of a healthy settings approach?

Recommended reading

Johnson and Paton (2007) *Health promotion and health services* provides a guide to the processes of change management relevant to all healthy settings projects. It contains a section on tools and techniques for change agents.

Edwards and Tsouros (2008) *A healthy city is an active city: A physical activity planning guide*. This useful planning guide provides a range of ideas, information and tools for developing a comprehensive plan for creating a healthy, active city by enhancing physical activity in the urban environment.

Useful websites

The Alliance for Healthy Cities provides a website www.alliance-healthycities.com with details of the Alliance Charter and activities.

The European Healthy Cities Network website www.euro.who.int/en/health-topics/environment-and-health/urban-health/activities/healthy-cities/who-european-healthy-cities-network describes the goals of Healthy Cities and gives examples of initiatives across Europe.

VicHealth has useful resources for establishing partnerships for health promotion www.vichealth.vic.gov.au/media-and-resources/publications/the-partnerships-analysis-tool

The Healthy Together Victoria website www.healthytogether.vic.gov.au provides an example of a state-wide initiative that incorporates a range of healthy settings including cities, schools and workplaces.

24

HEALTHY PUBLIC POLICY

Healthy public policy ... puts health on the agenda of policy makers in all sectors and at all levels, directing them to be aware of the health consequences of their decisions and to accept their responsibilities for health.

Ottawa Charter for Health Promotion, World Health Organization, 1986

KEY CONCEPTS

Introduction
What is policy?
What is healthy public policy?
Policy formulation
Phases in policy making
Approaches to policy formulation
Policies and power
Healthy public policy in a globalised world
Examples of healthy public policy
What makes for healthy public policy?
Conclusion

INTRODUCTION

The first strategy in the Ottawa Charter is Building Healthy Public Policy, which recognises the limitations of behavioural approaches to health promotion and puts emphasis on policies in all sectors to ensure protection from disease and injury and promotion of health. The main aim of healthy public policy is to create environments in which people can live healthy lives and make healthy choices. Public health policies can be implemented by local, state or federal governments, and organisations in the private, public and non-government sectors, and there have been some spectacular successes. Much government policy takes the form of legislation.

WHAT IS POLICY?

> Policy is rather like the elephant—you recognise it when you see it but cannot easily define it.
>
> *Cunningham, 1963, p. 229*

Political scientists have extensive debates about the definition of policy. Definitions stress that policy is about taking decisions, setting goals and ways of achieving them and taking action or not to achieve these goals. Hill (2005) provides an excellent discussion of these debates. He stresses that most commentators see policy as a course of action or a web of decisions rather than just one decision and that the values underlying policy are important. He notes several crucial factors stemming from this. First, action may result from a decision network of considerable complexity that extends over a long period of time far beyond the initial decision-making process. Second, there will be a series of decisions. Third, policies invariably change over time because the policy-making process is dynamic rather than static and changes in response to external events. Fourth, the policy process does not exist on a desert island. Most policy spaces are crowded and so any 'new' policies will be influenced by others and will have an impact on them. Hill then notes that it is important to recognise the importance of non-decision making to policy and to examine its impact on policy arenas. He also notes that policy writers have increasingly focused on the role of 'street level bureaucrats' (Lipsky, 1980) in formulating policy. Thus in a health system it is not only the minister of health and chief executive of the health department who determine policy but also other players in the system at different levels. These players will include doctors, nurses, unions and (hopefully) public health advocates. Marsh (2010) notes that policy results from an interaction between *structures* (the entities and rules within organisations or systems that influence policy making—for example, public health organisations or organisations for whom health is not core business but who nonetheless have a significant impact on health), *actors* (the stakeholders involved in policy making) and *ideas* (the content of policy making)—each of which have to be considered in understanding policy.

WHAT IS HEALTHY PUBLIC POLICY?

> Healthy public policy is characterised by an explicit concern for health and equity in all areas of policy and by accountability for health impact. The main aim of healthy public policy is to create a supportive environment to enable people to lead healthy lives.
>
> *World Health Organization, 1988*

Healthy public policy covers a broad range of activities in most sectors of society, and aims to alter the socioeconomic and physical environments in which we live, and ultimately to affect individual behaviours so that quality of life, well-being and health are enhanced. It is distinct from health policy, which is concerned with those policies that determine the financing and operation of sickness care services (Brown, 1992).

Pederson et al. (1988) note that healthy public policy seems to have been plagued by both conceptual ambiguity and terminology. They suggest the following definition: 'public policy for health using health in the broadest, ecological sense'. It is difficult to imagine policy areas that do not have implications for health. Their view of healthy public policy recognises the complex factors that affect health and illness, and the complex relationships between different sectors in society. Draper (1991) defined six features of healthy public policy:

• Public health issues are, invariably multi-sectoral and involve a range of interest groups. For instance, attempts to control drink-driving involve the police, hospital emergency departments, alcohol producers and retailers, schools and workplaces.
• Healthy public policy should involve commerce and industry, voluntary organisations, the community and all three tiers of government.
• Increasingly, risks to public health are international and not confined within regional or national boundaries. This is particularly true of environmental problems.
• The aim should be 'educational and persuasive rather than dictatorial or puritanical' (Draper, 1991, p. 18) and should aim to 'make the healthy choices the easy choices'. However, health legislation may be necessary in some circumstances.
• Action for healthy public policy takes many forms, through formally organised lobby groups or the actions of local community health initiatives.
• Healthy public policy is an intrinsically political activity.

The European Union has coined the term 'Health in All Policies' to promote healthy public policy and this notion has been taken up by the World Health Organization (WHO) (see box 24.1 for definition).

BOX 24.1   DEFINITION OF HEALTH IN ALL POLICIES

Health in All Policies is an approach to public policies across sectors that systematically takes into account the health implications of decisions, seeks synergies, and avoids harmful health impacts in order to improve population health and health equity. It improves accountability of policymakers for health impacts at all levels of policy-making. It includes an emphasis on the consequences of public policies on health systems, determinants of health and well-being.

Source: WHO, 2013b.

Finally, Kickbusch and Szabo (2014) note that healthy public policy is increasingly global. They define three arenas that are crucial to it: global health governance (institutions and processes of governance that are related to an explicit health mandate, such as WHO); global governance for health (institutions and processes of global governance that have a direct and indirect health impact, such as the United Nations, the World Trade Organization or the Human Rights Council); and governance for global health (institutions and mechanisms established at the national and regional level to contribute to global health governance and/or to governance for global health, such as national global health strategies or regional strategies for global health). They note that 'Health ministers must now be concerned with the priorities and activities of the security, trade, finance, agriculture, development, and employment industries if they are to effectively address health issues domestically and in global negotiations' and that global initiatives such as the post-2015 Sustainable Development Goals need to be translated into domestic policies.

POLICY FORMULATION

'Policy' is a nebulous term, used in many different contexts from general references to the foreign policy of a country to the particular policies of an organisation. Milio (2001, p. 622) defines policy as follows: 'Policy is a guide to action to change what would otherwise occur, a decision about amounts and allocations of resources: the overall amount is a statement of commitment to certain areas of concern; the distribution of the amount shows the priorities of decision-makers. Policy sets priorities and guides resource allocation.'

Inevitably, creating policy is complex, and can take years or decades. Legge et al. (1995) suggest that policy should be viewed as a narrative that provides guidelines for coordinated action across sectors and institutions. They state: 'The policy narrative tells of a set of problems and contextual issues; it tells of a world in which these problems could be resolved or ameliorated and it tells of actions that people will take that will lead to changes (or avoid changes) and that in doing so will help to create the better situation envisaged' (Legge et al., 1995, p. 7).

This fits well with Walt's (1994) assessment of policy as involving the decision to act on a particular problem, but as also including subsequent decisions on implementation and enforcement. She points out that a government's decision not to do something may represent policy, so policy 'must include what governments say they will do, what they actually do, and what they decide not to do' (Walt, 1994, p. 41).

Policy commentators note that the present developments in the nature of bureaucracies and how they relate to other parts of society have complicated the policy-making process. Some have characterised this as a shift from 'government' to 'governance' (Hill, 2005, p. 11). Richards and Smith (2002) refer to the emergence of a 'postmodern state', which is significantly different from the Weberian model of a bureaucratised state as shown in table 24.1.

TABLE 24.1   THE WEBERIAN STATE VERSUS THE POSTMODERN STATE

Weberian state	Postmodern state
Government	Governance
Hierarchy	Heterarchy (networks etc.)
Power zero sum game, concentrated	Power positive sum game, diffuse
Elitist	Pluralist
Unitary, centralised, monolithic state	Decentralised, fragmented, hollowed state
Strong central executive	Segmented executive

Weberian state	Postmodern state
Clear lines of accountability	Blurred/fuzzy lines of accountability
State central control	State central steering

Source: Richards and Smith, 2002, p. 36, table 2.2.

A final and crucial point about policy making is that the formulation of a policy issue is crucial to how the policy is determined. Tesh (1988) points out that there are many 'hidden arguments' behind public health policies. She comments that individualism is often prevalent so that policies will find solutions in terms of changing individuals' behaviours rather than in changing the structures that set the context for those behaviours. These hidden arguments underline the importance of values behind policy formulation. Table 18.1 provided an example of contrasting solutions to diabetes. It compared medical and behaviours approaches with those that tackle the underlying structural causes such as the restriction of advertising of high-energy foods and provision of active transport options in cities. Tesh (1988) argues that the structural solutions are 'hidden' and not immediately obvious in the way that medical and behavioural solutions are.

PHASES IN POLICY MAKING

Policy making usually involves a series of phases, such as this framework offered by Walt (1994, p. 45):

- *Problem identification and issue recognition*: analysis asking which issues do and do not get on the policy agenda, and why
- *Policy formulation*: determining who formulates policy and how, and where the initiatives come from
- *Policy implementation*: asking how policies are implemented, what resources are available, and how implementation is enforced
- *Policy evaluation*: asking how the policy is monitored, whether it achieves its objectives, and whether it has unintended outcomes.

It would be very rare for policy stages to follow such a rational or ordered linear process in reality. Policy making and implementation are usually more iterative, subjective, and are affected by the social environment.

Kingdon (2011) argues that policy agendas are shaped and change in response to a range of influences—ideas, interests and institutions. He examined policy-making processes in relation to health care services and transportation in the federal government of the USA and on that basis developed the idea of streams of influence that need to come together in order for policy ideas to be adopted and implemented. These streams are shown in figure 24.1. For him, these streams (problems, policies and politics) interact in the process of policy development and change in often fairly chaotic and unpredictable but nevertheless understandable ways to create policy. Kingdon argues that the policy landscape is peopled with policy entrepreneurs who wait for opportunities (he uses the metaphor of the open policy window) to engage in agenda setting. These entrepreneurs could be politicians, public servants or civil society actors and form a policy network that interacts and keeps the policy-making cycle dynamic and ripe with opportunities for change (Kingdon, 2011). Considine (2005, p. 140) says that networks are 'a kind of capillary system' through which policy work is done. J. Lewis (2005, p. 170) draws on the notion of policy networks in her detailed study of Victorian health policy networks and concludes: 'The policy process is better understood as a series of ongoing interactions between actors, ideas and structures, all of which affect each other and re-shape positions and connections in an interdependent network.'

FIGURE 24.1   KINGDON'S THREE-STREAM MODEL OF AGENDA SETTING

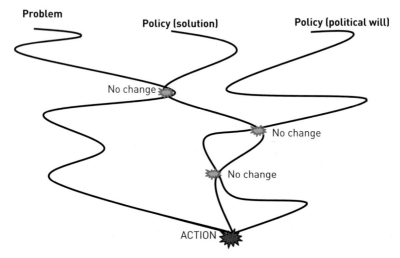

Source: Buse et al., 2005, p. 69 (adapted from Kingdon, 1984).

The notion of a network of policy actors, which can be influenced to varying degrees depending on the political and social climate, shows the ways in which public health actors need to understand this process in order to influence agenda opportunistically.

Kingdon helps us understand how policies come to be adopted and the role of different actors in this process. This agenda setting can be conceived of as a first stage in the policy process. Organisational theory provides frameworks to examine how innovative proposals move from the agenda through stages of implementation. Kaluzny and Hernandez (1988) describe four stages in the adoption of change: awareness, adoption, implementation and institutionalisation.

APPROACHES TO POLICY FORMULATION

Most literature on policy formulation identifies three main approaches:

- *Rational-deductive*: start with a problem and work through to its solution in a rational and linear way. The approach has its origins in the early part of the twentieth century when modernism held that scientific approaches and technology would lead to rational policy-making. The approach has been described as involving policy-makers in 'identifying the goals or objectives that should govern the choice of solutions to the problems, and in undertaking a comprehensive analysis of all possible alternatives and their consequences. On this basis, a solution is chosen as a master plan for maximising the objectives chosen' (Wiseman, quoted in Ziglio, 1987).

The main features of this approach are its rationality, following a logical sequence to arrive at decisions and making use of as much information as possible. Policy making tries as far as possible to follow a deductive approach to decision making, which is akin to the traditional scientific method.

- *Incremental*: an opportunistic approach that recognises that all implications are never known at the outset and so there is a constant need to reflect and amend. Lindblom (1959) characterised the approach as 'muddling through'. Policy decisions are never the definitive option but the one that makes sense at a particular point in time. Policies need to be adapted to suit new or changed circumstances. The approach relies heavily on the judgments of key players and reflects a pluralistic society in which there are multiple influences. Interest groups play a key role in shaping and reshaping the policy environment. In the health field J. Lewis's (2005) assessment of policy networks suggests they are powerful and underwrite medical power. Ziglio (1987) points out that incremental policy-making will tend to be reactive and likely to reinforce the status quo, rather than lead to innovation and change.

- *Mixed-scanning*: aims to combine the best features of the previous two. It is based on the understanding that the rational-deductive approach does not pay sufficient attention to the politics and values of any policy environment and that the incremental approach tends to be overly reactive. The main proponent of mixed-scanning is Etzioni (1967), who suggested that an overall scan of the policy environment is useful to identify those decisions that can be taken incrementally and those that are strategic. For the latter, information should be gathered and analysed in a way that takes into account prevailing political values and realities. By doing this, policy makers are able to retain a long-term vision and respond to immediate issues that require policy amendment. Hancock (1992) suggested that in the new public health, planning requires 'goal-directed muddling through'. This sits well with the mixed-scanning approach. After reviewing the three models, Ziglio (1987) recommends a mixed-scanning approach for healthy public policy. This approach is also suggested by Kingdon's (2011) policy streams and the opportunities that are created for policy entrepreneurs.

The process of policy formulation is complex and context specific, especially in relation to politics. Raphael (2012) and Navarro and Shi (2001) have noted that healthy public policies are more common under social democratic governments. The reasons for this relate back to the issues discussed in chapter 4, which show that more conservative and neo-liberal governments adhere to the values of individualism and the belief that society's role is not to become a 'nanny state'. Hence public health advocates for healthy public policy have to engage in the political process as is shown in chapter 22.

Assuming a political determination to take policy action, it is vital that there are mechanisms through which to formulate and enact policy. This can be done in municipal government through Healthy Cities processes (see chapter 23) and through the range of

Health in All Policies initiatives (see collection in Leppo et al., 2013). The 'health lens' mechanism used in South Australia is shown in box 24.2. Other useful mechanisms can be health impact assessment and health equity impact assessment. Health impact assessment typically focuses on interventions outside the health sector that do not have health improvement as their primary aim (Veerman, 2012). Health impact assessment offers a method to address the social and environmental determinants of health before implementing proposed policies, plans or projects designed to maximise future health benefits and minimise risks to health (Harris-Roxas, 2011). In this way it can make an important contribution to enacting healthy public policy.

BOX 24.2   THE ROLE OF POLICY ENTREPRENEURS IN THE ADOPTION OF HEALTH IN ALL POLICIES IN SOUTH AUSTRALIA

South Australia has a long history of engagement with the new public health. In the 1980s its health department established a Social Health Office, which produced a Social Health Strategy that drew directly on the Alma Ata Declaration on Primary Health Care and the WHO Ottawa Charter. The state hosted the 1988 WHO Conference on Healthy Public Policy and was one of the first in the world to enact tight tobacco control legislation. It also had a vibrant community health centre network and movement that lobbied for the new public health to be enacted. All this resulted by the twenty-first century in a local cadre of public servants and academics who were highly committed to the new public health values and had a good deal of experience in implementation and strategising about it. This resulted in an application to bring Ilona Kickbusch (an ex-WHO senior staff member who had been a driving force behind the Ottawa Charter) as a Thinker in Residence. The combination of the local policy entrepreneurs and the external Thinker in Residence enabled the new concept of Health in All Policies to be floated and then adopted. Work between Kickbusch and the local entrepreneurial public servants resulted in a method of health lens analysis.

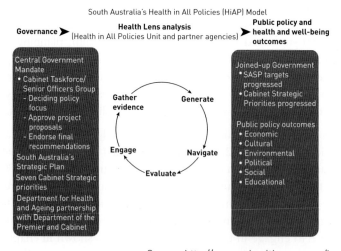

Source: http://www.sahealth.sa.gov.au/healthinallpolicies

Health lens analysis

Source: Baum et al., 2014b.

Professor Ilona Kickbusch working with the South Australia Health in All Policies team

POLICIES AND POWER

Milio (1983) defined the public policy environment in which policies for health come about. Key players are politicians, bureaucrats, media representatives and interest groups, and the process involves struggles between groups to ensure their desired policy ends are achieved in preference to those put forward by other groups. Inevitably, policy formulation is entangled with issues of power and influence. Different groups have different amounts of power and influence with which they can guide policy decisions.

Analysis of power is complex. Political science has presented an increasingly sophisticated understanding of how it operates in pluralist societies. Lukes (2005) pointed out that power does not just involve one person or group persuading another to act in a particular way, but also influencing the actual wants and desires of another person or group. When a carbon tax was introduced in Australia in 2013 the fossil fuel lobby developed a fund to fight against the tax and was successful in making it unpopular. People's views may be manipulated by those possessing power in such a way that their judgment about their real interests may become clouded. Foucault (1984) has also pointed out how power in modern societies is exercised through numerous interactions in everyday life. Medical knowledge, in particular, was seen as being very powerful in affecting people's behaviour. Bachrach and Baratz (1970) argue that 'non-decision' making can also influence policy processes. People with power may manipulate the decision-making process so that certain issues are not even raised. In Australia, it could, for instance, be argued that the power of parts of the medical profession and the private health insurance industry operate to ensure that the concept of an exclusively public health insurance scheme is not canvassed. Bacchi (2009) notes that power is exerted through the ways in which policy problems are represented. Often public health problems are constructed to highlight the role of individuals (for example, eating too much fast food rather than the increasing supply through the expansion of the fast food industry). The individualisation of the problem means the resulting policies are likely to stress behaviour change rather than regulation of industry. These perspectives on power make it imperative for policy analysts to go beyond obvious conflicts (or agreements) in any situation to determine how it is being manipulated. In analysing policy it is crucial to ask whose interests will be served or threatened by a policy change and their power to affect the policy formulation process. An equally important question is to consider involved groups that have no power or influence to affect policy. The move towards more consultative processes in bureaucracies was partly motivated by a desire to introduce less powerful voices. Themba-Nixon (2010, p. 138) notes that the 'best kind of policy initiative engages the community that shares the problem and ensures that the initiative is part of the solution'. Others have noted that participatory policy processes both reflect and encourage community empowerment (Burris et al., 2007). A review of the role of community organising in policy concluded that if community involvement is to be effective then community building should become 'a central consideration in how policies are shaped and what they look like' (Blackwell et al., 2012, p. 382). Baum (2007a) noted that equitable health policies result from a nutcracker effect between civil society and political and bureaucratic action (see figure 24.2).

FIGURE 24.2　HEALTH EQUITY NUTCRACKER

Source: Created by Simon Kneebone for Fran Baum; Baum, 2007a, p. 92.

Policy commentators point out that in the postmodern state (see table 24.1) policy is influenced by policy networks and policy communities. These networks and communities introduce a range of voices into the policy process and make it possible for public health advocates to play a role in the policy process. For the purposes of healthy public policy a consideration of the possible influence of these communities and networks is that it offers a way to study the actors that are likely to be arguing for or against a healthy public policy. They also underline the fact that the political system and the operation of the state are far from being unified and homogeneous systems but are in fact fluid systems that are open to influence at each stage of policy making (Kingdon, 2011; Hill, 2005; J. Lewis, 2005).

HEALTHY PUBLIC POLICY IN A GLOBALISED WORLD

When the Ottawa Charter was written in 1986 the focus of policy was on the nation state. By contrast the Bangkok Charter for Health Promotion (WHO, 2005) places much more emphasis on global relationships and approaches. Kickbusch and Seck (2007, p. 159) go so far as to say 'we are presently in a situation that all progress achieved so far towards health and well-being could be wasted unless effective global health policies are formulated ...' They suggest four factors that indicate a global governance crisis:

- *The lack of sustainable health systems* as shown by the increasing costs of health care in rich countries and the weakening health infrastructure in many poor countries. They stress this is especially dangerous at a time in which new disease challenges such as HIV/AIDS, SARS and Avian influenza threaten.
- *The consequences of global restructuring of economies*, which has led to a very different socioeconomic-political context of health. Inequities between rich and poor countries are increasing and public health is weakly protected under the new global trade agreements (see chapter 5).
- *Global health has no defined centre of action and is characterised by a growing and complex set of actors* including business (health is now one of the largest private markets in the world); international agencies like the World Bank, G8 and World Economic Forum; diverse new organisations, networks and alliances (such as the Global Fund on AIDS, Tuberculosis and Malaria), UNAIDS, Global Alliance in Vaccines and Immunization (GAVI) and the Gates Foundation making the scene more fragmented, harder to navigate and coordinate.
- *There are no obvious mechanisms for global accountability* for health as there is no systematic effort to build health globally (see chapter 5).

Thus advancing healthy public policy globally will require clear international commitment to health as a global public health good and to establishing mechanisms for the governance of health internationally. WHO is the obvious body to ensure policy coherence in global health and to play a brokering role in relation to the health impacts of the policies of other agencies (Kickbusch and Seck, 2007). The United Nations has recognised how crucial cross-sectoral policies are in its response to non-communicable diseases in its 2012 UN General Assembly resolution, which said that effective non-communicable disease prevention and control require 'leadership and multisectoral approaches for health at the government level, including, as appropriate, health in all policies and whole-of-government approaches across such sectors as health, education, energy, agriculture, sports, transport, communication, urban planning, environment, labour, employment, industry and trade, finance, and social and economic development'.

EXAMPLES OF HEALTHY PUBLIC POLICY

The examples given below range from those that provide for the universal provision of services and those that address a single issue. Both types are important to healthy and equitable societies.

UNIVERSAL POLICIES

Discussions of healthy public policy often overlook or take for granted the existence of universal provision of services that are beneficial for population health. Early theorists of the welfare state in the UK (Titmuss, 1958; Marshall, 1950) were clear that social spending was designed to alleviate poverty but was also part of a broader project of building citizenship and social solidarity. Key services include health, welfare and employment. A summary of these services is provided in table 24.2. Universal public policies mean that the service is available to the whole population and is publicly provided. Their benefits include the following:

- There is no stigmatisation associated with using the services because everyone uses them.
- Public provision is very efficient and effective for the whole population.
- Universal services can be funded by progressive taxation so funding can be equitable.
- Welfare state models do not eliminate inequities, but they do reduce them (Lundberg et al., 2008).
- The very existence of universalism underpins the development of citizen commitment to equity as a goal of government.

TABLE 24.2   EXAMPLES OF UNIVERSAL POLICIES AND HOW THEY PROMOTE HEALTH

Universal policy type	Examples	Evidence for health impact
Publicly funded health care	National Health Service in the UK, Medicare in Australia	Cost is a major barrier to health care and universal services ensure that this barrier is removed
Primary and secondary education	All OECD countries have free public education systems	Education is strongly associated with improved health outcomes
Cradle to grave welfare provision accessible to whole population in times of need	European welfare state, Australia (although increasingly becoming targeted)	Avoids catastrophic health impacts when families lose breadwinner support through unemployment or disability; provides for people in old age

Universal policy type	Examples	Evidence for health impact
Child allowances	European welfare states, Australia	Reduces likelihood that children will live in poverty
Water and sewerage services	Many high-income countries	Prevention of infectious diseases
Childcare	Finland	Reduces stress on parents and provides a safe place for children

As we saw in chapter 5 the introduction of neo-liberal thinking into government has been accompanied by a retreat from the universalism of welfare services in countries where they existed. This has been true of the National Health Service in the UK, welfare services in the Nordic countries, and in Australia. In Australia there has been increased targeting of a range of human services and increasing paternalism in their provision (Spies-Butcher, 2014). As market reforms have been introduced by both Labor and Liberal governments in Australia egalitarianism has been undermined (Spies-Butcher, 2014).

GUN CONTROL

One of the most spectacularly successful policy changes in Australia was the restriction on gun ownership. This contrasts strongly with the USA where a powerful gun lobby has ensured that legislation to control guns has not been introduced. This is despite ongoing mass gun murders including that at the Sandy Hook Primary School in December 2012, when 20 children and six teachers were killed by a young man. Gun ownership and policy is a major public health issue because the availability of guns is strongly associated with fatalities and injury from firearms (Chapman, 1998). Restricting the right of individuals to own firearms is thus an important public health measure. It is also very controversial, with sections of the community maintaining that it is their right to be able to defend themselves. The policy change in Australia, which introduced national uniform gun laws, was possible only because of the massacre of 35 people at Port Arthur at Easter in 1996. This provided a window of opportunity that was taken by Prime Minister John Howard and used to exert leadership and push the policy through, despite entrenched opposition to uniform gun laws from well-organised gun lobbies with strong links to the conservative side of politics. In each state this gun lobby had for many years successfully lobbied any government considering tougher laws. Chapman reports that the gun lobby was responsible for the defeat of the Unsworth NSW Labor government at the 1988 state election and following this 'the NSW Labor Party hierarchy proclaimed that any talk of serious gun control was a political no-go zone' (Chapman, 1998, p. 67). Beresford (2000) analyses the three things Howard had to achieve in order to push through the policy and capitalise on the public mood following Australia's worst-ever massacre:

- Demonstrate leadership. He did this despite significant opposition and managed to push through the law without too many special exemptions. He was supported by

Deputy Prime Minister Tim Fisher, who was able to quell dissent from the farming lobby.

- Develop a process to reach agreement. Howard called an emergency meeting of the state police ministers, chaired by the federal attorney-general. He addressed this meeting and from it came a proposal for national uniform laws; prohibition on the importation and sale of military-style weapons, pump-action shotguns and self-loading rimfire rifles. Licensing procedures were tightened. After this meeting further ones were necessary and Howard continued to address rallies of angry opponents. He also used the threat of a referendum on the issue to keep pressure on his state colleagues.

- Develop a method of implementation. The federal government reached agreement on a buy-back scheme funded by a temporary rise in the Medicare levy. This allowed for semi-automatic weapons to be taken out of circulation and for compensation to the owners.

The success of the policy in reducing firearm-related deaths was shown by Chapman and Alpers (2013). They note that the rate of firearm homicides reduced as well as the rate of firearms-related suicides (59.9 per cent decline between 1997 and 2005) with no indication of any method of substitution for suicide. By contrast the US population is 13.7 times larger than Australia but it has 134 times the number of total firearms-related deaths (31 672 in the US compared with 236 in Australia in 2010) and 27 times the rate of firearm homicides (Chapman and Alpers, 2013, p. 770).

POLICY AND LEGISLATION RELATING TO DRUGS

HISTORY OF DRUG POLICY

Our current policies surrounding drugs have emerged from a history in which drugs have been intimately linked with our culture and commerce. Plant (1999) provides a rich account of the ways in which drugs have been used in Western and other cultures. Mind-altering substances have played a role in the arts. Plant describes the ways in which they helped inspire many nineteenth-century poets including Wordsworth and Coleridge.

In the past two centuries attempts by the state to control the use of drugs have intensified. In the 1930s the USA made an attempt to control alcohol through prohibition. This was quite unsuccessful and the main result was the establishment of a firm link between the supply of illegal drugs and criminal activity, especially organised crime. In recent decades the links have been between a range of drugs declared illicit by the states, most prominently cocaine and heroin. President Nixon declared a 'war on drugs' in the 1970s and since that time the USA has seen drugs as a kind of Enemy Number One. Plant (1999, p. 243) sees the result of this war as 'a vast and complex alternative economy that positively thrives on the laws and attempts to enforce them'. She reports that this attempt at control has been as unsuccessful as Prohibition in the 1930s, and that the law enforcement agencies only seize a small proportion of the drugs available. In the countries where the drugs come from (for example Colombia, Afghanistan and Burma) the CIA has been involved in covert operations that have resulted in its becoming complicit in the trade (McCoy, 1991).

Control of legal drugs has generally been more successful than that of illegal drugs. Tobacco, alcohol and illegal drug policies are reviewed below.

TOBACCO CONTROL INTERNATIONALLY

Tobacco is the most widely available and harmful product on the market—more than one billion tobacco-related deaths are projected for the twenty-first century, especially afflicting low- and middle-income countries (Bettcher and da Costa e Silva, 2013). Tobacco is the second leading cause of death in the world, killing nearly six million people each year. More than five million of those deaths are the result of direct tobacco use, while more than 600 000 are the result of non-smokers being exposed to second-hand smoke. Unless urgent action is taken, the annual death toll could rise to more than eight million by 2030. Half the people who smoke today (about 650 million people) will eventually be killed by tobacco (WHO, 2014h).

One of the most significant international healthy public policies has been the Framework Convention on Tobacco Control (WHO, 2003b), which was passed by the World Health Assembly in May 2003. The core of the treaty calls for demand reduction through price, tax, and non-price measures. In 2014, 168 countries were listed as having signed the treaty (WHO, 2015c). It is the first global health treaty negotiated under the auspices of the World Health Organization and provides a model for international healthy public policy in the future. The core measures it calls for are listed in box 24.3.

BOX 24.3   DEMAND AND SUPPLY REDUCTION METHODS IN WHO's FRAMEWORK CONVENTION ON TOBACCO CONTROL

CORE DEMAND REDUCTION MEASURES

- Price and tax measures to reduce the demand for tobacco
- Protection from exposure to tobacco smoke
- Regulation of the contents of tobacco products
- Regulation of tobacco product disclosures
- Packaging and labelling of tobacco products
- Education, communication, training and public awareness
- Tobacco advertising, promotion and sponsorship
- Measures to reduce tobacco dependence and encourage cessation

CORE SUPPLY REDUCTION MEASURES

- Eliminating illicit trade in tobacco products
- Prohibiting sales to and by minors
- Provision of support for economically viable alternative activities

Source: adapted from Bettcher and da Costa e Silva, 2013, p. 207.

TOBACCO CONTROL IN AUSTRALIA

The Cancer Council Victoria reports that tobacco is the most harmful recreational drug used in Australia, and is estimated to be responsible for more than 15 000 premature deaths each year. It accounts for more than 90 per cent of all drug-related deaths. It is the largest single preventable cause of death in Australia (Scollo and Winstanley, 2012).

Prior to 1976, the only regulation of tobacco use was aimed at ensuring fire safety but in that year the federal government prohibited the advertising of cigarettes on television and radio. Between 1988 and 2013 most Australian states and the Commonwealth passed comprehensive tobacco control legislation restricting availability (particularly to children); requiring price increases, and larger and more explicit health warnings on labelling; restricting promotion via print, point-of-sale and outdoor advertising; prohibiting tobacco company sponsorship of most sporting and cultural events; and introducing plain packaging. Australian courts have accepted that passive smoking is harmful and the implications of this have encouraged increased prohibitions on smoking in workplaces and other public areas.

As Reynolds (1995) points out, change through legislation is only possible if there is political support, which in turn depends upon a groundswell of social support. That support requires the building of coalitions for change. Reynolds cites the forces that gathered to ensure the passage of the 1988 South Australian legislation as an example. The Australian Medical Association (AMA) and the anti-cancer societies portrayed the legislation as a critical response to a major public health problem. The Health Minister, Dr John Cornwall, was a passionate advocate. Members of Parliament received letters from the AMA advising them how many of their constituents were dying from tobacco-induced diseases. The state was also fortunate in not having a tobacco-growing industry to raise opposition. More recently the tobacco industry is taking steps to sue Australia because it has introduced plain packaging legislation (see chapter 5). Since the mid-1980s Australia has had considerable success in tobacco control. In 1964, 43 per cent of Australians aged 15 years or older were daily smokers (OECD, 2009, as cited in AIHW, 2011a), compared with 15 per cent in 2010 (AIHW, 2011a). Daily smoking rates have almost halved since 1991 (24.3%) (AIHW, 2014e). Smoking prevalence has declined overall, but rates for men declined more rapidly than those for women. In Australia daily smoking rates have dropped steadily from 40 per cent for men and 31 per cent for women in 1983 (Chapman and Wakefield, 2001) to 14.5 and 11.2 per cent for men and women, respectively, by 2013 (AIHW, 2014e). The prevalence of smoking is also inversely related to occupational and educational status. Working-class people are more likely to smoke than their better-off counterparts, most likely reflecting that smoking is a short-term relief and pleasure in difficult life circumstances (Graham, 1987). Smoking is also high among Aboriginal and Torres Strait Islander people and is even higher among the members of those communities who are members of the Stolen Generations (Thomas et al., 2008). This suggests a strong link between stressful life events and taking up addictive behaviours (Maté, 2012). The legislative and other changes that enabled the dramatic drop in smoking rates to occur are listed in box 24.4.

BOX 24.4   ACHIEVEMENTS IN AUSTRALIAN TOBACCO CONTROL

- Harm reduction: Australian advocates were the first to arrange for the tar and nicotine content of cigarettes to be tested.
- Advertising bans: Australia was one of the first democracies to ban all tobacco advertising and sponsorship.
- Packet warnings: Australia has among the world's largest packet warnings.
- Mass reach campaigns: In the 1970s Australia was one of the first countries to run these campaigns.
- Civil disobedience: Australian was one of the first nations to experience widespread civil society organised action against tobacco companies with campaigns such as BUGA UP.
- Smokeless tobacco: In 1986 the South Australian Government become the first government in the world to ban smokeless tobacco.
- Small packets banned: Small 'kiddies' packs were banned nationally.
- Tax: Australia has a relatively high tobacco tax, and among the most expensive cigarettes in the world.
- Replacement of sponsorship: Victoria pioneered the use of a dedicated 5 per cent rise in tobacco tax to enable the buyout of tobacco sponsorship.
- Clean indoor air: Australia has among the world's highest rates of smoke-free workplaces and domestic environments. Smoking is banned on all public transport. Most states have banned smoking in restaurants.
- Graphic packet warnings: In 2006 Australia upgraded warnings on cigarette packets with graphic depictions of end results of tobacco use such as gangrene toes and a diseased lung.
- Plain packaging was introduced in 2013 with the particular aim of making it unattractive to start smoking

Source: Chapman and Wakefield, 2001, pp. 275–6.

Litigation has been a powerful tool in the fight against tobacco. A WHO report noted that the law is usually seen as an instrument of justice rather than health, but that the experience from the early 1990s to the early twenty-first century has shown that it can help fight the global tobacco epidemic. The law has been mainly used in the USA, where there have been some dramatic successes in which juries have found for individual smokers and awarded large sums in punitive damages. In April 2002 the first Australian successful judgment against a tobacco company was made when $700 000 was awarded to a 52-year-old woman dying of lung cancer. The judge noted that the company had tried to hide evidence of its knowledge of the link between tobacco smoking and cancer. Thus recent experience indicates that use of the law can 'awaken public outrage, strengthen public policies and redress injuries' (Blanke and Mitchell, 2002, p. 867).

While legislation has played a part in this process, it is not possible to quantify or directly attribute its contribution. Rather, it has been one element in an interplay of social, political and scientific forces that together have resulted in a significant public health advance. Chapman and Wakefield (2001) analyse the success of the tobacco-control movement in Australia and point out that it resulted from two decades of effective lobbying by those employed within government bureaucracies and non-government organisations and from effective civil society organisations such as BUGA UP (Billboard Utilising Graffitists Against Unhealthy Promotions).

ALCOHOL

WHO has recognised that excessive alcohol causes significant harm (intoxication, dependence, violence and a wide range of other harms) in many countries of the world to the extent that in low mortality countries it ranks first as a risk factor responsible for the global disease burden (WHO, 2002). But it is also an important drug that people use for relaxation and makes a significant contribution to the culture and economy of many countries. Australia's public policy response to alcohol demonstrates a solid harm-minimisation approach. Australian per capita alcohol consumption is high by world standards, ranking 34 of 185 countries. Eighty-three per cent of Australian adults reported drinking alcohol in 2004 so it is a behaviour with wide social acceptance (contrasting with some countries, such as Iran and Saudi Arabia, where the use of alcohol is illegal). Each year approximately 3000 Australians die as a result of excessive alcohol consumption and around 65 000 people are hospitalised. Particular concern is expressed about young people and episodic risky drinking (colloquially called 'binge drinking') including the violence, injuries and harms to others associated with this behaviour; drinking during pregnancy; the high levels of health and other harms among Aboriginal and Torres Strait Islander peoples; and the longer term health harms from regularly drinking above the level of NHMRC recommended guidelines (Australian National Preventive Health Agency, 2013a). The total cost to society of alcohol-related problems in 2010 was estimated to be $14.352 billion. Of this, $2.958 billion (or 20.6 per cent) represents costs to the criminal justice system, $1.686 billion (or 11.7 per cent) comprises costs to the health system, $6.046 billion (or 42.1 per cent) involve costs to Australian productivity and $3.662 billion (or 25.5 per cent) are costs associated with traffic accidents. This estimate of total costs, however, does not incorporate the negative impacts on others ($6.807 billion estimated by Laslett et al., 2010) associated with someone else's drinking (Manning et al., 2013). Alcohol also brings benefits in that alcohol-related taxes produced $6.8 billion in 2009–10 (AIHW, 2011a) and the industry is estimated to have contributed $18.3 billion to the Australian economy in 2004–05.

Public policy on alcohol is influenced by the beliefs that people should be able to make personal choices about alcohol use, and that alcohol, unlike tobacco, can be used at levels that are not harmful and may even have some health benefits.

Australia had a National Alcohol Strategy until 2011 and that policy is under an overarching National Drug Strategy that includes legal and illicit drugs (see box 24.5). The very act of putting all drugs in one strategy supports harm minimisation. Alcohol's harm was extensively covered in the report of the National Preventive Health Taskforce and was the subject of a good deal of work by the Australian National Preventive

Health Agency before it was defunded by the Abbott Coalition government in June 2014. Collectively this work indicated that industry self-regulation of advertising was not sufficiently tight to protect children and young people, and alcohol was also consistently associated with sporting events. The agency was also examining binge drinking and its harmful effects on young people (see box 24.6). It was recommended that taxation be used more effectively to control the harmful effects of cheap alcohol, particularly wine (Australian National Preventive Health Agency, 2013a). Alcohol pricing through taxation is a powerful tool that governments can use to control the consumption of alcohol.

BOX 24.5   NATIONAL DRUG STRATEGY 2010–2015

MISSION

To build safe and healthy communities by minimising alcohol, tobacco and other drug-related health, social and economic harms among individuals, families and communities.

Since the National Drug Strategy began in 1985, harm minimisation has been its overarching approach. This encompasses the three equally important pillars of demand reduction, supply reduction and harm reduction being applied together in a balanced way:

- **Demand reduction** means strategies and actions that prevent the uptake and/or delay the onset of use of alcohol, tobacco and other drugs; reduce the misuse of alcohol and the use of tobacco and other drugs in the community; and support people to recover from dependence and reintegrate with the community.
- **Supply reduction** means strategies and actions that prevent, stop, disrupt or otherwise reduce the production and supply of illegal drugs; and control, manage and/or regulate the availability of legal drugs.
- **Harm reduction** means strategies and actions that primarily reduce the adverse health, social and economic consequences of the use of drugs.

Source: Australian Government, 2011.

BOX 24.6   THE NATIONAL BINGE DRINKING STRATEGY

This strategy aims to address Australia's harmful binge drinking culture, especially among young people. The strategy focuses on raising awareness of the short- and long-term impacts of harmful or 'risky' drinking among young people, and over time, contribute to the development of a more responsible drinking culture within Australian society. 'Be the Influence—Tackling Binge Drinking' has created a large presence at music festivals and sporting events across Australia, including Splendour in the Grass, Listen Out, Stereosonic, the Big Day Out and the Australian Boardriders Battle.

Source: Australian National Preventive Health Agency, 2013b.

ILLICIT DRUGS

Around the world there are very different public policies relating to drugs. At one extreme, for example in Singapore and Indonesia, drug possession can lead to a death sentence. Other countries operate on a harm-minimisation basis where the intent of policies is to reduce the harm associated with use of the drugs. Australia provides an example of a country whose policies have rested on harm minimisation but which is moving to one based more on zero tolerance. The debate between these two positions is often heated with strong views on both sides. Thus illicit drugs present a very contested area of policy characterised by much public debate. Between the mid-1980s and the late 1990s, Australia's public policy approach to illicit drugs was relatively moderate, though not permissive. Neal Blewett, Commonwealth Health Minister during the development of the National Campaign Against Drug Abuse, characterised the campaign's aim as 'to minimise the harmful effects of drugs on Australian society. Its ambition is thus moderate and circumscribed. No utopian claims to eliminate drugs, or drug abuse, or remove entirely the harmful effects of drugs, merely to "minimise" the effects of the abuse of drugs on a society permeated by drugs' (Blewett, 1985). He went on to acknowledge that the campaign's credibility partly rested on its ability to demonstrate that the problems associated with both illegal and legal drugs can be addressed.

The struggle has been to develop a conceptual and policy framework capable of doing just that. The Australian approach to illicit drugs has eschewed the prohibition model evident in the USA, where the addict is seen as a criminal rather than a health problem. Many of the ill effects attributed to drug use (corruption, crime and justice system costs) are, in fact, attributable to prohibition, rather than to drug abuse per se. This tends to focus on the abuser and sets up barriers to treatment and rehabilitation, while maintaining a restricted supply situation that creates high prices and attracts new suppliers. It is an approach that is hard on the individuals, but soft on the problem. The difference between the two approaches is shown in figure 24.3.

FIGURE 24.3   DRUG POLICY DILEMMAS

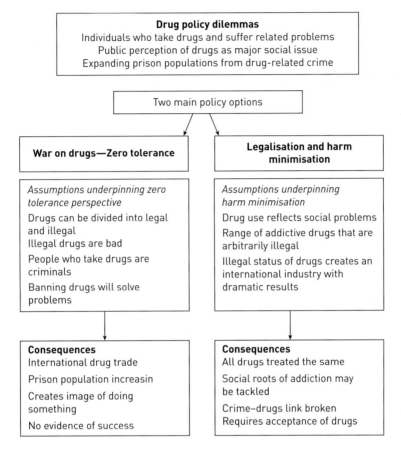

Drug policy dilemmas
Individuals who take drugs and suffer related problems
Public perception of drugs as major social issue
Expanding prison populations from drug-related crime

Two main policy options

War on drugs—Zero tolerance

Legalisation and harm minimisation

Assumptions underpinning zero tolerance perspective
Drugs can be divided into legal and illegal
Illegal drugs are bad
People who take drugs are criminals
Banning drugs will solve problems

Assumptions underpinning harm minimisation
Drug use reflects social problems
Range of addictive drugs that are arbitrarily illegal
Illegal status of drugs creates an international industry with dramatic results

Consequences
International drug trade
Prison population increasin
Creates image of doing something
No evidence of success

Consequences
All drugs treated the same
Social roots of addiction may be tackled
Crime–drugs link broken
Requires acceptance of drugs

Wardlaw (1992, pp. 1–37) compares policy in the USA, Australia, the Netherlands and the UK. The Netherlands approach is characterised by acknowledging that it is not practical to base policy on moral or ideological grounds, but that it is impossible to eliminate drugs. It accepts that different drugs have different effects and are used by different people for different reasons, and develops policies that normalise drug use so as to minimise harm and maximise treatment and support. Cannabis use is effectively decriminalised, but prohibition is still enforced for large amounts, for harder drugs, and high-level dealing. In terms of outcomes the policy seems successful—cannabis use is lower than in comparable nations. Heroin users are proportionately fewer and their contact with the treatment system very good. Marshall (1988, pp. 391–420) estimates that 70–80 per cent of the Dutch heroin user population has contact, compared with 10–15 per cent of the US heroin user population. This is enormously significant in relation to HIV control, where the effectiveness of prevention strategies using messages advocating safe sex and clean needles depends on access to, and credibility with, the particular communities. Researchers in the USA (Buchanan et al., 2004) have compared two communities that have adopted syringe exchange programs (SEPs) (a classical tool of harm minimisation), Hartford and New Haven, Connecticut, with one that did not—Springfield, Massachusetts. Their research suggested there was a significantly lower rate of HIV diagnosis in the communities with SEPs compared to the one without. People living in the community with SEPs were less likely to use risky sources to obtain their syringes and the needles were not in circulation as long so intravenous drug users were at less risk of blood-borne disease. Disposal of needles was also safer in the communities that provided facilities for this purpose.

Wardlaw (1992) concludes that Australia, like the UK, lies somewhere between the Dutch and American models—but is much closer to the Dutch. He notes the considerable influence that the USA has had on drug policy in other nations. Because the prohibition approach is focused on reducing supply, rather than demand, producer and trans-shipping countries find themselves under great pressure to conform with US requirements. Fortunately, Australia avoided this until 1997.

Over the past decades there have been some changes in the extent to which governments have embraced harm minimisation. Despite this, harm minimisation has remained as the driving force of the National Drug Strategy (see box 24.5). Thus in the late 1990s a new policy direction under the title 'Tough on Drugs' was announced (Commonwealth of Australia, 1997), aiming to totally prevent the entry of illicit drugs into Australia and a zero tolerance to the use of illicit drugs (including in schools). Government support went to non-government treatment agencies that adopt abstinence-based treatment and research on such programs. The Howard government famously refused to endorse a heroin trial in the ACT, which would have tested the effectiveness of controlled availability of heroin (Bammer et al., 1996). The proposed trial was based on Swiss research, which indicated that it offers many benefits for opiate-dependent people, including not being homeless, being more likely to be employed, reduced debts and less likelihood of committing criminal acts (Uchtenhagen et al., 1997). The trial received much media coverage in 1997 and was publicly supported by the federal minister for health and the state and territory ministers for health. Despite this, the prime minister would not support the trial, probably reflecting fear of a public backlash. Consequently, there has been no heroin trial in Australia.

The current National Drug Strategy does stress a harm-minimisation approach. It also recognises that drug use is associated with social determinants such as unemployment, homelessness, poverty and family breakdown. Some illicit drug use appears to be declining—the 2007 National Drug Strategy Household Survey showed the proportion of people reporting recent use of illegal drugs fell from 22 per cent in 1998 to 13.4 per cent in 2007. The recent use of cannabis—the most commonly used illegal drug—fell from 17.9 per cent in 1998 to 9.1 per cent in 2007. Since its introduction in September 2005, non-sniffable Opal fuel has contributed to a 70 per cent reduction in petrol sniffing across 20 regional and remote communities in Western Australia, South Australia, the Northern Territory and Queensland, showing the value of supply reduction.

Recent data suggest an increasing use of methamphetamine (also called meth, crystal, chalk, and ice) as measured by the number of police detainees testing positive for it—21 per cent in 2011 compared with 16 per cent in 2010 and 13 per cent in 2009

(MacGregor and Payne, 2011). 'Meth users' are presented as an increasing problem for hospital emergency departments because the behaviours associated with its use make dealing with the patients very difficult. In the USA from 2007 to 2011 there was a 51 per cent increase in emergency department visits relating to methamphetamine (Substance Abuse and Mental Health Services Administration, 2014). The drug has a high potential for abuse and addiction and can cause a wide array of potentially harmful intoxicating effects, including altered judgment and reduced inhibitions, leading to unsafe behaviours. The drug can also cause severe dental problems, anxiety, confusion, insomnia, mood disturbances and violent behaviour.

FOOD AND NUTRITION POLICIES

Food and nutrition are basic elements of human survival, and policies to influence them are crucial aspects of public health. We live in a world where food security continues to be a major problem, in which some suffer from undernutrition and in which an increasing proportion of the world's population is overweight or obese. The past decade has seen a massive increase in public health concern on this latter issue because increasing weight is seen to be driving a chronic disease epidemic, which is evident in countries at all levels of development.

Food policy highlights many of the dilemmas in devising healthy public policy, including the role of values reflected in either individualist or free-market approaches, or more collectivist approaches that rely on state intervention. Through these issues, food policies also demonstrate the ways in which the needs of industry often conflict with public health, and the importance of cross-sectoral action. First we review two food policy issues—undernutrition and healthy weight. Then we consider *laissez-faire* and interventionist policies and finally consider the ways in which industry tries to influence policies.

UNDERNUTRITION

Black et al. (2013) estimate that undernutrition—including foetal growth restriction, stunting, wasting, and deficiencies of vitamin A and zinc, along with suboptimum breastfeeding—is a cause of 3.1 million child deaths annually or 45 per cent of all child deaths in 2011. Undernutrition reduces cognitive attainment; substantially increases the likelihood of being poor throughout adulthood; and is closely linked with illness or death during pregnancy and childbirth for both mothers and their babies (Gillespie et al., 2013). Despite decades of concern about this issue, effective policies have not been devised to meet this public health tragedy and the Millennium Development Goals relating to improving this situation will not be met. The causes of undernutrition reflect systematic factors, as shown in figure 24.4. These range from inadequate breastfeeding to policies in a range of sectors including health, agriculture, trade and poverty reduction. To give one example, policies in the agricultural sector that encourage farming for cash crops may undermine the food supply of farming families.

FIGURE 24.4   CONCEPTUAL FRAMEWORK OF DRIVERS AND DETERMINANTS OF UNDERNUTRITION

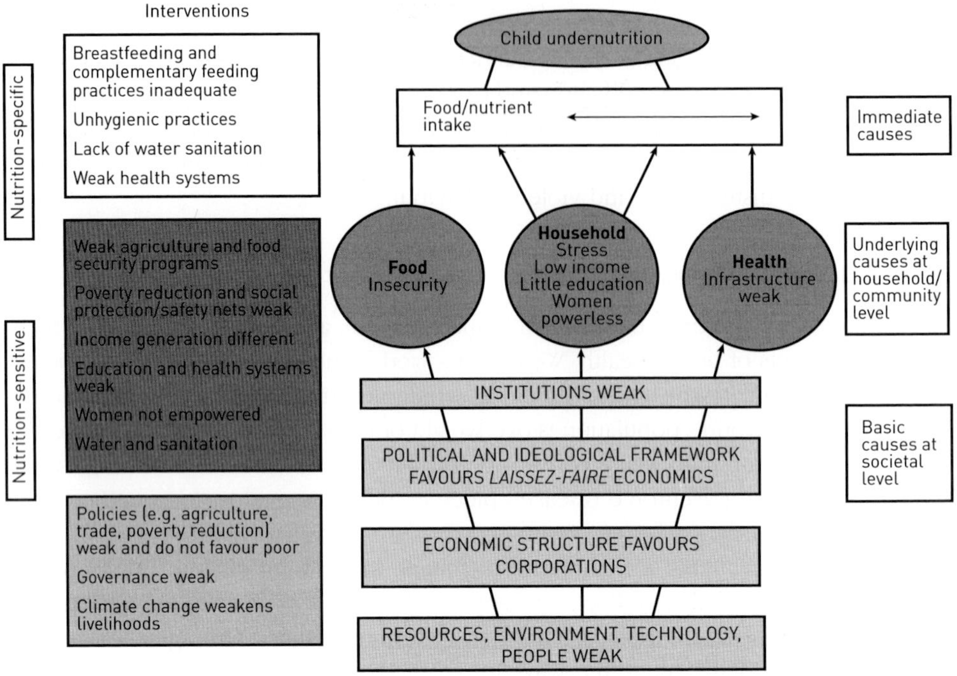

Source: Gillespie et al., 2013, p. 185, Figure 9.1.

Clearly the response to undernutrition needs to be as multisectoral as the causes, and will involve improving the income security of poor families, linking the health agenda with the agriculture agenda, and ensuring that economic policy does not undermine food security. The sorts of changes required in the agriculture sector are shown in box 24.7, demonstrating the importance of a Health in All Policies approach.

BOX 24.7   GUIDING PRINCIPLES FOR NUTRITION-SENSITIVE AGRICULTURE POLICY

NUTRITION-SENSITIVE AGRICULTURE POLICY:

1   increases incentives/reduces disincentives for production of nutrient-dense foods
2   increases incentives/reduces disincentives for production diversification
3   increases incentives/reduces disincentives for environmentally sustainable production
4   invests in research to increase productivity of nutrient-dense foods in low-resource conditions and diverse systems
5   invests in higher education that trains future leaders on agriculture–nutrition linkages

6   builds capacity among ministry staff and extension workers to understand linkages and communicate relevant behaviour-change information

7   improves gender equity in extension and training

8   provides nutrition information about foods and diets through schools, higher education and markets

9   improves smallholders' access to government-controlled markets such as food aid/ social protection, communal catering (e.g. school lunch programs, hospitals and workplace canteens)

10   improves infrastructure needed to provide market access for smallholders and other vulnerable groups, improves access to market price information, avoids trade policies that would preclude smallholders' market access

11   builds resilience against shocks through infrastructure and social safety-net programs

12   has institutional mechanisms and incentives to coordinate with other sectors relevant to nutrition (e.g. health, social protection, education).

Source: Gillespie et al., 2013, p. 185. Adapted from Food and Agriculture Organization, 2013, p. 20.

HEALTHY WEIGHT POLICIES

There has been increasing public health concern about the growing incidence of overweight and obesity (see box 24.8).

BOX 24.8   GLOBAL PICTURE OF OVERWEIGHT AND OBESITY

- Worldwide obesity has more than doubled since 1980.
- In 2014, 39 per cent of adults aged 18 and older (more than 1.9 billion people) were overweight. Of these, more than 600 million were obese. Overall, about 13 per cent of the world's adult population (11 per cent of men and 15 per cent of women) were obese in 2014.
- Most of the world's population live in countries where overweight and obesity kills more people than underweight.
- More than 42 million children under the age of five were overweight or obese in 2013.

Source: WHO, 2015b.

The health risks associated with obesity include type 2 diabetes, heart disease, hypertension, stroke and certain cancers, making it a leading cause of preventable disease and death. A recent report from WHO in Europe details the evidence that the social gradient for obesity is steepening: 'obese people in lower socioeconomic groups are getting heavier at a faster rate than people in higher socioeconomic groups' (Loring and Robertson, 2014, p. 4). Friel et al. (2007, p. 1242) indicate that these inequities reflect underlying structural inequities. The actual and future potential health impact of overweight and obesity has led to considerably increased policy attention to this area.

WHO's global strategy on diet, physical activity and health (WHO, 2004a) laid the basis for a series of strategy documents that note that regulatory interventions (such as restricting food marketing to children) are likely to lead to more equitable outcomes because they may have a greater impact on lower socioeconomic populations (WHO, 2012, p. 23). Caballero (2007) notes that 'political leaders still tend to regard obesity as a disorder of individual behavior, rather than highly conditioned by the socioeconomic environment' (see box 4.2 for examples of political expressions of individualism or collectivism). Yet in the public health literature there is growing consensus that it is indeed the socioeconomic environment that is the culprit. Egger and Swinburn (2010) talk of an obesogenic environment in which opportunities for exercise as part of everyday life are reduced, and low-cost, high-fat and high-sugar food are widely available. Freudenberg (2014) has detailed the role of transnational food corporations in encouraging the consumption of high-fat and high-sugar foods, showing how they advertise to children, including activities such as promoting Ronald McDonald as an ambassador for fast food, and opposing public health legislation. Others have linked obesity to economic policy. Wisman and Capehart (2010) argue that the increase in obesity reflects more than increasing access to unhealthy food and that neo-liberalism has led to a generation of insecurity, lack of control and increasing powerlessness, which has encouraged consumption of high-sugar and high-fat foods.

Policies likely to address these more structural causes of obesity are listed in box 24.9. These contrast with those that focus on changing the behaviours of individuals such as by providing information on the benefits of healthy eating or taking exercise, and the provision of advice from general practitioners or other health professionals.

BOX 24.9   EXAMPLES OF STRATEGIES THAT TACKLE STRUCTURAL IMPEDIMENTS TO HEALTHY WEIGHT

NATIONAL AND INTERNATIONAL STRATEGIES

- Discourage subsidies to industry that mean products undercut indigenous production in other countries or encourage ecologically unsustainable food production.
- Encourage local production and ensure international trade agreements do not impede this.
- Restrict the import of ultra-processed foods.
- Reduce taxes on healthy, fresh food.
- Legislate against the advertising and promotion of unhealthy foods that are high in fat and sugar.
- Ensure trade agreements do not restrict governments' abilities to regulate in favour of healthy diets.
- Develop national policies to encourage high-exercise environments, including active transport.

REGIONAL AND LOCAL STRATEGIES

- Develop model food service policies for canteens in schools, hospitals, childcare centres, clubs and other public eating places.
- Ensure that socioeconomically disadvantaged communities have public transport to low-cost shopping venues.
- Develop model workplace policies to support breastfeeding.
- Provide hot meals at community centres for older people or through Meals on Wheels.
- Provide lunches for all children in public schools.
- Encourage ecologically sustainable production of food—for example, through the establishment of community gardens. Community gardens are places where people of all ages, cultural backgrounds and abilities can find a space to rest, reflect, grow vegetables, fruit, herbs or flowers, and observe nature at work. Community gardens can also play an important role in community composting and community building.
- Develop schemes to promote backyard gardening and use of homegrown fruit and vegetables.
- Promote and subsidise local farmers' markets.
- Provide and maintain safe open space in all areas to encourage exercise.
- Encourage urban design that creates active transport options including public transport, walking and cycling.

The role of values in policy is well illustrated by Australia's response to growing obesity over the past decade. Australia has among the highest rates of adult obesity in the world, at 28.3 per cent (OECD, 2014b). In 2011–12, only 35.5 per cent of Australians were in the healthy weight range (ABS, 2013a). In response to this, the Council of Australian Governments (COAG) determined 'to increase the proportion of the population in the healthy weight range by 5 percentage points over the 2009 baseline, from 37 per cent to 42 per cent' by 2018 (National Health Performance Authority, 2013, p. 2). The Rudd–Gillard Labor government saw significant attention to the issue of obesity, most significantly through the work of the National Preventative Health Taskforce, which reported in 2009 (National Preventative Health Taskforce, 2009b), and the government's response (which included the establishment of the Australian National Preventive Health Agency, which had the reduction of obesity as one of its key objectives). Although the main emphasis remained on behavioural strategies, reflecting its limited terms of reference (Baum and Fisher, 2011), the National Preventative Health Taskforce did acknowledge that making healthy choices requires supportive environments: 'the healthy choice must be physically, financially and socially the easier and more desirable choice than the less easy option' (National Preventative Health Taskforce, 2009, p. 58). The National Preventative Health Taskforce paid some attention to the role of industry in addressing obesity, highlighting food production and food marketing to children.

The main strategies on healthy weight were part of Commonwealth–state agreements relating to programs for health promotion in communities, with children and in workplaces. These agreements were terminated in the Abbott Coalition government's 2014 Budget, and the Australian National Preventative Health Agency was defunded in June 2014, with its functions consolidated into the Department of Health. The Department of Health website now includes only behavioural lifestyle advice. This change in focus reflects the Coalition government's emphasis on individualist philosophies and personal responsibility, and a dislike of any policy that is perceived as reflecting a 'nanny state' philosophy. The Eat for Health website has been continued (www.eatforhealth.gov.au) and the dietary guidelines (NHMRC, 2013) it provides for Australia do recognise the constraints on people adopting healthy diets, but focus primarily on individual behaviour change rather than constraints on the food industry to improve the supply of food. The signs are that the Australian Government will be particularly reactive to the lobbying of the food industry, as shown by the issue of food labelling legislation (see box 24.10).

BOX 24.10   FOOD LABELLING IN AUSTRALIA

Labels on food products play an important role in providing information to consumers at the time when they are making crucial decisions about purchase and consumption. Using these labels to encourage healthier eating has been an important public health strategy. In 2009 the Australia and New Zealand Food Regulation Ministerial Council set up the National Review of Food Labelling to examine the structure, format, accuracy and appropriateness of information on food labels. The review saw food safety as the most crucial issue, followed by nutritional quality. Its recommendations included the need for a simple interpretative front-of-pack nutrition labelling system, and a clear preference for the use of a traffic-light system. The traffic-light system categorises the four key nutrients most associated with public health issues—fat, saturated fat, sugars and salt— as high, medium or low compared to the recommended level of intake of these nutrients. These ranks are portrayed as red, amber or green traffic lights on the package. Evidence from the UK suggests the system does encourage more healthy food choices, but the Australian Food and Grocery Council immediately rejected this recommendation as too simplistic.

The traffic-light system was rejected by health ministers in favour of watered down front-of-pack traffic-light labelling, which turned into a health star rating system that was approved in June 2013, implemented in 2014 and gives industry five years to comply. In February 2014 a website explaining the health star rating system briefly went live and then was removed after intervention from Senator Fiona Nash's then chief of staff, Alistair Furnival. It very soon become apparent that Mr Furnival had previously worked for the food industry, and he was forced to resign after extensive media coverage of the issue that pointed to his very direct conflicts of interest.

This example demonstrates the very significant power the food industry dictates over public health policy. Despite the evidence supporting other methods and advice from public health experts, we have a less effective system of food labelling, thanks to industry lobbying. The same thing happened in the European Union where the traffic-light system was rejected, and the industry is said to have spent €1 billion on its lobbying.

Sources: Gill, 2011; Sacks, 2014.

LAISSEZ-FAIRE OR INTERVENTIONIST FOOD POLICIES?

Milio (1989) contrasted two approaches to healthy food policies: the Norwegian interventionist, comprehensive public policy and the US *laissez-faire*, market-oriented strategy.

Overall, Milio determines that the Norwegian Food and Nutrition Policy made significant progress towards its goals. The food self-sufficiency goals were met and exceeded in some cases. Norway, through the 1970s and 1980s, made a significant contribution to world food security, particularly given the size of the country. Evaluation of the dietary objectives showed a more complex picture with some improvements but a continued high intake of saturated fats. There were declines in cardiovascular deaths, but these could also be linked to stringent smoking control. Nevertheless, Norwegian death rates from cardiovascular disease were more favourable than those in other Nordic countries that had not had comparable nutrition policies. Milio's (1989) analysis of the USA's market-driven approach to food and nutrition was not as favourable. She found the food and nutrition policies to 'be piecemeal, influenced by political and commercial interests, and consequently inconsistent, tending to neutralise or confound support for healthy nutritional patterns' (p. 420). Market interests took precedence over those of health. The effects of the absence of national policies were particularly clear in terms of equity. In the 1980s disparities grew between rich and poor in diet-related low birthweight babies and infant mortality, dietary risk factors and diet-related disease and deaths (Milio, 1989, p. 420). Overall, Milio concludes that the 'invisible hand' of the market is less effective than public policy in promoting healthy food choices for a population.

Her subsequent work for WHO (Milio and Helsing, 1998) stresses that a food and nutrition policy should address issues beyond the health sector and include agriculture and fisheries, the environment, rural development, food manufacture and foreign trade. She suggests that policy instruments should include economic subsidies and taxes, regulation of food standards, labelling and advertising, provision of direct services (such as nutrition counselling, meals on wheels and mass catering), training and education of personnel and the public, and research and evaluation of consumer and organisational behaviour. She stresses that while healthy food at the point of purchase is important, more impact can be gained by influencing or regulating farming organisations, food corporations, retailers, advertisers and educators, as their decisions create healthy choices for consumers in food shops, restaurants, workplace canteens and institutional catering. An example is shown in box 24.11. The importance of intersectoral approaches is also stressed in box 24.7 on the contribution that the agricultural sector can make to undernutrition.

BOX 24.11   SAN FRANCISCO'S HEALTHY MEALS INCENTIVE ORDINANCE

In 2010, San Francisco passed a groundbreaking ordinance that set nutritional standards for restaurant food accompanied by toys or other youth-focused incentive items. The model ordinance, created by public health law and policy and pioneered by Santa Clara

County, helps localities wanting to take a regulatory stand against unhealthy fast food by encouraging restaurants to develop healthier children's meals. The impact of these ordinances has spurred significant changes in industry practice. Organisers worked closely with San Francisco Supervisor Eric Mar, who sponsored the ordinance, and they mobilised health professionals, parents, school teachers and community organisations to advocate for its passage. According to Corporate Accountability International, the powerful coalition generated more than 5000 messages to the city's Board of Supervisors. A core group of 10–30 people regularly attended lobby visits and testified in favour of the legislation.

It was this level of grassroots organising that, in the end, secured the legislation's passage. Supervisor Bevan Dufty, who provided the swing vote in favour of the ordinance, explained in a hearing how the compelling testimony of residents finally convinced him to support the legislation, despite the hostile climate created by the industry.

Source: Gagnon et al., 2012.

FOOD POLICY AND COMPETING INTERESTS

Food and nutrition policy formulation and implementation in all settings demonstrate the importance of competing interests and agendas on policy. The food and agriculture industry is extremely powerful and dominated by multinational companies who have at their disposal powerful advertising mechanisms and access to governments, because of their ability to affect national economic issues. Inevitably, there are conflicts between the interests of these companies in maximising their profits and in contributing to national goals for healthy diets. Policy formulation and implementation are generally done in an environment in which the voices of these industries are clearly heard. Governments, including successive Australian ones, seek to make partnerships with industry when dealing with policy. Grossman and Webb (1991, p. 276) recognise the necessity of this in a pluralistic society, but also offer this warning:

> It is true that the industry is a power to be tapped … Nevertheless, what must be remembered in working with the food industry is that on many topics there is a straightforward conflict of interest between those of us who are (broadly speaking) working for the population as a whole and those who are paid to represent corporate interests. The syllogism which says that the public health lobby, once committed to intersectoral action, must court the food industry at all costs is foolish: it encourages the worst sort of appeasement.

Since those comments were written this issue has become more acute in that the food industry has become increasingly powerful in shaping the public health agenda. To give one example, Freudenberg (2014) details the power of McDonald's, the fast food chain. From 1989 to 2012 the company contributed nearly $10 million to candidates for public office in the USA. In the same period it paid 10 lobbyists more than $8 million, a figure that does not include the salaries of its own paid lobbyists. Much of its lobbying explicitly seeks to thwart public health regulation. The company also has its staff on six federal government advisory committees at the departments of Agriculture, Commerce and State. The power of the food lobby was seen when WHO came under blatant pressure from elements of the food industry in the USA, which objected to its recommendations for reducing the consumption of sugar (Beaglehole and Bonita, 2004). Another recent example was the extensive lobbying by Coca-Cola against the Mexican Government introduction of a tax on large bottles of soda. The tax was introduced in response to the fact that Mexico is the most overweight country in the world and suffering an epidemic of diabetes-related deaths. Coca-Cola lobbied extensively, including with full-page advertisements in the press (*The Economist*, 2013b). The power of the food lobby is also shown by the Australian example on food labelling in box 24.10, and it was in spite of pressure from industry that San Francisco introduced the Healthy Meals Incentive Ordinance (see box 24.11). Public health interests must be continually alert to the dangers of co-option by interests whose primary concerns are issues other than health. Such considerations underline that healthy public policy will remain complex, challenging and political.

ROAD SAFETY: EVIDENCE FOR THE EFFECTIVENESS OF LEGISLATION

The potential for healthy public policy is shown by the data on road accidents. These have shown a steady decline in the past four decades in Australia (see chapter 12 for data). In 2006 road crashes were estimated to cost $17.85 billion (1.7 per cent of gross domestic product). This was a real decrease of 7.5 per cent compared with 1996 (Bureau of Infrastructure, Transport and Regional Economics, 2009). The savings to the community in terms of the reduction in mortality are clearly considerable. It is less easy to demonstrate precisely what policy interventions brought about the change. However, figure 24.5 gives an indication of the impact of a range of safety measures.

FIGURE 24.5   THE IMPACT OF ROAD SAFETY MEASURES ON FATALITY RATES

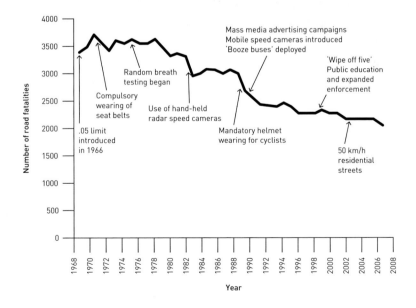

Source: Australian National Preventive Health Agency, 2013c, p. 25.

Australia has been a world leader in developing policy and legislation related to road accidents, and the combined effect of the policy measures has made a significant contribution. Powles and Gifford (1993, p. 127) point out that 'the temporal proximity of putative causes and their effects strongly suggests that a large part of the fall in mortality from traffic accidents is attributable to centrally coordinated action'. The measures included the introduction of compulsory seat-belt wearing, restrictions on the amount of alcohol a person can have in their blood when driving and policing of this limit with random breath testing, and speed limits in country and metropolitan areas. Wearing seat belts was made law in Victoria in 1970 and subsequent evidence indicates that this legislation made a significant impact on the decline in road accident deaths. The saving in lives as a result was estimated to be about 30 to 40 per cent (Joubert, 1979). Subsequently, similar legislation has been adopted in many other countries.

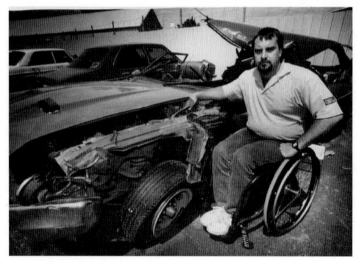

Motor vehicle accidents are the leading cause of death and injury among young Australian males. (*Advertiser* and *Sunday Mail*, Adelaide)

Between 1984 and 2013, fatalities have been more than halved. Over the past decade to 2013, total annual deaths decreased by 24.6 per cent. The estimated trend is an average reduction of 3.4 per cent per year. Passenger and pedestrian deaths have decreased at the fastest rates, with driver deaths also falling. In contrast, both motorcyclist and pedal cyclist deaths show non-decreasing trends (Bureau of Infrastructure Transport and Regional Economics, 2014). The data suggest that the introduction of legislation prohibiting driving after the consumption of alcohol has had a significant effect on the decline in fatalities. Random breath testing has been adopted in all Australian states, together with tough penalties for those found to be over the limit. The legislation has been accompanied by mass media campaigns highlighting the dangers of drink driving. A typical slogan is 'If you drink, then drive, you're a bloody idiot'. In recent years, advertising campaigns have become increasingly graphic in depicting the effects of road accidents. As a result of these campaigns drinking and driving has become largely unacceptable within the general Australian population. Road safety legislation is now relatively uncontroversial, but this was not the case when measures were first introduced. Seat-belt legislation in Victoria, for instance, needed a concerted campaign by the local newspaper, a professor of mechanical engineering, politicians and the Royal College of Surgeons. The lobbying process was typical of many other campaigns for healthy public policy. First, the number of road deaths was rising and involved many young people, so there was an undisputed public health problem. There was some evidence to suggest that legislation might be effective. The advantages to be gained from legislation for compulsory wearing had been proven by a 1961 Victorian law that compelled all motorcyclists to wear a crash helmet. This law produced a 50 per cent reduction in deaths and injuries to motorcyclists. Also, when the Snowy Mountains hydro-electric scheme was under construction, the authority insisted on its employees wearing seat belts. Despite some very serious accidents, there were no accident-related deaths or serious injuries in the six-year construction period (1961–67) (Joubert, 1979). An analysis of the process by which road safety legislation became acceptable by a Melbourne newspaper noted that prior to the legislation being introduced: 'There had been a tendency to blame poor driving skills and a belief that exhortation and slogans could change that' (Smith, 1993). An analysis of the Victorian success in reducing road deaths and injury concluded that the policies were successful because of persistent and persuasive lobbying from a wide range of community sectors; these community interests had access to decision-makers; there was an active all-party parliamentary committee dedicated to road safety; there was a secure government that was confident about making potentially contentious decisions; and there was a supportive and influential media that was willing to take on causes and promote them.

As a result of concerted policy reform and cooperation from a range of sectors, Australia has climbed from being one of the worst countries for vehicle accidents and mortality to one of the best in the world. This example of successful public policy for health has much to offer other countries where deaths from road accidents are increasing and there is little public policy to address the increase.

WHAT MAKES FOR HEALTHY PUBLIC POLICY?

These examples suggest that the processes needed for the successful adoption of healthy public policies are:

- an issue over which there is clear evidence about adverse effects on health
- effective lobby groups in favour of policy and legislation to control the source of the adverse effects on health
- winning of support for policy and legislation change from key opinion leaders, including the media and politicians, in spite of opposition from groups who favour the status quo. Healthy public policy will often mean challenging the power of groups who are influential and wish to protect their profits with unhealthy policies

- supportive bureaucratic players in key positions who are keen to advocate the public good over private interests
- a policy environment that supports government intervention to change social and economic structures in order to promote health
- ensuring that there are wins both for health and the other sectors involved.

CONCLUSION

This chapter has demonstrated the great potential of public policy and legislation as a public health tool. It indicates that an effective new public health relies on policies that do restrict the freedom of individuals, but which can be justified in terms of the gains in health status in the population as a whole. The extent to which the power of policy and legislation can continue to be harnessed to improve health depends on the acceptance of restrictions on the freedoms of individuals. The likelihood of such restrictions being acceptable is increased when community values are seen as a crucial component of a healthy society.

CRITICAL REFLECTION QUESTIONS

24.1   Why is the nutcracker approach (see figure 24.2) to health equity policy likely to be successful?

24.2   What are the hallmarks of a Health in All Policies approach?

24.3   What factors account for the decline in road deaths in Australia since the 1970s?

24.4   Why does the USA have a much higher rate of firearm-related deaths than Australia?

Recommended reading

Buse et al. (2005) *Making health policy* provides a good guide of how to get health issues on the public policy agenda, how evidence informs policy and why some initiatives are more likely to be implemented than others.

Clavier and de Leeuw (eds.) (2013) *Health promotion and the policy process: Practical and critical theories* has a series of chapters on the importance of theory to understanding healthy public policy and examples of the ways in which policy theory has been used to understand implementation processes.

Hill (2005) *The public policy process*. This book provides an excellent introduction to the process by which public policy is made. It makes the policy process understandable and combines and links theoretical and practical aspects of policy.

Leppo et al. (2013) *Health in all policies: Seizing opportunities, implementing policies*. This collection of essays provides an excellent guide to the different ways in which Health in All Policies has been implemented.

Loring and Robertson (2014) *Obesity and Inequities: Guidance for addressing inequities in overweight and obesity*. Good example of a social determinants approach to reducing inequities.

Useful website

The South Australian Government Health in All Policies website (http://www.sahealth.sa.gov.au/wps/wcm/connect/public+content/sa+health+internet/health+reform/health+in+all+policies) provides a comprehensive guide to the government's implementation of Health in All Policies and includes links to a Health in All Policies training manual.

PART 8

PUBLIC HEALTH IN THE TWENTY-FIRST CENTURY

> The only hope is that the mass of humanity will see the danger before the rot goes too far and the human and environmental damage becomes too great to repair.
>
> *Harvey, 2014, p. 293*

> Ultimately, we need to see the emergence of a new Weltanschauung: a new view of the world, a new framework of ideas within which to make choices and decisions.
>
> *Eckersley, 2005, p. 250*

Public health in the twenty-first century faces a world characterised by growing social unrest—economic austerity has seen economic inequities soar, and evidence is increasing that the scientific predictions of climate change are proving to be accurate. Responses to climate change are weak and not meeting the scale of the threats. There is also growing sectarian conflict in the Middle East, introducing new instabilities. Refugee numbers are increasing, while new xenophobias flourish in the wake of the uncertainties and insecurities this brings. Although many of these developments do not bode well for our collective health, there is also a growing realisation that only a radical change in our priorities and values will save the world from the dire predictions many commentators are making. Thus, this fourth edition of *The New Public Health* comes at a time when understanding the underlying social and economic causes of health is more crucial than ever.

Public health's central *raison d'être* is about shaping this future and working to ensure it is as healthy, sustainable and equitable as possible. New information technologies, rapid global transportation systems, the globalisation of trade and capital, rampant consumerism and the increasing importance of international agreements and treaties have already ensured that the twenty-first century will be one in which local, national, and global issues interrelate to a far greater degree than previously. Threats to the environment and health appear to be growing in number and intensity, yet life expectancy increased considerably in the twentieth century. Threats that were on the horizon during the twentieth century are now realities. Thus ideology-based conflicts and efforts to prevent them are centre stage on most political agenda and tangible evidence of climate change is apparent in our daily experience of unusual weather patterns.

In the first two decades of the twenty-first century life has proved to be uncertain, less equal, full of perceived risk and more threatening. For the first time for at least two centuries life expectancies are declining in some regions of the world. A future characterised by deteriorating natural environments, growing divides between rich and poor, declining health status for much of the world's population and a decline in the quality of civic and community life seems possible and not just a worst-case scenario.

The zeitgeist of the times is not conducive to progressive, visionary futures. Fear dominates so much political discourse and the global threats of terrorism and environmental catastrophe underline insecurities. Yet the creation of a more equal and liveable world will require considerable vision and commitment to experiment with political choices, ways of running economies, organising cities and communities, consuming and producing and relating across the differences between people.

Public health offers a means of assessing how well we are doing in creating a better global community. The visions presented in part 6 of this book suggest much of the architecture that will be required for this new world. Part 7 speaks directly of the strategies public health can offer to contribute to a better, healthier and more equitable world. Shaping creative alternatives will be a central task for our future and the new public health has a central role in this journey.

25

LINKING LOCAL, NATIONAL AND GLOBAL PUBLIC HEALTH

Equity, ecologically sustainable development and peace are at the heart of our vision of a better world—a world in which a healthy life for all is a reality; a world that respects, appreciates and celebrates life and diversity; a world that enables the flowering of people's talents and abilities to enrich each other; a world in which people's voices guide the decisions that shape our lives.

People's Charter for Health, People's Health Movement, 2000

It always seems impossible until it is done.

Nelson Mandela

KEY CONCEPTS

Introduction

Global issues of ecology

A just world?

Leadership for a healthy future

Public health for the brave-hearted

Reflective, flexible and eclectic

A vision for 2050

Conclusion

INTRODUCTION

This chapter provides a summary of the key arguments made in this book and provides a vision of what our global society and local communities could be like if the principles and ideals of the new public health were implemented. Health results from people's experiences in their everyday lives. Families, workplaces and community organisations shape people's lives. But these lives are also influenced by powerful social and economic forces nationally and globally, many of which are not health promoting. This book has demonstrated that, while medicine has been based on an understanding of health as an absence of illness, broader understandings have existed alongside it. Public health has come to embrace a broad view of health and increasingly its practitioners operate from the assumption that longer term and meaningful change in health will result only if the powerful structural factors that affect people are the focus of public health initiatives. This book has also stressed that our collective health and equity depends on effective stewardship of the natural environment.

Public health policy and practice, not surprisingly, reflect the economic, political and social climate in which they occur. Dominant political and social ideas have a crucial impact on health. Current ideas that are particularly crucial in shaping public health are the balance between individualism and collectivism, the dominance of economic considerations and market philosophy in public policy-making and the importance accorded to social solidarity and social participation. Each of these has a strong, if usually invisible, impact on public health.

Perhaps the influence of public policy factors on health is shown most starkly in health inequity. Australian and international evidence supports the view that inequalities in health status appear to be growing within and between countries, reflecting deep-rooted structural factors in societies. While behaviour might explain some differences between groups, factors such as income, wealth, housing quality, employment and educational opportunities appear to be far more significant. More than any other area of public health, differences in the health status of groups demonstrate that health reflects the experiences of people in their everyday lives. Some countries (including Cuba, Costa Rica and Sri Lanka) have managed to achieve a relatively high health status without moving into the league of rich countries. This is explained by the fact that these countries have public policies that provide educational opportunities, especially for women, a strong primary health care system and discourage large disparities in income. Differences in life expectancy between rich countries appear to reflect the equitability of income and wealth distribution rather than the overall level of wealth (Wilkinson and Pickett, 2009).

Evidence suggests that attempts to change behaviour without a parallel effort to change structures will only benefit those people who already have favourable living conditions such as employment, reasonable income, good housing and a safe environment (CSDH, 2008; Blaxter, 2010).

The very limited gains of the behavioural public health experiments have resulted in public health revisiting and extending the strategies used by earlier generations of public health reformers. As a result, public health concentrates more and more on reforming the operation and practices of organisations, institutions and communities. In this sense public health is being implemented differently to that in the past when reformers were more likely to impose solutions with little consultation. Participation in public health in the twenty-first century will continue to be as challenging as it has been in the past decade or so, and its effectiveness will require changes in the ways in which public health professionals and their organisations operate. But lessons are being learnt and organisations and professionals are changing, and communities are becoming more sophisticated in dealing with them.

Social issues are assuming more prominence in public health. Social support, high self-esteem and a sense of personal control are important determinants of health, best achieved in societies and communities that are relatively equal and that have reasonable levels of social solidarity. The development of supportive societies and communities should be at the heart of public health strategies for the future. The necessary strategies will include public policies, legislation, public sector interventions and local actions. Social capital and the trust it relies on are likely to be central to public health endeavours in the future.

If the focus of public health is to reflect the importance of social factors and achieving ecological sustainability, then policy will be the fundamental and essential tool. 'Making healthy choices the easy choices' became a catchcry of public health in the 1980s and 1990s. For the twenty-first century, it is 'making healthy, sustainable and convivial choices the easy choices'. This dictum applies to workplaces, schools, cities, communities and all places where people gather to play, learn and work, including social media such as Facebook and Twitter.

The overall mission of the new public health is to create a healthy and equitable society in which the natural environment is sustainable, political will for equity is articulated, policies are used to create health and equity in a proactive manner, and there are numerous opportunities for lifelong personal, intellectual, social and emotional development (figure 25.1). Stacking these factors up together in a coherent way will result in healthy populations with equitably distributed health.

FIGURE 25.1   THE NEW PUBLIC HEALTH SUSTAINABLE HEALTH EQUITY STACK

Source: Concept by Fran Baum, drawing by Simon Kneebone.

GLOBAL ISSUES OF ECOLOGY

Humanity is facing the enormity of the ecology crisis. Climate change and the resultant global warming are putting the environment at the centre of political and commercial debates. While some details of this crisis are disputed among experts, the overwhelming majority of scientists accept the evidence that the world's natural systems are out of balance, that the effects of this are unpredictable and that the cause is human activity. Climate change, air and water pollution, resource depletion, rapid urbanisation, loss of biodiversity, high energy use and lifestyles that take a heavy toll on the earth's environment combine to paint a bleak future. That public health is showing more concern with the environment has been evident in international and national public health policies and in community-based action to tackle local environmental problems. However, in the broader political community, action to reduce the use of carbon is timid and Australia has seen a reversal of the modest carbon tax introduced in 2013. Achieving an international treaty is proving extremely difficult and even if achieved, its likely goals appear unambitious.

The first public health movement was concerned with providing clean water and effective sanitation to fast-growing industrial cities. The preoccupation of this movement was to ensure that waste was taken away from areas of dense population. Now public health cannot simply ensure that waste is removed but must also be concerned about the impact of waste on the environment and the sustainability of solutions. The complexity of the problems faced today is far greater than those faced by earlier generations of public health reformers.

A decade or so ago there was confidence that the battle against infectious disease was being won. In developed countries chronic diseases were the big challenge and in developing countries, while infectious disease was still the major killer, solutions were felt to be at hand. Now there is much less certainty. Re-emergent and emergent infections are continuing to challenge public health. The increase in international travel appears to have assisted the spread of infectious diseases and makes global pandemics more likely. Most recently, Ebola in West Africa has claimed thousands of lives and required a concerted international response.

Public health will have to be global to be effective. Global inequities and inequities within countries show every sign of increasing and will be evident in the pattern of new diseases and burden of environmental problems. Social and economic justice will become more difficult to achieve, and public health will have to take on a stronger advocacy role. Redressing the disparity in wealth between rich and poor countries will become more pressing as the differences widen and fuel social, political and environmental problems. Since the Global Financial Crisis the existence and growth of inequities has become much more evident in popular discourse. Protests against these inequities are common, most recently in protests against austerity in Europe and against draconian budget measures in Australia that would have undermined the universal health and welfare system. An increasing number of academic texts document these inequities (see, for example, Stiglitz, 2012; Stuckler and Basu, 2013; Piketty, 2014) and call for systematic changes to our economic system to curb them. Public health practitioners should be at the forefront of advocacy for equity. Public health arguments need to be marshalled to support the importance of reducing the gaps between rich and poor (both within and between countries). These arguments should draw on the increasing evidence base that equitable societies appear to be both more cohesive and more healthy.

Public health will have the task of encouraging governments to make long-term investments in health and environmental enterprises. Doing this will not be easy. Short-term economic considerations often come first. But public health will have to decide which priorities will best promote human and environmental health in the future. Molecular and genetic approaches to the control of disease are gaining ground (Beaglehole and Bonita, 2004) (for instance, many millions of dollars are being spent globally on the human genome project). Proponents of these approaches argue that the money is well spent because of the potential that understanding genes and molecular structures has for promoting health and improving life expectancy. Meanwhile, millions die worldwide for want of basic public health measures that we know will promote health and extend life: clean water, sanitation, adequate food and housing and access to basic literacy. The environment continues to deteriorate because many governments will not slow development or invest in environmental protection. Petersen (1998) warns of the increasing 'geneticisation' of society, suggesting that it may lead to new insidious forms of surveillance and control. The new public health practitioners will need to take a critical and sceptical view of genetic technology, questioning its potential for impact on population health status and the impact its availability would have on equity. If genetics have an impact on health (other than that of promise) almost certainly the rich and powerful will benefit more than others.

Often decisions about development and the environment appear to be driven by the notion that there are no limits to growth and that one day we will entirely conquer death and disease. Yet the public health task is more to do with providing environments in which all people (not just the rich and those in wealthy countries) can lead healthy lives and then die with dignity in reasonable living environments without destroying the planet on which all health depends.

A JUST WORLD?

We live in an extremely unjust world. If you are born in Sierra Leone you face a life over 40 years shorter than someone born in Japan. If you are an Australian Aboriginal person you die, on average, some 12 years before other Australians. Such inequities are increasing globally and in many countries. Class, gender and ethnicity play a crucial role in determining health status. This book has argued that too often these inequalities are put down to some personal failure of the people within the group that suffer the worst health status. Thus in Australia many people will point to the fact that some Aboriginal people are heavy drinkers, that they are overweight or do not look after their houses. Such victim-blaming ignores the long history of dispossession of land and culture that goes back to the white invasion of 1788. A Royal Commission has documented the severe health effects of being part of the 'stolen generation'. Much other research has shown that the abuse of alcohol has complex causes concerning loss of a sense of cultural identity and lack of a hopeful future. Food choices in many poor communities, especially those in remote Australia, are limited and offer primarily unhealthy foods. Often housing and other infrastructure is not designed for harsh remote settings and not suited to typical lifestyles. In other words many Aboriginal people do not have the social, historical, cultural or physical infrastructure to lead a healthy life.

The same story is true of Africa. Centuries of colonialism have robbed Africa of its people (most notably through the slave trade) and resources, which have fuelled the wealth of countries that are rich today. Contemporary trading patterns reinforce these historic inequities and are heavily weighted in favour of already rich countries. 'Make poverty history' campaigns, which have led to some debt reduction, have made negligible changes to the lives and chances of Africans. An outsider to planet Earth is likely to marvel at the poverty we continue to tolerate alongside riches that, historically, are unheard of.

While poor countries suffer from a very uneven playing field, capitalism is producing larger and larger profits. The new 'barons' of the system, the chief executive officers of corporations, are paid huge salaries that, as a percentage of average salaries, have zoomed out of all proportion. The extent of these salaries commonly receives negative coverage in the media. Shareholders are receiving large dividends. The gap between the wealth of rich and poor continues to grow. Increasing attention is paid to the 0.1 per cent who have come to own an increasing share of the world's wealth in the past decade (Oxfam, 2015).

Despite this somewhat gloomy picture there are many people and social movements, including public health and health promotion activists, who believe the world should be more just. They recognise that the struggle for health equity is a struggle for social and economic justice.

LEADERSHIP FOR A HEALTHY FUTURE

Economic considerations have come to dominate public decision-making to the extent that this in itself has become a public health risk for society. The goal of public policy is being progressively narrowed to a preoccupation with reducing public sector costs and privatising public services, and a rejection of social ends as a goal of public policy. Political parties in most countries are adopting such views or having them thrust upon them by international monetary agencies. Public health is quintessentially a public service activity that has always challenged the logic of neo-liberalism. While part of the reason for public health activity may be to support a healthy workforce and protect the interests of trade and commerce, it also rests on a strong ethical argument that promoting health in communities and individuals is a social good in its own right.

For the future, public health has an investment in joining the voices that are arguing for a return to public policies that seek to promote civil society, encourage an investment in the social fabric of communities and protect the environment. Without a strong state that sees its role as leading societal efforts to balance social, environmental and economic concerns it is hard to imagine public health's goals of equity, sustainability and health being achieved. National governments and international public-good organisations around the world, Australia included, need to foster visionary national commitment to and leadership for public health.

This book has shown that public health problems and responses are becoming increasingly global. Consequently, the need for strong global leadership is growing, but the reverse appears to be happening. Since 1945 the United Nations World Health Organization has been the leading public health body. Other UN organisations have also played a role in health activities—the UN Children's Fund (UNICEF), the UN Population Fund, the UN Development Program (UNDP) and the World Bank—but until recently they have taken a secondary role to that of the World Health Organization (WHO). There is a serious lack of coordination and cooperation between the agencies (Beaglehole and Bonita, 2004; S. Lewis, 2005) and there is not the leadership from WHO that was seen in the 1970s and 1980s. Criticisms of WHO are becoming common (Howard, 1996; Clift, 2014) and nothing has captured the imagination of public health workers around the world in the way that Health for All by the Year 2000 and the Ottawa Charter did. Since Halfdan Mahler's term as director-general ended in 1988, WHO has been without a passionate leader who can convey a strong vision for the direction of health care services and health promotion. The strength of the Organization's advocacy for the poor and marginalised has been questioned. Werner and Sanders (1997, p. 171) comment that it is unfortunate that WHO has 'not stood up more firmly to pressure from governments, wealthy elites and multinational corporations. Thus the new public health needs a champion who will be visionary and brave in their advocacy of the needs of the world's poor and marginalised and prepared to 'speak honestly to power' about the impact that rampant pursuit of profit brings to human health and equity. Examples of such voices from formal institutions have been the Commission on Social Determinants of Health (2008), and the *The Lancet*–University of Oslo Commission on Global Governance for Health (Ottersen et al., 2014). Both promote a new public health understanding of the ways in which our society and economy need to change to promote health. The most positive signs of leadership for an alternative vision of economics and environment come from civil society, which provides a strong voice advocating progressive and visionary scenarios for the future. The People's Health Movement and the World Social Forums provide prime examples of how civil society is envisaging a healthier and fairer world. At the international and national level, policies and complementary strategies that set a clear direction and framework for a progressive new public health movement are important for meeting the challenges of the twenty-first century. The vision for 2050 in box 25.1 is intended as a discussion starter about how we might change the world for the better.

PUBLIC HEALTH FOR THE BRAVE-HEARTED

Tackling the social and economic aspects of health will be controversial and very likely to bring public health workers into conflict with those who have an interest in maintaining the status quo and opposing public policy that benefits the poorer and more vulnerable (and so less powerful) members of society. While some commentators suggest that the future will be less complicated because the fall of communist governments has signalled 'the end of history' (Fukuyama, 1992), present indications are that numerous complexities face us. Since 11 September 2001, terrorism has been a powerful reminder of the competing and complex forces that are shaping our future—recent acts of terrorism include the Sydney Martin Place attack in December 2014 and the Paris attack on the Charlie Hebdo office in January 2015. Market economies are still rife with inequities and very significant social and environmental problems. If public health engages in redressing these problems, the complexities and extent of powerful vested interests will be immediately evident. The threats to public health posed by climate change are of such a magnitude that senior Australian public health figures advocate that civil disobedience by public health professionals is required (Butler et al., 2015). Public health in the twenty-first century will not be for the faint-hearted.

REFLECTIVE, FLEXIBLE AND ECLECTIC

Hopefully this book has inspired a vision of public health as an enterprise that can contribute to the creation of a sustainable, convivial and productive future. Public health is exciting because it is continually changing. New problems, new solutions and new challenges are its heart and soul. What worked in one setting may not work in another. The great success of last year may not be this year's. Consequently, public health practitioners have to develop considerable flexibility and imagination. Public health has never been noted for its theoretical base. It has appealed to people who like to implement rather than theorise. Effective public health practitioners, however, are likely to be reflective in their practice and use theories from a variety of disciplines in an eclectic way. Evidence is essential to good public health practice but there will never be enough to provide complete certainty, hence the need for creativity and willingness to experiment. The methodologies available to public health are numerous and most areas of human enquiry (including the social sciences, medicine, epidemiology) can offer some insights to public health issues. None on their own will offer sufficient insight, however. Consequently, public health work is best done by multidisciplinary approaches (including community and lay knowledge) bringing a variety of complementary perspectives to work. Most crucially there must be a focus on populations and societies rather than the currently dominant thinking in health, which is on individuals and cure.

A VISION FOR 2050

I know from conversations with many public health students and workers that the threats to public health can seem overwhelming and as if there is little that can happen to ensure a healthy future. The last thing I would want this book to do is to leave people with such feelings. So in the apocryphal words: 'The situation is hopeless: we must take the next steps'. In this spirit, box 25.1 provides a vision of a much healthier world in 2050. This vision requires changes that are not that huge, is easily achievable within existing resources but needs political will to realise it. Another powerful means of overcoming feelings of hopelessness is to join others in collective action. For me this has meant being active in the Public Health Association of Australia (including a period as national president) and the People's Health Movement, which is a global network of health activists. Working with others for a better world provides comradeship and hope—essential ingredients for effective activism!

BOX 25.1   PUBLIC HEALTH DREAMING: 2050

LIMITS TO GROWTH ACCEPTED

Global citizens no longer look to economic growth as a measure of their progress. Rather, they consider human well-being, health, happiness and extent of equity. The rampant consumerism of the late twentieth and early twenty-first centuries is seen as a period of madness. In 2050 satisfaction is gained from such things as conviviality, visiting places of natural beauty, live art performances and slow food. Natural resources are used with care and recycling is 'cool' among young people and part of accepted norms. The excessive consumption of the early twenty-first century is viewed with disdain.

PHYSICAL ENVIRONMENT PROTECTED AND RESTORED

Ecological thinking has imbued governments' policies and practices around the world and so decisions are made to restore, protect and enhance the physical environment. The importance of physical ecology to human health and all life on planet Earth is well respected and this has resulted in areas such as the Amazon rainforest being protected for future generations. The Latin American concept of *buen vivir* (living well) is used to guide a more ecological way of thinking and being. A Global Biodiversity Fund and a Global Carbon Reduction Program developed by the United Nations and then endorsed by all nations have been crucial instruments in the new ecological regime.

CORPORATIONS TAMED

In 2050 the empires established by the transnational corporations (TNCs) have been dismantled. In the previous years there was increasing disquiet at the size of the TNCs' chief executive pay packets, the way in which the TNCs externalised the environmental and other costs of their activities, and their tax avoidance behaviour. By 2050 business and industry is on a much smaller scale, more locally controlled (although globally networked) and seen as a part of local communities to whom they give back through community projects, fair tax and skills.

NATIONAL AND GLOBAL GOVERNANCE AND REGULATION FOR HEALTH AND WELL-BEING THE NORM

Internationally there are many global treaties in 2050 to control the activities of the market when its activities impinge on health and well-being—for example, regulation of food and water to ensure equitable supply, good quality and local production. The World Bank is unrecognisable from the body it was at the turn of the century and now works to assist development and promote health equity. The privatisation of the late twentieth century and early twenty-first century is regarded as a mistake. In most countries, services judged as central to health and well-being are controlled or owned by the public. This includes water, power utilities, communications infrastructure, prisons and schools.

Taxation is now seen as a public good that people are willing to contribute to because of the benefits they see they gain. A Global Taxation treaty was concluded in 2030 that closed down tax loopholes and stemmed illicit financial flows from poor to rich countries. By 2050 the inequities between countries have been dramatically reduced and this has flowed through to reduced health inequities.

HEALTH FOR ALL

In 2050 the stark health inequities of the first decade of the twentieth century have been largely reduced. This is seen in Africa where life expectancies have increased dramatically, famine is a thing of the past, education is available to all and local enterprises provide plenty of employment opportunities. In Australia it was seen in 2032, when the life expectancy of Indigenous peoples reached parity with that of the non-Indigenous. This resulted from a national process of reconciliation and a massive investment into the Close the Gap strategy, which enabled action on all social determinants of health, including access to health services.

SHAPE OF HEALTH SERVICES

Primary health care services form the basis of health systems around the world in 2050. Most systems are publicly funded and run and they provide pretty much seamless integration. Local health centres are the heart of the system and provide a full range of services including nursing, medical, physiotherapy, psychology and social work, delivered through one-to-one encounters, groups and community development. These centres advocate for the health and environment of their local community. Hospitals are less prominent and powerful in the system but provide good care and make judgments about the value of treatment at the end of life.

Around the world strong public health departments conduct health equity impact assessments on major developments and on the operations of TNCs and work to protect the health of their community and promote it in a positive way.

CONCLUSION

Our collective health is threatened by ecological disasters, worsening health effects from ever-intensifying neo-liberal economic policies, increasing global tensions reflecting inequitable resource distribution, and communities that are less convivial and more stressed. There is no certainty that these obstacles to a healthier and more equitable society will be overcome. What is certain is that public health and other sectors driven by pursuit of the common good will have to play a central role if we are to meet these challenges. The task is huge, the journey will be tough, the barriers are many, but public health—with its long tradition of commitment to a just and good society—is well placed to make a major contribution to a sustainable, equitable and healthy future.

APPENDIX: PUBLIC HEALTH KEYWORDS

Definition is a vexed and confusing topic for public health. In this book I have used the terms 'public health', 'new public health', 'health promotion', 'primary health care', 'community health', and 'social health'. The boundaries between them are neither fixed nor constant; they have overlapping meanings that may change according to the context. Discussion of their meaning and comparison of different understandings are valuable exercises for students of public health or teams of public health work colleagues. The very process of discussion assists comprehension and helps to clarify understanding. I offer some definitions here but do not expect them to be cast in stone. They are a starting point for dialogue.

Some years ago a colleague and I tackled this issue of definition (Fry and Baum, 1992). In doing this we referred to the work of Raymond Williams (1983), *Keywords: A Vocabulary of Culture and Society*. His analysis of words is particularly relevant to public health as he starts from the position that:

- words usually have a range of meanings, and these meanings are influenced by both their historical beginnings and by people who use words in different ways
- words that involve important and strongly held values are particularly prone to transformation and different interpretations of meanings
- certain uses of words bind together certain ways of seeing and thinking about the world in general.

The discussion of the meaning of 'health' in chapter 1 illustrated Williams' points well. Here I offer some discussion on the keywords of public health.

PUBLIC HEALTH

Definitions of public health have changed and developed since the term first emerged in the nineteenth century. Armstrong (1988) sees that public health in the nineteenth century was concerned with the physical interface between the body and the environment, and substances that passed between these two were viewed as potential threats to human health. Public health was based on sanitation and hygiene and the interface between the body and the natural environment 'was constantly monitored, guarded and cleansed so as to prevent the transmission of disease' (p. 10). He sees a shift in the early twentieth century to a concern with personal hygiene and a greater focus on the individual. In 1920 Winslow (1920, p. 30) defined public health as '... the science and art of preventing disease, prolonging life and promoting health and efficiency through organized community effort for the sanitation of the environment, the control of communicable infections, the education of the individual in personal hygiene ... and for the development of the social machinery to ensure everyone a standard of living adequate for the maintenance of health ...' The rise of medical science in the second half of the twentieth century meant public health increasingly moved to define itself as different from medicine. Most influential in this respect was a monograph published by the Canadian federal minister of health entitled *A New Perspective on the Health of Canadians* (Lalonde, 1974) which asserted that health did not depend primarily on medical care but on non-medical factors such as socio-demographic, lifestyle and environmental influences.

The American Public Health Association (2007) defined public health as 'the practice of preventing disease and promoting good health within groups of people, from small communities to entire countries', which at its core is similar to the Winslow definition but lacks the sense of health being a right and responsibility of society. There may be confusion over the term 'public health' because it is often used to refer to the publicly funded health services.

NEW PUBLIC HEALTH

The 'new' public health developed from thinking about public health led by the Lalonde report and enshrined in the Ottawa Charter for Health Promotion (WHO, 1986), which is subtitled 'the move towards a new public health'. The Commission on Social Determinants of Health (2008) built on this charter and brought the social determinants of health equity to the fore. The following definition draws on both documents:

> The new public health is the totality of the activities organised by societies collectively (primarily led by governments) to protect people from disease and to promote their health. It seeks to do this in a way that promotes equity between different groups in society. New public health activities occur in all sectors and will include the adoption of policies which support health. They will also ensure that social, physical, economic and natural environments promote health. The new public health sees that the participation of communities in activities to promote health is as essential to the success of those activities as is the participation of experts. The new public health measures the practices of the government and private sector (including the health sector) to ensure that they do not detract from health and wherever possible promote health. The term the 'new public health' has been used in earlier periods when in 1916 H.W. Hill published a book entitled *The New Public Health* (Armstrong, 1988). The need to re-invent public health reflects the importance of reviewing public health activities in the light of changing social and economic circumstances.

POPULATION HEALTH

The term 'population health' has come to enjoy some currency since the mid 1990s in both Canada and Australia. It came to prominence following its use by Evans and Stoddart (1994) and has been adopted by the Canadian Institute for Advanced Research and the Australian Government Department of Health, which has a Population Health Division rather than a Public Health Division. The term has been criticised as being more politically neutral than 'public health' and 'health promotion', (Poland et al., 1998), which are associated with the values of equity, public provision of services and a social structural understanding of the determinants of health. Population health is distinct from clinical interventions in that it places emphasis on whole populations and takes actions that will change collective health. Chapter 1 explained the importance of public health practitioners working at a population health level and understanding that the determinants of population health are usually different from those of the health of individuals. The term is helpful to draw a distinction between clinical care and work that improved the health of a whole population.

HEALTH PROMOTION

The term 'health promotion' was virtually unknown until the late 1970s (Parish, 1995, p. 13). An early definition was offered in a WHO technical paper by Anderson (1984): 'Any combination of health education and related organisational, political and economic intervention designed to facilitate behavioural and environmental adaptations which will improve health.'

Health promotion was initially an activity that concentrated on changing individuals' behaviour (see chapter 19 for a detailed description) but following a series of WHO conferences that repositioned health promotion (see chapter 3), it now has a social and environmental focus too. The extent to which these new foci drive practice differs considerably from country to country and between sites of practice. This means health promotion remains a highly contested area of activity. Critical accounts of it have been published (Bunton et al., 1995). The key defining feature of health promotion is a focus on social determinants of health and empowerment (Kickbusch, 2007). Tones and Green (2004, p. 3), writing from a UK perspective, make the case for a 'critical health promotion', which they see as operating 'as a kind of militant wing of public health' that is concerned with social justice and achievement of equity. In 2015, progressive health promotion is very much in step with the new public health.

PRIMARY HEALTH CARE

Fry and Baum (1992, p. 305) note that the 'primary' part of primary health care has multiple meanings:

- first, as in the first place people go to seek help
- early stage, as in treating health problems at an early stage of their development
- basic, as in accessible, affordable care
- important or essential, as in the foundation for the rest of the health system.

These multiple meanings may be part of the reason this term's meaning is disputed. In many industrialised countries primary health care refers to the first level of medical care. A much broader definition was implied in the World Health Organization's Alma Ata Declaration. This document defined primary health care as both a level of service delivery and as an approach to health care. Five principles were incorporated:

- equitable distribution of resources
- community involvement
- emphasis on prevention
- use of appropriate technology
- an approach that involves a range of sectors (such as housing, agriculture, water supply).

The debate between comprehensive and selective primary health care was described in detail in chapter 3. In the 2008 World Health Report, WHO stressed that primary health care should provide integrated and coordinated care that provides for continuity. Primary health care teams were seen as the hub of other services including hospital, specialist medical, diagnostic and a range of social and support groups. Finally—and for the new public health, very importantly—primary health care was seen to have responsibility for a well-identified population and for promoting the health of the whole community and protecting it from health threats. Thus it can take a comprehensive approach to addressing the root causes of ill health and inequity and to structuring sustainable and fair health systems.

The sectors that may provide primary health care are:

- private fee-for-service practitioners including GPs, physiotherapists, pharmacists, nutritionists
- public sector agencies (e.g. community health centres)
- not-for-profit non-government agencies, which may receive a substantial amount of their funding from the government (Family Planning Associations)
- informal sector, including voluntary self-help groups, carers.

Unlike public health and health promotion, primary health care is concerned with treatment, cure and care of people with illness. It overlaps with public health in its focus on illness prevention and health promotion.

COMMUNITY HEALTH SERVICES

Internationally community health services generally refer to those that are based in the community but generally offer more than medical services. In Australia, 'community health' has referred to community-based health services funded largely by state governments or Aboriginal-controlled health services. In Canada they are provided by provincial governments and in the USA and the UK community health centres have developed as a response to service the needs of poor communities. Many poor countries also have such programs based on community health workers. More than any other sector of health systems they have been concerned with social determinants of health and empowerment of populations. Australia and Canada have had strong community health sectors since the 1970s (although in Australia there has been a withdrawal of government support in the past decade) and they are characterised as follows:

- responsibility to meet the main health needs of a defined community
- equity and accessibility; providing services close to where people live and work, without financial, geographic, cultural or other barriers
- comprehensive program content that includes primary health care, health promotion and the management of ongoing health problems
- the participation of people and communities in debate and decision-making about health issues and their own health care
- organisational structures that promote multidisciplinary teamwork among practitioners.

SOCIAL DETERMINANTS OF HEALTH

Social determinants of health are 'the causes of the causes' of ill health and emphasise social context, social stratification, the differential exposure people experience and their vulnerability to illness and injury. The Commission on Social Determinants of Health (2008) summarised the evidence on social determinants and stressed that health equity cannot be achieved without attention to the unequal power relations and unfair global distribution of wealth, income and other resources.

HEALTH EQUITY

Achieving health equity is a central aim of the new public health. Whitehead (1992) defined health inequities as 'differences in health that are unnecessary, avoidable, unfair and unjust'. The Commission on Social Determinants of Health defined health inequity as 'where systematic differences in health are judged to be avoidable by reasonable action they are, quite simply, unfair. It is this that we label health inequity' (CSDH, 2008, Executive Summary). Braveman and Gruskin (2003) define equity in health as 'the absence of systematic disparities in health (or in the major social determinants of health) between groups with different levels of underlying social advantage or disadvantage—that is, wealth, power, or prestige'.

CONCLUSION

The above discussion has demonstrated the overlapping meanings. I find the 'new public health' to be a useful all-embracing term. It signifies an interest in the broad range of social, economic and political activities implied by the Ottawa Charter and a commitment to community participation. 'Social health' neatly summarises the philosophy and distinguishes the new public health from approaches that have been dominated by medicine. Health promotion is one of the key aims of the new public health. Community health and primary health care are key strategic areas within the health sector.

REFERENCES

Abbott, D. (1990) *Regaining our senses: Conceptual frameworks for environmental health, Healthy environments in the 1990s: The community health approach*, Australian Community Health Association, Sydney, NSW.

Abbott, J. (1995) Community participation and its relationship to community development. *Community Development Journal*, 30, 158–168.

Abbott, T. (2006) Minister for Health and Ageing, Address to the Queensland Obesity Summit, 3 May, Brisbane, QLD.

Abdallah, S., Michaelson, J., Shah, S., Stoll, L. and Marks, N. (2012) *The Happy Planet Index, 2012 report: A global index of sustainable wellbeing*. The New Economics Foundation, London. From www.happyplanetindex.org/assets/happy-planet-index-report.pdf

Aboriginal and Torres Strait Islander Commission (1997) ATSIC submission to the National Inquiry into the separation of Aboriginal and Torres Strait Islander children from their Families. ATSIC, Canberra, ACT.

Aboriginal and Torres Strait Islander Social Justice Commissioner (2005) *Social Justice Report 2005*. HREOC, Sydney, NSW.

Aboriginal Health and Medical Research Council (AHMRC) (2015) Kick the habit social marketing campaign. http://www.ahmrc.org.au/index.php?option=com_content&view=article&id=240:tobacco-resistance-and-control&catid=99:tobacco-resistanceacontrol&Itemid=84 (accessed 17 March 2015).

Abramson, M. and Voigt, T. (1991) Ambient air pollution and respiratory disease. *Medical Journal of Australia*, 154, 543–553.

Acevedo-Garcia, D. and Lochner, K. A. (2003) Residential segregation and health. In Kawachi, I. and Berkman, L. F. (eds), *Neighborhoods and health*, Oxford University Press, Oxford, England. pp. 265–287.

Acheson, D. (1998) *Independent inquiry into inequalities in health*. UK Department of Health, London.

Adam, D. and Wintour, P. (2006) Most Britons willing to pay green taxes to save the environment. *The Guardian*, 23 February. From http://politics.guardian.co.uk/polls/story/0,,1717302,00.html.

Adamson, R., Nierenberg, D. and Arno, O. (2013) Valuing indigenous peoples. In Stark, L. (ed), *State of the World 2013: Is sustainability still possible?*, Worldwatch Institute/Island Press, Washington, DC. pp. 210–217.

Adler, N. E., Boyce, W. T., Chesney, M. A., Folkman, S. and Syme, S. L. (1993) Socioeconomic inequalities in health: No easy solution. *JAMA*, 269, 3140–3145.

Adler, N. E. and Stewart, J. (2009) Reducing obesity: Motivating action while not blaming the victim. *Milbank Quarterly*, 87, 49–70.

Adler, P. A. and Adler, P. (1994) Observational techniques. In Denzin, N. K. and Lincoln, Y. S. (eds), *Handbook of qualitative research*, SAGE Publications, Thousand Oaks, CA. pp. 377–392.

Advertising Age (2013) Marketing fact pack 2014: Annual guide to marketers, media and agencies. From http://gaia.adage.com/images/bin/pdf/MFPweb_spreadsv2.pdf

Ågren, G. (2003) *Sweden's new public health policy*. National Institute of Public Health, Sweden, Stockholm. From www.drugsandalcohol.ie/5868/1/Sweden_new_public_health_policy_2003.pdf

Aitkin, D. (1996) Address of welcome. In Furnass, B., Whyte, J., Harris, J. and Baker, A. (eds), Survival, health and wellbeing into the twenty-first century. Proceedings of a conference held at the Australian National University, 30 November–1 December, 1995, Nature and Society Forum, Canberra, ACT.

Ajzen, I. and Fishbein, M. (1980) *Understanding attitudes and predicting social behaviour*. Prentice-Hall, Englewood Cliffs, NJ.

Akaleephan, C., Wibulpolprasert, S., Sakulbumrungsil, R., Luangruangrong, P., Jitraknathee, A., Aeksaengsri, A., Udomaksorn, S., Tangcharoensathien, V. and Tantivess, S. (2009) Extension of market exclusivity and its impact on the accessibility to essential medicines, and drug expense in Thailand: Analysis of the effect of TRIPs-Plus proposal. *Health Policy*, 91, 174–182.

Akhter, F. (2001) Seeds in women's hands: A symbol of food security and solidarity. *Development*, 44, 52–55.

Alexander, C. (1967) The city as a mechanism for sustaining human contact. In Ewald, W. R. (ed), *Environment for man*, Indiana University Press, Bloomington.

Alexander, K. (1995) Community participation in hospitals. In Baum, F. (ed), *Health for All: The South Australian experience*, Wakefield Press, Adelaide, SA. pp. 107–124.

Alinsky, S. (1971) *Rules for radicals*. Random House, New York, NY.

Allende, S. (2006) Chile's medical-social reality. *Social Medicine*, 1, 151–155.

Alliance for Healthy Cities (2007) Our charter. From www.alliance-healthycities.com/htmls/charter/index_charter.html (accessed 21 January 2015).

Allison, K. R. and Rootman, I. (1996) Scientific rigor and community participation in health promotion research: are they compatible? *Health Promotion International*, 11, 333–340.

Almedom, A. M. (2005) Social capital and mental health: An interdisciplinary review of primary evidence. *Social Science and Medicine*, 61, 943–964.

Altman, D. G. (1986) A framework for evaluating community-based heart disease prevention programs. *Social Science and Medicine*, 22, 479–487.

Altman, D. G. (1991) Public policy: Community organisations and the new political challenges. *National AIDS Bulletin*, Sept, 13–16.

Altman, J. C. (2010) Income management and the rights of Indigenous Australians to equity, *CAEPR topical issue no. 03/2010*. Centre for Aboriginal Economic Policy Research, The Australian National University, Canberra, ACT.

American Public Health Association (APHA) (2007) What is Public Health? [Fact Sheet]. From www.apha.org/~/media/files/pdf/fact%20sheets/whatisph.ashx (accessed 25 March 2015).

Ampri, I. (2012) Fiscal support for the efficient and sustainable management of natural resources in Indonesia: Renewable energy and forestry resources, *Presentation by Vice Chairman of Fiscal Policy Agency for Climate Change Financing and Multilateral Policy Ministry of Finance, Indonesia*. OECD, Paris. From www.oecd.org/greengrowth/13%20Fiscal%20Support%20For%20the%20Efficient%20and%20Sustainable%20Management%20of%20Natural%20Resources%20In%20Indonesia%20OECD%20Bara%20191112%20rev%20ia%20231112%20grand%20final.pdf

Anaf, J., Newman, L., Baum, F., Ziersch, A. and Jolley, G. (2013) Policy environments and job loss: Lived experience of retrenched Australian automotive workers. *Critical Social Policy*, 33, 325–347.

Anderson, D. (1984) Health Promotion: An overview, *WHO Technical Paper*. WHO Regional Office for Europe, Copenhagen.

Anderson, I. (1996) Ethics and health research in Aboriginal communities. In Daly, J. (ed), *Ethical intersections: Health research, methods and researcher responsibility*, Allen & Unwin, Sydney, NSW. pp. 153–164.

Anderson, I. P. S. (2006) Mutual obligation, shared responsibility agreements and indigenous health strategy. *Australia and New Zealand health policy*, 3, 10.

Anderson, J., Miller, M. and McGuire, A. (2012) Organizing for health care reform. In Minkler, M. (ed), *Community organizing and community building for health and welfare*, 3rd edition, Chapter 21, Rutgers University Press, New Brunswick, NJ. pp. 386–406.

Anderson, T. (2006) Policy coherence and conflict of interest: The OECD guidelines on health and poverty. *Critical Public Health*, 16, 245–257.

Anderson, V. (1991) *Alternative economic indicators*. Routledge, London.

Anderson, V. and Draper, P. (1991a) Better economics and better economic policies. In Draper, P. (ed), *Health through public policy*, Green Print, London. pp. 240–243.

Anderson, V. and Draper, P. (1991b) Economics and hostile environments. In Draper, P. (ed), *Health through public policy*, Green Print, London. pp. 169–184.

Andre, F. E., Booy, R., Bock, H. L., Clemens, J., Datta, S. K., John, T. J., Lee, B. W., Lolekha, S., Peltola, H., Ruff, T. A., Santosham, M. and Schmitt, H. J. (2008) Vaccination greatly reduces disease, disability, death and inequity worldwide. *Bulletin of the World Health Organization*, 86, 81–160.

Anon. (1994) Population health looking upstream [Editorial]. *The Lancet*, 343, 429–430.

Anon. (1996) The economic totem pole. *New Internationalist*. April. From http://newint.org/features/1996/04/05/totem

Antonovsky, A. (1996) The salutogenic model as a theory to guide health promotion. *Health Promotion International*, 11, 11–18.

Arber, S., Gilbert, G. N. and Dale, A. (1985) Paid employment and women's health: a benefit or a source of role strain? *Sociology of Health and Illness*, 7, 375–400.

Armstrong, D. (1988) Historical origins of health behaviour. In Anderson, R., Davies, J., Kickbusch, I., McQueen, D. and Turner, R. (eds), *Health behaviour research and health promotion*, Oxford University Press, Oxford. pp. 8–21.

Arneil, B. (2006) *Diverse communities: The problem with social capital*. Cambridge University Press, Cambridge.

Arnstein, S. (1971) Eight rungs on the ladder of citizen participation. In Cahn, S. E. and Passett, B. A. (eds), *Citizen participation: Effecting community change*, Praeger, London. pp. 216–225.

Ashenden, R., Silagy, C. and Weller, D. (1997) A systematic review of the effectiveness of promoting lifestyle change in general practice. *Family Practice*, 14, 160–176.

Ashton, J. (1988) *Esmedune 2000: Vision or dream. A healthy Liverpool*. Department of Community Health, University of Liverpool.

Ashton, J. (ed) (1992) Healthy cities. Open University Press, Milton Keynes.

Ashton, J. and Seymour, H. (1988) *The new public health*. Open University Press, Milton Keynes.

Ashworth, P. D. (1995) The meaning of "participation" in participant observation. *Qualitative Health Research*, 5, 366–387.

Astbury, J. (2002) Mental health: Gender bias, social position, and depression. In Sen, G., George, A. and Ostlin, P. (eds), *Engendering international health: The challenge of equity*, MIT Press, Cambridge, MA. pp. 143–166.

Atkins, L. and Jarrett, D. (1979) The significance of "significance tests". In Miles, I. and Evans, J. (eds), *Demystifying social statistics*, Pluto Press, London. pp. 87–109.

Atkinson, P. and Hammersley, M. (1994) Ethnography and participant observation. In Denzin, N. K. and Lincoln, Y. S. (eds), *Handbook of qualitative research*, SAGE Publications, Thousand Oaks, CA. pp. 248–261.

Auer, J., Repin, Y. and Roe, M. (1993) *Just change: The cost-conscious manager's toolkit*. National Reference Centre for Continuing Education in Primary Health Care, Wollongong, NSW.

Auslander, G. K. (1988) Social networks and the functional health status of the poor: A secondary analysis of data from the national survey of personal health practices and consequences. *Journal of community health*, 13, 197–209.

Australian Broadcasting Corporation (ABC) (2005) Tony Abbott on childhood obesity, Generation 'O', 17 October, ABC Four Corners [Transcript]. From www.abc.net.au/4corners/content/2005/s1484310.htm

Australian Broadcasting Corporation (ABC) (2007) Bird flu: Risks, laws and rights, 21 January, ABC Background briefing program [Transcript]. From www.abc.net.au/rn/backgroundbriefing/stories/2007/1814815.htm

Australian Broadcasting Corporation (ABC) (2009) Nicola Roxon in Will to Live, 17 October, ABC AWAYE! program [Transcript]. From www.abc.net.au/radionational/programs/awaye/will-to-live-part-1/3670070#transcript

Australian Broadcasting Corporation (ABC) (2014) ISDS: The devil in the trade deal, 14 September [Transcript]. From www.abc.net.au/radionational/programs/backgroundbriefing/isds-the-devil-in-the-trade-deal/5734490

Australian Bureau of Statistics (ABS) (1997) Australian transport and the environment, 1997, *Cat no. 4605.0*. ABS, Canberra, ACT.

Australian Bureau of Statistics (ABS) (1999) Causes of Death Data 1970-90. ABS, Canberra, ACT.

Australian Bureau of Statistics (ABS) (2000) Suicides, 1921–1998, *Cat no. 3309.0*. ABS, Canberra, ACT.

Australian Bureau of Statistics (ABS) (2001) Australian Social Trends, 2001, *Cat no. 4102.0*. ABS, Canberra, ACT.

Australian Bureau of Statistics (ABS) (2003) Australian Social Trends, 2003, *Cat no. 4102.0*. ABS, Canberra, ACT.

Australian Bureau of Statistics (ABS) (2006) What do Australians think about protecting the environment? *Paper prepared for the 2006 Australian State of the Environment Committee.* Department of the Environment and Heritage, Canberra, ACT.

Australian Bureau of Statistics (ABS) (2008) Causes of Death, Australia, 2006, *Cat no. 3303.0*. ABS, Canberra, ACT.

Australian Bureau of Statistics (ABS) (2009) National Aboriginal and Torres Strait Islander Social Survey, 2008, *Cat no. 4714.0*. ABS, Canberra, ACT.

Australian Bureau of Statistics (ABS) (2010a) Australian Labour Market Statistics, Oct 2010. Trends in hours worked (feature article), *Cat no. 6105.0*. ABS, Canberra, ACT.

Australian Bureau of Statistics (ABS) (2010b) Causes of Death, Australia, 2008, *Cat no. 3303.0*. ABS, Canberra, ACT.

Australian Bureau of Statistics (ABS) (2012a) Australian Health Survey: First Results, 2011–12, *Cat no. 4364.0.55.001*. ABS, Canberra, ACT.

Australian Bureau of Statistics (ABS) (2012b) The Health and Welfare of Australia's Aboriginal and Torres Strait Islander Peoples, Oct 2010, *Cat no. 4704.0*. ABS, Canberra, ACT.

Australian Bureau of Statistics (ABS) (2012c) Profiles of Health, Australia, 2011–13, *Cat no. 4338.0*. ABS, Canberra, ACT.

Australian Bureau of Statistics (ABS) (2013a) Australian Health Survey: Updated Results, 2011–2012, *Cat no. 4364.0.55.003*. ABS, Canberra, ACT.

Australian Bureau of Statistics (ABS) (2013b) Australian Social Trends, Data Cube— Work. Table 1 Work, National Summary, 1999–2012, *Cat no. 4102.0*. ABS, Canberra, ACT.

Australian Bureau of Statistics (ABS) (2013c) Disability, Ageing and Carers, Australia: Summary of Findings, 2012, *Cat no. 4430.0*. ABS, Canberra, ACT.

Australian Bureau of Statistics (ABS) (2013d) Measures of Australia's Progress, 2013, *Cat no. 1370.0*. ABS, Canberra, ACT.

Australian Bureau of Statistics (ABS) (2014a) Australian Aboriginal and Torres Strait Islander Health Survey: First Results, Australia, 2012–13, *Cat no. 4727.0.55.001*. ABS, Canberra, ACT.

Australian Bureau of Statistics (ABS) (2014b) Australian Historical Population Statistics, 2014, *Cat no. 3105.0.65.001*. ABS, Canberra, ACT.

Australian Bureau of Statistics (ABS) (2014c) Causes of Death, Australia, 2012, *Cat no. 3303.0*. ABS, Canberra, ACT.

Australian Bureau of Statistics (ABS) (2014d) Deaths, Australia, 2013, *Cat no. 3302.0*. ABS, Canberra, ACT.

Australian Bureau of Statistics (ABS) (2014e) Estimates and Projections, Aboriginal and Torres Strait Islander Australians, 2001 to 2026, *Cat no. 3238.0*. ABS, Canberra, ACT.

Australian Bureau of Statistics (ABS) (n.d.) ABS Basic Community Profiles, 2011 Census. ABS, Canberra, ACT. From www.abs.gov.au/websitedbs/censushome.nsf/4a256353 001af3ed4b2562bb00121564/communityprofiles accessed 19 December 2014).

Australian Community Health Association (ACHA) (1986) *Review of the Community Health Program*. ACHA, Sydney, NSW.

Australian Council of Social Service (ACOSS) (2014) *Poverty in Australia 2014*. ACOSS, Strawberry Hills, NSW.

Australian Energy Market Operator (AEMO) (2013) *South Australian electricity report, 2014*. AEMO, Adelaide, SA.

Australian Government (2010) Taking preventative action—A response to Australia: The Healthiest Country by 2020—The Report of the National Preventative Health Taskforce. Commonwealth of Australia, Canberra, ACT.

Australian Government (2011) *The National Drug Strategy 2010–2015*. Commonwealth of Australia, Canberra, ACT.

Australian Health Ministers' Advisory Council (AHMAC) (2006) *Aboriginal and Torres Strait Islander Framework Report, 2006*. Commonwealth of Australia, Canberra, ACT.

Australian Health Ministers' Advisory Council (AHMAC) (2004) *Cultural Respect Framework for Aboriginal and Torres Strait Islander Health, 2004–2009*. Commonwealth of Australia, Canberra, ACT.

Australian Human Rights Commission (2015) *Close the Gap: Indigenous Health Campaign*. From www.humanrights.gov.au/close-gap-indigenous-health-campaign (accessed 11 February 2015).

Australian Indigenous Doctors' Association (AIDA) and Centre for Health Equity Training Research and Evaluation (CHETRE) (2010) *Health impact assessment of the Northern Territory Emergency Response*. AIDA, Canberra, ACT.

Australian Institute of Criminology (AIC) (2013) *Indigenous justice in focus*. From www.aic.gov.au/crime_types/in_focus/indigenousjustice.html (accessed 13 May 2013).

Australian Institute of Health and Welfare (AIHW) (1996) Australia's Health 1996, *Australia's health no. 5, Cat. no. AIHW 26*. AIHW, Canberra, ACT.

Australian Institute of Health and Welfare (AIHW) (2005) Health Expenditure Australia 2003–04, *AIHW Cat. no. HWE 32*. AIHW, Canberra, ACT.

Australian Institute of Health and Welfare (AIHW) (2006) Australia's Health 2006, *Australia's health no. 10, Cat. no. AUS 73*. AIHW, Canberra, ACT.

Australian Institute of Health and Welfare (AIHW) (2011a) 2010 National Drug Strategy Household Survey (NSDHS) report, *Drug statistics series no. 25, Cat. no. PHE 145*. AIHW, Canberra, ACT.

Australian Institute of Health and Welfare (AIHW) (2011b) Housing and homelessness services: Access for Aboriginal and Torres Strait Islander people, *Cat. no. HOU 237*. AIHW, Canberra, ACT.

Australian Institute of Health and Welfare (AIHW) (2011c) The health and welfare of Australia's Aboriginal and Torres Strait Islander people: An overview 2011, *Cat. no. IHW 42*. AIHW, Canberra, ACT.

Australian Institute of Health and Welfare (AIHW) (2011d) Young Australians. Their health and wellbeing 2011, *Cat. no. PHE 140*. AIHW, Canberra, ACT.

Australian Institute of Health and Welfare (AIHW) (2012) Australia's Health 2012, *Australia's health no. 13, Cat. no. AUS 156*. AIHW, Canberra, ACT.

Australian Institute of Health and Welfare (AIHW) (2013a) Aboriginal and Torres Strait Islander Health Performance Framework 2012: Detailed analyses, *Cat. no. IHW 94*. AIHW, Canberra, ACT.

Australian Institute of Health and Welfare (AIHW) (2013b) Australia's Welfare 2013. AIHW, Canberra, ACT.

Australian Institute of Health and Welfare (AIHW) (2014a) Australia's Health 2014, *Australia's health series no. 14, Cat. no. AUS 178*. AIHW, Canberra, ACT.

Australian Institute of Health and Welfare (AIHW) (2014b) Health expenditure in Australia, 2012–13, *Health and welfare expenditure series no. 52, Cat. no. HWE 61*. AIHW, Canberra, ACT.

Australian Institute of Health and Welfare (AIHW) (2014c) Mortality and life expectancy of Indigenous Australians, 2008 to 2012, *Cat. no. IHW 140*. AIHW, Canberra, ACT.

Australian Institute of Health and Welfare (AIHW) (2014d) Mortality inequalities in Australia, 2009–2011, *Bulletin no. 124, Cat. no. AUS 184*. AIHW, Canberra, ACT.

Australian Institute of Health and Welfare (AIHW) (2014e) Tobacco Smoking – NDSHS 2013 key findings and supplementary tables. From www.aihw.gov.au/alcohol-and-other-drugs/ndshs/2013/tobacco (accessed 4 February 2015).

Australian Institute of Health and Welfare (AIHW) (n.d.) Australian Cancer Incidence and Mortality (ACIM) Books. From www.aihw.gov.au/acim-books (accessed 4 February 2015).

Australian Institute of Health and Welfare (AIHW), Harrison, J. E. and Henley, G. (2014a) Suicide and hospitalised self-harm in Australia: trends and analysis, *Injury research and statistics series no. 93, Cat. no. INJCAT 169*. AIHW, Canberra, ACT.

Australian Institute of Health and Welfare (AIHW), Johnson, S., Bonello, M. R., Li, Z., Hilder, L. and Sullivan, E. A. (2014b) Maternal deaths in Australia 2006–2010, *Maternal deaths series no. 4, Cat. no. PER 61*. AIHW, Canberra, ACT.

Australian Medical Association (AMA) (2007) *AMA Indigenous health report card 2006*. AMA, Barton, ACT.

Australian National Preventive Health Agency (ANPHA) (2013a) *Exploring the public interest case for a minimum (floor) price for alcohol final report*. ANPHA, Canberra, ACT. From http://anpha.gov.au/internet/anpha/publishing.nsf/Content/minimum-price-final-report

Australian National Preventive Health Agency (ANPHA) (2013b) *National Binge Drinking Strategy*. From www.anpha.gov.au/internet/anpha/publishing.nsf/Content/NBDS (accessed 26 March 2015).

Australian National Preventive Health Agency (ANPHA) (2013c) State of Preventive Health 2013. Report to the Australian Government Minister for Health. ANPHA, Canberra, ACT.

Bacchi, C. (2009) *Analysing policy: What's the problem represented to be?* Pearson Education, Frenchs Forest, NSW.

Bachrach, P. and Baratz, M. S. (1970) *Power and poverty: Theory and practice*. Oxford University Press, Oxford.

Baekgaard, H. (1998) The distribution of household wealth in Australia: 1986 and 1993. NATSEM, Canberra, ACT.

Bailie, R. (2007) Housing. In Carson, B., Dunbar, T., Chenhall, R. D. and Bailie, R. (eds), *Social determinants of Indigenous health*, Allen & Unwin, Crows Nest, NSW. pp. 203–230.

Balabanova, D., McKee, M. and Mills, A. (eds) (2011) *'Good health at low cost' 25 years on. What makes a successful health system?* School of Hygiene and Tropical Medicine, London.

Balch, O. (2014) Water wars: a new reality for business and government, *The Guardian*, 7 October. From www.theguardian.com/sustainable-business/2014/oct/06/water-wars-business-governments-scarcity-pollution-access

Bammer, G., Dance, P., Stevens, A., Mugford, S., Ostini, R. and Crawford, D. (1996) Attitudes to a proposal for controlled availability of heroin in Australia: Is it time for a trial? *Addiction Research and Theory*, 4, 45–55.

Ban Ki-moon (2011) Prevention and control of non-communicable diseases. *Report of the Secretary-General, United Nations General Assembly, 19 May*. From www.un.org/en/ga/search/view_doc.asp?symbol=A/66/83&Lang=E, (accessed 26 March 2015).

Bandura, A. (1977) *Social learning theory*. Prentice Hall, Englewood Cliffs, NJ.

Banerji, D. (2002) A fundamental shift in the approach to international health by WHO, UNICEF, and the World Bank: Instances of the practice of "intellectual fascism" and totalitarianism in some Asian countries. In Navarro, V. (ed), *The political economy of social inequalities. Consequences for health and quality of life*, Baywood Publishing, New York, NY.

Bareja, C., Waring, J., Stapledon, R., Toms, C., Douglas, P. and National Tuberculosis Advisory Committee for the Communicable Diseases Network Australia (2014) Tuberculosis notifications in Australia, 2011 [Annual Report]. *Communicable Diseases Intelligence*, 38, E356–E368.

Barratt, A., Howard, K., Irwig, L., Salkeld, G. and Houssami, N. (2005) Model of outcomes of screening mammography: Information to support informed choices. *BMJ*, 330, 936.

Bartley, M. (1985) Coronary heart disease and the public health 1850–1983. *Sociology of Health and Illness*, 7, 289–313.

Bartley, M., Ferrie, J. and Montgomery, S. M. (1999) Living in a high-unemployment economy: Understanding the health consequences. In Marmot, M. and Wilkinson, R. G. (eds), *Social determinants of health*, Oxford University Press, Oxford. pp. 81–104.

Basch, E., Febbo, P. and Vickers, A. (2013) Untangling evidence and guidelines in prostate cancer screening. *ASCO Annual Meeting*. American Society of Clinical Oncology (ASCO). From http://am.asco.org/untangling-evidence-and-guidelines-prostate-cancer-screening

Bates, E. and Linder-Pelz, S. (1990) *Health care issues*. Allen & Unwin, Sydney, NSW.

Baum, F. (1988) Community-based research for promoting the new public health. *Health Promotion International*, 3, 259–268.

Baum, F. (1989) Community development and the new public health in Australia and New Zealand [Editorial]. *Community Health Studies*, 13, 1–4.

Baum, F. (1990) The new public health: Force for change or reaction? *Health Promotion International*, 5, 145–150.

Baum, F. (1992) Moving targets: Evaluating community development. *Health Promotion Journal of Australia*, 2, 10–15.

Baum, F. (1993) Healthy cities and change: Social movement or bureaucratic tool? *Health Promotion International*, 8, 31–40.

Baum, F. (ed) (1995a) *Health for All: The South Australian experience*. Wakefield Press, Kent Town, SA.

Baum, F. (1995b) Researching public health: Beyond the qualitative-quantitative methodological debate. *Social Science and Medicine*, 40, 459–468.

Baum, F. (1998) Measuring effectiveness in community-based health promotion. In Davies, J. K. and Macdonald, G. (eds), *Quality and effectiveness in health promotion*, Routledge, London, UK. pp. 68–89.

Baum, F. (2003) Primary health care: Can the dream be revived? *Development in Practice*, 13, 515–519.

Baum, F. (2007a) Cracking the nut of health equity: top down and bottom up pressure for action on the social determinants of health. *Promotion and Education*, 14, 90–95.

Baum, F. (2007b) Health for All Now! Reviving the spirit of Alma Ata in the twenty-first century: An introduction to the Alma Ata Declaration. *Social Medicine*, 2, 34–41.

Baum, F. (2011) From Norm to Eric: Avoiding lifestyle drift in Australian health policy. *Australian and New Zealand Journal of Public Health*, 35, 404–406.

Baum, F. (2014) Community health services in Australia. In Germov, J. (ed), *Second opinion: An introduction to health sociology*, 5th edition, Chapter 25, Oxford University Press, Melbourne, VIC.

Baum, F., Bégin, M., Houweling, T. A. and Taylor, S. (2009) Changes not for the faint-hearted: reorienting health care systems toward health equity through action on the social determinants of health. *American Journal of Public Health*, 99, 1967–1974.

Baum, F. and Brown, V. (1989) Healthy Cities (Australia) Project: Issues of evaluation for the new public health. *Community Health Studies*, 13, 140–149.

Baum, F. and Cooke, R. (1992) Healthy Cities Australia: Evaluation of the pilot project in Noarlunga, South Australia. *Health Promotion International*, 7, 181–193.

Baum, F., Cooke, R., Crowe, K., Traynor, M. and Clarke, B. (1990) *Healthy Cities Noarlunga pilot project evaluation*. Southern Community Health Research Unit, South Australian Health Commission, Adelaide, SA.

Baum, F. and Dwyer, J. (2014) The accidental logic of health policy in Australia. In Miller, C. and Orchard, L. (eds), *Australian public policy: Progressive ideas in the neo-liberal ascendency*, Policy Press, Bristol. pp. 187–208.

Baum, F. and Fisher, M. (2011) Are the national preventive health initiatives likely to reduce health inequities? *Australian Journal of Primary Health*, 17, 320–326.

Baum, F. and Fisher, M. (2014) Why behavioural health promotion endures despite its failure to reduce health inequities. *Sociology of Health and Illness*, 36, 213–225.

Baum, F., Fisher, M., Trewin, D., Duvnjak, A. and Members of the Public Health Advisory Group of the Public Health Association of Australia (2013a) Funding the 'H' in NHMRC [Editorial]. *Australian and New Zealand Journal of Public Health*, 37, 503–505.

Baum, F., Freeman, T., Jolley, G., Lawless, A., Bentley, M., Värttö, K., Boffa, J., Labonté, R. and Sanders, D. (2014a) Health promotion in Australian multi-disciplinary primary health care services: case studies from South Australia and the Northern Territory. *Health Promotion International*, 29, 705–719.

Baum, F., Freeman, T., Lawless, A. and Jolley, G. (2012) Community development: Improving patient safety by enhancing the use of health services. *Australian Family Physician*, 41, 424–428.

Baum, F., Fry, D. and Lennie, I. (eds) (1992) *Community health: Policy and practice in Australia*. Pluto Press, Sydney, NSW.

Baum, F., Jolley, G., Hicks, R., Saint, K. and Parker, S. (2006) What makes for sustainable Healthy Cities initiatives: A review of the evidence from Noarlunga, Australia after 18 years. *Health Promotion International*, 21, 259–265.

Baum, F. and Kahssay, H. M. (1999) Health development structures: An untapped resource. In Kahssay, H. M. and Oakley, P. (eds), *Community involvement in health development: A review of the concept and practice*, World Health Organization, Geneva.

Baum, F., Laris, P., Fisher, M., Newman, L. and MacDougall, C. (2013b) "Never mind the logic, give me the numbers": Former Australian health ministers' perspectives on the social determinants of health. *Social Science and Medicine*, 87, 138–146.

Baum, F., Lawless, A., Delany, T., MacDougall, C., Williams, C., Broderick, D., Wildgoose, D., Harris, E., McDermott, D., Kickbusch, I., Popay, J. and Marmot, M. (2014b) Evaluation of Health in All Policies: Concept, theory and application. *Health Promotion International*, 29, i130–i142.

Baum, F., Legge, D. G., Freeman, T., Lawless, A., Labonté, R. and Jolley, G. M. (2013c) The potential for multi-disciplinary primary health care services to take action on the social determinants of health: Actions and constraints. *BMC Public Health*, 13, 460.

Baum, F., MacDougall, C. and Smith, D. (2006) Participatory action research [Glossary]. *Journal of Epidemiology and Community Health*, 60, 854–857.

Baum, F., Newman, L. and Biedrzycki, K. (2014c) Vicious cycles: digital technologies and determinants of health in Australia. *Health Promotion International*, 29, 349–360.

Baum, F., Ollila, E. and Peña, S. (2013d) History of HiAP. In Leppo, K., Ollila, E., Peña, S., Wismar, M. and Cook, S. (eds), *Health in all policies: Seizing opportunities, implementing policies*, Ministry of Social Affairs and Health, Helsinki, Finland. pp. 25–42.

Baum, F. and Sanders, D. (1995) Can health promotion and primary health care achieve health for all without a return to their more radical agenda? *Health Promotion International*, 10, 149–160.

Baum, F., Sanderson, C. and Jolly, G. (1997) Community participation in action: An analysis of the South Australian Health and Social Welfare Councils. *Health Promotion International*, 12, 125–134.

Baum, F., Santich, B., Craig, B. and Murray, C. (1996) Evaluation of a national health promotion program in South Australia. *Australian and New Zealand Journal of Public Health*, 20, 41–49.

Baum, F. and Ziersch, A. (2003) Social Capital [Glossary]. *Journal of Epidemiology and Community Health*, 57, 320–323.

Bauman, A., Mant, A., Middleton, L., Mackertich, M. and Jane, E. (1989) Do general practitioners promote health? A needs assessment. *Medical Journal of Australia*, 151, 265–269.

Bauman, Z. (2007) Liquid times: Living in an age of uncertainty. Polity Press, Cambridge.

Baxter, J. (2011) The glossary of greed. *Pambazuka News*, 30 March. From http://www.pambazuka.org/en/category/features/72112.

Baybutt, M., Hayton, P. and Dooris, M. (2007) Prisons in England and Wales: An important public health opportunity? In Douglas, J., Earle, S., Handsley, S., Lloyd C. E., and Spurr S. (eds), *A reader in promoting public health*, SAGE Publications, London. pp. 134–142.

Bayer, R. (1986) AIDS, power, and reason. *Milbank Quarterly*, 64, 168–182.

Bazeley, P. and Jackson, K. (2013) *Qualitative data analysis with NVivo*. SAGE Publications, London.

BBC (2007) UK scientists' IPCC reaction. BBC News, 22 April. From http://news.bbc.co.uk/2/hi/science/nature/6324093.stm (accessed 2 February 2015).

BC Healthy Communities (BCHC) (2015) http://bchealthycommunities.ca/about (accessed 24 March 2015).

Beaglehole, R. and Bonita, R. (2004) *Public health at the crossroads*. Cambridge University Press, Cambridge.

Beall, R. and Kuhn, R. (2012) Trends in compulsory licensing of pharmaceuticals since the Doha Declaration: A database analysis. *PloS Medicine*, 9, e1001154.

Beatley, T. and Manning, K. (1997) *The ecology of place*. Island Press, Washington, DC.

Beauchamp, D. E. (1988) The health of the republic: Epidemics, medicine, and moralism as challenges to democracy. Temple University Press, Philadelphia, PA.

Beauchamp, T. L. and Childress, J. F. (2013) *Principles of biomedical ethics*. Oxford University Press, New York, NY.

Beaugrand, G., Reid, P. C., Ibanez, F., Lindley, J. A. and Edwards, M. (2002) Reorganization of North Atlantic marine copepod biodiversity and climate. *Science*, 296, 1692–1694.

Becker, H. S., Geer, B., Hughes, E. C. and Strauss, A. L. (1961) *Boys in white: Student culture in medical school*. University of Chicago Press, Chicago, IL.

Becker, M. H. (1974) The health belief model and personal health behaviour. *Education Monographs*, 2, 324–508.

Becker, M. H. and Rosenstock, I. M. (1987) Comparing social learning theory and the health belief model. In Ward, W. B. (ed), *Advances in health education and promotion*, JAT Press, Greenwich, CT. pp. 245–249.

Beder, S. (1993) *The nature of sustainable development*. Scribe Publications, Carlton North, VIC.

Beder, S. (2000) Costing the earth: equity, sustainable development and environmental economics. *New Zealand Journal of Environmental Law*, 4, 227–243.

Belfast Healthy Cities (2013) Health Equity in All Policies: Nine case studies of HEiAP in action. From http://belfasthealthycities.com/sites/default/files/publications/NineHEiAPCaseStudies.pdf

Belfast Healthy Cities (2014) Creating resilient communities and supportive environments. From www.belfasthealthycities.com/creating-resilient-communities-supportive-environments (accessed 21 January 2015).

Benach, J., Solar, O., Santana, V., Castedo, A., Chung, H. and Muntaner, C. (2010) A micro-level model of employment relations and health inequalities. *International Journal of Health Services*, 40, 223–227.

Beresford, Q. (2000) Governments, markets and globalisation: Australian public policy in context. Allen & Unwin, St Leonards, NSW.

Berg, M. and Medrich, E. A. (1980) Children in four neighbourhoods: The physical environment and its effects on play and play patterns. *Environment and Behaviour*, 12, 320–348.

Berkman, L. F. (1984) Assessing the physical health effects of social networks and social support. *Annual Review of Public Health*, 5, 413–432.

Berkman, L. F. (1986) Social networks, support and health: Taking the next step forward. *American Journal of Epidemiology*, 123, 559–562.

Berkman, L. F. and Breslow, L. (1983) *Health and ways of living: The Alameda County study*. Oxford University Press, New York, NY.

Berkman, L. F. and Glass, T. (2000) Social integration, social networks, social support and health. In Berkman, L. F. and Kawachi, I. (eds), *Social epidemiology*, Chapter 7, Oxford University Press, New York, NY. pp. 137–173.

Berkman, L. F. and Kawachi, I. (2000) A historical framework for social epidemiology. In Berkman, L. F. and Kawachi, I. (eds), *Social epidemiology*, Chapter 1, Oxford University Press, New York, NY. pp. 3–12.

Berkman, L. F. and Syme, S. L. (1979) Social networks, host resistance and mortality: A nine-year follow-up study of Alameda County residents. *American Journal of Epidemiology*, 109, 186–204.

Bernal, J. A. L., Gasparrini, A., Artundo, C. M. and McKee, M. (2013) The effect of the late 2000s financial crisis on suicides in Spain: An interrupted time-series analysis. *European Journal of Public Health*, 23, 732–736.

Bettcher, D. and da Costa e Silva, V. L. (2013) Tobacco or health. In Leppo, K., Ollila, E., Pena, S., Wismar, M. and Cook, S. (eds), *Health in All Policies: Seizing opportunities, implementing policies*, Ministry of Social Affairs and Health, Helsinki, Finland. pp. 203–224.

Better Health Commission (1986) Looking forward to better health. Commonwealth of Australia, Canberra, ACT.

Beveridge, W. (1944) *Full employment in a free society*. The New Statesman and Nation and Reynolds News, London.

Biddle, N. (2012) Paper 3: Indigenous housing need, *CAEPR Indigenous Population Project: 2011 Census papers*. Centre for Aboriginal Economic Policy Research, The Australian National University, Canberra.

Bidgood, J. and Philipps, D. (31 October 2014) Nurse pushes issue of Ebola quarantine, *International New York Times*. From http://ihtbd.com/ihtuser/print/old%20THT/31-10-2014/a3110x04xxxxxxxxx.pdf

Bijlmakers, L., Bassett, M. and Sanders, D. (1998) Socioeconomic stress, health and child nutritional status in Zimbabwe at a time of economic structural adjustment: A three-year longitudinal study, *Research Report No. 5*. Nordiska Afrikainstitutet, Uppsala.

Birdsall, N., Lusting, N. and McLeod, D. (2011) Declining inequality in Latin America: Some economics, some politics. US Center for Global Development, Washington, DC.

Biro, G. A., Ring, I. and Lawson, J. S. (1981) The Ryde Heart Disease Prevention Program: One-year follow-up of a controlled trial to lower coronary risk factors in self-selected employees by screening and intervention. *Community Health Studies*, 5, 275–282.

Bjārås, G., Haglund, B. J. and Rifkin, S. B. (1991) A new approach to community participation. *Health Promotion International*, 6, 199–206.

Black, D. (1996) The development of the Glasgow City Health Plan. In Price, C. and Tsouros, A. (eds), *Our cities, our future: Policies and action plans for health and sustainable development*, WHO Healthy Cities Project Office, Copenhagen. pp. 89–97.

Black, N. (1994) Why we need qualitative research. *Journal of Epidemiology and Community Health*, 48, 425–426.

Black, R. E., Victora, C. G., Walker, S. P., Bhutta, Z. A., Christian, P., De Onis, M., Ezzati, M., Grantham-McGregor, S., Katz, J. and Martorell, R. (2013) Maternal and child undernutrition and overweight in low-income and middle-income countries. *The Lancet*, 382, 427–451.

Blackburn, H., Luepker, R. V., Kline, F. G., Bracht, N., Carlaw, R., Jacobs, D., Mittelmark, M., Stauffer, L. and Taylor, H. L. (1984) The Minnesota Heart Health Program: A research and demonstration project in cardiovascular disease prevention. In Matarazzo, J. D., Weiss, S. M., Herd, J. A. and Miller, N. E. (eds), *Behavioral health: A handbook of health enhancement and disease prevention*, 3rd edition, John Wiley & Sons, New York, NY. pp. 1171–1178.

Blackwell, A. G., Thompson, M., Freudenberg, N., Ayers, J., Schrantz, D. and Minkler, M. (2012) Using community organizing and community building to influence public policy. In Minkler, M. (ed), *Community organizing and community building for health and welfare*, 3rd edition, Chapter 20, Rutgers University Press, New Brunswick, NJ. pp. 371–385.

Blanke, D. and Mitchell, W. (2002) Towards health with justice: Litigation and public inquiries as tools for tobacco control. World Health Organization, Geneva.

Blaxter, M. (1990) *Health and Lifestyles*. Routledge, London.

Blaxter, M. (1997) Whose fault is it? People's own conceptions of the reasons for health inequalities. *Social science and medicine*, 44, 747–756.

Blaxter, M. (2010) *Health*. Polity Press, Cambridge.

Blewett, N. (1985) National campaign against drug abuse: Assumptions, arguments and aspirations. Commonwealth of Australia, Canberra, ACT.

Blewett, N. (1996) Valuing the past … investing in the future. *Australian Journal of Public Health*, 20, 342–343.

Bloem, M., Biswas, D. and Adhikari, S. (1996) Towards a sustainable and participatory rural development: Recent experiences of an NGO in Bangladesh. In De Koning, K. and Martin, M. (eds), *Participatory research in health: Issues and experiences*, Zed Books, London.

Blondel, J. (1990) *Comparative government*. Philip Allan, New York.

Bloor, M., Samphier, M. and Prior, L. (1987) Artefact explanations of inequalities in health: An assessment of the evidence. *Sociology of Health and Illness*, 9, 231–264.

Bogdan, R. and Taylor, S. J. (1975) Introduction to qualitative research methods: A phenomenological approach to the social sciences. Wiley, New York, NY.

Bogdewic, S. P. (1992) Participant observation. In Crabtree, B. F. and Miller, W. I. (eds), *Doing qualitative research*, SAGE Publications, Newbury Park, CA. pp. 45–69.

Bohlmeijer, E., Prenger, R., Taal, E. and Cuijpers, P. (2010) The effects of mindfulness-based stress reduction therapy on mental health of adults with a chronic medical disease: A meta-analysis. *Journal of Psychosomatic Research*, 68, 539–544.

Bohme, S., Zorabedian, D. and Egilman, D. (2005) Maximising profit and endangering health: Corporate strategies to avoid litigation and regulation. *International Journal of Occupational and Environmental Health*, 11, 338–348.

Bonevski, B., Sanson-Fisher, R. W. and Campbell, E. M. (1996) Primary care practitioners and health promotion: a review of current practices. *Health Promotion Journal of Australia*, 6, 22–31.

Bonita, R., Beaglehole, R. and Kjellström, T. (2006) *Basic epidemiology*. World Health Organization, Geneva.

Boris, J. P. (2005) *Commerce inéquitable*. Hachette Littératures, Paris.

Borrell, C., Morrison, J., Burstrom, B., Pons-Vigués, M., Hoffmann, R., Gandarillas, A., Martikainen, P., Domínguez-Berjón, M. F., Tarkiainen, L. and Díez, E. (2013) Comparison of health policy documents of European cities: Are they oriented to reduce inequalities in health? *Journal of Public Health Policy*, 34, 100–120.

Bouma, M. J., Sondorp, H. E. and Van der Kaay, H. J. (1994) Climate change and periodic epidemic malaria. *The Lancet*, 343, 1440.

Bourdieu, P. (1986) The forms of capital. In Richardson, J. (ed), *Handbook of theory and research for sociology of education*, Greenwood Press, New York, NY.

Boutilier, M., Mason, R. and Rootman, I. (1997) Community action and reflective practice in health promotion research. *Health Promotion International*, 12, 69–78.

Boutilier, M., Cleverly, S. and Labonté, R. (2000) Community as a setting for health promotion. In Poland, B. D., Green, L. W. and Rootman, I. (eds), *Settings for health promotion: Linking theory and practice*, SAGE Publications, Thousand Oaks, CA. pp. 250–279.

Bowen, G. A. (2009) Document analysis as a qualitative research method. *Qualitative Research Journal*, 9, 27–40.

Bowles, D. C., Butler, C. D. and Friel, S. (2014) Climate change and health in Earth's future. *Earth's Future*, 2, 60–67.

Bowling, A. (1995) *Measuring disease*. Open University Press, Buckingham.

Bowling, A. (2005) Measuring health: A review of quality of life measurement scales. Open University Press, Maidenhead.

Bowman, D. (1997) Untrammelled power. *The Adelaide Review*, Adelaide, SA.

Boxall, A. and Leeder, S. (2006) The health system: What should our priorities be? *Health Promotion Journal of Australia*, 17, 200–205.

Boyden, S. (1996) Health of the biosphere. In Furnass, B., Whyte, J., Harris, J. and Baker, A. (eds), Survival, health and wellbeing into the twenty first century. Proceedings of a conference held at the Australian National University, November 30—December 1, 1995, Nature and Society Forum, Canberra, ACT. pp. 51–57.

Branson, C. (2014) Lecture: Business and human rights: The new global consensus? *Flinders Law Journal*, 16.

Braveman, P. and Gruskin, S. (2003) Defining equity in health. *Journal of Epidemiology and Community Health*, 57, 254–258.

Bray, J. R., Gray, M., Hand, K., Bradbury, B., Eastman, C. and Katz, I. (2012) *Evaluating new income management in the Northern Territory: First evaluation report.* Social Policy Research Centre, University of New South Wales, Sydney & Australian National University, Canberra.

Breilh, J. (2003) *Epidemiolgia critica.* Lugar Editorial, Buenos Aires.

Bright, C. (1996) Understanding the threats of bioinvasions. In Brown, L. R. (ed), *State of the World 1996*, Worldwatch Institute/Earthscan, London. pp. 95–113.

Briss, P. A. (2005) Evidence-based: US road and public-health side of the street. *The Lancet*, 365, 828–830.

Brito, N. H. and Noble, K. G. (2014) Socioeconomic status and structural brain development. *Frontiers in Neuroscience*, 8. 10.3389/fnins.2014.00276.

Broadhead, W. E., Kaplan, B. H., James, S. A., Wagner, E. H., Schoenbach, V. J., Grimson, R., Heyden, S., Tibblin, G. and Gehlbach, S. H. (1983) The epidemiologic evidence for a relationship between social support and health. *American Journal of Epidemiology*, 117, 521–537.

Brockington, C. F. (1975) The history of public health. In Hobson, W. (ed), *The theory and practice of public health*, Oxford University Press, Oxford. pp. 1–7.

Brotherhood of St Laurence (2003) Submission to the Senate Community Affairs References Committee Inquiry into Poverty and Financial Hardship in Australia. Brotherhood of St Laurence, Melbourne, VIC.

Brown, E. R. and Margo, G. E. (1978) Health education: Can the reformers be reformed? *International Journal of Health Services*, 8, 3–26.

Brown, G. and Harris, T. (1978) Social origins of depression: A study of psychiatric disorder in women. Tavistock, London.

Brown, J. S., Collins, A. and Duguid, P. (1989) Situated cognition and the culture of learning. *Educational Researcher*, 18, 32–42.

Brown, L. (2013) The real threat to our future is peak water. *The Guardian*, 6 July. From www.theguardian.com/global-development/2013/jul/06/water-supplies-shrinking-threat-to-food

Brown, L. R. (1996) *State of the World 1996.* Earthscan, London.

Brown, L. R. (2011) World on the edge: How to prevent environmental and economic collapse. Earthscan, London.

Brown, M. (1995) *Medical power of attorney: A benefit or burden for the well elderly?* (MSc Thesis). Department of Public Health, Flinders University, Adelaide: SA.

Brown, V. A. (1985) Towards an epidemiology of health: A basis for planning community health programmes. *Health Policy*, 4, 331–340.

Brown, V. A. (1992) Health care policies, health policies or policies for health? In Gardner, H. (ed), *Health policy development, implementation and evaluation in Australia*, Churchill Livingstone, Melbourne, VIC.

Brown, V. A., Grootjans, J., Ritchie, J., Townsend, M. and Verrinder, G. (eds) (2005) Sustainability and health: Supporting global ecological integrity in public health. Allen & Unwin, Crows Nest, NSW.

Browne, J. (2005) Survey design. In Grenne, J. and Browne, J. (eds), *Principles of social research*, Chapter 12, Open University Press, Maidenhead. pp. 116–127.

Bruce, N., Springett, J., Hotchkiss, J. and Scott-Samuel, A. (eds) (1995) *Research and change in urban community health.* Avebury Press, Aldershot.

Brulle, R. J. and Pellow, D. N. (2006) Environmental justice: Human health and environmental inequalities. *Annual Review of Public Health*, 27, 103–124.

Bryant, T. (2012) Applying the lessons from international experiences. In Raphael, D. (ed), *Tackling health inequalities: Lessons from international experiences*, Chapter 10, Canadian Scholars' Press, Toronto, Ontario. pp. 265–288.

Bryman, A. (1988) Quantity and quality in social research. Routledge, London.

Bryson, L. (1987) *A new iron cage? Experiences of managerial reform.* Flinders University, Adelaide, SA.

Buchanan, D., Singer, M., Shaw, S., Teng, W., Stopka, T., Khoshnood, K. and Heimer, R. (2004) Syringe access, HIV risk and AIDS in Massachusetts and Connecticut: The health implications of public policy. In Castro, A. and Singer, M. (eds), *Unhealthy health policy*, Altamira Press, Walnut Creek, CA. pp. 275–285.

Bulmer, M. (1987) *The social basis of community care.* Allen & Unwin, London.

Bunker, J. P., Frazier, H. S. and Mosteller, F. (1994) Improving health: Measuring effects of medical care. *Milbank Quarterly*, 72, 225–258.

Bunton, R., Nettleton, S. and Bunton, R. (eds) (1995) *The sociology of health promotion.* Routledge, London.

Bureau of Infrastructure Transport and Regional Economics (BITRE) (2009) Cost of road crashes in Australia 2006, *Research Report 118*. BITRE, Canberra, ACT.

Bureau of Infrastructure Transport and Regional Economics (BITRE) (2014) *Road deaths Australia, 2013, statistical summary*. BITRE, Canberra, ACT.

Bureau of Meteorology (BOM) (2013) Monthly climate summary for New South Wales. From www.bom.gov.au/climate/mwr/ (accessed 14 October 2015).

Burgess, P. and Morrison, J. (2007) Country. In Carson, B., Dunbar, T., Chenhall, R. D. and Bailie, R. (eds), *Social determinants of Indigenous health*, Allen & Unwin, Crows Nest, NSW. pp. 177–202.

Burgmann, V. (1993) Power and protest: Movements for change in Australian society. Allen & Unwin, St Leonards, NSW.

Burnham, G., Lafta, R., Doocy, S. and Roberts, L. (2006) Mortality after the 2003 invasion of Iraq: A cross-sectional cluster sample survey. *The Lancet*, 368, 1421–1428.

Burns, A., Robins, A., Hodge, M. and Holmes, A. (2009) Long-term homelessness in men with a psychosis: Limitation of services. *International journal of mental health nursing*, 18, 126–132.

Burris, S., Hancock, T., Lin, V. and Herzog, A. (2007) Emerging strategies for healthy urban governance. *Journal of Urban Health*, 84, 154–163.

Burton, B. (2014) Big coal flexes $100 million PR muscle on soft sell. *The Canberra Times*, 7 May. From www.canberratimes.com.au/comment/big-coal-flexes-100-million-pr-muscle-on-soft-sell-20140507-zr5kq.html

Buse, K., Mays, N. and Walt, G. (2005) *Making health policy.* Open University Press, Maidenhead.

Busfield, J. and Paddon, M. (1977) *Thinking about children: Sociology and fertility in post-war England*. Cambridge University Press, Cambridge.

Butler, C. D., Sainsbury, P. and Armstrong, F. (2015) Civil disobedience, the energy-climate nexus and Australian coal exports [Letters]. *Australian and New Zealand Journal of Public Health*, 39, 93.

Büttner, P. and Muller, R. (2011) *Epidemiology.* Oxford University Press, South Melbourne, VIC.

Caballero, B. (2007) The global epidemic of obesity: An overview. *Epidemiologic Reviews*, 29, 1–5. 10.1093/epirev/mxm012.

Cameron, D. (2011) David Cameron's speech on plans to improve services for troubled families. From www.gov.uk/government/speeches/troubled-families-speech (accessed 2 April 2013).

Campbell, M., Cleland, J., Ezeh, A. and Prata, N. (2007) Return of the population growth factor. *Science*, 315, 1501.

Campbell, M. A., Finlay, S., Lucas, K., Neal, N. and Williams, R. (2014) Kick the habit: A social marketing campaign by Aboriginal communities in NSW. *Australian Journal of Primary Health*, 20, 327–333.

Cape York Institute (CYI) (2007) Welfare reform. From www.cyi.org.au/welfarereform.aspx (accessed 31 March 2007).

Carley, M. and Christie, I. (1992) *Managing sustainable development*. Earthscan Publications, London.

Carpenter, M. (2000) Health for some: Global health and social development since Alma Ata. *Community Development Journal*, 35, 336–351.

Carroll, J. and Manne, R. (eds) (1992) *Shutdown: The failure of economic rationalism and how to rescue Australia*. Text Publishing, Melbourne, VIC.

Carson, B., Dunbar, T., Chenhall, R. D. and Bailie, R. (eds) (2007) *Social determinants of Indigenous health*. Allen & Unwin, Crows Nest, NSW.

Carson, R. (1963) *Silent spring*. Hamish Hamilton, London.

Center for Responsive Politics (n.d.) OpenSecrets.org – Lobbying database. From www .opensecrets.org/lobby (accessed 30 July 2014).

CentreForum Commission (2014) *The pursuit of happiness: A new ambition for our mental health*. CentreForum Mental Health Commission.

Chadwick, E. (1842) Report on the sanitary condition of the labouring population of Great Britain. HM Stationery Office, London.

Chadwick, P. (1998) Do media help or harm public health? *Australian and New Zealand Journal of Public Health*, 22, 155–158.

Chalmers, I. (1993) The Cochrane collaboration: Preparing, maintaining, and disseminating systematic reviews of the effects of health care. *Annals of the New York Academy of Sciences*, 703, 156–165.

Chalmers, I., Enkin, M. and Keirse, M. J. (eds) (1989) Effective care in pregnancy and childbirth. Oxford University Press, Oxford.

Chalmers, I., Sackett, D. and Silagy, C. (1997) The Cochrane Collaboration. In Maynard, A. and Chalmers, I. (eds), *Non-random reflections on health services research*, BMJ Publishing, London. pp. 231–249.

Chan, M. (2007) Health diplomacy in the 21st century. Address to Directorate for Health and Social Affairs, Oslo, Norway, 13 February. From www.who.int/dg/speeches/2007/130207_norway/en (accessed 19 March 2007).

Chan, M. (2013) WHO Director-General addresses the Sixty-sixth World Health Assembly, 20 May. From www.who.int/dg/speeches/2013/world_health_assembly_20130520/en (accessed 5 February 2015).

Chan, M. (2015) Report by the Director-General to the Special Session of the Executive Board on Ebola, 25 January. From www.who.int/dg/speeches/2015/executive-board-ebola/en (accessed 5 February 2015).

Change.org (n.d.) Hershey: Stop Exploiting Student Guestworkers. From www.change.org/en-AU/petitions/hershey-stop-exploiting-student-guestworkers (accessed 25 July 2014).

Change.org (n.d.) I've experienced homophobia in Aussie Rules Football first hand—now it's time to end it. From www.change.org/en-AU/petitions/i-ve-experienced-homophobia-in-aussie-rules-football-first-hand-now-it-s-time-to-end-it (accessed 25 July 2014).

Chapman, P. and Davey, P. (1997) Working 'with' communities, not 'on' them: A changing focus for local government health planning in Queensland. *Australian Journal of Primary Health*, 3, 82–91.

Chapman, S. (1985) Stop-smoking clinics: A case for their abandonment. *The Lancet*, 325, 918–920.

Chapman, S. (1994) What is public health advocacy? In Chapman, S. and Lupton, D. (eds), *The fight for public health*, BMJ Publishing, London.

Chapman, S. (1998) Over our dead bodies: Port Arthur and Australia's fight for gun control. Pluto Press, Annandale, NSW.

Chapman, S. (2008) Public health advocacy and tobacco control: Making smoking history. Blackwell Publishing.

Chapman, S. and Alpers, P. (2013) Gun-related deaths: How Australia stepped off "the American path". *Annals of Internal Medicine*, 158, 770–771.

Chapman, S., Haynes, A., Derrick, G., Sturk, H., Hall, W. D. and St George, A. (2014) Reaching "an audience that you would never dream of speaking to": influential public health researchers' views on the role of news media in influencing policy and public understanding. *Journal of Health Communication*, 19, 260–273.

Chapman, S. and Wakefield, M. (2001) Tobacco control advocacy in Australia: Reflections on 30 years of progress. *Health Education and Behavior*, 28, 274–289.

Charting Transport (2012) Trends in transport greenhouse gas emissions, *Charting transport – Looking at transport through graphs and maps.*

Chasin, B. (1997) Inequality and violence in the United States: Casualties of capitalism. Humanities Press, Atlantic Highlands, NJ.

Chaufan, C. (2004) Sugar blues: A social anatomy of the diabetes epidemic in the United States. In Castro, A. and Singer, M. (eds), *Unhealthy health policy*, Altamira Press, Walnut Creek, CA. pp. 257–274.

Cheek, J., Shoebridge, J., Willis, E. and Zadoroznyj, M. (1996) *Society and health: Social theory for health workers.* Longman, Melbourne, VIC.

Chen, S. and Ravallion, M. (2012) An update to the World Bank's estimates of consumption poverty in the developing world, *Briefing Note 03-01-12.* World Bank, Washington, DC.

Chen, Y.-Y., Subramanian, S. V., Acevedo-Garcia, D. and Kawachi, I. (2005) Women's status and depressive symptoms: A multilevel analysis. *Social science and medicine*, 60, 49–60.

Chin, S. (2007) Poet for the people. In Olson, A. (ed), *Word warriors: 35 women leaders in the spoken word revolution*, Seal Press, Emeryville, CA.

China Daily (2014) Traffic jams cost Beijing $11.3b a year. *ChinaDaily.com.cn*, 29 September. From www.chinadaily.com.cn/china/2014-09/29/content_18679171.htm

Chivian, E. (ed) (2003) *Biodiversity: Its importance to human health. Interim executive summary.* Center for Health and the Global Environment, Harvard Medical School, Boston, MA.

Chopra, M. (2005) The impact of globalisation on food. In Lee, K. and Collin, J. (eds), *Global change and health*, Open University Press, Maidenhead.

Christie, D., Gordon, I. and Heller, R. (1987) *Epidemiology: An introductory text for medical and other health science students.* University of NSW Press, Kensington, NSW.

Chu, C. and Forrester, C. A. (1992) *Workplace health promotion in Queensland.* Queensland Health, Brisbane, QLD.

City of New York (2010) Active design guidelines. Promoting physical activity and health in design. Center for Active Design, New York, NY.

Clarke, J. N. (1983) Sexism, feminism and medicalism: A decade review of literature on gender and illness. *Sociology of Health and Illness*, 5, 62–81.

Clavier, C. and de Leeuw, E. (eds) (2013) Health promotion and the policy process: Practical and critical theories. Oxford University Press, Oxford.

Clift, C. (2014) What's the World Health Organization for? Final report from the Centre on Global Health Security Working Group on Health Governance. Chatham House, London.

Coburn, D. (2004) Beyond the income inequality hypothesis: Class neo-liberalism and health inequalities. *Social Science and Medicine*, 58, 41–56.

Cochrane, R. (1983) *The social creation of mental illness.* Longman, London.

Cofala, J., Bertok, I., Borken-Kleefeld, J., Heyes, C., Kilmont, Z., Rafaj, P., Sander, R., Schopp, W. and Amann, M. (2012) *Emissions of air pollutants for the World Energy Outlook 2102 energy scenarios. Draft final report.* International Institute for Applied Systems Analysis, Laxenburg, Austria.

Coggan, C., Saunders, C. and Grenot, D. (2008) Art and safe communities: The role of Big hART in the regeneration of an inner city housing estate. *Health Promotion Journal of Australia*, 19, 4–9.

Cohen, J. M., Boniface, S. and Watkins, S. (2014) Health implications of transport planning, development and operations. *Journal of Transport and Health*, 1, 63–72.

Cole-Hamilton, I. and Lang, T. (1986) *Tightening belts: A report on the impact of poverty on food.* London Food Commission, London.

Coleman, A. (1985) Utopia on trial: Vision and reality in planned housing. Hilary Shipman, London.

Coleman, J. S. (1988) Social capital in the creation of human capital. *American Journal of Sociology*, 94, S95–S120.

Collier, P. (2007) The bottom billion: Why the poorest countries are failing and what can be done about it. Oxford University Press, New York, NY.

Colson, A. C. (1986) Aspects of isolation. *Community and institutional care for aged migrants in Australia*. Australian Institute of Management Affairs, Melbourne, VIC.

Commission on Social Determinants of Health (CSDH) (2008) Closing the gap in a generation: Health equity through action on the social determinants of health. *Final report of the Commission on Social Determinants of Health*. World Health Organization, Geneva.

Commonwealth of Australia (1997) *Tough on Drugs*. Commonwealth of Australia, Canberra, ACT.

Community Development in Health (CDIH) Project (1988) *Community development in health, a resources collection: A collection of resource materials for community workers in health*. Community Development in Health Project, District Health Councils Program, Northcote, VIC.

Coney, S. (1988) *The unfortunate experiment*. Penguin, Auckland, NZ.

Connell, B. (1994) Poverty and education. *Harvard Education Review*, 62, 125–149.

Connell, J. P. and Kubisch, A. C. (1998) Applying a theory of change approach to the evaluation of comprehensive community initiatives: Progress, prospects and problems. In Fulbright-Anderson, K., Kubisch, A. C. and Connell, J. P. (eds), *New approaches to evaluation community initiatives*, Aspen Institute, Washington, DC.

Connell, R. W. and Irving, T. H. (1991) Yes, Virginia, there is a ruling class. In Mayer, H. and Nelson, H. (eds), *Australian politics: A fourth reader*, Cheshire, Melbourne, VIC.

Considine, M. (1990) Managerialism strikes out. *Australian Journal of Public Administration*, 49, 166–178.

Considine, M. (1994) *Public policy: A critical approach*. Macmillan, South Melbourne, VIC.

Considine, M. (2005) Making public policy: Institutions, actors, strategies. Polity Press.

Convention on Biological Diversity (CBD) (2014) Global Biodiversity Outlook 4: A mid-term assessment of progress towards the implementation of the Strategic Plan for Biodiversity 2011–2020. Convention on Biological Diversity Secretariat, Montréal.

Coombs, H. C. (1990) *The return of scarcity: Strategies for an economic future*. Cambridge University Press, Melbourne, VIC.

CO-OPS Collaboration (n.d.). From www.co-ops.net.au (accessed 21 January 2015).

Cormack, S., Ali, R. and Pols, R. G. (1995) A public health approach to drug issues. In Baum, F. (ed), *Health for All: The South Australian experience*, Wakefield Press, Adelaide, SA. pp. 339–361.

Cornia, G. A., Jolly, R. and Stewart, F. (eds) (1988) *Adjustment with a human face: Protecting the vulnerable and promoting growth*. Oxford University Press, Oxford.

Cornwall, A. (1996) Towards participatory practice: Participatory rural appraisal (PRA) and the participatory process. In de Koning, K. and Martin, M. (eds), *Participatory research in health: Issues and experiences*, Zed Books, London. pp. 94–107.

Cornwall, J. (1988) Introduction, *A social health strategy for South Australia*. South Australian Health Commission, Adelaide, SA.

Cornwell, J. (1984) Hard earned lives: Accounts of health and illness from East London. Tavistock, London.

Cortie, B., Donovan, R. J. and Holman, C. D. (1996) Factors influencing the use of physical activity facilities: Results from qualitative research. *Health Promotion Journal of Australia*, 6, 16–21.

Costanza, R., Alperovitz, G., Daly, H., Farley, J., Franco, C., Jackson, T., Kubiszewski, I., Schor, J. and Victor, P. (2013) Building a sustainable and desirable economy-in-society-in-nature. In Stark, L. (ed), *State of the World 2013: Is sustainability still possible?*, Chapter 11, Worldwatch Institute/Island Press, Washington, DC. pp. 126–142.

Costanza, R., Daly, H. E. and Bartholomew, J. A. (1991) Goals, agenda and policy recommendations for ecological economics. In Costanza, R. (ed), *Ecological economics: The science and management of sustainability*, Columbia University Press, New York, NY.

Costello, A., Abbas, M., Allen, A., Ball, S., Bell, S., Bellamy, R., Friel, S., Groce, N., Johnson, A., Kett, M., Lee, M., Levy, C., Maslin, M., McCoy, D., McGuire, B., Montgomery, H., Napier, D., Pagel, C., Patel, J., de Oliveira, J. A. P., Redclift, N., Rees, H., Rogger, D., Scott, J., Stephenson, J., Twigg, J., Wolff, J. and Patterson, C. (2009) Managing the health effects of climate change: Lancet and University College London, Institute for Global Health Commission. *The Lancet*, 373, 1693–1733.

Costello, T. (1996) Kennett and the Casino led recovery. *In Touch*, 13, 13–14.

Costongs, C. and Springett, J. (1997) Towards a framework for the evaluation of health-related policies in cities. *Evaluation*, 3, 345–362.

Council of Australian Governments (COAG) (2008a) *National Indigenous Reform Agreement (Closing the Gap)*. COAG. From www.coag.gov.au/node/145

Council of Australian Governments (COAG) (2008b) *National Partnership Agreement on Preventive Health*. COAG. From www.federalfinancialrelations.gov.au/content/npa/health_preventive/national_partnership.pdf

Council of Australian Governments (COAG) (2008c) *National Partnership Agreement on Remote Indigenous Housing (NPARIH)*. COAG, Canberra. From www.federalfinancialrelations.gov.au/content/npa/housing/remote_indigenous_housing/national_partnership.pdf

Council of Australian Governments (COAG) Reform Council (2013) *Indigenous Reform 2011–12: Comparing performance across Australia*. COAG Reform Council, Sydney, NSW.

Cox, E. (1996) *A truly civil society*. Australian Broadcasting Corporation, Sydney, NSW.

Craig, B. (1989) Health costs and benefits of urban consolidation versus suburban expansion in Adelaide: A literature review. Southern Community Health Research Unit, Adelaide, SA.

Cramb, S. M., Garvey, G., Valery, P. C., Williamson, J. D. and Baade, P. D. (2012) The first year counts: cancer survival among Indigenous and non-Indigenous Queenslanders, 1997–2006. *Medical Journal of Australia*, 196, 270–274.

Crawford, R. (1977) You are dangerous to your health: The ideology and politics of victim blaming. *International Journal of Health Services*, 7, 663–680.

Crawford, R. (1984) A cultural account of "health": Control, release, and the social body. In McKinlay, J. B. (ed), *Issues in the political economy of health care*, Tavistock, New York, NY. pp. 60–103.

Credit Suisse (2012) *Global Wealth Databook 2012*. Credit Suisse.

Creese, A. (1991) User charges for health care: A review of recent experience. *Health Policy and Planning*, 6, 309–319.

Crockford, R. (1996) Culture jamming. *New Internationalist*, 278, 14.

Crombie, I. K., Irvine, L., Elliot, L. and Wallace, H. (2005) *Closing the health inequalities gap: An international perspective*. WHO Regional Office for Europe, Copenhagen.

Crotty, M. (1996) The ethics of ethics committees. In Willis, P. and Neville, B. (eds), *Qualitative research practice in adult education*, David Lovell, Ringwood, VIC. pp. 80–98.

Crotty, M. (1998) The foundations of social research: Meaning and perspective in the research process. Allen & Unwin, St Leonards, NSW.

CSIRO (2012) Understanding extreme weather changes. From www.csiro.au/Outcomes/Environment/Extreme-Events/Understanding-extreme-weather-changes.aspx (accessed 15 August 2014).

Cummings, T. G. and Worley, C. G. (2004) *Organisational development and change*. South Western College Publishing, Mason, Ohio.

Cunningham, G. (1963) Policy and practice. *Public Administration*, 41, 229–238.

Cunningham, J. (2002) Diagnostic and therapeutic procedures among Australian hospital patients identified as Indigenous. *Medical Journal of Australia*, 176, 58–62.

Curson, P. and McCracken, K. (1989) *Plague in Sydney: The anatomy of an epidemic*. New South Wales University Press, Sydney, NSW.

Curtis, S. and Taket, A. R. (1996) *Health and societies: Changing perspectives*. Arnold, London.

Cutler, D. and Meara, E. (2001) Changes in the age distribution of mortality over the 20th century. NBER, Cambridge, MA.

Dahl, R. (1961) *Who governs?* Yale University Press, New Haven, CT.

Dahlgren, G. and Whitehead, M. (2006) European strategies for tackling social inequities in health: Levelling up (Part 2). WHO Regional Office for Europe, Copenhagen.

Daly, H. E. (1992) *Steady state economics.* Earthscan, London.

Daly, H. E. and Cobb, J. B. (1990) *For the common good.* Green Print, London.

Daly, J. (ed) (1996) Ethical intersections: Health research, methods and researcher responsibility. Allen & Unwin, St Leonards, NSW.

Daly, J. and McDonald, I. (1992) Introduction. The problem as we saw it. In Daly, J. (ed), *Researching health care: Designs, dilemmas, disciplines*, Tavistock/Routledge, London. pp. 1–11.

Dark, E. P. (1939) Property and health. *Medical Journal of Australia*, March, 345–352.

Dark, E. P. (1941) Letter. *Medical Journal of Australia*, November, 526–527.

Darnton-Hill, I., Mandryk, J. A., Mock, P. A., Lewis, J. M. and Kerr, C. B. (1990) Sociodemographic and health factors in the well-being of homeless men in Sydney, Australia. *Social Science and Medicine*, 31, 537–544.

Davey Smith, G. (ed) (2003) *Health inequalities: Lifecourse approaches.* The Policy Press, Bristol.

Davey Smith, G., Neaton, J. D., Wentworth, D., Stamler, R. and Stamler, J. (1996) Socioeconomic differentials in mortality risk among men screened for the Multiple Risk Factor Intervention Trial: I. White men. *American Journal of Public Health*, 86, 486–496.

Davidson, R., Kitzinger, J. and Hunt, K. (2006) The wealthy get healthy, the poor get poorly? Lay perceptions of health inequalities. *Social Science and Medicine*, 62, 2171–2182.

Davies, J. K. and Kelly, M. P. (eds) (1993) *Healthy Cities: Research and practice.* Routledge, London.

Dávila, A. L. (2011) Global pharmaceutical development and access: Critical issues of ethics and equity. *MEDICC Review*, 13, 16–22.

Davis, A. (1995) Managerialised health care. In Rees, S. and Rodley, S. (eds), *The human costs of managerialism*, Pluto Press, Leichhardt, NSW.

Davis, M. (2006) *Planet of slums.* Verso, London.

Davis, M. (2014) Neoliberalism, the culture wars and public policy. In Miller, C. and Orchard, L. (eds), *Australian public policy: Progressive ideas in the neo-liberal ascendency*, Chapter 2, Policy Press, Bristol. pp. 27–42.

Davison, C., Frankel, S. and Smith, G. D. (1992) The limits of lifestyle: Re-assessing "fatalism" in the popular culture of illness prevention. *Social Science and Medicine*, 34, 675–685.

Davison, G. (1994) The past and future of the Australian suburb. In Johnson, L. C. (ed), *Suburban dreaming: An interdisciplinary approach to Australian cities*, Deakin University Press, Geelong, VIC. pp. 99–113.

Davoren, S., Mills, C. and Jones, A. (2014) National and regional perspectives of NCD risk regulation: Australia. In Voon, T., Mitchell, A. D. and Liberman, J. (eds), *Regulating tobacco, alcohol and unhealthy foods: The legal issues*, Routledge, UK.

de Koning, K. and Martin, M. (eds) (1996) *Participatory research in health: Issues and experiences.* Zed Books, London.

de Laine, M. D. (1997) *Ethnography: Theory and application in health research.* MacLennan & Petty, Sydney, NSW.

de Looper, M. and Magnus, P. (2005) Australian health inequalities 2: Trends in male mortality by broad occupational group. *Bulletin No. 25, AIHW Cat. no. AUS 58.* AIHW, Canberra, ACT.

De Silva, M. J. (2006) A systematic review of the methods used in studies of social capital and mental health. In McKenzie, K. and Harpham, T. (eds), *Social capital and mental health*, Kingsley, London.

de Snyder, V. N. S., Friel, S., Fotso, J. C., Khadr, Z., Meresman, S., Monge, P. and Patil-Deshmukh, A. (2011) Social conditions and urban health inequities: realities, challenges and opportunities to transform the urban landscape through research and action. *Journal of Urban Health*, 88, 1183–1193.

Dean, K. (ed) (1993) Population health research: Linking theory and methods. SAGE Publications, London

Dean, K., Kreiner, S. and McQueen, D. V. (1993) Researching population health: New directions. In Dean, K. (ed), *Population health research: Linking theory and methods*, SAGE Publications, London. pp. 227–237.

Deary, I. (2008) Why do intelligent people live longer? *Nature*, 456, 175–176.

D'Elia, A., Newstead, S. and Cameron, M. (2007) Overall impact during 2001–2004 of Victorian speed-related package, *Report No. 267*. Monash University Accident Research Centre, Clayton, VIC.

Denholm, M. (2013) Companies to get protection from activists' boycotts. *The Australian*, 23 September. From www.theaustralian.com.au/national-affairs/companies-to-get-protection-from-activists-boycotts/story-fn59niix-1226724817535

Denzin, N. K. (1978) The research act: A theoretical introduction to sociological methods. McGraw Hill, New York, NY.

Denzin, N. K. (1989) *Interpretive interactionism*. SAGE Publications, Newbury Park, CA.

Denzin, N. K. and Lincoln, Y. S. (2000) *Handbook of qualitative research*. SAGE Publications, Thousand Oaks, CA.

Denzin, N. K. and Lincoln, Y. S. (2011) *Handbook of qualitative research*. SAGE Publications, Thousand Oaks, CA.

Deparment of Families Housing Community Services and Indigenous Affairs (FaHCSIA) (2009) Policy statement: Landmark reform to the welfare system, reinstatement of the Racial Discrimination Act and strengthening of the Northern Territory Emergency Response. Commonwealth of Australia, Canberra, ACT.

Department of Families Housing Community Services and Indigenous Affairs (FaHCSIA) (2008) *The road home: A national approach to reducing homelessness*. Commonwealth of Australia, Canberra, ACT.

Department of Families Housing Community Services and Indigenous Affairs (FaHCSIA) (2011) *Northern Territory Emergency Response evaluation report 2011*. Commonwealth of Australia, Canberra, ACT. From www.fahcsia.gov.au/our-responsibilities/indigenous-australians/publications-articles/northern-territoryemergency-response-evaluation-report-2011

Department of Families Housing Community Services and Indigenous Affairs (FaHCSIA) (2013) *Stronger Futures in the Northern Territory*. Commonwealth of Australia, Canberra, ACT.

Department of Health (DoH) (2009) *National mental health policy 2008, Canberra*, Commonwealth of Australia. From www.health.gov.au/internet/publications/publishing.nsf/Content/mental-pubs-n-pol08-toc

Department of Health (DoH) (2012) National Strategic Framework for Rural and Remote Health. Commonwealth of Australia, Canberra, ACT. From www.ruralhealthaustralia.gov.au/internet/rha/publishing.nsf/Content/NSFRRH

Department of Health (DoH) (2013a) The Australian Immunisation Handbook 10th Edition (undated January 2014). Commonwealth of Australia, Canberra, ACT. From www.health.gov.au/internet/immunise/publishing.nsf/Content/Handbook10-home

Department of Health (DoH) (2013b) Healthy workplace initiative. From www.healthyworkers.gov.au (accessed 21 January 2015).

Department of Health (DoH) (2013c) National Aboriginal and Torres Strait Islander Health Plan (NATSIHP) 2013–2023. Commonwealth of Australia, Canberra, ACT. From www.health.gov.au/internet/publications/publishing.nsf/Content/oatsih-healthplan-toc

Department of Health (DoH) (2014a) Establishment of Primary Health Networks: Information session. [Presentation] From www.health.gov.au/internet/main/publishing.nsf/Content/phn_presentation (accessed 27 November 2014).

Department of Health (DoH) (2014b) Guidelines for healthy foods and drinks supplied in school canteens. Commonwealth of Australia, Canberra, ACT. From www.health.gov .au/internet/main/publishing.nsf/Content/phd-nutrition-canteens

Department of Health and Aged Care (1998) National Action Plan for Suicide Prevention. Commonwealth of Australia, Canberra, ACT.

Department of Health and Ageing (DoHA) (2012) Background Paper: Medicare Locals Health Needs Assessment and Planning. Commonwealth of Australia, Canberra, ACT. From www.thesandsingpframework.com/documents/ml/20120106_BRF_Background-Paper-Medicare-Locals-Health-Needs-Assessment-and-Planning.pdf

Department of Public Health, Flinders University and South Australian Community Health Research Unit (SACHRU) (2000) *Improving health services through consumer participation: A resource guide for organisations.* Australian Government, Canberra, ACT.

Department of Social Services (DSS) (2013) National Partnership on Remote Indigenous Housing – Progress review (2008–2013). Commonwealth of Australia, Canberra, ACT. From www.dss.gov.au/sites/default/files/files/indigenous/Final%20NPARIH%20Review%20 May%2020132.pdf

Department of the Environment (2010) The Hon Dr Mike Kelly AM MP Parliamentary Secretary for Water – $51.7 million to improve water and wastewater services in 17 indigenous communities [Media release]. From www.environment.gov.au/minister/archive/ps/2010/ mr20100423.html (accessed 9 January 2015).

Diamond, J. (2008) What's your consumption factor? *New York Times*, 2 January. From www .thomashylton.org/pdfs/Recommended%20Articles/2008-1-2%20Diamond.pdf

Diderichsen, F. (2002) Income maintenance policies: Determining their potential impact on socioeconomic inequalities in health. In Mackenbach, J. and Bakker, M. (eds), *Reducing inequalities in health: A European perspective*, Routledge, London. pp. 55–66.

Diderichsen, F., Evans, T. and Whitehead, M. (2001) The social basis of disparities in health. In Whitehead, M., Evans, T., Diderichsen, F., Bhuiya, A. and Wirth, M. (eds), *Challenging inequities in health: From ethics to action*, Oxford University Press, New York, NY. pp. 13–23.

Dillman, D. A. (1983) Mail and other self-administered questionnaires. In Rossi, P. H., Wright, J. D. and Anderson, A. B. (eds), *Handbook of survey research: Quantitative studies in social relations*, Academic Press, London.

Dillman, D. A., Smyth, J. D. and Christian, L. M. (2014) *Internet, phone, mail, and mixed-mode surveys: The tailored design method.* John Wiley & Sons, Hokoken, NJ.

Dimick, D. (2014) As world's population booms, will its resources be enough for us? *National Geographic*, 20 September.

Dixon, J. (1989) The limits and potential of community development for personal and social change. *Community Health Studies*, 13, 82–92.

Dixon, J. and Broom, D. (eds) (2007) *The 7 deadly sins of obesity: How the modern world is making us fat.* University of NSW Press, Sydney, NSW.

Dixon-Woods, M., Agarwal, S., Young, B., Jones, D. and Sutton, A. (2004) Integrative approaches to qualitative and quantitative evidence. Health Development Agency, London.

Dockery, D. W., Pope, C. A., Xu, X., Spengler, J. D., Ware, J. H., Fay, M. E., Ferris, B. G. and Speizer, F. E. (1993) An association between air pollution and mortality in six US cities. *New England Journal of Medicine*, 329, 1753–1759.

Doll, R. and Hill, A. B. (1954) The mortality of doctors in relation to their smoking habits. *British Medical Journal*, 1, 1451.

Donovan, R. J. and Spark, R. (1997) Towards guidelines for survey research in remote Aboriginal communities. *Australian and New Zealand Journal of Public Health*, 21, 89–95.

Dorfman, L. and Krasnow, I. D. (2014) Public health and media advocacy. *Annual Review of Public Health*, 35, 293–306.

Doyal, L. (1979) *The political economy of health*. Pluto Press, London.

Drahos, P. and Braithwaite, J. (2004) Who owns the knowledge economy? Political organising behind TRIPS. *Corner House Briefing 32*. From www.thecornerhouse.org.uk/sites/thecornerhouse.org.uk/files/32trips.pdf

Draper, G., Turrell, G. and Oldenburg, B. (2004) Health inequalities in Australia: Mortality. *Health inequalities monitoring series no. 1, AIHW Cat. no. PHE 55*. Queensland University of Technology, and Australian Institute of Health and Welfare, Canberra, ACT.

Draper, P. (ed) (1991) *Health through public policy*. Green Print, London.

Draper, R., Curtice, L., Hooper, J. and Goumans, M. (1993) WHO healthy cities project: Review of the first five years (1987–1992) – A working tool and a reference framework for evaluating the project. WHO Regional Office for Europe, Copenhagen.

Drohan, M. (1998) How the net killed the MAI. Grassroots used their own globalisation to derail deal, *The Globe and Mail*. 29 April. From www.gwb.com.au/gwb/news/mai/3004.html

Duckett, S. and Willcox, S. (2011) *The Australian health care system*. Oxford University Press, South Melbourne, VIC.

Duhl, L. (1992) Healthy cities: Myths or reality. In Ashton, J. (ed), *Healthy cities*, Open University Press, Milton Keynes. pp. 15–21.

Dunbar, T. and Scrimgeour, M. (2007) Education. In Carson, B., Dunbar, T., Chenhall, R. D. and Bailie, R. (eds), *Social determinants of Indigenous health*, Allen & Unwin, Crows Nest, NSW. pp. 135–152.

Dunn, S. and Flavin, C. (2002) Moving the climate change agenda forward. In Stark, L. (ed), *State of the World 2002*, Worldwatch Institute/WW Norton & Co, New York, NY.

Dunphy, D. and Stace, D. (1992) *Under new management*. McGraw Hill, Sydney, NSW.

Durham, G. (1997) *WHO and health promotion: Streets ahead or gone aground?* Paper presented at the VicHealth Conference: Health Promotion in the Twenty-first Century—Jakarta and Beyond, Melbourne, VIC.

Durkheim, E. (1979 [1897]) *Suicide: A study in sociology*. Free Press, New York, NY.

Dwyer, J. (1989) The politics of participation. *Community Health Studies*, 13, 59–65.

Dwyer, J. (2004) Australian health system restructuring – what problem is being solved? *Australia and New Zealand health policy*, 1, 6. 10.1186/1743–8462-1-6.

Dwyer, J., Liang, Z., Thiessen, V. and Martini, A. (2013) *Project management in health and community services*. Allen & Unwin, Melbourne, VIC.

Dwyer, J., Silburn, K. W. and Wilson, G. (2004a) National strategies for improving Indigenous health and health care, *Aboriginal and Torres Strait Islander Primary Health Care Review: Consultant Report No 1*, Commonwealth of Australia, Canberra, ACT.

Dwyer, J., Stanton, P. and Thiessen, V. (2004b) Project management in health and community services: Getting good ideas to work. Allen & Unwin, Sydney, NSW.

Eckersley, R. (2005) *Well and good: Morality, meaning and happiness*. Text Publishing, Melbourne, VIC.

Ecological Footprint (2011) Earth overshoot day. From www.footprintnetwork.org/en/index.php/GFN/blog/today_is_earth_overshoot_day1 (accessed 27 September 2012).

Edgar, D. (2005) *The war over work. The future of work and family*. Melbourne University Press, Melbourne, VIC.

Editorial. (1995) Public health advocacy: Unpalatable truths. *The Lancet*, 345, 597–598.

Edward, P. (2006) Examining inequality: Who really benefits from global growth? *World Development*, 34, 1667–1695.

Edwards, P. and Tsouros, A. D. (2008) *A healthy city is an active city: a physical activity planning guide*. WHO Regional Office for Europe, Copenhagen.

Egger, G., Fitzgerald, W., Frape, G., Monaem, A., Rubinstein, P., Tyler, C. and McKay, B. (1983) Results of large scale media antismoking campaign in Australia: North Coast "Quit for Life" programme. *British Medical Journal*, 287, 1125–1128.

Egger, G., Spark, R. and Donovan, R. J. (1990) *Health promotion strategies and methods*. McGraw-Hill, Sydney, NSW.

Egger, G. and Swinburn, B. (2010) *Planet obesity: How we're eating ourselves and the planet to death*. Allen & Unwin, Crows Nest, NSW.

Eisenberg, L. (1979) A friend, not an apple, a day will keep the doctor away. *American Journal of Medicine*, 66, 551–553.

Elder, J. P. (1986) Organisational and community approaches to communitywide prevention of heart disease: The first two years of the Pawtucket Heart Health program. *American Journal of Preventive Medicine*, 15, 107–117.

Elder, J. P., Schmid, T. I., Dower, P. and Hedlund, S. (1993) Community heart health programs: Components, rationale and strategies for effective interventions. *Journal of Public Health Policy*, 14, 463–479.

Elder, R. W., Shults, R. A., Sleet, D. A., Nichols, J. L., Thompson, R. S. and Rajab, W. (2004) Effectiveness of mass media campaigns for reducing drinking and driving and alcohol-involved crashes: A systematic review. *American Journal of Preventive Medicine*, 27, 57–65.

Ell, K. (1996) Social networks, social support and coping with serious illness: The family connection. *Social Science and Medicine*, 42, 173–183.

Ellis, G. and Walton, S. (2012) Building partnerships between local health departments and communities: Case studies in capacity building and cultural humility. In Minkler, M. (ed), *Community organizing and community building for health and welfare*, 3rd edition, Rutgers University Press, New Brunswick, NJ.

Engelman, R. (2013) Beyond sustainability. In Stark, L. (ed), *State of the World 2013: Is sustainability still possible?* Chapter 1, Worldwatch Institute/Island Press, Washington, DC. pp. 2–16.

Engelman, R., Halweil, B. and Nierenberg, D. (2002) Rethinking population, improving lives. In Starke, L. (ed), *State of the World 2002*, Worldwatch Institute/WW Norton & Co, New York, NY.

Engels, F. (1993 [1845]) *The conditions of the working class in England.* Oxford University Press, New York, NY.

Epstein, P. R., Ford, T. E. and Colwell, R. R. (1993) Marine ecosystems. *The Lancet*, 342, 1216–1219.

Erben, R., Franzkowiak, P. and Wenzel, E. (1992) Assessment of the outcomes of health intervention. *Social Science and Medicine*, 35, 359–365.

Erdmenger, C. and Führ, V. (2005) Hidden Subsidies for Urban Car Transportation: Public Funds for Private Transport. International Council for Local Environmental Initiatives (ICLEI) European Secretariat, Freiburg, Germany. From http://www.increase-public-transport.net/fileadmin/user_upload/Procurement/SIPTRAM/Hidden_subsidies_final.pdf

Erunke, C. E. and Hafsat, K. (2012) Globalization, multinational corporation and the Nigerian economy. *International Journal of Social Sciences Tomorrow*, 1, 1–8.

Estes, R. (1996) Tyranny of the bottom line: Why corporations make good people do bad things. Berrett-Koehler Publishers, San Francisco, CA.

Etzioni, A. (1967) Mixed-scanning: A third approach to decision-making. *Public Administrative Review*, 27, 385–392.

European Environment Agency (EEA) (2011) Mapping the impacts of natural hazards and technological accidents in Europe: An overview of the last decade, *EEA Technical report no. 13/2010*. European Environment Agency, Copenhagen.

Evans, G. and Newnham, J. (1992) The dictionary of world politics: A reference guide to concepts, ideas and institutions. Harvester Wheatsheaf, London.

Evans, R. G. (1994) Introduction. In Evans, R. G., Barer, M. L. and Marmor, T. R. (eds), *Why are some people healthy and others not? The determinants of health of populations*, Walter de Gruyter, New York, NY. pp. 3–26.

Evans, R. G., Barer, M. L. and Marmor, T. R. (eds) (1994a) *Why are some people healthy and others not? The determinants of health of populations.* Walter de Gruyter, New York, NY.

Evans, R. G., Hodge, M. and Pless, I. B. (1994b) If not genetics, then what? Biological pathways and population health. In Evans, R. G., Barer, M. L. and Marmor, T. R. (eds), *Why are some people healthy and others not? The determinants of health of populations*, Walter de Gruyter, New York, NY. pp. 161–188.

Evans, R. G. and Stoddart, G. L. (1994) Producing health, consuming health care. In Evans, R. G., Barer, M. L. and Marmor, T. R. (eds), *Why are some people healthy and others not? The determinants of health of populations*, Walter de Gruyter, New York, NY. pp. 27–64.

Evans, W., Wolfe, B. and Adler, N. (2012) The SES and health gradient: A brief review of the literature. In Wolfe, B., Evans, W. and Seeman, T. E. (eds), *The biological consequences of socioeconomic inequalities*, Russell Sage, New York, NY. pp. 1–37.

Eversley, D. (1978) A question of numbers? In Bulmer, M. (ed), *Social policy research*, Macmillan, London. pp. 271–301.

Fair Trade International (2009) *A charter of fair trade principles*. World Fair Trade Organization and Fairtrade Labelling Organizations International. From http://fairtrade-advocacy.org/images/Charter_of_Fair_Trade_principles_EN_v1.2.pdf

Farquhar, J. W. (1984) The Stanford Five City Project: An overview. In Matarazzo, J. D. (ed), *Behavioural health: A handbook of health enhancement and disease prevention*, John Wiley, New York, NY.

Farquhar, J. W., Maccoby, N., Wood, P. D., Alexander, J. K., Breitrose, H., Brown, B. W., Haskell, W. L., McAlister, A. L., Meyer, A. J., Nash, J. D. and Stern, M. P. (1977) Community education for cardiovascular health. *The Lancet*, 1, 1192–1195.

Feacham, R. (1995) Valuing the past … Investing in the future: Evaluation of the National HIV/Aids Strategy 1993–94 to 1995–96. Commonwealth of Australia, Canberra, ACT.

Feacham, R. (2001) Globalisation is good for your health, mostly. *British Medical Journal*, 323, 504–506.

Feinstein, A. R. (1983) An additional basic science for clinical medicine: 11. The limitations of randomised trials. *Annals of Internal Medicine*, 99, 544–550.

Feldbaum, H., Lee, K. and Michaud, J. (2010) Global health and foreign policy. *Epidemiologic Reviews*, 32, 82–92.

Fenner, F. (1984) Smallpox, "the most dreadful scourge of the human species". Its global spread and eradication (second of two parts). *Medical Journal of Australia*, 141, 841–846.

Ferreira, F. H. G. (1997) *Economic transition and the distribution of income and wealth*. Office of the Chief Economist for East Asia and Pacific, Washington, DC.

Ferrie, J. E., Shipley, M. J., Marmot, M. G., Stansfeld, S. A. and Davey Smith, G. (1998) The health effects of major organisational change and job insecurity. *Social Science and Medicine*, 46, 243–254.

Ferrie, J. E., Westerlund, H., Virtanen, M., Vahtera, J. and Kivimäki, M. (2008) Flexible labor markets and employee health. *SJWEH Supplements*, 98–110.

Feuerstein, M. T. (1986) Partners in evaluation: Evaluating development and community programmes with participants. Macmillan, London.

Filmer, P., Phillipson, M., Silverman, D. and Walsh, D. (1972) *New directions in sociological theory*. Collier-Macmillan, London.

Fincher, R. (1990) Women in the city. *Australian Geographical Studies*, 28, 29–37.

Fincher, R. (2007) Space, gender and institutions in processes creating difference. *Gender, Place & Culture*, 14, 5–27.

Fischbach, R. L. and Herbert, B. (1997) Domestic violence and mental health: Correlates and conundrums within and across cultures. *Social Science and Medicine*, 45, 1161–1176.

Fisher, M. and Baum, F. (2010) The social determinants of mental health: implications for research and health promotion. *Australian and New Zealand Journal of Psychiatry*, 44, 1057–1063.

Fisher, M., Baum, F., MacDougall, C., Newman, L. and McDermott, D. (2014a) A qualitative methodological framework to assess uptake of evidence on SDH in health policy. *Evidence and Policy*. From http://dx.doi.org/10.1332/174426414X14170264741073

Fisher, M., Milos, D., Baum, F. and Friel, S. (2014b) Social determinants in an Australian urban region: A 'complexity' lens. *Health Promotion International*. 10.1093/heapro/dau071.

Flannery, T. (2005) The weather makers. The history and future impact of climate change. Text Publishing, Melbourne, VIC.

Flavin, C. (1996) Facing up to the risks of climate change. In Brown, L. R. (ed), *State of the World 1996*, Worldwatch Institute/Earthscan, London. pp. 21–39.

Flitcroft, K., Gillespie, J., Salkeld, G., Carter, S. and Trevena, L. (2011) Getting evidence into policy: The need for deliberative strategies? *Social Science and Medicine*, 72, 1039–1046.

Fogelman, K., Fox, A. J. and Power, C. (1989) Class and tenure mobility: Do they explain social inequalities in health among young adults in Britain? In Fox, J. (ed), *Health inequalities in European countries*, Gower, Aldershot, UK. pp. 333–352.

Foley, G. (1997) Trends over recent years, *Australian Occupational Health and Safety Statistics Bulletin No. 1*. National Workplace Statistics and Epidemiology Team, National Occupational Health and Safety Commission, Canberra, ACT.

Fontana, A. and Frey, J. H. (1994) Interviewing: The art of science. In Lincoln, Y. S. and Denzin, N. K. (eds), *Handbook of qualitative research*, SAGE Publications, Thousand Oaks, CA. pp. 361–376.

Food and Agriculture Organization (FOA) (2013) Synthesis of guiding principles on agriculture programming for nutrition. United Nations.

Fook, J. (ed) (1996) The reflective researcher: Social workers' theories of practice research. Allen & Unwin, Sydney, NSW.

Forster, C. A. (2000) Rising levels of disadvantage in Adelaide's outer south: A study of four postcodes 1991–1996. South Australian Department of Human Services, Adelaide, SA.

Foucault, M. (1973) The birth of the clinic: An archeology of medical perception. Pantheon, New York, NY.

Foucault, M. (1979) Discipline and punish: The birth of the prison. Penguin, London.

Foucault, M. (1984) Space knowledge and power. In Rabinow, P. (ed), *The Foucault reader: An introduction to Foucault's thought*, Pantheon Books, New York, NY.

Framework Convention on Global Health (FCGH) (2014) *Platform for a Framework Convention on Global Health (FCGH)*. From www.globalhealthtreaty.org/ (accessed 8 October 2014).

Franks, P. and Campbell, T. L. (1992) Social relationships and health: The relative role of family functioning and social support. *Social Science and Medicine*, 34, 779–788.

Freeman, H. (1992) The environment and mental health, *Streetwise—The Magazine of Urban Studies*, 11, 22–28.

Freidson, E. (1970) Professional dominance: The social structure of medical care. Aldine, Chicago.

Freire, P. (1972) *Pedagogy of the oppressed*. Penguin, Harmondsworth, UK.

French, H. (2000) Coping with ecological globalization. In Starke, L. (ed), *State of the World 2000*, Worldwatch Institute/WW Norton & Co, New York, NY.

French, J. and Adams, L. (1986) From analysis to synthesis. *Health Education Journal*, 45.

Freudenberg, N. (2014) Lethal but legal: Corporations, consumption, and protecting public health. Oxford University Press, New York, NY.

Fried, M. (1994) Life and death in the free zone. *New Internationalist*, July 16–18.

Friel, S. (2014) Climate change will widen the social and health gap. http://theconversation.com/climate-change-will-widen-the-social-and-health-gap-30105 (accessed 15 August 2014).

Friel, S., Akerman, M., Hancock, T., Kumaresan, J., Marmot, M., Melin, T., Vlahov, D. and GRNUHE members (2011) Addressing the social and environmental determinants of urban health equity: Evidence for action and a research agenda. *Journal of Urban Health*, 88, 860–874.

Friel, S., Chopra, M. and Satcher, D. (2007) Unequal weight: Equity oriented policy responses to the global obesity epidemic. *BMJ*, 335, 1241–1243.

Friel, S., Hattersley, L. and Townsend, R. (2014) Trade policy and public health. *Annual Review of Public Health*. 10.1146/annurev-publhealth-031914-122739.

Frumkin, H. (2002) Urban sprawl and health. *Public Health Reports*, 117.

Fry, D. and Baum, F. (1992) Keywords in community health. In Baum, F., Fry, D. and Lennie, I. (eds), *Community health: Policy and practice in Australia*, Pluto Press, Sydney, NSW. pp. 296–309.

Fry, D. and Furler, J. (2000) General practice, primary health care and population health interface, *General Practice in Australia 2000*. Commonwealth Department of Health and Aged Care, Canberra, ACT.

Fukuyama, F. (1992) *The end of history and the last man*. Avon Books, New York, NY.

Gagnon, M., Freudenberg, N. and Corporate Accountability International (2012) *Slowing down fast food: A policy guide for healthier kids and families*. Corporate Accountability International, Boston, MA.

Galbraith, J. K. (1994) *The world economy since the wars*. Mandarin, London.

Galea, S. and Tracy, M. (2007) Participation rates in epidemiologic studies. *Annals of Epidemiology*, 17, 643–653.

Galea, S. and Vlahov, D. (2005) Urban health: Evidence, challenges, and directions. *Annual Review of Public Health*, 26, 341–365.

Gallus, C. (1989) *Marion, Brighton and Glenelg community health needs assessment: Youth report*. South Australian Health Commission. Southern Community Health Services Research Unit, Morphett Vale, SA.

Ganz, M. (2009) Why stories matter. *Sojourners Magazine*, March. From www.sojo.net/magazine/2009/03/why-stories-matter

Gardner, M. J., Snee, M. P., Hall, A. J., Powell, C. A., Downes, S. and Terrell, J. D. (1990) Results of case-control study of leukaemia and lymphoma among young people near Sellafield nuclear plant in West Cumbria. *British Medical Journal*, 300, 423–429.

Garrett, L. (1994) The coming plague: Newly emerging diseases in a world out of balance. Penguin Books, New York.

Garvey, D. (2008) *Review of the social and emotional wellbeing of Indigenous Australian peoples – considerations, challenges and opportunities*. Australian Indigenous HealthInfoNet, Edith Cowan University, Perth, WA. From www.healthinfonet.ecu.edu.au/sewb_review

Gasfield Free Northern Rivers (n.d.) About the Gasfield Free Campaign. http://csgfreenorthernrivers.org/about-the-csg-free-campaign (accessed 28 July 2014).

Gaventa, J. (1988) Participatory research in North America. *Convergence*, 21, 17–46.

Geertz, C. (1973) *The interpretation of culture*. Basic Books, New York, NY.

Gelbspan, R. (1997) The heat is on: The high stakes battle over Earth's threatened climate. Addison Wesley Longman, Reading, MA.

Geneva Declaration Secretariat (2011) *Global Burden of Armed Violence (GBAV) 2011*. Geneva Declaration on Armed Violence and Development, Geneva.

Gerland, P., Raftery, A. E., Ševčíková, H., Li, N., Gu, D., Spoorenberg, T., Alkema, L., Foskick, B. K., Chunn, J., Lalic, N., Bay, G., Buettner, T., Heilig, G. K. and Wilmoth, J. (2014) World population stabilization unlikely this century. *Science*, 346, 234–237.

Germov, J. (2014) *Second opinion: An introduction to health sociology*. Oxford University Press, South Melbourne, VIC.

Geronimus, A. T., Bound, J. and Colen, C. G. (2011) Excess black mortality in the United States and in selected black and white high-poverty areas, 1980–2000. *American Journal of Public Health*, 101, 720–729.

GetUp! Australia (n.d.) Help stop Adani from destroying the Great Barrier Reef. www.getup.org.au/adani (accessed 30 July 2014).

Giddens, A. (1999) Runaway world: How globalisation is reshaping our lives. Profile, London.

Giles-Corti, B., Ryan, K. and Foster, S. (2012) *Increasing density in Australia: Maximising the health benefits and minimising harm*. National Heart Foundation of Australia, Melbourne, VIC.

Gill, T. (2011) Food industry digs in heels over traffic light labels. *The Conversation*, 30 March 2011. From http://theconversation.com/food-industry-digs-in-heels-over-traffic-light-labels-311

Gillespie, S., Egal, F. and Park, M. (2013) Agriculture, food and nutrition. In Leppo, K., Ollila, E., Pena, S., Wismar, M. and Cook, S. (eds), *Health in All Policies: Seizing opportunities, implementing policies*, Ministry of Social Affairs and Health, Helsinki, Finland. pp. 183–202.

Gillies, P. (1997) Social capital: Recognising the value of society, *Healthlines*, 15–17.

Gilson, L., Doherty, J., Loewenson, R. and Francis, V. (2007) Challenging inequity through health systems, *Final report of the Knowledge Network on Health Systems*. Commission on Social Determinants of Health, World Health Organization, Geneva.

Glaser, B. G. and Strauss, A. L. (1967) The discovery of grounded theory: Strategies for qualitative research. Aldine, Chicago.

Glasziou, P., Vandenbroucke, J. and Chalmers, I. (2004) Assessing the quality of research. *BMJ*, 328, 39–41.

Gleeson, D. (2013) What you need to know about the Trans Pacific Partnership. *The Conversation*, 9 December. From http://theconversation.com/what-you-need-to-know-about-the-trans-pacific-partnership-21168

Gleeson, D. and Friel, S. (2013) Emerging threats to public health from regional trade agreements. *The Lancet*, 381, 1507–1509.

Glesne, C. and Peshkin, A. (1992) *Becoming qualitative researchers: An introduction*. Longman, White Plains, NY.

Glibert, P. M., Anderson, D. M., Gentien, P., Graneli, E. and Sellner, K. G. (2005) The global, complex phenomena of harmful algal blooms. *Oceanography*, 18, 12.

Global Alliance for the Rights of Nature (2014) Universal Declaration of Rights of Mother Earth. From http://therightsofnature.org/universal-declaration (accessed 3 March 2015).

Global Burden of Disease 2013 Mortality and Causes of Death Collaborators (2015) Global, regional, and national age–sex specific all-cause and cause-specific mortality for 240 causes of death, 1990–2013: A systematic analysis for the Global Burden of Disease Study 2013. *The Lancet*, 385, 117–171.

Global Forum for Health Research (2004) *10/90 Report on Health Research 2003–2004*. In Davey, S. (ed), Global Forum for Health Research, Geneva.

Global Greengrants Fund (2010) Grassroots activism blocks GM eggplant in India, 25 February. From www.greengrants.org/2010/02/25/grassroots-activism-blocks-gm-eggplant-in-india

Global Health Watch (GHW) (n.d.) From www.ghwatch.org/about (accessed 24 July 2014).

Global Policy Forum (2010) Comparison of the world's 25 largest corporations with the GDP of selected countries. From www.globalpolicy.org/component/content/article/150-general/50950-comparison-of-the-worlds-25-largest-corporations-with-the-gdp-of-selected-countries.html (accessed 5 April 2014).

Global Research Network on Urban Health Equity (GRNUHE) (2010) Improving urban health equity through action on the social and environmental determinants of health: *Final Report of the GRNUHE*. University College London and the Rockefeller Foundation.

Global Wind Energy Council (GWEC) (2013) *Global Wind Report 2013*. GWEC, Brussels.

Globalization Knowledge Network (2007) Towards health-equitable globalisation: Rights, regulation and redistribution. *Final Report to the Commission on Social Determinants of Health*. Institute of Population Health, University of Ottawa, Canada.

Glover, J., Harris, K. and Tennant, S. (1999) *A Social Health Atlas of Australia, 2nd Edition*. Public Health Information Development Unit, University of Adelaide, Adelaide, SA.

Glover, J., Hetzel, D., Glover, L., Tennant, S. and Page, A. (2006) *A Social Health Atlas of South Australia, 3rd Edition*. Public Health Information Development Unit, University of Adelaide, Adelaide, SA.

Goffman, E. (1961) Asylums: Essays on the social situation of mental patients and other inmates. Penguin, Harmondsworth, UK.

Gold, R. L. (1958) Roles in sociological field observations. *Social Forces*, 36, 217–223.

Goldblatt, P. (1990) Mortality and alternative social classification. In Goldblatt, P. (ed), *Longitudinal study: Mortality and social organisation 1971–1981*, HMSO, London.

Goldenberg, S. (2013) America's first climate refugees. *The Guardian*, 13 May. From www.theguardian.com/environment/interactive/2013/may/13/newtok-alaska-climate-change-refugees

Goldman, E. and Galea, S. (2014) Mental health consequences of disasters. *Annual Review of Public Health*, 35, 169–183.

Goldsmith, O. (1770) *The deserted village*. W. Griffin, London.

Goodland, R. J. A., Daly, H. E. and El Serafy, S. (1992) *Population, technology, and lifestyle: the transition to sustainability*. Island Press, Washington, DC.

Goodwin, D. M., Cummins, S., Sautkina, E., Ogilvie, D., Petticrew, M., Jones, A., Wheeler, K. and White, M. (2013) The role and status of evidence and innovation in the healthy towns programme in England: A qualitative stakeholder interview study. *Journal of Epidemiology and Community Health*, 67, 106–112.

Gove, W. R. and Hughes, M. (1979) Possible causes of the apparent sex differences in physical health: an empirical investigation. *American Sociological Review*, 44, 126–146.

Grace, V. M. (1991) The marketing of empowerment and the construction of the health consumer: A critique of health promotion. *International Journal of Health Services*, 21, 329–343.

Graham, H. (1984) *Women, health and family*. Wheatsheaf Books, Brighton, UK.

Graham, H. (1987) Women's smoking and family health. *Social Science and Medicine*, 25, 47–56.

Graham, H. (1994) Gender and class as dimensions of smoking behaviour in Britain: Insights from a survey of mothers. *Social science and medicine*, 38, 691–698.

Graham, H. (ed) (2000) *Understanding health inequalities*. Open University Press, Buckingham.

Graham, H. (2007) *Unequal lives*: Health and socioeconomic inequalities. Open University Press, Maidenhead.

Graham, S., Guy, R. J., Cowie, B., Wand, H. C., Donovan, B., Akre, S. P. and Ward, J. S. (2013) Chronic hepatitis B prevalence among Aboriginal and Torres Strait Islander Australians since universal vaccination: A systematic review and meta-analysis. *BMC infectious diseases*, 13, 1–10.

Gramsci, A. (1978) Selections from the prison notebooks of Antonio Gramsci: Ed. and transl. by Quintin Hoare and Geoffrey Nowell Smith, International Publishers, New York, NY.

Gray, J. (2001) The era of globalisation is over. *New Statesman*, 25–27.

Grbich, C. (1999) *Qualitative research in health. An introduction*. Allen & Unwin, St Leonards, NSW.

Green, J. (2005) *Nuclear power: No solution to climate change*. Australian Conservation Foundation, Carlton, VIC.

Green, L. W. and Raeburn, J. M. (1988) Health promotion. What is it? What will it become? *Health Promotion International*, 3, 151–159.

Groome, H. (1995) Towards improved understandings of Aboriginal young people. *Youth Studies Australia*, Summer, 17–21.

Grossman, J. and Webb, K. (1991) Local food and nutrition policy. *Australian Journal of Public Health*, 15, 271–276.

Grossman, P., Niemann, L., Schmidt, S. and Walach, H. (2004) Mindfulness-based stress reduction and health benefits: A meta-analysis. *Journal of Psychosomatic Research*, 57, 35–43.

Guba, E. G. and Lincoln, Y. S. (1994) Competing paradigms in qualitative research. In Denzin, N. K. and Lincoln, Y. S. (eds), *Handbook of qualitative research*, SAGE Publications, Thousand Oaks, CA. pp. 105–117.

Haavio-Mannila, E. (1986) Inequalities in health and gender. *Social Science and Medicine*, 22, 141–149.

Hall, S. (1990) Cultural identity and diaspora. In Rutherford, J. (ed), *Identity: Community, culture, difference*, Lawrence & Wishart, London.

Halpern, D. (2005) *Social capital*. Polity Press, Cambridge.

Halstead, S. B., Walsh, J. A. and Warren, K. S. (1985) *Good health at low cost*. The Rockefeller Foundation, New York, NY.

Hamilton, C. (2002) Overconsumption in Australia: The rise of the middle-class battler. Discussion Paper Number 49: The Australia Institute.

Hamilton, C. (2003) *Growth fetish*. Allen & Unwin, Sydney, NSW.

Hamilton, C. (2007) *Scorcher. The dirty politics of climate change*. Black Inc, Melbourne, VIC.

Hamilton, C. and Denniss, R. (2005) *Affluenza*. Allen & Unwin, Sydney, NSW.

Hancock, T. (1986) Lalonde and beyond: Looking back at "A New Perspective on the Health of Canadians". *Health Promotion International*, 1, 93–100.

Hancock, T. (1992) The healthy city: Utopias and realities. In Ashton, J. (ed), *Healthy cities*, Open University Press, Buckingham. pp. 22–29.

Hancock, T. (1994) A healthy and sustainable community: The view from 2020. In Chu, C. and Simpson, R. (eds), *Ecological public health: From vision to practice*, Institute of Applied Environmental Research, Griffith University, Nathan, QLD. pp. 245–253.

Hancock, T. and Duhl, L. (1986) Healthy cities: Promoting health in the urban context, *WHO Healthy Cities Paper No.1*. FADL Publishers, Copenhagen.

Hannes, K. and Lockwood, C. (2011) *Synthesizing qualitative research: Choosing the right approach*. John Wiley & Sons, Hoboken.

Hansen, J. and Sato, M. (2012) Update of greenhouse ice sheet mass loss: Exponential?

Harden, A. (2001) The fine detail: Conducting a systematic review. In Oliver, S. and Peersman, G. (eds), *Using research for effective health promotion*, Open University Press, Buckingham. pp. 111–122.

Harding, A. (2005) *Recent trends in income inequality in Australia*, Presentation to the Conference on Sustaining Prosperity: New Reform Opportunities for Australia, 31 March. NATSEM, University of Canberra, ACT.

Harding, M. (1987) *The relationship between economic status and health status: A synthesis*. Ontario Social Assistance Review Committee, Toronto, Canada.

Hare, W. L., Marlow, J. P., Australian Conservation Foundation, Greenpeace Australia, Wilderness Society (Australia) and World Wide Fund for Nature (Australia) (1991 [1990]) *Ecologically sustainable development: a submission*. Australian Conservation Foundation, Fitzroy, VIC.

Harman, E. (1988) Capitalism, patriarchy and the city. In Baldock, C. and Cass, B. (eds), *Women, social welfare and the state in Australia*, Allen & Unwin, Sydney, NSW.

Harris, M. and Lloyd, J. (2012) The role of Australian primary health care in the prevention of chronic disease. Australian National Preventive Health Agency, Canberra, ACT.

Harrison, J. (2015) *Deaths from motor vehicle accidents, 1940–2014*. Research Centre for Injury Studies, Flinders University, Adelaide, SA.

Harris-Roxas, B. (2011) Health impact assessment in the Asia Pacific. *Environmental Impact Assessment Review*, 31, 393–395.

Hart, N. (1991) Social, economic, and cultural environment and human health. In Detels, R., McEwan, J., Beaglehole, R. and Tanaka, H. (eds), *Oxford textbook of public health*, 2nd edition, Oxford University Press, London.

Harvey, D. (2005) *A brief history of neoliberalism*. Oxford University Press, Oxford.

Harvey, D. (2010) The enigma of capital and the crises of capitalism. Profile Books, London.

Harvey, D. (2014) Seventeen contradictions and the end of capitalism. Profile Books, London.

Hawe, P. (1994) Measles control: A best practice challenge in public health. *Australian Journal of Public Health*, 18, 241–243.

Hays, G. C., Richardson, A. J. and Robinson, C. (2005) Climate change and marine plankton. *Trends in Ecology and Evolution*, 20, 337–344.

Healthy Alberta Canada (2015) Creating a sense of place in your community. From www .healthyalberta.com/688.htm (accessed 8 January 2015).

Healthy Policies (n.d.). www.healthypolicies.com (accessed 26 July 2014).

Healthy Skepticism (n.d.) www.healthyskepticism.org/global (accessed 26 July 2014).

Held, D. and McGrew, A. (2007) *Globalization/Anti-globalization: Beyond the great divide*. Polity Press, Cambridge.

Henry, J. S. (2012) The price of offshore revisited: New estimates for "missing" global private wealth, income, inequality, and lost taxes. Tax Justice Network.

Herzlich, C. (1973) *Health and illness*. Academic Press, London.

Hetzel, B. S. (1976) *Health and Australian society*. Penguin, Ringwood, VIC.

Hetzel, B. S. and McMichael, T. (1989) *The L S factor: Lifestyle and health*. Penguin, Ringwood, VIC.

Heymann, J. (2006) Forgotten families: Ending the crisis confronting children and working parents in the global economy. Oxford University Press, New York, NY.

Heywood, M. and Altman, D. G. (2000) Confronting AIDS: Human rights, law, and social transformation. *Health and Human Rights*, 5, 149–179.

Hibbs, J. R., Benner, L., Klugman, L., Spencer, R., Macchia, I., Mellinger, A. K. and Fife, D. (1994) Mortality in a cohort of homeless adults in Philadelphia. *New England Journal of Medicine*, 331, 304–309.

Higgins, D. and Scott, D. (2012) Child abuse and neglect in Australia's Northern Territory: The Northern Territory Emergency Response. In Dubowitz, H. (ed), *World perspectives on child abuse*, 10th edition, International Society for the Prevention of Child Abuse and Neglect, Aurora, Colorado.

Hill, D. (2005) *The public policy process*. Pearson Longman, Harlow, Essex.

Hill, D. and Marks, R. (2008) Health promotion programs for melanoma prevention: Screw or spring? *Archives of Dermatology*, 144, 538–540.

Hillman, M. (1991) Healthy transport policy. In Draper, P. (ed), *Health through public policy*, Green Print, London. pp. 82–91.

Hinde, S. (2007) The vehicle that drives obesity. In Dixon, J. and Broom, D. (eds), *The 7 deadly sins of obesity: How the modern world is making us fat*, University of NSW Press, Sydney. pp. 82–100.

Hinscliff, G. (2006) Ten years to save the planet from mankind, *Observer*. p. 22.

Hocking, B., Gordon, I. R., Grain, H. L. and Hatfield, G. E. (1996) Cancer incidence and mortality and proximity to TV towers. *Medical Journal of Australia*, 165, 601–605.

Hofrichter, R. (ed) (2003) Health and social justice: A reader on politics, ideology and inequity in the distribution of disease. Jossey-Bass, San Francisco, CA.

Holman, R. (1991) *The ethics of social research*. Longman, London.

Holmes, N. and Gifford, S. M. (1997) Narratives of risk in occupational health and safety: Why the "good" boss blames his tradesman and the "good" tradesman blames his tools. *Australian and New Zealand Journal of Public Health*, 21, 11–16.

Homedes, N. and Ugalde, A. (2005) Why neoliberal health reforms have failed in Latin America. *Health Policy*, 71, 83–96.

Hospitals and Health Services Commission (1973) *A community health program for Australia*. Commonwealth of Australia, Canberra, ACT.

Hospitals and Health Services Commission (1976) *Review of the community health program*. Commonwealth of Australia, Canberra, ACT.

House, J. S., Landis, K. R. and Umberson, D. (1988) Social relationships and health. *Science*, 214, 540–545.

House, J. S., Robbins, C. and Metzner, H. L. (1982) The association of social relationships and activities with mortality: Prospective evidence from the Tecumseh Community Health Study. *American Journal of Epidemiology*, 116, 123–140.

House of Representatives Standing Committee on Family and Community Affairs (2000) *Health is life: Report on the inquiry into Indigenous health*. Parliament of the Commonwealth of Australia, Canberra, ACT.

Howard, B. (1996) The World Health Organization and its critics. *Current Affairs Bulletin*, June/July, 23–25.

Howard-Grabman, L. (1996) Planning together: Developing community plans to address priority maternal and neonatal health problems in rural Bolivia. In de Koning, K. and Martin, M. (eds), *Participatory research in health: Issues and experiences*, Zed Books, London. pp. 153–163.

Hoy, W., Norman, R. J., Hayhurst, B. G. and Pugsley, D. J. (1997) A health profile of adults in a Northern Territory Aboriginal community, with an emphasis on preventable morbidities. *Australian and New Zealand Journal of Public Health*, 21, 121–126.

Hseih, Y. H. and Chen, C. W. S. (2009) Turning points, reproduction number, and impact of climatological events for multi-wave dengue outbreaks. *Tropical Medicine and International Health*, 14, 628–629.

Hughes, D. (1996) Coping with contracting: The implications of the contract culture on community service organisations. *Community Quarterly*, 41, 37–41.

Hughes, M. (2013) Cuts to community-based health services short-sighted, *Croakey*. Crikey.

Hughes, P., Bellamy, J. and Black, A. (2000) Building social trust through education. In Winter, I. (ed), *Social capital and public policy in Australia*, Australian Institute of Family Studies, Melbourne, VIC.

Human Rights and Equal Opportunity Commission (HREOC) (1989) *Our Homeless Children: Report of the National Inquiry into Homeless Children*. Australian Government Publishing Service, Canberra, ACT.

Human Rights and Equal Opportunity Commission (HREOC) (1997) Bringing Them Home: National Inquiry into the Separation of Aboriginal and Torres Strait Islander Children from Their Families. HREOC, Sydney, NSW.

Humphrey, L. (1970) *Tea-room trade*. Aldine, Chicago.

Hunt, L. M., Jordan, B., Irwin, S. and Browner, C. H. (1989) Compliance and the patient's perspective: Controlling symptoms in everyday life. *Culture, Medicine and Society*, 13, 315–334.

Hunt, S. M. (1987) Evaluating a community development project: Issues of acceptability. *British Journal of Social Work*, 17, 661–667.

Hunt, S. M. (1989) The public health implications of private cars. In Martin, C. J. and McQueen, D. V. (eds), *New public health*, Edinburgh University Press, Edinburgh. pp. 100–115.

Hunt, S. M., McEwan, J. and McKenna, S. P. (1986) *Measuring health status*. Croom Helm, London.

Hunter, D. J., Popay, J., Tannahill, C. and Whitehead, M. (2010) Getting to grips with health inequalities at last? *British Medical Journal*, 340, 323–324.

Hunter, E. (1993) *Aboriginal health and history*. Cambridge University Press, Cambridge.

Hunter, P., Mayers, N., Couzos, S., Daniels, J., Murray, R., Bell, K., Kehoe, H., Brice, G. and Tynan, M. (2005) Aboriginal community controlled health services. In *General Practice in Australia 2004*. Commonwealth Department of Health and Ageing, Canberra, ACT. pp. 337–356.

Hurley, S. F. and Matthews, J. P. (2008) Cost-effectiveness of the Australian national tobacco campaign. *Tobacco Control*, 17, 379–384. 10.1136/tc.2008.025213.

Hutton, W. (1995) *The state we're in*. Jonathan Cape, London.

Huxley, M. (1994) Space, knowledge, power and gender. In Johnson, L. C. (ed), *Suburban dreaming: An interdisciplinary approach to Australian cities*, Deakin University Press, Geelong, VIC. pp. 181–192.

Ifanti, A. A., Argyriou, A. A., Kalofonou, F. H. and Kalofonos, H. P. (2013) Financial crisis and austerity measures in Greece: their impact on health promotion policies and public health care. *Health Policy*, 113, 8–12.

Ife, J. (1995) Community development. Creating community alternatives: Vision, analysis and practice. Longman, Melbourne, VIC.

Ife, J. (2001) Human rights and social work: Towards rights-based practice. Cambridge University Press, Cambridge.

Ife, J. and Tesoriero, F. (2006) Community development: Community-based alternatives in an age of globalisation. Pearson, Melbourne, VIC.

Illsley, R. (1986) Occupational class, selection and the production of inequalities in health. *Quarterly Journal of Social Affairs*, 2, 151–165.

Inner Activists (n.d.) From www.inneractivist.com (accessed 27 July 2014).

Intergovernmental Panel on Climate Change (IPCC) (2007) Summary for policymakers. In Solomon, S., Qin, D., Manning, M., Chen, Z., Marquis, M., Averyt, K. B., Tignor, M. and Miller, H. L. (eds), *Climate change 2007: The physical science basis. Contribution of Working Group I to the Fourth Assessment Report of the Intergovernmental Panel on Climate Change*. Cambridge University Press, Cambridge.

Intergovernmental Panel on Climate Change (IPCC) (2012) Managing the risks of extreme events and disasters to advance climate change adaption. A Special Report of Working Groups

I and II of the Intergovernmental Panel on Climate Change. In Field, C. B., Barros, V., Stocker, T. F., Qin, D., Dokken, D. J., Ebi, K. L., Mastrandrea, M. D., Mach, K. J., Plattner, G.-K., Allen, S.K., Tignor, M. and Midgley, P. M. (eds). Cambridge University Press, New York, NY.

Intergovernmental Panel on Climate Change (IPCC) (2013) Climate change 2013: The physical science basis. Contribution of Working Group I to the Fifth Assessment Report of the Intergovernmental Panel on Climate Change. Cambridge University Press, New York.

Intergovernmental Panel on Climate Change (IPCC) (2014) Climate change 2014: Synthesis report, Contribution of Working Groups I, II and III to the Fifth Assessment Report of the Intergovernmental Panel on Climate Change [Core Writing Team, RK Pachauri and LA Meyer (eds.)]. IPCC, Geneva.

International Council for Local Environmental Initiatives (ICLEI). (1996) Planning elements for Local Agenda 21. *Local Agenda 21 Network News*, 4, 1–4.

International Labour Organization (ILO) (2015) Decent work agenda: Promoting decent work for all. From www.ilo.org/global/about-the-ilo/decent-work-agenda/lang--en/index.htm (accessed 23 January 2015).

International Monetary Fund (IMF) (2014) Debt relief under the Heavily Indebted Poor Countries (HIPC) initiative. *IMF Factsheet*. From www.imf.org/external/np/exr/facts/pdf/hipc.pdf

Irwin, A. and Scali, E. (2005) Action on the social determinants of health: learning from previous experiences. World Health Organization, Geneva.

Irwin, L. G., Siddiqi, A. and Hertzman, C. (2007) Early child development: A powerful equalizer. HELP, University of British Columbia, Vancouver, BC.

Islam, M. K., Merlo, J., Kawachi, I., Lindstrom, M. and Gerdtham, U.-G. (2006) Social capital and health: Does egalitarianism matter? A literature review. *International Journal of Equity in Health*, 5, 28.

Israel, B. A., Checkoway, B., Schulz, A. and Zimmerman, M. A. (1994) Health education and community empowerment: Conceptualizing and measuring perceptions of individual, organizational and community control. *Health Education Quarterly*, 21, 149–170.

Ivanova, M. (2014) Assessing the outcomes of Rio+20. In Mastny, L. (ed), *State of the World 2014: Governing for sustainability*, Chapter 13, Worldwatch Institute/Island Press, Washington, DC. pp. 138–151.

Jackson, T. (2009) Prosperity without growth: Economics for a finite planet. Earthscan, London.

Jackson, T., Mitchell, S. and Wright, M. (1989) The community development continuum. *Community health studies*, 13, 66–73.

Jacobs, J. (1961) *The death and life of great American cities*. Random House, New York, NY.

Jacobs, M. (1991) *The green economy*. Pluto Press, London.

James, C. (2009) Economic insights: Australian homes are biggest in the world. CommSec.

James, O. (2008) *The selfish capitalist*. Vermilion, London.

Janesick, V. (1994) The dance of qualitative research design: Metaphor, methodolatry, and meaning. In Denzin, N. K. and Lincoln, Y. S. (eds), *Handbook of qualitative research*, SAGE Publications, Thousand Oaks, CA. pp. 209–219.

Jarlais, D. C. D., Simtson, G. V., Hagan, H., Perlman, D., Choopanya, K., Bastos, F. I. P. M. and Friedman, S. R. (1996) Emerging infectious diseases and the injection of illicit psychoactive drugs. *Current Issues in Public Health*, 2, 130–137.

Jelinek, M. (1993) The clinician and the randomised controlled trial. In Daly, J. (ed), *Researching health care: Designs, dilemmas, disciplines*, Tavistock/Routledge, London. pp. 76–89.

Jervis-Bardy, J., Sanchez, L. and Carney, A. (2014) Otitis media in Indigenous Australian children: Review of epidemiology and risk factors. *The Journal of Laryngology and Otology*, 128, S16–S27.

Jezet, Z. (1987) *Ten years without smallpox*. WPRO Information Unit, WHO Western Pacific Regional Office, Manila.

Johnson, A. and Baum, F. (2001) Health promoting hospitals: A typology of different organizational approaches to health promotion. *Health Promotion International*, 16, 281–287.

Johnson, A. and Paton K. (2007) *Health promotion and health services*. Oxford University Press, Melbourne, VIC.

Joint United Nations Programme on HIV/AIDS (UNAIDS) (2006) *Aids Epidemic Update: Global summary*. UNAIDS, Geneva.

Joint United Nations Programme on HIV/AIDS (UNAIDS) (2013) *Global report: UNAIDS report on the global AIDS epidemic, 2013*. UNAIDS, Geneva.

Joint United Nations Programme on HIV/AIDS (UNAIDS) (2014) *Fast track: Ending the AIDS epidemic by 2030*. UNAIDS, Geneva.

Jones, M. (1995–96) *Healthy settings—healthy scepticism*. Health for All Network (UK), Liverpool.

Jorgensen, D. I. (1989) *Participant observation*. SAGE Publications, Newbury Park, CA.

Joubert, P. N. (1979) *Development and effects of seat belt laws in Australia*. Department of Mechanical Engineering, University of Melbourne, Melbourne, VIC.

Judge, K. and Bauld, L. (2001) Strong theory, flexible methods: Evaluating complex community-based initiative. *Critical Public Health*, 11, 19–38.

Kabat-Zinn, J. (1990) Full catastrophe living: Using the wisdom of your body and mind to face stress, pain and illness. Delacorte, New York, NY.

Kaldor, J. (1996) Feachem's report on Australia's National HIV/AIDS Strategy [Editorial]. *Australian Journal of Public Health*, 20, 342–343.

Kale, D. (2012) Transnational corporations: Significance and impacts. In Papaioannou, T. and Butcher, M. (eds), *International development in a changing world*, Bloomsbury Academic.

Kalucy, E. and Baum, F. (1992) The epidemiology of caring: The pattern in a South Australian suburban population. *Australian Journal of Ageing*, 11, 3–8.

Kaluzny, A. D. and Hernandez, S. R. (1988) Organization change and function. In Shortell, S. M. and Kaluzny, A. D. (eds), *Health care management: A text in organization theory and behavior*, 2nd edition, Wiley, New York, NY. pp. 379–417.

Kane, I. (1991) *Women's health*. Macmillan, London.

Kaplan, G. A., Salonen, J. T., Cohen, R. D., Brand, R. J., Syme, S. L. and Puska, P. (1988) Social connections and mortality from all causes and from cardiovascular disease: prospective evidence from eastern Finland. *American Journal of Epidemiology*, 128, 370–380.

Karmel, T., Misko, J., Blomberg, D., Bednarz, A. and Atkinson, G. (2014) Improving labour market outcomes through education and training, *Issues paper no. 9. Produced for the Closing the Gap Clearinghouse*. Australian Institute of Health and Welfare and Australian Institute of Family Studies, Canberra, ACT.

Katz, A. (2008) Chernobyl: The great cover-up. *Le Monde diplomatique*, April. From http://mondediplo.com/2008/04/14who#nh7

Kawachi, I. (2007) Social capital and cohesion as community health assets, *Health assets and the social determinants of health*. WHO European Office for Health Investment, Vienna.

Kawachi, I. and Berkman, L. F. (2003) Introduction. In Kawachi, I. and Berkman, L. F. (eds), *Neighbourhoods and health*, Oxford University Press, Oxford. pp. 1–19.

Kawachi, I., Colditz, G. A., Ascherio, A., Rimm, E. B., Giovannucci, E., Stampfer, M. J. and Willett, W. C. (1996) A prospective study of social networks in relation to total mortality and cardiovascular disease in men in the USA. *Journal of Epidemiology and Community Health*, 50, 245–251.

Kawachi, I., Kennedy, B. P., Lochner, K. and Prothrow-Stith, D. (1997) Social capital, income inequality, and mortality. *American Journal of Public Health*, 87, 1491–1498.

Kearns, G. (1988) Private property and public health reform in England 1830–70. *Social Science and Medicine*, 26, 187–199.

Keirse, M. J. N. C. (1988) Amniotomy or oxytocin for induction of labor: Re-analysis of a randomised controlled trial. *Acta Obstetricia et Gynecologica Scandinavica*, 67, 731–735.

Keirse, M. J. N. C. (1994) Electronic monitoring: Who needs a trojan horse? *Birth*, 21, 111–113.

Keleher, H. (2013) Policy scorecard for gender mainstreaming: Gender equity in health policy. *Australian and New Zealand Journal of Public Health*, 37, 111–117.

Kellehear, A. (1993) *The unobtrusive researcher: A guide to method*. Allen & Unwin, St Leonards, NSW.

Keller, E. F. and Longino, H. E. (eds) (1996) Feminism and science. Oxford University Press, Oxford.

Kelly, M. and Swann, C. (2004) Foreword. In Dixon-Woods, M., Agarwal, S., Young, B., Jones, D. and Sutton, A. (eds), *Integrative approaches to qualitative and quantitative evidence*, Health Development Agency, London.

Kelly, M. P., Bonnejoy, J., Morgan, A. and Florenzano, F. (2006) *The development of the evidence base about the social determinants of health*. WHO Commission on Social Determinants of Health (CSDH) Measurement and Evidence Knowledge Network (MEKN), Geneva.

Kelsey, J. (1995) *Economic fundamentalism*. Pluto Press, London.

Kemmis, S. and McTaggart, R. (1988) *The action research planner*. Deakin University, Melbourne, VIC.

Kennedy, A. (1995) Measuring health for all: A feasibility study in a Glasgow community. In Bruce, N., Springett, J., Hotchkiss, J. and Scott-Samuel, A. (eds), *Research and change in urban community health*, Avebury, Aldershot. pp. 199–217.

Kenny, S. (1999) Developing communities for the future: Community development in Australia. Nelson, Melbourne, VIC.

Kenworthy, J. and Inbakaran, C. (2011) Differences in transport and land use in thirteen comparable Australian, American, Canadian and European cities between 1995/6 to 2005/6 and their implications for more sustainable transport, *34th Australasian Transport Research Forum (ATRF), 28–30 September 2011*, Adelaide, SA.

Kenworthy, J. R., Laube, F. B., Raad, T., Poboon, C. and Guia, B. (1999) *An international sourcebook of automobile dependence in cities 1960–1990*. University Press of Colorado, Boulder, CO.

Kiama Municipal Council (2011) *Kiama Health Plan 2011–2017 (updated June 2013)*. Kiama Municipal Council, Kiama, NSW.

Kickbusch, I. (1996) Tribute to Aaron Antonovsky, "What creates health?" *Health Promotion International*, 11, 5–6.

Kickbusch, I. (2003) The contribution of the World Health Organization to a New Public Health and Health Promotion. *American Journal of Public Health*, 93, 383–388.

Kickbusch, I. (2006) Mapping the future of public health: Action on global health. *Canadian Journal of Public Health/Revue Canadienne de Sante'e Publique*, 97, 6–8.

Kickbusch, I. (2007) The move towards a new public health. *Promotion and Education*, 14, 9.

Kickbusch, I. (2015) Healthy cities and settings for health. From www.ilonakickbusch.com/kickbusch/healthy-cities-and-settings-for-health/index.php (accessed 21 January 2015).

Kickbusch, I. and Cassar Szabo, M. M. (2014) Acta Obstetricia et Gynecologica Scandinavica. *Global Health Action*, 7. http://dx.doi.org/10.3402/gha.v7.23507.

Kickbusch, I. and Seck, B. (2007) Global public health. In Douglas, S., Earle, S., Handsley, S., Lloyd, C. E. and Spurr, S. (eds), *A reader in promoting public health*, SAGE Publications, London.

Kim, I.-H., Muntaner, C., Vahid Shahidi, F., Vives, A., Vanroelen, C. and Benach, J. (2012) Welfare states, flexible employment, and health: A critical review. *Health Policy*, 104, 99–127.

Kingdom, J. E. (1992) *No such thing as society? Individualism and community*. Open University Press, Buckingham.

Kingdon, J. W. (2011) *Agendas, alternatives, and public policies*. Longman, New York, NY.

Kingsley, G. T. (2003) Housing, health, and the neighbourhood context. *American Journal of Preventive Medicine*, 24, 6–7.

Kingsley, P. (2012) Participatory democracy in Porto Alegre. *The Guardian*, 11 September. From www.theguardian.com/world/2012/sep/10/participatory-democracy-in-porto-alegre

Kingsolver, B. (1998) *The Poisonwood Bible*. Faber & Faber, London.

Kirby, M. (2014) AIDS 2014 Opening Addresses, Jonathan Mann Memorial Lecture, 20th International AIDS Conference, Melbourne, 20–25 July. From www.aids2014.org/WebContent/File/AIDS2014_Opening_Addresses_Michael_Kirby.pdf

Kirke, J. and Miller, M. (1986) *Reliability and validity in qualitative research*. SAGE Publications, Newbury Park, CA.

Kirke, K. (1995) A state public and environmental health authority. In Baum, F. (ed), *Health for All: The South Australian experience*, Wakefield Press, Adelaide, SA. pp. 242–252.

Kjellstrom, T. and McMichael, A. J. (2013) Climate change threats to population health and well-being: The imperative of protective solutions that will last. *Global Health Action*, 6.

Klein, N. (2001) *No logo*. Flamingo, London.

Klein, N. (2011) Capitalism vs. the Climate. *The Nation*, 28 November. From www.thenation.com/article/164497/capitalism-vs-climate

Klos, D. M. and Rosenstock, I. M. (1982) Some lessons from the North Karelia Project. *American Journal of Public Health*, 72, 53–54.

Koivusalo, M., Labonté, R., Wibulpolprasert, S. and Kanchanachitra, C. (2013) Globalization and national policy space for health and a HiAP approach. In Leppo, K., Ollila, E., Peña, S., Wismar, M. and Cook, S. (eds), *Health in All Policies: Seizing opportunities, implementing policies*, Ministry of Social Affairs and Health, Helsinki, Finland. pp. 81–102.

Koivusalo, M. and Tritter, J. (2014) "Trade Creep" and implications of the Transatlantic Trade and Investment Partnership Agreement for the United Kingdom National Health Service. *International Journal of Health Services*, 44, 93–111.

Korpi, W. and Palme, J. (1998) The paradox of redistribution and strategies of equality: Welfare state institutions, inequality, and poverty in the Western countries. *American Sociological Review*, 63, 661–687.

Korten, D. (1995) *When corporations rule the world*. Earthscan, London.

Korten, D. (1996) "Development" is a sham. *New Internationalist*, 278, 12–13.

Korten, D. (2000) The post-corporate world: Life after capitalism. Pluto Press, Sydney, NSW.

Korten, D. (2006) *The great turning: From empire to earth community*. Berrett-Koehler Publishers, San Francisco, CA.

Korten, D. (2015) Change the story, change the future: A living economy for a living earth. Berrett-Koehler Publishers, Oaklands, CA.

Koutroulis, G. (1990) The orifice revisited: Women in gynaecological texts. *Community health studies*, 14, 73–84.

Koutsoukis, J. (2014) Delhi dirtiest city on earth, says World Health Organisation. *The Sydney Morning Herald*, 10 May. From www.smh.com.au/world/delhi-dirtiest-city-on-earth-says-world-health-organisation-20140509-zr8py.html

Krieger, N. (1994) Epidemiology and the web of causation: Has anyone seen the spider? *Social Science and Medicine*, 39, 887–903.

Krieger, N. (2000) Passionate epistemology, critical advocacy and public health: Doing our profession proud. *Critical Public Health*, 10, 287–294.

Krieger, N. (2011) *Epidemiology and the people's health: Theory and context*. Oxford University Press, New York, NY.

Kroeger, A. and Franken, H. P. (1981) The educational value of participatory evaluation of primary health care programmes: An experience with four indigenous populations in Ecuador. *Social science and medicine*, 15, 535–539.

Krueger, R. A. and Casey, M. A. (2009) *Focus groups: A practical guide for applied research*. SAGE Publications.

Krug, E., Dahlberg, L. L., Mercy, J. A., Zwi, A. B. and Lozano, R. (2002) *World Report on Violence and Health*. World Health Organization, Geneva.

Kubiszewski, I., Costanza, R., Franco, C., Lawn, P., Talberth, J., Jackson, T. and Aylmer, C. (2013) Beyond GDP: Measuring and achieving global genuine progress. *Ecological Economics*, 93, 57–68.

Kunitz, S. (2001) Accounts of social capital. In Leon, D. and Walt, G. (eds), *Poverty, inequality and health: An international perspective*, Oxford University Press, Oxford.

Kuzel, A. J. (1992) Sampling in qualitative inquiry. In Miller, B. F. C. W. L. (ed), *Doing qualitative research*, SAGE Publications, Newbury Park, CA. pp. 31–44.

Labonté, R. (1990) Empowerment: Notes on professional and community dimensions. *Canadian Review of Social Policy*, 26, 1–12.

Labonté, R. (1992) Heart health inequalities in Canada: Models, theory and planning. *Health Promotion International*, 7, 119–127.

Labonté, R. (1997) Power, participation and partnerships for health promotion. VicHealth, Melbourne, VIC.

Labonté, R. (1999) Brief to the World Trade Organization: World trade and population health. *Promotion and Education*, 6, 24–32.

Labonté, R. (2001) Amended brief to the Genoa Non-Governmental (GNG) Initiative on International Governance and World Trade Organization (WTO) reform. International Union for Health Promotion and Education.

Labonté, R. (2012) The austerity agenda: How did we get here and where do we go next? *Critical Public Health*, 22, 257–265.

Labonté, R., Baum, F. and Sanders, D. (2015) Poverty, justice and health. In Detels, R., Tan, C. C. and Karim, Q. A. (eds), *Oxford textbook of public health*, 6th edition, Oxford University Press, Oxford.

Labonté, R. and Laverack, G. (2008) *Health promotion in action: From local to global empowerment*. Palgrave Macmillan, London.

Labonté, R., Mohindra, K. and Schrecker, T. (2011) The growing impact of globalization for health and public health practice. *Annual Review of Public Health*, 32, 263–283.

Labonté, R. and Penfold, S. (1981) Canadian perspectives in health promotion: A critique. *Health Education*, 19, 4–9.

Labonté, R. and Schrecker, T. (2006) Globalization and social determinants of health: Analytic and strategic review paper, *Globalization Knowledge Network*. University of Ottawa, Ottawa, Canada.

Labonté, R. and Schrecker, T. (2007) Globalization and social determinants of health: Introduction and methodological background. *Globalization and Health*, 3, 5. 10.1186/1744-8603-3-5.

Lagarde, C. (2013) Managing Director IMF: A New Global Economy for a New Generation. January 23. From www.imf.org/external/np/speeches/2013/012313.htm (accessed 4 November 2014).

Laing, R. D. (1982) The voice of experience: Experience, science and psychiatry. Allen Lane, London.

Lalonde, M. (1974) *A new perspective on the health of Canadians*. Ministry of National Health and Welfare, Ottawa, Canada.

Lang, J. (1987) Creating architectural theory: The role of behavioural science in environmental design. Van Nostrand Reinhold, New York, NY.

Lang, T. (1999) The new GATT round: Whose development? Whose health? *Journal of Epidemiology and Community Health*, 53, 681–682.

Langlieb, A. M., Cooper, C. P. and Gielen, A. (1999) Linking health promotion with entertainment television. *American Journal of Public Health*, 89, 1116–1117.

Laris, P. (1995) Boards of directors of community health centres. In Baum, F. (ed), *Health for All: The South Australian experience*, Wakefield Press, Adelaide, SA. pp. 82–92.

Laris, P., Baum, F., Schaay, N., Sanders, D. and Kahssay, H. (2001) *Tapping into civil society: Guidelines for linking health systems with civil society*. South Australian Community Health Research Unit, Adelaide, SA.

Laslett, A.-M., Catalano, P., Chikritzhs, Y., Dale, C., Doran, C., Ferris, J., Jainullabudeen, T., Livingston, M., Matthews, S., Mugavin, J., Room, R., Schlotterlein, M. and Wilkinson, C. (2010) *The range and magnitude of alcohol's harm to others*. AER Centre for Alcohol Policy Research, Turning Point Alcohol and Drug Centre, Eastern Health, Fitzroy, VIC.

Last, J. (1987) *Public health and human ecology*. Prentice Hall, Upper Saddle River, NJ.

Last, J. (ed) (1995) A dictionary of epidemiology. Oxford University Press, New York, NY.

Last, J. (ed) (1998) Public health and human ecology. McGraw-Hill.

Laurie, V. (2011) Home improvement. Indigenous housing [Essay]. *The Monthly*, June, From http://mnth.ly/1mnL5Zq

Laverack, G. (2004) *Health promotion practice: Power and empowerment*. SAGE Publications, London.

Laverack, G. (2013) *Health activism: Foundations and strategies*. SAGE Publications, London.

Laycock, A., Walker, D., Harrison, N. and Brands, J. (2011) *Researching Indigenous health: A practical guide for researchers*. Lowitja Publishing, Carlton South, VIC.

Le Quéré, C., Peters, G. P., Andres, R. J., Andrew, R. M., Boden, T., Ciais, P., Friedlingstein, P., Houghton, R. A., Marland, G. and Moriarty, R. (2013) Global carbon budget 2013. *Earth System Science Data Discussions*, 6, 689–760.

Leaning, J. (2014) Guatemalan Farmers Stop a Mining Operation. *Grassroots International*, 28 March. From www.grassrootsonline.org/news/articles/guatemalan-farmers-stop-mining-operation

Lederberg, J. (1996) Emerging infectious disease threats, *American Public Health Association Annual Conference*, New York.

Lee, A. C. K. and Maheswaran, R. (2011) The health benefits of urban green spaces: A review of the evidence. *Journal of Public Health*, 33, 212–222.

Lee, A. J., Crombie, I. K., Smith, W. and Tunstall, H. D. (1991) Cigarette smoking and employment status. *Social Science and Medicine*, 33, 1309–1312.

Lee, K. (2005) Global social change and health. In Lee, K. and Collin, J. (eds), *Global change and health*, Open University Press, Maidenhead. pp. 13–27.

Lefebvre, R. and Flora, J. (1988) Social marketing and public health interventions. *Health Education Quarterly*, 15, 299–315.

Legge, D. (1990) Community participation: Models and dilemmas, Making the connections—People, communities and the environment. First National Conference of Healthy Cities Australia. ACHA, Wollongong, NSW.

Legge, D. (1992) Community management: Open letter to a new committee member. In Baum, F., Fry, D. and Lennie, I. (eds), *Community health: Policy and practice in Australia*, Pluto Press, Sydney, NSW. pp. 95–114.

Legge, D., Butler, P. and Scott, J. (1995) *Policies for a healthy Australia*. Commonwealth Department of Human Services and Health, Canberra, ACT.

Legge, D., Wilson, G., Butler, P., Wright, M., McBride, T. and Attewell, R. (1996) Best practice in primary health care. *Australian Journal of Primary Health*, 2, 12–26.

Leigh, A. (2013) *Battlers and billionaires: The story of inequality in Australia*. Black Inc., Collingwood, VIC.

Leppo, K., Ollila, E., Peña, S., Wismar, M. and Cook, S. (eds) (2013) *Health in All Policies: Seizing opportunities, implementing policies*. Ministry of Social Affairs and Health, Helsinki, Finland.

Lerner, M. (1986) *Surplus powerlessness*. Institute for Labor and Mental Health, Oakland, CA.

Lewin, K. (1946) Action research and minority problems. *Journal of Social Issues*, 2, 34–46.

Lewis, B. and Walker, R. (1997) *Changing central-local relationships in health service provision: Final report*. School of Health Systems Science, La Trobe University, Melbourne, VIC.

Lewis, J. M. (2005) *Health policy and politics: Networks, ideas and power*. IP Communications, Melbourne, VIC.

Lewis, M. J. (2003) The people's health. Public health in Australia 1799–1950. Preader, Westport, CT.

Lewis, S. (2005) *Race against time*. House of Anansi Press, Toronto, Canada.

Li, C. (ed) (2010) *China's emerging middle class: Beyond economic transformation*. Brookings Institution Press, Washington, DC.

Lillie-Blanton, M., Parsons, P. E., Gayle, H. and Dievler, A. (1996) Racial differences in health: Not just black and white, but shades of gray. *Annual Review of Public Health*, 17, 411–448.

Lincoln, Y. S. and Guba, E. G. (1985) *Naturalistic inquiry*. SAGE Publications, Beverly Hills, CA.

Lindblom, C. E. (1959) The science of muddling through. *Public Administration Review*, 19, 79–88.

Lindheim, R. and Syme, L. (1983) Environments, people and health. *Annual Review of Public Health*, 4, 335–359.

Lipsky, M. (1980) *Street-level bureaucracy*. Russell Sage, New York, NY.

Lipson, D. J. (2001) The World Trade Organization's health agenda. *BMJ*, 323, 1139–1140.

Litva, A. and Eyles, J. (1994) Health or healthy: Why people are not sick in a southern Ontarian town. *Social science and medicine*, 39, 1083–1091.

Lock the Gate Alliance (n.d.) From www.lockthegate.org.au (accessed 28 July 2014).

Locker, D. (1981) Symptoms and illness: The cognitive organization of disorder. Tavistock, London.

Loevinsohn, M. E. (1994) Climatic warming and increased malaria incidence in Rwanda. *The Lancet*, 343, 714–718.

Lopez, A. (1983) The sex mortality differential in developing countries. In Lopez, A. D. and Ruzicka, L. T. (eds), *Sex differentials in mortality: Trends, determinants and consequences*, ANU Press, Canberra, ACT.

Loring, B. and Robertson, A. (2014) Obesity and inequities: Guidance for addressing inequities in overweight and obesity. WHO Regional Office for Europe, Copenhagen.

Lovelock, J. (2004) Nuclear power is the only green solution. *Independent*, 24 May. From www.ecolo.org/media/articles/articles.in.english/love-indep-24-05-04.htm

Low, C. and Rinaudo, J. (1993) More than a bit of shadecloth: Healthy cities and shires in action in Queensland. Community Health Association, Brisbane, QLD.

Lowe, I. (2005) Living in the hothouse: How global warming affects Australia. Scribe Publications, Melbourne, VIC.

Lowe, I. (2012) *Bigger or better? Australia's population debate*. Penguin Group, Melbourne, VIC.

Lukes, S. (2005) *Power: A radical view*. Macmillan, London.

Lumley, J. (1996) Ethics and epidemiology: Problems for the researcher. In Daly, J. (ed), *Ethical intersections. Health research-methods and researcher responsibility*, Text Publishing, Melbourne, VIC. pp. 24-33.

Lund, R., Christensen, U., Nilsson, C. J., Kriegbaum, M. and Rod, N. H. (2014) Stressful social relations and mortality: A prospective cohort study. *Journal of Epidemiology and Community Health*, 68, 720–727.

Lundberg, O., Yngwe, M. Å., Stjärne, M. K., Elstad, J. I., Ferrarini, T., Kangas, O., Norström, T., Palme, J. and Fritzell, J. (2008) The role of welfare state principles and generosity in social policy programmes for public health: an international comparative study. *The Lancet*, 372, 1633–1640.

Lundy, P. (1996) Limitations of quantitative research in the study of structural adjustment. *Social Science and Medicine*, 42, 313–324.

Lupton, D. (1995) The imperative of health: Public health and the regulated body. SAGE Publications, London.

Lwanga, S. K. and Lemeshow, S. (1991) *Sample size determination in health studies*. World Health Organization, Geneva.

Lynch, J. W. and Davey Smith, G. (2005) A life course approach to chronic disease epidemiology. *Annual Review of Public Health*, 26, 1–35.

Lynch, J. W., Davey Smith, G., Kaplan, G. A. and House, J. S. (2000) Income inequality and mortality: Importance to health of individual income, psychosocial environment or material conditions. *BMJ*, 320, 1200–1204.

Lynch, J. W., Kaplan, G. A. and Salonen, J. T. (1997) Why do poor people behave poorly? Variation in adult health behaviours and psychosocial characteristics by stages of the socioeconomic lifecourse. *Social Science and Medicine*, 44, 809–819.

Lynch, K. (1977) *Growing up in cities*. UNESCO, Paris.

Lyons, G., Moore, E. and Smith, J. W. (1995) Is the end nigh? Internationalism, global chaos and the destruction of the Earth. Avebury, Aldershot.

MacDonald, T. H. (2005) Third World Health: Hostage to First World Wealth. Radcliffe, Oxon, UK.

MacDougall, C. (2007) Reframing physical activity. In Keleher, H., MacDougall, C. and Murphy, B. (eds), *Understanding health promotion*, Oxford University Press, Melbourne, VIC. pp. 326–342.

MacDougall, C. and Baum, F. (1997) The devil's advocate: A strategy to avoid groupthink and stimulate discussion in focus groups. *Qualitative Health Research*, 7, 532–541.

MacDougall, C., Wright, C. and Atkinson, R. (2002) Supportive environments for physical activity and the local government agenda: A South Australian example. *Australian Health Review*, 24, 178–184.

MacGregor, S. and Payne, J. (2011) *Findings from the DUMA program*. Australian Institute of Criminology, Canberra, ACT.

Macintyre, S. (1986) The patterning of health by social position in contemporary Britain: Directions for sociological research. *Social Science and Medicine*, 23, 393–415.

Macintyre, S. and Ellaway, A. (2000) Ecological approaches: Rediscovering the role of the physical and social environment. In Berkman, L. F. and Kawachi, I. (eds), *Social epidemiology*, Oxford University Press, New York, NY.

Macintyre, S., Hiscock, R., Keans, A. and Ellaway, A. (2000) Housing tenure and health inequalities: a three-dimensional perspective on people, homes and neighbourhoods. In Graham, H. (ed), *Understanding health inequalities*, Chapter 8, Open University Press, Buckingham. pp. 129–142.

Mackenbach, J. P. (1996) The contribution of medical care to mortality decline: McKeown revisited. *Journal of clinical epidemiology*, 49, 1207–1213.

Mackenbach, J. P. (2005) Health inequalities: European in profile, *Report for UK Presidency of the European Union*. UK Presidency of the European Union, London.

Mackenbach, J. P., Bouvier-Colle, M. and Jougla, E. (1990) "Avoidable" mortality and health services: A review of aggregate data studies. *Journal of Epidemiology and Community Health*, 44, 106–111.

Maclean, U. (1988) Ethnographic approaches to health. In Anderson, R. (ed), *Health behaviour research and health promotion*, Oxford University Press, Oxford. pp. 41–44.

Maddison, S. and Hamilton, C. (2007) Non-government organisations. In Hamilton, C. and Maddison, S. (eds), *Silencing dissent: How the Australian government is controlling public opinion and stifling debate*, Chapter 5, Allen & Unwin, Crows Nest, NSW. pp. 78–100.

Maggi, S., Irwin, L. G., Siddiqi, A., Poureslami, I., Hertzman, E. and Hertzman, C. (2005) *Analytic and strategic review paper: International perspectives on early child development*. Commission on Social Determinants of Health, World Health Organization, Geneva.

Maher, C. and Tilton, E. (1994) *Health promotion or self promotion? A Central Australian Alcohol Media Strategy*. Central Australian Aboriginal Congress, Alice Springs, NT.

Mahler, H. (1988) Opening Address, 2nd International Conference on Health Promotion, 5–9 April, Adelaide, SA.

Mahler, H. (2008) Dr Halfdan Mahler, Former Director-General of WHO. Address at the 61st World Health Assembly. From www.who.int/mediacentre/events/2008/wha61/hafdan_mahler_speech/en (accessed 5 January 2015).

Maibach, E. and Holtgrave, D. R. (1995) Advances in public health communication. *Annual Review of Public Health*, 16, 219–238.

Make Roads Safe (2012) Road injury ranks high in Global Burden of Disease, 17 December. From www.makeroadssafe.org/news/2012/Pages/RoadinjuryrankshighinGlobalBurdenof Disease.aspx (accessed 2 January 2015).

Manandhar, D. S., Osrin, D., Shrestha, B. P., Mesko, N., Morrison, J., Tumbahangphe, K. M., Tamang, S., Thapa, S., Shrestha, D. and Thapa, B. (2004) Effect of a participatory intervention with women's groups on birth outcomes in Nepal: Cluster-randomised controlled trial. *The Lancet*, 364, 970–979.

Manderson, L. and Aaby, P. (1992) An epidemic in the field? Rapid assessment procedures and health research. *Social Science and Medicine*, 35, 839–850.

Mann, A. (2014) *Global activism in food politics: Power shift*. Palgrave Macmillan, New York, NY.

Manne, R. and McKnight, D. (eds) (2010) Goodbye to all that? On the failure of neo-liberalism and the urgency of change. Black Inc. Agenda, Melbourne, VIC.

Manning, M., Smith, C. and Mazerolle, P. (2013) The societal costs of alcohol misuse in Australia, *Trends and issues in crime and criminal justice, No. 454.* Australian Institute of Criminology, Canberra, ACT.

Mares, P. (2001) Borderline: Australia's treatment of refugees and asylum seekers. University of NSW Press, Sydney, NSW.

Mark, M. M., Henry, G. T. and Julnes, G. (2000) Evaluation: An integrated framework for understanding, guiding and improving policies and programs. Jossey-Bass, San Francisco, CA.

Marley, J. E. and McMichael, A. J. (1991) Disease causation: The role of epidemiological evidence. *Medical Journal of Australia,* 155, 95–101.

Marmot, M. (2001) Economic and social determinants of disease. *Bulletin of the World Health Organization,* 79, 988–989.

Marmot, M. (2004) The status syndrome: How social standing affects our health and longevity. Times Books, New York, NY.

Marmot, M. (2006) Health in an unequal world. *The Lancet,* 368, 2081–2094.

Marmot, M., Allen, J., Goldblatt, P., Boyce, T., McNeish, D., Grady, M. and Geddes, I. (2010) *Fair society, healthy lives, the Marmot review, executive summary: Strategic review of health inequalities in England post-2010.* UK Department of Health, London.

Marmot, M. and Consortium members (2013a) Health inequalities in the EU. *Final report of a Consortium led by Sir Michael Marmot.* European Commission Directorate-General for Health and Consumers.

Marmot, M. and Kivimäki, M. (2009) Social inequalities in mortality: A problem of cognitive function? *European heart journal,* 30, 1819–1820. 10.1093/eurheartj/ehp264.

Marmot, M., Pellegrini Filho, A., Vega, J., Solar, O. and Fortune, K. (2013b) Action on social determinants of health in the Americas [Editorial]. *Pan American Journal of Public Health,* 34, 379–381.

Marmot, M. and Wilkinson, R. G. (eds) (1999) *Social determinants of health.* Oxford University Press, Oxford.

Marmot, M. G., Bobak, M. and Davey Smith, G. (1995) Explanations for social inequalities in health. In Amick, B. C., Levine, S., Tarlov, A. R. and Walsh, D. C. (eds), *Society and health,* Oxford University Press, New York, NY. pp. 172–210.

Marmot, M. G., Shipley, M. J. and Rose, G. (1984) Inequalities in death-specific explanations of a general pattern. *The Lancet,* 1, 1003–1006.

Marsh, D. (2010) Meta-theoretical issues. In Marsh, D. and Stoker, G. (eds), *Theory and methods in political science,* Palgrave, Basingstoke, UK. pp. 212–231.

Marshall, I. H. (1988) Trends in crime rates, certainty of punishment and severity of punishment in the Netherlands. *Criminal Justice Policy Review,* 2, 21–52.

Marshall, T. H. (1950) *Citizenship and social class.* Cambridge University Press, Cambridge.

Marston, G., McDonald, C. and Bryson, L. (2013) *The Australian welfare state: Who benefits now?* Palgrave Macmillan, Melbourne, VIC.

Martin, G. and Davis, C. (1995) Mental health promotion: From rhetoric to reality. In Baum, F. (ed), *Health for All: The South Australian experience,* Wakefield Press, Adelaide, SA. pp. 406–425.

Maté, G. (2012) Realm of hungry ghosts: Close encounters with addiction. Vintage Canada, Toronto.

Mathews, R. (2012) The Mondragon model: How a Basque cooperative defied Spain's economic crisis. *The Conversation,* 19 October. From http://theconversation.com/the-mondragon-model-how-a-basque-cooperative-defied-spains-economic-crisis-10193

Maulia, E. (2014) Jakarta's air quality takes a toxic turn for the worse. *Jakarta Globe,* 9 May. From http://thejakartaglobe.beritasatu.com/news/jakarta/jakartas-air-quality-takes-toxic-turn-worse

Mayo, M. and Craig, G. (1995) Community participation and empowerment: The human face of structural adjustment or tools for democratic transformation? In Craig, G. and Mayo, M. (eds), *Community empowerment: A reader in participation and development*, Zed Books, London. pp. 1–11.

McAuley, I. (2013) Taxes—our payment for civilization. *Paper to accompany presentation to South Australia Council of Social Services AGM, 25 November.* Centre for Policy Development and University of Canberra.

McCaffery, K. J. and Barratt, A. L. (2004) Assessing psychosocial/quality of life outcomes in screening: how do we do it better? *Journal of Epidemiology and Community Health*, 58, 968–970.

McColl, M. (1985) *The high cost of home care: The carer's perspective.* Southern Community Health Research Unit, Adelaide, SA.

McCord, C. and Freeman, H. P. (1990) Excess mortality in Harlem. *New England Journal of Medicine*, 322, 173–177.

McCoy, A. (1991) The politics of heroin: CIA complicity in the global drug trade. Lawrence Hill Books, Brooklyn, NY.

McCoy, D., Narayan, R., Baum, F., Sanders, D., Serag, H., Salvage, J., Rowson, M., Schrecker, T., Woodward, D., Labonté, R., Sengupta, A., Qizphe, A. and Schuftan, A. (2006) A new Director General for WHO: An opportunity for bold and inspirational leadership. *The Lancet*, 368, 2179.

McCoy, D., Sanders, D., Baum, F., Narayan, T. and Legge, D. (2004) Pushing the international health research agenda towards equity and effectiveness. *The Lancet*, 364, 1630–1631.

McDaid, D. and Oliver, A. (2005) Inequalities in health: International patterns and trends. In Scriven, A. and Garman, S. (eds), *Promoting health: Global perspectives*, Palgrave Macmillan, Basingstoke.

McDonald, E., Slavin, N., Bailie, R. and Schobben, X. (2011) No germs on me: A social marketing campaign to promote hand-washing with soap in remote Australian Aboriginal communities. *Global health promotion*, 18, 62–65.

McGinn, A. P. (2000) Phasing out persistent organic pollutants. In Starke, L. (ed), *State of the World 2000*, WW Norton & Co, New York.

McGlade, C. and Ekins, P. (2015) The geographical distribution of fossil fuels unused when limiting global warming to 2°C. *Nature*, 517, 187–190.

McGuiness, M. and Wadsworth, Y. (1992) *Understanding, anytime: A consumer evaluation of an acute psychiatric hospital.* Victorian Mental Illness Awareness Council, Melbourne, VIC.

McKeown, T. (1979) The role of medicine: Dream, mirage or nemesis? Basil Blackwell, Oxford.

McKinlay, J. (1984) Introduction. In McKinlay, J. (ed), *Issues in the political economy of health care*, Tavistock, New York, NY. pp. 1–19.

McKinlay, J. B. and McKinlay, S. M. (1997) Medical measures and the decline of mortality. In Conrad, P. (ed), *The sociology of health and illness: Critical perspectives*, 5th edition, St. Martin's Press, New York, NY. pp. 10–23.

McKnight, J. L. (1985) Health and empowerment. *Canadian Journal of Public Health*, 76, 37–38.

McLean, C., Carmona, C., Francis, S., Wohlgemuth, C. and Mulvihill, C. (2005) *Worklessness and health—what do we know about the casual relationship? Evidence review.* Health Development Agency, London.

McMichael, A. J. (1993) *Planetary overload.* Cambridge University Press, Cambridge.

McMichael, A. J. (2001) *Human frontiers, environments and disease.* Cambridge University Press, Cambridge.

McMichael, A. J. (2005) Global environmental changes, climate change and human health. In Lee, K. and Collin, J. (eds), *Global change and health*, Open University Press, Maidenhead, Berkshire.

McMichael, A. J. (2013) Globalization, climate change, and human health. *New England Journal of Medicine*, 368, 1335–1343.

McMichael, A. J. and Butler, C. D. (2011) Promoting global population health while constraining the environmental footprint. *Annual Review of Public Health*, 32, 179–197.

McMichael, A. J., Woodruff, R., Whetton, P., Hennessy, K., Nicholls, N., Hales, S. and Woodward, A. (2003) *Human health and climate change in Oceania: A risk assessment.* Commonwealth of Australia, Canberra, ACT.

McMurtry, J. (1997) The Multilateral Agreement on Investment: The plan to replace democratically responsible government. Interdisciplinary Conference on the Evolution of World Order: Building a Foundation of Peace in the Third Millenium Toronto: June. From http://islandnet.com/plethora/mai/maiplan.html

McNeill, P. M., Berglund, C. A. and Webster, I. W. (1992) Do Australian researchers accept committee review and conduct ethical research? *Social science and medicine*, 35, 317–322.

McPherson, P. D. (1992) Health for all Australians. In Gardner, H. (ed), *Health policy: Development, implementation and evaluation in Australia*, Churchill Livingstone, Melbourne, VIC. pp. 119–135.

McQueen, D. V. (1993) A methodological approach for assessing the stability of variables used in population research on health. In Dean, K. (ed), *Population health research: Linking theory and methods*, SAGE Publications, London. pp. 95–115.

McTaggart, R. (1991) *Action research: A short modern history.* Deakin University, Melbourne, VIC.

McWaters, N., Hurwood, C. and Morton, D. (1989) Step by step on a piece of string: An illustration of community work as a social health strategy. *Community health studies*, 13, 23–33.

Meadows, D., Randers, J. and Meadows, D. (2004) *Limits to growth: The 30-year update.* Chelsea Green Publishing.

Mechanic, D. (1976) Sex, illness, illness behaviour and the use of services. *Social Science and Medicine*, 12, 207.

Medalie, J. and Goldbourt, U. (1976) Angine pectoris among 10 000 men. Psychosocial and other risk factors as evidenced by a multivariate analysis of a five year incidence study. *American Journal of Medicine*, 60, 910–921.

Médecins Sans Frontières (MSF), World Health Organization (WHO) and Joint United Nations Programme on HIV/AIDS (UNAIDS) (2003) *Surmounting challenges: Procurement of antiretroviral medicines in low- and middle-income countries.* MSF, WHO, UNAIDS.

Meedeniya, J., Smith, A. and Carter, P. (2000) Food supply in rural South Australia: A survey on food cost, quality and variety. Eat Well SA, Adelaide, SA.

Mehta, K. (2007) Food advertising to children: The battle for children's dollars or health? *Public Health Bulletin SA*, 4, 20–22.

Meltzer, M., Cox, N. and Fukuda, K. (1999) The economic impact of pandemic influenza in the United States; priorities for intervention. *Emerging Infectious Diseases*, 5, 659–671.

Mertens, D. M. (2005) Research and evaluation in education and psychology: Integrating diversity with quantitative, qualitative and mixed methods. SAGE Publications, Thousand Oaks, CA.

Merton, R. M., Fiske, M. and Kendall, P. L. (1956) *The focused interview.* Free Press, New York, NY.

Middleton, N. and O'Keefe, P. (2003) *Rio plus ten. Politics, poverty and the environment.* Pluto Press, London.

Miles, I. and Evans, J. (eds) (1979) Demystifying social statistics. Pluto Press, London.

Miles, M. B. and Huberman, A. M. (1994) *Qualitative data analysis.* SAGE Publications, Thousand Oaks, CA.

Milio, N. (1983) *Promoting health through public policy.* F. A. Davis, Philadelphia.

Milio, N. (1989) Nutrition and health: Patterns and policy perspectives in food-rich countries. *Social science and medicine*, 29, 413–423.

Milio, N. (2001) Healthy public policy [Glossary]. *Journal of Epidemiology and Community Health*, 55, 622–623.

Milio, N. and Helsing, E. (eds) (1998) *European food and nutrition policies in action.* WHO Regional Office for Europe, Copenhagen.

Miller, C. (2014) Citizen engagement in Australian policy-making In Miller, C. and Orchard, L. (eds), *Australian public policy: Progressive ideas in the neo-liberal ascendency*, Policy Press, Bristol.

Miller, C. and Orchard, L. (eds) (2014) Australian public policy: Progressive ideas in the neo-liberal ascendency. Policy Press, Bristol.

Miller, P. and Rainow, S. (1997) Don't forget the plumber: research in remote Aboriginal communities [Commentary]. *Australian and New Zealand Journal of Public Health*, 21, 96–97.

Mills, A., Kanters, S., Hagopian, A., Bansback, N., Nachega, J., Alberton, M., Au-Yeung, C. G., Mtambo, A., Luboga, S., Hogg, R. S. and Ford, N. (2011) The financial cost of doctors emigrating from sub-Saharan Africa: human capital analysis. *BMJ*, 343, 13. 10.1136/bmj. d7031.

Minichiello, V., Aroni, R. and Hays, T. (eds) (2008) *In-depth interviewing*. Pearson, Australia.

Minkler, M. (ed) (2005) *Community organizing and community building for health*. Rutgers University Press, New Brunswick, NJ.

Minkler, M. and Wallerstein, N. (eds) (2003) *Community-based participatory research for health*. Jossey-Bass, San Francisco, CA.

Minkler, M. and Wallerstein, N. (2012) Improving health through community organization and community building: Perspectives from health education and social work. In Minkler, M. (ed), *Community organizing and community building for health and welfare*, 3rd edition, Chapter 3, Rutgers University Press, New Brunswick, NJ. pp. 37–58.

Mitchell, A. D. and Studdert, D. M. (2012) Plain packaging of tobacco products in Australia: a novel regulation faces legal challenge. *JAMA*, 307, 261–262.

Mitchell, J. (2006) History. In Carson, B., Dunbar, T., Chenhall, R. D. and Bailie, R. (eds), *Social determinants of Indigenous health*, Allen & Unwin, Crows Nest, NSW. pp. 41–64.

Mitchell, K. (2004) Geographies of identity: Multiculturalism unplugged. *Progress in Human Geography*, 28, 641–651.

Mitlin, D. and Satterthwaite, D. (2004) Empowering squatter citizen. Local government, civil society and urban poverty reduction. Earthscan, London.

Mittlemark, M., Hunt, M., Heath, G. W. and Schmid, T. I. (1993) Realistic outcomes: Lessons from community based research and demonstration programs for the prevention of cardiovascular diseases. *Journal of Public Health Policy*, 14, 437–462.

Modrek, S., Stuckler, D., McKee, M., Cullen, M. R. and Basu, S. (2013) A review of health consequences of recessions internationally and a synthesis of the US response during the Great Recession. *Public Health Reviews*, 35.

Monaem, A., Tyler, C., McPhee, L., Egger, G. and NSW North Coast Health Region (1985) *Healthy lifestyle: Survey data*. NSW Department of Health, Lismore, NSW.

Monbiot, G. (2003) Universal fair trade, 8 September. From www.monbiot.com/archives/2003/09/08/universal-fair-trade

Monbiot, G. (2013) Feral: Searching for enchantment on the frontiers of rewilding. Allen Lane.

Monbiot, G. (2014) Growth: The destructive god that can never be appeased. *The Guardian*, 19 November. From www.theguardian.com/commentisfree/2014/nov/18/growth-destructive-economic-expansion-financial-crisis

Mondragon Corporation (n.d.) About us. From www.mondragon-corporation.com/eng/about-us (accessed 5 January 2015).

Montague, M., Borland, R. and Sinclair, C. (2001) Slip! Slop! Slap! And SunSmart, 1980–2000: Skin cancer control and 20 years of population-based campaigning. *Health Education and Behaviour*, 28, 290-305.

Moodie, R. (2006) Why I like paying my rates! *VicHealth*, Winter, 3.

Moodie, R., Stuckler, D., Monteiro, C., Sheron, N., Neal, B., Thamarangsi, T., Lincoln, P. and Casswell, S. (2013) Profits and pandemics: Prevention of harmful effects of tobacco, alcohol, and ultra-processed food and drink industries. *The Lancet*, 381, 670–679.

Mooney, G. (2012) The health of nations: Towards a new political economy. Zed Books, London.

Moore, D. (1992) Beyond the bottle: Introducing anthropological debate to research into Aboriginal alcohol use. *Australian Journal of Social Issues*, 27, 173–193.

Morgan, D. L. (1988) *Focus groups as qualitative research*. SAGE Publications, Portland.

Morgan, D. L., Slade, M. D. and Morgan, C. M. (1997) Aboriginal philosophy and its impact on health care outcomes. *Australian and New Zealand Journal of Public Health*, 21, 597–601.

Morgan, D. L. and Spanish, M. T. (1985) Social interaction and the cognitive organisation of health-relevant knowledge. *Sociology of Health and Illness*, 7, 401–422.

Morrison, D. E. and Henkel, R. E. (eds) (1970) The significance test. Butterworth, London.

Morrison, D. S. (2009) Homelessness as an independent risk factor for mortality: Results from a retrospective cohort study. *International Journal of Epidemiology*, 38, 877–883.

Morton, S. M., Bandara, D. K., Robinson, E. M. and Carr, P. E. (2012) In the 21st century, what is an acceptable response rate? *Australian and New Zealand Journal of Public Health*, 36, 106–108.

Moser, K., Goldblatt, P. and Pugh, H. (1990) Occupational mortality of women in employment. In Goldblatt, P. (ed), *Longitudinal study: Mortality and social organisation 1971–1981*, HMSO, London. pp. 129–144.

Moser, K. A., Fox, A. J. and Jones, D. R. (1984) Unemployment and mortality in the OPCS longitudinal study. *The Lancet*, 324, 1324–1329.

Moss, I. (1994) Water: A report on the provision of water and sanitation in remote Aboriginal and Torres Strait Islander communities. Commonwealth of Australia, Canberra, ACT.

Moss, N. E. (2002) Gender equity and socioeconomic inequality: A framework for the patterning of women's health. *Social science and medicine*, 54, 649–661.

Moynihan, R. and Henry, D. (2006) The fight against disease mongering: Generating knowledge for action. *PloS Medicine*, 3, e191.

Moynihan, R. and Murphy, K. (2002) Doctors causing a drug costs blowout, *Australian Financial Review*.

Muller, H. J. and Ventriss, C. (1985) *Public health in a retrenchment era: An alternative to managerialism*. State University of New York Press, Albany, NY.

Municipal Services Project (MSP) (n.d.) From www.municipalservicesproject.org (accessed 26 July 2014).

Murray, C. J., King, G., Lopez, A. D., Tomijima, N. and Krug, E. G. (2002) Armed conflict as a public health problem. *British Medical Journal*, 324, 346–349.

Naidoo, J. (1986) Limits to individualism. In Rodmell, S. and Watt, A. (eds), *The politics of health*, Routledge & Kegan Paul, London. pp. 17–37.

Naidoo, J. and Wills, J. (1994) *Health promotion: Foundations for practice*. Bailliere Tindall, London.

Naidoo, J. and Wills, J. (2001) *Health studies: An introduction*. Palgrave, Basingstoke.

Nathanson, C. A. (1975) Illness and the feminine role: A theoretical review. *Social Science and Medicine*, 9, 57–62.

National Aboriginal Community Controlled Health Organisation (NACCHO) (2014) Definitions. From www.naccho.org.au/aboriginal-health/definitions (accessed 2 October 2014).

National Aboriginal Community Controlled Health Organisation (NACCHO) and Royal Australian College of General Practitioners (RACGP) (2012) *National guide to a preventive health assessment for Aboriginal and Torres Strait Islander people*. 2nd edition, RACGP, South Melbourne, VIC.

National Council for Voluntary Organisations (NCVO) (n.d.) *NCVO UK Civil Society Almanac 2014*—What is civil society? From http://data.ncvo.org.uk/a/almanac14/what-is-civil-society-2 (accessed 23 July 2014).

National Health and Hospitals Reform Commission (2009) *A healthier future for all Australians*. Commonwealth of Australia, Canberra, ACT.

National Health and Medical Research Council (NHMRC) (1989) *Health effects of ozone layer depletion*. Commonwealth of Australia, Canberra, ACT.

National Health and Medical Research Council (NHMRC) (1991) *Ecologically sustainable development: The health perspective*. Commonwealth of Australia, Canberra, ACT.

National Health and Medical Research Council (NHMRC) (1996) Ethical aspects of qualitative methods in health research: An information paper for institutional ethics committees. Commonwealth of Australia, Canberra, ACT.

National Health and Medical Research Council (NHMRC) (2003) *Values and ethics: Guidelines for the ethical conduct in Aboriginal and Torres Strait Islander research*. Commonwealth of Australia, Canberra, ACT. From www.nhmrc.gov.au/guidelines/publications/e52

National Health and Medical Research Council (NHMRC) (2013) *Eat for Health—Australian Dietary Guidelines 2013*. NHMRC, Canberra, ACT. From www.eatforhealth.gov.au/sites/default/files/files/the_guidelines/n55_australian_dietary_guidelines.pdf

National Health and Medical Research Council (NHMRC), Australian Research Council (ARC) and Australian Vice-Chancellors' Committee (2007 [updated March 2014]) National Statement on Ethical Conduct in Human Research 2007. Commonwealth of Australia, Canberra, ACT. From www.nhmrc.gov.au/guidelines/publications/e72

National Health Performance Authority (NHPA) (2013) Overweight and obesity rates across Australia, 2011–12. *inFocus—Healthy Communities*, October 2013. From www.myhealthycommunities.gov.au/Content/publications/downloads/NHPA_HC_Report_Overweight_and_Obesity_Report_October_2013.pdf

National Health Strategy (1992) Enough to make you sick: How income and environment affect health. National Health Strategy Unit, Melbourne, VIC.

National Preventative Health Taskforce (NPHT) (2009a) *Australia: The Healthiest Country by 2020. National Preventative Health Strategy – the roadmap for action*. Commonwealth of Australia, Barton, ACT. From www.preventativehealth.org.au/internet/preventativehealth/publishing.nsf/Content/nphs-roadmap-toc

National Preventative Health Taskforce (NPHT) (2009b) Australia: The healthiest country by 2020. Technical Report 1, Obesity in Australia: a need for urgent action. Including addendum for October 2008 to June 2009. Commonwealth of Australia, Barton, ACT. From www.preventativehealth.org.au/internet/preventativehealth/publishing.nsf/Content/tech-obesity-toc

Navarro, V. (ed) (1979) *Imperialism, health and medicine*. Baywood Publishing, New York, NY.

Navarro, V. (ed) (2002) The political economy of social inequalities: Consequences for health and quality of life. Baywood, New York, NY.

Navarro, V. and Shi, L. (2001) The political context of social inequalities and health. *Social Science and Medicine*, 52, 481–491.

Newell, D. J. (1993) Randomised controlled trials in health care research. In Daly, J. (ed), *Researching health care: Designs, dilemmas, disciplines*, Tavistock/Routledge, London. pp. 47–61.

Newman, L., Ludford, I., Williams, C. and Herriot, M. (2014) Applying Health in All Policies to obesity in South Australia. *Health Promotion International*. 10.1093/heapro/dau064.

Newman, O. (1972) Defensible space: Crime prevention through urban design. Macmillan, New York, NY.

Newman, P. (1993) Planning in an age of uncertainty, *Urban Planning Seminar*. Hobart Metropolitan Councils Association, Hobart, TAS.

Newman, P. (2006) Beyond peak oil: Will our cities and regions collapse? *Res Publica*, 15, 1–7.

Newman, P., Beatley, T. and Boyer, H. (2009) *Resilient Cities. Responding to peak oil and climate change*. Island Press, Washington, DC.

Newman, P. and Hogan, T. (1981) A review of urban density models: Towards a resolution of the conflict between populace and planner. *Human Ecology*, 9, 269–303.

Newman, P., Kenworthy, J. and Robinson, L. (1992) *Winning back the cities*. Australian Consumers Association and Pluto Press, Sydney, NSW.

Newman, P. and Kenworthy, J. R. (1999) *Sustainability and cities: Overcoming automobile dependence*. Island Press, Washington DC.

Newman, P. and Matan, A. (2013) *Green urbanism in Asia: The emerging green tigers.* World Scientific Publishing.

Nisbet, E. G. (1991) *Leaving Eden: To protect and manage the Earth.* Cambridge University Press, New York, NY.

Nose, T. and Oiwa, Y. (2013) More suspected and confirmed cases of thyroid cancer diagnosed in Fukushima children. *The Asahi Shimbun,* 13 November. From https://ajw.asahi.com/article/0311disaster/fukushima/AJ201311130066

Novotny, P. (1994) Popular epidemiology and the struggle for community health: Alternative perspectives from the environmental justice movement. *Capitalism, Nature and Socialism,* 5, 29–42.

NSW Nurses and Midwives' Association (2014) Street theatre during the G20 Trade Ministers meeting in Sydney 19 July 2014. From www.nswnma.asn.au/street-theatre-during-the-g20-trade-ministers-meeting-in-sydney-19-july-2014

Nutbeam, D. (1986) *Health promotion glossary.* World Health Organization, Geneva.

Nutbeam, D. (2006) Using theory to guide changing individual behaviour. In Davies, M. and Macdowall, W. (eds), *Health promotion theory,* Open University Press, Maidenhead.

Nutbeam, D., Macaskill, P., Smith, C., Simpson, J. M. and Catford, J. (1993a) Evaluation of two smoking education programmes under normal classroom conditions. *British Medical Journal,* 306, 102–107.

Nutbeam, D., Smith, C., Murphy, S. and Catford, J. (1993b) Maintaining evaluation designs in long-term community-based health promotion programs: The Heartbeat Wales experience. *Epidemiology and Community Health,* 47, 127–133.

O'Brien, M., Hartwig, F., Schanes, K., Kammerlander, M., Omann, I., Wilts, H., Bleischwitz, R. and Jäger, J. (2014) Living within the safe operating space: a vision for a resource efficient Europe. *European Journal of Futures Research,* 2. 10.1007/s40309-014-0048-3.

O'Keefe, E. (2000) Equity, democracy and globalization. *Critical Public Health,* 10, 167–177.

O'Loughlin, J., Elliot, S. J., Cameron, R., Eyles, J., Harvey, G., Robinson, K. and Hanusaik, N. (2001) From diversity comes understanding: Health promotion capacity building and dissemination research in Canada. *Promotion and Education,* 1, 4–8.

O'Meara, M. (1999) Exploring a new vision for cities. In Starke, L. (ed), *State of the World 1999,* Worldwatch Institute/WW Norton & Co, New York, NY.

O'Sullivan, G., Sharman, E. and Short, S. (1999) *Goodbye normal gene: Confronting the genetics revolution.* Pluto Press, Sydney, NSW.

Oakley, A. (1981) Interviewing women: A contradiction in terms. In Roberts, H. (ed), *Doing feminist research,* Routledge & Kegan Paul, London. pp. 30–61.

Oakley, A. (1985) Social support in pregnancy: The soft way to increase birthweight. *Social Science and Medicine,* 21, 1259–1268.

Oakley, A. (1989) Who's afraid of the randomised controlled trial? Some dilemmas of the scientific method and "good" research practice. *Women and Health,* 15, 25–59.

Oakley, A. (2001) Evaluating health promotion: Methodological diversity. In Oliver, S. and Peersman, G. (eds), *Using research for effective health promotion,* Open University Press, Buckingham. pp. 16–31.

Oakley, P. (1989) Community involvement in health development: An examination of the critical issues. World Health Organization, Geneva.

Oakley, P. (1991) *Projects with people.* International Labor Organization, Geneva.

Obama, B. (2012) Why organize? Problems and promise in the inner city. In Minkler, M. (ed), *Community organizing and community building for health and welfare,* Rutgers University Press, New Brunswick, NJ. pp. 27–31.

Oke, K. (1997) Removing the community from health centre boards. *Health Issues,* June, 5.

Oldenburg, R. (1997) Our vanishing "third places". *Planning Commissioners Journal,* 25, 8–10.

Olesen, V. (1994) Feminisms and models of qualitative research. In Denzin, N. K. and Lincoln, Y. S. (eds), *Handbook of qualitative research,* SAGE Publications, Thousand Oaks, CA. pp. 158–174.

Olivier, J. G. J., Janssens-Maenhout, G., Muntean, M. and Peters, J. A. H. W. (2013) *Trends in global CO_2 emissions: 2013 Report*. PBL Netherlands Environmental Assessment Agency, The Hague.

Ong, B. N. (1996) *Rapid appraisal and health policy*. Chapman & Hall, London.

Organisation for Economic Co-operation and Development (OECD). (2014a) Life expectancy at birth, total population. *Health: Key Tables from OECD, No. 11*. 10.1787/lifexpy-total-table-2014-1-en.

Organisation for Economic Co-operation and Development (OECD) (2014b) Obesity update—June 2014. From www.oecd.org/els/health-systems/Obesity-Update-2014.pdf

Organisation for Economic Co-operation and Development (OECD) (2014c) Public expenditure on health. As a percentage of total expenditure on health. *Health: Key Tables from OECD, No. 3*. From http://dx.doi.org/10.1787/hlthxp-pub-table-2014-1-en

Organisation for Economic Co-operation and Development (OECD) (2014d) Total expenditure on health. As a percentage of gross domestic product. *Health: Key tables from OECD No.1*. From http://dx.doi.org/10.1787/hlthxp-total-table-2014-1-en

Organisation for Economic Co-operation and Development (OECD). (2014e) Total expenditure on health per capita. At current prices and PPPs, US$. *Health: Key Tables from OECD, No. 2*. From http://dx.doi.org/10.1787/hlthxp-cap-table-2014-1-en

Organisation for Economic Co-operation and Development (OECD) and International Transport Forum (ITF) (2010) *Reducing transport greenhouse gas emissions: Trends and data, 2010*. OECD, Paris.

Orr, D. W. (2013) Governance in the long emergency. In Stark, L. (ed), *State of the World 2013: Is sustainability still possible?* Chapter 26, Worldwatch Institute/Island Press, Washington, DC. pp. 279–291.

Ortiz, I. and Cummins, M. (2013) The age of austerity: A review of public expenditures and adjustment measures in 181 countries, *Working paper*. Initiative for Policy Dialogue and the South Centre New York, NY.

Osborne, D. and Gaebler, T. (1992) Re-inventing government: How the entrepreneurial spirit is transforming the public sector. Addison-Wesley, Reading, MA.

Osborne, K., Baum, F. and Brown, L. (2013) What works? A review of actions addressing the social and economic determinants of Indigenous health, *Issues paper no. 7 produced for the Closing the Gap Clearinghouse*. Australian Institute of Health and Welfare, Canberra, and Australian Institute of Family Studies, Melbourne.

Ottersen, O. P., Dasgupta, J., Blouin, C., Buss, P., Chongsuvivatwong, V., Frenk, J., Fukuda-Parr, S., Gawanas, B. P., Giacaman, R., Gyapong, J., Leaning, J., Marmot, M., McNeill, D., Mongella, G. I., Moyo, N., Møgedal, S., Ntsaluba, A., Ooms, G., Bjertness, E., Lie, A. L., Moon, S., Roalkvam, S., Sandberg, K. I. and Scheel, I. B. (2014) The Lancet–University of Oslo Commission on Global Governance for Health: The political origins of health inequity: prospects for change. *The Lancet*, 383, 630–667.

Ovretveit, J. (1995) *Purchasing for health*. Open University Press, Buckingham.

Owen, A. and Lennie, I. (1992) Health for all and community health. In Baum, F., Fry, D. and Lennie, I. (eds), *Community health: Policy and practice in Australia*, Pluto Press, Sydney, NSW. pp. 6–27.

Owen, D. (2004) Green Manhattan, *New Yorker*.

Owens, J. (2013) Emergency state in battle against 300km firestorm in NSW, *The Australian*. 21 October. From www.theaustralian.com.au/in-depth/bushfires/emergency-state-in-battle-against-300km-firestorm-in-nsw/story-fngw0i02-1226743494282

Oxfam (2002) Rigged rules and double standards: Trade globalisation and the fight against poverty. Oxfam International, Oxford.

Oxfam (2014a) Risk of reversal in progress on world hunger as climate change threatens food security, *Oxfam media briefing, July*. Oxfam International, Oxford.

Oxfam (2014b) Working for the few: Political capture and economic inequality, *178 Oxfam briefing paper, 20 January*. Oxfam International, Oxford.

Oxfam (2015) Wealth: Having it all and wanting more, *Oxfam issue briefing, January*. Oxfam International, Oxford.

Paddon, M. (1996) The private world of market politics. *Australian options*, 2.

Palmer, G. and Short, S. (1994) *Health care and public policy*. Macmillan Education Australia, Melbourne, VIC.

Panaretto, K. S., Wenitong, M., Button, S. and Ring, I. T. (2014) Aboriginal community controlled health services: Leading the way in primary care. *Medical Journal of Australia*, 200, 649–652.

Paradies, Y. (2007) Racism. In Carson, B., Dunbar, T., Chenhall, R. D. and Bailie, R. (eds), *Social determinants of Indigenous health*, Allen & Unwin, Crows Nest, NSW.

Parenti, M. (1978) *Power and the powerless*. St Martin's Press, New York, NY.

Parish, R. (1995) Health promotion: Rhetoric and reality. In Bunton, R., Nettleton, S. and Burrows, R. (eds), *The sociology of health promotion*, Routledge, London. pp. 13–23.

Parnell, B., Lie, G., Hernandez, J. J. and Robins, C. (1996) Development and the HIV epidemic: A forward looking evaluation of the approach of the UNDP HIV and Development Programme. UNDP HIV and Development Programme, New York, NY.

Patterson, M. and Johnston, J. (2012) Theorizing the obesity epidemic: Health crisis, moral panic and emerging hybrids. *Social Theory and Health*, 10, 265–291.

Patton, M. Q. (1990) *Qualitative evaluation and research methods*. SAGE Publications, Newbury Park, CA.

Patton, M. Q. (2011) Developmental evaluation: Applying complexity concepts to enhance innovation and use. Guildford Press, New York, NY.

Patton, M. Q. (2015) *Qualitative research and evaluation methods*. SAGE Publications, London.

Pearce, F. (1992) Why it's cheaper to poison the poor. *New Scientist*, issue 1806, 1 February.

Pearse, W. (1997) Occupational health and safety: Model for public health? *Australian and New Zealand Journal of Public Health*, 21, 9–10.

Peberdy, A. (1993) Observing. In Shakespeare, P., Atkinson, D. and French, S. (eds), *Reflecting on research practice: Issues in health and social welfare*, Open University Press, Buckingham. pp. 47–57.

Peckham, S., Hann, A. and Boyce, T. (2011) Health promotion and ill-health prevention: the role of general practice. *Quality in Primary Care*, 19, 317–323.

Pederson, A., Edwards, R. K., Kelner, M., Marshal, V. M. and Allison, K. R. (1988) *Co-ordinating healthy public policy: An analytic literature review and bibliography*. Department of Behavioural Science, University of Toronto, Toronto, Canada.

Peersman, G., Oliver, S. and Oakley, A. (2001) Systematic reviews of effectiveness. In Oliver, S. and Peersman, G. (eds), *Using research for effective health promotion*, Open University Press, Buckingham. pp. 96–108.

People's Health Movement (PHM) (2000) The People's Charter for Health. From www .phmovement.org/en/resources/charters/peopleshealth?destination=home   (accessed 17 September 2014).

People's Health Movement (PHM) (2005) The Cuenca Declaration. From www.phmovement .org/en/node/798 (accessed 17 September 2014).

People's Health Movement (PHM) (2012) The Final Cape Town Call to Action. 13 July. From www.phmovement.org/en/pha3/final_cape_town_call_to_action

People's Health Movement (PHM), Medact and Global Equity Gauge Alliance (GAGE) (2005) *Global Health Watch 1 (2005–2006): An alternative world health report*. Zed Books, London.

People's Health Movement (PHM) (2014) Ebola epidemic exposes the pathology of the global economic and political system. *PHM position paper*, 26 September 2014. From www .phmovement.org/sites/www.phmovement.org/files/phm_ebola_23_09_2014final_0.pdf

People's Health Movement (PHM) Medact, Health Action International, Medicos International and Third World Network (2011) *Global Health Watch 3: An alternative world health report*. Zed Books, London.

People's Health Movement (PHM), Medact, Health Action International, Medicos International and Third World Network (2014) *Global Health Watch 4: An alternative world health report*. Zed Books, London.

Pepper, D. (1996) Modern environmentalism: An introduction. Routledge, London.

Petersen, A. (1994) In a critical condition: Health and power relations in Australia. Allen & Unwin, Sydney, NSW.

Petersen, A. (1998) The new genetics and the politics of public health. *Critical Public Health*, 8, 59–72.

Petersen, A. and Bunton, R. (2002) *The new genetics and the public's health*. Routledge, London.

Petersen, A. R. and Lupton, D. (1996) *The new public health: Health and self in the age of risk*. Allen & Unwin, St Leonards, NSW.

Phillimore, P. and Moffatt, S. (1994) Discounted knowledge: Local experience, environmental pollution and health. In Popay, J. and Williams, G. (eds), *Researching the people's health*, Routledge, London.

Phillips, R. (2003) The Future Role of the Divisions Network: report of the review of the role of divisions of general practice. Commonwealth of Australia, Canberra, ACT.

Pholeros, P., Rainow, S. and Torzillo, P. (1993) Housing for health: Towards a healthy living environment for Aboriginal Australia. Health Habitat, Sydney, NSW.

Pierson, C. (1994) *Beyond the welfare state*. Polity Press, London.

Piketty, T. (2014) *Capital in the twenty-first century*. Harvard University Press, Cambridge, MA.

Pill, R. (1988) Health beliefs and behaviours in the home. In Anderson, R. (ed), *Health behaviour research and health promotion*, Oxford University Press, Oxford. pp. 140–153.

Pilotto, L. S., Burch, M. D., Douglas, R. M., Cameron, S., Beers, M., Rouch, G. J., Robinson, P., Kirk, M., Cowie, C. T., Hardiman, S., Moore, C. and Attewell, R. G. (1997) Health effects of exposure to cyanobacteria (blue–green algae) during recreational water–related activities. *Australian and New Zealand Journal of Public Health*, 21, 562–566.

Pinner, R. W., Teutsch, S. M., Simonsen, L., Klug, L. A., Graber, J. M., Clarke, M. J. and Berkelman, R. L. (1996) Trends in infectious diseases mortality in the United States. *JAMA*, 275, 189–193.

Pinto, A. D., Manson, H., Pauly, B., Thanos, J., Parks, A. and Cox, A. (2012) Equity in public health standards: A qualitative document analysis of policies from two Canadian provinces. *International Journal for Equity in Health*, 11. From www.equityhealthj.com/content/11/1/28

Pirkis, J., Hickie, I., Young, L., Burns, J., Highet, N. and Davenport, T. (2005) An evaluation of beyondblue, Australia's national depression initiative. *International Journal of Mental Health Promotion*, 7, 35–53.

Plant, A. (1995) Emerging infectious diseases: What should Australia do? *Australian Journal of Public Health*, 19, 541–542.

Plant, S. (1999) *Writing on drugs*. Faber & Faber, London.

Platt, A. E. (1996) Confronting infectious diseases. In Brown, L. R. (ed), *State of the World 1996*, Worldwatch Institute/Earthscan, London. pp. 114–132.

Pocock, B. (2003) *The work/life collision*. Federation Press, Leichhardt, NSW.

Pocock, B., Williams, P. and Skinner, N. (2012) *Time bomb: Work, rest and play in Australia today*. NewSouth, Sydney, NSW.

Poland, B., Coburn, D., Robertson, A. and Eakin, J. (1998) Wealth, equity and health care: A critique of a "population health" perspective on the determinants of health. *Social Science and Medicine*, 46, 785–798.

Pollock, A. M. and Price, D. (2013) From cradle to grave. In Davis, J. and Tallis, R. (eds), *NHS SOS: how the NHS was betrayed—and how we can save it*, Oneworld, London. pp. 174–203.

Pomagalska, D., Putland, C., Ziersch, A., Baum, F., Arthurson, K., Orchard, L. and House, T. (2009) *Practical social capital: A guide to creating health and wellbeing*. Finders University, Adelaide, SA.

Popay, J., Bennett, S., Thomas, C., Williams, G., Gatrell, A. and Bostock, L. (2003) Beyond "beer, fags, egg and chips"? Exploring lay understanding of social inequalities in health. *Sociology of Health and Illness*, 25, 1–23.

Popay, J., Whitehead, M. and Hunter, D. J. (2010) Injustice is killing people on a large scale—but what is to be done about it? *Journal of Public Health*, 32, 148–149.

Popay, J. and Williams, G. (1996) Public health research and lay knowledge. *Social Science and Medicine*, 42, 759–768.

Popper, K. (1972) *The logic of scientific discovery*. Hutchinson, London.

Portes, A. and Landolt, P. (1996) The downside of social capital. *American Prospect*, 26, 18–21.

Potter, G. (1988) *Dialogue on debt: Alternative analyses and solutions*. Center of Concern, New York, NY.

Potts, L. K. (2004) An epidemiology of women's lives: The environmental risk of breast cancer. *Critical Public Health*, 14, 133–147.

Powell-Davies, G. and Fry, D. (2004) General practice in the health system, *General Practice in Australia 2004*. Primary Care Division, Australian Department of Health and Ageing, Canberra, ACT.

Power, C., S., M. and Manor, O. (1996) Inequalities in self-rated health in the 1958 birth cohort: Lifetime, social circumstances or social mobility? *BMJ*, 313, 449–453.

Powles, J. (1973) On the limitations of modern medicine. *Science, Medicine and Man*, 1, 1–30.

Powles, J. (1988) Professional hygienists and the health of the nation. In MacLeod, R. (ed), *The Commonwealth of Science: ANZAAS and the Scientific Enterprise in Australasia 1888–1988*, Oxford University Press, Melbourne, VIC. pp. 292–307.

Powles, J. and Gifford, S. (1990) How healthy are Australia's immigrants? In Reid, J. and Trompf, P. (eds), *The health of immigrant Australia: A social perspective*, Harcourt Brace Jovanovich, Sydney, NSW.

Powles, J. and Gifford, S. (1993) Health of nations: Lessons from Victoria, Australia. *British Medical Journal*, 306 125–127.

Powles, J. and Salzberg, M. (1989) Work, class or lifestyle? Explaining inequalities in health. In Lupton, G. M. and Najman, J. M. (eds), *Sociology of Health and Illness: Australian Readings*, Macmillan, Melbourne, VIC. pp. 135–168.

Press, I. and Smith, M. E. (1980) Urban places and process: Readings in the anthropology of cities. Macmillan, New York, NY.

Price, D., Pollock, A. M. and Shaoul, J. (1999) How the World Trade Organization is shaping domestic policies in health care. *The Lancet*, 354, 1889–1892.

Prochaska, J. O. and DiClemente, C. (1984) *The transtheoretical approach: Crossing traditional foundations of change*. Don Jones/Irwin, Hanrewood, IL.

Prochaska, J. O. and DiClemente, C. (1992) In search of how people change. *American Psychologist*, 47, 1102–1114.

Productivity Commission (PC) (2014) *Overcoming Indigenous disadvantage: Key indicators 2014*. Productivity Commission, Canberra, ACT.

Prugh, T., Flavin, C. and Sawin, J. L. (2005) Changing the oil economy. In Stark, L. (ed), *State of the World 2005: Redefining global security*, Worldwatch Institute/WW Norton & Co, New York, NY.

Prüss-Üstün, A. and Corvalán, C. (2006) Preventing disease through healthy environments. Towards an estimate of the environmental burden of disease. World Health Organization, Geneva.

Public Health Association of Australia (PHAA) (2012) Policy-at-a-glance – Immunisation Policy. From www.phaa.net.au/documents/130201_Immunisation%20Policy%20FINAL.pdf

Public Health Information Development Unit (PHIDU) (2014) *Data Archive: Social Health Atlases of Australia—Released Online: 2008 to 2013*. PHIDU, University of Adelaide, Adelaide, SA.

Punch, M. (1994) Politics and ethics in qualitative research. In Denzin, N. K. and Guba, Y. S. (eds), *Handbook of qualitative research*, SAGE Publications, Thousand Oaks, CA. pp. 83–97.

Pusey, M. (1991) Economic rationalism in Canberra: A nation building state changes its mind. Cambridge University Press, Cambridge.

Pusey, M. (2010) 25 years of neo-liberalism in Australia. In McKnight, D. and Manne, R. (eds), *Goodbye to all that? On the failure of neo-liberalism and the urgency of change*, Black Inc. Agenda, Melbourne, VIC. pp. 124–146.

Puska, P., Nissinen, A. and Tuomilehto, J. (1985) The community-based strategy to prevent coronary heart disease: Conclusions from the ten years of the North Karelia project. *Annual Review of Public Health*, 6, 147–193.

Puska, P., Vartiainen, E., Laatikainen, T., Jousilahti, P. and Paavola, M. (2009) *The North Karelia Project: from North Karelia to national action*. National Institute for Health and Welfare, Helsinki, Finland.

Putland, C., Baum, F. and MacDougall, C. (1997) How can health bureaucracies consult effectively about their policies and practices? Some lessons from an Australian study. *Health Promotion International*, 12, 299–309.

Putnam, R. D. (1993a) *Making democracy work: Civic traditions in modern Italy*. Princeton University Press, Princeton, NJ.

Putnam, R. D. (1993b) Social capital and public life. *New Prospect*, 13, 35–42.

Putnam, R. D. (1996) The strange disappearance of civic America. *American Prospect*, 24, 34–48.

Putnam, R. D. (2000) Bowling alone: The collapse and revival of American community. Simon & Schuster, New York, NY.

Quiggin, J. (2012) *Zombie Economics: How dead ideas still walk among us*. Princeton University Press, Princeton, NJ.

Raftery, J. (1995) The social and historical context. In Baum, F. (ed), *Health for All: The South Australian experience*, Wakefield Press, Adelaide, SA. pp. 19–37.

Raloff, J. (2012) Insurance payouts point to climate change. *Science News*, 4 January. From www.sciencenews.org/blog/science-public/insurance-payouts-point-climate-change

Randolph, W. and Viswanath, K. (2004) Lessons learned from public health mass media campaigns: Marketing health in a crowded media world. *Annual Review of Public Health*, 25, 419–437.

Rao, M., Afshin, A., Singh, G. and Mozaffarian, D. (2013) Do healthier foods and diet patterns cost more than less healthy options? A systematic review and meta-analysis. *BMJ open*, 3, e004277. 10.1136/bmjopen-2013-004277.

Raphael, D. (2012) An analysis of international experiences in tackling health inequalities. In Raphael, D. (ed), *Tackling health inequalities. Lessons from international experiences*, Chapter 9, Canadian Scholars' Press, Toronto, Ontario, Canada. pp. 229–264.

Rappaport, J. (1987) Terms of empowerment/exemplars or prevention: Towards a theory for community psychology. *American Journal of Community Psychology*, 15, 121–148.

Rawles, J. (1971) *A theory of justice*. Harvard University Press, Cambridge, MA.

Reason, P. (ed) (1988) Human inquiry in action: Developments in new paradigm research. SAGE Publications, London.

Reason, P. (1994) Three approaches to participative inquiry. In Denzin, N. K. and Lincoln, Y. S. (eds), *Handbook of qualitative research*, SAGE Publications, Thousand Oaks, CA. pp. 324–339.

Reconciliation Australia (2010) About us. From www.reconciliation.org.au/home/about-us (accessed 21 August 2013).

Redland City Council (2010) Redlands 2030 Community Plan. From www.redland.qld.gov.au/AboutCouncil/CommunityPlan/Pages/default.aspx (accessed 12 December 2014).

Rees, S. and Rodley, G. (1995) *The human cost of managerialism*. Pluto Press, Leichhardt, NSW.

Rees, W. E. and Wackernagel, M. (1994) Ecological footprints and appropriated carrying capacity: Measuring the natural capital requirement of the human economy. In Jannson, A., Hammer, M., Folke, C. and Costanza, R. (eds), *Investing in natural capital: The ecological economics approach to sustainability*, Island Press, Washington, DC.

Reid, E. (1997) Power, participation and partnerships for health promotion. In Labonté, R. (ed), *Power, participation and partnerships for health*, VicHealth, Melbourne, VIC. pp. 1–11.

Renner, M. (1999) Ending violent conflict. In Brown, L. R. (ed), *State of the World 1999*, W. W. Norton & Co, New York.

Renner, M. and Prugh, T. (2014) Failing governance, unsustainable planet. In Mastny, L. (ed), *State of the World 2014: Governing for sustainability*, Worldwatch Institute/Island Press, Washington, DC. pp. 3–19.

Repower Port Augusta Alliance (n.d.). From http://repowerportaugusta.org (accessed 11 June 2014).

Research Centre for Injury Studies (RCIS) (2000) *Deaths from motor vehicle accidents data 1999*. RCIS, Flinders University, Adelaide, SA.

Revicki, D. A. and Mitchell, J. P. (1990) Strain, social support and mental health in rural elderly individuals. *Journal of Gerontology: Social Sciences*, 45, s267–274.

Reynolds, C. (1989) Editorial: Legislation and the new public health: Introduction. *Community health studies*, 13, 397–402.

Reynolds, C. (1995) Health and public policy: The tobacco laws. In Baum, F. (ed), *Health for All: The South Australian experience*, Wakefield Press, Adelaide, SA. pp. 215–229.

Reynolds, C. (2011) *Public and environmental law*. Sydney Federation Press, Sydney, NSW.

Rice, M. and Rasmusson, E. (1992) Healthy cities in developing countries. In Ashton, J. (ed), *Healthy cities*, Open University Press, Milton Keynes. pp. 70–81.

Richards, D. and Smith, M. J. (2002) *Governance and public policy in the United Kingdom*. Oxford University Press, Oxford.

Richardson, D. and Denniss, R. (2014) Income and wealth inequality in Australia, *Policy Brief No. 64*. The Australia Institute, Canberra, ACT.

Rider, M. and Flynn, B. (1992) Indiana. In Ashton, J. (ed), *Healthy cities*, Open University Press, Milton Keynes. pp. 195–204.

Rifkin, S. and Walt, G. (1986) Why health improves: Defining the issues concerning 'comprehensive primary health care' and 'selective primary health care'. *Social Science and Medicine*, 23, 559–566.

Rissel, C. (1994) Empowerment: The holy grail of health promotion. *Health Promotion International*, 9, 39–47.

Rissel, C., Curac, N., Greenaway, M. and Bauman, A. (2012) Key health benefits associated with public transport, *An evidence check review brokered by the Sax Institute for the NSW Ministry of Health*. Sax Institute, Haymarket, NSW.

Rissel, C. and McCue, P. (2014) Healthy places and spaces: The impact of the built environment and active transport on physical activity and population health. *Health Promotion Journal of Australia*, 25, 155–156.

Ritchie, J. and Spencer, L. (1994) Qualitative data analysis for applied policy research. In Bryman, A. and Burgess, R. G. (eds), *Analyzing qualitative data*, Routledge, London. pp. 173–194.

Ritchie, J. E. (1996) Using participatory research to enhance health in the work setting: An Australian experience. In de Koning, K. and Martin, M. (eds), *Participatory research in health: Issues and experiences*, Zed Books, London.

Robert, C. F., Bouvier, S. and Rougemont, A. (1989) Epidemiology, anthropology and health education. *World Health Forum*, 10, 355–364.

Roberts, I., Owen, H., Lumb, P. and MacDougall, C. (1995) *Pedalling health: Health benefits of a modal transport shift*. SA Department of Transport, Adelaide, SA.

Robertson, J. (1989) Future wealth: A new economics for the 21st century. Cassell, London.

Robson, C. (2000) *Small-scale evaluation*. SAGE Publications, London.

Rocan, C. (2009) Multi-level collaborative governance: The case of the Canadian Heart Health Initiative. University of Ottawa, Canada.

Roddick, A. (ed) (2001) *Take it personally*. Thorsons, London.

Rodmell, S. W., A. (1986) *The politics of health education*. Routledge & Kegan Paul, London.

Rogers, R. G., Everett, B. G., Saint Onge, J. M. and Krueger, P. M. (2010) Social, behavioral, and biological factors, and sex differences in mortality. *Demography*, 47, 555–578.

Romm, J. J. and Ervin, C. A. (1996) How energy policy affects public health. *Public Health Reports*, 111, 391–399.

Rose, G. (1985) Sick individuals and sick populations. *International Journal of Epidemiology*, 14, 32–38.

Rose, G. (1992) *The strategy of preventive medicine*. Oxford University Press, Oxford.

Rose, G. and Marmot, M. (1981) Social class and coronary heart disease. *British Heart Journal*, 45, 13–19.

Rosella, L. C., Wilson, K., Crowcroft, N. S., Chu, A., Upshur, R., Willison, D., Deeks, S., Schwartz, B., Tustin, J., Sider, D. and Goel, V. (2013) Pandemic H1N1 in Canada and the use of evidence in developing public health policies: A policy analysis. *Social Science and Medicine*, 83, 1–9.

Rosen, G. (1958) *A history of public health*. MD Publications, New York, NY.

Roser, M. (2014) 'Trust'. Our world in data [Online Resource]. From www.ourworldindata .org/data/culture-values-and-society/trust (accessed 27 December 2014).

Ross, C. E. and Huber, J. (1985) Hardship and depression. *Journal of Health and Social Behavior*, 26, 312–327.

Ross, E. (1991) The origins of public health: Concepts and contradictions. In Draper, P. (ed), *Health through public policy*, Green Print, London. pp. 26–40.

Ross, M., Toohey, D., Peinemann, M. and Ross, P. (2009) Limits on the space launch market related to stratospheric ozone depletion. *Astropolitics*, 7, 50–82.

Royal Australian College of General Practitioners (RACGP) (2012) *Guidelines for preventive activities in general practice (the red book)*. Royal Australian College of General Practitioners, East Melbourne, VIC. From www.racgp.org.au/redbook

Ruberman, W., Weinblatt, E., Goldberg, J. D. and Chaudhary, B. S. (1984) Psychosocial influences on mortality after myocardial infarction. *New England Journal of Medicine*, 311, 552–559.

Rudd, K. (2008) Prime Minister Kevin Rudd—Apology to Australia's Indigenous peoples. 13 February. From http://australia.gov.au/about-australia/our-country/our-people/apology-to-australias-indigenous-peoples (accessed 23 May 2014).

Rukmana, D. (2014) The megacity of Jakarta: Problems, challenges and planning efforts, *Indonesia's Urban Studies*. 29 March. From http://indonesiaurbanstudies.blogspot.com. au/2014/03/the-megacity-of-jakarta-problems.html

Ryan, W. (1972) *Blaming the victim*. Vintage Press, New York, NY.

Sachs, A. (1996) Upholding human rights and environmental justice. In Brown, L. R. (ed), *State of the World 1996*, Worldwatch Institute/Earthscan, London. pp. 133–151.

Sacks, G. (2014) Big Food lobbying: tip of the iceberg exposed *The Conversation*, 19 February. From http://theconversation.com/big-food-lobbying-tip-of-the-iceberg-exposed-23232

Saegert, S. (1985) The androgynous city: From critique to practice. *Sociological Focus*, 18, 161–176.

Safe Work Australia (2012a) *Australian Work Health and Safety Strategy 2012–2022*. Safe Work Australia, Canberra, ACT.

Safe Work Australia (2012b) The cost of work-related injury and illness for Australian employers, workers and the community, 2008–09. Safe Work Australia, Canberra, ACT.

Safe Work Australia (2014) *Work-related traumatic injury fatalities Australia, 2013*. Safe Work Australia, Canberra, ACT.

Saggers, S. and Gray, D. (1991) Aboriginal health and society: The traditional and contemporary Aboriginal struggle for better health. Allen & Unwin, Sydney, NSW.

Saint-Arnaud, S. and Bernard, P. (2003) Convergence or resilience? A hierarchical cluster analysis of the welfare regimes in advanced countries. *Current Sociology*, 51, 499–527.

Saltman, D. (1991) *Women's health: An introduction to issues*. Harcourt Brace Jovanovich, Sydney, NSW.

Sample, I. (2013) Big pharma mobilising patients in battle over drugs trials data. *The Guardian*, 22 July. From www.theguardian.com/business/2013/jul/21/big-pharma-secret-drugs-trials?CMP=EMCNEWEML6619I2

Sampson, R., Raudenbush, S. and Earls, F. (1997) Neighborhoods and violent crime: a multilevel study of collective efficacy. *Science*, 277, 918–924.

Sandel, M. J. (2012) What money can't buy: The moral limits of markets. Allen Lane, London.

Sandelowski, M. (1986) The problem of rigor in qualitative research. *Advanced Nursing Studies*, 8, 27–37.

Sanders, D. (1985) The struggle for health: Medicine and the politics of underdevelopment. Macmillan, London.

Sanders, D. (2006) A global perspective on health promotion and the social determinants of health. *Health Promotion Journal of Australia*, 17, 165–167.

Sanders, D., Baum, F., Benos, A. and Legge, D. (2011) Revitalising primary health care requires an equitable global economic system: Now more than ever. *Journal of Epidemiology and Community Health*, 65, 661–665.

Sanders, D., Labonté, R., Baum, F. and Chopra, M. (2004) Making research matter: A civil society perspective on health research. *Bulletin of the World Health Organization*, 82, 757–763.

Sanders, D., Todd, C. and Chopra, M. (2005) Education and debate. Confronting Africa's health crisis: More of the same will not be enough. *British Medical Journal*, 331, 755–758.

Sanders, R. (1993) Is Bruntland's model of sustainable development a case of having our cake and eating it too? *Ecopolitics VII Conference*, Griffith University.

Sanderson, C. and Baum, F. (1995) Health and social welfare councils. In Baum, F. (ed), *Health for All: The South Australian experience*, Wakefield Press, Adelaide, SA. pp. 67–81.

Sarantakos, S. (2005) *Social research*. Palgrave Macmillan, New York, NY.

Sarkissian, W. and Heine, W. (1978) *Social mix: The Bournville experience*. Bournville Village Trust and the South Australian Housing Trust, Adelaide, SA.

Sassen, S. (1999) *Guests and aliens*. New Press, New York, NY.

Sassen, S. (2014) *Expulsions: Brutality and complexity in the global economy*. Harvard University Press, Harvard, MA.

Saul, J. R. (1997) *The unconscious civilisation*. Penguin Books, Ringwood, VIC.

Sax, S. (1984) A strife of interests: Politics and policies in Australian health services. George Allen & Unwin, Sydney, NSW.

Sax, S. (1990) *Health care choices and the public purse*. Allen & Unwin, Sydney, NSW.

Saywell, T., Fowler, G. A. and Crispin, S. W. (2003) SARS deals blow to Asian economies analysis shows virus could drain $10.6 billion from 2003's forecast GDP, *Wall Street Journal (Eastern Edition)*, 21 April, p.10.

Schlosser, C. A., Strzepek, K., Gao, X., Fant, C., Blanc, É., Paltsev, S., Jacoby, H., Reilly, J. and Gueneau, A. (2014) The future of global water stress: An integrated assessment. *Earth's Future*, 2. 10.1002/2014EF000238.

Schneider, M. (2004) *The distribution of wealth*. Edward Elgar, Cheltenham, UK.

Schoenbach, V. J., Kaplan, B. H., Fredman, L. and Kleinbaum, D. G. (1985) Social ties and mortality in Evans County, Georgia. *American Journal of Epidemiology*, 123, 577–591.

Schram, S. F. and Silverman, B. (2012) The end of social work: Neoliberalizing social policy implementation. *Critical Policy Studies*, 6, 128–145.

Schrecker, T. (2012) How not to think about social determinants of health: A cautionary tale from Canada. *Healthy Policies for a healthier world*. From www.healthypolicies.com/2012/04/how-not-to-think-about-social-determinants-of-health-a-cautionary-tale-from-canada

Schumacher, E. F. (1973) Small is beautiful: Economics as if people mattered. Abacus, London.

Schwartz, S. (1994) The fallacy of the ecological fallacy: The potential misuse of a concept and the consequence. *American Journal of Public Health*, 84, 819–824.

Scollo, M. M. and Winstanley, M. H. (2012) *Tobacco in Australia: Facts and issues*. 4th Edition. Cancer Council Victoria, Melbourne, VIC.

Scott-Samuel, A. (1995) A new synthesis: Population health research for the 21st century. In Bruce, N., Springett, J., Hotchkiss, J. and Scott-Samuel, A. (eds), *Research and change in urban community health*, Avebury, Aldershot. pp. 49–55.

Seabrook, J. (1984) *The idea of neighbourhood*. Pluto Press, London.

Seabrook, J. (1995) The population humbug. In Bissio, R. R. (ed), *The World: A Third World Guide 1995/6*, Instituto del Tercer Mundo (Third World Institute), Montevideo, Uruguay. pp. 10–11.

Sebastián, M. S. and Hurtig, A. K. (2005) Oil development and health in the Amazon basin of Ecuador: The popular epidemiology process. *Social Science and Medicine*, 60, 799–807.

Seedhouse, D. (2002) Total health promotion: Mental health, rational fields and the request for autonomy. John Wiley, Hoboken, NJ.

Seidman, I. (2005) Interviewing as qualitative research: A guide for researchers in education and the social sciences. Teachers College Press, New York, NY.

Self, P. (1997) Governments and the cult of economic rationalism. *Public Service Association Review*, 13–14.

Seligman, M. (2012) Flourish: A visionary new understanding of happiness and well-being. Free Press, New York, NY.

Sen, A. (1992) *Inequality reexamined*. Harvard University Press, Boston, MA.

Sen, A. (1999) *Development as freedom*. Oxford University Press, Oxford.

Sen, A. (2001a) Economic progress and health. In Leon, D. and Walt, G. (eds), *Poverty, inequality and health: An international perspective*, Oxford University Press, Oxford.

Sen, A. (2001b) Many faces of gender inequality. *Frontline*, 18, 4–14.

Sen, A. (2013) Universal health care in India: Making it public, making it a reality. In McDonald, D. A. and Ruiters, G. (eds), *Occasional Paper No. 19*, Municipal Services Project.

Senate Community Affairs Committee (2013) Australia's domestic response to the World Health Organisation's (WHO) Commission on Social Determinants of Health report "Closing the gap within a generation". Commonwealth of Australia, Canberra, ACT.

Senate Select Committee on Regional and Remote Indigenous Communities (SSCRRIC) (2010) *Indigenous Australians, incarceration and the criminal justice system.* Commonwealth of Australia, Canberra, ACT.

Sendziuk, P. (2002) Denying the grim reaper: Australian responses to AIDS. *Eureka Street*. From www.eurekastreet.com.au/articles/0310sendziuk.html

Shakespeare, P., Atkinson, D. and Franch, S. (eds) (1993) Reflecting on research practice: Issues in health and social welfare. Open University Press, Buckingham.

Shan, H., Muhajarine, N., Loptson, K. and Jeffery, B. (2014) Building social capital as a pathway to success: Community development practices of an early childhood intervention program in Canada. *Health Promotion International*, 29, 244–255.

Shannon, P. and Worsley, A. (1991) A long-term investment: Report on the evaluation of the health and social welfare councils of South Australia. South Australian Health Commission, Adelaide, SA.

Shapcott, M. (2004) Housing. In Raphael, D. (ed), *Social determinants of health: Canadian perspectives*, Canadian Scholars' Press, Toronto, Ontario.

Shaw, M. (2004) Housing and public health. *Annual Review of Public Health*, 25, 1–22.

Sheills, A. and Hawe, P. (1996) Health promotion community development and the tyranny of individualism. *Health Economics*, 5, 241–247.

Shields, K. (1991) In the tiger's mouth: An empowerment guide for social action. Millennium Books, Sydney, NSW.

Shiva, V. (1988) Staying alive: Women, ecology and development. Zed Books, London.

Shiva, V. (1996) Economic globalisation, ecological feminism and sustainable development, *6th International Interdisciplinary Congress on Women*, Adelaide, SA.

Shiva, V. (1998) *Biopiracy: The plunder of nature and knowledge*. Green Books, Totnes, Devon.

Shiva, V. (2014) Seeds of Freedom. *The Asian Age*, 23 April. From http://seedfreedom.in/seeds-of-freedom

Shiva, V. and Shiva, M. (1995) Third world women denied right to development. *Impact*, 30, 19–27.

Short, J. R. (1989) *The humane city: Cities as if people matter*. Basil Blackwell, Oxford.

Shuttleworth, C. and Auer, J. (1995) Women's health centres in Adelaide. In Baum, F. (ed), *Health for All: The South Australian experience*, Wakefield Press, Adelaide, SA. pp. 253–267.

Shy, C. M. (1997) The failure of academic epidemiology: Witness for the prosecution. *American Journal of Epidemiology*, 145, 479–484.

Siegel, D. J. (2012) *Pocket guide to interpersonal neurobiology*. WW Norton & Co., New York.

Sigerist, H. (1941) *Medicine and human welfare*. Yale University Press, New Haven.

Silagy, C. (1993) Developing a register of randomised controlled trials in primary care. *British Journal of Medicine*, 306, 897–900.

Simberkoff, M. (1994) Drug-resistant pneumococcal infections in the United States. *JAMA*, 271, 1875–1876.

Simnett, I. (1995) Managing health promotion: Developing health promotion. John Wiley & Sons, Chichester.

Sivasakthivel, T. and Reddy, K. K. (2011) Ozone layer depletion and its effects: a review. *International Journal of Environmental Science and Development*, 2, 30–37.

Skolbekken, J. A. (1995) The risk epidemic in medical journals. *Social science and medicine*, 40, 291–305.

Smith, A. B. and Katz, R. W. (2013) US billion-dollar weather and climate disasters: Data sources, trends, accuracy and biases. *Natural Hazards*, 67, 387–410.

Smith, B. (1995) Promoting healthy eating. In Baum, F. (ed), *Health for All: The South Australian experience*, Wakefield Press, Adelaide, SA. pp. 317–338.

Smith, C., Moore, L., Roberts, C. and Catford, J. (1994) Health-related behaviours in Wales, 1985–1990. *Health Trends*, 26, 18–21.

Smith, J. W. and Maddern, G. J. (2011) Surgical implications of global warming. *Medical Journal of Australia*, 195, 32.

Smith, R. (1993) How the road toll war was won. *Sunday Age*, Melbourne.

Smith, S. E., Pyrch, T. and Lizardi, A. O. (1993) Participatory action-research for health. *World Health Forum*, 14, 319–324.

Smith, W., Mitchell, P., Attebo, K. and Leeder, S. (1997) Selection bias from sampling frames: Telephone directory and electoral roll compared with door-to-door populations census: Results from the Blue Mountain eye study. *Australian and New Zealand Journal of Public Health*, 21, 127–133.

Social Determinants of Health Alliance (n.d.) From www.socialdeterminants.org.au/ (accessed 29 July 2014).

Solar, O. and Irwin, A. (2005) Towards a conceptual framework for analysis and action on the social determinants of health. World Health Organization, Geneva.

Sontag, S. (1979) *Illness as metaphor*. Allen Lane, London.

Soul City Institute (2007) Soul City evaluation report: Series 7. Soul City Institute, South Africa.

South Australian Community Health Research Unit (SACHRU) (1991) Planning healthy communities. SACHRU, Flinders University, Adelaide, SA.

South Australian Community Health Research Unit (SACHRU) (1996) Changing times: Planning, evaluation and outcomes in metropolitan community and women's health services. SACHRU, Flinders University, Adelaide, SA.

South Australian Council of Social Service (SACOSS) (n.d.) A fair and sustainable tax base. From http://sacoss.org.au/fair-and-sustainable-tax-base (accessed 5 January 2015).

South Australian Health Commission (SAHC) (1988) A Social Health Strategy for South Australia. SAHC, Adelaide, SA.

South Australian Health Commission (SAHC) (1989) Policy on primary health care. SAHC, Adelaide, SA.

Spies-Butcher, B. (2014) Marketisation and the dual welfare state: Neoliberalism and inequality in Australia. *The Economic and Labour Relations Review*, 25, 185–201. 10.1177/1035304614530076.

Spradley, J. (1980) *Participant observation*. Holt, Rinehart & Winston, New York, NY.

Springett, J. (2003) Issues in participatory evaluation. In Minkler, M. and Wallerstein, N. (eds), *Community-based participatory research for health*, Jossey-Bass, San Francisco, CA.

Stainton-Rogers, W. (1991) *Explaining health and illness: An exploration of diversity*. Harvester Wheatsheaf, London.

Stake, R. E. (1995) *The art of case study research*. SAGE Publications, Thousand Oaks, CA.

Standing Committee on Agriculture and Resource Management and Agriculture and Resource Management Council of Australia and New Zealand (1997) Evaluation report on the decade of Landcare plan: National overview. Department of Agriculture, Fisheries and Forestry, Australian Government, Canberra, ACT.

Standing Senate Committee on Social Affairs, Science and Technology (2001) The health of Canadians—The federal role. Volume 1—The story so far, *Interim Report on the state of health care system in Canada*. Parliament of Canada.

Stanley, L. and Wise, S. (1990) Method, methodology and epistemology in feminist research processes. Feminist praxis. Research theory and epistemology. In Stanley, L. (ed), *Feminist sociology*, Chapter 2, Routledge & Kegan Paul, London.

Stansfeld, S. A. (1999) Social support and social cohesion. In Marmot, M. and Wilkinson, R. G. (eds), *Social determinants of health*, Oxford University Press, Oxford.

Starfield, B., Shi, L. and Macinko, J. (2005) Contribution of primary care to health systems and health. *Milbank Quarterly*, 83, 457–502.

State of the Environment Advisory Council (1996) State of the Environment (SOE) 1996. An Independent Report Presented to the Commonwealth Minister for the Environment. CSIRO Publishing, Collingwood, VIC.

State of the Environment 2011 Committee (2011). Australia state of the environment 2011. Independent report to the Australian Government Minister for Sustainability, Environment, Water, Population and Communities. Department of Sustainability, Environment, Water, Population and Communities, Canberra, ACT.

Stavropoulos, P. (2008) Living under liberalism: The politics of depression in western democracies. Florida Universal Publishers, Boca Raton, FL.

Steffen, W., Grinevald, J., Crutzen, P. and McNeill, J. (2011) The anthropocene: Conceptual and historical perspectives. *Philosophical Transactions of the Royal Society A: Mathematical, Physical and Engineering Sciences*, 369, 842–867.

Stendahl, K. (2014) Statement by Kerstin Stendahl, Executive Secretary ad interim of the Basel, Rotterdam and Stockholm Conventions, to the fifth meeting of the Assembly of the GEF, Cancun, Mexico, 28 May. From http://chm.pops.int/Implementation/PublicAwareness/Speeches/ExecutiveSecretary5thAssemblyGEF/tabid/3773/Default.aspx

Stern, N. (2007) *The economics of climate change: The Stern Review*. Cambridge University Press, Cambridge.

Stiglitz, J. (2002) *Globalization and its discontents*. WW Norton & Co, New York, NY.

Stiglitz, J. (2007) *Making globalization work*. WW Norton & Co, New York, NY.

Stiglitz, J. (2012) The price of inequality: How today's divided society endangers our future. WW Norton & Co, New York, NY.

Stock, C. and Ellaway, A. (eds) (2013) Neighbourhood structure and health promotion. Springer, New York, NY.

Stone, W. and Hughes, J. (2000) What role for social capital in family policy? *Family Matters*, 56, 20–25.

Stott, R. (2000) The Ecology of Health. Green Books, Totnes, Devon.

Stretcher, V. J., DeVellis, M., Cevoy, B., Becker, M. H. and Rosenstock, I. M. (1986) The role of self-efficacy in achieving health behaviour change. *Health Education Quarterly*, 13, 74–92.

Stretton, H. (1987) *Political essays*. Georgian House, Melbourne, VIC.

Stretton, H. (2000) *Economics: A new introduction*. University of NSW Press, Sydney, NSW.

Stretton, H. (2005) *Australia fair*. University of NSW Press, Sydney, NSW.

Strong, K., Mathers, C., Leeder, S. and Beaglehole, R. (2005) Preventing chronic diseases: how many lives can we save? *The Lancet*, 366, 1578–1582.

Stuckler, D. and Basu, S. (2013) *The body economic: Why austerity kills*. Basic Books, New York, NY.

Substance Abuse and Mental Health Services Administration (SAMHSA) (2014) Report shows rise in methamphetamine-related hospital emergency department visits, 19 June. From www.samhsa.gov/newsroom/press-announcements/201406191200

Suzuki, D. (2010) The legacy: An elder's vision for our sustainable future. Greystone Books, Vancouver, BC.

Suzuki, D. and Knudtson, P. (1992) *Wisdom of the elders*. Bantam Books, New York, NY.

Swan, P. and Raphael, B. (1995) Ways forward: National consultancy report on Aboriginal and Torres Strait Islander mental health. Parts 1 & 2. Commonwealth of Australia, Canberra, ACT.

Sweet, M. (2012) Overview of Qld Health changes, including the "historic" dismantling of public and preventative health services, 11 September, *Croakey*. From http://blogs.crikey. com.au/croakey/2012/09/11/overview-of-qld-health-changes-including-the-historic-dismantling-of-public-and-preventative-health-services

Swerissen, H. and Duckett, S. (1997) Health policy and financing. In Gardner, H. (ed), *Health policy in Australia*, Oxford University Press, Melbourne, VIC. pp. 13–45.

Syme, S. L. (1996) To prevent disease: The need for a new approach. In Blane, D., Brunner, E. and Wilkinson, R. (eds), *Health and social organisation: Towards a health policy for the 21st century*, Routledge, London.

Szreter, S. (1988) The importance of social intervention in Britain's mortality decline c. 1850–1914: A reinterpretation of the role of public health. *Society for the Social History of Medicine*, 1, 1–37.

Szreter, S. (1992) Mortality and public health 1815–1914. In Digby, A., Feinstein, C. and Jenkins, D. T. (eds), *New directions in economic and social history*, Macmillan, London.

Szreter, S. (1995) Rapid population growth and security: Urbanisation and economic growth in Britain in the nineteenth century. Centre for History and Economics, King's College, Cambridge.

Szreter, S. (2002) The state of social capital: Bringing back in power, politics, and history. *Theory and Society*, 31, 573–621.

Szreter, S. and Woolcock, M. (2004) Health by association? Social capital, social theory, and the political economy of public health. *International Journal of Epidemiology*, 33, 650–667.

Talbot-Smith, A. and Pollock, A. M. (2006) *The new NHS: A guide*. Routledge, London.

Tarasuk, V. (2004) Health implications of food insecurity. In Raphael, D. (ed), *Social determinants of health: Canadian perspectives*, Chapter 13, Canadian Scholars' Press, Toronto, Ontario.

Tarimo, E. and Webster, E. G. (1994) *Primary health care concepts and challenges in a changing world. Alma Ata revisited*. Division of Strengthening of Health Services, World Health Organization, Geneva.

Tausig, M. (2013) The sociology of work and well-being. In Aneshensel, C. S., Phelan, J. C. and Bierman, A. (eds), *Handbook of the sociology of mental health*, 2nd edition, Chapter 21, Springer Science and Business Media, Dordrecht. pp. 433–455.

Tax Justice Network (2013) Beyond BEPS. TJN briefing on the OECD's "BEPS" project on corporate tax avoidance. Tax Justice Network.

Taylor, C. and Jolly, R. (1988) The straw men of primary health care. *Social Science and Medicine*, 26, 971–977.

Telphia, J., Menzies, R. and McIntyre, P. (2006) *Vaccination for our mob*. Commonwealth Department of Health and Ageing, Canberra, ACT.

Tenover, F. and Hughes, J. (1996) The challenges of emerging infectious diseases. *JAMA*, 275, 300–304.

Tesh, S. (1982) Political ideology and public health in the nineteenth century. *International Journal of Health Services*, 12, 321–342.

Tesh, S. (1988) Hidden arguments: Political ideology and disease prevention policy. Rutgers University Press, New Brunswick, NJ.

Tesoriero, F. (1995) Community development and health promotion. In Baum, F. (ed), *Health for All: The South Australian experience*, Wakefield Press, Adelaide, SA. pp. 268–280.

Teves, O. (2012) Protesters picket key tobacco show in Philippines. *Yahoo News*, 15 March. From http://news.yahoo.com/protesters-picket-key-tobacco-show-philippines-073230482.html

The Economist (2013a) Crash course. *The Economist*, 7 September. From www.economist.com/news/schoolsbrief/21584534-effects-financial-crisis-are-still-being-felt-five-years-article

The Economist (2013b) Soft drinks in Mexico: Fizzing with rage. A once-omnipotent industry fights what may be a losing battle. *The Economist*, 19 October. From www.economist.com/news/business/21588088-once-omnipotent-industry-fights-what-may-be-losing-battle-fizzing-rage

The Haven (n.d.) From www.haven.ca (accessed 31 January 2015).

The Lancet (2014) Health and health care in Australia at a crossroads [Editorial]. *The Lancet*, 383, 2100.

Themba-Nixon, M. (2010) The power of local communities to foster policy. In Cohen, L., Chavez, V. and Chehimi, S. (eds), *Prevention is primary: Strategies for community well being*, Jossey-Bass, San Francisco, CA.

Thomas, D. P., Briggs, V., Anderson, I. P. and Cunningham, J. (2008) The social determinants of being an Indigenous non-smoker. *Australian and New Zealand Journal of Public Health*, 32, 110–116.

Thomson, H., Petticrew, M. and Douglas, M. (2003) Health impact assessment of housing improvement: Incorporating research evidence. *Journal of Epidemiology and Community Health*, 57, 11–16.

Tilden, J. (1996) Bunyas and bladoy grass. *New Internationalist*, 278, 22–23.

Titmuss, R. M. (1958) *Essays on "The Welfare State"*. George Allen & Unwin, London.

Tones, B. K. (1986) Health education and the ideology of health promotion: A review of alternative approaches. *Health Education Research*, 1, 3–12.

Tones, K. (1992) Health promotion, self-empowerment and the concept of control. In Colquhoun, D. (ed), *Health education: Politics and practice*, Deakin University, Geelong, VIC.

Tones, K. and Green, J. (2004) *Health Promotion: Planning and Strategies*. SAGE Publications, London.

Toole, M. J. and Waldman, R. J. (1997) The public health aspects of complex emergencies and refugees situations. *Annual Review of Public Health*, 18, 283–312.

Townsend, P. (1979) *Poverty in the United Kingdom*. Penguin, Harmondsworth, UK.

Townsend, P. (2004) From universalism to safety nets: The rise and fall of Keynesian influence on social development. In Mkandawire, T. (ed), *Social policy in a development context*, Palgrave, Houndsmills, UK.

Townsend, P., Davidson, N. and Whitehead, M. (eds) (1992) *Inequalities in health*. Penguin Books, London.

Travers, K. D. (1996) The social organisation of nutritional inequities. *Social Science and Medicine*, 43, 543–553.

Travers, P. and Richardson, S. (1993) *Living decently: Material well-being in Australia*. Oxford University Press, Melbourne, VIC.

Traynor, M. (1989) *Measuring the health of the city: The invisible Christies Downs*. SACHRU, Flinders University, Adelaide, SA.

Troy, P. (1996) *The perils of urban consolidation*. Federation Press, Sydney, NSW.

Tsey, K. (1996) Aboriginal health workers: Agents of change? *Australian and New Zealand Journal of Public Health*, 20, 227–228.

Tsouros, A. D. (1995) The WHO Healthy Cities project: State of the art and future plans. *Health Promotion International*, 10, 133–141.

Tsouros, A. D. (1996) A nine-year investment. *World Health*, 49, 7–9.

Tuomilehto, J. and Puska, P. (1987) The changing role and legitimate boundaries of epidemiology: Community-based prevention programmes. *Social Science and Medicine*, 25, 589–598.

Turner, B. S. (1984) *The body and society*. Basil Blackwell, Oxford.

Turrell, G., Kavanagh, A., Draper, G. and Subramanian, S. V. (2007) Do places affect the probability of death in Australia? A multilevel study of area-level disadvantage, individual-level socio-economic position and all-cause mortality, 1998–2000. *Journal of Epidemiology and Community Health*, 61, 13–19.

Turrell, G., Oldenburg, B., McGuffog, I. and Dent, R. (1999) *Socioeconomic determinants of health: Towards a national research program and a policy and intervention agenda*. School of Public Health, Queensland University of Technology, Brisbane, QLD.

Turrell, G., Western, J. S. and Najman, J. M. (1994) The measurement of social class in health research: Problems and prospects. In Waddell, C. and Petersen, A. R. (eds), *Just health: Inequality in illness care and prevention*, Churchill Livingstone, Melbourne, VIC. pp. 87–104.

Twenge, J. M., Campbell, W. K. and Carter, N. T. (2014) Declines in trust in others and confidence in institutions among American adults and late adolescents, 1972–2012. *Psychological science*, 25, 1914–1923.

Twiss, J., Duma, S., Look, V., Shaffer, G. S. and Watkins, A. C. (2000) Twelve years and counting: California's experience with a Statewide Healthy Cities and Communities Program. *Public Health Reports*, 115, 125–133.

Uchtenhagen, A., Gutzwiller, F., Dobler-Mikola, A. D. and Steffen, T. (1997) *Programme for a medical prescription of narcotics: Synthesis report*. Institute for Social and Preventive Medicine, University of Zurich.

UK Department of Health (2002) *Health promoting prisons: A shared approach*. UK Government, London.

Underwood, P., Owen, A. and Winkler, R. (1986) Replacing the clockwork model of medicine. *Community health studies*, 10, 275–283.

Unger, A. and Riley, L. W. (2007) Slum health: From understanding to action. *PloS Medicine*, 4, e295.

United for a Fair Economy (2011) Responsible Wealth. From http://faireconomy.org/responsible_wealth (accessed 5 January 2015).

United Nations (UN) (2012) Resolution adopted by the General Assembly. 66/2. Political Declaration of the High-level Meeting of the General Assembly on the Prevention and Control of Non-communicable Diseases. United Nations. From www.who.int/nmh/events/un_ncd_summit2011/political_declaration_en.pdf?ua=1

United Nations (UN) (2013) *The Millennium Development Goals Report 2013*. United Nations, New York, NY.

United Nations (UN) (2014) Water for Life Decade—Water scarcity. 24 November. From www.un.org/waterforlifedecade/scarcity.shtml

United Nations Department of Economic and Social Affairs (n.d.) Open Working Group proposal for Sustainable Development Goals. From https://sustainabledevelopment.un.org/focussdgs.html (accessed 8 January 2015).

United Nations Department of Economic and Social Affairs (UN DESA) (2013) *World Population Prospects: The 2012 Revision*. United Nations New York, NY.

United Nations Department of Economic and Social Affairs (UN DESA) (2014) *World Urbanization Prospects: The 2014 Revision, Highlights*. United Nations, New York, NY.

United Nations Development Programme (UNDP) (2000) *Human Development Report 2000. Human rights and human development*. United Nations, New York, NY.

United Nations Development Programme (UNDP) (2010) Human Development Report 2010. The real wealth of nations: Pathways to human development. United Nations, New York, NY.

United Nations Development Programme (UNDP) (2014) Human Development Report 2014. Sustaining human progress: Reducing vulnerabilities and building resilience. United Nations, New York, NY.

United Nations Environment Programme (UNEP) (1993) *Transport and the environment: Facts and figures*. United Nations, New York, NY.

United Nations High Commissioner for Refugees (UNHCR) (1995) *The State of the World's Refugees*. UNHCR, Geneva.

United Nations High Commissioner for Refugees (UNHCR) (2014) *UNHCR Global Trends 2013.* UNHCR, Geneva.

United Nations Research Institute for Social Development (UNRISD) (1995) *States of disarray: The social effects of globalization.* UNRISD, Geneva.

Ura, K. and Galay, K. (2004) *Gross national happiness and development.* Centre for Bhutan Studies, Thimphu, Bhutan.

US Centers for Disease Control and Prevention (CDC) (2013) Table 24 (page 1 of 3). Age-adjusted death rates, by race, sex, region, and urbanization level: United States, average annual, selected years 1996–1998 through 2008–2010, *Health, United States, 2013.* US CDC.

US Centers for Disease Control and Prevention (CDC) (2014a) Economic facts about U.S. tobacco production and use, 6 February. From www.cdc.gov/tobacco/data_statistics/ fact_sheets/economics/econ_facts

US Centers for Disease Control and Prevention (CDC) (2014b) Malaria Facts, 26 March. From www.cdc.gov/malaria/about/facts.html

US Energy Information Administration (EIA) (n.d.) International Energy Statistics. From www .eia.gov/countries/data.cfm (accessed 8 July 2014).

Utell, M. J., Warren, J. and Sawyer, R. F. (1994) Public health risks from motor vehicle emissions. *Annual Review of Public Health*, 15, 157–178.

Van der Kolk, B. A., McFarlane, A. C. and Weisaeth, L. (eds) (1996) *Traumatic stress: The effects of overwhelming experience on mind-body and society.* Guildford Press, New York, NY.

Van Eyk, H. (1996) Overcoming isolation: Ethnicity, ageing and the provision of health services (MSc thesis). Department of Public Health. Flinders University, Adelaide, SA.

Veerman, J. L. (2012) Health Impact Assessment. In Chadwick, R. (ed), *Encyclopedia of Applied Ethics*. 2nd Edition. Academic Press, San Diego. pp. 551–555.

Verbrugge, L. M. (1977) Sex differences in morbidity and mortality in the United States. *Biodemography and Social Biology*, 23, 275–296.

Verbrugge, L. M. (1979) Female illness rates and illness behavior: Testing hypothesis about sex differences in health. *Women and Health*, 4, 61–79.

Verbrugge, L. M. (1985) Gender and health: An update on hypotheses and evidence. *Journal of Health and Social Behavior*, 26, 156–182.

Verheul, E. and Cooper, G. (2001) Poverty Reduction Strategy Papers: What is at stake for health? Wemos, Amsterdam.

VicHealth (1999) Mental Health Promotion Plan. Foundation Document 1999–2002. VicHealth, Melbourne, VIC.

VicHealth (2011) Resources for Organisations: The partnerships analysis tool. From www .vichealth.vic.gov.au/media-and-resources/publications/the-partnerships-analysis-tool (accessed 15 January 2015).

VicHealth (2013) Fair Foundations: The VicHealth framework for health equity. From www .vichealth.vic.gov.au/Publications/Health-Inequalities/The-VicHealth-framework-for-health-equity.aspx (accessed 6 October 2014).

VicHealth (2014) Pilot Projects—Creating Healthy Workplaces program. From www .vichealth.vic.gov.au/programs-and-projects/creating-healthy-workplaces-program (accessed 21 January 2015).

Victorian Government (2015) Healthy Together Victoria. From www.health.vic.gov.au/ prevention/healthytogether.htm (accessed 21 January 2015).

Victorian Government Department of Health (2010) *Primary care partnerships: Achievements 2000 to 2010.* Integrated Care Branch of the Wellbeing, Integrated Care and Aged Division, Victorian Government Department of Health, Melbourne, VIC.

von Hippel, F. N. (2011) The radiological and psychological consequences of the Fukushima Daiichi accident. *Bulletin of the Atomic Scientists*, 67, 27–36.

Wade, R. H. (2001) The rising inequality of world income distribution. *Finance and Development*, 38, 1–6.

Wadsworth, M. E. J. (1997) Health inequalities in the life course perspective. *Social science and medicine*, 44, 859–869.

Wadsworth, Y. (1984) *Do it yourself social research*. Victorian Council of Social Services, Melbourne, VIC.

Wadsworth, Y. (1991) *Everyday evaluation on the run*. Action Research Issues Association, Melbourne, VIC.

Waitzkin, H. (1981) The social origins of illness: A neglected history. *International Journal of Health Sciences*, 11, 77–103.

Waitzkin, H. (2006) One and a half centuries of forgetting and rediscovering: Virchow's lasting contributions to social medicine. *Social Medicine*, 1, 5–10.

Wakefield, M., Freeman, J. and Inglis, G. (2004) Changes associated with the National Tobacco Campaign: Results of the third and fourth follow-up surveys, 1997–2000. *Australia's National Tobacco Campaign Evaluation Report Volume Three*. Commonwealth Department of Health and Ageing, Canberra, ACT.

Wakefield, M., Loken, B. and Hornik, R. C. (2010) Use of mass media campaigns to change health behaviour. *The Lancet*, 376, 1261–1271.

Waldron, I. (1976) Why do women live longer than men? *Social Science and Medicine*, 10, 349–362.

Waldron, I. (1991) Patterns and causes of gender difference in smoking. *Social Science and Medicine*, 32, 989–1005.

Wallace, R. and Wallace, D. (1993) The coming crisis of public health in the suburbs. *Milbank Quarterly*, 71, 543–563.

Wallack, L., Dorfman, L., Jernigan, D. H. and Themba-Nixon, M. (1993) *Media advocacy and public health: Power for prevention*. SAGE Publications, Newbury Park, CA.

Wallerstein, N. (1992) Powerlessness, empowerment and health: Implications for health promotion programs. *American Journal of Health Promotion*, 6, 197–205.

Wallerstein, N. and Bernstein, E. (1994) Introduction to community empowerment, participatory education, and health. *Health Education Quarterly*, 21, 141–148.

Walsh, D. C., Sorenson, G. and Leonard, L. (1995) Gender, health and cigarette smoking. In Amick, B., Levine, S., Tarlov, A. R. and Walsh, D. C. (eds), *Society and health*, Oxford University Press, New York, NY. pp. 131–171.

Walsh, J. A. and Warren, K. S. (1979) Selective primary health care: an interim strategy for disease control in developing countries. *New England Journal of Medicine*, 301, 967–974.

Walt, G. (1994) Health policy: An introduction to process and power. Zed Books, London.

Walter, D. (1996) Gnomes go to work in trams, *New Statesman*, 40–41.

Walter, M. and Mooney, G. (2007) Employment and welfare. In Carson, B., Dunbar, T., Chenhall, R. D. and Bailie, R. (eds), *Social determinants of Indigenous health*, Allen & Unwin, Crows Nest, NSW.

Wang, H., Deeks, S., Glasswell, A. and McIntyre, P. (2008) Trends in invasive Haemophilus influenzae type b disease in Australia, 1995–2005. *Communicable Diseases Intelligence*, 32, 316–325.

Ward, J., Gordon, J. and Sanson-Fisher, R. W. (1991) Strategies to increase preventive care in general practice. *Medical Journal of Australia*, 154, 523–531.

Wardlaw, G. (1992) Overview of National Drug Control Strategies. Comparative Analysis of Illicit Drug Strategy, *NCDA Monograph Series No. 18*. Commonwealth of Australia, Canberra, ACT.

Ware, J. E., Kosinski, M. and Keller, S. (1994) *SF-36 physical and mental health summary measures: A user's manual*. The Health Institute, Boston, MA.

Waring, M. (1988) Counting for nothing: What men value and what women are worth. Allen & Unwin, Wellington.

Warren, M. and Francis, H. (eds) (1987) Recalling the Medical Officer of Health: Writing by Sydney Chave. King's Fund Publishing, London.

Wass, A. (2000) Promoting health: The primary health care approach. Harcourt Brace, Sydney, NSW.

Watts, J. (2005a) Korean farmers take lemming—like plunge into Hong Kong harbour. *The Guardian*, 14 December. From www.theguardian.com/business/2005/dec/14/koreanews.wto

Watts, J. (2005b) Satellite data reveals Beijing as air pollution capital of the world. *The Guardian*, 31 October. From www.theguardian.com/news/2005/oct/31/china.pollution

Watts, N., Campbell-Lendrum, D., Maiero, M., Montoya, L. F. and Lao, K. (2014) Strengthening health resilience to climate change. Technical Briefing for the World Health Organization Conference on Health and Climate.

Weatherill, J. (2014) Letter dated 8 July, 2014 from the Premier Jay Weatherill RE reform of S.A. Boards and Committees. From http://dpc.sa.gov.au/sites/default/files/pubimages/documents/boards-committees/letters/SA%20Medical%20Education%20and%20Training%20HAC%20-%201.pdf

Webster, I. W. (1997) Health and tuberculosis in Sydney's homeless. *Australian and New Zealand Journal of Public Health*, 21, 444–446.

Wedes, J. (2013) Occupy Wall Street, two years on: We're still the 99%. *The Guardian*, 17 September. From www.theguardian.com/commentisfree/2013/sep/17/occupy-wall-street-99-percent

Weeramanthri, T. (1996) Knowledge, language and mortality: Communicating health information in Aboriginal communities in the Northern Territory. *Australian Journal of Primary Care*, 2, 3–11.

Welin, L., Tibblin, G., Svärdsudd, K., Ander-Peciva, S., Tibblin, B., Larsson, B. and Wilhelmsen, L. (1985) Prospective study of social influences on mortality: The study of men born in 1913 and 1923. *The Lancet*, 325, 915–918.

Weller, D. (1997) Cancer screening in general practice. *Australian Family Physician*, 26, 517–519.

Weller, D. and Dunbar, J. (2005) History, policy and context, *General Practice in Australia: 2004*. Commonwealth Department of Health and Ageing, Canberra, ACT.

Weller, D. P. and Campbell, C. (2009) Uptake in cancer screening programmes: a priority in cancer control. *British Journal of Cancer*, 101, S55–S59.

Welsh Government (2014) Mark Drakeford, Health Minister—People must take responsibility for their own lifestyle choices. From wales.gov.uk/newsroom/healthandsocialcare/2014/140521health-survey-wales/?lang=en (accessed 4 December 2014).

Werna, E. and Harpham, T. (1996) The implementation of the Healthy Cities project in developing countries: Lessons from Chittagong. *Habitat International*, 20, 221–228.

Werna, E., Harpham, T., Blue, I. and Goldstein, G. (1998) Healthy city projects in developing countries: An international approach to local problems. Earthscan, London.

Werner, D. (1997) Nothing with us without us: Developing innovative technologies for, by and with disabled persons. HealthWrights, Palo Alto, CA.

Werner, D. (2005) PHA-Exchange> Re: Bangkok Charter—Action Needed! From http://phm.phmovement.org/pipermail/phm-exchange-phmovement.org/20050706/002231.html (accessed 27 November 2014).

Werner, D. and Sanders, D. (1997) Questioning the solution: The politics of primary health care and child survival. Health Wrights, Palo Alto, CA.

Westaway, J. (2012) Globalization, transnational corporations and human rights: A new paradigm. *International Law Research*, 1. 10.5539/ilr.v1n1p63.

Weston, D. (2014) The political economy of global warming: The terminal crisis. Routledge, Abingdon, Oxon.

Weston, H. and Putland, C. (1995) Public health and local government. In Baum, F. (ed), *Health for All: The South Australian experience*, Wakefield Press, Adelaide, SA. pp. 281–306.

WHAM ABC 13 (2014) RIT Professor saw ebola crisis firsthand, 4 October [Transcript]. From http://13wham.com/template/cgi-bin/archived.pl?type=basic&file=/news/features/top-stories/stories/archive/2014/10/2R7jKXNG.xml (accessed 26 March 2015).

White, S. K. (2010) Corporations, public health, and the historical landscape that defines our challenge. In Wiist, W. (ed), The bottom line or public health: Tactics corporations use to influence health and health policy, and what we can do to counter them, Chapter 2, Oxford University Press, New York, NY. pp. 72–96.

Whitehead, M. (1992) The health divide. In Townsend, P., Davidson, N. and Whitehead, M. (eds), *Inequalities in health*, Penguin Books, London. pp. 219–437.

Whitehead, M. and Dahlgren, G. (2006) Levelling up, part 1: Concepts and principles for tackling social inequalities in health. World Health Organization, Copenhagen.

WHO Kobe Centre (2005) A billion voices: Listening and responding to the health needs of slum dwellers and informal settlers in new urban settings, *An analytic and strategic review paper for the Knowledge Network on Urban Settings, WHO Commission on Social Determinants of Health*. WHO Kobe Centre, Japan.

WHO Kobe Centre (2014) Best Practices: Healthy Urban Planning in New York City. From www.who.int/kobe_centre/interventions/urban_planning/HUP_NYC/en   (accessed 9 January 2015).

WHO Kobe Centre (2015a) One billion people more in urban areas since 2000. Tokyo remains the largest city in the world. Measuring urban health. From www.who.int/kobe_centre/measuring/WUP_2014/en (accessed 16 March 2015).

WHO Kobe Centre (2015b) Urban health equity assessment and response tool (HEART). From www.who.int/kobe_centre/measuring/urbanheart/en (accessed 7 December 2014).

WHO Regional Office for Europe (1997) *Twenty steps for developing a healthy cities project*. WHO Regional Office for Europe, Copenhagen.

WHO Regional Office for Europe (2005) *Health impact assessment toolkit for cities*. WHO Regional Office for Europe, Copenhagen.

WHO Regional Office for Europe (2011) Impact of economic crises on mental health. WHO Regional Office for Europe, Copenhagen.

WHO Regional Office for Europe (2013) Review of evidence on health aspects of air pollution—REVIHAAP project: Final technical report. WHO Regional Office for Europe, Copenhagen.

WHO Regional Office for Europe (2014b) Strasbourg conclusions on prisons and health. WHO Regional Office for Europe, Copenhagen.

WHO Regional Office for Europe (2015a) Healthy city checklist. From www.euro.who .int/en/health-topics/environment-and-health/urban-health/activities/healthy-cities/who-european-healthy-cities-network/what-is-a-healthy-city/healthy-city-checklist   (accessed 26 March 2015).

WHO Regional Office for Europe (2015b) Health Promoting Hospitals Network (HPH). From www.euro.who.int/en/health-topics/Health-systems/public-health-services/activities/health-promoting-hospitals-network-hph (accessed 26 March 2015).

WHO Regional Office for Europe (2015c) What is a healthy city? From www.euro.who.int/en/health-topics/environment-and-health/urban-health/activities/healthy-cities/who-european-healthy-cities-network/what-is-a-healthy-city (accessed 21 January 2015).

WHO Regional Office for Europe (2015d) WHO European Healthy Cities Network—Six strategic goals. From www.euro.who.int/en/health-topics/environment-and-health/urban-health/activities/healthy-cities/who-european-healthy-cities-network/six-strategic-goals (accessed 21 January 2015).

WHO Regional Office for the Western Pacific (2002) *The vision of healthy islands for the 21st century*. WHO Regional Office for the Western Pacific, Manila.

Whyte, W. F. (1943) *Street corner society: The social structure of an Italian slum*. University of Chicago Press, Chicago, IL.

Wigg, N. (1995) Promoting health with children, adolescents and their families. In Baum, F. (ed), *Health for All: The South Australian experience*, Wakefield Press, Adelaide, SA. pp. 393–405.

Wiist, W. (ed) (2010) The bottom line or public health: Tactics corporations use to influence health and health policy, and what we can do to counter them. Oxford University Press, New York, NY.

Wilcox, S. (2014) Chronic diseases in Australia: the case for changing course, *Background and policy paper No. 02/2014*. Mitchell Institute, Melbourne, VIC.

Wild, R. and Anderson, P. (2007) Ampe Akelyernemane Meke Mekarle "Little children are sacred": *Report of the Northern Territory Board of Inquiry into the Protection of Aboriginal Children from Sexual Abuse*. NT Government, Darwin, NT.

Wilkinson, R. G. (1996) Unhealthy societies: The afflictions of inequality. Routledge, London.

Wilkinson, R. G. (1999) Putting the picture together: Prosperity, re-distribution, health and welfare. In Marmot, M. and Wilkinson, R. G. (eds), *Social determinants of health*, Oxford University Press, Oxford.

Wilkinson, R. G. (2005) The impact of inequality. How to make sick societies healthier. The New Press, New York, NY.

Wilkinson, R. G. and Marmot, M. (eds) (2003) *Social determinants of health: The solid facts*. World Health Organization, Geneva.

Wilkinson, R. G. and Pickett, K. (2009) The spirit level: Why more equal societies almost always do better. Allen Lane, London.

Wilkinson, W. and Sidel, V. W. (1991) Social applications and interventions in public health. In Detels, R., McEwan, J., Beaglehole, R. and Tanaka, H. (eds), *Oxford textbook of public health*, Oxford University Press, Oxford. pp. 47–70.

Willaby, H. (2014) Why do people not vaccinate? *The Conversation*, 27 March. From https://theconversation.com/why-do-people-not-vaccinate-24882

Williams, C. (1981) Open cut: The working class in an Australian mining town. Allen & Unwin, Sydney, NSW.

Williams, J. (2004) *50 facts that should change the world*. Icon Books Inc, Cambridge.

Williams, R. (1983) Keywords: A vocabulary of culture and society. Fontana Press, London.

Williams, R., Neighbours, H. and Jackson, J. (2003) Racial/ethnic discrimination and health: Findings from community studies. *American Journal of Public Health*, 93, 200–208.

Willis, E. (1983) *Medical dominance*. Allen & Unwin, Sydney, NSW.

Willis, P. (1977) *Learning to labour*. Gower, Farnborough, Hants.

Wilson, S. and Breusch, T. (2004) After the tax revolt: Why Medicare matters more to middle Australia than lower taxes. *Australian Journal of Social Issues*, 39, 99–116.

Winkelstein, W. and Marmot, M. (1981) Primary prevention of ischemic heart disease: Evaluation of community interventions. *Annual Review of Public Health*, 2, 253–276.

Winslow, C. E. A. (1920) The Untilled Fields of Public Health (PDF). *Science*, 51, 23–33.

Winter, I. (ed) (2000) *Social capital and public policy in Australia*. Australian Institute of Family Studies, Melbourne, VIC.

Wisman, J. D. and Capehart, K. W. (2010) Creative destruction, economic insecurity, stress, and epidemic obesity. *American Journal of Economics and Sociology*, 69, 936–982.

Wolff, J. (2014) Paying people to act in their own interests: Incentives versus rationalization in public health. *Public Health Ethics*. 10.1093/phe/phu035.

Wolfson, M., Rowe, G., Gentleman, J. F. and Tomiak, M. (1993) Career earnings and death: A longitudinal analysis of older Canadian men. *Journal of Gerontology: Social Sciences*, 48, S167–S179.

Wong, B. and McKeen, J. (2013a) *Being: A manual for life*. Haven Institute Press, Gabriola Island, BC.

Wong, B. and McKeen, J. (2013b) *Joining: The relationship garden*. Haven Institute Press, Gabriola Island, BC.

Wong, E. (2013) Air pollution linked to 1.2 million premature deaths in China. *New York Times*, 1 April. From www.nytimes.com/2013/04/02/world/asia/air-pollution-linked-to-1-2-million-deaths-in-china.html?_r=0

Woodruff, P. (1984) *Two million South Australians*. Peacock, Adelaide, SA.

Woodward, A. (2014) Climate change and health: recent progress [editorial]. *Bull World Health Organization*, 92, 774. From http://dx.doi.org/10.2471/BLT.14.148130.

Woodward, A., Guest, C., Steer, K., Harman, A., Scicchitano, R., Pisaniello, D., Calder, I. and McMichael, A. J. (1995) Tropospheric ozone: respiratory effects and Australian air quality goals. *Journal of Epidemiology and Community Health*, 49, 401–407.

Woodward, D., Drager, N., Beaglehole, R. and Lipson, D. J. (2001) Globalization and health: A framework for analysis and action. *Bulletin of the World Health Organization*, 79, 875–881.

Woollard, K. (1996) Aboriginal problem will get worse not better. *Australian Medicine*, 7.

World Bank (1993) World Bank Development Report, 1993—Investing in health. Oxford University Press.

World Bank (2011) Violence in the city: Understanding and supporting community responses to urban violence. World Bank Social Development Department Conflict, Crime and Violence Team, Washington, DC.

World Bank (2014) *World Development Indicators*. From http://data.worldbank.org/data-catalog/world-development-indicators (accessed 18 December 2014).

World Commission on Environment and Development (WCED) (1987) *Our Common Future [Brundtland Report]*. Oxford University Press, Oxford.

World Commission on the Social Dimension of Globalization (2004) *A fair globalization. Creating opportunity for all*. International Labour Office, Geneva.

World Health Organization (WHO) (1948) Constitution of the World Health Organization, 1946. From http://whqlibdoc.who.int/hist/official_records/constitution.pdf

World Health Organization (WHO) (1978) Declaration of Alma-Ata. International Conference on Primary Health Care, 6–12 September, Alma-Ata, USSR. From www.who.int/publications/almaata_declaration_en.pdf?ua=1

World Health Organization (WHO) (1981) *Global Strategy for Health for All by the Year 2000*. WHO, Geneva.

World Health Organization (WHO) (1986) Ottawa Charter for Health Promotion, First International Conference on Health Promotion, 21 November, Canada. From www.who.int/healthpromotion/conferences/previous/ottawa/en

World Health Organization (WHO) (1988) Healthy public policy: The Adelaide recommendations. International Conference on Health Promotion. WHO, Geneva.

World Health Organization (WHO) (1989) *Health principles of housing*. WHO, Geneva.

World Health Organization (WHO) (1993) The urban health crisis: Strategies for health for all in the face of rapid urbanisation. *Report of the technical discussions at the forty-fourth World Health Assembly*. WHO, Geneva.

World Health Organization (WHO) (1994) Health development structures: A hidden resource for health. WHO, Geneva.

World Health Organization (WHO) (1995a) *Health consequences of the Chernobyl accident*. WHO, Geneva.

World Health Organization (WHO) (1995b) *World Health Report 1995—Bridging the Gaps*. WHO, Geneva.

World Health Organization (WHO) (1996) *Creating Healthy Cities in the 21st Century*. WHO, Geneva.

World Health Organization (WHO) (1998) *Health Promotion Glossary*. WHO, Geneva.

World Health Organization (WHO) (2000) World Health Report 2000—Health systems: improving performance. WHO, Geneva.

World Health Organization (WHO) (2002) *World Report on Violence and Health*. WHO, Geneva.

World Health Organization (WHO) (2003a) Summary table of SARS cases by country, November 1 2002–August 7, 2003. From www.who.int/csr/sars/country/2003_08_15/en (accessed 5 January 2015).

World Health Organization (WHO) (2003b) World Health Assembly Resolution 56.1: Framework Convention on Tobacco Control. From www.who.int/tobacco/framework/final_text/en/index.html (accessed 8 April 2007).

World Health Organization (WHO) (2004a) Global Strategy on Diet, Physical Activity & Health—Process for developing the Global Strategy—May 2004: final strategy document and resolution. From www.who.int/dietphysicalactivity/strategy/eb11344/en (accessed 27 January 2015).

World Health Organization (WHO) (2004b) World Report on Knowledge for Better Health—Strengthening Health Systems. WHO, Geneva.

World Health Organization (WHO) (2005) The Bangkok Charter for Health Promotion. From www.who.int/healthpromotion/conferences/6gchp/bangkok_charter/en

World Health Organization (WHO) (2006) World Health Report 2006—Working together for health. WHO, Geneva.

World Health Organization (WHO) (2007) Everybody's business. Strengthening health systems to improve health outcomes: WHO's framework for action. WHO, Geneva.

World Health Organization (WHO) (2008) World Health Report 2008—Primary health care (Now more than ever). WHO, Geneva.

World Health Organization (WHO) (2010a) *Urban HEART—Urban health equity assessment and response tool: User manual.* WHO Centre for Health Development, Kobe, Japan.

World Health Organization (WHO) (2010b) Urbanization and health. *Bulletin of the World Health Organization*, 88, 241–320.

World Health Organization (WHO) (2010c) WHO Global Code of Practice on the International Recruitment of Health Personnel, 63rd World Health Assembly, 21 May. WHO, Geneva.

World Health Organization (WHO) (2011) Rio Political Declaration on Social Determinants of Health. World Conference on Social Determinants of Health, 21 October, Rio de Janeiro, Brazil. From www.who.int/sdhconference/declaration/en

World Health Organization (WHO) (2012) Prioritizing areas for action in the field of population-based prevention of childhood obesity. WHO, Geneva.

World Health Organization (WHO) (2013a) Global Health Estimates (GHE)—Projections of mortality and causes of death, 2015 and 2030. From www.who.int/healthinfo/global_burden_disease/projections/en

World Health Organization (WHO) (2013b) The Helsinki Statement on Health in All Policies. 8th Global Conference on Health Promotion, 10–14 June, Helsinki, Finland. From www.who.int/healthpromotion/conferences/8gchp/statement_2013/en

World Health Organization (WHO) (2013c) Road traffic injuries. Fact sheet N°358, updated March 2013. From www.who.int/mediacentre/factsheets/fs358/en (accessed 2 January 2015).

World Health Organization (WHO) (2014a) Air quality declining in many of the world's cities, 7 May [News Release]. From www.who.int/mediacentre/news/releases/2014/air-quality/en (accessed 2 October 2014).

World Health Organization (WHO) (2014b) Frequently Asked Questions—Ambient and Household Air Pollution and Health, Update 2014. From www.who.int/phe/health_topics/outdoorair/databases/faqs_air_pollution.pdf?ua=1 (accessed 2 October 2014).

World Health Organization (WHO) (2014c) Global Health Estimates (GHE)—Deaths By Cause, Age And Sex, By World Bank Income Category, 2000–2012. WHO, Geneva.

World Health Organization (WHO) (2014d) *Global Status Report on Violence Prevention 2014.* WHO, Geneva.

World Health Organization (WHO) (2014e) Poliomyelitis. Fact sheet N°114, updated October 2014. From www.who.int/mediacentre/factsheets/fs114/en (accessed 19 December 2014).

World Health Organization (WHO) (2014f) *Preventing suicide: A global imperative.* WHO, Geneva.

World Health Organization (WHO) (2014g) Quantitative risk assessment of the effects of climate change on selected causes of death, 2030s and 2050s. WHO, Geneva.

World Health Organization (WHO) (2014h) Tobacco. Fact sheet N°339, updated May 2014. From www.who.int/mediacentre/factsheets/fs339/en (accessed 19 December 2014).

World Health Organization (WHO) (2014i) The top 10 causes of death, Fact sheet N°310, updated May 2014. From www.who.int/mediacentre/factsheets/fs310/en (accessed 19 December 2014).

World Health Organization (WHO) (2014j) World Health Statistics 2014. WHO, Geneva.

World Health Organization (WHO) (2014k) WHO guidance to protect health from climate change through health adaptation planning. WHO, Geneva.

World Health Organization (WHO) (2015a) Ebola situation report, 22 April 2015. From http://apps.who.int/ebola/current-situation/ebola-situation-report-22-april-2015-0 (accessed 24 May 2015).

World Health Organization (WHO) (2015b) Obesity and overweight. Fact sheet N°311, updated January 2015. From www.who.int/mediacentre/factsheets/fs311/en (accessed 25 January 2015).

World Health Organization (WHO) (2015c) Parties to the WHO Framework Convention on Tobacco Control. From www.who.int/fctc/signatories_parties/en (last accessed 5 January 2015).

World Health Organization (WHO) (2015d) Global Health Observatory (GHO) data—Life Expectancy, situation. From www.who.int/gho/mortality_burden_disease/life_tables/situation_trends_text/en (accessed 25 March 2014).

World Health Organization (WHO) (2015e) Global Health Observatory (GHO) data—Under-five mortality situation. From www.who.int/gho/child_health/mortality/mortality_under_five_text/en (accessed 25 March 2015).

World Health Organization (WHO), Department of Reproductive Health and Research, London School of Hygiene and Tropical Medicine and South African Medical Research Council (2013a) *Global and regional estimates of violence against women: Prevalence and health effects of intimate partner violence and non-partner sexual violence*. WHO, Geneva.

World Health Organization (WHO) and United Nations Children Fund (UNICEF) (2013b) *Progress on sanitation and drinking water: 2013 update. WHO and UNICEF*, Geneva.

World Health Organization (WHO) and United Nations Children Fund (UNICEF) (2014a) *Global Polio Eradication Initiative. Financial Resource Requirements 2013–2018*. WHO, Geneva.

World Health Organization (WHO) and United Nations Children Fund (UNICEF) (2014b) *Progress on drinking-water and sanitation—2014 update*. WHO and UNICEF, Geneva.

World Health Organization (WHO) and United Nations Human Settlements Programme (UN-HABITAT) (2010) *Hidden cities: Unmasking and overcoming health inequities in urban settings*. WHO, Geneva.

World Health Organization (WHO) and World Bank (2011) *World Report on Disability 2011*. WHO, Geneva.

Worldwatch Institute (ed) (2013) *State of the World 2013—Is sustainability still possible?* Island Press, Washington, DC.

Worsley, T. S. (1990) *National Evaluation of Healthy Cities Australia Pilot Project*. Australian Community Health Association, Bondi Junction, NSW.

Wright, P., Davies, C., Haseman, B., Down, B., White, M. and Rankin, S. (2013) Arts practice and disconnected youth in Australia: impact and domains of change. *Arts and Health*, 5, 190–203.

Yeatman, A. (1987) The concept of public management and the Australian state in the 1980s. *Australian Journal of Public Administration*, 156, 339–356.

Yeatman, A. (1990) Bureaucrats, technocrats, femocrats: Essays on the contemporary Australian state. Allen & Unwin, Sydney, NSW.

Yen, I. H. and Syme, S. L. (1999) The social environment and health: A discussion of the epidemiologic literature. *Annual Review of Public Health*, 20, 287–308.

Yin, R.K (1989) *Case study research: design and method*. SAGE Publications, Newbury Park, CA.

Yin, R. K. (2014) *Case Study Research: Design and Methods*. SAGE Publications, Newbury Park, CA.

Young, I. M. (1990) The ideal of community and the politics of difference. In Nicholson, L. J. (ed), *Feminism/Postmodernism*, Routledge, New York. pp. 300–323.

Young, M. and Willmott, P. (1957) *Family and kinship in East London*. Penguin Books, Harmondsworth, UK.

Zacune, J. (2012) Combatting Monsanto Grassroots resistance to the corporate power of agribusiness in the era of the 'green economy' and a changing climate. La Via Campensina, Friends of the Earth International, Combat Monsanto.

Ziersch, A. and Baum, F. (2004) Involvement in civil society groups: Is it good for your health? *Journal of Epidemiology and Community Health*, 58, 493–500.

Ziersch, A., Baum, F., Woodman, R. J., Newman, L. and Jolley, G. (2014) A longitudinal study of the mental health impacts of job loss: The role of socioeconomic, sociodemographic, and social capital factors. *Journal of Occupational and Environmental Medicine*, 56, 714–720.

Ziersch, A., Gallaher, G., Baum, F. and Bentley, M. (2011a) Racism, social resources and mental health for Aboriginal people living in Adelaide. *Australian and New Zealand Journal of Public Health*, 35, 231–237.

Ziersch, A., Gallaher, G., Baum, F. and Bentley, M. (2011b) Responding to racism: Insights on how racism can damage health from an urban study of Australian Aboriginal people. *Social science and medicine*, 73, 1045–1053.

Ziglio, E. (1987) Policy making and planning in conditions of uncertainty: Theoretical considerations for health promotion policy. Research Unit in Health and Behavioural Change, University of Edinburgh, Edinburgh.

Zimmerman, M. A. (1990) Taking aim on empowerment research: On the distinction between individual and psychological conceptions. *American Journal of Community Psychology*, 18, 169–177.

Zubrick, S. R., Silburn, S. R., Lawrence, D. M., Mitrou, F. G., Dalby, R. B., Blair, E. M., Griffin, J., Milroy, H., De Maio, J. A., Cox, A. and Li, J. (2005) *The Western Australian Aboriginal Child Health Survey: The social and emotional wellbeing of Aboriginal children and young people*. Curtin University of Technology and Telethon Institute for Child Health Research, Perth, WA.

Zuppa, J. A., Morton, H. and Mehta, K. P. (2003) TV food advertising: Counterproductive to children's health? A content analysis using the Australian Guide to Healthy Eating. *Nutrition and Dietetics*, 60, 78–84.

Zwi, A. B., Grove, N. J., Kelly, P., Gayer, M., Ramos-Jimenez, P. and Sommerfield, J. (2006) Child health in armed conflict: time to rethink. *The Lancet*, 367, 1886–1888.

INDEX